MACHINE DESIGN

Machine Design

Third Edition

PAUL H. BLACK
Professor of Mechanical Engineering
Ohio University

O. EUGENE ADAMS, JR.
Associate Professor of Mechanical Engineering
Ohio University

McGRAW-HILL BOOK COMPANY
New York St. Louis San Francisco Toronto London Sydney

Machine Design

Library of Congress Catalog Card Number 68-13623

07-005524-6

4567890KPKP754321

Preface

This book is intended to serve as a textbook for courses in general machine design and as a reference book in mechanical-design offices. It is assumed that the reader has a knowledge of mechanics, strength of materials, kinematics, mechanical processes, and engineering materials, including their properties.

The analytical method combined with results of experimental investigation has been used whenever possible, and the empirical method has been used only when it is presently impractical to use the analytical approach. In the latter class of problems, the direction of rationalization is indicated by references.

As in earlier editions, we assume that the application of principles is the best way to solve a problem. An effort is made to base designs on the likely methods of failure, such as static and fatigue failures, lack of rigidity, and lack of wear resistance. In this edition, emphasis is further placed on other methods of failure such as corrosion, creep, and failure due to improper processing.

The problems at the back of the book are representative of the field covered; most of them are based on actual installations, with emphasis on both design and analysis. In this revision, the number of short problems has been increased. Also, there have been added several longer problems which require the consideration of related items such as application requirements, selection of material, units of power transmission, motor selection, and other phases of comprehensive problems. Additional problems are aimed at giving the designer experience in making assumptions, choices, compromises, and decisions—in short, providing him with an opportunity to exercise his ingenuity and creative ability.

Special appreciation is extended to Irene Cunningham Black for her assistance in preparation of the manuscript.

Paul H. Black
O. Eugene Adams, Jr.

Contents

Preface v

CHAPTER 1 Introduction 1

2 Machine-design Computations 7

3 Stress and Deflection Analysis 16

4 Stress Concentration in Machine Members 59

5 Design Criteria 81

6 Residual Stresses 109

7 Engineering Materials 122

8 Mechanical Fabrication and Processes 144

9 Detachable Fastenings 179

10 Springs 206

11 Pressure Cylinders 234

12 Translation Screws 249

13 Shafting 263

14 Belt Drives and Hoists 273
15 Power-transmission Chains 300
16 Shaft Couplings 313
17 Clutches and Brakes 322
18 Spur and Parallel Helical Gears 345
19 Gears for Nonparallel Shafts 378
20 Surface Finish, Friction, and Wear 399
21 Sliding-contact Bearings 419
22 Rolling-contact Bearings 459
23 Metal Fits and Tolerances 481
24 Vibration and Vibration Control 498
25 Motor Selection 514
APPENDIX I Abbreviations 523
II Elements of Sections 525
III Bending-moment and Deflection Formulas 526
IV Values of $\pi d^3/16$ and $\pi d^3/32$ 527
V Properties of Materials 528
VI Tensile Strength and Yield Point versus Brinell Hardness for Steel 529
VII Endurance Limit versus Tensile Strength 530
VIII Yield Points for Steel 531
IX Hardness Numbers Conversion Chart 533
X Curves for Stress-concentration Factors 534
XI WR^2 for Rotating Bodies 550
XII Method for Determining Lewis Factor 551
XIII-A Graphical Integration 552
XIII-B Shaft Deflection by Digital Computer 556
XIII-C Rupture-stress Diagram Curve for Elevated-temperature Systems 560
XIV Table of Wire Sizes 561
XV-A Decimal Equivalents of Fractions 562
XV-B Conversion Factors, U.S. and Metric Units 563
XVI Horsepower Ratings of USAS Standard Roller Chain, Single Strand 564
XVII Case Histories 573
Problems 589
Name Index 667
Subject Index 671

1
Introduction

Mechanical engineering is the branch of engineering that is concerned with the conversion of energy and with the design of machines. The designer of machines is a versatile person, not a specialist, because he must have knowledge of science, "to know," and a knowledge of engineering, "to produce." For instance, the design of the machinery for missile launching requires, in part, considerable knowledge of physics, mechanics, aerodynamics, performance of materials in high-intensity, short-time use, electronics, and safety requirements.

The engineering design process[1] begins with the awareness of a human need or want. In Fig. 1-1, this is indicated as the GOAL (1). The goal should be stated in words as clearly as possible at this stage of the design process. When the goal is stated in general terms, it is then necessary to develop more specific and quantitative information which determines the TASK SPECIFICATION (2) necessary to achieve the goal. The next step is the framing of a CONCEPT (3) with which the final form

[1] R. J. McCrory, The Design Method, *Mech. Eng.*, May, 1963, p. 30.

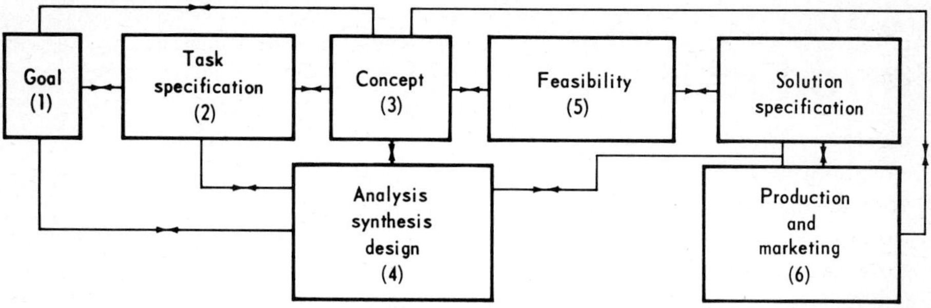

Fig. 1-1 The engineering design process. (*From Robert W. Mann, Mechanical Engineering Department, Massachusetts Institute of Technology.*)

of the machine is concerned. The step between task specification and the concept is *synthesis*. This is an art, a creative ability,[1,2] which involves a study of the problem, recording and evaluating a number of possible solutions, and the synthesizing of the best ones.

At this stage, it may be necessary to revise the task specification if it becomes apparent a solution is impossible. This backtracking is indicated by the reverse arrowhead on the task-specification–concept interconnection.

At this junction, there are two interrelated and dependent avenues, one to proceed to the specification of the solution to the enterprise, or, a second, to make a critical ANALYSIS (4) of the concept. If the latter is necessary one must reconsider and reevaluate the concept before proceeding to FEASIBILITY (5).

The step from concept to feasibility is the *design* province, and the inevitable question is: Will it work? This province includes organizing, making decisions, optimizing, compromising, standardizing, and detailing.

When the design has proved feasible, it may be ready for the PRODUCTION-MARKETING (6) stage. Between the feasibility-production stages, technical and manufacturing *development* and possibly a reconsideration of one or more of the previous stages may be necessary.

This book is concerned mainly with the concept, analysis, and feasibility stages of design in which the components and units of machinery are described. Descriptions, applications, and limitations of standard units are discussed, and necessary equations are derived. As far as

[1] H. R. Buhl, "Creative Engineering Design," The Iowa State University Press, Ames, Iowa, 1960.

[2] Edward V. Krick, "An Introduction to Engineering and Engineering Design," John Wiley & Sons, Inc., New York, 1965.

possible, these equations are presented in the design form rather than in the analysis form.

As one stands before any machine which is operating—such as a plastic-molding machine turning out, say, a hundred switch-box covers each hour, or a punch press performing a dozen operations simultaneously at each stroke on a steel part for a computing machine, or the main reduction gears of the propulsion machinery of a ship which change the speed from the turbine speed of 3,600 rpm to propeller-shaft speed of 225 rpm while transmitting 8,500 hp, or a turboalternator running at 1,800 rpm supplying light and power to a community at a peak load of 100,000 kw at 10 A.M. on a February day—as one watches such machines in operation, he is bound to have feelings of awe and reverence for their performance and their power, both mechanical and economic. He will no doubt notice the form of the machine and of the visible parts. He may note that many of the parts appear complicated and that their operation contributes to the operation of the machine as a whole in a synchronized manner. Watching such a machine is an inspiration to a mechanically minded person to understand its construction, operation, and performance.

The operation and performance of the machine as a whole depends on the individual parts, each of which has individual characteristics, just as the machine as a whole has its characteristics. These individual parts may be shafts, couplings, bearings, gears, brakes, controls, or many other units.

The production of such a machine is the result of an enormous amount of directed work: theoretical, experimental, and manual. The theoretical work includes the application of the laws of physics, including thermodynamics, mechanics, and strength of materials, chemistry and, invariably, of the mathematical methods. The experimental work involves the verification of the laws and relations that were developed from theoretical considerations; the determination of the properties of materials; the analyses of stress, strain, and vibrations in complicated structures; and the performance of complete machines and units of machines, such as power-transmission units. The manual work involves actual production in the shop, which makes use of highly developed techniques. Other factors that must be considered are safety and economics. Each of these, as well as many others that are not mentioned, is in the province of experts who confine their activities mainly to the fields of their specialization.

In the production of every machine, someone must have the overall responsibility of deciding on the mechanical arrangements to be used, of making use of the information provided by the specialized fields mentioned above, and of combining them by engineering judgment so that the machine can be manufactured economically, will perform its specified function properly, and will have a satisfactory life and maintenance. The man who has this overall responsibility is the mechanical designer, or machine designer.

In each design, a number of operations are involved.

1. It may be necessary to decide on the kinematic arrangement of the parts of the machine. In a new machine, this may call on the creative ability of the designer, that is, on his ingenuity, inventiveness, resourcefulness, originality, and intuition. In this procedure it is always well to begin by setting down a clear statement of the problem.[1] Creative ability applies not only to the kinematic arrangement of the parts but also to other phases of design, such as devising new operating processes, new materials, and new production methods. It may be an individual or a group operation. After the kinematic arrangement of the parts has been established, their motion may require detailed analysis. This analysis may involve the use of tentative dimensions that may require alteration as the design progresses and the dynamic characteristics become established.

2. The maximum forces on the parts and the nature of the forces must be determined. The forces may be due to either static or impact loading, and they may or may not be applied repeatedly. Inertia forces may also be induced, and vibration forces which are either locally induced or transmitted from another member may be present. Frequently it is not possible to determine the forces exactly, and in these cases the judgment of the designer is required to estimate the maximum forces, his decision being based on factors which occasionally may be somewhat indeterminate.

3. The ability of the member to function properly must be considered. This involves both the material of the member and its size and shape. The selection of the material requires a consideration of mechanical properties, of heat treatment, and or manufacturing processes. The load-carrying ability, which will be discussed in detail later, may depend on the yield point of the material, its modulus of rigidity, ductility, heat-treating characteristics, endurance limit, strength, or creep limit. In addition, the shape and finish of the member may have a considerable effect on its strength.

[1] For instance, in telephone equipment the soldered connection has proved unsatisfactory. The Bell Laboratories set out to devise a method of making connections that would have the following characteristics: high conductivity, long life, speed in manufacturing, quality independent of maker, small space, no necessity for clipping ends of wire, and low cost. That was the statement of the problem. The solution was the development of the solderless wrapped connection made by an electrically driven wrapping tool. This connection has replaced soldered connections in Western Electric communication equipment.

4. The manufacturing process by which the member is to be made must be selected. This involves consideration of the effects of the process on the properties of the material and on the cost of the processing. The latter item is obviously of great importance in any design intended for production in large numbers.

After the part has been satisfactorily designed to operate properly (*functional design*), it may be necessary to make a *production design* in order to reduce the cost to a minimum.

5. In most designs it is necessary to consider additional items, such as safety, appearance, cost of operation, standard codes, thermal effects, contamination, corrosion, and maintenance.

Chapters 3 to 8 cover essentially the design of parts for strength and rigidity. This portion of the book includes the selection of material for making the parts. Chapters 9 to 12 cover fastenings, springs, cylinders, and translation screws. Chapters 13 to 19 may be termed the power-transmission section. This is a most important section for industrial applications. Chapters 20 to 22 cover surface finish, friction, wear, lubrication, and journal-bearing design and selection of rolling-contact bearings. Chapter 24 is a brief discussion of mechanical vibrations. The last chapter, Chap. 25, is on motor selection and is included since many machines include a motor in the drive and the characteristics of the motor may have a marked effect on the design of the entire equipment.

The problems in the back of the book are based in most cases on data from actual installations. They are mainly of the design type rather than analysis problems. The comprehensive problems at the end of the problem section are intended to illustrate the design procedure, the integration of work covered in the various chapters, and the necessary correlation between the five steps stated earlier in this chapter. This correlation is made necessary by the related functioning and performance of each part of the machine. Experience in the field gives the machine designer an appreciation of this interesting process.

Numerous references are given throughout the text as footnotes. Although some readers may not have the opportunity to consult these references or may not have them readily available, they are given to indicate the scope and sources of further developments and for the direct use of interested designers. Most of these references carry selected bibliographies for further reference.[1]

Just as the field of machine design today involves requirements of

[1] Two excellent general sources of additional information are Baumeister and Marks, "Standard Handbook for Mechanical Engineers," 7th ed., McGraw-Hill Book Company, New York, 1967, and Kent, "Mechanical Engineers' Handbook, Design and Production," 12th ed., John Wiley & Sons, Inc., New York, 1950.

manufacture and performance which were unknown or unappreciated a generation ago, the coming generation must deal with problems which today are just coming to light. High- and low-temperature and high-pressure applications of materials are two of the pressing developments. The advent of the gas turbine and of atomic power production and utilization is requiring performances which were considered impossible a few years ago. And, finally, the trend today is to require smaller and smaller things to do bigger and bigger jobs and to do them better. The machine designer who appreciates each of the five items listed earlier in this chapter and also a sixth item, *to make each of the other five work together effectively,* is in an excellent position to handle the solution of the problems that will be demanded tomorrow by industry.

2 Machine-design Computations

2-1 IMPORTANCE OF COMPUTATIONS

Design procedure in engineering offices usually requires the making of computations and notes that become the property of the firm or company rather than of the individual who makes them. As a rule the computations are checked by someone other than the designer; hence they should be made and recorded so that they can be followed readily by the reviewer; in other words they should, so far as possible, be self-explanatory.

In order that a designer may acquire the habit of making computations in good form and of keeping them orderly, he should constantly keep in mind that they are a record of that phase of the design and that others may have occasion to refer to them.

It is interesting to look back over a sample of one's own computations which were made a year or so previously. If they are readily understood, it indicates that they have good form. If not, one can easily imagine that another person would not be impressed with the ability or the courtesy of their originator.

2-2 FORM OF COMPUTATIONS

To aid the designer in developing good form, some suggestions are made here. Many of them are undoubtedly already followed by the reader, some of them may be new, but they are generally characteristic of a presentable record and it is suggested that all of them be tried.

1. Each sheet should carry the designer's name, date, and consecutive numbers if more than one sheet is used.
2. One side only of the paper should be used and a 1-in. strip on the left edge of the sheet should be reserved for binding.
3. If possible, a sketch should be drawn for every problem. The sketch should be neat and correctly drawn and may be used to show the arrangement of the parts and such data as dimensions, speeds, materials, etc. For data that cannot be shown on the sketch, a list should be used.
4. Each step should be set off, labeled or lettered, so that divisions of the work can be readily identified. This corresponds to using paragraphs in ordinary written matter and is helpful in making computations understandable.
5. All pertinent assumptions should be stated and references given.
6. Equations should be solved for the unknown quantity before numerical substitutions are made.
7. The numerical substitutions should follow the literal equation and be in order, term for term.
8. Units should not be placed in numerical equations. If a check on units is desirable, a separate dimensional equation should be used.
9. Units should be indicated for every separate value or result that has units, whether it be data, intermediate calculated value, or final result. The United States of America Standards Institute (USAS) abbreviations are recommended (see Appendix I).
10. Long equations should be avoided if possible; instead, intermediate values or particular physical quantities should be calculated separately before being inserted into the parent equation.
11. An unjustified number of significant figures should not be used.
12. Calculated dimensions of parts should be rounded out to nominal sizes and indicated, as, for example, $d = 2.95$; use 3 in.
13. Values in an equation in formal computations should not be canceled. Since erasures should be avoided, a change is indicated by drawing a line through the old value and writing the new value over or under the old. Canceling values would be confusing. Do any canceling off to one side.
14. All results should be considered from the standpoint of their reasonability.

2-3 GENERAL SOLUTION

In solving problems involving several quantities, it is usually desirable to make a general solution first, using symbols for the quantities involved, and, after the general solution is completed, to make numerical substitutions for the particular solution that is required. This method of making a general solution first has several advantages, some of which are as follows:

1. It provides the possibility of making a dimensional check on the final expression.
2. It makes it possible to determine the number of significant figures to which the quantities must be determined in order to obtain a desired degree of accuracy of the final result.
3. It facilitates making an alternative solution with changed data. This procedure, which is discussed later, is invariably necessary in engineering offices.
4. A general solution can be used for making subsequent solutions of the same problem. If one mixes particular numerical values in a derivation, it becomes necessary to repeat the entire derivation if a new set of data is specified.

2-4 ABSOLUTE AND RELATIVE ERRORS

The difference between a measured value and its exact value is termed the absolute error. Thus, if the measured value is too large, the error is positive, and if too small, it is a negative error. If the absolute error is divided by the exact value, the result is the relative error and is usually expressed as a percentage. In engineering, the relative error frequently has more significance than the absolute error. The following comparison will illustrate the difference.

1. An absolute error of 30 mph in determining the speed of an airplane traveling 300 mph is a relative error of 10 percent. This illustrates poor accuracy.
2. A measurement of the distance from New York to San Francisco with an absolute error of 30 miles is a relative error of 1.2 percent. This is good accuracy.
3. Michelson determined that the commonly used velocity of light was 30 mps in absolute error. This is a relative error of 0.016 percent, which is highly accurate.

2-5 ERRORS IN COMPUTATIONS[1]

Let A and B represent the correct values of two quantities and m and n be, respectively, the relative errors involved in the measurement of the two quantities.

Addition of two numbers when one contains an error

$$A + B = C \qquad \text{the correct result}$$
$$(A + mA) + B = D \qquad \text{the result containing the error}$$

The absolute error in the result is equal to

$$D - C = (A + mA) + B - (A + B) = mA$$

The relative error in the result is equal to

$$\frac{mA}{A + B} = \frac{m}{1 + (B/A)}$$

The last result is less than m. Therefore, in adding quantities, a relative error in one of the quantites does not affect the result by as great a relative error as in the inexact quantity.

Subtraction of two numbers when one contains an error

$$A - B = C \qquad \text{the correct result}$$
$$(A + mA) - B = D \qquad \text{the result containing the error}$$

The absolute error in the result is equal to

$$D - C = (A + mA) - B - (A - B) = mA$$

The relative error in the result is equal to

$$\frac{mA}{A - B} = \frac{m}{1 - (B/A)}$$

The last result is greater than m. Therefore, in subtracting two quantities, an error in one of the quantities affects the result by a larger relative error than that in the original quantity. In case the two quantities A and B have nearly the same value, the relative error in the result due to an error in A or B may be considerable. This magnification of error is shown by Fig. 2-1, which illustrates the effect of a 1 percent error in the larger of two numbers in subtraction.

Addition and subtraction of two numbers when both contain errors It may be shown that, in addition and subtraction of numbers containing

[1] J. B. Scarborough, "Numerical Mathematical Analysis," 3d ed., Johns Hopkins Press, Baltimore, 1955.

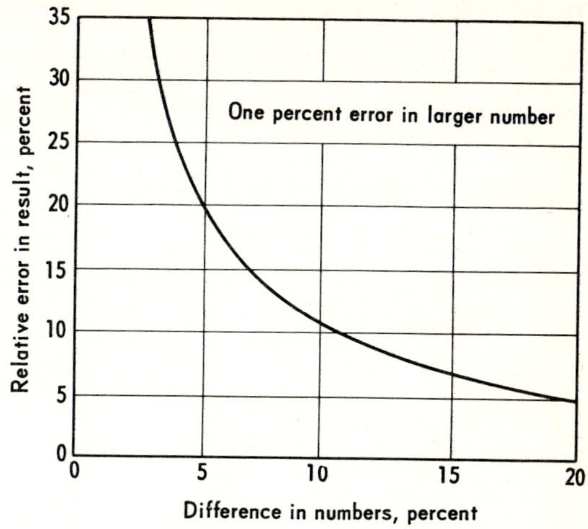

Fig. 2-1 Effect of errors in subtraction of numbers.

errors, the result contains an absolute error equal to the *sum* of the absolute errors of the individual quantities.

Multiplication and division of two numbers when one contains an error

$$A \times B = C \qquad \text{the correct result}$$
$$(A + mA) \times B = D \qquad \text{the result containing the error}$$

The absolute error in the result is equal to

$$D - C = (A + mA) \times B - AB = mAB$$

The relative error in the result is equal to

$$\frac{mAB}{AB} = m$$

Therefore, in multiplication as well as in division of quantities, a relative error in one of the quantities affects the result by the same relative error.

Multiplication of two numbers when both contain errors

$$A \times B = C \qquad \text{the correct result}$$
$$(A + mA)(B + nB) = D \qquad \text{the result containing the error}$$

The absolute error in the result is

$$\begin{aligned} D - C &= (A + mA)(B + nB) - AB \\ &= AB(m + n + mn) \end{aligned}$$

The relative error is equal to

$$\frac{AB(m + n + mn)}{AB} = m + n + mn$$

For small values of m and n, the term mn can be dropped to obtain an approximate result. Therefore, for relative errors of small value in two (or more) numbers, the relative error in their product is approximately equal to the algebraic sum of the relative errors in the numbers. Thus, if two numbers each have 3 percent errors, their product will be in error by 6 percent, using the approximate method. The actual percentage error is 6.09.

Raising a number to a power In this case, the relative error in the result is approximately equal to the relative error in the number *multiplied* by the power when the relative error of the number is not large.

Extracting a root of a number In this case, the relative error in the result equals the relative error in the number *divided* by the root where the relative error is not large.

2-6 REVISING COMPUTATIONS

The methods discussed above may be used to revise computations as required by small changes in specifications or data. Here, the change in the quantity may be considered as a *difference* between a new and an old value just the same as an error is a *difference* between a measured value and the correct value. In changed data, the effect of the change on the quantity being solved for depends on the mathematical operations involved (addition, multiplication, etc.), and the methods worked out above may be used to alter an old result without re-solving the equations involved.

As an example, the diameter of a solid circular shaft subjected to a torsional moment T may be determined by the use of the equation

$$d = \sqrt[3]{\frac{16T}{\pi s}}$$

where s is the allowable shearing stress. Assume that a diameter d has been calculated for a particular value of T and s. If, for example, the data are changed so that the torque T is increased 10 percent, the corresponding diameter of the shaft will be increased approximately $3\frac{1}{3}$ percent over that indicated by the original calculation.

2-7 THE DIGITAL COMPUTER

This is at once a high-speed calculator, a data-processing machine, a device that accepts thousands of instructions and executes them in correct sequence, and a device that can make decisions and check its own work.

Instead of using wheels and gears as in mechanical calculators, digital computers use electrons to accomplish the job and in short time. It has been stated that this type of computer can multiply two 10-digit numbers in the time it takes a jet plane to fly 1 in. Whether computers will eventually become highly creative is something that cannot be decided a priori, because we do not know the nature of creative thought.[1]

Any mathematical problem can be related to addition. Multiplication is a rapid form of addition, for instance, 3×4 is equivalent to $3 + 3 + 3 + 3$. Subtraction is the converse of addition, and division is the converse of multiplication.

The fundamental units of the digital computer are:[2]

Input To insert information into the computer
Memory To store information and make it available at the proper time
Arithmetic To perform the calculations
Output To deliver information
Control To coordinate all functions of the computer
Program To instruct the computer what to do

Examples of uses of digital computers in the mechanical engineering design field follow:

Computation As examples, a designer may wish to determine the deflections of a stepped shaft with complex loading; to determine contact pressure in a cam drive with roller follower and spring return; to determine bearing reactions in a bevel-gear drive; or to work out the design of a nozzle for a jet engine. Some computers have available a "library" of generalized mathematical routines, including solutions of simultaneous equations, matrix inversions, additions and multiplications, linear and nonlinear programming, and curve fitting. An addition is the algebraic language Fortran by which a designer can state a specific problem in

[1] D. M. Michael, Peace Research Institute, "The Challenge of Cybernation," Automation and Technological Change, Battelle Memorial Institute, Columbus, Ohio, The American Assembly Report, May 9–11, 1963.

[2] P. Siegel, "Understanding Digital Computers," John Wiley & Sons, Inc., New York, 1961.

algebraic language which the computer can readily understand and to which it can respond.[1]

Preparation of tables As an example, a manufacturer may use stock gears to make up complex gear trains to suit a customer's order for a definite horsepower and overall gear ratio. By a long process of trial and error the designer can select a suitable set of gears to correspond to the desired input and output speeds. However, by the use of a digital computer, calculations can be quickly made for all possible gear combinations and sorted by the computer to obtain values for a tabular form from which a suitable selection of gears can be made to suit any set of speeds specified by the customer.

Equation generation The computer can be programmed to produce an equation relating the variables to the optimum results for use in the design of mechanical units for power production or utilization, or for chemical processes, for programming maintenance of manufacturing or chemical plants, for the layout of factories, ships, missiles, mechanical units or assemblies of units, and so forth without end. This has been called *spacial data processing.*

Simulation This technique enables engineers to evaluate alternative designs without the expense, time, and often risk of loss of life required to develop and test systems in real situations.

Two types of simulation currently used are the analog and digital. In analog simulation, electrical pulses are arranged to behave in a manner analogous to the behavior of the components in the real machine, so that the designer may experiment by analogy with different arrangements of parts, materials, pressures, angles, and other components, all without the necessity of making hardware.

In digital simulation, an electronic digital computer performs a series of step-by-step numerical manipulations according to a "program" of instructions. Any reasonable degree of accuracy can be secured and with great saving in labor and time.

Iterative design Most mechanical-design problems do not have a unique solution and require the trial-and-error method of solution. Such designs are usually carried out by making initial assumptions based on results of past experience. Frequently there are more quantities to be calculated than there are equations. Also, for each initial assumption

[1] D. N. Leeson and D. L. Dimitry, "Basic Programming Concepts and the IBM 1620 Computer," Holt, Rinehart and Winston, Inc., New York, 1962; C. B. Germain, "Programming the IBM 1620," Prentice-Hall, Inc., Englewood Cliffs., N.J., 1962; R. H. Pennington, "Introductory Computer Methods and Numerical Analysis," The Macmillan Company, New York, 1965.

there must be an equation or a check in terms of standard values, proportions, or range in values.

Such problems can be programmed for a digital computer, which will yield all the possible solutions. The selection of the best design is then an easy task. Examples are the design of a gear train and the design of a journal bearing.

The computer has taken a firm place in global academic, industrial, technical, humanistic, cultural, sociological, arts, and science areas. In science and engineering its future is virtually unlimited. Its extensive use will continue strongly in both small computer areas, in quasi-creative engineering, and in large computer centers which are being widely developed, with intercommunication with the small areas and with each other, and including time-sharing computers. Not only is manual drudgery reduced and time saved, but also the horizons of learning are being extended.[1,2] The authors feel, however, that, although the design engineer will derive untold benefit from the computer, he will never become its slave.

[1] Christopher Rand, Profiles—Center of a New World II, *The New Yorker*, April 18, 1964, p. 58 et seq.

[2] *The Scientific American*, vol. 215, no. 3 (whole issue on computers), September, 1966.

3 Stress and Deflection Analysis

3-1 THE DESIGN PROBLEM

A machine or machine member is usually designed to perform its function for a specified length of time (operating life). In many machines, the members must be able to resist external forces, called *applied loads,* and, in addition, they must satisfy rigidity requirements. The machine designer must reach decisions concerning the nature of the action in the member that may cause it to fail, in order to determine what quantity or quantities should be expressed in terms of the loads and dimensions of the member. This generally entails the consideration of the loads applied to a member and some quantity, such as stress,[1] strain, deflection, or energy which characterizes the action that may cause its failure. The action that initiates failure frequently is referred to as the *mode of failure* (discussed further in Chap. 5).

[1] Stress, in this book, means force per unit area, an artificial concept which is seldom measured. Strain means change in length per unit length and is, in general, the quantity measured experimentally.

To prevent failure of critical parts, the following design procedure is usually followed:[1]

1. Develop a suitable shape. This may require a kinematic analysis of the system.
2. Anticipate probable loads and environmental effects to be encountered by the part.
3. Establish and evaluate *criteria of behavior* related to the expected modes of failure, such as measures of the effects of loading and/or environment of the part (stress, strain, deflection, corrosion, etc.).
4. Select materials on the basis of their mechanical, physical, and chemical properties and their economy. The selection involves a comparison of the anticipated effects of loading and environment, expressed in terms of the appropriate criteria of behavior, and corresponding limiting characteristics or design criteria of the material for the particular condition of loading. The choice of design criteria may be influenced by the type of material (brittle or ductile), the nature of stress, or accepted design codes or standards.
5. Select final dimensions, tolerances, surface finish, etc.
6. Construct a prototype, test, evaluate the performance, and, if necessary, redesign the part in the light of test results.

The designer should realize that many of the quantities with which he must deal are not known accurately, such as induced stresses and deflections, properties of materials, and the inevitable inaccuracies in manufacture and assembly. Therefore, he must make assumptions, compromises, and decisions. This requires creative ability and ingenuity. In situations such as these, he learns that in the success or failure of his products "nature is the final arbiter."[2]

The design of a camshaft may serve as an example of a design problem where the induced stresses and deflections cannot be accurately determined. In this case, the forces on the shaft due to the action of a cam and its follower depend on the mass of the follower and frequently also on the spring force; on the speed of operation and cam profile; on the clearance between the cam and the follower, which may or may not be properly adjusted in service; and on the deflection and vibration characteristics of the shaft itself.

The stresses induced in the shaft may be kept within limits by using a large shaft, but space and weight limitations may require a small shaft;

[1] J. B. Hartman and R. E. Benner, Stress Analysis in Design, *Machine Design*, March, 1954; F. B. Seely and J. O. Smith, "Advanced Mechanics of Materials," pp. 3–7, John Wiley & Sons, Inc., New York, 1952.

[2] Quote from Mechanical Engineering staff at Massachusetts Institute of Technology.

also heat treatment is more effective on small parts and usually small parts cost less. On the other hand, allowable deflections and vibration characteristics may require a larger shaft. Then the bearings for the shaft should be considered, since their selection is influenced by the size of the shaft which in turn affects the loads on the shaft and bearings. To save weight, a hollow shaft may be considered but this may be too expensive for the price class of the machine appliance.[1]

If the shaft is too large for space requirements, the designer may consider using alloy steel instead of carbon steel, but the alloy steel and the required heat treatment may unduly increase the cost; also alloy steel and carbon steel have the same deflection properties and if deflection is important, which is frequently the case since it affects vibration characteristics and bearing performance, alloy steel is no better than carbon steel.

Thus it can be appreciated that the solution of this problem is more than substituting given values in an equation and using a slide rule for the result; instead it becomes an interesting experience of balancing several interrelated and generally conflicting characteristics and influences.

In the following articles in this chapter, the various types of stresses and deflections are discussed. Stresses may be divided into three classifications based upon time of load application: namely, static stresses, cyclic or fatigue stresses, and impact stresses. Static stresses are stresses that do not vary in magnitude, direction, or point of application. These stresses are induced by static loads. Some examples are static stresses due to dead weights, tightening-up loads on bolts, and constant pressure on the wall of a vessel.

Cyclic or fatigue stresses are stresses that vary in magnitude, direction, or both, and that are induced by varying loads acting on the member. Examples of cyclic stresses are those induced in gear teeth, in rotating shafts subjected to bending moments, and in vehicle suspension springs.

Impact or shock stresses are induced when the time of load application is less than about half the natural period of vibration of the member.[2] The loading may be caused by an impact, by a displacement motion of one part of the member relative to other parts, or by a force pulse. An impulse generally is caused by an impact that acts for a very short period of time—just sufficient to allow for a stress wave to travel at the velocity of sound through the material of the member and back to the source of the excitation. However, in the case of a force pulse, the force acts for an appreciable length of time.

In addition to classification according to the time of application of

[1] For an appliance recently placed on the market by a large manufacturing company, the first item in the specifications was: Retail selling price shall be $14.95.

[2] J. H. Faupel, "Engineering Design," pp. 556–557, John Wiley & Sons, Inc., New York, 1964.

the load, stresses may be classified in relation to the nature of loading: for example, axial stresses, bending stresses, torsional stresses, etc.

3-2 DEVELOPMENT OF STRESS FORMULAS

In order to apply stress formulas in a rational manner, it is important that we achieve a basic understanding of the methods for determining the relation between the loads on the member and the stresses at any given point in the member. Both the *strength of materials* and the *classical theory of elasticity* are available methods; however, the former method will be emphasized in this book. From either method, formulas are obtained which relate stresses in the member to the load acting on the member, the size of the member, and, to a certain degree, the shape of the member. The major steps in the strength-of-materials method are as follows.

1. Develop equations of equilibrium or equations of motion involving the external forces acting on some portion of the member and the internal forces at critical sections through the body. At this section, usually a plane is passed through the body and the portion of the body lying on one side of this plane is removed and replaced by the force system exerted at the cut section. Draw a free-body diagram showing all the original external forces on the portion and the forces at the cut section. The latter forces usually are divided into normal forces (acting normal or perpendicular to the section) and shearing forces (acting in the plane of the cut section). These forces are expressed as the product of the stress at a point times a differential area of the section; thus a normal force at the section is expressed as $\sigma\, da$† and a shearing force as $\tau\, da$. The equations of equilibrium will contain expressions of the form $\int \sigma\, da$, $\int \tau\, da$, or both, and $\int \sigma y\, da$ (equilibrium of moments); however, they cannot be evaluated until the stress distribution over the area of the cut section is known or assumed.

2. Determine the stress distribution in the vicinity of the cut section. Obtain the variation of strain of the elements of the body by experimental technique (strain gauges), or assume the strain distribution is based upon the deformations of the body as a whole. In general, the strain distribution for the body as a whole is considered to be the same

† In this book, symbols for stresses and strains are used as follows: In derivations of equations, σ represents normal stress, τ represents shearing stress, and ϵ represents normal strains. This nomenclature corresponds with usual practice in theoretical mechanics. However, in mechanical-design calculations in engineering offices, the symbols usually used are s for normal stress (with subscript t for tension or c for compression) and s_s for shearing stress. This nomenclature is used in this book in the design form of equations.

as that for the cut section (unless stress concentration is present; see Chap. 4). Although the distribution of stresses can be determined by the photoelastic method, the stresses in metallic parts seldom can be measured directly; therefore, one must rely on the measurements of displacement (or deformation) and strains and then express strain as a function of stress. This implies the application of Hooke's law, namely, $\sigma = E\epsilon$ (on the assumption that there is only one principal stress, the elastic limit is not exceeded, and the material is isotropic). The modulus of elasticity, E, is a constant for the material. It may be necessary in certain cases to test specimens of the material under increasing load in order to obtain the relation between stress and strain for the material.

3. Substitute the expression for stress distribution into the equilibrium equations obtained in the first step. Solve the resulting equations for stress in terms of external loads and dimensions of the body.

The method of the classical theory of elasticity follows a similar sequence of steps and differs from the strength-of-materials method only in the manner in which these steps are carried out. In the theory-of-elasticity method, equilibrium equations (in the form of partial differential equations) are written for an elemental cube from the interior of the body.[1] In addition, equations expressing the boundary conditions must be satisfied. These boundary conditions are mathematical statements that the components of stress at the boundary must equal the components of the surface forces at the boundary (external loads per unit area of boundary surface).

In addition to these equations, a series of *compatibility equations* are developed which interrelate the various strains. By satisfying these equations, one ensures that the displacement components can be expressed by a set of single-valued continuous functions and that the body remains a continuous one (no discontinuities or voids) after deformation. The compatibility equations, which are general in nature, apply to many types of problems and, unlike the strength-of-materials method, the prediction or measurement of strain distribution is not necessary in the theory of elasticity. The principal drawback to the application of the theory-of-elasticity method is the mathematical complexity encountered in the solution of real problems.

In the application of either of these methods, the following assumptions are made regarding the properties of the material:

1. *Perfect elasticity.* The material returns to its original form immediately upon removal of the external loads.

[1] S. P. Timoshenko and J. N. Goodier, "Theory of Elasticity," 2d ed., McGraw-Hill Book Company, New York, 1951.

2. *Isotropy.* The elastic properties of materials are the same for every direction and position in the body.

3. *Homogeneity.* The material is continuously distributed through its volume.

4. *Adherence to Hooke's law.* The stress-strain relations have the linearity determined from Hooke's law. As a result of its application, the method of "superposition" may be employed in determining stresses and deflections in elastic bodies. According to this method, the actual loading may be considered to be composed of two (or more) systems of loads. Each system of load results in stresses and deflections independent of the other system of load. The actual effect is then the summation of the effects of the two (or more) systems of loads.

3-3 STRESSES DUE TO AXIAL LOADING

The strength-of-materials method discussed in Art. 3-2 is employed in obtaining the stress formula for axially loaded members as follows:

Tensile stress The average tensile stress induced in the body of a simple tension member as shown in Fig. 3-1 is

$$s = \frac{P}{A} \tag{3-1}$$

where s = average stress, psi
P = axial load, lb
A = cross-sectional area, in.^{2}

Compressive stress By reversing the directions of the loads in Fig. 3-1, a compressive stress will be induced in the member. Equation (3-1) may be used to determine this stress, unless the member is so slender that it is classified as a column, in which case stability considerations will determine failure. This type of failure is treated in Art. 3-7.

In the derivation of the equation $s = P/A$ it is assumed that the axial stresses as well as the axial strains are equal for every fiber at the section being considered. In order to fulfill this assumption the following conditions or restrictions must be satisfied:

1. The longitudinal part of the member under consideration must be straight.

2. The member must be axially and centrally loaded.

3. That portion of the member being considered must have a uniform cross section since an abrupt change in section, such as a notch, would induce large localized strains at the roots of the notch or groove.

4. The stress must be computed for a section well removed from

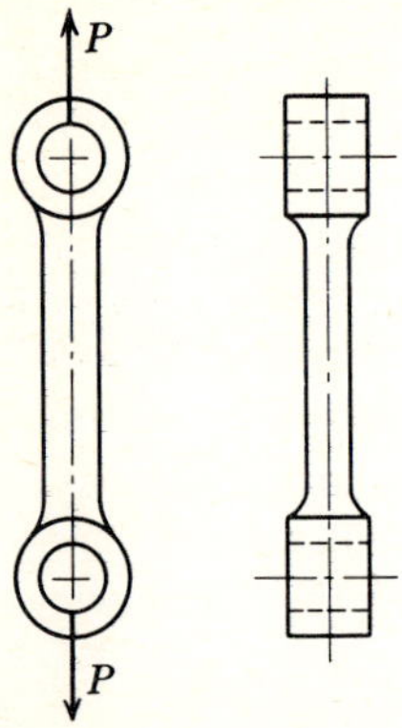

Fig. 3-1 Tension member.

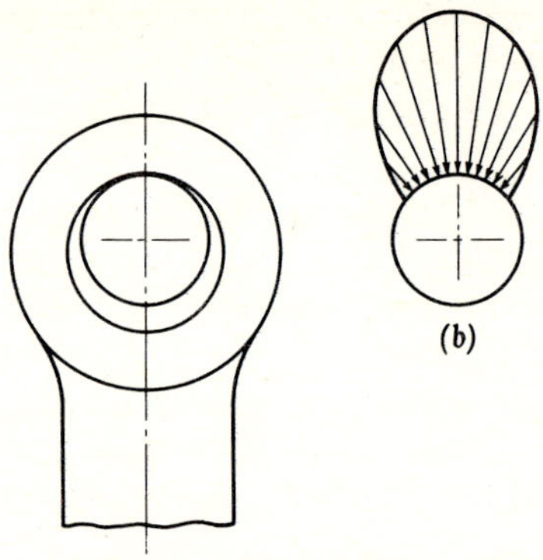

Fig. 3-2 Distribution of bearing pressure by pin connection.

the section on which the load is applied since the stresses in the vicinity of the contact areas are highly concentrated. This topic will be discussed further in Chap. 4.

5. The load must be a gradually applied load, with no impact.

6. The stress computed by the formula P/A does not represent the actual stress present if the member contains residual stress. See Chap. 6 for further discussion.

Bearing pressure There is a particular type of local compression that requires special attention. This is known as "bearing pressure," or "contact pressure," and occurs between two members held in contact. If the load is applied to the member in Fig. 3-1 by pins, the distribution of pin pressure will not be uniform but will be in accordance with the shape of the surfaces in contact and the deformation characteristics of the two materials. The distribution of pressure will be similar to that shown in Fig. 3-2(*b*).

Since the actual distribution of pin pressure is difficult to determine accurately, the calculation of bearing pressure usually consists of dividing the load by the "projected bearing area" of the pin, i.e.,

$$s = \frac{P}{td} \tag{3-2}$$

where P = load on pin, lb
t = length of pin in contact, or thickness of link, in.
d = diameter of pin, in.

The pressures as calculated by Eq. (3-2) from results of tests on links and pins or on riveted connections may be used to determine allowable pressures in design.

3-4 NORMAL STRESSES DUE TO BENDING

The normal stresses induced in straight beams by bending are a particular distribution of tensile and compressive stresses. For a straight beam the following equation applies:

$$s = \frac{My}{I}$$

where s = stress at distance y from neutral axis, psi
I = moment of inertia of cross section about neutral axis, in.^{4}
M = applied bending moment, lb-in.

The above equation gives the maximum tensile stress when it is evaluated for the maximum value of y on the tension side of the specimen. This value of y may be denoted as c_t. The maximum compressive stress is similarly found by using the maximum value of y on the compressive side as c_c. For symmetrical beam sections, $c_t = c_c$, which may be denoted as c (see Fig. 3-3). Hence, in design nomenclature,

$$s_t = s_c = \frac{Mc}{I} = \frac{M}{I/c} \qquad (3\text{-}3)$$

The quantity I/c is known as the "section modulus" and has the units of in.^{3}

During bending, a straight beam is considered to behave so that plane sections remain plane and hence unit fiber strains and stresses are proportional to the distance of the fiber from the neutral surface. The following assumptions must be satisfied.

1. The beam is straight or nearly so. If this were not the case, the fiber strains would not be proportional to distances from the neutral surface.
2. The cross section is uniform. Plane sections do not remain plane at or near changes in section.
3. Those forces acting on the beam must be perpendicular to the

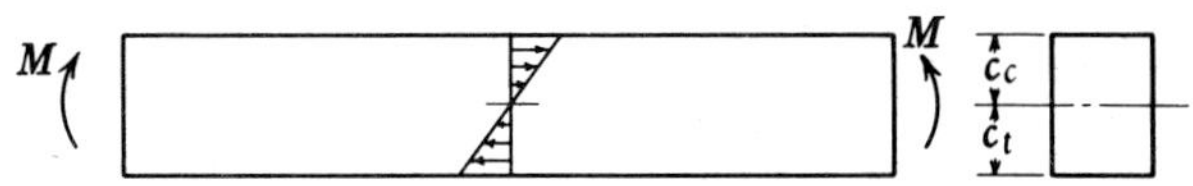

Fig. 3-3 Distribution of bending stress in straight beam.

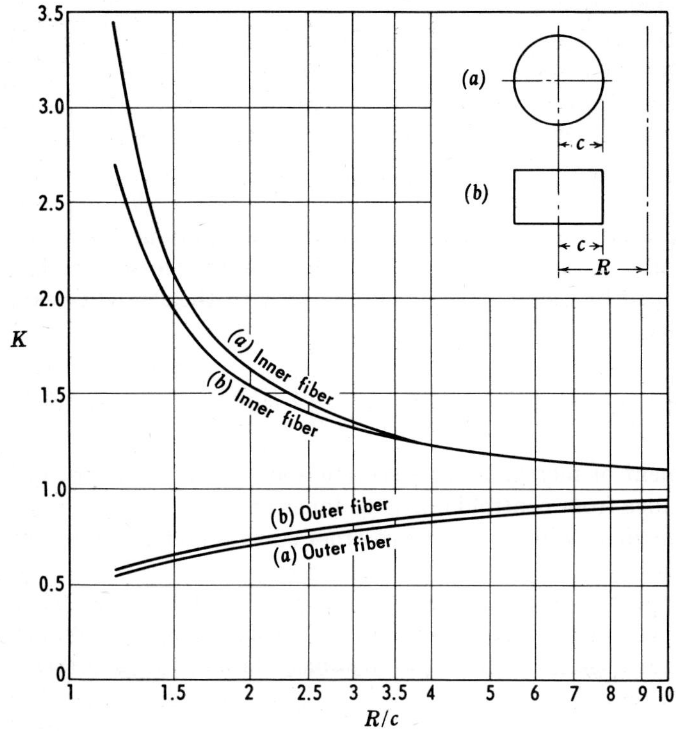

Fig. 3-5 Factor K for curved beam of circular and rectangular cross section.

Torsion Torsional moments induce shear stresses on cross sections normal to the axis of bars and shafts (see Fig. 3-7). For circular shafts, the shear stress at any point a distance r from the center is given by

$$\tau = \frac{Tr}{J} \tag{3-7}$$

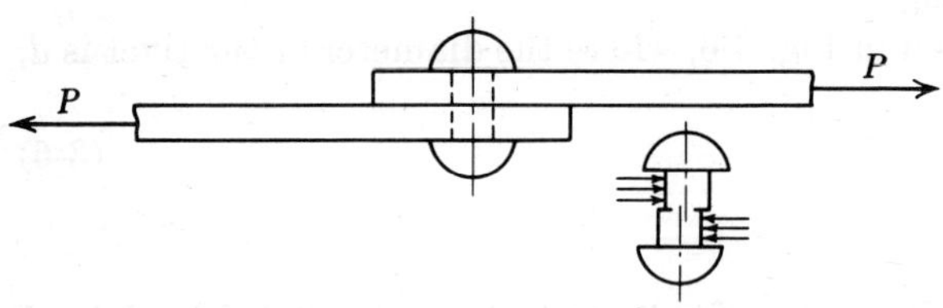

Fig. 3-6 Rivet connection and shear deformation.

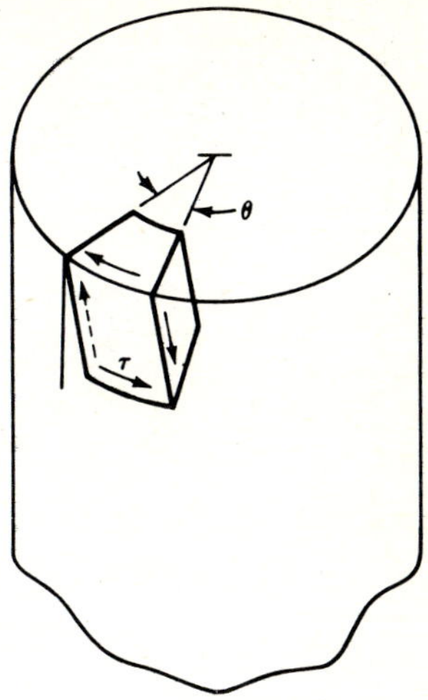

Fig. 3-7 Shear stress due to torsion.

where τ = shear stress, psi
r = distance from center of shaft to point of stress, in.
J = polar moment of inertia, in.^{4}
= $\pi d^4/32$ for solid circular shaft where d is diameter, in.
= $\pi(d_o^4 - d_i^4)/32$ for hollow circular shaft where d_o and d_i are outside and inside diameters, respectively, in.

It follows that the maximum shear stress for a solid circular shaft occurs at the outside fiber and is given by the formula

$$s_s = \frac{16T}{\pi d^3} \tag{3-8}$$

Where weight must be a minimum, it is common practice to use hollow shafts. This removes material from the center of the shaft where the stresses are low, and it also improves the heat-treating characteristics.

In applying Eq. (3-7), the following limitations must be considered:

I. The shearing stress on any cross section normal to the axis varies directly as the distance from the center of the section. Therefore, a plane transverse section must remain plane. This implies:
   A. The transverse cross sections of a bar or shaft must be circular.
   B. That portion of the shaft being considered must be free from abrupt changes in cross section.
   C. The sections under consideration must be well removed from the point of load application.

II. The shaft must have no residual stresses.

III. Impact loads are excluded.

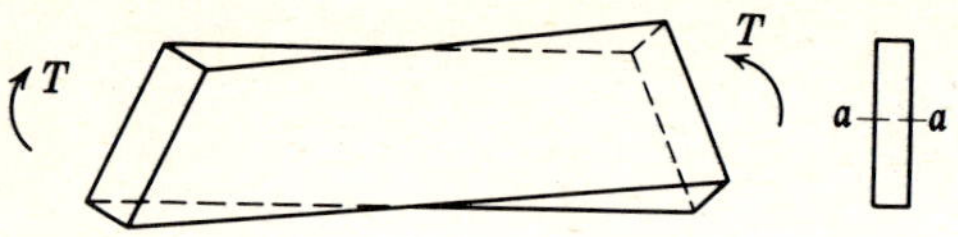

Fig. 3-8 Rectangular bar in torsion. Maximum shear stress occurs at *a*.

Bars in torsion with noncircular cross sections have a complicated stress distribution for whose determination the theory-of-elasticity method is essential.[1] The fact that the cross section is noncircular means that a plane transverse section before twisting becomes a warped surface after twisting. Accompanying the warping is an increase of shearing stress at some locations on the section and a decrease at other points, as contrasted to the stress distribution on the section if no warping occurs.

The theory of elasticity shows that the maximum stresses in a bar of rectangular cross section occur at the midpoints of the long sides, as indicated at a in Fig. 3-8.

The maximum shear stress for bars with a rectangular cross section may be determined by means of the formula[2]

$$s_s = \frac{T}{\alpha b t^2} \tag{3-9}$$

where T = torque on bar, lb-in.
b = breadth of rectangular section, in.
t = thickness of rectangular section, in.
α = coefficient depending on ratio b/t (see Table 3-1)

For long and narrow cross sections, the value of α is approximately

[1] Timoshenko and Goodier, *op. cit.*, p. 258.
[2] *Ibid.*, p. 278.

Table 3-1 Stress coefficients for rectangular bars

b/t	α
1.00	0.208
1.50	0.231
2.00	0.246
3.00	0.267
4.00	0.282
6.00	0.299
8.00	0.307
10.00	0.313
∞	0.333

⅓; hence for $b/t \geqq 10$

$$s_s = \frac{3T}{bt^2}$$

Shear stresses due to bending In addition to normal stresses induced by bending of a beam, transverse shearing stresses are induced between the elements or fibers, provided the bending moment varies along the length of the beam. According to the strength-of-materials method,

$$\tau = \frac{V}{Ib}\int_z^c y\,dA = \frac{VQ}{Ib} \tag{3-10}$$

where τ = shear stress, psi
I = moment of inertia of cross section about neutral axis, in.^{4}
V = shearing force at beam section under consideration, lb
b = beam width at section, in.
$Q = \int_z^c y\,dA$ = moment of area of element about neutral axis, in.^{3}
z = location where shear stress is desired, in.
(See Fig. 3-9.)

From Eq. (3-10), it is concluded that the shearing stress is zero at the upper surface of the beam ($z = c$) and is maximum at the neutral axis ($z = 0$). The maximum shearing stresses for some common shapes of beams are as follows:

$$\tau = \frac{3V}{2A} \qquad \text{for rectangular cross section}$$

$$\tau_{\max} = 2\frac{V}{A} \qquad \text{for solid circular cross section}$$

$$\tau_{\max} = 2\frac{V}{A} \qquad \text{for thin-walled circular-tube cross section}$$

where A is the cross-sectional area of the beam, in square inches.

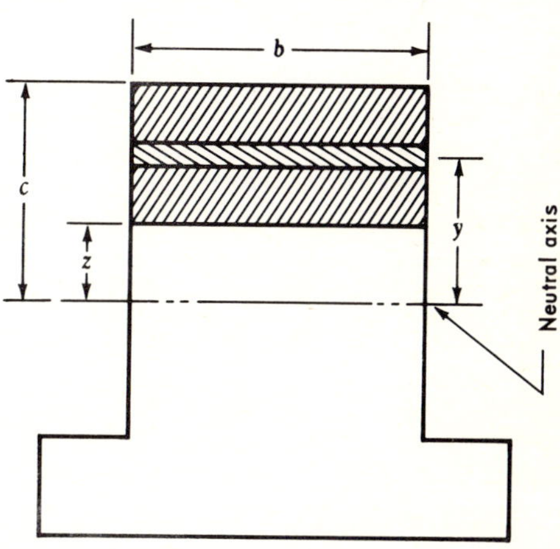

Fig. 3-9 Beam cross section.

The derivations of the above formulas have been based upon the assumptions that the shear stresses are uniformly distributed across the beam width and that the direction of shear stresses throughout the cross section is the same as the direction of the shear force.

3-6 COMBINED STRESSES

Stresses due to eccentric loading If the short prismatic bar in Fig. 3-10 is loaded by a compressive force collinear with its longitudinal axis, the stress on cross sections of the bar remote from the points of application of the loads will be uniformly distributed and Eq. (3-1) will apply.

When the point of application of the load is displaced a distance e along a principal axis of the bar, as shown in Fig. 3-10(*b*), the stress on cross sections of the bar will not be uniformly distributed. As shown at (*c*), two forces P_1 and P_2 equal in magnitude to P may be introduced without altering the equilibrium of the forces on the bar. It may be noted that the force P_1 will induce a uniformly distributed stress whose value is equal to P_1 (or its equal P) divided by the cross-sectional area A. The couple equal to Pe will bend the bar, and this will induce a compressive stress at n and a tensile stress at m. By using the principle of superposition, the direct compressive and bending stress may be added at n and subtracted at m. Thus

$$s_c = \frac{Pec_c}{I} + \frac{P}{A} \tag{3-11}$$

$$s_t = \frac{Pec_t}{I} - \frac{P}{A} \tag{3-12}$$

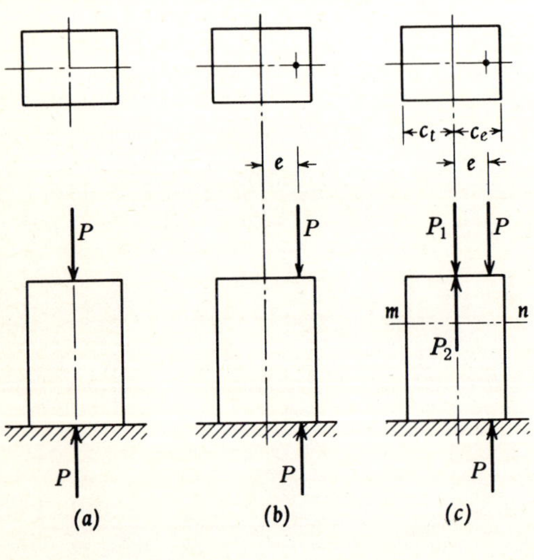

Fig. 3-10 Axial and concentric loading.

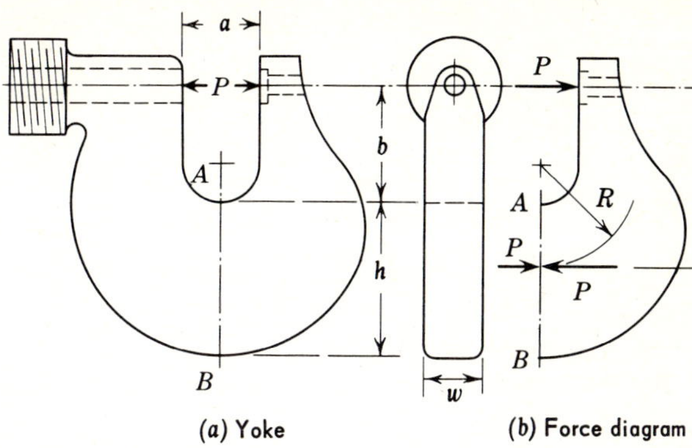

Fig. 3-11 Yoke for hydraulic riveter.

where s_c = maximum compressive stress, psi
s_t = maximum tensile stress, psi

When the loading on the members is *tension*, the equations above may be used by interchanging the subscripts c and t.

Example 3-1 *Curved beam design.* Figure 3-11 shows the yoke of a portable hydraulic riveter. The yoke screws on a riveting head (not shown), which carries the hydraulic cylinder, handle, and manual control valve and is connected to the pressure pump by a hydraulic hose. These units have capacities of 10 to 30 tons and weigh 10 to 15 lb. To keep the size of the unit small for accessibility to the work and its weight low for ease of handling, it is necessary to make the yoke of alloy steel. Since each head should accommodate yokes of several sizes, it is customary for convenience in mounting the yokes on the head to make all yokes the same width, dimension w.

Data: Capacity P = 10 tons; gap a = 2 in.; reach b = 2 in.; width of yoke w = 1½ in. Assume SAE 2340 steel. This steel in 1½-in. thickness and oil-quenched and tempered at 800°F has an ultimate strength of 180,000 psi and a yield point of 165,000 psi. Let us assume an allowable stress equal to 75,000 psi. This value compares favorably with calculated induced stresses in existing riveters. The design problem is to determine the depth of the yoke h.

Solution By introducing at the gravity axis of the section two forces each equal in magnitude to P and opposite in direction, as shown in Fig. 3-11(*b*), one of these forces and the original force form a bending moment on the section equal to $P \times e$. This moment induces tensile stress in bending at A and compression at B. The remaining force induces a uniform tensile stress over the section and adds to the tensile stress due to bending at A to produce a total tensile stress at A, as given by Eq. (3-11) for tension loading with allowance for the effect of curvature from Eq. (3-5) as follows:

$$s_t = \frac{P}{A} + \frac{KPe}{I/c} = \frac{P}{wh} + \frac{6KPe}{wh^2}$$

In this equation, s_t can be made equal to the allowable stress, P and w are known, and h is the unknown quantity on which K and e depend. The equation is in quadratic form so that a direct solution for h is not convenient.

Let us simplify the above equation as much as possible, as follows:

$$\frac{s_t}{P} = \frac{1}{wh} + \frac{6Ke}{wh^2}$$

By substituting values for s, P, and w, and simplifying, the following form is reached:

$$h^2 - 0.178h = 1.07Ke$$

This equation could be solved by assuming values for h and solving by trial but this method would be laborious and not enlightening. It is helpful to note from Fig. 3-5 that, if we obtain an approximate value for h, K will not change markedly for new values of h. Let us assume a reasonable value for h and use that value to determine K and e, then solve the equation for the unknown quantity h. One might expect h to be, say, 2 in. The following values may then be found: $e = 3$ in., $R = 2$ in., $R/c = 2$, and, from Fig. 3-5, $K = 1.6$. Therefore

$$h^2 - 0.178h = 1.07Ke = 1.07 \times 1.6 \times 3 = 5.14$$

Solving this equation directly for h, we find that

$$h = \frac{0.178 \pm \sqrt{(0.178)^2 + 4(5.14)}}{2}$$

$$h = 2.358 \text{ in.}$$

This value is larger than the assumed value of h; therefore, let us increase the value to, say, $2\frac{1}{2}$ in. and solve again.

For $h = 2\frac{1}{2}$ in., $e = 3.25$, $R = 2.25$, $R/c = 1.80$, and $K = 1.65$. Then

$$1.07Ke = 1.07 \times 1.65 \times 3.25 = 5.74$$

$$h = \frac{0.175 + \sqrt{(0.178)^2 + 4(5.74)}}{2}$$

$$h = 2.48 \text{ in.}$$

This value is slightly less than the assumed value; hence, h will be specified as $2\frac{1}{2}$ in.

From the initial equation for the maximum stress in the yoke, it is evident that, if the dimensions of the section are known, it is relatively easy to solve for the stress. This is an analysis problem. But the design problem is one of determining the dimensions of the member, using a reasonable value for the allowable stress. In many cases, a direct solution is not possible, but the problem can be solved readily by devising a procedure similar to that used in Example 3-1 or by the use of a digital computer.[1]

[1] Daniel D. McCracken, "A Guide to Fortran IV Programming," John Wiley & Sons, Inc., New York, 1965.

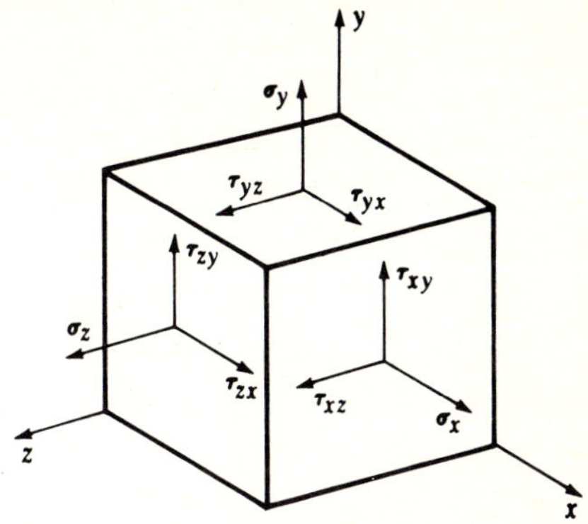

Fig. 3-12 Stress components on an element.

Determination of principal stresses—Mohr's circle diagram The several induced stresses, such as axial, bending, torsion, and direct shear, are either (1) normal stress, i.e., the stress normal to the area on which it acts, or (2) the shear stress, which acts along the area. Whatever the aspect of the stress at a point may be, it can always be expressed in terms of normal stresses and the shear stresses, as shown in Fig. 3-12, in which

$\sigma_x, \sigma_y, \sigma_z$	normal stresses
$\tau_{yx} = \tau_{xy}$	
$\tau_{yz} = \tau_{zy}$	shear stresses
$\tau_{zx} = \tau_{xz}$	

Many design problems deal with systems in which several types of loading exist simultaneously, such as bending and torsion stresses in a shaft.

In general, most stress systems encountered in design fall within the designation of plane stress, which is defined as stress distributions such that, of the six possible components of stress indicated in Fig. 3-12, all components in one direction (that is, τ_{xz}, τ_{yz}, and σ_z) are equal to zero. Since many design criteria are expressed in terms of the maximum and minimum values of normal stress, also designated as principal stresses, it is necessary to consider methods of determining these stresses from the combinations of stresses induced by the separate loads.

Equations will be derived that express the principal stresses and the maximum shear stress at critical points in terms of known stresses on orthogonal planes through any point. These are the equations for the plane-stress situation shown in Fig. 3-13(*a*) in which the plane *MN* represents the trace of any plane through the point, all planes being normal to the *plane* of zero stress, that is, the paper. The following sign convention is used in the derivations: The angle θ is considered positive in a counterclockwise direction, tensile stresses are positive, and shearing

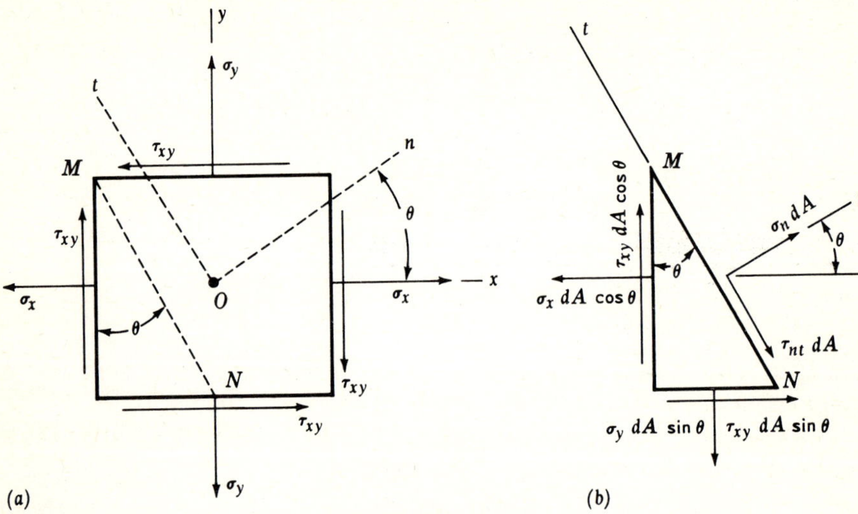

Fig. 3-13 Stress and force components on an element.

stresses that form a clockwise couple (the vertical shearing stresses in this case) are positive.

Figure 3-13(*b*) is a free-body diagram of a wedge-shaped element whose inclined face MN has an area dA, the vertical and horizontal faces of which have areas of $dA \cos\theta$ and $dA \sin\theta$, respectively.

By writing a force equation for equilibrium in the n direction, and dividing by dA, we obtain

$$\sigma_n = \sigma_x \cos^2\theta + \sigma_y \sin^2\theta - \tau_{xy} \sin 2\theta$$

or

$$\sigma_n = \frac{\sigma_x + \sigma_y}{2} + \frac{\sigma_x - \sigma_y}{2} \cos 2\theta - \tau_{xy} \sin 2\theta \tag{3-13}$$

In a similar manner, the following equation can be obtained for the shear stress in the t direction:

$$\tau_{nt} = (\sigma_x - \sigma_y) \sin\theta \cos\theta + \tau_{xy}(\cos^2\theta - \sin^2\theta)$$

$$\tau_{nt} = \frac{\sigma_x - \sigma_y}{2} \sin 2\theta + \tau_{xy} \cos 2\theta \tag{3-14}$$

The direction of the principal stresses (maximum and minimum values) may be found by differentiating σ_n with respect to θ, setting the derivative equal to zero, and solving for θ. This gives

$$\tan 2\theta_{1,2} = -\frac{\tau_{xy}}{(\sigma_x - \sigma_y)/2} \tag{3-15}$$

(Note that θ_1 and θ_2 are 180° apart.)

Substituting expressions for the sine and cosine of the angle $2\theta_{1,2}$ into Eq. (3-14) gives the principal stresses:

$$\sigma_1 = \frac{\sigma_x + \sigma_y}{2} + \sqrt{\left(\frac{\sigma_x - \sigma_y}{2}\right)^2 + \tau_{xy}^2} \tag{3-16}$$

and $$\sigma_2 = \frac{\sigma_x + \sigma_y}{2} - \sqrt{\left(\frac{\sigma_x - \sigma_y}{2}\right)^2 + \tau_{xq}^2} \tag{3-17}$$

in which σ_1 is the maximum principal stress corresponding to θ_1, and σ_2 is the minimum principal stress corresponding to θ_2.

In a similar manner, the maximum shear stresses may be obtained from Eq. (3-14). The result is

$$\tau_{\max} = \pm\sqrt{\left(\frac{\sigma_x - \sigma_y}{2}\right)^2 + \tau_{xy}^2} \tag{3-18}$$

These equations can be represented graphically by a diagram known as "Mohr's circle," shown in Fig. 3-14. This diagram is constructed so that the coordinates of each point on the circle represent the normal and shearing stresses on one plane through a stressed point. Normal stresses are plotted as horizontal coordinates, with tensile stress considered positive and compressive stress negative. Shearing stresses are plotted vertically, with shearing stresses that produce a clockwise couple considered positive and those that give a counterclockwise couple considered negative.

To draw the diagram for the stress situation of Fig. 3-14(b) (in which the stresses were calculated from the loading on the member), the points representing the given stresses are plotted [see points A and B in Fig. 3-14(a)]. The coordinates of point A are the stresses on the vertical plane (x face, perpendicular to the x axis), and the coordinates of point B are the stresses on the horizontal plane of the element. Draw the line AB (diameter of Mohr's circle) and use the point where AB intersects the σ axis (point C) as the center of a circle of radius AC.

The stresses on an element oriented at an angle θ may be determined by laying off an angle 2θ from CA in the same direction as the angle θ is measured from the x axis of the element, as in Fig. 3-14(a) and (c). The diameter DE is then determined. The coordinates of point D represent the stresses σ_n and τ_{nt}. The stresses at point D, as calculated by means of Mohr's circle, are the same stresses as given by Eqs. (3-13) and (3-14), and the stresses at point E correspond to those stresses for the surface of the element whose normal lies at an angle of $90° + \theta$ from the x axis.

The maximum normal stress σ_1 will be equal to OF, the minimum normal stress σ_2 will be equal to OG, and the maximum shearing stress τ_{max} will be equal to CH. These relations can be readily verified by comparison with Eqs. (3-16) to (3-18).

Example 3-2 Determine the stresses acting on an element at A in Fig. 3-15(a). Calculate the principal stresses at A. Show the stresses acting on the element in each case.

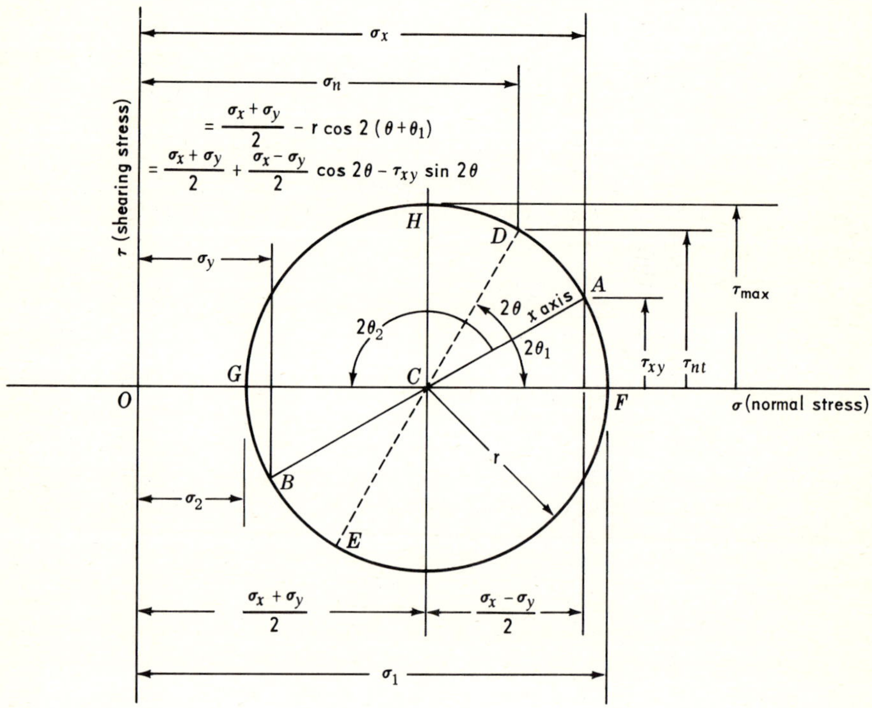

(a) Mohr's circle

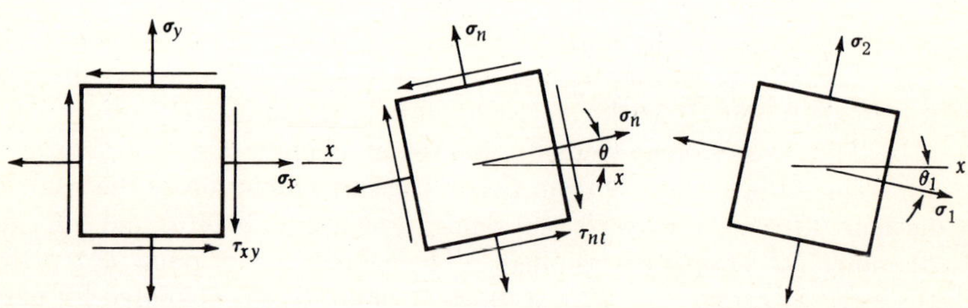

(b) Given state of stress ($\sigma_x > \sigma_y$) (c) Stresses on element at angle θ (d) Principal stresses

Fig. 3-14 Mohr's circle.

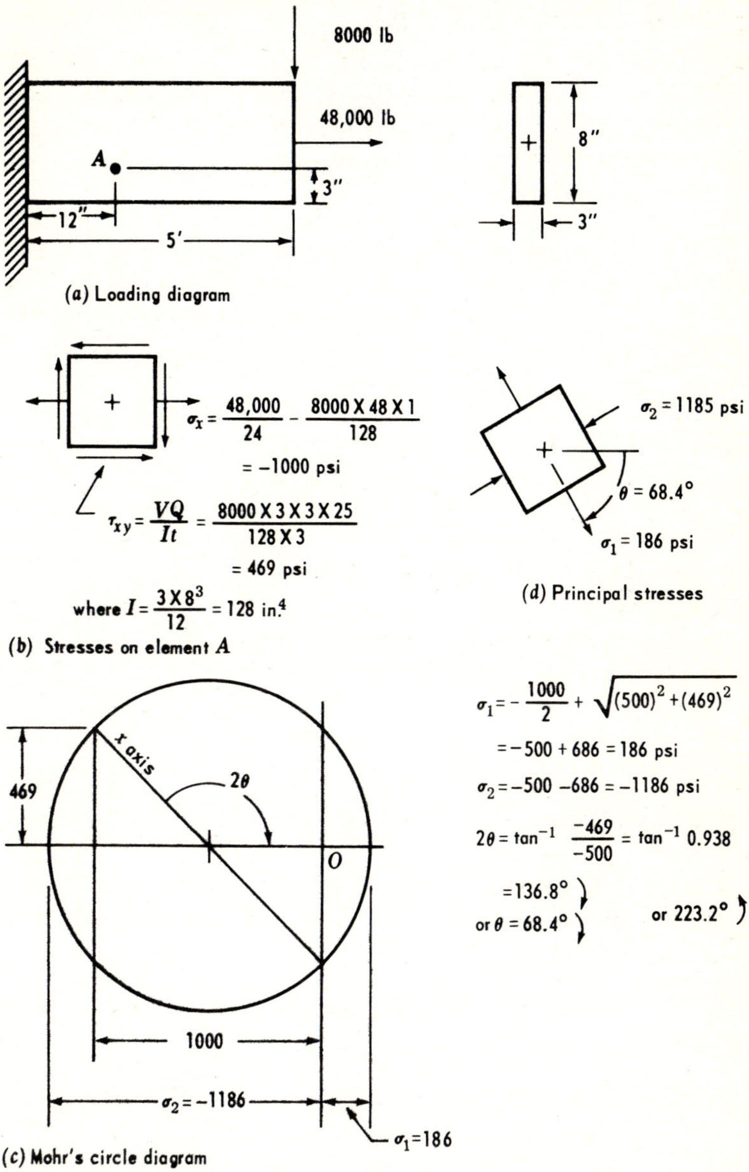

Fig. 3-15 Stresses in a cantilever.

3-7 INSTABILITY CONSIDERATIONS

Short compressive members subjected to centrally applied loads may be analyzed or designed on the basis of direct compression. If, however, the member has a length greater than four to six times the least dimension perpendicular to its axis, it is classed as a column or strut, and failure may be caused by *buckling*.

Theoretical equations for the determination of the buckling load of columns were first developed by Euler. The Euler formula for the buckling load, or critical load, of a long column was derived on the assumption that the column bows sideways while the stresses are within the elastic limit. This type of failure is the result of *elastic instability*.

If the column is of less slender proportions, the maximum stress may reach the yield point before sideways bowing occurs; hence, the Euler formula does not predict the critical load. This type of failure is the result of *plastic instability*. A rational equation for the critical load has not been developed for this type of failure; hence, formulas based on experimental results must be used for columns that fail by this method. A formula that gives a reasonable check on test results for columns of this type is the J. B. Johnson formula.

A third type of failure for columns made of thin sections, such as tubes, or for built-up columns is *local buckling*.

Columns The Euler and the J. B. Johnson formulas are as follows:

The Euler formula:

$$F_{\text{crit}} = \frac{n\pi^2 EI}{L^2}$$

The J. B. Johnson formula:

$$F_{\text{crit}} = As_y\left(1 - \frac{s_y L^2}{4n\pi^2 E\rho^2}\right)$$

where F_{crit} = critical load causing failure, lb
A = cross-sectional area, in.^{2}
I = moment of inertia of area, in.^{4}
L = length of column, in.
ρ = least radius of gyration of cross section, in.
n = end-fixity coefficient (see Fig. 3-16)
E = modulus of elasticity of material, psi
s_y = yield point of material, psi
$B = s_y L^2/n\pi^2 E$

In the design of a column or strut, the length is usually known but, since the dimensions of the cross section are not known initially, one cannot determine which equation, Euler's or J. B. Johnson's, should be used.

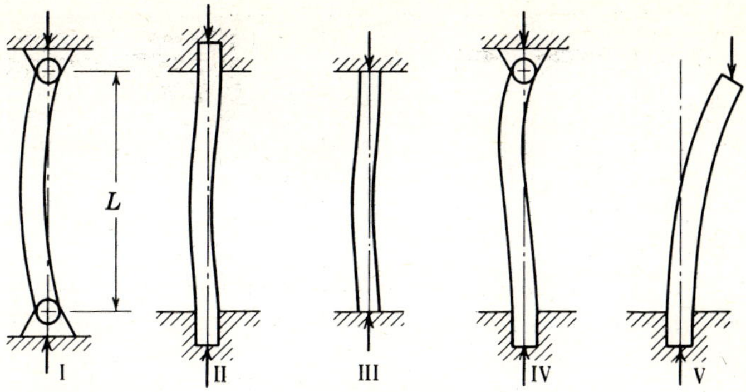

CASE:	n
I Round ends or pin ends, free but guided	1
II Fixed ends	4
III Flat ends	1 to 4
IV One end fixed, other free but guided	2
V One end fixed, other free	0.25

Fig. 3-16 End-fixity coefficient for columns.

As an aid in expressing the range in which the equations apply, and in checking the results, the dimensionless quantity B/ρ^2 is useful in which

$$B = \frac{s_y L^2}{n\pi^2 E}$$

The Euler and J. B. Johnson equations may be rewritten by introducing B and ρ:

The Euler formula (for $B/\rho^2 > 2$):

$$F_{\text{crit}} = \frac{n\pi^2 AE}{(L/\rho)^2} = \frac{s_y A \rho^2}{B} \tag{3-19}$$

The J. B. Johnson formula (for $B/\rho^2 < 2$):

$$F_{\text{crit}} = A s_y \left(1 - \frac{B}{4\rho^2}\right) \tag{3-20}$$

If these two equations for F_{crit} are equated, it will be found that $B/\rho^2 = 2$. For that condition, the column would be at the point of simultaneous failure by elastic and plastic instability; that is, for values of B/ρ^2 less than 2, failure would be predicted by the J. B. Johnson equation, whereas for B/ρ^2 greater than 2, the Euler formula applies. These ranges are shown in Fig. 3-17.

Most struts used in machinery are of proportions in the J. B. Johnson range rather than the Euler range, so that, unless it is evident that

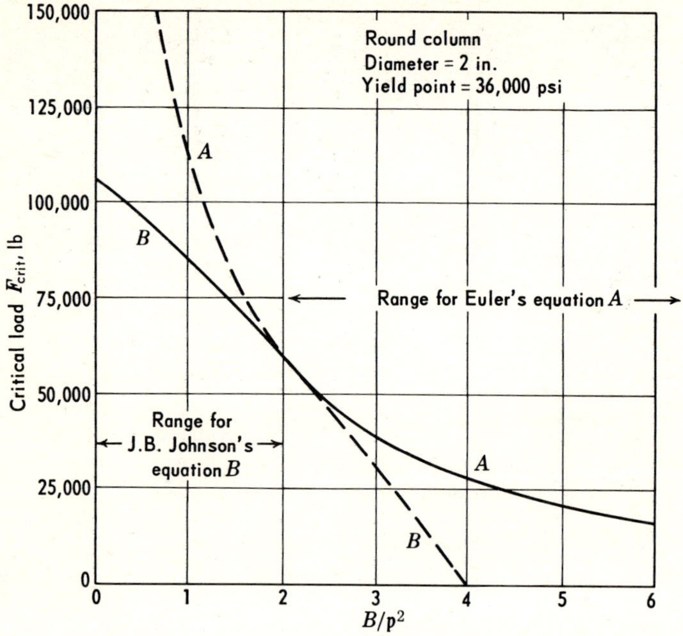

Fig. 3-17 Ranges for Euler's and J. B. Johnson's equations.

the column will be long and of small cross section, the J. B. Johnson equation should be tried first and the assumption checked by the criterion of B/ρ^2 after the cross-sectional dimensions have been determined. The design of columns will be treated in Chap. 5.

Beam columns Members required to withstand axial loads and transverse loads simultaneously are called *beam columns.* Examples of beam columns are an eccentrically loaded strut, as shown in Fig. 3-18, and an end-supported beam carrying a weight, as shown in Fig. 3-19.

For the eccentrically loaded strut shown in Fig. 3-18, it is apparent

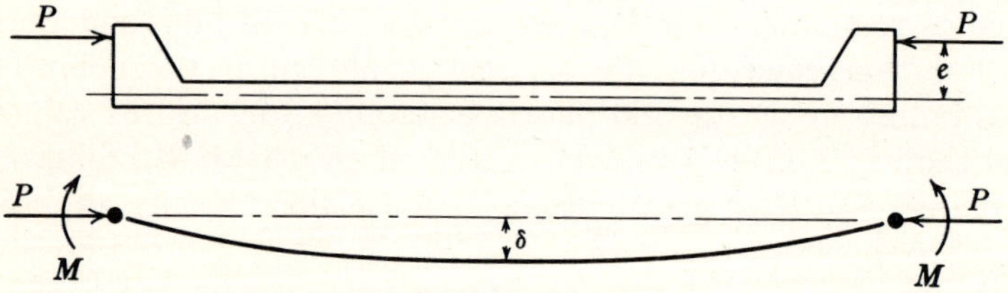

Fig. 3-18 Eccentrically loaded strut.

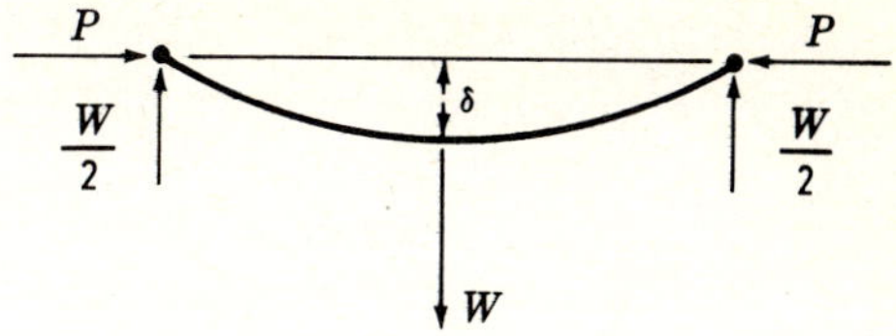

Fig. 3-19 Beam with transverse and axial loading.

that the moment at the ends, Pe, will cause transverse deflection of the strut. The axial compressive load P will then have a line of action eccentric to the strut by an amount δ at the center. This force and distance in turn induce an additional bending moment on the beam. The magnitude of the moment therefore depends on the deflection of the structure, which in turn is related to the loading. This interrelation makes the determination of the moment difficult.[1]

The maximum stress may be determined from

$$s = \frac{P}{A} + \frac{Mc}{I} \tag{3-21}$$

where P = axial load, lb
A = area of cross section, in.^{2}
I/c = section modulus, in.^{3}
M = maximum bending moment due to combined axial and transverse loads acting simultaneously, in.-lb

Usually one assumes a value for e, for example, ec/ρ^2 [see Eq. (3-23)], and employs the maximum deflection δ.

It is usually difficult to determine the bending moment M analytically. Tabulations of some of these bending moments are available for various combinations of bending.[2]

One case of particular interest is the eccentrically loaded strut in Fig. 3-18. From the consideration of the equilibrium of a strut with axial loads applied eccentrically, the deflection at the center can be determined, and the moment will be[3]

$$M = Pe \sec \frac{L}{2\rho} \sqrt{\frac{P}{EA}} \tag{3-22}$$

[1] Archie Higdon, E. H. Ohlsen, W. B. Stiles, and John A. Weese, "Mechanics of Materials," p. 430, John Wiley & Sons, Inc., New York, 1967.

[2] William Griffel, "Handbook of Formulas for Stress and Strain," Frederick Ungar Publishing Co., New York, 1966.

[3] Stephen P. Timoshenko and Gleason H. MacCullough, "Elements of Strength of Materials," 4th ed., D. Van Nostrand Company, Inc., Princeton, N.J., 1962.

where P = axial compressive load, lb
e = eccentricity of axial load from centroid of section, in.
L = length of strut, in.
ρ = radius of gyration of cross section, in.
A = area of cross section, in.^{2}
E = modulus of elasticity of material, psi

The stress at the center of the beam is, from Eqs. (3-21) and (3-22),

$$s = \frac{P}{A} + \frac{Pec}{I} \sec \frac{L}{2\rho}\sqrt{\frac{P}{EA}}$$

$$= \frac{P}{A}\left(1 + \frac{ec}{\rho^2} \sec \frac{L}{2\rho}\sqrt{\frac{P}{EA}}\right) \tag{3-23}$$

for $P < F_{\text{crit}}$ [see Eq. (3-19)].

For the beam with a central transverse load and an axial compressive load, the bending moment at the center is[1]

$$M = \frac{WL}{4}\frac{\tan U}{U}$$

where $U = \frac{\pi}{2}\sqrt{\frac{P}{F_{\text{crit}}}}$
F_{crit} = critical load from Euler's equation, lb
L = length of beam, in.
W = transverse load, lb

The stress in this case becomes, from Eq. (3-21),

$$s = \frac{P}{A} + \frac{WcL}{4I}\frac{\tan U}{U} \tag{3-24}$$

The design of beam columns is discussed in Chap. 5.

3-8 CONTACT STRESSES

Contact stresses are due to pressure of one elastic solid upon another, where the pressure is distributed over a small area relative to the size of the parts. In many cases, the contact stresses that are induced when surfaces of two curved bodies are pressed together by external loads are more significant than those occurring in portions of the body well removed from the contact area. Contact stresses frequently cause failure of one or both of the bodies. For example, these stresses may be encountered in the contact between a locomotive wheel and a railroad rail, between a roller or ball and its race in a rolling-contact bearing, between a cam and its follower, or between mating gear teeth.

[1] S. P. Timoshenko and J. M. Gere, "Theory of Elastic Stability," 2d ed , p. 7, McGraw-Hill Book Company, New York, 1961.

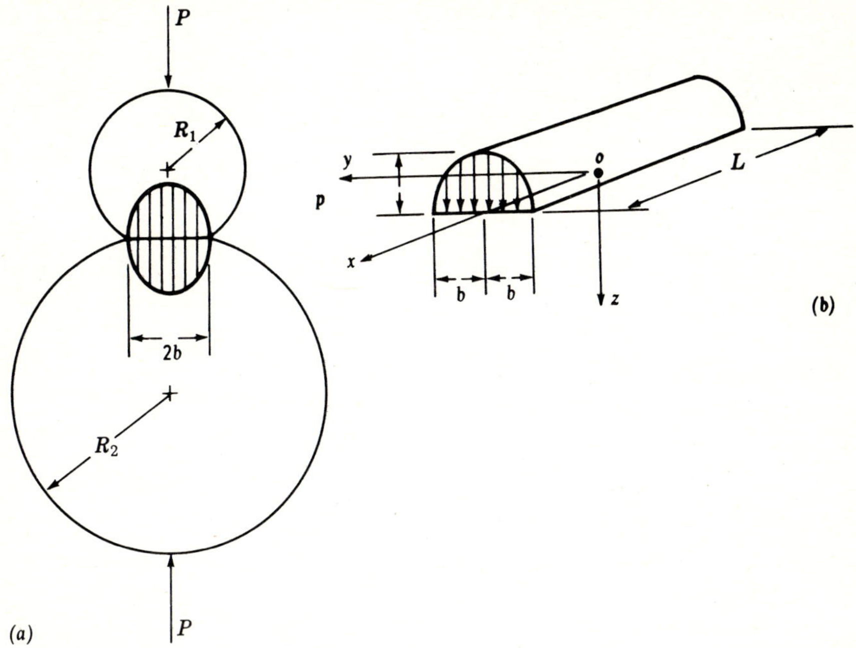

Fig. 3-20 Pressure distribution across contact surfaces of two cylinders.

In all these examples, the contact stresses are repeated a large number of times and frequently cause fatigue failure, beginning as a crack. Sometimes, ball bearings and gear teeth fail as a result of contact stresses that produce pitting near the surface of contact. For fatigue failures in rolling-contact bearings, see Art. 22-9(*a*) and (*b*).

A common case of contact stress is that between two parallel cylinders in contact along an axial element, such as in roller bearings. The usual method for determining stresses in the contact zone of two elastic bodies was first established by H. Hertz (hence the name "Hertz stress" for contact stress); the theory-of-elasticity approach is used. Solutions for this stress problem are also presented in several references.[1]

When two cylinders are in axial contact, the area of contact is a narrow rectangle (half width b and length L) as shown in Fig. 3-20(*b*). The distribution of pressure on the contact area is an elliptical distribution shown in Fig. 3-20(*a*) and (*b*).

The half width of the contact band b for two cylinders with parallel

[1] Timoshenko and Goodier, *op. cit.*, pp. 377–382; Seely and Smith, *op. cit.*, pp. 342–378; E. I. Radzimovsky, Stress Distribution and Strength Condition of Two Rolling Cylinders Pressed Together, *Univ. Illinois, Bull.* 408, vol. 50, no. 44, February, 1953.

axes can be computed from the expression

$$b = 2\sqrt{\frac{1-\mu^2}{\pi}\frac{P(1/E_1 + 1/E_2)}{L(1/R_1 + 1/R_2)}} \tag{3-25}$$

The maximum pressure (and also the maximum compressive stress) on the contact strip is equal to

$$p = \frac{2P}{\pi b L} \tag{3-26}$$

where μ = Poisson's ratio for two materials
P = external force pressing two cylinders together, lb
E_1, E_2 = moduli of elasticity for cylinders 1 and 2, psi
R_1, R_2 = radii of two cylinders (for plane surface, $R = \infty$ and is negative for internal cylindrical surface), in.

For this analysis, the following assumptions apply:

1. The proportional limit of the materials is not exceeded.
2. The loading acts perpendicular to the surface.
3. The dimensions of the contact area are small compared with the surfaces of the two bodies.

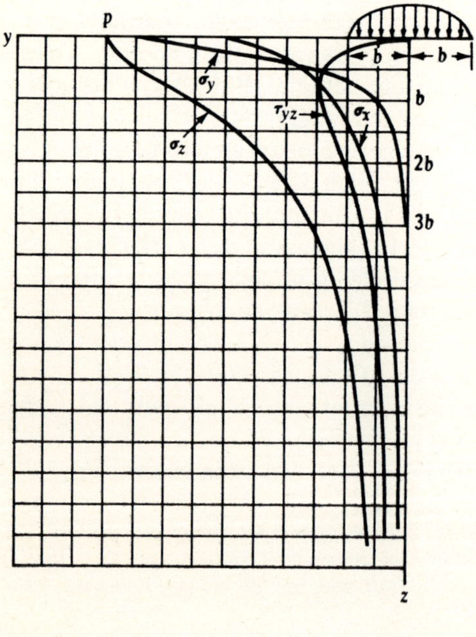

Fig. 3-21 Distribution of stresses in contact zone along line of symmetry.

4. The radii of the contact areas are large in comparison with the dimensions of these areas.

Figure 3-21 shows the distribution of stress components along the line of symmetry for two cylinders pressed together. The distance of the points from the contact surface is given in terms of b. Only the value of the larger shear stress, τ_{yz}, is shown; it is in the yz plane at 45° to the y axis, and its value is[1]

$$\tau_{yz} = \frac{\sigma_z - \sigma_y}{2} \tag{3-27}$$

The maximum shearing stress is located at approximately $z = 0.78b$ from the contact surface and is equal to $0.304p_{\max}$ (for $\mu = 0.3$).

3-9 CYCLIC OR FATIGUE STRESSES

In previous articles in this chapter, the external loads and induced stresses were considered as static. However, many machines and structural members are subjected to loads that are not static but vary in magnitude. This type of loading induces cyclic or fatigue stresses in the members. With the increase in mechanization, design to resist fatigue failure has become of increasing importance in many areas of engineering. For example, many factories employ heavy machinery operating at high speeds, new means of rapid transportation are being developed with corresponding sources of vibration and cyclic stresses, and improvements in design efficiency require component parts to operate at higher static- and dynamic-stress levels. In particular, the design of aircraft structural members to resist fatigue failure becomes of paramount importance in view of increased speeds and efficiencies. Authorities[2] believe that between 80 and 90 percent of all service *fractures* are caused by fatigue; therefore, it is necessary to consider fatigue failure in the design of machine members.

In order to design a member for fatigue loading, the stress variation with time must be either known or approximated. Frequently, such stresses vary in an irregular and sometimes unpredictable manner. This is true particularly when random loads, such as those due to wind, are encountered. Nevertheless, a pattern of variation must be constructed as a basis for design; however, the exact shape of the curve "does not appear to be of particular significance."[3] The most common stress-time

[1] Radzimovsky, *loc. cit.*

[2] V. M. Faires, "Design of Machine Elements," 4th ed., pp. 99, 104, The Macmillan Company, New York, 1965; Faupel, *op. cit.*, p. 628.

[3] J. E. Shigley, "Mechanical Engineering Design," p. 177, McGraw-Hill Book Company, New York, 1963.

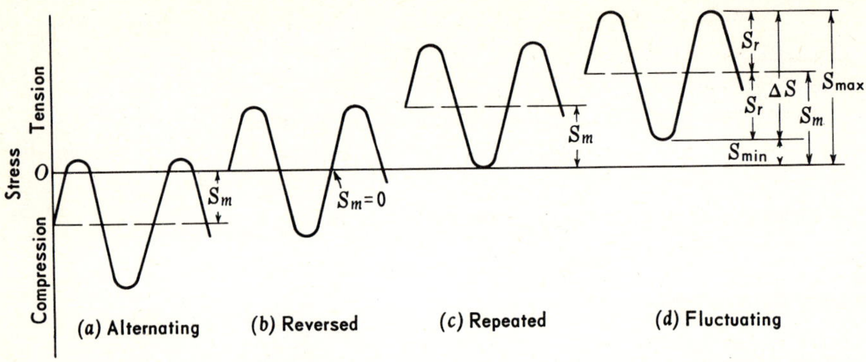

Fig. 3-22 Fatigue-stress variations.

variations are sinusoidal and are shown in Fig. 3-22; therefore, for any of these particular models, $\sigma = \sigma_m + \sigma_r \sin (2\pi t/T)$, where T is the period of the load application. The following expressions also apply:

$$\sigma_m = \frac{\sigma_{\max} + \sigma_{\min}}{2}$$

$$\sigma_r = \frac{\sigma_{\max} - \sigma_{\min}}{2}$$

$$R = 2\sigma_r$$

where σ_m = mean stress
$\sigma_{\max}$ = maximum stress
$\sigma_{\min}$ = minimum stress
R = range of stress

In general, the stresses corresponding to the maximum, minimum, and mean values of loads, or to bending moments and torque, are computed by means of the static formulas discussed earlier in the chapter.

The application of cyclic stresses to fatigue design, the prediction of operating life, and the nature of fatigue will be more thoroughly discussed in Chap. 5.

3-10 STRESSES DUE TO SHOCK AND IMPACT LOADING

Impact is defined in the "Shock and Vibration Handbook"[1] as "a single collision of one mass in motion with a second mass which may be either in motion or at rest." If the time of load application is less than about half the natural period of vibration of the member being struck, calculations of stress and strain should be based upon wave analysis, because

[1] Cyril M. Harris and Charles E. Crede (eds.), "Shock and Vibration Handbook," pp. 1–20, McGraw-Hill Book Company, New York, 1961.

of the predominance of the propagation, reflection, and interference of elastic and plastic waves.[1]

If the end of an elastic bar is struck axially by a hammer, a compressive wave develops that travels at the speed of sound in the material:[2]

$$c = \sqrt{\frac{E}{\rho}} \tag{3-28}$$

where c = velocity, in. per sec

E = modulus of elasticity, psi

ρ = mass density, lb-sec^2 per in.^{4}

In the case of an unsupported uniform-cross-section elastic bar subjected to a longitudinal impact from a rigid body moving with velocity v (in. per sec), a compressive stress wave[3] of intensity σ is propagated through the bar as follows:

$$\sigma = \frac{v}{c} E = v \sqrt{\rho E} \tag{3-29}$$

In the event that the time of load application is in the range of one-half to three times the natural period of vibration, the wave phenomena are of minor importance.[4] For this loading situation, generally an energy method is employed to calculate the impact stresses, or the response of the body receiving the impact load is determined from vibration theory.[5,6] Application of either of these methods depends upon the following conditions:

1. An undamped single-degree-of-freedom system (or one with a small degree of damping)
2. The impact load distributed over the cross section of the member in nearly a uniform manner and the induced stresses distributed throughout the member, as in the case of static loading
3. The fundamental mode of vibration uncoupled from the higher modes

To investigate the effect of shock, an elastic system loaded by a falling weight W, as shown in Fig. 3-23, will be used as an example. The system shown is one whose elastic displacement δ of the point of

[1] Faupel, *op. cit.*, p. 557.

[2] *Ibid.*, p. 570.

[3] R. J. Roark, "Formulas from Stress and Strain," McGraw-Hill Book Company, New York, 1965.

[4] Harris and Crede, *op. cit.*, p. 9-8.

[5] Faires, *op. cit.*, p. 148.

[6] J. M. Frankland, Effects of Impact on Simple Elastic Structures, *Proc. Soc. Exptl. Stress Anal.*, vol. 6, no. 2, 1948.

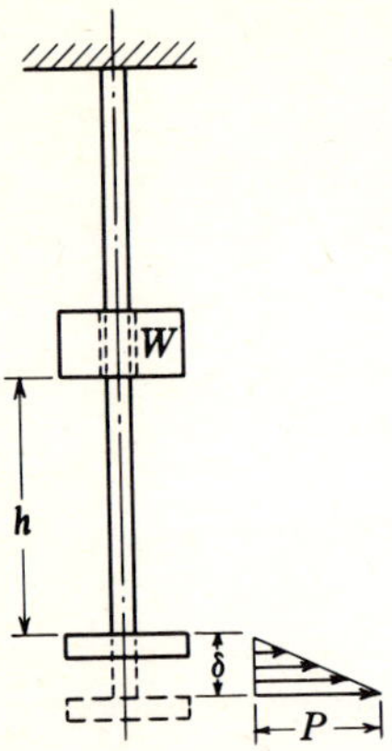

Fig. 3-23 Shock loading.

load application is proportional to the load producing it, that is, $P = C\delta$, where C is a constant for the system.

Let W = falling weight, lb
h = height of free fall, in.
δ = displacement of point of load application, in.
P = impact load, lb
$C = P/\delta$ = lb per in. of deflection

By assuming that the support is unyielding, the energy given up by the falling weight will be absorbed wholly by the system. If it is also assumed that the mass of the system is small so that its inertia forces may be neglected and that the stresses induced are within the elastic limit, the energy absorbed by the system will be equal to the product of the average load and the displacement, or $\frac{1}{2}P\delta$. By equating this energy to that given up by the falling weight, the following equation is obtained:

$$\tfrac{1}{2}P\delta = W(h + \delta) \tag{3-30}$$

But $\delta = P/C$, where C is a constant, lb per in., for a system. Substituting this value of δ into Eq. (3-30) and solving for P give

$$P = W\left(1 + \sqrt{1 + \frac{2Ch}{W}}\right) \tag{3-31}$$

If Eq. (3-31) is written as

$$\frac{P}{W} = 1 + \sqrt{1 + \frac{2Ch}{W}} \tag{3-32}$$

the quantity P/W may be termed a "shock factor" and represents the magnification of load caused by impact.

Note that the shock factor can be decreased by lowering C, that is, by reducing the number of pounds required to deflect the member 1 in., that is, make the system more "springy" or "soft."

In the example above, for a bar in tension, $\delta = PL/AE$, or

$$C = \frac{P}{\delta} = \frac{AE}{L}$$

By decreasing the area A, or by decreasing the modulus of elasticity E (using, for instance, rubber instead of steel), or by increasing L, then the value of C is decreased and the magnitude of the shock load P would be decreased.

The equation for P/δ depends on the system, whether it be a bar in tension, a beam in bending, or a helical spring in tension or compression; in the case of a bar in torsion, the equivalent of C is radians per inch-pounds of torque. In the latter case, a new equation may be derived for the shock factor in which the equivalent value for C may be determined from $\theta = TL/JG$, that is, $C = T/\theta$.

If the weight W in Eq. (3-31) were applied instantaneously and without initial velocity, h in the equation would be zero and the equation would reduce to

$$P = 2W \tag{3-33}$$

This equation indicates that the effect of applying a load suddenly is twice as severe as applying it statically. This is termed a suddenly applied load without velocity of approach. An example approximating this type of loading is the pressure caused by the explosion of gases in an internal-combustion-engine cylinder which is caused by a rapid building up of pressure in a time of the order of 0.001 sec.

In the above example, the effects of the mass of the falling load W and the mass of the bar were neglected in determining the maximum stress in the bar. However, if a load W without velocity of approach is suddenly transferred to the free end of the bar, the induced stress in the bar is characterized by a wave of stress as in impact loading, but, because of the effects of inertia, the space-average value of the induced tension in the bar is greater than one-half the average value so that the maximum stress is less than twice that due to gradual loading. It has been shown[1] that, if the ratio r of the mass of the load to that of the bar is unity, sudden loading induces a stress 1.63 times that due to static loading. For r equals 2, the ratio is 1.68; for r equal to 4, the ratio is 1.84; and, as the ratio r increases, the ratio approaches 2 as a limit. Thus, the assumption that sudden loading without velocity of approach induces a stress twice the value due to static loading is on the side of safety.

In case the time interval of application of the load is appreciable as

[1] A. E. H. Love, "Mathematical Theory of Elasticity," 2d ed., Cambridge University Press, London, 1906.

compared with the natural frequency of the system, the shock factor would be less than indicated by the above equation.[1]

In the example of Fig. 3-23, the energy of the falling weight was transferred to the bar as strain energy so that when the weight reached its lowest position the bar would be in maximum tension. The tensile force in the bar would then accelerate the weight upward and in the absence of friction and windage the weight would be returned to its original position.

If the falling weight, on striking the end of the bar, became locked to it, the bar in tension would then accelerate the weight upward until the stress in the bar became zero. Then the moving weight, being locked to the bar, would place the bar in compression until the weight came to rest. At this point, consider a negative sign for the radical in Eq. (3-31).

In the above-described manner the bar would be subjected to a repeated load which would cause vibration of the support, granted that actually the support would not be rigid as assumed. The vibration would continue until the energy was dissipated by damping.

Example 3-3 A rigidly built-in steel cantilever has a cross section ½ in. deep, ⅜ in. wide, and is loaded 12 in. from the support. Determine the maximum bending stress in the beam, assuming (*a*) that a load of 8 lb is applied gradually and (*b*) that an 8-lb weight is dropped on the beam through a distance of $\frac{3}{16}$ in.

Solution The section modulus is

$$\frac{I}{c} = \frac{bh^2}{6} = \frac{0.375 \times 0.5^2}{6} = 0.0156 \text{ in.}^3$$

(*a*) Static stress:

$$\frac{M}{I/c} = \frac{8 \times 12}{0.0156} = 6{,}150 \text{ psi}$$

(*b*) Impact stress:

The moment of inertia is

$$I = \frac{bh^3}{12} = \frac{0.375 \times 0.5^3}{12} = 0.00390 \text{ in.}^4$$

From the equation for deflection of a cantilever,

$$\delta = \frac{PL^3}{3EI} \qquad \text{(Appendix III)}$$

$$C = \frac{P}{\delta} = \frac{3EI}{L^3} = \frac{3 \times 30{,}000{,}000 \times 0.00390}{12^3} = 204 \text{ lb per in.}$$

The shock factor, Eq. (3-32), is

$$\frac{P}{W} = 1 + \sqrt{1 + \frac{2Ch}{W}} = 1 + \sqrt{1 + \frac{2 \times 204 \times 0.1875}{8}} = 4.33$$

The impact stress then is

$$4.33 \times 6{,}150 = 26{,}600 \text{ psi}$$

[1] See W. M. Murray, Effects of Shock Loading, *Prod. Eng.*, December, 1954, p. 171.

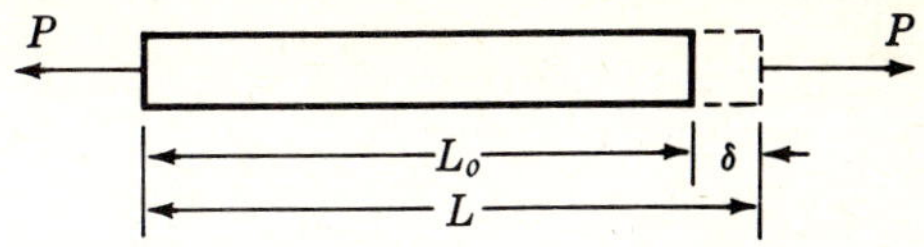

Fig. 3-24 Strain energy in bar in tension.

The actual value of the stress would be less than this value because of the inevitable nonrigidity of the support, the inertia of the beam, and the local deformations of the weight and the beam at the region of contact at the impact. These influences could be taken into account if a more exact analysis were required. Stress concentration at the built-in end of the beam may be taken into account separately.

3-11 BEAM DEFLECTIONS

In a basic strength-of-materials course, the double-integration method and the area-moment method generally are studied. These methods will not be discussed further in this text.

One of the most versatile methods of determining deflections is provided by Castigliano's theorem. Since this theorem is based upon the elastic strain energy stored within an elastic body, it is necessary that methods of computing this energy be discussed first before discussing this theorem.

In accordance with the principle of conservation of energy, external work on a body or structure is stored internally as *strain energy*. Strain energy (a form of potential energy) represents the total work done against the mutual actions between the particles of a body.[1] In the case of a bar that has elongated an amount δ by a gradually applied load P, as shown in Fig. 3-24, the external work and the corresponding internal strain energy equal $P\delta/2$. The bar is assumed to be perfectly elastic, and strains or displacements are proportional to loads.

Strain energy may also be expressed in terms of stress and strain. Thus, if the element shown in Fig. 3-25 is subjected to a uniaxial normal

[1] Timoshenko and Goodier, *op. cit.*, pp. 146–153; Faupel, *op. cit.*, p. 446.

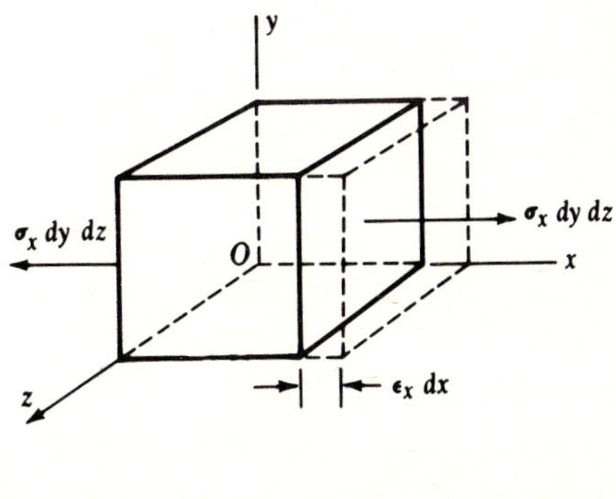

Fig. 3-25 Strain energy in an element.

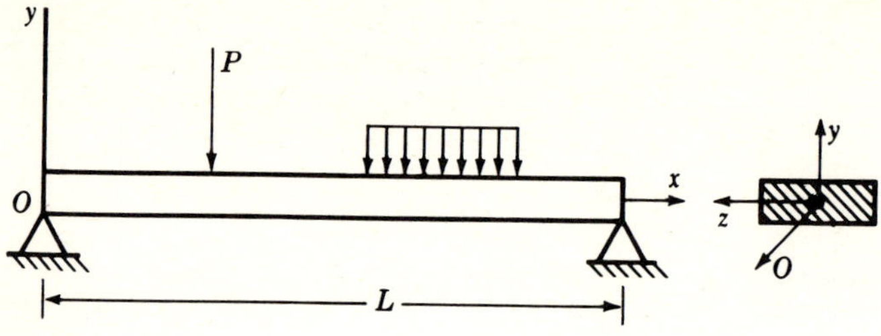

Fig. 3-26 Beam loading for strain-energy determination.

stress σ_x, a force $\sigma_x\, dy\, dz$ does work over an elongation $\epsilon_x\, dx$. Therefore, the strain energy $dU = \frac{1}{2}\sigma_x\epsilon_x\, dx\, dy\, dz$, where the strain energy per unit volume is

$$du = \frac{1}{2}\sigma_x\epsilon_x = \frac{\sigma_x^2}{2E} \qquad \text{uniaxial stress} \tag{3-34}$$

By the use of Eq. (3-34), an expression for the strain energy of a beam in uniaxial bending, as shown in Fig. 3-26, may be obtained as

$$U = \frac{1}{2}\int_{\text{length}}\int_{\text{area}} \frac{1}{E}\left(\frac{My}{I}\right)^2 dA\, dx = \frac{1}{2}\int_{\text{length}} \frac{1}{E}\frac{M^2}{I^2}\, dx \int_{\text{area}} y^2\, dA$$

$$U = \frac{1}{2}\int_0^L \frac{M^2}{EI}\, dx \tag{3-35}$$

where $I = \int y^2\, dA$ is the moment of inertia about the z axis.

Theorem of Castigliano[1,2,3] This theorem states that *the displacement corresponding to any force of a system of forces acting on an elastic*

[1] R. V. Southwell, "An Introduction to the Theory of Elasticity for Engineers and Physicists," p. 15, Oxford University Press, Fair Lawn, N.J., 1941.
[2] G. Murphy, "Advanced Mechanics of Materials," pp. 226–227, McGraw-Hill Book Company, New York, 1946.
[3] Faupel, *op. cit.*, p. 454.

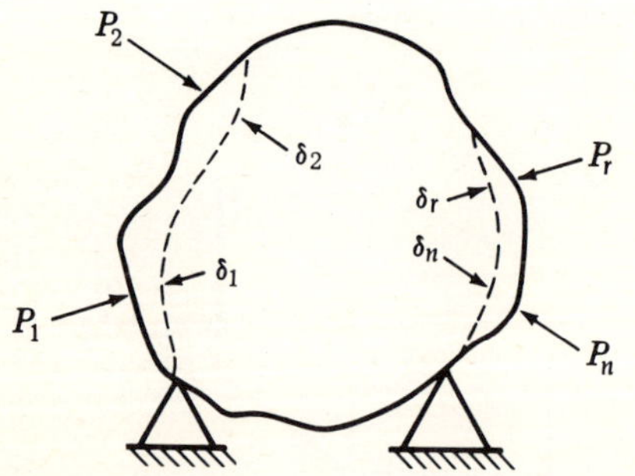

Fig. 3-27 Body subjected to system of external forces.

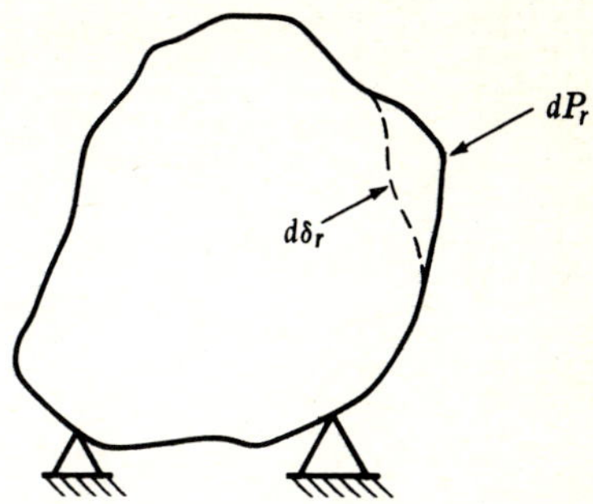

Fig. 3-28 Body subjected to differential load.

body can be determined by taking the partial derivative of the elastic strain energy (expressed in terms of the forces) with respect to that force.

$$\frac{\partial U}{\partial P_r} = \delta_r \tag{3-36}$$

where δ_r = deflection in direction of force P_r.

The theorem may be derived as follows: Figure 3-27 shows an elastic body subjected to a system of forces, together with their corresponding displacements. The resulting elastic strain energy due to this system of loading is

$$U = \frac{P_1\delta_1}{2} + \frac{P_2\delta_2}{2} + \frac{P_r\delta_r}{2} + \frac{P_n\delta_n}{2}$$

If the force P_r is increased a differential amount, as shown in Fig. 3-28, the total strain energy is

$$U_t = U + \frac{\partial U}{\partial P_r}\,dP_r \tag{3-37}$$

If the loading sequence is reversed so that dP_r is applied first and then followed by the other systems of loads, the strain energy associated with dP_r is

$$U_1 = \frac{dP_r\,d\delta_r}{2} \tag{3-38}$$

When the force system of Fig. 3-27 is applied, the force dP_r acts through a distance δ_r and is considered to remain constant through distance δ_r so that the strain energy corresponding to this force system becomes

$$U_2 = U + dP_r\,\delta_r \tag{3-39}$$

Therefore, the total strain energy for this sequence of loading is

$$U_1 + U_2 = U + \frac{dP_r\,d\delta_r}{2} + dP_r\,\delta_r \tag{3-40}$$

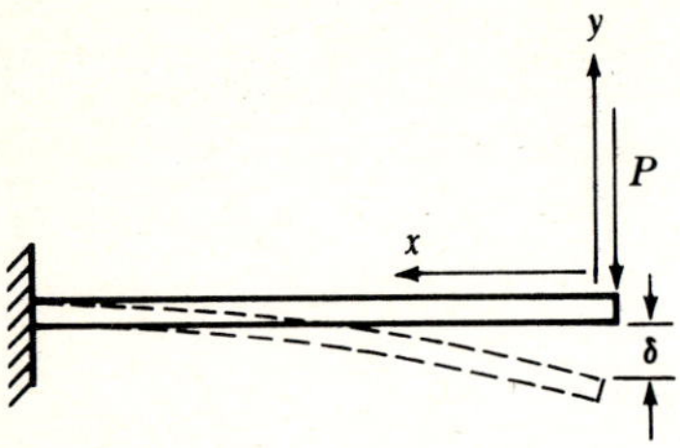

Fig. 3-29 Cantilever concentrated load.

Since the total strain energy stored in the body must be independent of the order of application of the forces, the value of total energy from Eq. (3-37) equals that given in Eq. (3-40). Hence, neglecting products of differentials,

$$U + \frac{\partial U}{\partial P_r} dP_r = U + dP_r \, \delta_r \tag{3-41}$$

from which $$\frac{\partial U}{\partial P_r} = \delta_r \tag{3-42}$$

Example 3-4 Consider the determination of the end deflection of a cantilever as shown in Fig. 3-29. At point x along the beam, the moment is $M = -Px$ (negative moment because P causes the beam to be convex upwards).

Using Eq. (3-35),

$$U = \frac{1}{2} \int_0^L \frac{M^2}{EI} dx$$

By differentiating U with respect to P, the deflection becomes

$$\delta = \frac{\partial U}{\partial P} = \int_0^L \frac{M \, \partial M/\partial P \, dx}{EI}$$

$$= \int_0^L \frac{(-Px)(-x)}{EI} dx = \frac{PL^3}{3EI}$$

The deflection is in the direction of the force with respect to which the derivative was taken—downward.

Example 3-5 Determine the end deflection of a uniformly loaded cantilever beam as shown in Fig. 3-30.

If an applied force does not exist at the point where the deflection is to be determined, then a fictitious force Q must be applied. After the strain-energy equation has been differentiated with respect to Q, the force Q is set equal to zero. The resulting expression is the displacement at the point of application

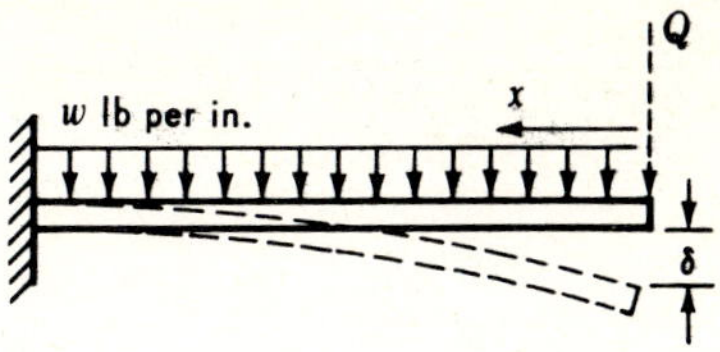

Fig. 3-30 Cantilever uniform load.

of Q and is in the same direction as Q was assumed to be acting:

$$M = -Qx - \frac{wx^2}{2}$$

$$\frac{\partial M}{\partial Q} = -x$$

$$\delta = \frac{\partial U}{\partial Q} = \int_0^L \frac{M(\partial M/\partial Q)}{EI}\,dx$$

$$\delta = \int_0^L \frac{(-Qx - wx^2/2)(-x)}{EI}\,dx$$

$$\text{or } \delta = \frac{wL^4}{8EI} \qquad \textit{Ans.}$$

Example 3-6 Determine the deflection of the semicircular ring, at load P, in Fig. 3-31(a).

$$M = -PR\sin\theta$$

$$\delta = \frac{\partial U}{\partial P} = \int_0^{\pi/2} \frac{M(\partial M/\partial P)}{EI}\,R\,d\theta$$

$$\delta = \int_0^{\pi/2} \frac{(-PR\sin\theta)(-R\sin\theta)}{EI}\,R\,d\theta$$

$$= \frac{PR^3}{EI}\int_0^{\pi/2} \sin^2\theta\,d\theta$$

$$\text{or } \delta = \frac{\pi PR^3}{4EI}$$

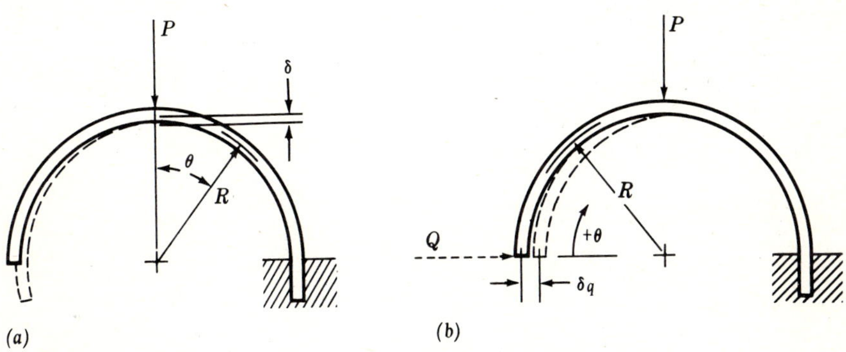

Fig. 3-31 (a) Vertical deflection due to load P. (b) Horizontal deflection of the free end.

Determine the horizontal deflection of the free end [Fig. 3-31(*b*)].

$$M = QR\sin\theta \qquad \text{for } 0 \leqq \theta \leqq \pi/2$$
$$M = QR\sin\theta + PR\sin(\theta - 90) \qquad \text{for } +\pi/2 \leqq \theta \leqq \pi$$
$$= QR\sin\theta - PR\cos\theta$$

By taking the partial derivative of the moment with respect to the load Q in each section of the member, we find

$$\frac{\partial M}{\partial Q} = R\sin\theta \qquad \frac{\partial M}{\partial Q} = R\sin\theta$$

$$\delta q = \frac{\partial U}{\partial Q} = \int_0^{\pi/2} = \frac{(QR\sin\theta)R\sin\theta\, R\, d\theta}{EI} + \int_{\pi/2}^{\pi} \frac{QR\sin\theta - PR\cos\theta}{EI} R\sin\theta\, R\, d\theta$$

$$= -\frac{PR^3}{EI}\int_{\pi/2}^{\pi} \sin\theta\cos\theta\, d\theta$$

$$\delta q = -\frac{PR^3}{EI}\frac{\sin^2\theta}{2}\Big|_{\pi/2}^{\pi} = \frac{PR^3}{2EI}$$

A useful method for calculating deflections is to program the double-integration method so that digital-computer solutions can be obtained. Another approach is the graphical-integration method discussed in Appendix XIII-A. These methods are particularly well adapted to the calculation of deflections of beams having variable cross sections, such as stepped shafts.

3-12 ANGULAR DEFORMATIONS

Castigliano's theorem also may be employed to calculate the angle of twist in members subject to torsion. By employing reasoning similar to that used in Art. 3-11, it can be shown that the strain energy per unit volume is

$$u = \frac{1}{2}\tau\gamma = \frac{\tau^2}{2G}$$

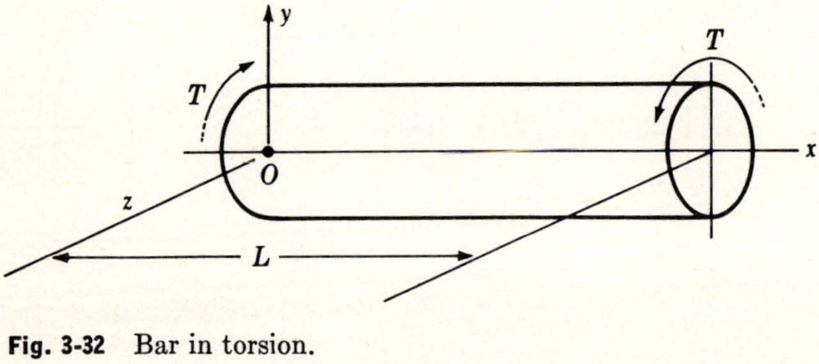

Fig. 3-32 Bar in torsion.

The total strain energy for a cylindrical bar in torsion (Fig. 3-32) is

$$U = \int_{\text{length}} \int_{\text{area}} \frac{\tau^2}{2G}\, dx\, dA = \int_{\text{length}} \int_{\text{area}} \frac{T^2 r^2}{2GJ^2}\, dx\, dA$$

or $$U = \int_0^L \frac{T^2}{2JG}\, dx$$

where τ = shear stress, psi
γ = shear strain, in. per in.
A = cross-sectional area, in.^{2}
T = torque, in.-lb
J = polar moment of inertia (for circular shaft), in.^{4} $= \int_{\text{area}} r^2\, dA$

By differentiating the strain-energy expression for torsion with respect to the torque, the angle of twist, θ, may be obtained for that portion of the shaft where the torque acts.

$$\frac{\partial U}{\partial T} = \theta = \int_0^L \frac{2T}{2JG}\, dx$$

$$\theta = \int_0^L \frac{T\, dx}{JG}$$

If the torque is uniform throughout the length of the shaft, the total angle of twist is

$$\theta = \frac{TL}{JG}$$

which is the expression for the angle of twist of a cylindrical shaft in torsion.

It should be noted that the slope, or the rotation of a section, of a beam at a particular location also may be determined by a similar approach, namely, taking the derivative of the strain energy in bending with respect to the moment (or couple) at the pertinent section:

$$\theta = \frac{\partial U}{\partial C} = \int_0^L \frac{M(\partial M/\partial C)}{EI}\, dx$$

where C is the couple at the section of interest.

In the event that the beam is under pure bending, where $M = C$ (throughout the length of the beam),

$$\theta = \frac{ML}{EI}$$

where E and I are assumed to be constant.

Example 3-7 Determine the rotation θ of the free end of a tube in the plane of a torque T_0; see Fig. 3-33. Both portions of the tube lie in the same plane. Neglect the effect of deflection of the radius of the quarter bend.

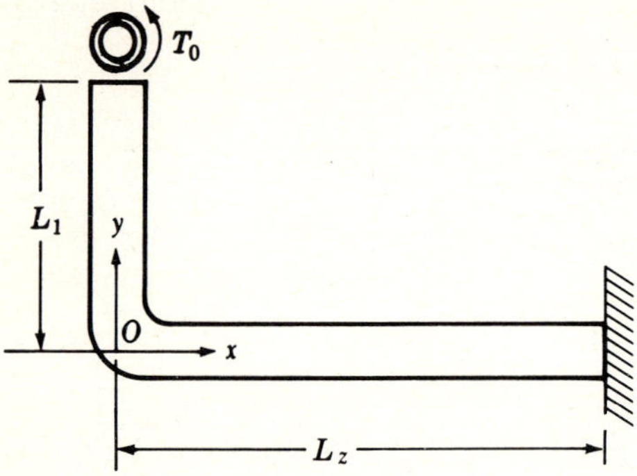

Fig. 3-33 Bent tube in torsion.

Length L_1 of pipe is subjected to torque T_0.
Length L_2 of pipe is subjected to bending moment T_0 (about the y axis).

$$U = \int_0^{L_1} \frac{T^2\,dy}{2JG} + \int_0^{L_2} \frac{M^2\,dx}{2EI}$$

$$= \int_0^{L_1} \frac{T_0^2\,dy}{2JG} + \int_0^{L_2} \frac{T_0^2\,dx}{2EI}$$

Therefore

$$\theta = \frac{\partial U}{\partial T_0} = \int_0^{L_1} \frac{T_0\,dy}{JG} + \int_0^{L_2} \frac{T_0\,dx}{EI} = \frac{T_0 L_1}{JG} + \frac{T_0 L_2}{EI}$$

4
Stress Concentration in Machine Members

4-1 GENERAL DISCUSSION

In local regions of machine members, the stresses may not be distributed as indicated by the elementary theory of strength of materials. Such localized variations in stress distribution are caused by sharp discontinuities in shape of the member or by surface roughness or scratches.

These localized stresses are frequently of large magnitude and, under the action of loads in service, may give rise to a crack at the region of the localized stresses. Such a crack usually leads to failure of the member by rupture. It is evident, therefore, that it is necessary for the mechanical designer to have a knowledge of stress concentrations, especially in regard to their causes and the means for avoiding them or lessening their seriousness.

4-2 CAUSES OF STRESS CONCENTRATION

Variation in properties of materials from point to point in a member Some examples of this type of stress concentration are (1) internal cracks and flaws, (2) cavities in welds, (3) air holes in steel and concrete, and (4) nonmetallic inclusions.[1]

Pressures at points or areas at which loads on a member are applied Some examples of this type of stress concentration are (1) contact between a wheel and rail, (2) contact between the balls and races of a ball bearing, (3) contact between a beam and its supports, and (4) contact between gear teeth (see Fig. 4-12 and Art. 3-8).

Abrupt changes of section There are two classes of this type of stress concentration: The first involves surface condition, for example, scratches due to improper machining or to roughness in handling. In machining, the dimensions of the scratches or irregularities vary in size from large in roughing operations to microscopic in fine finishing. Stress concentration may also be caused by abrupt changes in form of a member. Stress concentration of this kind may be extremely serious, and its prevention lies squarely within the designer's responsibility. Some common examples of this kind of stress concentration are considered in Art. 4-4.

4-3 STRESS CONCENTRATION DEFINED

The stress at a point in a member when influenced by one or more of the causes discussed in Art. 4-2 is, in general, greater than that determined by elementary strength of materials and is referred to as *stress concentration.*

4-4 STRESS CONCENTRATION DUE TO HOLES AND NOTCHES

Tension member with a transverse hole In a prismatic bar subjected to a tensile load as shown in Fig. 4-1(a), the stress is uniformly distributed over a cross section AA of the bar.

If the bar has an elliptical hole, as shown in Fig. 4-1(b), the stress at a section CC remote from the hole will be uniformly distributed over the section, but the stress over the cross section BB through the hole will not be uniformly distributed over the section carrying the load. The maximum stress will be induced at the edge of the hole and it may have a value several times that of the stress at the section CC.† The value of the maximum stress in terms of the stress at the section CC is given by the expression

$$s_{\max} = s_0\left(1 + 2\frac{b}{c}\right) \tag{4-1}$$

[1] J. N. Goodier, Concentration of Stress around Spherical and Cylindrical Inclusions and Flaws, *Trans. ASME*, vol. 55, p. APM 55-7, 1933.

† In this discussion, the hole is assumed to be so small that it does not appreciably reduce the cross section of the bar.

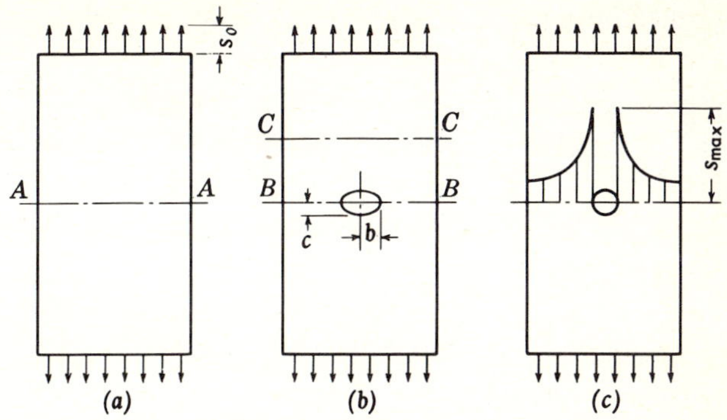

Fig. 4-1 Effects of holes on stress distribution in tension members.

in which s_0 is the stress at the section CC, and b and c are the semiaxes of the ellipse perpendicular and parallel, respectively, to the line of the load as shown in the figure.[1]

It may be seen from Eq. (4-1) that for large values of b/c, which represents an ellipse approaching a transverse slit, s_{max} reaches a very high value.

For small values of b/c, which represents an ellipse approaching a longitudinal slit, s_{max} is not markedly increased over s_0.

For the case of a circular hole, $b = c$, and $s_{max} = 3s_0$. The stress distribution for this case is shown in Fig. 4-1(c).

Tension member with small notches The stress concentration in the notched tension member in Fig. 4-2 is influenced by the depth b of the notch and the radius r at the bottom of the notch. The maximum stress may be calculated by the following equation,[2] which applies to members having notches that are small in comparison with the width of the bar.

$$s_{max} = s_0 \left(1 + 2\frac{b}{r}\right) \tag{4-2}$$

where s_0 is the uniformly distributed stress at a section remote from the notch.

[1] S. P. Timoshenko and J. N. Goodier, "Theory of Elasticity," 2d ed., p. 84, McGraw-Hill Book Company, New York, 1951. In the derivation of Eq. (4-1), it was assumed that the bar is of infinite width. However, for bars of finite width and with holes not larger than one-sixth the width of the bar, the equation may be used without introducing large errors. For circular holes larger than one-sixth the width of the bar, the experimental data in Art. 4-8 should be used to determine the value of the maximum stress.

[2] C. E. Inglis, Stresses in a Plate Due to the Presence of Cracks and Sharp Corners, *Trans. Inst. Naval Architects*, 1913.

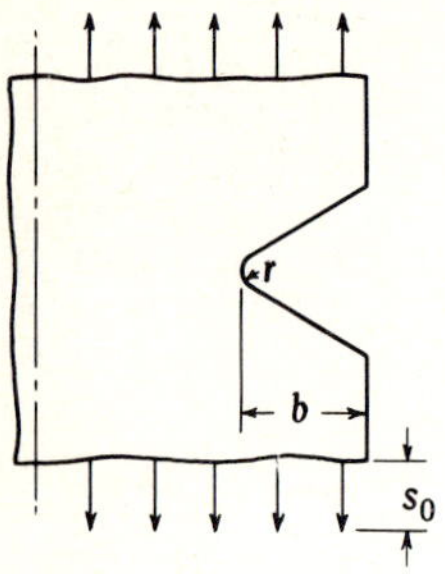

Fig. 4-2 Notch in edge of tension member.

4-5 VISUALIZATION OF STRESS CONCENTRATION

The stress distribution in tension members having transverse holes or notches, as discussed in Art. 4-4, are representative of the few cases that have been solved by analytical methods. For determining stress concentration in the majority of members of the forms found in machinery, it is necessary to resort to experimental methods, some of which are discussed in Art. 4-7. However, there are qualitative methods that are useful to the designer in visualizing probable stress concentrations in members. Two of these methods of visualizing stress concentration are discussed below.

Shafts subjected to torsion A qualitative representation of stress concentration in circular shafts of variable cross section subjected to torsion considers that the shaft is divided into a number of concentric tubes, each of which carries an equal share of the twisting moment on the shaft.[1] On the section of a typical shaft shown in Fig. 4-3, the left-hand end is shown divided into five concentric tubes, including the solid central one.[2] The right-hand end of the shaft is divided into five concentric tubes in a similar manner. The sets of tubes for the left-hand and the right-hand ends of the shafts are joined by smooth curves, as indicated in the figure.

It may now be noted that the thickness of the outer tube in the vicinity of the fillet is very small, which indicates high stresses at the thin section. It becomes apparent that means for maintaining the uniformity of thicknesses of the equitorque tubes are identical with means for reducing stress concentration. Some of these means are illustrated in

[1] L. S. Jacobsen, Torsional Stress Concentration in Shafts of Circular Cross Section and Variable Diameter, *Trans. ASME*, vol. 47, p. 619, 1925.

[2] The thicknesses of the tubes have been chosen so that the angle of twist per unit length of each tube has the same value. This division results in a wall thickness h, which decreases as the tubes become larger in diameter, as shown; it is necessary in order that the assembly of concentric tubes will twist without relative displacement and thus, by fulfilling the condition of continuity, simulate the twisting of the solid shaft.

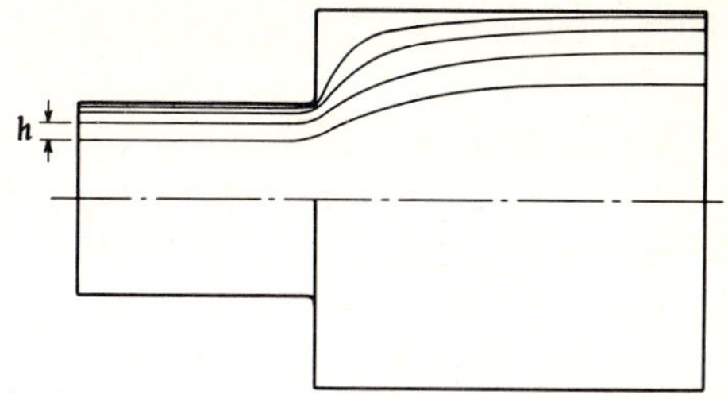

Fig. 4-3 Circular stepped shaft divided into equitorque tubes.

Fig. 4-4, in which it may be seen that low stress concentration is indicated by *gradual* variation of the distances between the equitorque-tube curves. An interesting case is Fig. 4-4(j) in which the strength of a crank under repeated loading was increased 40 percent by cutting an eccentric circular groove as indicated in the improved section, although the groove lowered the moment of inertia of the crankpin by 50 percent.[1]

Tension members A similar method of representing stress concentration in tension members makes use of lines indicating the direction of the principal stresses. Figure 4-5 shows an axially loaded tension member with the direction lines of the principal stresses. In the upper end

[1] *Zeit. Ver. deut. Ing.*, vol. 76, p. 981, 1934.

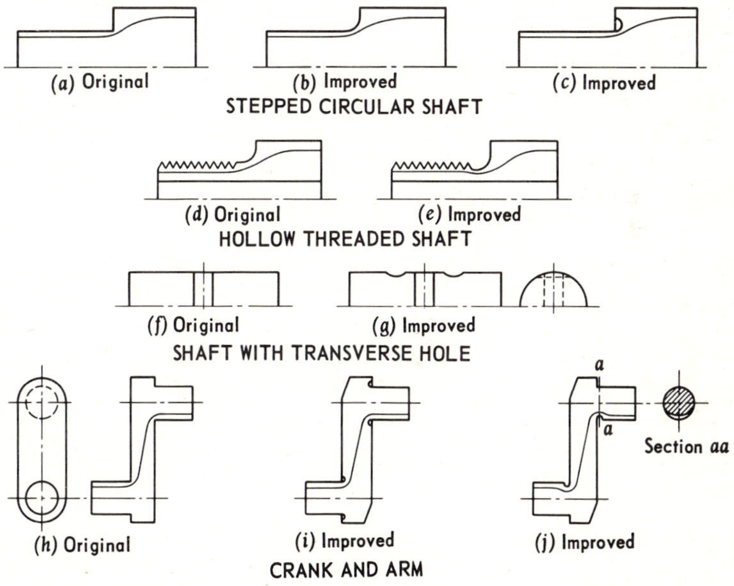

Fig. 4-4 Examples of stress reduction in members subjected to torsion.

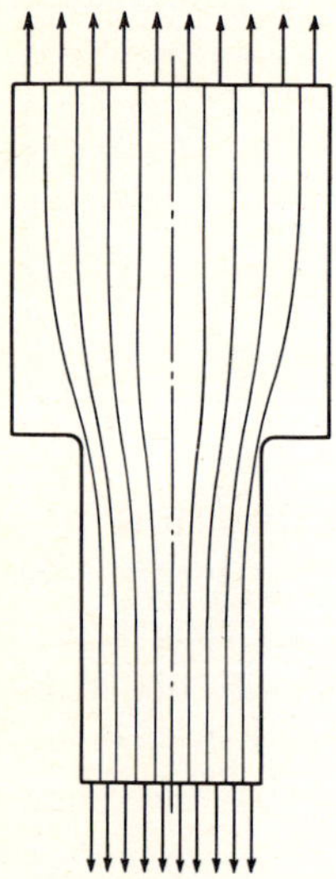

Fig. 4-5 Stress-direction lines for tension member.

of the member, the lines are parallel and are drawn equal distances apart to indicate uniform stress distribution. These lines are then extended to join with the corresponding lines for the lower end of the member, as indicated.

In the vicinity of the fillet, it may be noted that the changes in slopes of the stress-direction lines are pronounced and the lines are close together, both of which may be shown to be an indication of stress concentration.[1]

As in the case of the equitorque-tube visualization of torsional stress concentration, the stress-direction lines for tension members may be used to suggest means for reducing stress concentration by adopting shapes of the member to favor equispaced and low-slope lines of stress direction. Some of these methods are illustrated in Fig. 4-6.

The existence and location of stress concentration in a nonuniform axially loaded member, as shown in Fig. 4-7(*a*), can be determined in a qualitative manner.[2] Figure 4-7(*b*) shows a free-body diagram of the portion $EBB'E'$. Since the side EB is an unloaded surface and $E'B'$ is subjected to tension, a shearing force Q_n must be developed on the side EE' for equilibrium. This force is accompanied by a shearing force Q_t along $E'B'$, which must be balanced by a tensile force F_t on the side EE'.

[1] R. V. Baud, Avoiding Stress Concentration by Using Less Material, *Prod. Eng.*, May, 1934.

[2] Glen Murphy, "Advanced Mechanics of Materials," p. 91, McGraw-Hill Book Company, New York, 1946.

Since the small element at the corner E' is in a state of biaxial stress, there exists a principal tensile stress that is greater than the average stress on the side $E'B'$. In general, the same situation exists at all points in the vicinity of E.

Further reason for the existence of the shearing stress may be found by considering the geometry of the member in the vicinity of the change in cross section. If the line $A'B'$ is considered to be fixed, the tensile load will cause the line EE' to deform more than BB' so that the line $ADCEB$ will be deformed as shown in Fig. 4-7(c). The resulting change in shape is indicative of the presence of shearing strains[1] which increase in magnitude toward point E. These shearing strains are accompanied by shearing stresses that are also a maximum near point E; the principal stress is a maximum at E. The approximate stress distribution across the section DE is shown in Fig. 4-7(d), where the maximum stresses are dependent on the ratio of the widths DE and AB as well as on the radius of the fillets at D and E.

[1] Shearing strain is defined as the change in a right angle resulting from deformation.

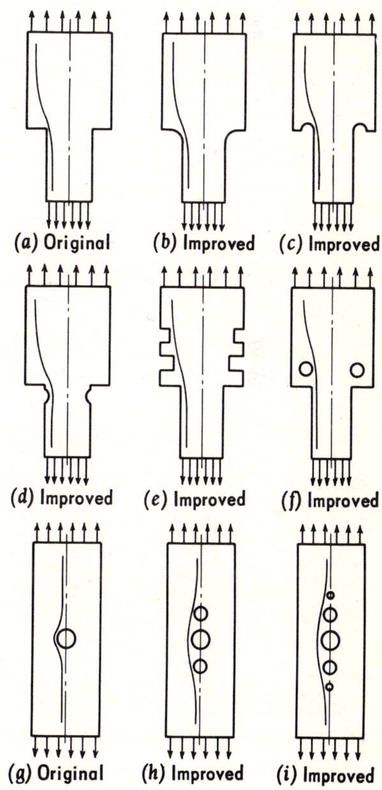

Fig. 4-6 Examples of reduction of stress concentration in members subjected to tension.

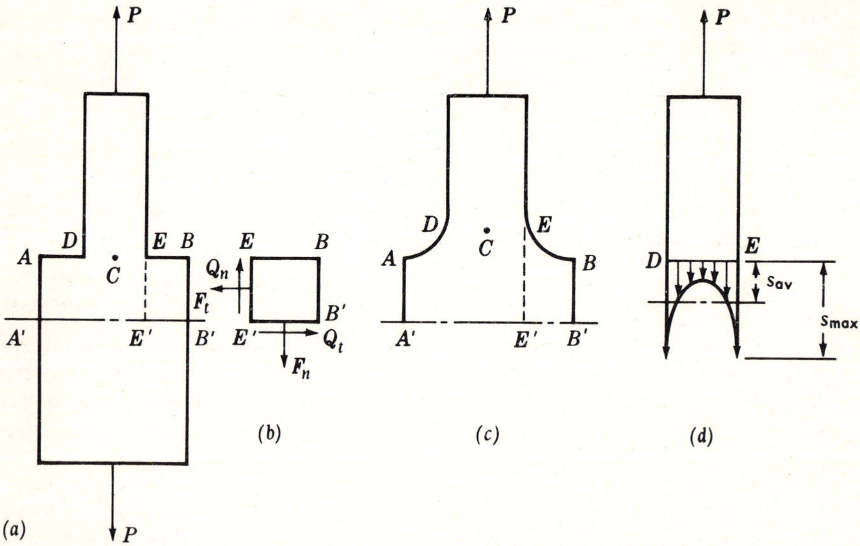

Fig. 4-7 Stress concentration at change in cross section.

4-6 SERIOUSNESS AND MITIGATION OF STRESS CONCENTRATION

It is evident from the preceding discussion that high concentration of stress may exist in certain regions in machine and structural members. Since a crack which may form at a region of stress concentration invariably leads to failure of the member, it is important that the designer reduce the concentration to the practicable minimum. The designer can effectively reduce stress concentration in most cases by simple expedients, such as the use of gradual transition curves and generous fillets.

The seriousness of stress concentration depends on the properties of the material and on the type of loading, namely, whether the load is static or cyclic.

Static loading It was noted in Art. 4-4 that the stress at the edge of a small transverse circular hole in a tension member has a value three times that in the sections remote from the hole. The distribution of stresses across the section including the hole is indicated in Fig. 4-1(*c*).

If the material of which the member is made has a stress-strain diagram that is a straight line up to rupture, the proportions of the stress-distribution diagram in Fig. 4-1(*c*) will not be altered as the load on the member is increased. This is true because the stresses would increase proportionately until the maximum stress in the member reached the breaking stress of the material. At this stage, a crack would form at the edge of the hole. The crack would have the effect of introducing additional stress concentration and of decreasing the section carrying the load. Both of these effects would cause very rapid failure of the member.

Brittle materials, such as cast iron, undergo relatively little yielding; therefore a member which has stress concentration and which is made of brittle material will fail substantially as described above, since the concentration of stress, as indicated by Fig. 4-1(c), remains in the member until failure occurs.

In considering the failure of a member made of a *ductile material*, such as wrought steel, and with stress concentration, it is necessary to note that the stress-strain diagram for the material indicates a region which involves plastic flow. The plastic flow occurs beyond the yield point and allows considerable strain to take place before failure occurs. The stress-distribution diagram in Fig. 4-1(c) will be valid if the stresses are below the proportional limit; however, beyond the yield point, plastic flow at the region of stress concentration will cause the stresses to be redistributed. The redistribution of stresses will be one that will approach a uniform distribution. To illustrate this change in distribution, one-half of the section in Fig. 4-1(c) is shown in Fig. 4-8. Figure 4-8(a) shows an

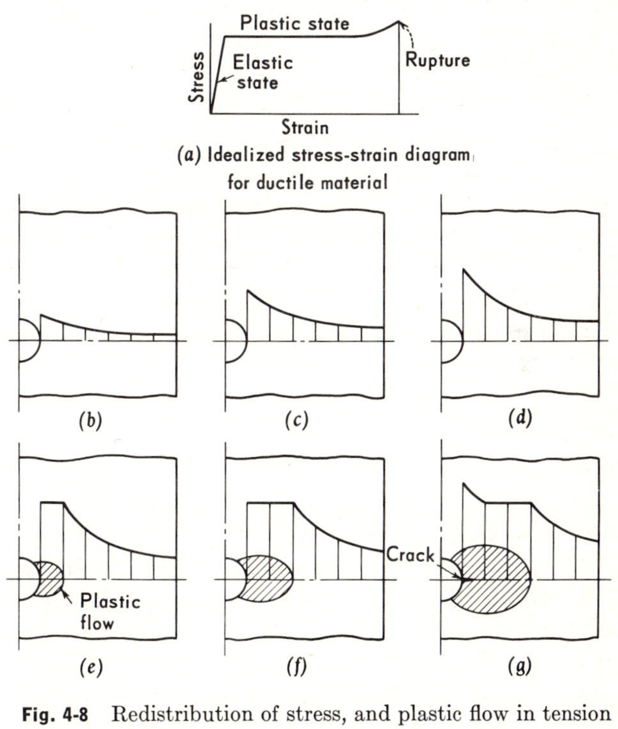

Fig. 4-8 Redistribution of stress, and plastic flow in tension member made of ductile material [see Fig. 4-1(c)].

idealized stress-strain diagram for a ductile material, and (*b*) represents a portion of the bar adjacent to the hole and the stress-distribution diagram for low stresses in the member. Figure 4-8(*c*) represents the diagram for increased stresses, and (*d*) shows the stage when the maximum or peak stress is just equal to the yield point. Figure 4-8(*e*) shows the diagram for an increased load on the member, and the corresponding crosshatched area indicates the region that has undergone plastic flow, leading to the next stage (*f*). Finally (*g*) indicates the diagram at the time the peak stress reaches the breaking stress of the material, when a crack forms and the member fails.

From the accompanying sequence diagrams, it is apparent that plastic flow in ductile materials will lessen the seriousness of stress concentration. Frequently this plastic flow takes place in machine and structural members when the working loads are applied. Structural members with rivet holes, keyways in shafting, and riveting or welding in ships that are "broken in" in calm waters before being subjected to service on the high seas are examples of this kind of reduction of stress concentration.

At times an intentional overload is applied to a member to cause plastic flow and corresponding stress redistribution. Examples are found in overstressing steel link chains and stressing generator rotors by overspeed test operation. See Art. 6-3.

It should be noted that, when a crack develops, the *real breaking stress* of the material has been reached. In a ductile material, the real breaking stress is appreciably higher than that indicated by the conventional stress-strain diagram, which is expressed in terms of the original area of the test specimen.

In concluding the discussion of the effects of the properties of material on static stress concentration, it may be stated that stress concentration in *static loading* is very serious in brittle materials, and it is less serious in ductile materials owing to the relief of stress concentration by plastic flow. However, the designer should reduce stress concentration wherever possible without regard to the class of material of which the member is made.

Cyclic loading In the preceding discussion it was stated that in static loading stress concentration is very serious in members made of brittle materials and is somewhat less serious in members made of ductile materials. In the latter case it was noted that the seriousness of stress concentration was lessened by the local plastic flow, which resulted in a more favorable distribution of stresses.

In cyclic loading, however, stress concentration is always serious since the ductility of the material is not effective in relieving the concentration of stress caused by cracks, flaws, surface roughness, or any sharp discontinuity in the geometrical form of the member. If the stress at *any* point in a member is above the endurance limit of the material, a crack may develop under the action of the repeated load, and the crack will, in all probability, lead to failure of the member. It is important to realize that, even though the region subjected to the peak stress is extremely small, the crack is liable to form and, once formed, will lead quickly to failure of the member.

The last statement is emphasized by the report[1] that approximately 90 percent or more of the number of cycles of application of load to produce failure of a member are undergone before the crack may be detected by laboratory means. It is evident then that the opportunity for an inspector to detect a crack under service conditions in time to prevent failure is relatively slight.

Figure 4-9 illustrates the effect of surface condition on the endurance limit of test specimens, i.e., the marked reduction in endurance limit due to surface condition. The figure shows also that high-strength alloy steels are more sensitive to stress concentration. The lower ductility of the high-strength alloy steels is partly responsible for the increased sensitivity, but it is not generally true that ductility is insurance against stress concentration in fatigue.

[1] Battelle Memorial Institute, "Prevention of Fatigue of Metals," appendix 27, John Wiley & Sons, Inc., New York, 1941.

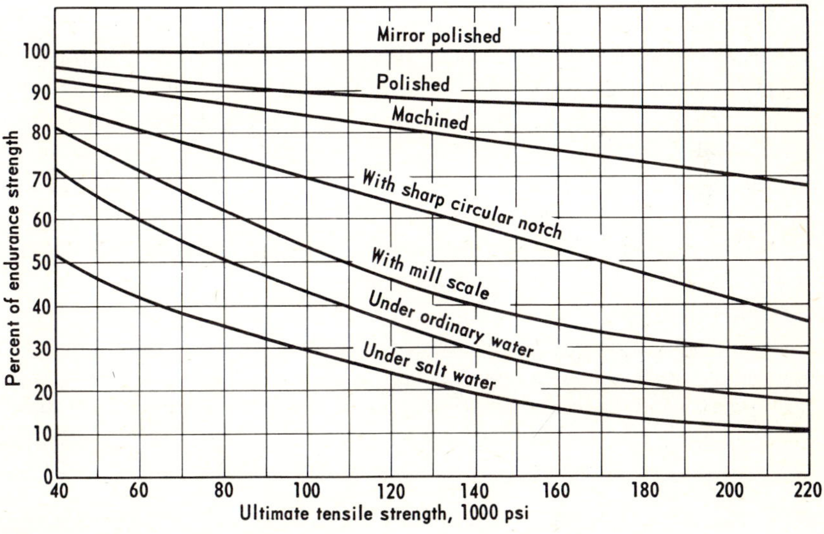

Fig. 4-9 Effect of surface condition on endurance strength.

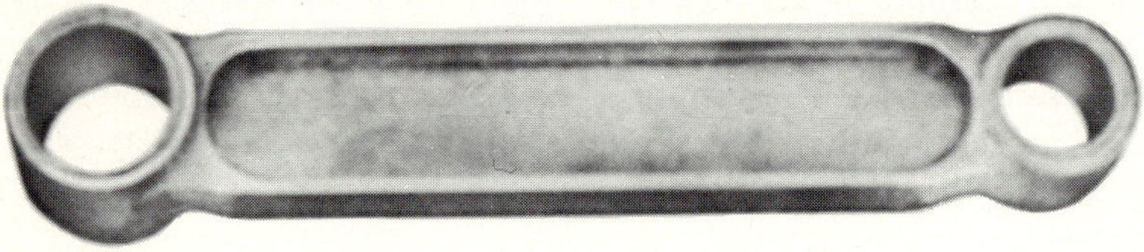

Fig. 4-10 Link rod for aircraft engine.

The gain in endurance limit is the reason for polishing aircraft-engine connecting rods and link rods. The overall polish is expensive, but it is justified by the increase in safety. Figure 4-10 shows a polished link rod for a radial aircraft engine.

The *endurance limit*, or *fatigue limit*, is the maximum stress that a member subjected to a completely reversed load can withstand for an infinite number of load applications without failure. This is shown in Fig. 5-4 by the stress at the horizontal part of the plots, which, for SAE 1050 steel, is approximately 41,000 psi. To save weight and space, many machine members are designed for *finite life*. For 10^4 repeated applications of the load, the stress for SAE 1050 steel is approximately 52,000 psi. The statistical nature of fatigue tests is discussed in Art. 5-3.

The endurance limit is not strictly a property of a material although it is frequently referred to as such, because it is affected by size, form, and finish of the part as well as by the stress range. In the laboratory the specimens are usually tested on a rotating-beam testing machine which gives results for repeated, completely reversed bending. The specimens are machined, according to specifications for one of the standard sizes,[1] and polished, to avoid stress concentration and effect of surface roughness.

Although it is necessary in machinery to use threads, oilholes, keyways, and other forms that introduce concentration of stress, it is generally possible for the designer to devise means for markedly reducing stress concentration. As has been discussed, this is especially important in members subjected to cyclic loading. When the designer realizes the seriousness of stress concentration in cyclic loading, coupled with the frequent difficulty of estimating service loads, the chance of occasional or accidental overloads, and the possibility of large forces due to resonant

[1] H. F. Moore and G. N. Krouse, Repeated Stress (Fatigue) Testing Machines Used in the Materials Testing Laboratory of the University of Illinois, *Univ. Illinois Eng. Expt. Sta. Circ.* 23, 1934.

vibration, he will take steps to reduce the loads to a minimum and to avoid the use of geometrical forms that invite stress concentration. He will also endeavor to select proper material, to specify proper heat treatments and surface finish, and to guard against careless machining, handling, and inspection.[1]

Some means for reducing stress concentration that apply especially to members subjected to repeated loading are illustrated in Fig. 4-11.

Figure 4-11(b) shows a circular fillet used to reduce stress concen-

[1] Inspection marks and numeral stampings have been known to be the cause of failure. See *Prod. Eng.* January, 1941, p. 25.

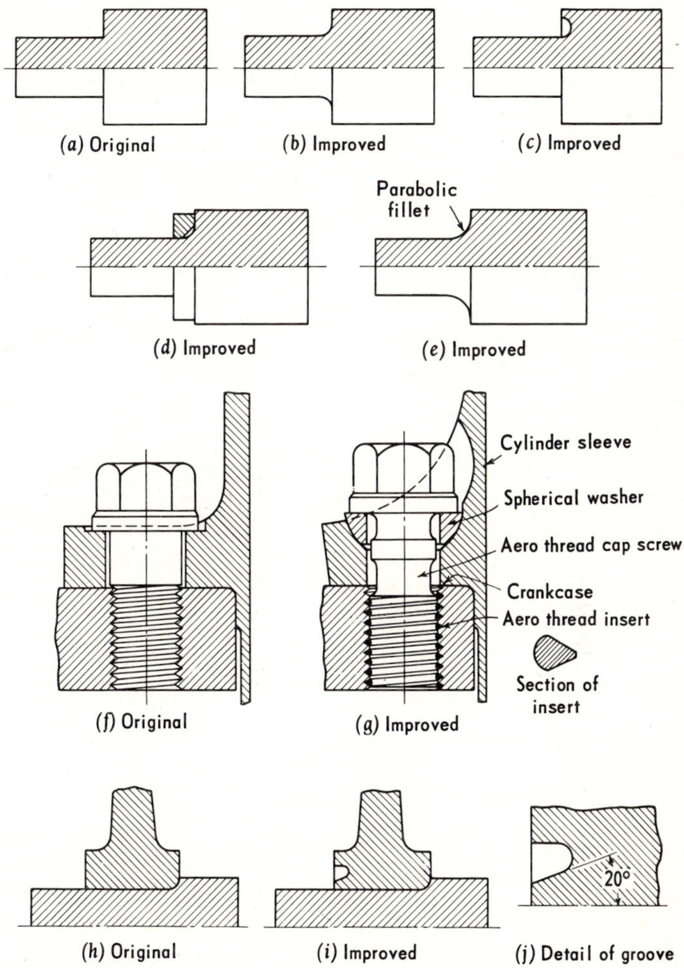

Fig. 4-11 Examples of reducing stress concentration.

tration. The circular fillet is the usual form, although it has been shown that a spiral or parabolic fillet, as shown at (*e*), results in slightly less concentration.[1] The gain in the use of the noncircular-arc fillets is generally not justified in view of the increased manufacturing cost. Figure 4-11(*d*) illustrates a separate collar used to provide a flat thrust cheek. The design in (*c*) is preferred, however, to that in (*d*) where the flat cheek is desired.

Figure 4-11(*f*) shows a cylinder flange bolted to the crankcase of an internal-combustion engine. It is evident that stress concentration will exist at the fillet joining the cylinder wall to the flange, at the corner of the counterbore, and at the roots of the threads of the screw. Figure 4-11(*g*) illustrates an improved design using large-radii fillets and the Aero thread employing an insert that reduces the stress concentration at the roots of the threads on the screw (see Art. 9-3 for discussion of the Aero thread).

Figure 4-11(*h*) shows a hub of a gear, wheel, or sheave pressed on a shaft. The press fit and loading will produce contact stresses in the shaft.[2] The improvement at (*i*) employs a "relieving groove" cut in the face of the hub. The groove increases the flexibility of the hub at its end, thereby cushioning the contact pressures. A desirable form of the groove is indicated in the sketch at (*j*).[3]

The groups of examples shown in Figs. 4-4, 4-6, and 4-11 should be regarded as commonly used methods for relieving stress concentration for the particular cases shown. It is necessary for the designer to use his ingenuity to devise appropriate methods for each particular case. An acquaintance with the literature on the general subject of stress concentration is valuable for this purpose.[4]

In addition to avoiding forms that cause severe concentration of stress, and selecting suitable material, there are means at the disposal of the designer for increasing the fatigue strength of members in special applications. Some of these are discussed below.

Cold working,[5] especially rolling at fillets on shafts, may materially increase the fatigue strength. In this process the shape of roller, the

[1] E. E. Weibel, Studies in Photoelastic Stress Determinations, *Trans. ASME*, vol. 56, p. APM-56-13, 1934.

[2] M. V. Barton, The Circular Cylinder with a Band of Uniform Pressure on a Finite Length of the Surface, *J. Appl. Mech.*, vol. 8, no. 3, p. A-97, 1941.

[3] Peterson and Wahl, Fatigue of Shafts at Fitted Members with a Related Photoelastic Analysis, *J. Appl. Mech.*, vol. 2, no. 1, p. A-1, 1935; Buckwalter and Peterson, Locomotive Axle Failures and Wheel Press Fits, *Railway Mech. Eng.*, April, 1935, p. 127.

[4] "Three Keys to Satisfaction," Climax Molybdenum Company, New York.

[5] O. J. Horger, Effect of Surface Rolling on the Fatigue Strength of Steel, *Trans. ASME*, vol. 2, no. 4, p. A-128, 1935.

pressure, and the axial feed are important for best results. Good results have been secured in railway axles by surface rolling.

Shot peening[1] is widely used, particularly in the automotive and aircraft industries, to increase the fatigue life of highly stressed parts subjected to fatigue loading. It is known that fatigue cracks spread only in regions of tensile stress and that by shot peening cyclically loaded parts a layer of compressive stress is introduced. Since surface material is weaker than sound subsurface material, and since load stresses in many cases are of greatest magnitude at the surfaces, the compressive stress at the surface induced by shot peening will prevent or delay the formation and spreading of fatigue cracks.[1] Substantial economy of manufacture, increased safety, and saving of weight can be achieved in cyclically loaded parts by the proper use of shot peening.

Understressing occurs when a metal is subjected to a stress somewhat less than the endurance limit, whereas *overstressing* occurs when a metal is subjected to a stress exceeding the endurance limit. Both understressing and overstressing[2] may increase the fatigue strength of members. Overstressing must be conducted very cautiously, preferably by successively increasing loadings. This procedure has been called "coaxing."

Improper heat treatment should be guarded against. Surface decarburization results in lowering the fatigue life of members. Quenching may cause surface cracks. An excessively deep layer of surface residual compressive stress induced by too deep nitriding or carburizing may induce in the core residual tensile stresses so high that subsurface fatigue cracks may occur in service and lead to fracture of the member.[3] See Art. 6-3.

4-7 EXPERIMENTAL METHODS FOR INVESTIGATING STRESS CONCENTRATION

The stress-concentration factor has been defined as the ratio of the maximum (localized) stress to the nominal stress on the section (often based on net section) as calculated from the ordinary formulas of mechanics.

In determining the maximum stress, it is generally necessary to resort to one of the experimental methods. A knowledge of these methods is of great value to the designer since such knowledge enables him to appreciate stress concentration and the applications and limitations of the factors, and also since a firsthand acquaintance with the experimental methods aids the designer in visualizing the effects of various design

[1] See Chap. 6; also J. O. Almen and P. H. Black, "Residual Stresses and Fatigue in Metals," McGraw-Hill Book Company, New York, 1963.

[2] J. B. Kommers, Effect of Understressing and Overstressing in Fatigue, *Proc. Soc. Exptl. Stress Anal.*, vol. 3, no. 2, 1946.

[3] See Battelle Memorial Institute, *op. cit.*, p. 97.

Fig. 4-12 Photoelastic photograph of loaded gear teeth. (*Courtesy of University of Illinois Engineering Experiment Station.*)

expedients in reducing stress concentration. The photoelastic method is especially useful in this respect, and laboratory experience with this method is invaluable to the mechanical engineer.

The photoelastic method This method uses polarized light and transparent plastic models whose contours are geometrically similar to those of the prototype in which the stress distribution is desired. In general, the models are made from such materials as Catalin 61-893, Castolite, Columbia resin CR-39, exposy resin, or urethane rubber.[1]

An interference of light rays in a stressed photoelastic model produces isochromatic fringes or bands of constant principal stress difference. Figure 4-12 shows a photoelastic model of a loaded gear tooth; the region of stress concentration is indicated by the closely spaced fringes at the fillets of the gear tooth.[2] In three-dimensional photoelasticity a three-dimensional model is constructed. Slices are removed from the model

[1] J. W. Dally and W. F. Riley, "Experimental Stress Analysis," pp. 165–221, McGraw-Hill Book Company, New York, 1965.

[2] M. M. Frocht, "Photoelasticity," John Wiley & Sons, Inc., New York, 1941.

and analyzed after the deformations due to the applied loading have been locked into the model (generally by the stress-freezing technique).

Two-dimensional photoelasticity is based upon a phenomenon whereby certain transparent plastics become doubly refracting when subjected to stress. The plastic material experiences a change in optical characteristics which is directly proportional to the difference in principal stresses and is observed, by means of polarized light, in an instrument called a *polariscope*.

During the past few years the birefringent-coating method of photoelasticity has become increasingly popular. In this technique, a thin layer of photoelastic plastic either is bonded to the surface of a metallic specimen by means of a reflective cement or is sprayed or brushed on the surface and hardened by heating. When the specimen is loaded and strained, the photoelastic coating responds in the same manner as the transparent model described earlier, and the resulting fringe pattern is observed in a reflected-light polariscope. This fringe pattern is related to the surface strains of the metallic specimen.

The advantages of the birefringent-coating method are that the investigation can be carried out on the actual metal part, rather than on models, and the reflective polariscope is portable, thus enabling it to be moved to the actual machine or structure.

A disadvantage is that only surface stresses can be investigated. This is not a serious disadvantage in most cases, however. Surface stresses, rather than subsurface stresses, are frequently the critical ones, because stress concentration is usually pronounced at surfaces and because surface material is weaker than subsurface material. See Art. 6-1.

The electric strain gauge This gauge has found extensive use in industry on account of its accuracy, its adaptability to all kinds of structures and parts to indicate local surface conditions of strain, and its ability to indicate either static or dynamic strains. In the latter use, an oscilloscope or an oscillograph is necessary. The most widely used types of electric strain gauges are the resistance-wire type (the SR-4 gauge made by Baldwin-Lima-Hamilton Corporation) and the metal-foil gauge, which are bonded to the part under investigation. Their operation is based on the change in electric resistance of either a fine wire or foil element as it changes in length when the part is strained. The gauges are small and light (so-called "postage-stamp gauges") and have been used to measure strains in many applications, for instance, in fillets of engine crankshafts of aircraft in flight, in drive rods and crankpins of locomotives, and in concrete structures such as dams. These gauges are used in many fields, for example, for measuring torques in rotating shafts and also as internal-combustion-engine pressure indicators.

The use of *brittle coatings* on machine parts has extensive use for a qualitative measure of strains and in indicating the directions of principal stresses. In this method the part is sprayed with a lacquer which becomes brittle on drying. When the part is loaded, cracks form at the region of high stress. The pattern of the cracks indicates the distribution of principal tensile stresses, and the closeness of the cracks may be used to estimate the magnitude of the stress. This method is frequently used to indicate the direction of the principal stresses prior to installing electric strain gauges so that they may be orientated most advantageously. Brittle coating and equipment are made by the Magnaflux Corporation under the trade name Stresscoat.

Brittle-material method Some brittle materials have an almost constant stress-strain characteristic to rupture. Stress concentration in a member made of such a material will exist until rupture occurs, as discussed in Art. 4-6. Since the load at rupture of the member can be determined by test, this method enables one to compare the peak stress at the point of stress concentration with the calculated stress and thus to determine values for stress-concentration factors.

Pottery plaster is one of the brittle materials that is convenient to use for models of machine members. Figure 4-13 is a stress-strain diagram for a high-grade pottery plaster. In determining the stress-concentration factor for a shaft with a shoulder, for instance, a plaster model of the shaft is cast and machined. The model is then gradually loaded in a manner that simulates the actual loading of the member, such as torsion or bending, and it is closely observed as the load is applied in order to detect the formation of a crack. The crack will form at the most highly

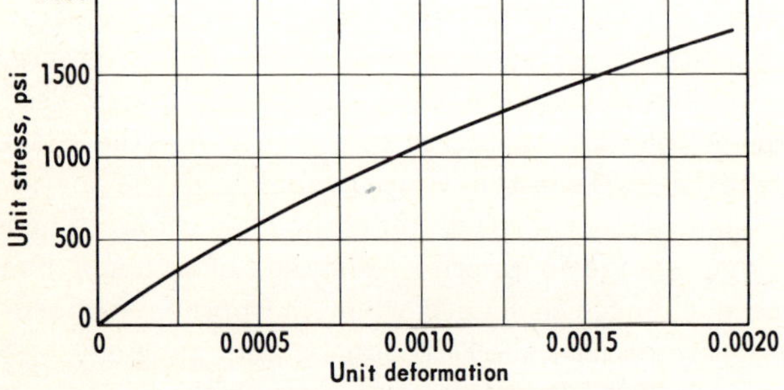

Fig. 4-13 Stress-strain diagram for cured plaster in compression. (*From Roark and Hartenberg, Predicting the Strength of Structures from Tests of Plaster Molds, Univ. Wisconsin Bull.* 81, 1935.)

stressed region, and when it forms it is apparent that the stress at that region is the breaking stress of the plaster and is the maximum stress in the model. By the use of the loading at the instant the crack appears, the nominal stress at the region of the crack is calculated by means of the appropriate equation. The ratio of the actual maximum stress to the calculated stress is the static-stress-concentration factor. The brittle-material method may be used for one-, two-, or three-dimensional problems.[1] The results are qualitative.

Short-gauge-length method This method employs a delicate extensometer that has a short gauge length (0.1 in., by R. E. Peterson). The strain is measured at critical points on the surface of the machine member. The method requires a highly developed technique but gives very accurate and informative results.[2]

4-8 STRESS-CONCENTRATION DESIGN FACTORS

Three common types of loading in machine members are tension, bending, and torsion. Tension and bending induce normal stresses s, and torsion (as well as direct shear) induces shear stress s_s.

The *theoretical stress-concentration factor* is a design factor that is used to take into account the concentration of stress. This factor is termed a "theoretical factor" because it is determined mathematically or by the photoelastic method.

The types of simple loading on members without and with stress concentration are shown in Fig. 4-14 and illustrate what the theoretical stress-concentration factor is and how it can be used in design equations. There are, of course, other kinds of loading and many other forms of members; those shown are samples. In design, each case is a special one from the standpoint of loading (tension, bending, and torsion or a combination of them); shape of the member (bar with transverse hole, shaft with circumferential groove, stepped shaft, etc.); type of material (brittle or ductile); and kind of loading (static or cyclic). This article is intended to acquaint the designer with the appropriate methods for selecting the factor and properly using it and also to help him gain an appreciation of its importance.

The theoretical stress-concentration factor may be defined as the maximum stress in a member (at, for instance, a notch or a fillet) divided

[1] F. B. Seely and J. O. Smith, "Advanced Mechanics of Materials," 2d ed., John Wiley & Sons, Inc., New York, 1952; Brewer and Glassco, Determination of Strain Distribution by the Photogrid Process, *J. Aeronaut. Sci.*, vol. 9, no. 1, November, 1941.

[2] Peterson and Wahl, Two- and Three-dimensional Cases of Stress Concentration, and Comparison with Fatigue Tests, *J. Appl. Mech.*, vol. 3, no. 1, p. A-1, 1936.

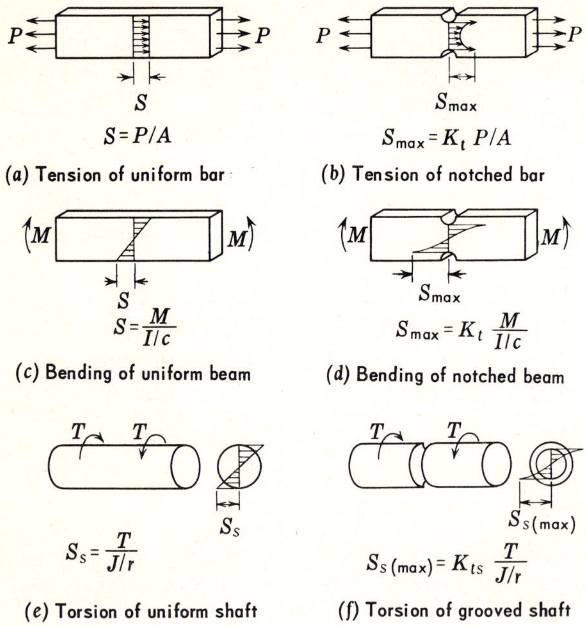

Fig. 4-14 Simple cases of loading without and with stress concentration.

by the nominal stress at the same section as calculated by a simple equation from strength of materials. The maximum stress is determined mathematically or photoelastically. The nominal stress is usually calculated across the net or minimum section.

The notch factor is defined as the strength of an unnotched member divided by the strength of a notched member of the same size and material.

Stress-concentration design-factor charts for several common shapes of members are given in Appendix X.[1]

Figure 4-14(*a*) shows an example of tension on a uniform bar in which the tensile stress is uniformly distributed over cross sections and is equal to P/A, where A is the cross-sectional area. At (*b*) is shown a similar bar with notches. As discussed in Art. 4-4, the notches cause a con-

[1] These charts as well as symbols, nomenclature, and methods are taken by permission of the author and publisher of "Stress-concentration Design Factors," by R. E. Peterson, John Wiley & Sons, Inc., New York, 1953. This source contains discussions of the determination of the factors, an extensive bibliography, and some 94 charts for factors for a wide variety of types of loading and shapes of members. Sixteen of these charts have been selected for inclusion in Appendix X.

centration of stress so that the maximum stress s_{max} is greater than P/A and is equal to $K_t \times P/A$. In this case A is the *net area* across the notches. By definition, then, K_t is the ratio of the maximum stress to the nominal stress P/A. Its value is always greater than unity.

For design purposes, we may solve the equation for the area

$$A = K_t \times \frac{P}{s_{max}}$$

In the design of a machine member, the applied load can be determined from the loading data. The value of s_{max} can be limited to the allowable stress as discussed in the following chapter. The value for K_t may be found or estimated from available data such as the charts in Appendix X.

Figure 4-14(*c*) shows a uniform beam in bending for which the equation $s = Mc/I$ applies, and (*d*) is a beam with notches for which $s_{max} = K_t \times Mc/I$. At (*e*) and (*f*) are shown cases of torsion. Here the stress-concentration factor is K_{ts}, the s in the subscript referring to shear stress, so that $s_{s(max)} = K_{ts} \times Tr/J$.

In all cases, in computing the area or section modulus, the *net* section at the notch, fillet, hole, etc., should be used (along with the stress-concentration factor) to determine the quantity desired.

In notched flat bars and stepped bars, the maximum stress at the notch or fillet is a uniaxial system. In a grooved or stepped shaft in tension or bending, however, in addition to the maximum stress as given by the equation containing the stress-concentration factor as discussed above, there is also a stress at right angles to the maximum stress; i.e., the stress system is biaxial. When there are two stresses at right angles, we need a theory of strength in order to predict failure. The Mises[1] criterion gives values which are in good agreement with results of failure tests of members made of ductile material. By taking into account both stress concentration and the Mises criterion, a new factor has been determined and labeled $K_{t'}$, the prime indicating the combined factor. For members in which the highest stressed element is subjected to biaxial tension, the value of $K_{t'}$ is always less than K_t, so that if the former is not available the latter may be used. The greatest difference is not over 15 percent. In the charts, values for the case of shear ($K_{t's}$) are not given for torsion of round shafts because with notches or fillets they are the same as K_{ts} values.

In cyclic loading, the effect of the notch or fillet is usually less than predicted by the use of the theoretical factors as discussed above, except for very small notches, although for high-strength steels the full theo-

[1] R. von Mises, Mechanik der festen Korper im plastisch deformablen Zustand, *Nach. Ges. Wiss. Göttingen Jahresber, Geschäftsjahr, Math-physik. Kl.*, 1913, p. 582; also A. Nádai, Theories of Strength, *Trans. ASME*, vol. 56, p. APM-55-15, 1933.

retical factor is generally obtained. The difference depends on the *stress gradient* in the region of stress concentration and on the hardness of the material. The term *notch sensitivity* has been applied to this behavior. Notch sensitivity may be defined as the degree to which the theoretical effect of stress concentration is actually reached. The stress gradient depends mainly on the radius of the notch, hole, or fillet and on the grain size of the material.[1]

[1] "Metals Handbook," 8th ed., vols. I, II (edited by Taylor Lyman), American Society for Metals, Cleveland, 1964.

5
Design Criteria

5-1 GENERAL

In Chaps. 3 and 4, relations were discussed for determining the values of stresses that are induced by applied loads on machine members. The effects of impact loading and of sharp changes in section and other "stress raisers" were considered so that the designer can calculate or estimate the probable value of the maximum stress[1] in the part being designed. One of the major decisions confronting the designer is the selection of appropriate design (or failure-prevention) criteria. This is largely influenced by the *mode of failure* of the machine or structural member. The common modes of failure are as follows:

[1] In case residual stresses are expected in the part because of casting, fabrication, or heat-treatment, these may also be included in the maximum stress; however, residual stresses are difficult to estimate so that, if it is not feasible to relieve them by annealing or other means, their influence may more properly be dealt with by using an increased value of the factor of safety or by using methods as discussed in Chap. 6.

I. Yielding
   A. Maximum stress exceeding the yield strength of the material, causing it to deform plastically
   B. Creep deformation, whereby the member deforms under a constant load, usually at an elevated temperature

II. Fracture
   A. Due to static loads
   B. Due to fatigue loads
   C. Due to impact loads

III. Excessive elastic deflection

IV. Wear

V. Buckling

VI. Corrosion fatigue and caustic embrittlement

The mode of failure in turn is influenced by such factors as the type and duration of the load, the shape of the part (effect of stress concentration), and the nature of the material (ductile or brittle). There are a number of conditions that cause a normally ductile material to act like a brittle material. They are:[1]

1. Repeated or fatigue loading
2. Impact, particularly at low temperature
3. Creep
4. Triaxial state of stress
5. Severe quenching without tempering
6. Strain hardening accompanying yielding

When the designer has determined the expected maximum induced stress in the member to be designed and has selected a suitable material, he is in a position to apply the design criteria in calculating the cross-sectional dimensions of the member. These criteria are usually in the form of an equation. In the selection of the material to be used, the choice is based partly upon its mechanical properties so that its *allowable stress* will be at least equal to and frequently greater than the maximum induced stress. The allowable stress is thus the upper limit to the induced stress. In order to provide a margin against failure, it is common practice in machine design to determine the allowable stress by dividing the failure stress for the member by a *factor of safety* (fs). The failure stress depends squarely on the method of failure and is discussed in the following articles.

[1] J. B. Hartman and R. E. Benner, Stress Analysis in Design, *Machine Design*, March, 1954, pp. 3–4.

An expression relating allowable stress to failure stress (or strength) is as follows:

$$s_{all} = \frac{s_{crit}}{fs} \tag{5-1}$$

where s_{all} = allowable stress, psi
s_{crit} = failure stress (or strength), psi

It must be cautioned that in the case of column design it is necessary to obtain an allowable *load* by applying a factor of safety to a failure load. This approach is necessary since load and stress are not linearly proportional. Whenever failure is due to deformation, strain, or some other quantity, the allowable value of this quantity is determined by dividing the quantity at failure by a suitable factor of safety.

The factor of safety is a number used to divide into the appropriate property of the material to obtain the allowable stress. In design for strength to avoid fracture or permanent deformation of parts subject to static loads, the appropriate property for static loading of brittle material is the ultimate strength, and for ductile material it is the yield point, whereas in cyclic loading for infinite life the appropriate property is the endurance limit; see Fig. 3-22. The factor of safety, particularly in control systems and in space-craft requirements, becomes an index for expressing the reliability of the product. If the product is part of a missile going to the moon, a reliability of 99.99 percent would be appropriate, however expensive. If the product requires less reliability, such as a part of a jukebox, 85 percent reliability would be appropriate and much more economical. Many design engineers pay for reliability that they do not need.

The determination of reliability is a statistical problem. There is little information available on the relation of endurance limits as found in the laboratory and the probability of failures of actual mechanical parts in service. Among the several influences for the difference between laboratory test results and service behavior are residual stresses, as discussed in Chap. 6. Because of the difference between expected life as determined by laboratory tests and actual life in service, the usual practice is to make the best use of the available information and data in the design of the part and then test the parts under conditions simulating service conditions as closely as possible. This procedure will indicate any need for redesigning the part.

In one instance, namely, rolling-contact bearings, the variables in operating conditions of the various components of the bearing kindly limit statistical variations so that the performance of the bearings can be closely predicted. This is discussed in Art. 22-4.

5-2 THEORIES OF FAILURE BY YIELDING AND BY FRACTURE UNDER STATIC LOADING

In the design of machine or structural members to resist failure by yielding or fracture, the question arises as to what conditions of stress will cause these modes of failure. The ability of a material to undergo an appreciable amount of permanent deformation before fracturing is characteristic of a "ductile" material. If the amount of deformation prior to fracture is small, the material is considered "brittle."[1] In general, *ductile* materials have an elongation in simple-tension tests greater than 5 percent,[2] whereas brittle materials have an elongation less than 5 percent.

The strength of ductile materials in simple tension is defined by the yield strength or by an offset yield stress when a definite yield point does not exist. Yield strengths of a ductile material in tension or in compression may be assumed to be equal. In the case of brittle materials, the strength is usually assumed to be the ultimate stress in tension or compression.

Over many years, several failure theories have been developed. However, only three currently used theories will be considered here. According to each theory of failure (whether a stress theory or an energy theory), the selected criteria of failure for the complex stress state are equated to the value of the same criteria corresponding to failure in a simple tensile test.

Frequently, the question arises concerning the significance of the stress-concentration factor. The failure load for a statically loaded member made of ductile material is not markedly affected by stress concentration because of the redistribution of stress that takes place beyond the yield point. Therefore, in the design of a member made of a *ductile* material to support a *static load*, it is permissible, generally, to modify the value of the stress-concentration factor or to neglect it entirely. The following general rules may be applied: If it is desired to prevent plastic flow entirely, use the full value of the stress-concentration factor. If plastic flow at the regions of stress concentration is not objectionable, as is usually the case, neglect the effect of stress concentration, that is, use $K_t = 1$. In some cases it may be desirable to use values of K_t between its full value and unity.

Figure 5-1 shows a steel bolt that failed in static loading caused by simultaneous tension and bending. This is failure as a ductile material, and necking down is apparent. Stress concentration caused by the threads had little or no effect on failure.

[1] N. H. Polakowski and E. J. Ripling, "Strength of Engineering Materials," pp. 133–318, 227–330, Prentice-Hall, Inc., Englewood Cliffs, N.J., 1966.

[2] Joseph Marin, "Mechanical Behavior of Engineering Materials," p. 8, Prentice-Hall, Inc., Englewood Cliffs, N.J., 1962.

Fig. 5-1 Failure of a ductile material under static load.

For materials that do not have a well-defined yield point, as cast iron, the allowable stress for static loading may be based on the ultimate strength, with a factor of safety larger than for a ductile material. In members made of *brittle* material there is practically no plastic flow of the material to relieve stress concentration; hence the full value of the factor should be used.

Maximum-normal-stress theory According to this theory, failure (yielding or fracture) occurs at a point in a body when one of the principal stresses at that point equals a critical stress for the material (yield stress or ultimate strength). If s_1 is the numerically greatest principal stress ($|s_1| > |s_2| > |s_3|$), failure for combined stresses occurs when

$$s_1 = \pm s_{\text{crit}} \tag{5-2}$$

where s_{crit} is the critical stress (yield stress or ultimate strength), assuming that the critical stress is equal in tension or compression.

The allowable stress based upon this theory of failure may be determined from Eq. (5-2) by dividing the critical stress by a suitable factor of safety fs:

$$s_1(\text{all}) = \frac{\pm s_{\text{crit}}}{\text{fs}} \tag{5-3}$$

This theory generally applies to the failure of brittle materials; however, the maximum-stress theory for yielding does not agree with test results.[1] In the latter case, the differences between the theory and the results are fairly high, particularly in the case of biaxial stresses of opposite sign. Even in the case of yielding, this theory is sometimes used in design and is employed when the influence of combined stresses is ignored.

The maximum-normal-stress theory, as applied to a biaxial stress state, can be represented graphically by plotting Eq. (5-2) for the case

[1] Marin, *op. cit.*, pp. 112–115.

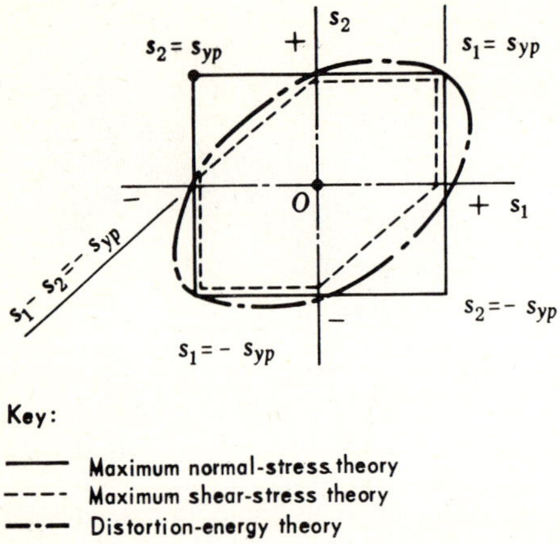

Fig. 5-2 Theories of failure.

when the critical stress equals the yield stress s_{yp} (see Fig. 5-2). A square is formed by the vertical lines passing through the points $s_1 = \pm s_{yp}$ and the horizontal lines passing through the points $s_2 = \pm s_{yp}$ (when $|s_2| > |s_1|$). These lines represent a boundary denoting the beginning of yielding, according to this theory.

Maximum-shear-stress theory This theory assumes that failure (yielding or fracture) occurs for a combined stress condition when the maximum shear stress equals the value of a critical shear stress (yield shear stress or ultimate shear stress) produced in an element subjected to simple tension. For a three-dimensional-stress state, the maximum shear stresses are given by one of the following expressions:

$$\frac{s_1 - s_2}{2} \qquad \frac{s_2 - s_3}{2} \qquad \text{or} \qquad \frac{s_3 - s_1}{2} \tag{5-4}$$

In the case of an element subjected to simple tension at the yield condition, it can be shown that

$$(s_s)_{yp} = \frac{s_{yp}}{2}$$

as shown in Art. 3-6. In accordance with the maximum-shear-stress theory, yielding will occur where the largest shear stress in the element [given by one of the terms in (5-4)] reaches a value equal to the shear yield stress in simple tension. If it is assumed that the shear stress at yield for simple tension is equal to that for simple compression, then, from (5-4),

$$\begin{aligned} s_1 - s_2 &= \pm s_{yp} \\ s_2 - s_3 &= \pm s_{yp} \\ s_3 - s_1 &= \pm s_{yp} \end{aligned} \tag{5-5}$$

where the $\pm$ sign means that the maximum shear stress may be positive or negative.

For two-dimensional-stress states, $s_3 = 0$, and the controlling equation depends upon the sign of the principal stresses. If the principal stresses are of opposite sign, then yielding occurs when $s_1 - s_2 = \pm s_{yp}$; however, when the principal stresses are of the same sign, yielding is defined by the second or third of Eqs. (5-5). In the latter situation, yielding will depend upon whether s_2 or s_1 is greater in magnitude.

Equations for the biaxial-stress state ($s_3 = 0$) are shown graphically in Fig. 5-2. For the case when both principal stresses are of like sign, Fig. 5-2 shows that the maximum-normal-stress theory and the maximum-shear theories coincide (for failure by yielding). When the principal stresses are of opposite sign, there is considerable difference between these two theories, with the maximum-shear-stress theory being more conservative.

The maximum-shear-stress theory is widely used by designers for predicting failure of ductile materials by yielding.[1,2,3] Its wide acceptance is indicated by its inclusion in such codes as the American Society of Mechanical Engineers (ASME) Pressure Vessel Code for Nuclear Vessels and the ASME Code for Design of Transmission Shafting. In addition, the maximum-shear-stress theory is in good agreement with experimental results, being slightly on the safe side.[1]

Equations (5-5) can be used to predict failure by fracture provided the yield stress s_{yp} is replaced by the ultimate stress or tensile strength. For two-dimensional stresses, the maximum-shear-stress theory is expressed by one of the following equations:[2]

$$s_1 = \pm s_{\text{ult}} \qquad \text{or} \qquad s_2 = \pm s_{\text{ult}} \tag{5-6}$$

where principal stresses are of the same sign, and

$$s_1 - s_2 = \pm s_{\text{ult}} \tag{5-7}$$

where principal stresses are of the opposite sign.

The term s_{ult} is the ultimate strength in tension or compression, whichever is smaller in magnitude; however, equal strengths in tension and compression are assumed when applying Eqs. (5-6) and (5-7).

Distortion-energy theory This theory is concerned mainly with predicting the beginning of yielding and seldom is used in connection with

[1] George E. Dieter, "Mechanical Metallurgy," pp. 58–59, McGraw-Hill Book Company, New York, 1961.

[2] Marin, *op. cit.*, pp. 115–116.

[3] Polakowski and Ripling, *op. cit.*, pp. 134–136.

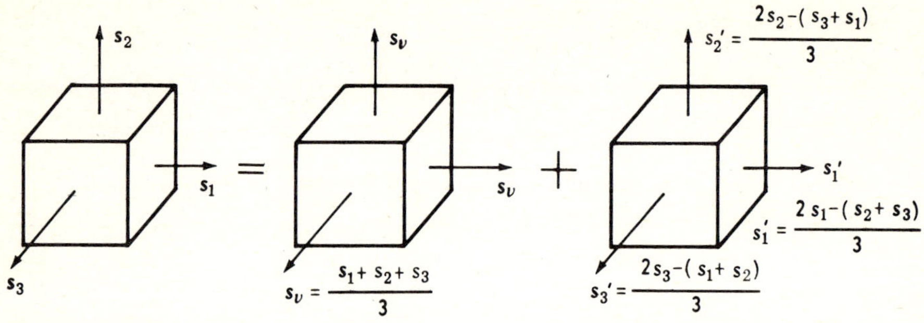

Fig. 5-3 Resolution of three-dimensional-stress state.

fracture. However, in accordance with this criterion, yielding will occur when the strain energy of distortion per unit volume equals the strain energy of distortion per unit volume for a specimen in uniaxial tension or compression (strained to the yield stress).

The strain energy of distortion per unit volume for a body subjected to a triaxial state of stress is given by[1,2,3]

$$u_s = \frac{1+\mu}{6E}[(s_1 - s_2)^2 + (s_2 - s_3)^2 + (s_3 - s_1)^2] \tag{5-8}$$

Yielding would occur when

$$\frac{1+\mu}{6E}[(s_1 - s_2)^2 + (s_2 - s_3)^2 + (s_3 - s_1)^2] = \frac{1+\mu}{6E} \times 2s_{yp}^2$$

or $$(s_1 - s_2)^2 + (s_2 - s_3)^2 + (s_3 - s_1)^2 = 2s_{yp}^2 \tag{5-9}$$

where s_{yp} in tension and in compression are assumed to be equal.

The expression for distortion energy is obtained from the total strain energy of deformation, u, by subtracting from u the strain energy of volume change, u_v. That is,

$$u_s = u - u_v$$

Figure 5-3 shows a general three-dimensional state of stress (in terms of principal stresses) and how it can be expressed in terms of a hydrostatic component of stress, s_v, and a stress deviator, s'.

The hydrostatic stresses s_v account for the entire change of volume, and the deviator stresses account for the distortion of the element. Experiments have shown that, even at fairly large values of hydrostatic

[1] Dieter, *op. cit.*, pp. 60–62.
[2] Polakowski and Ripling, *op. cit.*, pp. 136–139.
[3] Marin, *op. cit.*, pp. 117–120.

pressure, the hydrostatic state of stress has no effect on yielding. A complete derivation of Eqs. (5-9) is given in the references.[1,2]

For a biaxial-stress state ($s_3 = 0$), the distortion-energy criterion for yielding can be expressed by

$$s_1^2 - s_1 s_2 + s_2^2 = s_{yp}^2 \tag{5-10}$$

This equation is that of an ellipse whose major semiaxis is $\sqrt{2}\, s_{yp}$ and minor semiaxis is $\sqrt{2/3}\, s_{yp}$. Figure 5-2 shows a plot of Eq. (5-10). It is interesting to note that the maximum-normal-stress theory, the maximum-shear-stress theory, and the distortion-energy theory predict the same yield stress for conditions of uniaxial stress and when $s_1 = s_2$. The greatest discrepancy between the latter two theories occurs for a state of pure shear ($s_1 = -s_2$). According to the distortion-energy theory,

$$s_1^2 - s_1(-s_1) + s_1^2 = s_{yp}^2$$
$$3s_1^2 = s_{yp}$$
$$s_1 = \frac{s_{yp}}{\sqrt{3}} = 0.577 s_{yp}$$

According to the maximum-shear-stress theory [Eq. (5-5)], $s_1 = s_{yp}/2$. Therefore, the shear-stress law predicts a yield stress that is 15 percent lower than the value obtained by the distortion-energy theory.

Experiments have shown that the distortion-energy theory is in better agreement with data from the yielding of bodies under combined stress than any of the other theories discussed.[3,4] In general, the distortion-energy theory is recommended for defining yielding of ductile materials; however, the maximum-shear theory is a good approximation. In applying the distortion-energy theory for design purposes, the maximum allowable stress, s_{yp}/fs, is substituted for s_{yp} in Eq. (5-9) or Eq. (5-10). Thus,

$$(s_1 - s_2)^2 + (s_2 - s_3) + (s_3 - s_1)^2 = 2\,\frac{s_{yp}^2}{\text{fs}}$$

or $$s_1^2 - s_1 s_2 + s_2^2 = \left(\frac{s_{yp}}{\text{fs}}\right)^2$$

5-3 FATIGUE

The importance of fatigue failures was discussed in Art. 3-9. In this article, the nature of fatigue failures, fatigue design procedures, and cumulative fatigue damage are discussed.

[1] Marin, *op. cit.*, pp. 118–119.
[2] Dieter, *op. cit.*, pp. 60–62.
[3] Dieter, *op. cit.*, pp. 60, 62–64.
[4] Marin, *op. cit.*, pp. 120, 122.

Fatigue failures It has been realized for at least a hundred years that a metal subjected to an alternating stress will fail at a stress considerably lower than that required to induce fracture by a single application of the load.[1,2] A fatigue failure may be described as a progressive fracture leading from a point of stress concentration such as may occur in some machine and structural members. Fatigue failures are insidious since they frequently result from small cracks that are difficult to detect by Magnaflux or x-ray examination.

There is a marked difference between the type of fracture produced in ductile metals under fatigue loading and that occurring in ductile metals under static loading.[3] Since plastic flow of a ductile metal precedes fracture under static loading, the ruptured surface has a fibrous structure. However, in the case of fatigue fractures, a fatigue crack spreads from a point of stress concentration in a progressive manner until the stressed region is so small that it cannot sustain the load and fracture occurs suddenly. The fractured surface shows two regions of failure, one caused by the rubbing action as the crack propagates and the other by the sudden fracture. The latter zone is a rough surface resembling the fractured surface of a brittle material under static tensile loading.

There are three key factors in fatigue failures: i.e., a maximum tensile stress of sufficient magnitude, an applied stress fluctuation of large enough magnitude, and a sufficient number of cycles of the applied stress. Other variables such as stress concentration, residual stresses, corrosion, temperature, metallurgical structure, and stress state affect the conditions for fatigue.

With regard to the mechanism of fatigue, microscopic examinations of fatigue specimens show the presence of small irregular cracks oriented randomly. In general, the cracks tend to follow the directions of weaker crystallographic planes.

During the initial stages of fatigue loading, inelastic action occurs, resulting in the formation of submicroscopic permanent slip lines.[4] As repeated loading continues, submicroscopic voids form within the slip bands, eventually growing to microscopic size. Under further repeated loading, these voids join together to form cracks.

Fatigue design procedures The most common method of presenting engineering fatigue data is by means of the *S-N* curve, which represents a graph of the maximum applied stress *S* versus the number of cycles to

[1] John O. Almen and Paul H. Black, Basic Theories Relating to Fatigue in Metals, paper presented at National Metals/Materials Congress, Detroit, Mich., Oct. 21, 1965.

[2] Dieter, *op. cit.*, pp. 296–297.

[3] Marin, *op. cit.*, pp. 174–178.

[4] Marin, *loc. cit.*

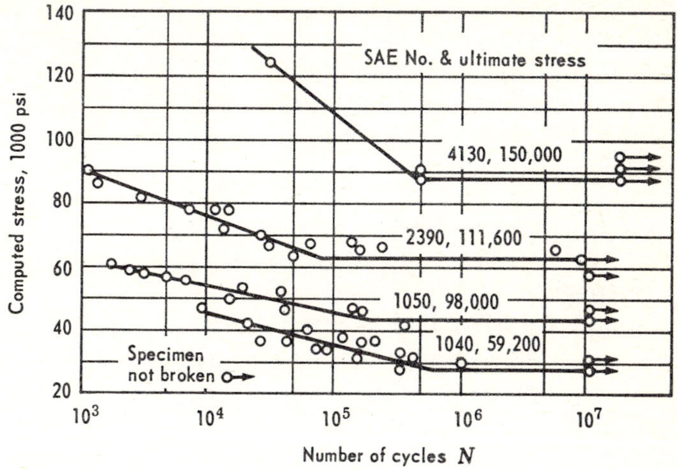

Fig. 5-4 Fatigue-test results for endurance-limit determination.

failure, N.[1] In the majority of fatigue tests, the rotating-beam testing machine has been used, where the mean stress is zero [see Fig. 3-22(*b*)]. Typical *S-N* curves for certain ferrous metals are shown in Fig. 5-4. According to this figure, the number of cycles of stress that a metal can withstand before complete failure increases with a decrease in stress. For ferrous metals, the *S-N* curve becomes asymptotic to a horizontal line which corresponds to failure at an infinite number of cycles. This stress is called the *fatigue* or *endurance limit*. For most nonferrous metals, the *S-N* curve slopes gradually downward with an increasing number of cycles. A true fatigue limit is not observed for these metals. For such metals it is accepted practice to specify a *fatigue strength*, which is defined as the fracture stress value representing the ordinate to the *S-N* curve for a selected number of stress cycles, N, for example, 10^8 cycles. This is referred to as "finite life."

The *S-N* curve is only an approximate curve,[2] since it is an average curve drawn to represent test data from a limited number of specimens. This curve may be in error, owing to the scatter of test data for the material being tested. During recent years it has become necessary to use statistical procedures to interpret test results.[3] It can be assumed that the percentage of failure for each selected life can be approximated by a *normal-distribution curve* (see Art. 23-7) corresponding to each stress

[1] Dieter, *op. cit.*, pp. 299–300.

[2] Marin, *loc. cit.*

[3] Guide for Fatigue Testing and the Statistical Analysis of Fatigue Data, *ASTM Spec. Tech. Publ.* 91-A, 1958.

level. With the aid of these normal-distribution curves, families of S-N-P curves can be drawn for each specific probability $P = 0.50$ (mean or median value); this means that only for 50 percent of the time will the fatigue-strength values be less than defined by the $P = 0.50$ curve.

Although most fatigue data presented in the technical literature have been determined for conditions of completely reversed stress (zero mean stress), conditions are often encountered where the stress situation consists of a mean stress s_m and a superimposed alternating stress s_r (see Fig. 3-22). In Art. 3-9 various expressions between s_m and s_r (or σ_m and σ_r) were given.

For the latter type of stress variation, it is necessary to obtain a series of S-N diagrams. One such diagram may present strength values s_r for failure corresponding to a particular value of s_m. A series of such diagrams are drawn, corresponding to different values of s_m. From these results a set of values of s_r and s_m can be determined which correspond to a particular life N_1 (number of cycles) whereby the s_r values are the ordinates of the S-N curves for $N = N_1$ (see Fig. 5-5).

One method of plotting the failure data (s_r and s_m) at fracture corresponding to the selected life N_1 is shown in Fig. 5-6. The alternating stress is plotted against the mean stress. The point s_e represents the fatigue strength corresponding to the case of complete reversal ($s_m = 0$), and the point s_{ult} corresponds to the static ultimate strength ($s_r = 0$). Various empirical laws defining the relationship between the variable stress and mean stress are presented in Fig. 5-6. A straight line drawn between s_e and s_{ult} follows the suggestion of Goodman, whereas a parabolic curve drawn between the same two points was proposed by Gerber. In general, test data for ductile metals fall closer to the Gerber curve, but because of scatter in the test results, the Goodman line is

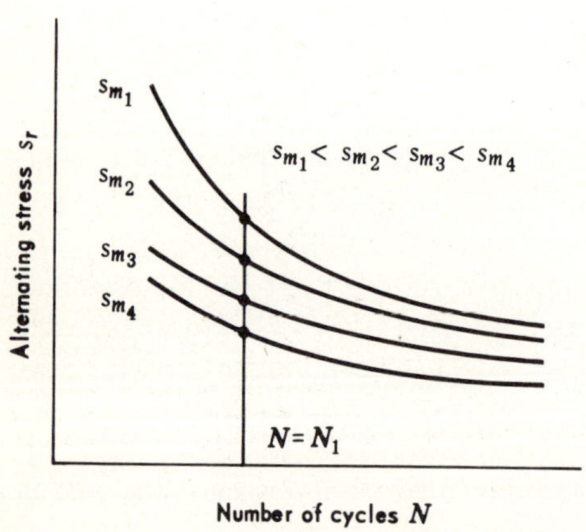

Fig. 5-5 S-N curves for various values of mean stress s_m.

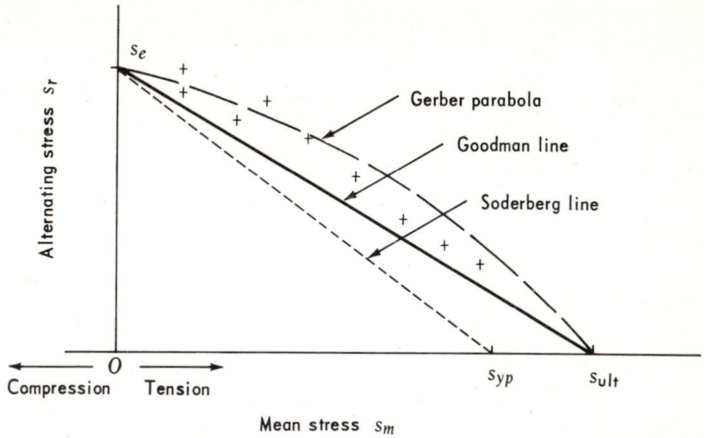

Fig. 5-6 Failure as a function of mean stress s_m and alternating stress s_r.

usually preferred in engineering design. These expressions may be put in the form of an equation:[1]

$$s_r = s_e\left[1 - \left(\frac{s_m}{s_{\text{ult}}}\right)^p\right] \tag{5-11}$$

where $p = 1$ for the Goodman line and $p = 2$ for the Gerber curve.

A third relationship, known as the Soderberg law, is used for the case where design is based on the yield strength. This expression may be obtained from Eq. (5-11) by substituting s_{yp} for s_{ult} and $p = 1$. With these substitutions, the resulting Soderberg expression plots as a straight line drawn between s_e and s_{yp}.

When a factor of safety is required in a design, both s_e and s_{ult} in Eq. (5-11) should be divided by fs. If the design is based on the Soderberg law, Eq. (5-11) becomes

$$s_r = \frac{s_e}{\text{fs}} - \frac{s_e}{s_{yp}} \times s_m \tag{5-12}$$

Since many machine and structural parts that are subjected to fatigue loads contain regions of high stress concentration, it is necessary that Eq. (5-12) be altered to include this effect. In the discussion of failure of brittle materials due to static loading, the stress-concentration factor was included in design calculations, but not for ductile materials. However, *fatigue* design calculations for ductile materials as well as for brittle materials must include the effect of stress concentra-

[1] Dieter, *op. cit.*, p. 325.

tion. Some authors[1,2] make use of the fatigue-strength reduction factor or fatigue notch factor, which is defined as the ratio of the fatigue limit of unnotched specimens to the fatigue limit of notched specimens. Since values of the fatigue notch factor published in the literature are subject to considerable scatter,[3] this factor will not be used in this book. Instead, the static theoretical stress-concentration factor K_t will be used. This procedure is actually somewhat more conservative than using the fatigue notch factor.

For ductile materials, and when design is based upon s_{yp} rather than s_{ult}, the stress-concentration factor is used to multiply the alternating stress s_r in Eq. (5-12); thus,

$$K_t s_r = \frac{s_e}{\text{fs}} - \frac{s_e}{s_{yp}} \times s_m \tag{5-13}$$

If the calculations are to be based upon the ultimate strength, then s_{ult} is substituted for s_{yp}. For fatigue design of brittle materials, both the alternating component of stress and the mean stress must be multiplied by the stress-concentration factor. In addition, the latter calculation uses the ultimate strength s_{ult} rather than the yield strength. Another approach to the design of brittle materials is to determine approximately a failure curve for the material and to design from the curve.[4]

Example 5-1 Determine the diameter of a circular rod made of ductile material with a fatigue strength (complete stress reversal) $s_e = 38{,}000$ psi and a tensile yield strength of 50,000 psi. The member is subjected to a varying axial load from $P_{\min} = -60{,}000$ lb to $P_{\max} = 140{,}000$ lb and has a stress-concentration factor $= 1.8$. Use a safety factor $= 2.0$.

$$P_m = \frac{140{,}000 + (-60{,}000)}{2} = 40{,}000 \text{ lb}$$

$$P_r = P_{\max} - P_m = 100{,}000 \text{ lb}$$

$$s_m = \frac{P_m}{A} = \frac{40{,}000}{A} \qquad s_r = \frac{P_r}{A} = \frac{100{,}000}{A}$$

By Eq. (5-13),

$$K_t s_r = \frac{s_e}{\text{fs}} - \frac{s_e}{s_{yp}} s_m$$

$$\frac{K_t P_r}{A} + \frac{s_e}{s_{yp}} \frac{P_m}{A} = \frac{s_e}{\text{fs}}$$

[1] Marin, *op. cit.*, p. 311.

[2] Joseph E. Shigley, "Mechanical Engineering Design," p. 181, McGraw-Hill Book Company, New York, 1963.

[3] Dieter, *op. cit.*, p. 311.

[4] Shigley, *op. cit.*, pp. 182–183.

$$A = \frac{K_t P_r + (s_e/s_{yp})P_m}{s_e/\text{fs}}$$

$$= \frac{(1.8)(100{,}000) + (38{,}000/50{,}000)(40{,}000)}{38{,}000/2.0}$$

$$A = 11.10 \text{ in.}^2$$

$$d = \sqrt{\frac{4A}{\pi}}$$

$$d = 3.76 \text{ in.} \qquad \text{Use } 3\tfrac{3}{4} \text{ in.}$$

Many machine members experience complex fatigue loading that induces three-dimensional stresses. A prediction of the fatigue performance of a combined stress system can be made by substituting the fatigue strength for simple types of loading into the equation of failure in the same manner as the yield strength in tension is used with the distortion-energy theory of failure to predict static yielding under a complex-stress state. On the basis of a rather limited amount of data, fatigue tests with different combinations of bending and torsion show that the distortion-energy theory provides the best agreement for ductile metals. The maximum-principal-stress theory appears to be the best criterion of failure for brittle materials.

A failure theory that considers the effect of combined stresses and the influence of a static mean stress is due to Sines:[1,2]

$$\frac{1}{\sqrt{2}}[(s_1 - s_2)^2 + (s_2 - s_3)^2 + (s_3 - s_1)^2]^{\frac{1}{2}} = s_e - C_2(s_x + s_y + s_z) \qquad (5\text{-}14)$$

where s_1, s_2, s_3 = alternating principal stresses
s_x, s_y, s_z = static stresses
C_2 = material constant giving variation of s_r versus s_m (slope of Goodman line in Fig. 5-6)

According to this criterion, failure occurs when the left side of Eq. (5-14) is greater than the right side. Constants s_e and C_2 can be evaluated for any number of cycles to failure. In addition, the effect of residual stresses can be included by adding these terms to the static stresses on the right side of the equation. It is interesting to note that Eq. (5-14) demonstrates that compressive residual stress will allow a greater magnitude of alternating stress for the same fatigue life.

Example 5-2 (*a*) Determine the thickness of a 5-in.-wide uniform plate for safe continuous operation if the plate is to be subjected to a tensile load that has a

[1] G. Sines, Failure of Materials under Combined Repeated Stresses with Superimposed Static Stresses, *NACA Tech. Note* 3495, 1955.
[2] Dieter, *op. cit.*, p. 326.

maximum value of 50,000 lb and a minimum value of 20,000 lb. The necessary properties of the material are $s_{yp} = 60{,}000$ psi and $s_e = 45{,}000$ psi. Assume a factor of safety equal to 1.5 based on the yield point.

Solution

$$s_m = \frac{50{,}000 + 20{,}000}{2 \times 5t} = \frac{7{,}000}{t} \qquad \text{See Fig. 5-6}$$

$$s_r = \frac{50{,}000 - 20{,}000}{2 \times 5t} = \frac{3{,}000}{t}$$

$$s_r = \frac{s_e}{\text{fs}} - \frac{s_e}{s_{yp}} s_m \qquad \text{See Eq. (5-12)}$$

$$\text{or } \frac{3{,}000}{t} = \frac{45{,}000}{1.5} - \frac{45{,}000}{60{,}000} \times \frac{7{,}000}{t}$$

and $t = 0.275$ in.
Use $\frac{5}{16}$-in. plate.

(*b*) Repeat Example 5-2(*a*) except that the plate is shaped as shown in Fig. X-3 in Appendix X, where $D = 5$ in., $d = 2\frac{1}{2}$ in., and the fillet radius $r = \frac{3}{4}$ in.

Solution

$$\frac{r}{d} = \frac{0.75}{2.5} = 0.3$$

$$\frac{D}{d} = \frac{5}{2.5} = 2$$

$$K_t = 1.63 \qquad \text{(from Fig. X-3)}$$

$$\text{Then } K_t s_r = \frac{s_e}{\text{fs}} - \frac{s_e}{s_{yp}} s_m$$

$$\text{or } 1.63 \times \frac{6{,}000}{t} = \frac{45{,}000}{1.5} - \frac{45{,}000}{60{,}000} \frac{14{,}000}{t}$$

From this equation, t is found to be 0.676 in. Use $\frac{3}{4}$-in. plate.

(*c*) Determine the diameter of a solid circular shaft that is to be subjected to an alternating torque varying from 16,000 to 20,000 in.-lb. The shaft is to be made of hot-rolled SAE 1020 wrought steel which has $s_{yp} = 35{,}000$ psi, $s_e = 22{,}000$ psi. Use a factor of safety equal to 2.

In a circular shaft in pure torsion, the maximum shear stress (in transverse sections) is equal to the maximum tensile stress (in 45° helical regions). The design of this shaft is to be based on an equivalent maximum-shear-stress equation for alternating stress. The calculated diameter of the shaft equals 2.44 in. Use $2\frac{7}{16}$ in.

Cumulative fatigue damage Certain equipment exposed to large intermittent loads or to high-frequency transient loads is frequently designed for finite fatigue life. Aircraft, missiles, automobiles, and earth-moving machines are typical examples. For purposes of design, it is customary to assume a typical loading sequence and to anticipate the number of cycles to be encountered at each stress level during the useful life of the member. Useful data that can be employed in the above estimation are frequently available from actual measurements on prototypes or previously built equipment.

There has been developed a criterion, known as Miner's rule,[1,2] that relates the degree of damage to the number of cycles, n, at a particular fatigue strength. This criterion assumes a linear relation between the degree of damage and the number of cycles to failure, N. According to this rule, if the fatigue lives at fatigue-strength amplitudes $s_1, s_2, \ldots, s_k$ are $N_1, N_2, \ldots, N_k$ (as obtained from an S-N curve using constant-stress amplitudes), respectively, fatigue failure will occur when

$$\frac{n_1}{N_1} + \frac{n_2}{N_2} + \cdots + \frac{n_k}{N_k} = \sum_{i=1}^{k} \frac{n_i}{N_i} = 1$$

Although this rule is not accurate, it represents a useful means of arriving at a reasonable prediction of life.

It must be cautioned that, although these procedures of analysis have been accepted in some quarters, they are still subject to review and reevaluation. In all likelihood, some of them will be changed in the future.

5-4 DEFLECTION

Many situations are encountered in practice where excessive deflection of machine members prevents the machine from functioning properly. An example of this type of failure is a shaft that had sufficient strength to support gears but lacked necessary stiffness. The gears mounted on the shaft were not held in alignment, and failure of the gears was the result. This shaft should be considered to have failed as completely as though it had broken. Lack of necessary stiffness may also cause failure of bearings or of other parts because of excessive vibration of the shaft. Since steel, which is used for most severely loaded machine parts, has practically uniform modulus of elasticity, the control over stiffness lies in sufficient size of parts and, for steel, is independent of composition or heat treatment.

Other examples of parts for which rigidity is a critical characteristic are housings for gear drives; supports for bearings of rolls where controlled thickness is necessary, as in making photographic film; printing-press parts; and crankcases that support crankshaft bearings.

The only design criterion in these situations is that the deflection of the element must be held within certain limits that are governed by clearance between moving parts, failure of bearings, environment, exces-

[1] Polakowski and Ripling, *op. cit.*, p. 511.

[2] N. A. Miner, Cumulative Damage in Fatigue, *Trans. ASME*, vol. 67, p. A-159, 1945.

sive vibration, or improper operation of the machine. Therefore, design calculations are made to determine a size for the member such that the deflection limit is not exceeded. Methods of calculating deflections were discussed in Arts. 3-11 and 3-12.

5-5 WEAR

As stated in Art. 20-6, wear is the predominant cause of failure of mechanical parts. Wear may be defined as "the removal of material from solid surfaces as the result of mechanical action."[1] Design criteria for wear involve consideration of the following: effect of hardness and roughness of surfaces; type of surface material (ductile or brittle); contact pressure; type of contact, such as sliding or rolling (there is a sliding component in rolling resistance; see Art. 20-9); size of abrasive particles; environment (moisture content of atmosphere); corrosive situations; velocity; temperature; and conditions of lubrication.

In early days, the mechanism of wear did not receive much attention. This was due to its being a slow process, to the difficulty of its measurement, and to the fact that other types of failure, such as fracture of parts, frequently terminated the useful life of mechanical equipment before it was "worn out." The prevention of early failure by wear has not kept pace with prevention of failure by other means, such as fracture, so that the proportion of failures by wear in modern machines and equipment in all areas, including appliances, has increased.

There are two main reasons for this situation. The first is the complications in the wear process. To date, considerable background work has been accomplished,[2] and qualitative equations have been derived to express the phenomena of wear, but they cannot be used directly by the designer in the same manner that he uses equations from mechanics of materials with test results for properties of materials from handbooks. Therefore the designer, like a surgeon, must use considerable judgment in doing his work to prolong life.

The second reason that wear of machines is the cause of many mechanical failures is the lack of concern and interest in the mechanism of wear. The approach in mechanical design has been frequently to rely casually on past experience in similar situations and to forgo experimental investigations for the actual conditions of operation for the new parts being designed. The qualitative equations referred to cannot be used directly in the design of new equipment, but they can be valuable in indicating parameters for consideration.

[1] Ernest Rabinowicz, "Friction and Wear of Materials," John Wiley & Sons, Inc., New York, 1965.
[2] *Ibid.*

The design criteria discussed and applied in Chap. 20 should be helpful as illustrations of designing mechanical parts for satisfactory wear life.

5-6 BUCKLING

In this article, the design of columns and beam columns to avoid failure by buckling is discussed. The Euler and the J. B. Johnson formulas presented in Art. 3-7 are used in the design of columns. For ease of reference these formulas are presented again, as follows:

The Euler formula:

$$F_{\text{crit}} = \frac{n\pi^2 EI}{L^2} \tag{5-15}$$

The J. B. Johnson formula:

$$F_{\text{crit}} = As_y\left(1 - \frac{s_y L^2}{4n\pi^2 E\rho^2}\right) \tag{5-16}$$

The variables in these formulas and the range of application are discussed in Art. 3-7.

In the design of columns, the criterion used is that the load to be supported by the column must be equal to or less than the critical or failure load. Since the maximum stress that exists in a column at failure is indeterminate, it is necessary to specify the factor of safety in column design in terms of loads rather than stresses:

$$F_{\text{crit}} = \text{fs} \times F_{\text{all}}$$

where F_{crit} = critical or failure load, lb
F_{all} = allowable or safe load, lb

The value for the factor of safety to be used in design depends on the installation and the service. Where weight must be kept to an absolute minimum and the conditions of loading and quality of material are accurately known, a value as low as 1.25 may be used. In the usual application, a value of from 2 to 3 or 4 may be suitable.

In arriving at the design value for the end-fixity coefficient n, it is necessary to consider the degree to which the actual support approaches one of the ideal cases in Fig. 3-16. For instance, the fixed-end case is seldom realized in an actual installation because of the deflection of the support so that the design value of n may well be limited to 3 or possibly 3.5 instead of the ideal value of 4.

In determining the radius of gyration, it is necessary to consider the plane in which failure occurs. The plane of failure will be that for which the combination of bending resistance and end fixity is the least, i.e., the plane for which the product nI is the least.

Example 5-3 A pin-ended strut 6 in. long is to be made of SAE 1030 steel, of circular cross section, and is to support a compressive load of 2,000 lb. The diameter of the loading pins at the ends of the strut is ½ in. See Fig. 3-1 for the appearance of the strut. Assuming a factor of safety of 1.5 for the strut and an allowable bearing pressure at the pins of 10,000 psi, determine the dimensions for the strut.

Solution For the strut,

$$B = \frac{s_y L^2}{n\pi^2 E} = \frac{42{,}000 \times 6^2}{1 \times \pi^2 \times 30{,}000{,}000} = 0.0051 \text{ in.}$$

$$F_{\text{crit}} = \text{fs} \times F_{\text{all}} = 1.5 \times 2{,}000 = 3{,}000 \text{ lb}$$

From the J. B. Johnson equation,

$$F_{\text{crit}} = As_y\left(1 - \frac{B}{4\rho^2}\right)$$

If substitutions are made for $A = \pi d^2/4$ and $\rho = d/4$ and the equation is solved for d^2, the result is

$$d^2 = \frac{4F_{\text{crit}}}{\pi s_y} + 4B = \frac{4 \times 3{,}000}{\pi \times 42{,}000} + 4 \times 0.0051$$

$$= 0.091 + 0.020 = 0.111 \text{ in.}^2$$

$$d = 0.333 \qquad \text{Use } \tfrac{3}{8} \text{ in. in diameter for the strut.}$$

As a check on the choice of the J. B. Johnson equation,

$$\frac{B}{\rho^2} = \frac{16B}{d^2} = \frac{16 \times 0.0051}{(0.375)^2} = 0.580$$

Since the value of B/ρ^2 is less than 2, the assumption of the J. B. Johnson equation was correct.

For the dimensions of the eye, the projected bearing area of the pins may be calculated as

$$\frac{2{,}000}{10{,}000} = 0.2 \text{ in.}^2$$

The minimum width of the eye therefore would be

$$\frac{0.2}{0.5} = 0.4 \text{ in.}$$

By using ½ in. for the width of the eye, clearance would be provided for machining the faces of the eye.

The dimensions of the strut would be as follows:

Diameter of body of strut	⅜ in.
For the eye:	
Inside diameter	½ in.
Outside diameter	1 in.
Width	½ in.

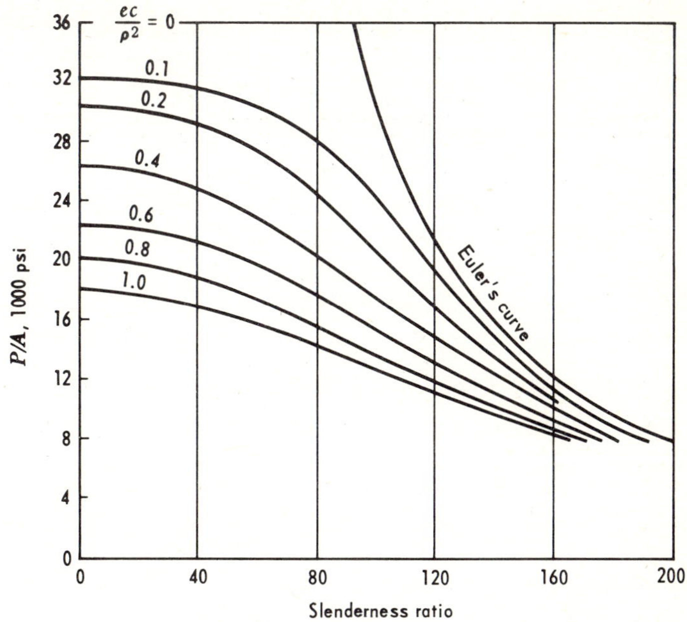

Fig. 5-7 Typical secant-formula curves. *(From Singer's "Strength of Materials," p. 419, Harper & Row, New York,* 1962.)

In Art. 3-7, the secant formula [Eq. (3-23)] for a *beam column* was discussed. This formula may be used for design purposes by substituting the yield strength S_{yp} for S so that the equation becomes

$$S_{yp} = \frac{P}{A}\left(1 + \frac{ec}{\rho^2} \sec \frac{L}{2\rho}\sqrt{\frac{P}{EA}}\right) \tag{5-17}$$

where P is the axial load at failure (corresponding to yielding) and the other variables are discussed in Art. 3-7.

A typical plot of the secant formula for various values of the eccentricity ratio ec/ρ^2 is shown in Fig. 5-7.[1] A steel with a yield strength of 36,000 psi and a modulus of elasticity of 30,000,000 psi was assumed. For purposes of comparison, a Euler curve is included.

Care must be exercised when using a factor of safety[1] with Eq. (5-17) because the maximum stress (yield stress) is not a linear function of load. In this case, the maximum stress increases at a greater rate than the lateral loads. To achieve a particular factor of safety equal to fs, the cross-sectional dimensions of the beam must be chosen so that the maximum fiber stress becomes equal to the yield stress when all loads

[1] Stephen P. Timoshenko and James M. Gere, "Theory of Elastic Stability," 2d ed., pp. 37–41, McGraw-Hill Book Company, New York, 1961.

acting on the beam are taken as the factor of safety times the allowable load (or safe load).[1] Substituting $P = \text{fs} \cdot P_{\text{all}}$ in Eq. (5-17) yields

$$S_{yp} = \frac{\text{fs} \cdot P_{\text{all}}}{A}\left(1 + \frac{ec}{\rho^2} \sec \frac{L}{2\rho}\sqrt{\frac{\text{fs} \cdot P_{\text{all}}}{EA}}\right) \tag{5-18}$$

where P_{all} is the allowable or safe load in pounds.

Equation (5-18) can always be solved for P_{all}/A by using a trial-and-error method but it is far more satisfactory to program the formula on a digital computer to obtain an optimum set of dimensions with regard to weight considerations as well as load-carrying capacity.

5-7 CREEP AND STRESS RUPTURE

It has often been noted that the available materials of construction restrict or limit a rapid advancement in technology. For example, the operating temperatures of such equipment as nuclear-reactor core components, rocket engines, and missile nose cones impose some severe problems in the selection of materials. In these cases, one of the major design considerations is that a metal subjected to a constant tensile load at an elevated temperature undergoes a time-dependent increase in length, known as *creep*. This behavior is attributed to the following three deformation processes that occur at elevated temperatures:

Slip
Subgrain formation
Grain-boundary sliding[2]

[1] *Ibid.*
[2] Dieter, *op. cit.*, pp. 342–345.

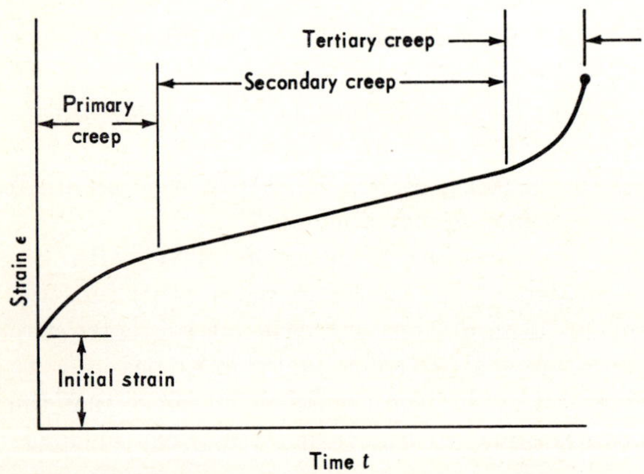

Fig. 5-8 Typical creep curve.

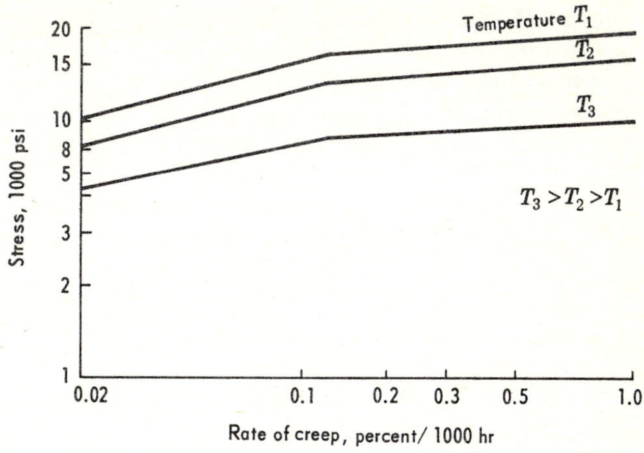

Fig. 5-9 Typical stress as a function of minimum creep rate.

Creep is most commonly evaluated by measuring the extension of a specimen subjected to a constant tensile load and a constant temperature. A typical creep curve is shown in Fig. 5-8, in which the slope $(d\epsilon/dt)$ is referred to as the "creep rate." After an initial rapid elongation of the specimen, the creep rate decreases with time and attains an essentially constant value in the secondary-creep region. During tertiary or third-stage creep, the creep rate increases rapidly and terminates in fracture of the specimen. However, the three well-defined stages of creep are found for only certain combinations of stress and temperature.

At present, theoretical methods are not sufficiently developed to predict creep behavior accurately for design purposes. However, there are available in the technical literature[1] considerable reliable creep test data which may be employed to advantage. One common manner of presenting creep data is by plotting the logarithm of the stress against the logarithm of the minimum creep rate (see Fig. 5-9). With these curves, the designer can determine the *creep strength* of the material. The creep strength is defined as the stress at a specified temperature that produces a minimum creep rate of a certain amount, such as 0.0001 percent per hr or 0.001 percent per hr.[2]

Another important design criterion is *rupture strength*, which is defined as the stress at an operating temperature to produce a specified life to rupture, such as 1,000 hr, 10,000 hr, 10 years, or 20 years.

[1] Ian Finnie and W. R. Heller, "Creep of Engineering Materials," pp. 232–239, McGraw-Hill Book Company, New York, 1959.

[2] Dieter, *op. cit.*, p. 355.

Rupture-strength properties are determined from stress or creep-rupture tests. These tests are basically identical to creep tests except that they are carried out to the failure of the material. The primary information obtained from the stress-rupture tests is the time to fracture at a specified nominal stress for a particular constant temperature. In addition, elongation and reduction of area at fracture are also determined. Simpler strain-measuring equipment can be used with the stress-rupture tests (as compared with that used for creep tests) since larger total strains are encountered.[1]

Stress-rupture data may be presented in a variety of ways. Two of them are shown in Fig. 5-10. Frequently, both creep and rupture data are shown in the same diagram.

Situations often arise when high-temperature stress-rupture data are needed for conditions of service life for which there are no experimental data. This is particularly true when equipment is being designed for a 10- to 20-year service life, and the time-to-rupture tests extend to only 1 year. It then becomes necessary to extrapolate the existing data to longer times. The predictions of rupture times at various stresses and temperatures are accomplished by the use of temperature-time parameters whose main function is to "buy" time with temperature.[2] However, as a note of caution, this procedure is reliable only when no structural changes occur for the material within the region of extrapolation.[3]

One of the most commonly used parameters is the *Larson-Miller parameter* (see Appendix XIII-C for data to construct a rupture-stress versus temperature curve for a definite rupture life), which has the form[1]

$$(T + 460)(\log t + C) = \text{LMP} = \text{const} \tag{5-19}$$

where t = time, hr
T = temperature, °F
C = material constant determined experimentally, usually from 10 to 30

By using Eq. (5-19) as the relation between temperature and time corresponding to a particular rupture stress, a graph is plotted of the rupture stress versus the Larson-Miller parameter for each combination of T and t that led to failure. A typical master curve for the Larson-Miller parameter is shown in Fig. 5-11.

When the master curve has been obtained, design curves of the type shown in Fig. 5-10 may be drawn. For a particular rupture stress, there will be a corresponding value of the Larson-Miller parameter. To plot a curve similar to Fig. 5-10(*a*)—for example, the rupture curve for a tem-

[1] Dieter, *op. cit.*, p. 341.
[2] Polakowski and Ripling, *op. cit.*, pp. 444–451.
[3] Dieter, *op. cit.*, p. 357.

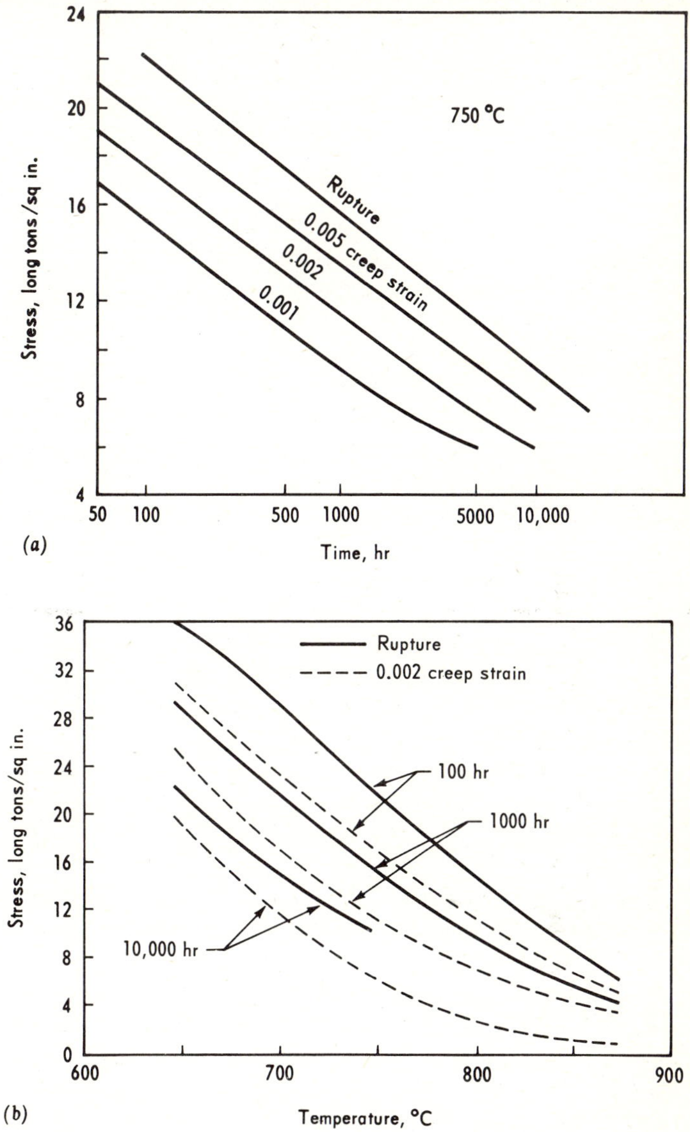

Fig. 5-10 Combined rupture and creep-strain data for steel. *(From I. Finnie and W. R. Heller, "Creep of Engineering Materials," p. 243, McGraw-Hill Book Company, New York, 1959.)*

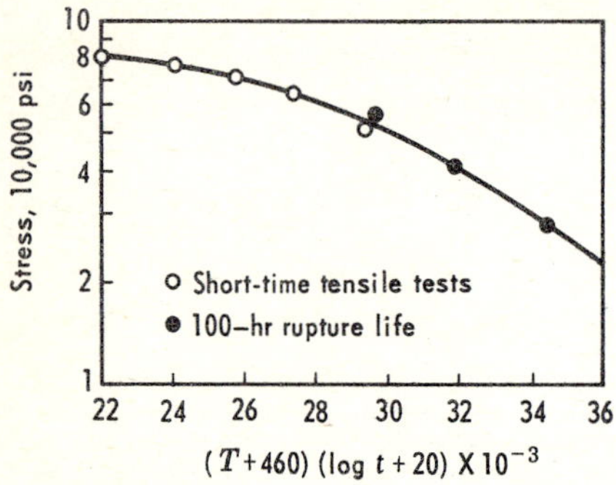

Fig. 5-11 Larson-Miller parameter as a function of stress for 1 Cr, 1 Mo + V steel normalized. (*From American Society for Testing Materials.*)

perature of 750°C—this temperature in degrees Fahrenheit is substituted into Eq. (5-19), and the equation is solved for the time. This gives a value of time corresponding to the particular rupture stress. The rupture curve in Fig. 5-10(*b*) may be obtained from the master curve in a similar manner, only, in this case, the time (for example, $t = 100$ hr) is substituted into Eq. (5-19) and the temperature in degrees Fahrenheit can be determined.

The creep curves in Fig. 5-10 also may be obtained by the Larson-Miller parametric method. Rather than defining a master curve for rupture, in this case the curve represents a specific total strain, such as 0.2 percent. This value will generally correspond to the percent elongation at the beginning of tertiary creep.[1]

5-8 IMPACT

Although the yield and ultimate impact strengths are higher than the respective static strengths, it is recommended that static design stresses be employed for impact conditions.[2] There are several reasons for this recommendation. In many situations, the yield or ultimate strength cannot be determined accurately. Frequently, overloads resulting from impact loads are more serious than for static loads, because of the tendency for permanent deformations to increase with further impacts. In addition, it is more difficult to determine impact loads (see Art. 3-10) than static loads.

In designing for impact loading, some important considerations should be kept in mind. It is important that the materials selected should have a high yield strength and that they be reasonably ductile to allow for stress relieving at regions of high stress concentration. If

[1] Finnie and Heller, *op. cit.*, pp. 241–242.
[2] Marin, *op. cit.*, pp. 265–269.

Eq. (3-31) for an impact load of an axial member is divided by the cross-sectional area A, the impact stress is obtained

$$S = \frac{W}{A}\left(1 + \sqrt{1 + \frac{2Ch}{W}}\right) \tag{5-20}$$

where $C = AE/L$, lb per in.
L = length, in.
E = modulus of elasticity, psi

Examination of this equation shows that a low value of C is desirable, since the impact stress varies with this constant. Another consideration is that stress concentration should be avoided if possible, but when it is present the stress-concentration factor must be used in the design calculations, as in the case of static design of a *brittle* material.

5-9 CORROSION FATIGUE AND CAUSTIC EMBRITTLEMENT

Corrosion fatigue[1] When stress concentration in cyclic loading occurs in a corrosive environment, the seriousness of stress concentration is greatly increased. Even the presence of air reduces the endurance limit of a member from its value in a vacuum. Ordinary tap water, salt water, and other corrosive substances markedly reduce the endurance limit. In 1915 the problem of corrosion fatigue presented itself in the steel cables of mine sweepers, which were acted on by vibration forces owing to the flow of sea water. Values are often quoted for endurance limits in certain substances, such as the brine endurance limit for steel. It is interesting to note that the corrosion effect on a member may not be great enough to "eat away" the material visibly. In fact, polished test specimens that have failed in corrosion fatigue in salt-water spray appear as polished as the original test pieces, even though the test may indicate an endurance limit that is 40 percent of its dry-air value.

Although there is extensive information on this type of failure in the literature, no generally valid rule can be expressed.[2] To help arrest corrosion-fatigue surface residual compressive stresses, corrosion-resistant coatings or special alloys may be used.

Caustic embrittlement This type of failure involves stress concentration in the presence of a caustic corrosive substance and has appeared in pressure vessels. Stress concentration is produced by rivet holes,

[1] H. H. Uhlig, "Corrosion and Corrosion Control," John Wiley & Sons, Inc., New York, 1963; B. B. Wescott, Corrosion Fatigue, "Corrosion Handbook," p. 578, John Wiley & Sons, Inc., New York, 1961.

[2] Polakowski and Ripling, *op. cit.*, p. 516.

unannealed welded joints, or sharp reentrant corners of flanges. The loading has been of the fatigue type because of the repeated expansion and contraction of the vessel. The fatigue stress concentration in the presence of certain chemicals, such as caustic soda in feedwater, has led to embrittlement that has caused many boiler failures.[1] Some means used to combat caustic embrittlement are (1) boiler feedwater treatments to reduce the effect of caustic soda on the metal and (2) design of the vessels to reduce stress concentration.

[1] F. G. Straub, Embrittlement in Boilers, *Univ. Illinois Bull.* 216, 1930; A. E. White, Changes in a High-pressure Drum to Eliminate Recurrence of Cracks Due to Corrosion Fatigue, *Trans. ASME*, vol. 61, 1929.

6
Residual Stresses

6-1 FATIGUE FAILURES

In formal courses in strength of materials, the major emphasis, which is based on simplifying assumptions, is on the static strength of machine and structural members. Failures in static loading form only a small proportion of the total number of failures of machine parts. The predominant number of failures are due, not to static failures, but to excessive deflection, fatigue fractures, corrosion, and wear.

Residual stresses have a profound effect on fatigue life and, since the simplifying assumptions referred to above deviate so widely from real circumstances, it has become necessary to revise them in accordance with the following conclusions:

1. Fatigue fractures as well as other brittle-type fractures occur only as a result of tensile stresses.
2. Metal surfaces are weaker in fatigue loading than sound subsurface material.

3. The yield point in fatigue loading is less than the static yield point.

4. Practically all machine-part fractures occurring in normal service are fatigue fractures.

5. Fully 90 percent of all fatigue failures occurring in normal service are traceable to design and fabrication errors.

It is the purpose of this chapter to present the fundamental theory of the effect of residual stresses on the fatigue strength of machine parts to serve as a guide for better designs and better use of materials, that is, for better performance at less weight and cost.[1]

6-2 CLASSIFICATION OF STRESSES

For our purpose, stresses can be classified as load stresses and residual stresses. *Load stresses* are those within the elastic limit and induced by external loads, such as tension, compression, bending, and torsion. These stresses return to zero when the external loads are removed, and they can be calculated by the use of equations from strength of materials. *Residual stresses* can be defined simply as stresses that are independent of external loads. They may be harmful or beneficial and affect markedly the fatigue strength of machine members.

6-3 SOURCES OF RESIDUAL STRESSES

Common sources of residual stresses are discussed in this article. Some of them are harmful, some beneficial, some are incident to the manufacturing process by which the part is made, and some are intentionally induced to improve the fatigue strength of the part.

Plastic deformation Assume an annealed bar made of material that has a stress-strain diagram in tension and compression as shown in Fig. 6-1(*a*). If the bar is subjected to a bending moment as shown in Fig. 6-1(*b*), so that some parts of the bar are stressed beyond the elastic limits, the tensile-stress distribution in the upper part of the bar will be as represented by *oa* and the compressive-stress distribution in the lower part as *oa′*. Note that the curves *oa* and *oa′* are miniature stress-strain diagrams.

If the maximum stress in the bar is represented by point *a* in Fig. 6-1(*a*), then, if the bending moment in the bar is released, the stress will not return along the heavy line from *a* to *o* in Fig. 6-1(*a*) but will return along the dotted line which is parallel to the line from *o* to the elastic

[1] For a more complete discussion of this subject and for an extensive bibliography, see J. O. Almen and P. H. Black, "Residual Stresses and Fatigue in Metals," McGraw-Hill Book Company, New York, 1963.

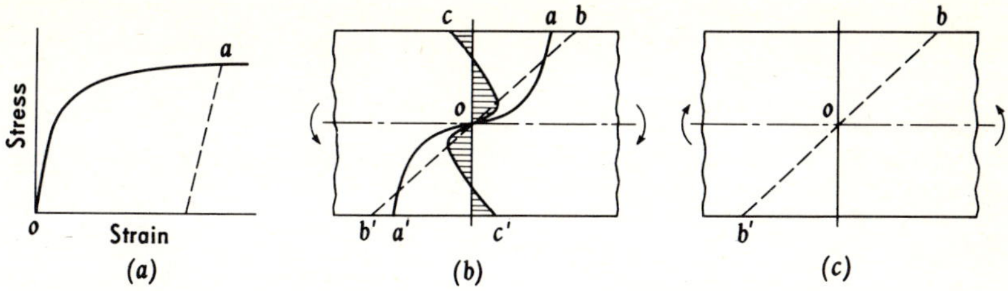

Fig. 6-1 Residual stresses in bending. (*a*) Stress-strain diagram. (*b*) Composite diagram. (*c*) Unloading diagram. (*From J. O. Almen and P. H. Black, "Residual Stresses and Fatigue in Metals," McGraw-Hill Book Company, New York,* 1963.)

limit. The dotted line thus represents the unloading path, and there will be initial set as shown. The unloading distribution is represented in Fig. 6-1(*b*) by *bo* and *b′o*.

Note that the moment required to bend the bar as represented by the stress-distribution diagram *aoa′* at (*b*) in the figure is equal to the moment represented by the unloading diagram *bob′*.

The method for determining the resultant or residual stresses, as developed by Wöhler,[1] is to subtract the unloading stress distribution from the loading distribution; the result is the residual stress as shown by the shaded areas in Fig. 6-1(*b*).

The top of the bar, in which inelastic tensile stress was introduced by loading, now has residual compressive stress, and the bottom originally stressed inelastically in compression has residual tensile stress. Note that, over the section of the bar shown, the tensile stresses are in equilibrium, that is, $\Sigma F = 0$, and the moments produced by the stresses are also in equilibrium, that is, $\Sigma M = 0$.

There is a principle in this example: A residual stress caused by plastic flow is of the opposite sign to the stress producing the plastic flow; that is, as shown in Fig. 6-1(*b*), plastic flow in *tension* represented by point *a* on loading causes a *compressive* residual stress *c*, and plastic flow in *compression a′* causes a *tensile* residual stress *c′*. This principle will be evident later in this chapter in the discussion of residual stresses produced by shot peening and surface rolling. The principle also applies to subsurface microresidual stresses in rolling bodies.[2]

Casting and forging During the cooling of a casting or a forging, the first parts to cool and become solid are the surface layers. Later,

[1] S. P. Timoshenko, "History of Strength of Materials," McGraw-Hill Book Company, New York, 1953.

[2] J. O. Almen, Effect of Residual Stresses on Rolling Bodies, Automotive Engineering Congress, Detroit, Society of Automotive Engineers, 1962.

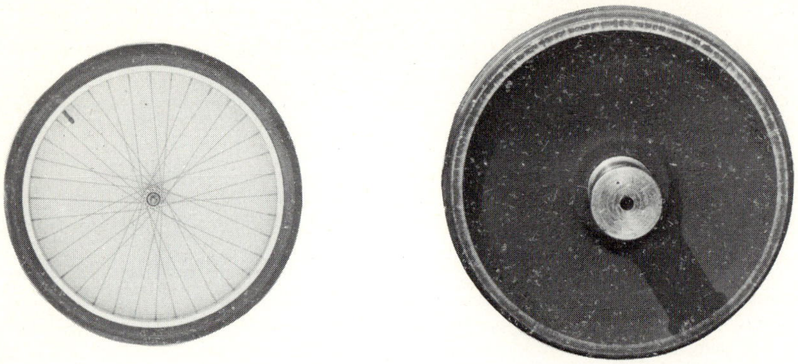

Fig. 6-2 Thermally induced residual-stress pattern in railway-car wheels closely resembles spoke and rim stresses in bicycle wheels. *(From J. O. Almen and P. H. Black, "Residual Stresses and Fatigue in Metals," McGraw-Hill Book Company, New York,* 1963.)

as the core cools, it cannot contract freely owing to the restraint of the solid surface layers (the case). Therefore, the core cools with a residual tensile stress whereas the case, being acted on by the inward tensile stresses in the core, becomes residually stressed in compression. The compressive stresses in the surface layers will prevent the spreading of fatigue cracks and thus increase the fatigue strength of the part and improve its wear characteristics. The *principle* is that the part to cool earliest will acquire residual compressive stresses, and the part to cool last will acquire residual tensile stresses.

The surface residual compressive stresses in castings and forgings improve the fatigue life because fatigue cracks cannot spread in regions of compressive stress. Also, the surface compressive stress gives improved resistance to wear because the surface particles offer greater resistance to being torn away by friction in sliding by a contacting body.

An example of intentionally creating residual compressive stresses in castings and forgings to prevent the propagation of cracks is the processing of railway-car wheels. For cast-iron wheels "chills" are used in the mold to cool the rim quickly, and for forged-steel wheels the rim is water-quenched. Both quick-cooling methods cause contraction of the rim on the still plastic plate. Later, as the plate cools, its thermal contraction places the rim in compression and the plate becomes stressed in tension. This action is analogous to tightening the spokes of a bicycle wheel because the spokes in tension pull inward so that the rim becomes compressively stressed. See Fig. 6-2.

If the rims of railway-car wheels are not deeply compressively stressed, the high temperature of the surface layer of the rim during severe braking expands the surface layer which, on cooling after the brakes are released, acquires residual tensile stress. These may lead to thermal cracks. However, if the rim has a sufficiently deep layer of residual compressive stress, the cracks are prevented from spreading so that they do not lead to catastrophic failure.

Quenching Stresses due to quenching may reach considerable magnitudes and may be harmful or beneficial to fatigue durability, depending on whether the resulting residual stresses increase or decrease the maximum tensile stresses due to the applied loads. In quenching a solid member from high temperature, the case is cooled first; then, as the core cools and contracts, it develops a tensile stress which "pulls in" the case, introducing therein a compressive residual stress.

In hollow castings to be subject to internal pressure, the fatigue durability is increased by *internal* quenching; this produces on the inner surfaces a compressive-residual-stress pattern that is of the opposite sign to the load-stress pattern.

Excessive quenching, like excessive carburizing and nitriding, should be avoided, because high residual compressive stresses in surface layers mean their counterpart of high residual tensile stresses in the core. The latter may lead to fatigue cracks originating in subsurface material.

Grinding and polishing Because of the high temperature reached in normal commercial grinding practice, the surface layer of the workpiece becomes plastic and expands as it passes under the wheel. As the workpiece cools, the contraction of the surface layer is constrained by the rigid subsurface material so that high tensile stresses remain in the surface layer. Figure 6-3 shows the residual-stress pattern in a specimen of annealed spring stock $\frac{1}{16}$ in. thick, 1 in. wide, and 7 in. long, which was ground to a depth of 0.002 in. After grinding, the previously flat specimen was curved concave on the ground side, which indicated tensile stresses. By the "dissection method" (see Art. 6-5), the magnitude of the stresses was found to be as shown in Fig. 6-3. Such high surface

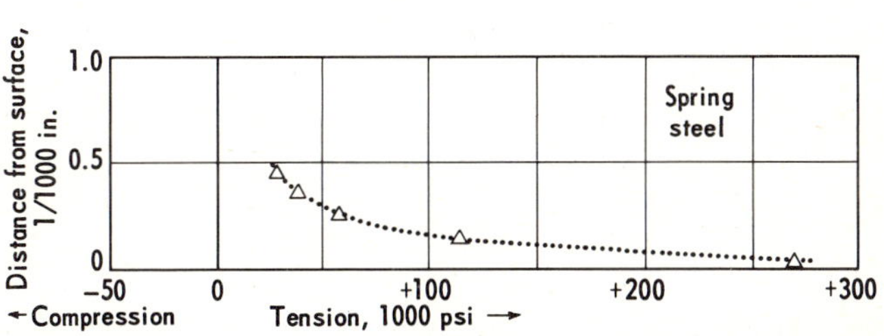

Fig. 6-3 Residual stresses resulting from grinding. (*From J. O. Almen and P. H. Black, "Residual Stresses and Fatigue in Metals," McGraw-Hill Book Company, New York,* 1963.)

(a)

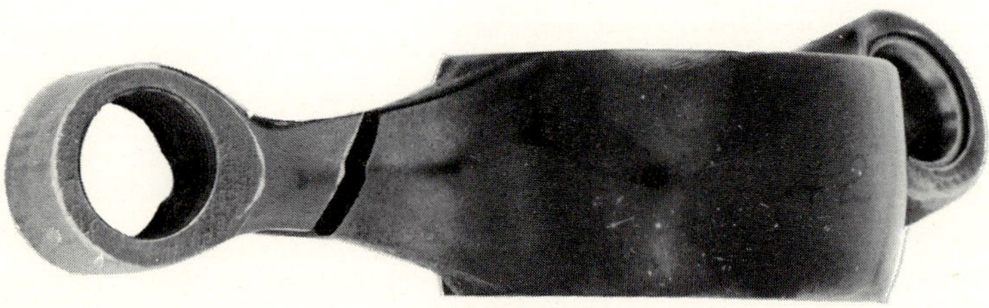

(b)

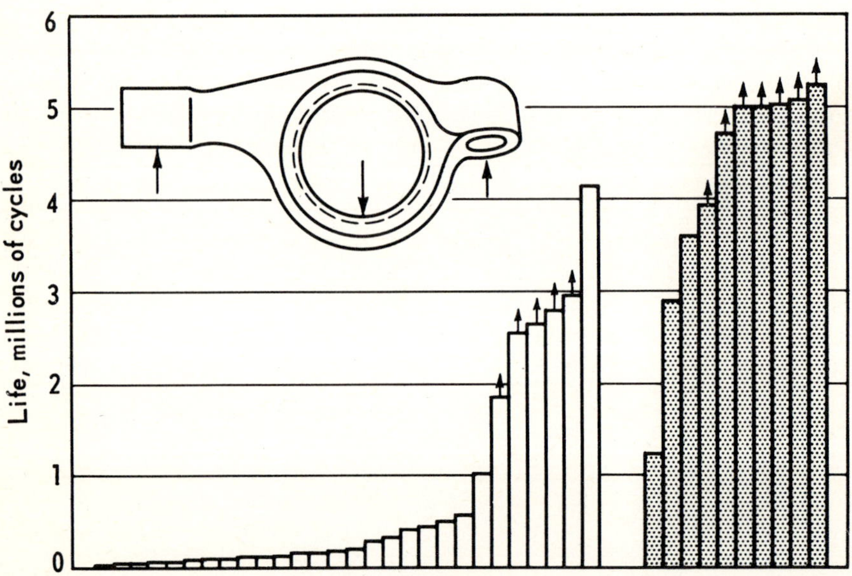

(c)

Fig. 6-4 Aircraft-engine rocker arm. Polishing destroys 96 percent of potential fatigue life. (*a*) Surface: forged, flash-sheared, shot-peened; fatigue durability 100 percent. (*b*) Surface: finished and polished; fatigue durability 4 percent. (*c*) Polished versus shot-peened aircraft-engine rocker arms; approximate stress + 3,100 + 83,100 psi. (*From J. O. Almen and P. H. Black, "Residual Stresses and Fatigue in Metals," McGraw-Hill Book Company, New York,* 1963.)

tensile stresses lower markedly the fatigue strength of ground parts. Indeed, cracks due to grinding may appear before the part is put in service. Even when parts are finished by machine polishing or by the use of abrasive paper, abrasive cloth, or abrasive-covered felt wheels, sufficient heat is often generated to introduce serious surface tensile stresses.

An instructive example is that of an aircraft-engine rocker arm shown in Fig. 6-4. Figure 6-4(*b*) shows a production specimen that was surface-finished, polished, and carefully heat-treated. The fatigue life of several of these specimens is represented by the left-hand bar chart in Fig. 6-4(*c*). The specimens in (*a*) were forged, flash-sheared, and shot-peened, with heat treatment omitted. The specimens in (*b*) were ground and polished. Their fatigue durability is represented by the speckled bar chart shown at the right in Fig. 6-4(*c*). The minimum life of the two groups shows that finishing and polishing reduced the fatigue durability to about 4 percent of that of the shot-peened specimens.

Carburizing and nitriding Machine parts are carburized or nitrided to increase their fatigue strength and/or wear resistance. The parts are placed in a carbon or nitrogen atmosphere and heated at suitable temperature and time conditions. Because of the diffusion of the carbon or nitrogen into the surface of the parts and the volume increase due to phase transformation, a residual compressive surface stress is introduced, as shown in Fig. 6-5. The surface compressive stress increases the fatigue strength because fatigue cracks cannot spread in a region of compressive

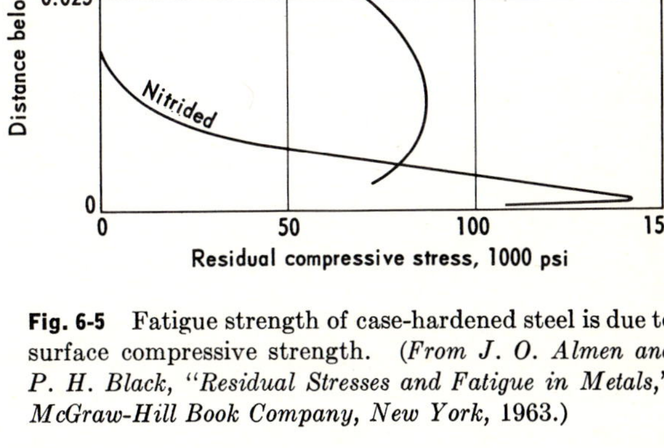

Fig. 6-5 Fatigue strength of case-hardened steel is due to surface compressive strength. (*From J. O. Almen and P. H. Black, "Residual Stresses and Fatigue in Metals," McGraw-Hill Book Company, New York,* 1963.)

stress; also, it increases the wear resistance by the more dense surface structure. It also decreases the tensile average stress.

Although the usual experience with carburized and nitrided parts is that fatigue strength is improved, it is possible to overdo the treatment because the surface compressive stress is balanced by an internal tensile stress of equal total value, so that subsurface fatigue cracks may develop in a region of high tensile stress and lead to early fatigue failure.

Induction and flame hardening The surface compressive stress in a carburized layer is the result of thermal contraction of the outer layer upon quenching and of the increased volume due to phase transformation. In induction and flame heating, the sequence of events is reversed from that of carburizing because the outer material becomes plastic and the inner material is cool and rigid. The hot outer layer tends to expand but, being restrained by the cold inner material, it is plastically deformed. Upon cooling, the thermal contraction of the deformed outer layer on the cool unyielding inner core results in tension in the case, and from this sequence the part is therefore weak in fatigue.

However, when hardenable steel is quenched from a temperature above the hardening temperature, the surface layer, upon cooling, experiences a phase change to martensite of greater volume. This volume increase, by itself, produces compressive stress in the transformed layer.

It follows that the net surface residual stress in surface-hardened steel is the algebraic sum of the tensile stress produced by the thermal contraction and the compressive stress produced by the transformation. In specimens that have been measured, it was found that the surface was compressively stressed to about 30,000 to 40,000 psi.

Rolling and peening For many years residual compressive stresses have been produced by cold-rolling cylindrical shafts or the fillets of stepped shafts.[1] In this process, a hard roller is pressed against the rotating shaft, and at the contacting regions, the elements of the shaft are stretched beyond the elastic limit, but the deeper layers are stretched elastically. As the roller passes on, the elastic recovery of the deeper material then "draws" the plastically deformed surface elements to a state of compressive stress. This procedure conforms to the principle stated earlier in this article. The compressively stressed surface layer increases the fatigue durability of the shaft, particularly in regions of stress concentrations such as fillets.

In peening by a hand or pneumatic ball-point tool, or by shot peening, the action is the same as described above.

In shot peening, steel or chilled cast-iron shot $\frac{1}{64}$ to $\frac{3}{32}$ in. in

[1] W. M. Murray et al., "Surface Stressing of Metals," American Society for Metals, Cleveland, 1947.

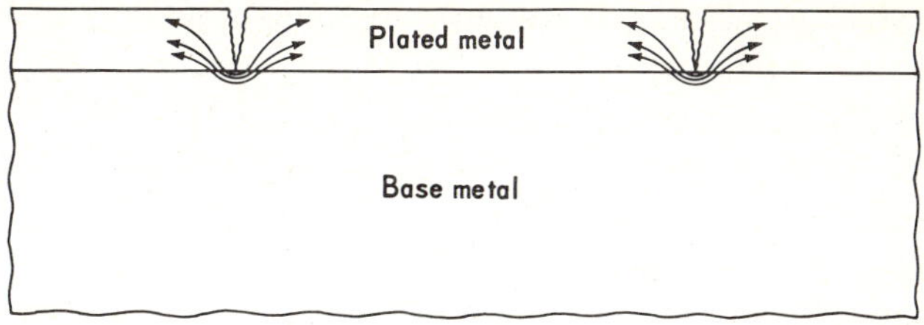

Fig. 6-6 Effects of cracks in plated members. *(From J. O. Almen and P. H. Black, "Residual Stresses and Fatigue in Metals," McGraw-Hill Book Company, New York,* 1963.)

diameter are impinged on the surface by an air blast or by centrifugal means. By the action described above, a residual compressive stress is induced in the surface layer. The maximum magnitude of the stress is about one-half the nominal yield point of the material. The depth of the compressively stressed layer depends on the hardness and ductility of the metal, the size of the shot, the velocity of impingement, and the length of time of application.

Electroplating The building up of worn and undersized machine parts by electrodeposits is extremely hazardous when applied to parts that are to be subjected to repeated loads. Nickel and chromium are usually selected for rebuilding operations because of their better wear resistance and presumed structural strength. However, these two materials are deposited in a state of residual tensile stress, from which it follows that, for equilibrium, the base metal is residually stressed in compression. When a crack forms in a strongly adhering deposit, the residual stress is transferred to the base metal, as shown in Fig. 6-6, thereby

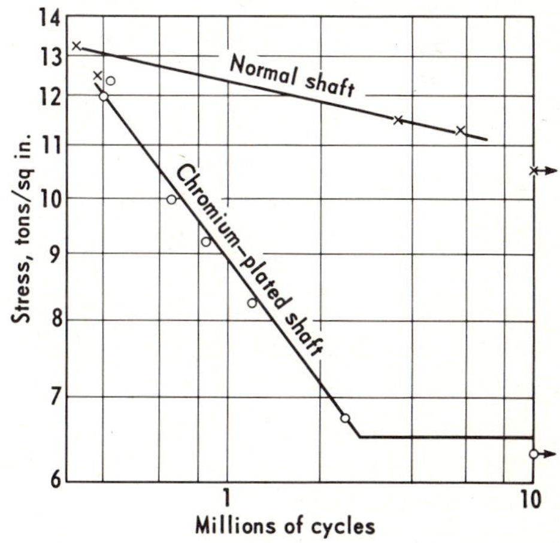

Fig. 6-7 Fatigue strength decreased by chromium plate. *(From J. O. Almen and P. H. Black, "Residual Stresses and Fatigue in Metals," McGraw-Hill Book Company, New York,* 1963.)

reducing the fatigue strength of the base metal. A further loss results because the crack introduces stress concentration which greatly decreases the fatigue strength of the plated part. This reduction is shown in Fig. 6-7.[1]

6-4 STRESS DIAGRAMS

Two examples are discussed in this article: a nitrided steel beam and a glass beam. The method for plotting and combining load stresses and residual stresses in these examples holds for other types of members and their loading and for residual stresses produced by the methods discussed in the preceding article.

For the purpose of determining the notch sensitivity of hard nitrided specimens, comparative fatigue tests were conducted[2] on notched and unnotched specimens in completely reversed bending (rotating-beam testing machine). The residual-stress pattern in the nitrided specimens is shown in Fig. 6-8, in which the compressively stressed case is approximately 0.02 in. Note the residual tensile stress in the core, which exists to balance the residual compressive stress in the case. In this series of tests it was found that notches of a depth less than that of the case did not affect adversely the fatigue strength of the specimens. The reason is that fatigue cracks at the roots of the notches were prevented from forming by the residual compressive stress in the case.

When the depth of the notches exceeded about 0.012 in., approximately 70 percent of the thickness of the case, the fractures originated at the roots of the notches.

Figure 6-8 shows also the bending stress induced by the external load and the sum of the load stress and the residual stress. Note that the maximum resultant stress occurs at a depth of approximately 0.03 in. below the surface.

In a second part of the experimental work referred to above, three groups of specimens were tested to determine their endurance limits.

Group A, not nitrided, unnotched	62,000 psi
Group B, nitrided, unnotched	82,000 psi
Group C, nitrided, notched	82,000 psi

In the nitrided specimens, the origins of the fatigue cracks were within the core because the residual compressive stress in the case prevented the spreading of fatigue cracks in that location; that is, the nitrided layer was an immunizing barrier.

[1] C. G. Williams and J. S. Brown, Fatigue Strength of Crankshafts, *Engineering (London)*, vol. 154, 1945.

[2] J. B. Johnson and T. T. Oberg, Effects of Notches in Nitrided Steel, *Metals and Alloys*, vol. 5, June, 1934.

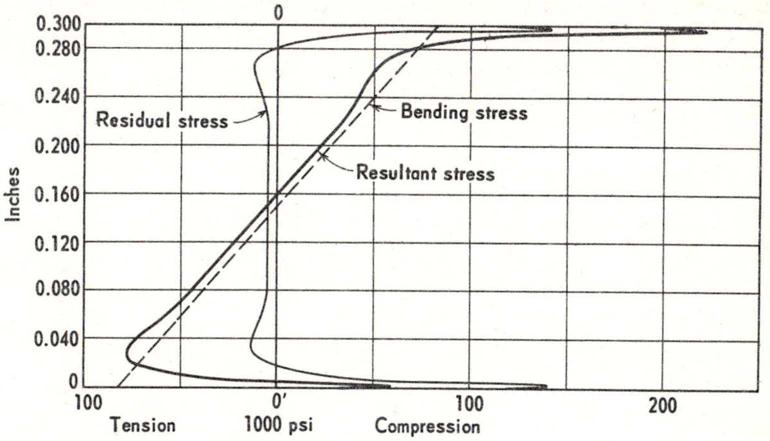

Fig. 6-8 Fatigue fractures cannot originate at surface of nitrided specimens. *(From J. O. Almen and P. H. Black, "Residual Stresses and Fatigue in Metals," McGraw-Hill Book Company, New York,* 1963.)

Glass is an interesting material to study because glass, a brittle material, fractures in static loading in the same manner as low-carbon steel in fatigue loading. Both materials exhibit "brittle fractures."

Figure 6-9 shows the stress distribution in a quenched glass plate in which the residual-stress pattern is indicated by the long-dashed line. The calculated external stress, or load stress, is indicated by the dashed

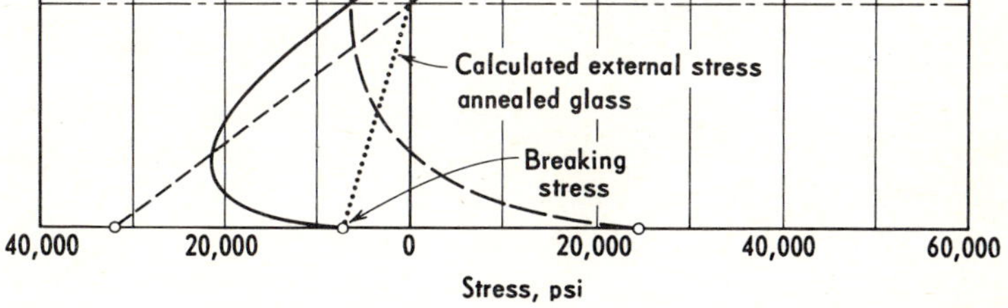

Fig. 6-9 Stress distribution in glass when loaded to failure. *(From J. O. Almen and P. H. Black, "Residual Stresses and Fatigue in Metals," McGraw-Hill Book Company, New York,* 1963.)

line and shows the nominal stress at fracture equal to 32,000 psi; however, the resultant stress (load stress plus residual stress) at fracture is 7,000 psi. Thus it is seen that the quenched-glass specimen failed at the same resultant surface stress as the annealed specimen, but the former supported a load over four times as great as the latter. It may also be seen from the diagram that the maximum tensile stress in the quenched glass beam reached a value of approximately 21,000 psi but, because of surface weakness, the fracture presumably originated at the surface from a tensile stress of only 7,000 psi.

6-5 MEASUREMENT

Although the magnitudes of residual stresses cannot be determined by calculations, they can be determined by experimental methods. The two most common laboratory methods are x-ray and dissection. The x-ray method, which is nondestructive, is limited presently to determining residual stresses to a depth of two or three atoms below the surface. The required laboratory equipment is expensive.

The dissection method, which is destructive and requires a careful and experienced technique but not elaborate equipment, is generally used at present. The residual stresses can be determined almost completely through the part. In this method, surface layers are successively removed by fine grinding, and the resulting change in curvature of the specimen after each layer has been removed can be used to calculate the sign and magnitude of the residual stress in the removed layer. The change in curvature of the remaining specimen is due to the release of the residual stress contained in the removed layer.

Each layer is usually removed to a depth of 0.0005 in., in steps of 0.0001 in. This step procedure is used to avoid introducing additional stresses by the grinding operation. As each layer is removed, the remaining residual stress in the specimen changes its curvature; this is measured by determining the "arc height" over a standard gauge length. During the grinding, the specimen must be held flat in a clamp in order to maintain uniform thickness of the layer removed.

In using the dissection method to determine residual stresses in cylindrical bars, as in prestressed torsion-bar springs, successive layers are removed by grinding. The stresses relieved by the removal of each layer alters the remaining stress system, which changes the angular deflection of one end of the bar with respect to the other. The change in angular deflection is then used to calculate the direction and magnitude of the residual stress.

As an alternative to removing layers of the specimen by dissection (grinding), the layers may be removed by acid etching.

6-6 SHOT-PEENING SPECIFICATION

The *intensity* of shot peening to give the desired surface residual compressive stress depends on the size and material of shot, velocity of shot stream, and length of time of peening. In a development department, a series of machine members are shot-peened with varying intensities; then by testing the members under real or simulated service conditions, the optimum intensity is selected. The optimum intensity should be sufficient to produce the required surface residual compressive stress to increase satisfactorily the endurance life of the part but not so heavy an intensity as to create excessive subsurface residual tensile stresses. The latter may cause subsurface fatigue failures.

In order to specify the optimum intensity, Almen test strips and gauges are used. The test strips are standardized as steel, Rockwell C 44 to 50, in size 3 in. long and $\frac{3}{4}$ in. wide, and are available in three thicknesses as follows:

N strips, thickness 0.031 in., for light intensity
A strips, thickness 0.051 in., for medium intensity
C strips, thickness 0.094 in., for heavy intensity

During each experimental peening of a machine part, a test strip is mounted on a holder to keep it flat, and the assembly is placed in the shot-peening cabinet so that it will receive the same peening intensity as the machine part.

After peening, the test strip is removed from the holder, and it is found that the residual compressive stress that was induced by peening causes convex curvature on the peened side. The curvature of the strip is measured on the unpeened side in terms of arc height over a standard chord by a peening intensity gauge.[1]

The production drawing of the part should carry the instructions to shot-peen to an intensity of Almen strip N, A, or C, and arc height in inches; for example, shot-peen, Almen strip N; arc height 0.10 in.

In production, the peening operator would then select the size of shot, adjust the air nozzle, and determine the time required to yield the specified arc height. This procedure establishes the correct peening intensity in the manufacture of the part.

[1] Almen and Black, *op. cit.*, pp. 64–70.

7 Engineering Materials

7-1 INTRODUCTION

In designing a machine member, the selection of material and the manufacturing process by which the part is to be made should be considered together. For this reason, the mechanical designer should be familiar with such interdependent factors as the adaptability of materials to the various processes, the effects of the processes on the properties of the materials, and the design details involved in the processes. Since the selection of material and the process for each machine member represent a special case involving a large number of variables, comprehensive rules cannot be stated. However, some important general considerations are discussed in this chapter.

The most economical design is arrived at by considering the total cost of labor, materials, and overhead for each of a number of proposed designs that are satisfactory from a technical standpoint.

7-2 FACTORS AFFECTING THE SELECTION OF MATERIALS

Some of the more important economic factors and physical and mechanical properties that are involved in material and process selection are discussed briefly in the following paragraphs.

Availability and *cost* of materials vary continually, and, as the change is toward favorable or unfavorable conditions, designs will necessarily undergo corresponding alterations for economic reasons. At times certain materials may become unavailable for general industrial use, and the necessity may arise for substitute designs based on procurable materials. A forceful example of this expediency occurs during national emergencies when alternative designs are required to avoid the use of strategic materials as well as processes that are affected by the unavailability of equipment.

Strength and *rigidity* are important properties of materials that are used for machine members. Strength, as measured by the ultimate strength, is necessary to prevent failure of the member by rupture. However, some steels have the desirable property of high ultimate strength coupled with low ductility, which may be undesirable in members subject to stress concentration. The use of high-ultimate-strength steels should imply carefully controlled heat treatment to avoid harmful effects, such as those caused by concentration of mass, surface decarburization, and quenching cracks. To guard against permanent deformation of the member, the *elastic limit* should be considered in design. For ductile materials, the yield point ordinarily may be used instead of the elastic limit.

Rigidity is of importance in members whose deflections are limited by service requirements. For instance, the transverse deflection of a shaft of a speed reducer may be limited by the following requirements: (1) clearance in film-lubricated bearings which support the shaft, for satisfactory bearing life as related to wear; (2) alignment, for instance, of gears mounted on the shaft, as related to gear-teeth wear, and vibration characteristics of the shaft, which depend mostly on the *deflection* of the shaft. Rigidity depends upon the modulus of elasticity. It should be noted that all steels have practically the same value for the modulus of elasticity. It follows that a change from a soft low-strength steel to a hard high-strength steel will not materially alter the rigidity of the part. It may also be emphasized that a member made of cast iron will generally be more rigid than a member of *equal load-carrying ability* made of steel, since the larger size required for the cast-iron member will more than compensate for its lower modulus of elasticity.

Resistance to fatigue should be the basis for the design of members that are subjected to cyclic loading. This property is measured by the *endurance limit*. *Notch sensitivity* and *damping capacity* of the material

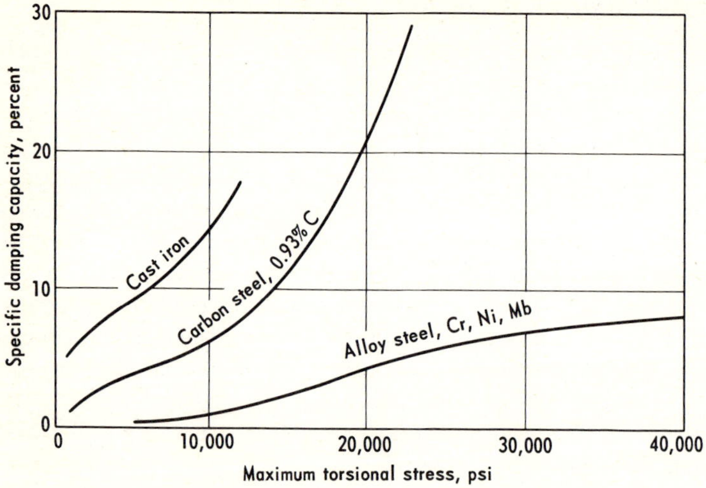

Fig. 7-1 Comparison of damping characteristics for cast iron and two steels. *(From G. S. von Heydenkampt, Damping Capacity of Materials, Proc. ASTM, vol.* 31, II, *p.* 157, 1931.)

should also be considered. Carefully controlled heat treatment should be applied to members that are subjected to fatigue, in order to avoid harmful surface effects. The surface residual stress is very influential in members subjected to fatigue loading. See Chap. 6.

Damping capacity is defined as the energy dissipated as heat by a unit volume of the material during a completely reversed cycle of stress.[1] The damping capacity is related to internal friction in the material and is represented by a mechanical-hysteresis diagram. Figure 7-1 shows the specific damping capacities of cast iron, of a carbon steel, and of an alloy steel. It may be noted that the damping capacity depends on the maximum stress. It has been suggested[2] that the significant critical damping capacity for use in design is the value at the endurance limit. The damping capacity is markedly affected by magnetic fields.[3] Low damping capacity is of value when resonance is desired, as in the case of tuning forks, bells, or gongs, but high damping capacity is desirable in most

[1] George E. Dieter, "Mechanical Metallurgy," p. 221 et seq., McGraw-Hill Book Company, New York, 1961.

[2] Battelle Memorial Institute, "Prevention of Fatigue of Metals," appendix 23, John Wiley & Sons, Inc., New York, 1940.

[3] E. R. Parker, The Influence of Magnetic Fields on Damping Capacity, *Trans. Am. Soc. Metals,* vol. 28, p. 661, 1940.

machine members to prevent an accumulation of serious resonant stresses, to decrease vibration, to decrease chatter, as in machine tools, or to decrease noise.

Resilience should be considered when the material is subjected to shock loading. A material with a suitable yield point should be selected, and the member should be designed so as to secure a desirable resilience of the part.

Hardness and *ductility* are important in many members. In bearing surfaces which have relative motion and in which fluid lubrication does not exist, hardness is of importance to limit wear. Hardness and ductility generally have a reciprocal relation in ferrous materials. The latter property is frequently desirable in order to relieve concentration of stress and it is effective in this respect in static loading but not in cyclic loading.

Weight may be important; often it is desirable, as in the case of foundations and flywheels; but it is undesirable in other cases, such as aircraft parts, where light metals, such as aluminum, magnesium alloys, or titanium may be used.

Electrical properties and *thermal properties* may be important. The latter property may be involved in expansion or contraction conditions that call for a consideration of the coefficients of expansion.

Resistance to wear or corrosion may be determining properties, the former in unlubricated or poorly lubricated friction surfaces. Generally, like materials in contact are not satisfactory for friction surfaces, although there are exceptions. *Corrosion resistance* is important in members subjected to a corrosive environment. The effect of corrosion is especially serious in the presence of stress concentration in cyclic loading.

Casting and *forging characteristics* are important for materials that are used for castings or forgings. Some materials, especially alloy steels, are difficult to cast because of effects of high temperature. *Fluidity* is of importance for good-quality thin sections in castings.

Machinability is frequently a critical factor, for instance, in stock for parts made by automatic machine tools. Often an expensive material which is readily machinable is a more economical selection than a lower-priced material which may be difficult to machine.

Low friction is of importance in bearing materials. The friction conditions are affected by the materials in contact and the surface finish. Certain combinations of materials that are in contact in bearing surfaces produce satisfactory results. This subject is considered in the discussion on bearings.

Friction is desirable in many instances: in friction surfaces of brakes, clutches, and power-transmission belts.

7-3 MATERIALS

The more common materials of which machine parts are made are discussed briefly in the following paragraphs. Detailed properties of these materials may be found in publications pertaining to engineering materials and to mechanical design.[1] Appendix V contains data for some of these materials.

Cast iron is the cheapest of the cast metals, and in the plain or alloyed state it can be cast into almost any size and form that is desired. The three classes of cast iron are as follows: (1) Gray iron, so called from the gray fracture as cast, is easily machinable but is not as abrasive-wear-resistant as chilled iron; it will resist galling wear, as in the case of brake shoes, cylinders, and pistons. (2) White iron exhibits a white fracture; it is hard and difficult to machine, and it is usually produced to be made into malleable cast iron. (3) Chilled cast iron is white iron composing parts of castings that are cooled rapidly to provide hard wear-resistant surfaces. The depth of chill can be controlled by composition and heat treatment. Chilled cast iron is used in such members as dies, car wheels, plow points, and cam followers.

The damping capacity of cast iron is high, as shown by Fig. 7-1. This property makes it suitable for beds, bases, and frames of machinery that are subject to vibration and for crankshafts and camshafts. The effects of mass and cooling rates on the structure and properties of the casting should be considered. In high-temperature applications, change in the physical dimensions known as "creep" and "growth" may be important to consider.

Cast iron is used for cylinder blocks, heads, crankcases, flywheels, crankshafts and camshafts, beds, bases, frames and housings, car wheels, brake drums and shoes, dies, wheels, gears, sprockets, pulleys, rolls, and many other machine parts of irregular form.[2]

Malleable iron is made by transforming white-iron castings, which are originally hard and brittle, by a heat treatment that converts the material into a matrix of ferrite containing nodules of temper carbon, so that it becomes tough and ductile, resistant to impact, and easily machinable.[3]

Wrought iron is a mechanical mixture of highly refined metallic iron and a small quantity of slag. It is resistant to corrosion in ordinary atmospheres and welds readily. The endurance limit is low as compared with other ferrous materials.

[1] "ASME Handbook, Metals Engineering—Design," Oscar J. Horger (ed.), McGraw-Hill Book Company, New York, 1953.

[2] "Cast Metals Handbook," and "Alloy Cast Iron," American Foundrymen's Society, Chicago.

[3] See "American Malleable Iron Handbook," Malleable Founders' Society, Cleveland.

Cast steel, either plain or alloyed, is used where castings of improved properties over iron or malleable castings are desired. Cast steel is stronger and tougher, weighs less for the same strength or stiffness, and has a higher endurance limit than cast iron. Corrosion resistance can be improved by the use of alloys or protective coatings. Some of the uses of steel castings are as bases for heavy-duty machinery, highly stressed castings, frames for transportation and construction equipment, gears, wheels, and many small and intricate highly stressed machine parts.[1]

Wrought steel in its various forms is commonly used for machine parts. Its applications range from members subjected to mild service, where low-carbon steel having relatively low strength is satisfactory, to members subjected to severe conditions of loading and environment where high-carbon steel or alloy steel is used. It is important to consider that the necessity for proper heat treatment increases with the carbon and alloy content of steel. The mechanical properties may be markedly controlled by the composition and heat treatment, and the selection problem is usually one of choosing from the large number of available steels the one that is most satisfactory for the member being designed.

The common alloying elements used in steel are manganese, nickel, chromium, molybdenum, vanadium, and wolfram (tungsten). Nickel increases the strength, yield point, endurance limit, shock resistance, and hardness without a corresponding loss in ductility.

Wrought steel may be forged, swaged, rolled, drawn, and welded, and it is especially susceptible to machining and finishing. It finds innumerable applications in members subjected to high stresses due to external loads either static or cyclic, to shock loading, or to stress concentration, and in parts subjected to wear, high or low temperature, or corrosion; it is also used in applications where special properties, such as electrical or thermal requirements, are involved and where space and weight must be conserved.

Some applications of steel in machinery parts are in crankshafts, connecting rods, piston rods, fastenings, keys, pins, rivets and bolts, ball and roller bearings, springs, shafting, gears, valves, frames of heavy stationary and transportation equipment, tubes, levers, arms, dies, and rolls.[2]

Brass and *bronze* are used in machinery parts in both cast and wrought form. Brass is used where moderate strength and ductility, resistance to corrosion, or good wearing qualities are desired. Bronze is

[1] See "Steel Castings Handbook," Steel Founders' Society of America, and "Cast Metals Handbook," American Foundrymen's Society, for information regarding cast steel.

[2] Two sources of information on metallic materials are the "Metals Handbook," American Society for Metals, Cleveland, and "SAE Handbook," Society of Automotive Engineers, New York.

superior to brass in these properties but is more expensive. Phosphor bronze has relatively high tensile strength, yield point, endurance limit, and shock-resistant properties.

Aluminum and *aluminum alloys* have a variety of uses. Pure aluminum resists oxidation, is highly ductile, and has good forming properties, but it has poor casting characteristics and machining properties. The effects of copper as an alloying element are to raise the ultimate strength and endurance limit and to improve the casting characteristics and machinability, but to lower its resistance to corrosion. Aluminum-copper alloys are used in such applications as crankcases, transmission housings, and fittings. The aluminum-silicon alloys have better corrosion resistance and mechanical properties but poorer machinability than aluminum-copper alloys. They are suitable for marine castings, water jackets, housings, and castings where a minimum of machining is required.

The aluminum-copper-magnesium-manganese alloy known as "duralumin" is suitable for parts that require severe working, and it has good corrosion resistance and strength. It is obtainable in sheets, plates, tubes, rods, wire, extruded sections, bolts, and rivets. The plates are well adapted to spot welding and are widely used in aircraft construction.

Magnesium alloys are corrosion-resistant in salt-free atmosphere, and galvanic corrosion can be prevented by coatings. Their weight is about one-fifth that of steel. Their modulus of elasticity is 6.5×10^6 psi. They can be cast, wrought, and extruded.

Titanium alloys are developed for useful strength and corrosion resistance at temperature ranges within -400 to 1000°F.[1]

Rubber and similar synthetic materials such as neoprene have a variety of applications in machinery. Rubber should be protected from high temperature, oil, and sunlight. It is an excellent material for seals and diaphragms, for water-lubricated bearings, for parts subjected to vibratory forces (such as vibration mountings, flexible couplings, and flexible bearings), and for tubes and hose. Rubber may be bonded to steel by the use of an intermediate nickel plating.

Nonmetallic materials have come into extensive use. In comparison with metals, the good features of plastic materials include low cost; light weight; good resistance to shock and vibration; self-lubrication, which means low friction and high wear resistance; and ease of fabrication. On their unfavorable side are low strength, high thermal expansion, low heat resistance, high creep and deformation under load, and embrittlement at low temperature. Plastics may be filled or unfilled and, if

[1] "Nonferrous Metals Book," Machine Design, Penton Publishing Company, Cleveland, Sept., 1963.

properly chosen, may be very satisfactory.[1] Examples of applications are telephone handsets (used before World War I), gears, bearing liners, and tubing.

Laminated Bakelite, Micarta, and Celoron are used extensively for lightly loaded gears where quietness is desired. The plastic gears are usually mated with cast-iron gears. This type of plastic is also used as bushings for bearings. Molded nylon is finding increased use for small gears, bearings, friction inserts, and many applications where low wear and friction are desirable.

A combination of glass, filler, and resin known as Fiberglas has replaced sheet metal in both flat plate form and in matched-metal-die processes where the parts are molded in the preformed shape. Its tensile strength varies between 10,000 and 24,000 psi. Corrosion resistance is high, and the useful maximum temperature is about 200°F.

7-4 MECHANICAL PROPERTIES OF MATERIALS

SAE numbering system The numbers for materials are given in Table 7-1.

Properties of steel[2] The selection of a definite steel for a particular application involves the consideration of the service requirements for the part, availability of the material, and the combined cost of the material and processing of the part. Normally the designer may select for many parts plain carbon steel on account of its low cost, ease of fabrication, and any necessary heat treatment. If the service requirements are severe, it may be necessary to select alloy steel to secure higher strength, greater hardenability, greater ductility, resistance to shock loading, and/or better machinability. The choice of alloy steel, however, will have the disadvantages of higher cost of material, more exacting heat treatment, and the possibility of temper brittleness in some grades.

The relative costs of steel are roughly as follows, taking hot-rolled plain-carbon steel bars as a basis: hot-rolled alloy steel, 60 percent greater; cold-finished plain-carbon steel, 30 percent greater; and cold-finished alloy steel, 85 percent greater.

Steel with less than 0.30 to 0.35 percent carbon will not have its properties altered materially by heat treating. With higher carbon content, the effects of heat treating are illustrated by the sample chart (Fig.

[1] "Plastics Book," Machine Design, Penton Publishing Company, Cleveland, Sept. 20, 1962; "Designing with DuPont Plastics," E. J. du Pont de Nemours & Company, Wilmington, Del.

[2] For a general discussion, see L. H. Van Vlack, "Elements of Materials Science," 2d ed., Addison-Wesley Publishing Company, Inc., Reading, Mass., 1964.

Table 7-1

Material	*Number*
Carbon steels	1XXX
Plain carbon	10XX
Free machining	11XX
Manganese (intermediate)	13XX
Nickel	2XXX
3.5% Ni	23XX
5.0% Ni	25XX
Nickel-chromium	3XXX
1.25% Ni, 0.75% Cr	31XX
3.5% Ni, 1.50% Cr	33XX
Corrosion and scale-resistant	30XXX
Molybdenum	4XXX
C-Mo	40XX
Cr-Mo	41XX
Ni-Cr-Mo	43XX
Ni-Mo (1.75% Ni)	46XX
Ni-Mo (3.50% Ni)	48XX
Chromium	5XXX
Low Cr (0.5% Cr)	50XX
Med Cr (1.0% Cr)	51XX
Chromium-vanadium	6XXX
Cr 1%	61XX
Triple-alloy steels	
0.55% Ni, 0.50% Cr, 0.20% Mo	86XX
0.55% Ni, 0.50% Cr, 0.25% Mo	87XX
3.25% Ni, 1.20% Cr, 0.12% Mo	93XX
1.00% Ni, 0.80% Cr, 0.25% Mo	98XX
Silicon manganese	
2% Si	92XX
Boron	
0.0005% B (min)	XXBXX

7-2) which shows that, with low tempering temperature, the tensile strength, yield point, and hardness are high but ductility and resistance to impact are low.[1] On the basis of service requirements the designer can select the optimum tempering temperature.

Oil quenching is not as drastic as water quenching and gives lower strength and hardness but with less distortion of the part (Fig. 7-3).

The size of the part affects markedly its tensile properties, as shown

[1] In comparing impact values between the Charpy keyhole notch and the Izod V notch, it is well to remember that the former usually lies between 15 and 45, whereas for the Izod test values are frequently over 100.

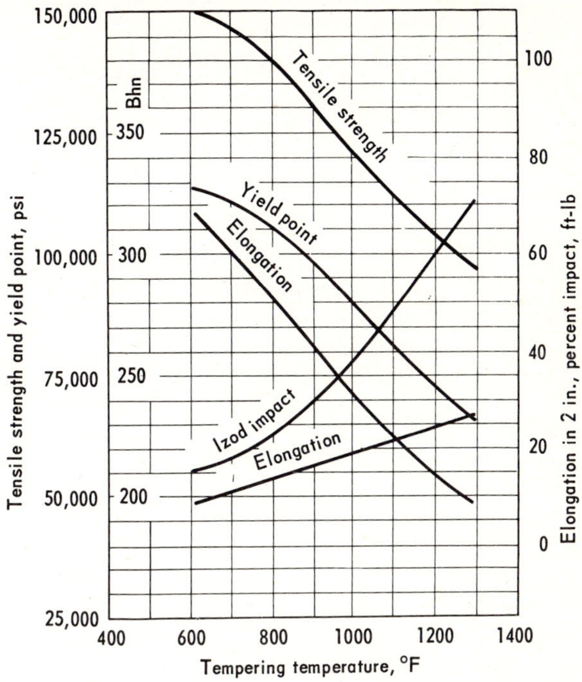

Fig. 7-2 Mechanical properties of SAE 1045 steel, water-quenched. *(Courtesy of The International Nickel Company, Inc.)*

by the sample chart of Fig. 7-4. For the yield point of several carbon and alloy steels, see Appendix VIII.

If hardenability is a factor for wear resistance, plain carbon steel with case hardening may be suitable, or if surface requirements are severe for resistance to shock and high pressure, a steel of the hardenable class called *H* steels may be necessary. Hardenability curves for selecting these steels are available which give results of standard end-quench tests and are expressed as shown in the sample graph (Fig. 7-5). These curves are presently available for about a hundred steels from 1320H to 9445H.[1]

In specifying hardenability, the maximum and minimum hardness values at a definite distance from the end of the test bar may be stated. The distance chosen would be that corresponding to the section used in making the part. For example, in Fig. 7-5, the specifications $J_{\frac{42}{53}} = 4/16$ in. means that at 4/16 in. from the end of the control specimen the hardness

[1] For a set of compositions and hardenability for the H steels, refer to the "SAE Handbook," "AISI Steel Products Manual," or "Metals Handbook," *op. cit.*

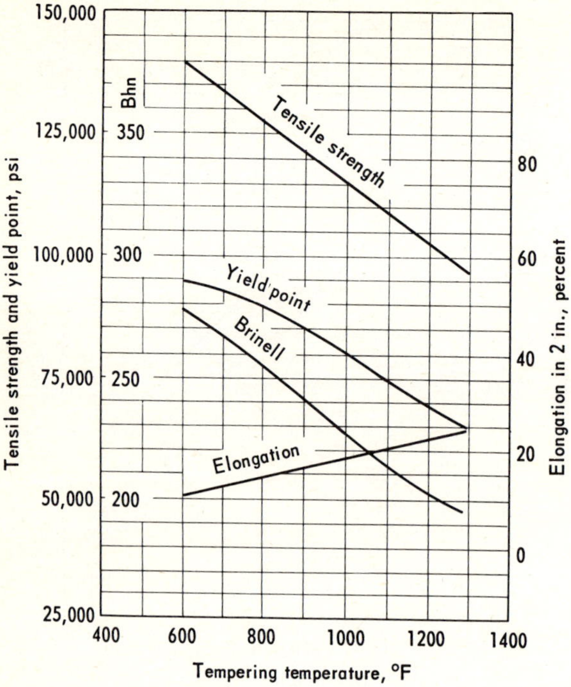

Fig. 7-3 Mechanical properties of SAE 1045 steel, oil-quenched. (*Courtesy of The International Nickel Company, Inc.*)

by a Jominy test for SAE 2330H steel would lie between the limits of 42 and 53 Rockwell C.

In determining strength properties for steel to use in design calculations, the yield point can be determined from Appendix VIII. The tensile strength can be estimated from the diagram in Appendix VI, and the endurance limit from the curves in Appendix VII for the required surface finish. See Example 7-1 for the procedure.

For steel, the ultimate strength in shear is approximately two-thirds the ultimate strength in tension. There is no general relation to determine the yield point in shear but it may be approximated from the values given in Table 7-2. The shear values in this table were determined from torsion tests.

In the absence of more specific data, the endurance limit in shear may be estimated for design purposes by multiplying the endurance limit in flexure by 0.55 for carbon steels and by 0.58 for alloy steels. Appendix V gives some properties of steel and nonferrous materials.

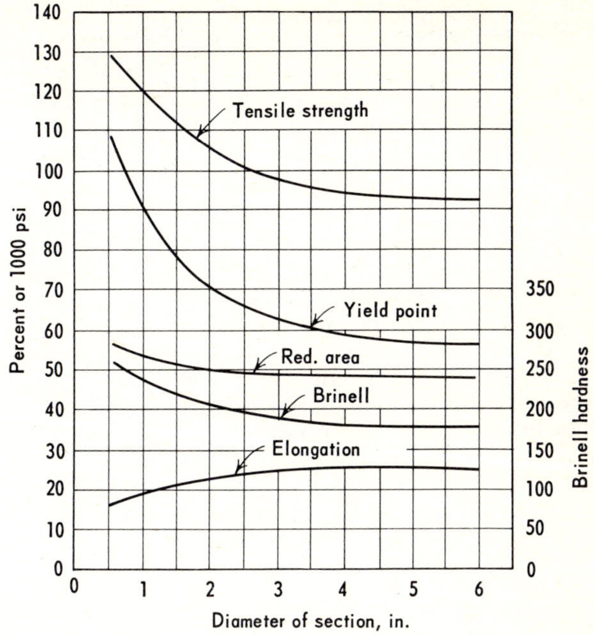

Fig. 7-4 Effect of size on properties of SAE 1045 steel. *(Courtesy of The International Nickel Company, Inc.)*

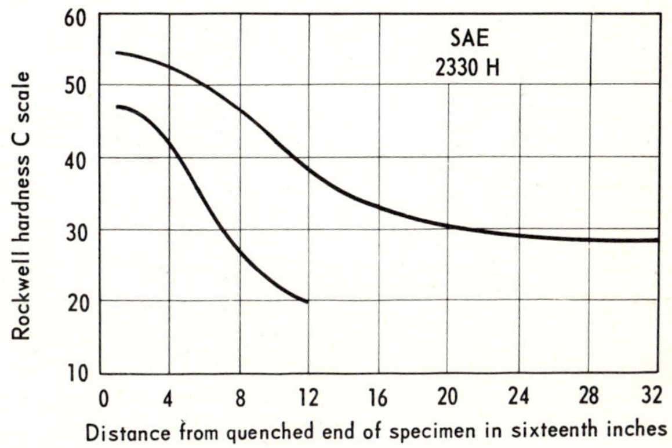

Fig. 7-5 Hardenability band for SAE 2330H steel. *(Courtesy of Society of Automotive Engineers.)*

Table 7-2 Yield ratios of some steels*

SAE no.	*Condition*	*Ratio* $\dfrac{\text{yield point in torsion}}{\text{yield point in tension}}$
1010	Normalized	0.70
1020	Normalized	0.71
1030	Normalized	0.62
1040	Normalized	0.62
1095	Normalized	0.62
2350	Annealed	0.78
3140	Quenched and tempered	0.68

* From "The Torsional and Shear Properties of Quenched and Tempered Nickel Alloy Steels in Different Sizes," The International Nickel Company, Inc., New York.

Example 7-1 Determine the yield points and endurance limits for SAE 1045 steel, oil-quenched and tempered at 900°F, assuming that the parts are in the size range ½ to 1½ in. with machined surfaces.

From Appendix VIII,

$s_y = 86{,}000$ psi

From Appendix VI,

$s_t = 115{,}000$ psi

From Table 7-2,

s_y in shear $= 0.62 \times 86{,}000 = 53{,}000$ psi

From Appendix VII,

s_e in tension $= 41{,}000$ psi
s_e in shear $= 0.55 \times 41{,}000 = 22{,}500$ psi

Machinability From the standpoint of production cost, a paramount property of the material is its machinability. This property at present cannot be directly measured in units;[1] but it can be satisfactorily expressed as a combined result of three factors: (1) tool life, (2) quality of the machined surface, and (3) required cutting power. It turns out that the most significant of these factors is tool life, since long tool life usually goes hand in hand with good surface quality and low cutting

[1] The state of our present knowledge of machinability recalls the statement: "I often say that when you can measure what you are speaking about, and can express it in numbers, you know something about it; but when you cannot measure it, when you cannot express it in numbers, your knowledge is of a meagre and unsatisfactory kind; it may be the beginning of knowledge, but you have scarcely, in your thoughts, advanced to the stage of *Science*, whatever the matter may be." Lord Kelvin (1824-1907).

power. A useful measure of tool life is the volume of material that is removed from the parts being machined before the tool becomes dull and must be reground. It becomes necessary then to establish a criterion for a dull tool.

In a cutting tool, wear occurs chiefly along the tool flank at the side cutting edge. This wears away the side relief adjacent to the cutting edge and produces what is called a *wear land.* When the wear land becomes large, the power required increases, the surface finish becomes poor, and tool wear is accelerated. Thus, an arbitrary value can be assigned to the *width* of the wear land (perpendicular to the side cutting edge) and the volume of metal removed from the parts being machined can be used as an expression of "machinability."

Cutting tools made of high-speed steels have been run satisfactorily until the wear land was as much as 0.060 in., and carbide tools to 0.030 in. To obtain comparative values of machinability for different metals, a standard wear land may be set for a series of tests and the volume of metal removed may be used as a basis for comparing the machinability of various metals. In the U.S. Air Force Machinability Report,[1] a wear land of 0.015 in. was selected for most of the tests. In production, carbide tools can be run with over twice this amount of wear land.

To pass on to the part being machined, it has been demonstrated that machinability of steel is governed more by its microstructure than by any other characteristic. The chart in Fig. 7-6 was prepared from U.S. Air Force test data, using 78B carbide cutting tools, a widely used type, with a limiting wear land of 0.0150 in. These data had little scatter, but similar results secured by using tools made of high-speed steel show more spread than with carbide tools.

In design of heavily loaded parts for production in large quantities, the problem is to select a material and heat treatment to satisfy the loading requirements and then to determine the probable microstructure of the part to use in determining the machining requirements. Often the machining requirements and strength requirements call for a compromise in selection of material and its heat treatment.

There are also available in tables relative machinability values for various materials in terms of their chemical composition and Brinell hardness numbers.[2] These values are usually stated in terms of 100 for SAE 1112 (free machining), with lower values for poorer machining materials. However, machinability cannot be predicted accurately by chemical composition and Brinell hardness numbers alone. When production cost is

[1] United States Air Force Machinability Report, Curtiss-Wright Corp., Wood Ridge, N.J., vol. 1, 1950; vol. 2, 1951.

[2] Kent's "Mechanical Engineers' Handbook, Design and Production," John Wiley & Sons., Inc., New York.

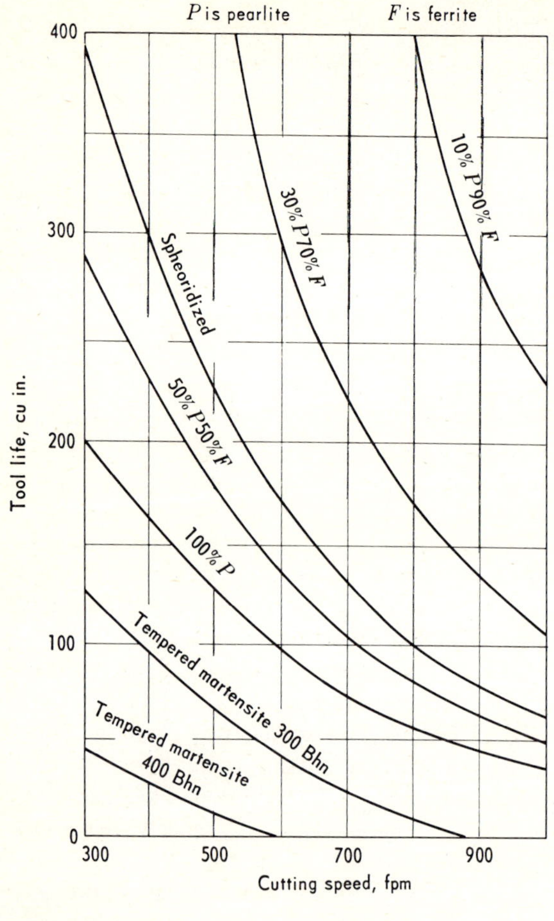

Fig. 7-6 Tool life versus cutting speed. Tool, 78B carbide; wear land, 0.015 in.; depth of cut, 0.100 in. (*From United States Air Force Machinability Report, vol.* 2, *Curtiss-Wright Corp., Wood Ridge, N.J.,* 1951.)

critical, efforts should be made to select cutting tools and speeds on the basis of microstructure. The Air Force report referred to includes a large number of charts of tool life and cutting time versus cutting speed for a variety of carbon and alloy steels represented by their composition, heat treatment, and microstructures.

High-temperature applications Because of the increase in efficiency at high operating temperatures of heat engines as steam and gas turbines and turbosuperchargers and of processing equipment as cracking stills, it has become necessary to develop materials that will have the required properties at these temperatures. The loss of strength in a carbon steel at high temperatures is shown in Fig. 7-7, which indicates the inadequacy of this steel for high-temperature service. Although considerable progress has been made in recent years in developing heat-resistant alloys and refractory metals, the frontiers of applications in the high-temperature, high-efficiency field are limited by available materials to retain strength and dimensions at the high temperatures. At present the maximum temperature in steam- and gas-turbine castings is limited to about 1000°F, although experimental work is in the 1400 to 1600°F and higher areas.

It is significant that, at the operating temperatures of high-temperature steam turbines and of gas turbines, the blades are of red-white heat and their strength, if made of ordinary steel, would be less than of cast iron at room temperature.

In addition to the loss of strength at high temperatures, steel and other metals exhibit the phenomenon of *creep*, which is the gradual elongation of the entire member at high temperatures over a long period of time. High temperatures and high stresses increase the creep rate so that at high temperatures we must limit the stress so that the part will not elongate beyond a certain amount over a desired life of the machine or equipment. For instance, a turbine blade may be designed with an

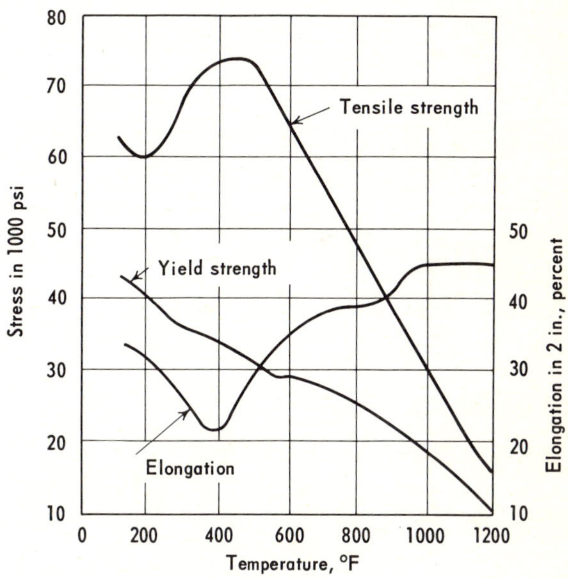

Fig. 7-7 Short-time tensile properties of hot-rolled steel. Effect of temperature on the properties of metals. (*ASME-ASTM Symposium,* 1931.)

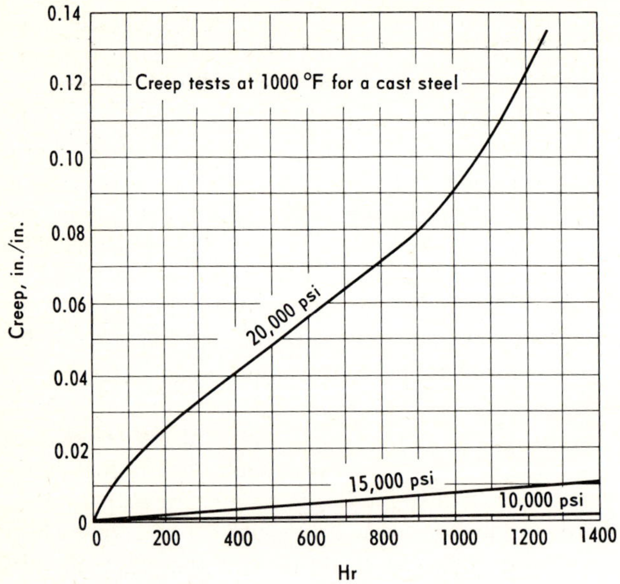

Fig. 7-8 Tensile creep curves for high and low stress.

allowable stress which will limit the total creep of the blade to not over 0.01 in. in 25 years. In Fig. 7-8 is shown the form of creep curves. In the use of creep data, it should be remembered that extrapolation of the curves may be unreliable. This poses a problem since if we are designing a unit for a life of 10 years, or 87,600 hr, we cannot wait that long to con-

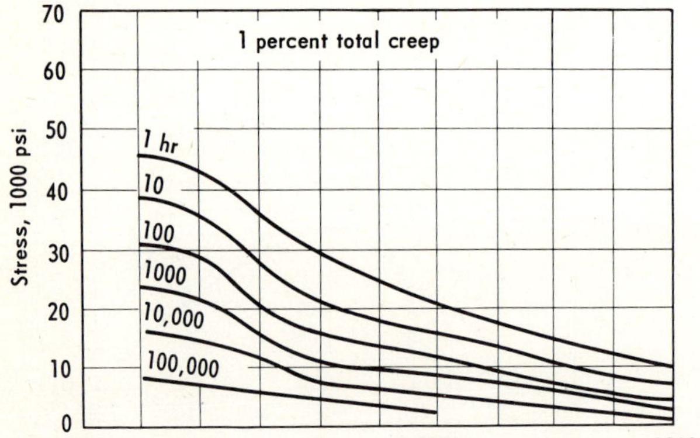

Fig. 7-9 Creep data for annealed cast carbon steel. *(Courtesy of The Crane Company.)*

duct long-time creep tests; instead we may be forced to make judicious use of short-time tests.

In Fig. 7-9 is shown a series of creep-design curves for a carbon steel. For extensive data on the properties of metal at elevated temperatures, see the reference below.[1]

A classic example of creep is that of a sheet-lead roof originally ¼ in. thick and weighing 16 lb per ft^2 which was placed as a cover on the roof of the Cathedral of St. Maurice at Angers, France, completed in 1247. The thickness of the lead at the ridge is now $\frac{1}{32}$ in., whereas at the eaves it has gathered into folds 10 in. high.[2]

[1] "Metals Handbook," *op. cit.*; "The Nickel Alloy Steel Handbook," The International Nickel Company, Inc., New York.

[2] *Prod. Eng.*, June, 1941, p. 293.

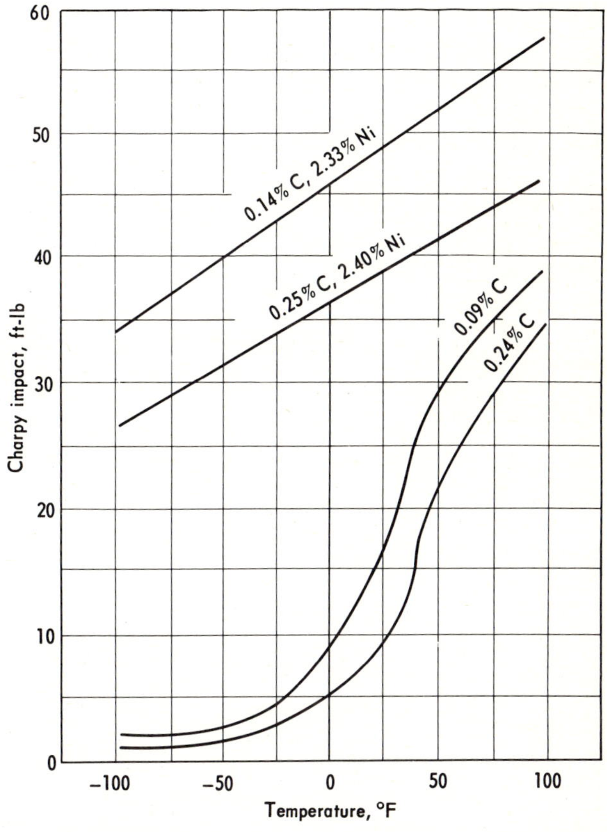

Fig. 7-10 Effect of temperature on impact resistance of carbon and nickel steel.

At *low temperatures*, down to at least $-125°F$, the strength, elastic properties, and endurance limit of steel are increased and the ductility changes little, but the notched-bar impact values usually decrease markedly, which indicates brittleness. This change in property is important in the design of low-temperature processing and refrigerating equipment and also in low ambient temperature applications as railroad and ordnance equipment and in airplanes. Maximum resistance to this embrittlement is secured with low carbon content but this of course reduces the strength. The addition of nickel as an alloying element toughens steel markedly at low temperatures in both cast and wrought steels (see Fig. 7-10).

7-5 SELECTION OF STEEL

Low-carbon steel Steel with less than 0.20 percent carbon has high ductility, is rolled or stamped, and is suitable for sheet or strip products. Structural steel and boiler plate are usually of 0.15 to 0.25 percent carbon. SAE 1015 to 1020 steel is known as carburizing grade; it may be formed cold and when carburized has good wear characteristics. SAE 1020 steel finds wide use in cold-worked parts as bolts and rivets and cold-drawn parts as bars and shafting. It provides a good finish for this class of parts. SAE 1022 to 1024 steel is used for heavier sections, such as forged parts, and may be readily welded and brazed.

Medium-carbon steel This steel is of 0.30 to 0.83 percent carbon. SAE 1030 to 1035 steel has lower ductility than the low-carbon grade but it responds to heat treatment. It may be forged cold for small forgings for connecting rods, shafts, and bar stock. SAE 1040 steel finds extensive use for tubing, automotive bolts, connecting rods, and crankshafts. SAE 1050 steel is harder and is usually oil-quenched. It is used for gears and heavy steel forgings.

Steels with 0.55 to 0.83 percent carbon have high strength and good wear properties. They are generally forged hot for spring wire and for parts having cutting edges. Steel with 0.65 to 0.75 percent carbon is used for railroad rails. SAE 1065 to 1085 steel is used for coil springs and flat springs, and SAE 1070 to 1083 is used for plow shares, toothed diggers and scrapers, and other parts subjected to shock and wear.

High-carbon steel In the high-carbon steel group, SAE 1090 to 1095 is widely used for springs, punches and dies, chisels, saws, cutting tools, steel wire, and cables. Steel with 1.0 to 1.2 percent carbon is used for saws, files, and razor blades, and alloy steel with 1.2 to 1.5 percent carbon is used for balls, rollers, and races for rolling bearings.

Alloy steels Nickel steel is the most widely used of the alloy group on account of its high strength, high ductility, and good impact properties. SAE 2335 steel is used for highly stressed gears, axles, castings, nuts and bolts, and SAE 2335 to 2350 is used for many other highly stressed machine parts. SAE 2515 steel is used for more severe service where high fatigue strength or good carburizing qualities are necessary; this steel, however, is costly. It finds use in wrist pins, kingpins, transmission gears, ring gears, pinions, and cams.

The chromium steels, SAE 50XX and 51XX, have improved hardening properties and higher strength and wear resistance but may be temper brittle. These steels are used in applications as ball bearings.

Nickel-chromium steel has further increased strength and hardenability but may be temper brittle. SAE 31XX is used for drive shafts and axles, transmission gears, and connecting rods.

Carbon-molybdenum steel, SAE 40XX, is a carburizing grade used for spline shafts, transmission gears, and pinions where the service is not too severe, and SAE 41XX, chromium-molybdenum, has good deep-hardening characteristics, high ductility, and weldability and is relatively inexpensive.

7-6 RELATION OF COOLING RATES TO DESIGN[1]

The serviceability of a machine part may be greatly affected by the interrelation between heat treatment and several factors that are involved in the design of the part, such as its form and mass.

The nonuniformity of physical properties of steel throughout a member may be due to unequal cooling rates in various sections and to the resulting effect on the decomposition of austenite. For example, in quenching a solid sphere heated to a uniform temperature throughout, the surface will be cooled more rapidly than the interior, and the difference in cooling rates will vary with the type of coolant. Also, a small sphere will have a greater difference in cooling rates than a larger sphere, since the ratio of surface to volume is inversely proportional to the diameter.

It is evident that points in an irregularly shaped member may have widely different cooling rates, which may result in a corresponding difference in the effect of heat treatment. For instance, the point of a tapered pin will be cooled more rapidly during quenching than the larger end. Likewise, a corner of a rectangular bar will be cooled faster than an edge, and the edge faster than a flat surface, and the surface faster than the interior. Thus, if uniform properties are required, the designer should

[1] See discussion on design and heat treatment in "Metals Handbook," American Society for Metals.

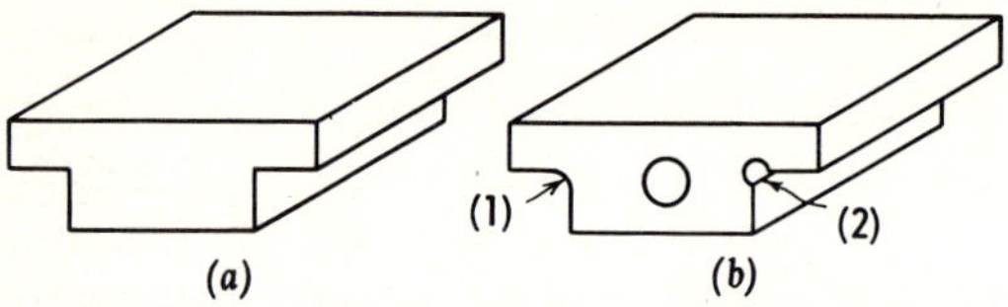

Fig. 7-11 (*a*) Original design of block to be heat-treated. (*b*) Improved design, showing reduction of concentration of mass, and substitution of fillets for sharp reentrant corners, by methods 1 and 2.

attempt to shape the part so that the cooling rates in the various sections will be as nearly uniform as possible. An approach to this condition is found in the distribution of the material as evenly as possible, i.e., in the avoidance of large masses of metal. The procedure is illustrated in Fig. 7-11 where material has been removed from heavy sections.

The Effect of Mass charts[1] given in handbooks indicate the effect of size or mass and the quenching operations on the physical properties of steels. For instance, one of these charts indicates that the tensile strength of SAE 1045 steel quenched in water at 1500°F and drawn at 1000°F has a value for a 4-in.-diameter specimen which is approximately 15 percent less than that for a 1-in.-diameter specimen. Other properties show corresponding differences.

An example of the mass effect is shown by the comparison between the strengths of a solid shaft and a hollow shaft. An 8-in. shaft with a 4-in. axial hole will have its torsional section modulus reduced by $6\frac{1}{4}$ percent and its weight reduced by 25 percent, as compared with an 8-in. solid shaft. Tests on shafts show that the bored shaft is more than 40 percent stronger than the solid shaft with the same heat treatment.

Internal stresses may be serious in that they may lead to cracks forming during heat treatment or under service loads. A bar or plate of section as shown in Fig. 7-12(*a*) may have cracks developed at the reentrant corners because of unequal cooling of the wide and narrow

[1] See "SAE Handbook," Society of Automotive Engineers.

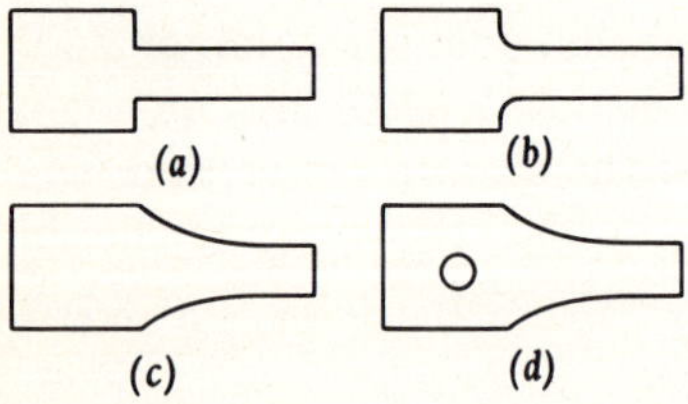

Fig. 7-12 Progressive improvement in section of bar to be heat-treated.

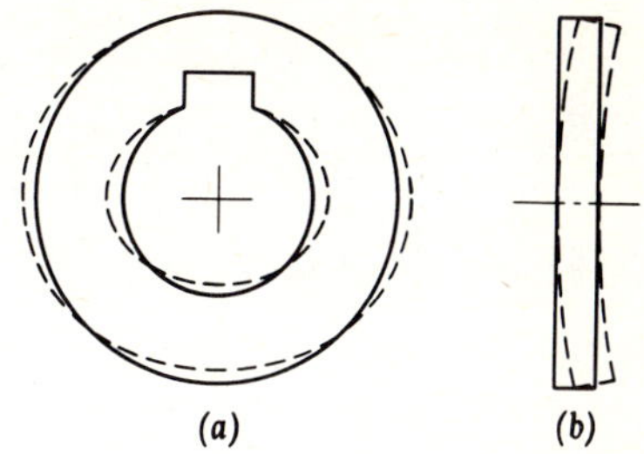

Fig. 7-13 Distortion and warping of slotted ring.

portions. An improvement results by introducing fillets, as shown in (*b*), or gradual transition slopes, as shown in (*c*). Further improvement may result from the removal of some material, as shown in (*d*), for reasons discussed in the preceding paragraphs. In some cases, however, the use of a large fillet, as required by considerations of stress concentration, may have its advantages reduced by reason of the mass effect, so that a medium-sized fillet may be a compromise.

Distortion and warping may be caused by the unsymmetrical shape of the part or by nonuniform application of the coolant during quenching. Thus the slotted ring in Fig. 7-13 may be distorted during cooling, as shown by the dotted lines in (*a*). Such distortion may be decreased by the use of two or more symmetrical slots. Distortion of gear teeth in heat treatment may lead to noise and excessive wear. Special heat treatment, such as induction hardening or flame hardening, may be used to give a minimum of distortion in this case.

The dotted lines in Fig. 7-13(*b*) indicate warping that may be caused by the sudden application of the coolant from one side of the ring.

8 Mechanical Fabrication and Processes

8-1 INTRODUCTION

In this discussion mechanical fabrication is intended to mean the permanent joining of structural or machine parts that are made as separate units for manufacturing, assembling, or transportation reasons. Thus large plates are fabricated by riveting or welding together small plates of sizes that are easily manufactured. Similarly, a drum may be fabricated from a cylinder and two circular plates by welding or riveting. Fabrication is intended to represent a process different from the *assembly* of units that are essentially different in functions, such as a gear and a shaft assembled by a force fit or a valve seat in a cylinder block assembled by a shrink fit.

8-2 RIVETING—GENERAL REMARKS

The function of rivets in a joint is to make a connection that has *strength* and *tightness*. Strength is necessary to prevent failure of the joint.

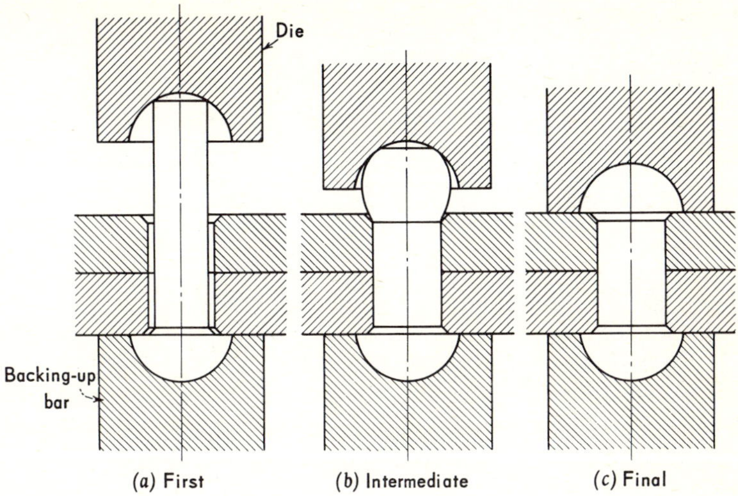

Fig. 8-1 Progressive stages in riveting.

Tightness is necessary in order to contribute to strength and to prevent leakage; for example, it is required in a boiler or in a ship hull.

Rivets may be driven by hand or by riveting machines. In hand riveting, the original rivethead is backed up by a hammer or heavy bar, and then the *die* or *set*, as shown in Fig. 8-1, is placed against the end to be headed and the blows are applied by a hammer. This forms the second head called the *point*. In machine riveting, the die is part of the hammer, which is operated by air, hydraulic, or steam pressure.

In structural and pressure-vessel riveting, the diameter of the rivet hole is usually $\frac{1}{16}$ in. larger than the nominal diameter of the rivet. After the rivet is driven, it will contract on cooling. The lateral contraction will be slight, but there will be a longitudinal tension introduced in the rivet that will draw the plates tightly together in well-driven rivets. This tension will cause frictional resistance between the plates that will resist sliding of the plates under load. The resistance to sliding due to interplate friction is seldom considered in calculations of the strength of riveted joints. The procedure of neglecting interplate friction is followed because the pressures caused by the driving and cooling of the rivets and the coefficient of friction between the plates are difficult to determine, and neglecting those factors is on the side of safety. Also in high-temperature installations, relaxation may cause the tension in the rivets to be reduced.

Although in some applications rivets carry an external tensile load,

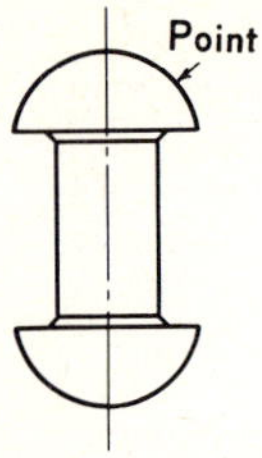

Fig. 8-2 Fillets in rivets.

their general use should be limited to shearing loads. If it is necessary for a tensile load to be carried by a connection, a bolt should be used.

Stress concentration exists in the parts riveted and in the rivets themselves; however, since the materials used in most riveted parts are ductile, stress concentration may not be serious. In ductile material, plastic flow around the rivet hole may be relied upon to relieve stress concentration if the loads on the joint are static. Stress concentration exists also in the rivet, and it may be serious at the junctures of the shank of the rivet and the heads. This concentration of stress may be reduced by countersinking the rivet holes $\frac{1}{16}$ in. so that a fillet is formed as shown in Fig. 8-2. Since loads on the rivets may be variable because

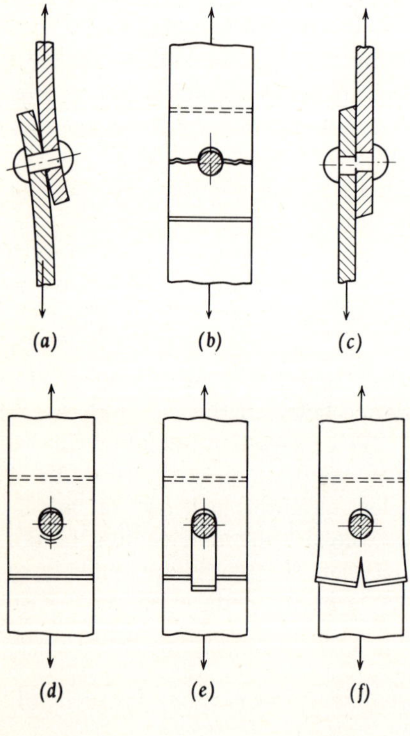

Fig. 8-3 Modes of failure of riveted joint: (*a*) bending of plate; (*b*) rupturing of plate by tension; (*c*) shearing of rivets; (*d*) crushing of rivets or plates; (*e*) shearing of margin; (*f*) tearing of margin.

of the character of the external load, the vibration, or the temperature change, the rivet may fail owing to stress concentration in fatigue, even though the rivets are made of ductile material.

The ways in which a single riveted lap joint may fail are illustrated in Fig. 8-3.

The maximum stresses in a riveted joint at failure are difficult to calculate. Instead of attempting to use exact stresses, a safe joint may be designed on the basis of simple tension of the plates, shear of the rivets, and bearing at the rivets and plates, using allowable stresses determined from similar existing and satisfactory installations. The allowable stresses quoted in the following article have been determined on this basis and are recommended by the structural standards.

8-3 STRUCTURAL AND MACHINE-MEMBER RIVETING

In the riveting of structural and machine members, the rivet holes may be punched, drilled, and reamed, or some combinations of these operations may be used. If punched holes are used without being reamed, as in structural work, it is assumed in calculating the strength of the plate that the punching operation may damage the plate to the extent of a $\frac{1}{32}$-in.-thick ring around the hole. Thus, in determining the net area of the plate between rivet holes in a tension plate, the rivet hole should be considered as having a diameter that is $\frac{1}{8}$ in. larger than the *nominal* diameter of the rivet. Furthermore, in calculating the strength of a structural riveted joint, the nominal diameter of the rivet should be used in determining the strength of the rivet, since it is assumed here that the rivet may not completely fill the hole throughout the length of the rivet. These two assumptions are illustrated in Fig. 8-4.

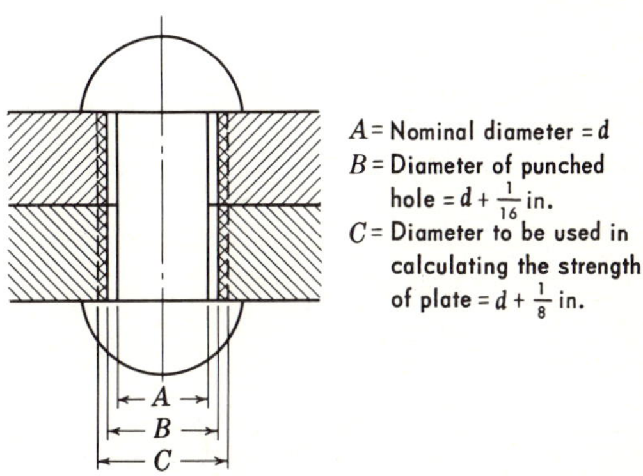

Fig. 8-4 Structural riveting.

In structural riveting, spacing of the rivets should be governed by the following considerations: (1) stability of the joint, (2) the existence of a margin between the rivet hole and the edge of the plate that is not less than the rivet diameter, and (3) clearance for the die during riveting.

For structural steel, the following allowable stresses may be used:

For plates in tension	20,000 psi
For rivets in shear	15,000 psi
For bearing of rivets and plates	
Single shear	32,000 psi
Double shear	40,000 psi

The determination of the strength of a simple structural joint is illustrated by the analysis of the connection shown in Fig. 8-5.

The problem is to determine the safe load that may be carried by the joint illustrated in the sketch. The angles and the plate are structural steel, and the rivets are $\frac{3}{4}$ in. in nominal diameter.

1. The strength of the plate along the section AA is equal to

$$[3\tfrac{1}{2} - (\tfrac{3}{4} + \tfrac{1}{8})] \times \tfrac{1}{2} \times 20{,}000 = 26{,}250 \text{ lb}$$

2. The strength of all the rivets in shear is equal to

$$6 \times \frac{\pi}{4}\left(\frac{3}{4}\right)^2 \times 15{,}000 = 39{,}700 \text{ lb}$$

3. The strength of all the rivets in bearing is equal to

$$3(\tfrac{3}{4} \times \tfrac{1}{2}) \times 40{,}000 = 45{,}000 \text{ lb}$$

4. The strength in failure of the plate in tension along section BB combined with failure of the lower rivet (a) in shear and (b) in bearing

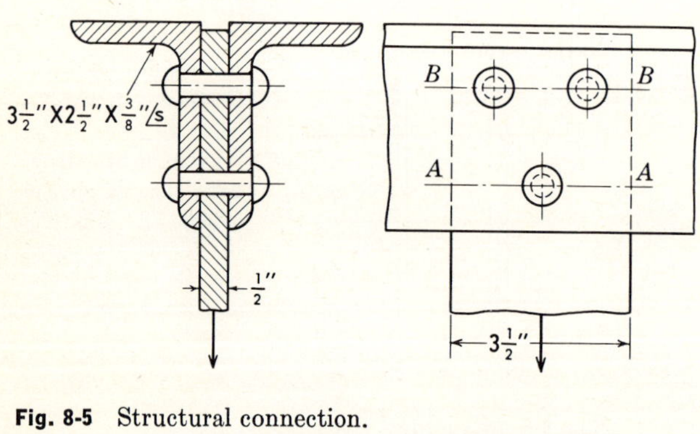

Fig. 8-5 Structural connection.

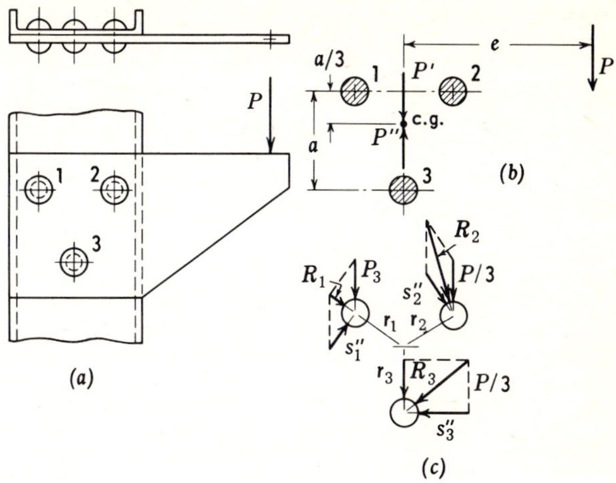

Fig. 8-6 Eccentrically loaded connection and rivet loading

is equal to

$$(a)\quad \{[3\tfrac{1}{2} - 2(\tfrac{3}{4} + \tfrac{1}{8})] \times \tfrac{1}{2} \times 20{,}000\} + 2 \times \pi/4\,(\tfrac{3}{4})^2 \times 15{,}000 = 30{,}750 \text{ lb}$$

$$(b)\quad \{[3\tfrac{1}{2} - 2(\tfrac{3}{4} + \tfrac{1}{8})] \times \tfrac{1}{2} \times 20{,}000\} + \tfrac{3}{4} \times \tfrac{1}{2} \times 40{,}000 = 32{,}500 \text{ lb}$$

The above calculations indicate that the joint would fail by rupture of the plate along section AA and that the safe load would be 26,250 lb.

Note that stress concentration has been neglected, a procedure that is justified by the lack of seriousness of stress concentration because of the ductility of the structural steel and the relatively high factor of safety used in determining the allowable stresses.

Eccentrically loaded connection A connection of this type is illustrated in Fig. 8-6(a). The *group* of three rivets supports the loads. The first step in the solution of the problem is to locate the center of gravity of the rivet group. For convenience, the rivet group is redrawn in Fig. 8-6(b). If the rivets are of the same size and the upper two rivets are equally spaced with regard to the vertical centerline, the center of gravity of the rivet group will be at the point indicated by c.g.[1]

The next step is to introduce at the center of gravity of the rivet

[1] If the rivets are of different diameters or are unsymmetrically spaced, the center of gravity of the rivet group may be located by the use of a well-known principle of mechanics, $X = \Sigma AX/\Sigma A$, as illustrated in Fig. 8-7.

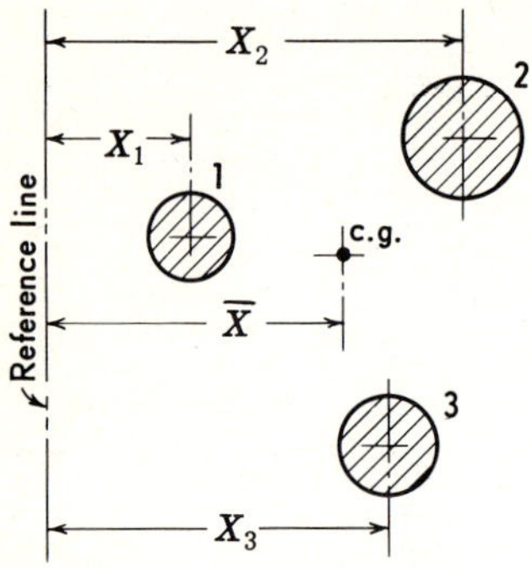

Fig. 8-7 Center of gravity of rivet group.

group two forces that are equal and opposite to P, as shown by P' and P'' in Fig. 8-6(*b*), and then to regard P' as inducing a direct shearing load, known as "primary shear," on the rivets, and to consider P and P'' with their moment arm e as producing a moment equal to Pe, which tends to rotate the rivets about the center of gravity of the group. The loads induced by the turning moment are known as "secondary shear."

Now the primary shear on each rivet is equal to $P/3$, and these loads are shown in Fig. 8-6(*c*), where the rivet group has been redrawn.

The turning moment will induce loads on rivets 1, 2, and 3, which are of different magnitudes and directions. The direction of the secondary shear on each rivet is at right angles to the direction of its moment arm r. Thus s_1'', or the secondary shear on rivet 1, will act at right angles to r_1, and s_2'' and s_3'' will act at right angles to r_2 and r_3, respectively. To determine the magnitudes of s_1'', s_2'', and s_3'', it is necessary to make an assumption.

A reasonable assumption is that the two members riveted together, i.e., the plate and the channel in Fig. 8-6(*a*), are rigid and that the deformation takes place by the rivets. This assumption is valid for the proportions of plates and the number and size of rivets that are met with in the usual structural or machine members. However, it may not be valid for very small, thin plates and for relatively large rivets.

If the assumption stated above is used, it follows that as a result of the turning of the plate the deformation of each rivet will be proportional to its distance r from the center of gravity of the group. Therefore

$$\frac{s_1''}{r_1} = \frac{s_2''}{r_2} = \frac{s_3''}{r_3}$$

From the condition that the sum of the external turning moment and of the internal resisting moment equals zero,

$$Pe = s_1''r_1 + s_2''r_2 + s_3''r_3$$

The equations given above are sufficient to determine the secondary shear on each rivet and, in general, it may be found that

$$s_n'' = \frac{Per_n}{r_1^2 + r_2^2 + r_3^2}$$

where the subscript n refers to any rivet.

The secondary shear loads are drawn in Fig. 8-6(c), which shows for both primary and secondary shear the loads produced by the *plate on the rivets.*

The primary and secondary shear loads may be added vectorially to determine the resultant load R on each rivet.

In the solution of a problem, the primary and secondary shear loads may be laid off approximately to scale; generally the rivet having the *maximum* total shear load will be apparent by inspection. The values of the resultant load for that rivet may then be calculated. The maximum loaded rivet becomes the critical one for determining the strength of an existing connection if all the rivets are the same size, or for selecting the rivet size or sizes in a design.

Draw rivets have been used widely for riveting in locations that are accessible from one side only. Figure 8-8(a) shows a rivet of this type

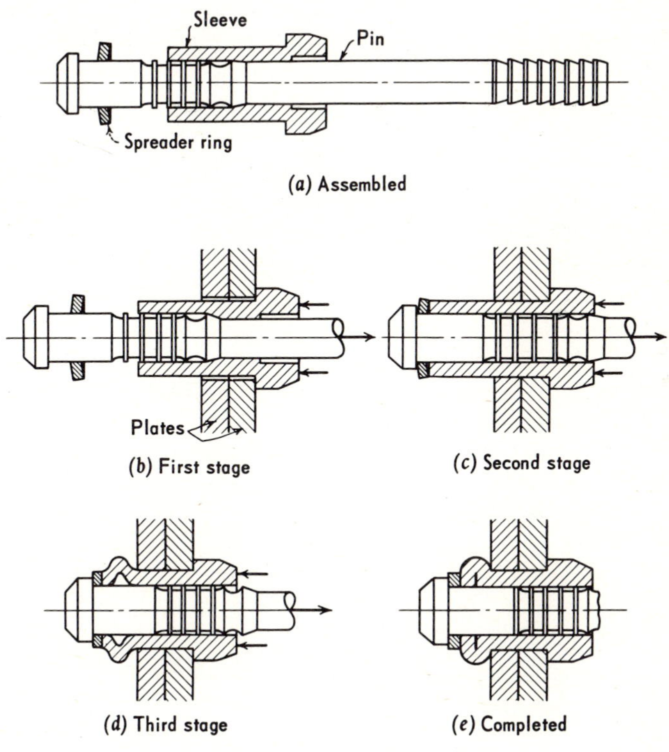

Fig. 8-8 Draw rivets for riveting from one side of plate.

known as the "Huck rivet," which is composed of three essential parts: (1) the hardened steel *pin*, (2) the soft steel *sleeve*, and (3) the *spreader ring*.

The complete rivet is supplied to the user assembled as shown in Fig. 8-8(*a*). At (*b*) is shown the rivet inserted from the right into the holes of the plates. The sleeve is held against the plates while the pin is pulled toward the right. This operation spreads the sleeve as shown in the diagrams at (*c*) and (*d*) and finally the pin breaks off, leaving the completed rivet as shown at (*e*). Another and simpler draw rivet is the Cherry rivet.

8-4 RIVETING THIN PLATES

The same general requirements of strength and tightness that apply to general riveting apply also to thin plates; however, because of the possibility of failure of thin plates carrying compression or shear loads by buckling of the plates, special requirements are necessary in thin-plate riveting.

A frequent cause of weakness of thin-plate construction is buckling of the plate because of the inaccurate spacing of rivet holes and/or improper riveting. This wrinkling of the plate is illustrated in Fig. 8-9; it may be serious in built-up structural members and in columns or beams or in plates carrying shear loads, such as aircraft structures.

Flush riveting may be used in the wetted surfaces of ship hulls or aircraft structures in order to reduce skin friction. Special procedures are necessary, especially in aircraft construction, to prevent wrinkling of the plate between rivets and "dimpling" of the plate around rivets. Figure 8-10 illustrates a flush rivet that has been developed for aircraft construction.[1]

Explosive rivets are used in aircraft structures in quantity production and are used also for locations accessible from one side only of the plate. Figure 8-11(*a*) shows a section of an unexploded aluminum rivet with the powder, or chemical, charge *c* in the shank. The rivet that has been

[1] Berlin and Rossman, Flush Riveting Considerations for Quantity Production, *S.A.E. Journal*, August, 1939, p. 328.

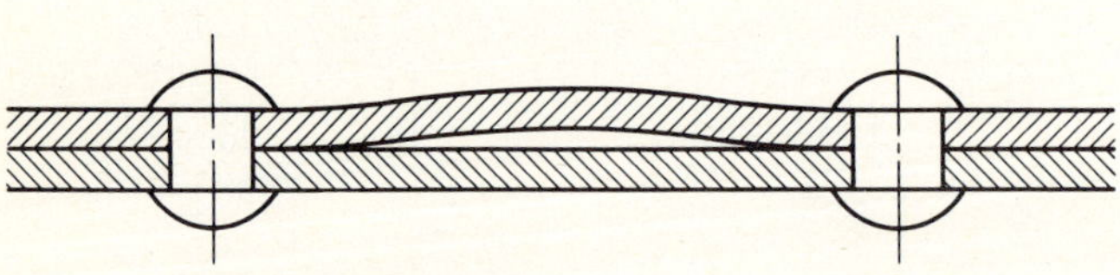

Fig. 8-9 Buckling of riveted thin plates.

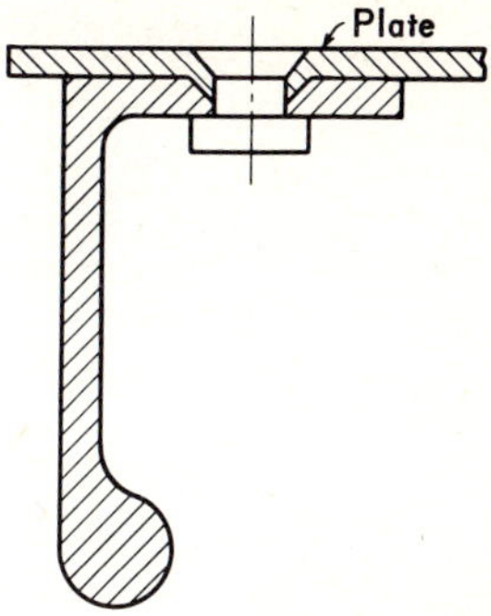

Fig. 8-10 Flush rivet developed for aircraft construction.

exploded by a heated-tip riveting gun placed in contact with the rivet-head is shown at (*b*).

8-5 PRESSURE-VESSEL RIVETING

The riveting of drums and tanks intended to withstand high pressure requires special consideration in design, materials, and construction. This is especially true for fired pressure vessels, such as a steam generator drum. The design of pressure-vessel joints should be based on strength and tightness.

The strength of the joint should be as near as possible to the strength of the unpunched plate. It is possible to produce joints of high efficiency, i.e., more than 90 percent.

Tightness of the joint is necessary to prevent leakage, and to ensure tightness the spacing of the rivets must be limited. Calking of the joint may be necessary to increase the resistance to leakage. Figure 8-12 illustrates the calking operation.

Joints for vessels subjected to high pressure are usually made of the double-strap construction shown in Fig. 8-13(*a*) in order to minimize bending of the plate, as illustrated in Fig. 8-3(*a*). The plates are generally of unequal width in order to increase the joint efficiency and to decrease stress concentration in the plate. This stress concentration is

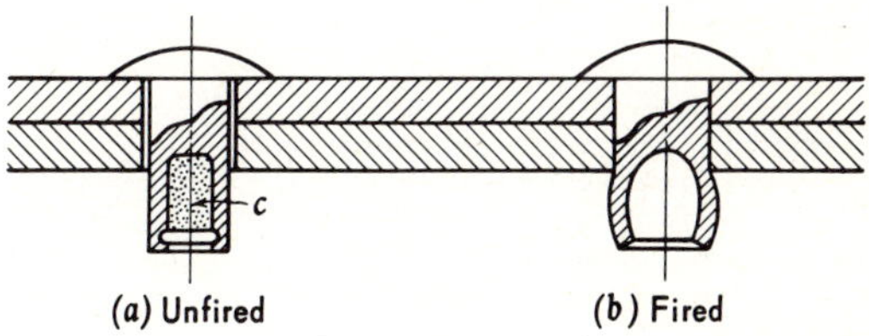

Fig. 8-11 Explosive rivets: (*a*) unfired; (*b*) fired.

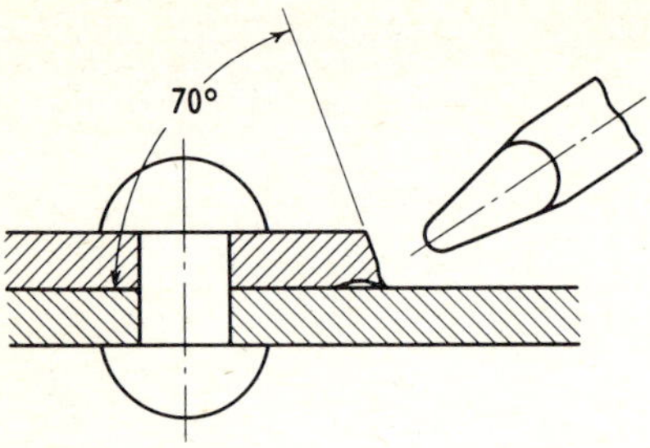

Fig. 8-12 Calking of riveted joint.

caused by flexure of the plate due to the expansion and contraction of the shell. A consideration of the stress-flow lines, as discussed in Chap. 4, indicates that a joint with unequal-width cover plates, as in Fig. 8-13(*a*), would have less stress concentration than one with equal-width cover plates, such as the one in (*b*). Placing the narrow strip on the outside of the drum facilitates calking of the joint adjacent to closely spaced rivets. This latter consideration will be evident from Fig. 8-14.

In order to design a joint for a pressure vessel, it is necessary first to determine the type of joint to be used. The choice involves the cost of construction and its relation to the efficiency of the joint, since the higher-efficiency joints have a larger number of rows of rivets. On this basis, a type of joint may be selected that will be satisfactory from an economic standpoint. After the type of joint has been decided upon, the design of the joint involves the determining of the size of the rivets and the pitch in order to secure approximately equal strengths of the joint for all methods of failure. It is evident that with large rivets spaced close

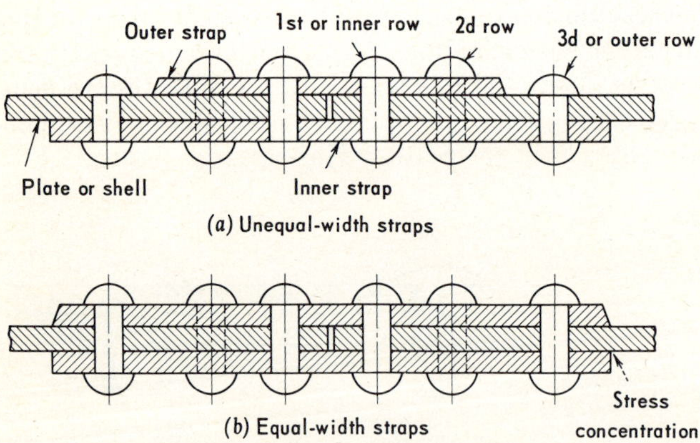

Fig. 8-13 Double-strap butt joints.

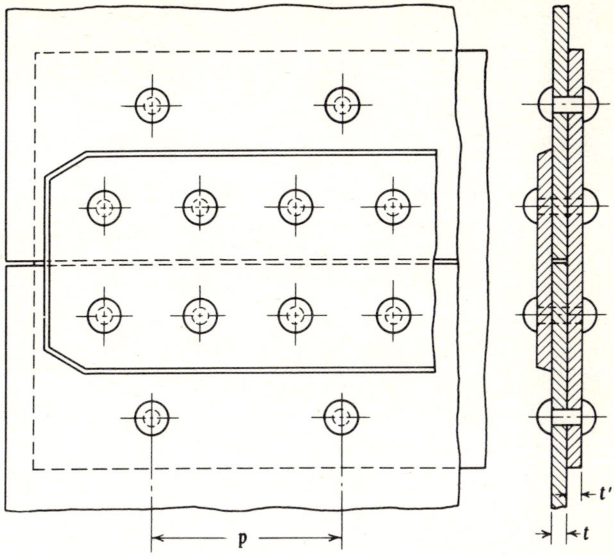

Fig. 8-14 Double-riveted double-strap butt joint.

together the rivets will have a high strength and will favor a tight joint, but the plate will be unduly weakened by the large, closely spaced holes.

In the treatment of pressure-vessel joints given here a typical standard joint as specified by the ASME Code[1] will be investigated for strengths under various probable methods of failure.

The following assumptions are made:

1. Each rivet will take an equal share of the load. This assumption implies that the shell and plate are rigid and that all the deformation of the joint takes place in the rivets themselves. Because of the actual deformation of the shell and plates, however, the rivets in the outer row in a multirow joint are subjected to more than an equal share of the load.

2. The plates are undamaged by the hole-forming operation, and the rivet fills the hole after it is driven. These assumptions are valid in high-grade construction using drilled and reamed holes and well-driven rivets. Since the diameter of the hole is $\frac{1}{16}$ in. larger than the nominal diameter of the rivet, the nominal diameter plus $\frac{1}{16}$ in. should be used in all calculations.

3. Failure of the margin by shear or tearing will not take place. This is true for ASME Boiler Code joints.

4. Finally, stress concentration is neglected. The stress concen-

[1] ASME Boiler Construction Code, Sec. I, Power Boilers.

tration in the shell, plates, and rivets may not be serious because of the ductility of the material; however, in the presence of corrosive substances, such as those found in certain kinds of boiler-feed water, corrosion fatigue may be serious (see Art. 5-9).

The strengths of a double-riveted double-strap butt joint, shown in Fig. 8-14, used as a longitudinal joint of a drum will be determined for various methods of failure.

In the figure, p indicates the pitch of rivets in the row having the greatest pitch, and indicates the length of a repeating group whose efficiency is typical of the efficiency of the entire length of joint.

Let p = pitch of rivets in outer row, in.
d = diameter of rivet after driving, or diameter of rivet hole, in.
t = thickness of plate, in.
t' = thickness of cover plate, in.
s_t = tensile strength of plate = 55,000 psi
s_s = shearing strength of rivet in single shear = 44,000 psi
s_c = crushing strength of rivet or plate = 95,000 psi

Assume that

$$p = 4\tfrac{7}{8} \text{ in.} = 4.875 \text{ in.}$$
$$d = \tfrac{7}{8} \text{ in.} = 0.875 \text{ in.}$$
$$t = \tfrac{3}{8} \text{ in.} = 0.375 \text{ in.}$$
$$t' = \tfrac{5}{16} \text{ in.} = 0.3125 \text{ in.}$$

1. The strength of the solid plate is equal to

$$pts_t = 4.875 \times 0.375 \times 55{,}000 = 100{,}500 \text{ lb}$$

2. The strength of the plate between rivet holes in the outer row is equal to

$$(p - d)ts_t = (4.875 - 0.875) \times 0.375 \times 55{,}000 = 82{,}500 \text{ lb}$$

3. The strength of all rivets in shear is equal to

$$5\frac{\pi d^2}{4} s_s = 5\left(\pi \frac{0.875^2}{4}\right) \times 44{,}000 = 132{,}290 \text{ lb}$$

4. The strength of the plate between rivet holes in the second row plus (*a*) shearing strength of rivets in the outer row or (*b*) crushing strength of rivets in the outer row is equal to

$$(a)\quad (p - 2d)ts_t + \frac{\pi d^2}{4} s_s = (4.875 - 2 \times 0.875) \times 0.375 \times 55{,}000 + \left(\pi \frac{0.875^2}{4}\right) \times 44{,}000 = 90{,}910 \text{ lb}$$

$$(b)\quad (p - 2d)ts_t + dt's_c = (4.875 - 2 \times 0.875) \times 0.375 \times 55{,}000 + 0.875 \times 0.3125 \times 95{,}000 = 90{,}429 \text{ lb}$$

5. Crushing strength of all rivets is equal to

$$2dts_c + dt's_c = d(2t + t')s_c = 0.875(2 \times 0.375 + 0.3125) \times 95{,}000 = 88{,}320 \text{ lb}$$

6. Crushing strength of rivets in second row plus shearing strength of rivets in outer row is equal to

$$2dts_c + \frac{\pi d^2}{4} s_s = 2 \times 0.875 \times 0.375 \times 95{,}000 + \left(\pi \frac{0.875^2}{4}\right) \times 44{,}000 = 88{,}800 \text{ lb}$$

The above analysis indicates that the joint will fail by method 2. The efficiency of the joint is

$$\frac{82{,}500}{100{,}500} \times 100 = 82.0 \text{ percent}$$

The efficiencies of some commercial boiler joints are given in Table 8-1.

Table 8-1 Efficiencies of commercial boiler joints

Type of joint	*Efficiency, percent*
Lap joints:	
Single riveted	45–60
Double riveted	63–70
Triple riveted	72–80
Butt joints:	
Single riveted	55–60
Double riveted	70–83
Triple riveted	80–90
Quadruple riveted	85–94

8-6 WELDING—GENERAL REMARKS

There are three general fields of application of welding:

1. *Fabrication* in which welding is used as an alternative method for casting or forging. In this application, *cost of construction* and *service-*

ability are important in making the choice between welding and the alternative methods. *Cost of construction* involves a consideration of the following items:

1. Welding:
   Cost of materials, such as structural steel
   Cost of cutting and forming parts
   Cost of positioning parts, welding, and machining
2. Casting:
   Cost of materials, as iron and steel
   Cost of patterns and molds
   Cost of casting and machining
3. Forging:
   Cost of materials
   Cost of dies
   Cost of forging and machining

Serviceability involves a consideration of strength and stiffness of the parts, and, in parts subjected to variable loads and/or vibration, of the damping capacity of the structure. In general, strength requirements are favored by the welding of structural steel parts, deflection requirements are favored by the larger sections of castings, and damping capacity is favored by larger volumes of cast and forged parts, especially in the use of cast iron.

In many instances where relatively large deflections are not objectionable and high damping capacity is not essential, the use of welded

Table 8-2 Comparison of fabrication methods for bearing support

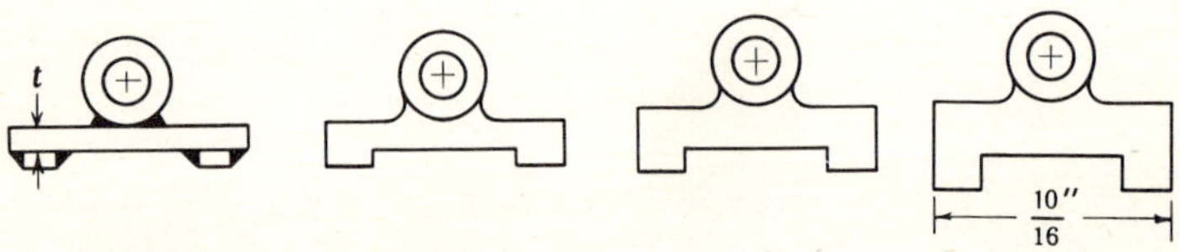

	Welded, percent	*Forged steel, percent*	*Cast steel, percent*	*Cast iron, percent*
Static strength	100	100	100	100
Deflection	259	259	191	100
Damping capacity	26.1	26.1	27.8	100
Weight	71	71	76	100
Cost	45	300	110	100
Thickness t	54	54	60	100

structural steel results in a substantial saving in production cost; however, where deflections are limited, or where impact loads and vibration require high damping capacity, forged or cast construction is more satisfactory. Approximate comparisons for a bearing support made in small numbers by different methods are given in Table 8-2.

2. *Fabrication* in which welding is substituted for riveting. In this application, welding is generally favored from the standpoints of strength and saving in weight. In welding, however, the cost of holding the parts during welding and the cost of welding itself, including materials, equipment, and operators, may favor riveting in some applications.

3. Welding used as a *repair* medium. Welding can be employed to reunite metal at a crack, to build up a small part that has broken off, such as a gear tooth, or to repair a worn surface, such as a bearing surface. Metal spraying also can be used in the latter example.

8-7 METHODS OF WELDING

The common methods of welding are forge welding, electric-resistance welding, and fusion welding.

In *forge welding* the parts must be heated to the plastic state at the regions where they are to be joined, and then impact, which is produced by a hand hammer or a press, causes the parts to unite. Wrought iron and low-carbon steel may be forge-welded. The process has limited use; it is employed in the manufacture of wrought-iron pipe and also as a repair medium.

In *electric-resistance welding*, the parts to be joined are pressed together and an electric current is passed from one part to the other until the metal is heated to the fusion temperature at the joint. A butt joint may be formed in this manner from relatively thick plates, or if the plates are thin a lap joint may be formed. In the latter case, if the pressure is applied by the two electrodes, one on each side of the overlapped plates, a *spot weld* is produced, or if two rollers are substituted for the point electrodes and the plates are pulled between the rollers a *seam weld* is produced.

In *fusion welding*, the parts to be welded must be held in position while molten metal is placed at the joints, thus fusing the metal of the parts, called the *parent metal*. The molten metal then solidifies to form the welded joint. The weld formed by the welding metal when it cools is essentially cast metal.

The heat to melt the weld metal and to make the parent metal plastic is generated in a number of ways, each of which defines the method of fusion welding as (1) Thermit welding, (2) gas welding, and (3) electric-arc welding.

In *Thermit welding,* a mold is built around the joints at the region where it is desired to confine the molten metal. Then the Thermit, which is composed of a mixture of finely divided iron oxide and aluminum, is placed in a reservoir in the mold. The Thermit is ignited and the aluminum reduces the iron oxide to molten steel, at a high temperature, which flows in the mold, melts the parts, and forms the joint on solidifying. Thermit welding is used principally to repair heavy cast-iron and steel parts and to weld heavy sections, such as rails, in the field where other welding equipment is not available.

Gas welding, often called *autogenous welding,* uses an oxygen-hydrogen or oxygen-acetylene gas, which is burned in a torch providing a pointed flame. The flame heats the parts to be welded and melts the welding metal, or filler rod, which on cooling forms the joint.

The oxygen-hydrogen process is generally used for welding nonferrous metals of low melting points, whereas the oxygen-acetylene process is used for welding ferrous and some nonferrous metals and alloys in thin sheets, and as a repair medium. Gas-welding equipment may be modified for use in flame cutting of plates.

Arc welding employs either a carbon-rod electrode with a separate rod used for the source of weld metal or a metallic-rod electrode in which the welding rod itself serves the dual purpose of electrode and source of welding metal. The metallic-rod arc-welding process is adapted to both structural and machine welding because of the consistently high quality of the welding that it is possible to obtain by the use of the proper welding procedure by trained operators of the welding equipment.

The making of high-quality welds requires specially designed equipment in order to produce a welding arc that has the necessary qualities for the proper control of the welding in producing joints of desired properties.

If a bare electrode or filler rod is used, the deposited weld metal while it is hot will absorb oxygen and nitrogen from the atmosphere, which will decrease the strength of the weld metal and lower its ductility and its resistance to corrosion. Shielded-arc welding is employed to prevent oxygen and nitrogen from coming in contact with the weld metal while it is molten. The shielded arc is produced by using welding rods coated with solid material which melts and vaporizes when heated by the arc and which forms a shield of inert gas around the molten weld metal; it also furnishes a flux which floats out impurities to form a slag on the top of the weld which further prevents harmful oxidizing effect on the weld while it is cooling. This slag is brushed off after the joint has cooled, leaving a good-appearing bead.

Shielded-arc welds are used in high-grade construction where strength is paramount. Figure 8-15 shows details of the shielded arc.

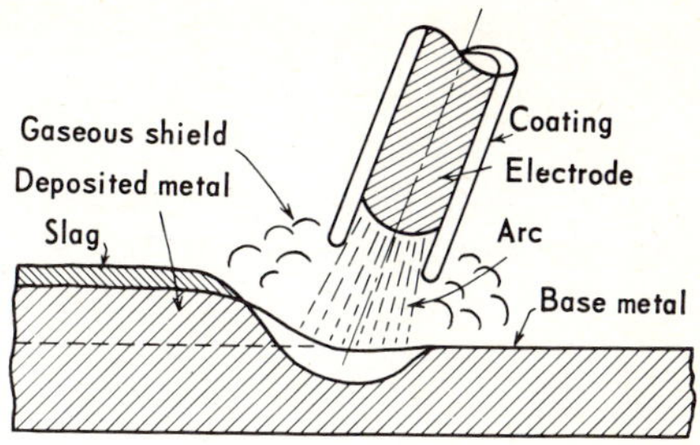

Fig. 8-15 Shielded arc.

8-8 FORMS OF WELDED JOINTS

Some common forms of welded joints are shown in Fig. 8-16. When butt welds are used, it is not necessary to bevel the edges of plates that are $\frac{1}{4}$ in. or less in thickness. For heavier plates, the edges should be beveled. For plates $\frac{3}{4}$ in. or more in thickness, it is desirable to weld from both sides of the plate in order to minimize distortion of the welded part and to reduce the amount of deposited metal required.

8-9 STRUCTURAL AND MACHINE WELDING

In welding of structural members or machine members, the primary purpose is to reduce the cost over other fabrication methods and to equal or improve the serviceability of the structure. Structural-member welding

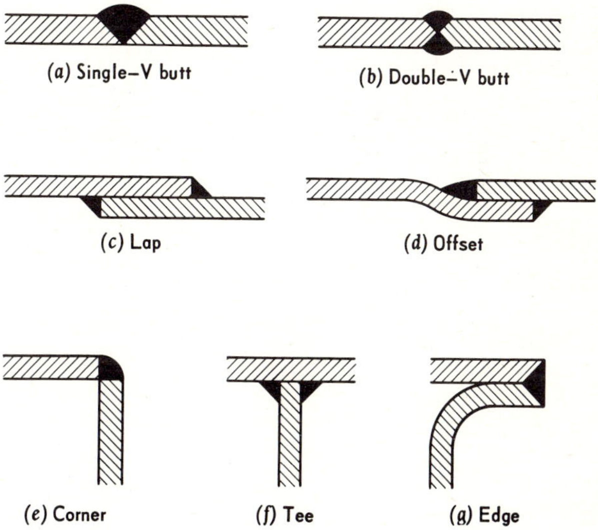

Fig. 8-16 Common forms of welded joints.

is employed generally as an alternative to riveting. Machine-member welding may be used as a substitute for rivets or bolts, for casting or forging, or for the purpose of permitting the joining of small cast units to form a large part that would be inconvenient to cast as a single unit.

In both structural and machine welding, the welds should be symmetrical with respect to the axis of the welded member unless the loading is unsymmetrical. The latter case is represented in Fig. 8-17, in which, assuming that the stresses at the welds are uniformly distributed,

$$L_a \times a = L_b \times b$$

which represents equilibrium of moments of resistance offered by the fillet welds.

The assumption that the stresses in the welds are uniformly distributed is not true because of the elongation of the welded member under load. The deviation from uniform distribution of stress is small for short lengths of fillet welds but becomes considerable for long weld lengths; hence the design of long welds should take into account the concentration of stress at the load-input end of the bead.

In the design of a welded part, an attempt should be made to arrive at the most economical combination of structural material, of cutting and bending, and of deposited welded metal. It should be realized that deposited weld metal costs per pound from 25 to 50 times as much as structural steel. From this standpoint welded joints should be reduced to a minimum in favor of the use of formed parts. Figure 8-18 illustrates alternative designs for a support.

Overwelding also should be avoided. In Fig. 8-18(b) and (c), the top of the support need not be held by a continuous weld. Likewise, in welding three strips to form a beam similar in section to an I beam, intermittent welding of the flanges and web is entirely satisfactory and results in considerable reduction in cost.

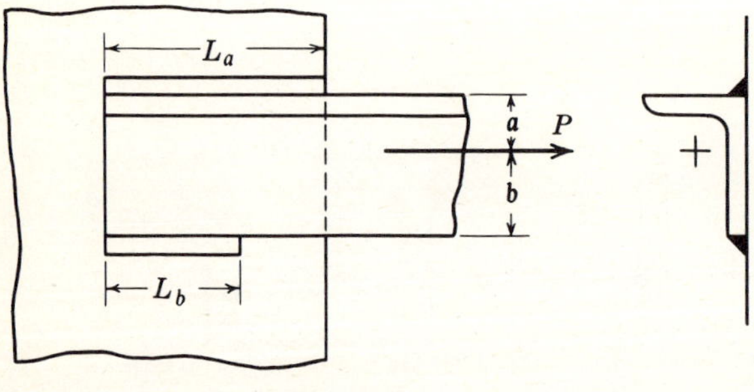

Fig. 8-17 Unsymmetrical welds.

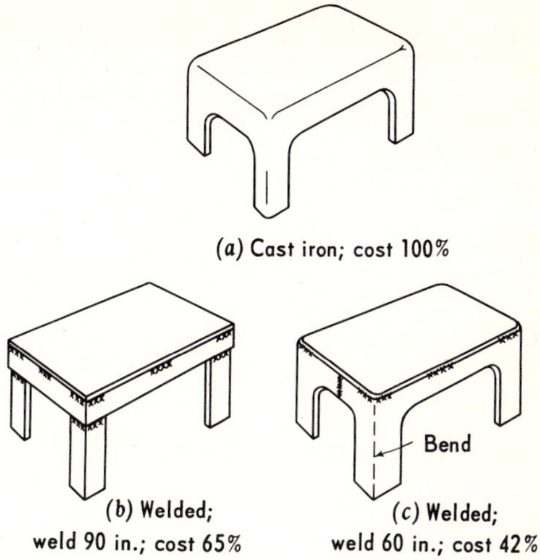

Fig. 8-18 Designs for a support.

8-10 WELDING DESIGN[1]

There are three types of welding design, as follows: (1) substitution method, (2) approximate calculation, and (3) precise calculation.

Substitution method In the changeover from a cast, forged, or riveted machine member to a welded design by the use of the substitution method, the welded part is fashioned from plates and structural and formed parts to simulate the shape of the original member. This method is illustrated in Fig. 8-19 in the changeover from a cast-iron to a welded link. If a program calls for changing to welding of a complicated machine, the changeover is generally made on one part at a time in order that any trouble that develops because of the substitution of welding may be localized and dealt with independently.

[1] See O. W. Blodgett, "Design of Weldments," James F. Lincoln Arc Welding Foundation, Cleveland, 1963.

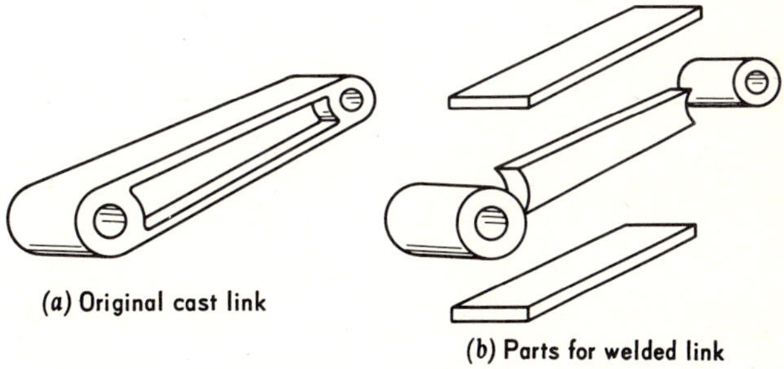

Fig. 8-19 Substitution method in welding design.

Table 8-3 Allowable static loads on mild-steel fillet welds

	Allowable static load per linear inch of weld, lb			
	Bare welding rod		*Shielded arc*	
Size of weld, in.	*Normal weld*	*Parallel weld*	*Normal weld*	*Parallel weld*
1/8	1,000	800	1,250	1,000
3/16	1,500	1,200	1,875	1,500
1/4	2,000	1,600	2,500	2,000
5/16	2,500	2,000	3,125	2,500
3/8	3,000	2,400	3,750	3,000
1/2	4,000	3,200	5,000	4,000
5/8	5,000	4,000	6,250	5,000
3/4	6,000	4,800	7,500	6,000

Approximate calculation This method is generally applied to a new design or to a moderately stressed existing machine member. The sizes of parts and the strengths of the welded joints are determined in accordance with allowable values that have been found satisfactory in practice. Allowable loads for normal and parallel fillet welds are given in Table 8-3. It may be noted in the table that the values for normal welds are higher than those for parallel welds. This is due to the fact that there is a more uniform distribution of stress for the normal weld than there is along the length of the parallel weld. The values of shielded-arc welds are higher than those for bare welding-rod welds for reasons discussed in Art. 8-7.

In determining the length of weld required, ½ in. should be added to the length of each weld to allow for starting and stopping of the bead.

For welded joints that are subjected to cyclic loading, special precautions are necessary to avoid harmful effects of stress concentration. The precautions should be in accordance with the principles discussed in Chap. 4. The use of stress-flow lines in the design of a welded connection

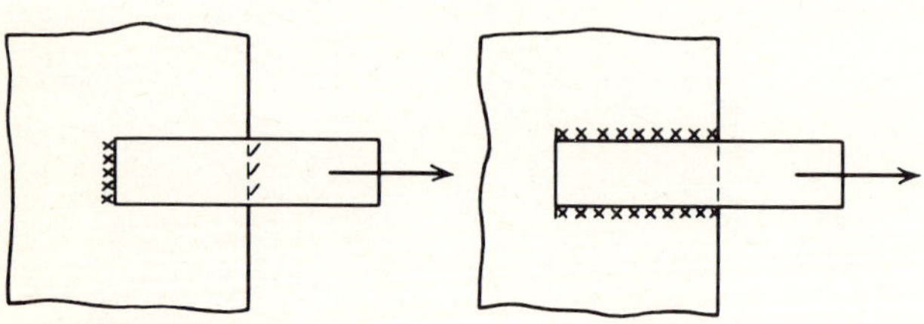

(a) Normal (b) Parallel

Fig. 8-20 Fillet welds.

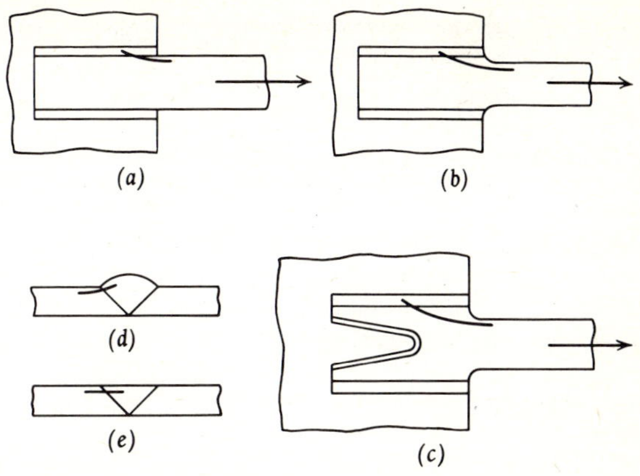

Fig. 8-21 Stress flow lines in welded joints.

indicates that the joint in Fig. 8-21(*b*) is stronger in fatigue loading than that shown at (*a*). Further increases[1] in endurance limit may be realized by the use of the joint similar to that in Fig. 8-21(*c*). The butt weld shown at (*d*) and (*e*) illustrates improvement in endurance limit secured by grinding to form a flush joint.

Example 8-1 A structural steel plate 3 in. wide and ⅜ in. thick loaded statically in tension by a load of 10,000 lb is welded to a plate with normal welds, as in Fig. 8-20(*a*). Determine the size of weld required.

[1] Martin J. Siegel, L. Vladimir, V. L. Maleev, and James B. Hartman, "Machine Design," International Textbook Company, Scranton, Pa., 1965.

Table 8-4 Minimum size of fillet welds for different thicknesses of plate*

Plate thickness, in.	*Minimum weld size, in.*
⅛ to 3⁄16	⅛
¼ to 5⁄16	3⁄16
⅜ to ⅝	¼
¾ to 1	⅜
1⅛ to 1⅜	½
Over 1½	¾

* From C. H. Jennings, Welding Design, *Trans. ASME*, vol. 58, p. 497, 1936.

Solution The net length of each weld is 3 − ½ = 2½ in. The load on the two welds would then be

$$\frac{10{,}000}{2 \times 2.5} = 2{,}000 \text{ lb per in.}$$

From Table 8-3, ¼-in. shielded-arc welds would be satisfactory.

Example 8-2 A ½-in. plate 2 in. wide carries a static tensile load of 8,000 lb and is to be welded to a plate by parallel fillet welds, as in Fig. 8-20(*b*). Determine the length of each weld, using ¼-in. shielded-arc welds.

Solution The net length of welds (Table 8-3) is

$$\frac{8{,}000}{2 \times 2{,}000} = 2 \text{ in.} \qquad \text{Use welds } 2\tfrac{1}{2} \text{ in. long.}$$

Precise calculations This method is used in exceptional cases where the geometry of the joint is unusual or where safety is of prime importance, and where the savings in cost justifies the additional design expenses. In this method an attempt is made to approximate the value of the stresses in the weld and in the adjacent parts of the joint as well as in the structure as a whole. See Table 8-5 for allowable stresses for

Table 8-5 Allowable stresses for coated-electrode welds on low-carbon steels*

Type of weld	*Static loads, psi*	*Cyclic loads, psi*
Butt welds:		
Tension	16,000	8,000
Compression	18,000	8,000
Shear	10,000	5,000
Fillet welds:		
Transverse and parallel welds	14,000	5,000

* From C. H. Jennings, Welding Design, *Trans. ASME,* vol. 58, p. 497, 1936.

Table 8-6 Stress-concentration factors for welds*

Type of weld	*Stress-concentration factor*
Reinforced butt weld	1.2
Toe of transverse fillet weld	1.5
End of parallel fillet weld	2.7
T butt joint with sharp corners	2.0

* From C. H. Jennings, Welding Design, *Trans. ASME,* vol. 58, p. 497, 1936.

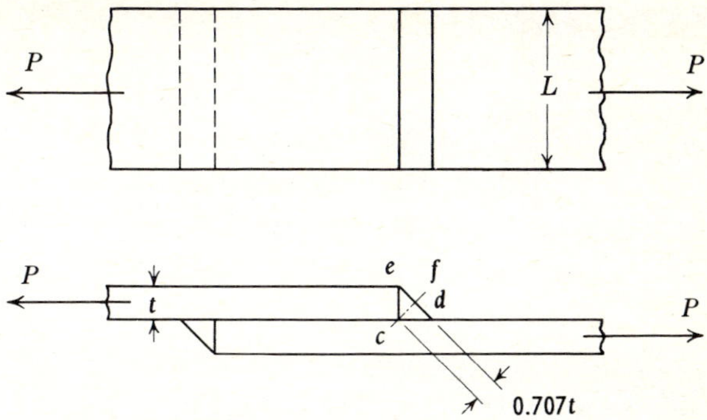

Fig. 8-22 Lap weld.

coated-electrode welds and Table 8-6 for stress-concentration factors for maximum stresses in these welds. Stress concentration should be considered for cyclic loading but may be neglected for static loading for reasons given in Art. 4-6.

In a normal fillet-welded lap joint (Fig. 8-22), the size of the weld is specified by dimension *ec* or *cd*. These dimensions are equal. The size of the weld is not greater than the thickness of the plate, and it may be less. Minimum sizes of welds are recommended in Table 8-4. In arriving at the size of welds to be specified, it is well to realize that doubling the size of a weld doubles the strength but requires four times the metal to be deposited.

In the weld, maximum shearing stresses occur along the face *cd* but are concentrated at *c* (the heel of the weld) on account of the plate being more rigid than the weld. In normal fillet welds, the stresses may be assumed as uniformly distributed along the length of the weld if the load is centrally applied. In Fig. 8-22 it may be assumed that each of the two welds carries one-half of the load P, so that the normal stress on the section *ec* will be equal to $P/2Lt$. The component of that load which is normal to the throat section *cf* is 0.707 times its value across *ec*. Since the area of the throat section *cf* is 0.707*ec*, it follows that the normal stress across the throat is equal to the tensile stress at *ec*. The 0.707 value above is the function of the 45° angle of the throat section.

Single-weld lap joint This type of joint should be avoided whenever possible because of the opening of the joint on the unwelded side and because the stresses under load are high and difficult to estimate. The stresses are high because of the sharp bending of the plate. Figure 8-23 shows a rubber model of such a joint in the unloaded and the loaded states and also, for comparison, a model of a double-weld lap joint.

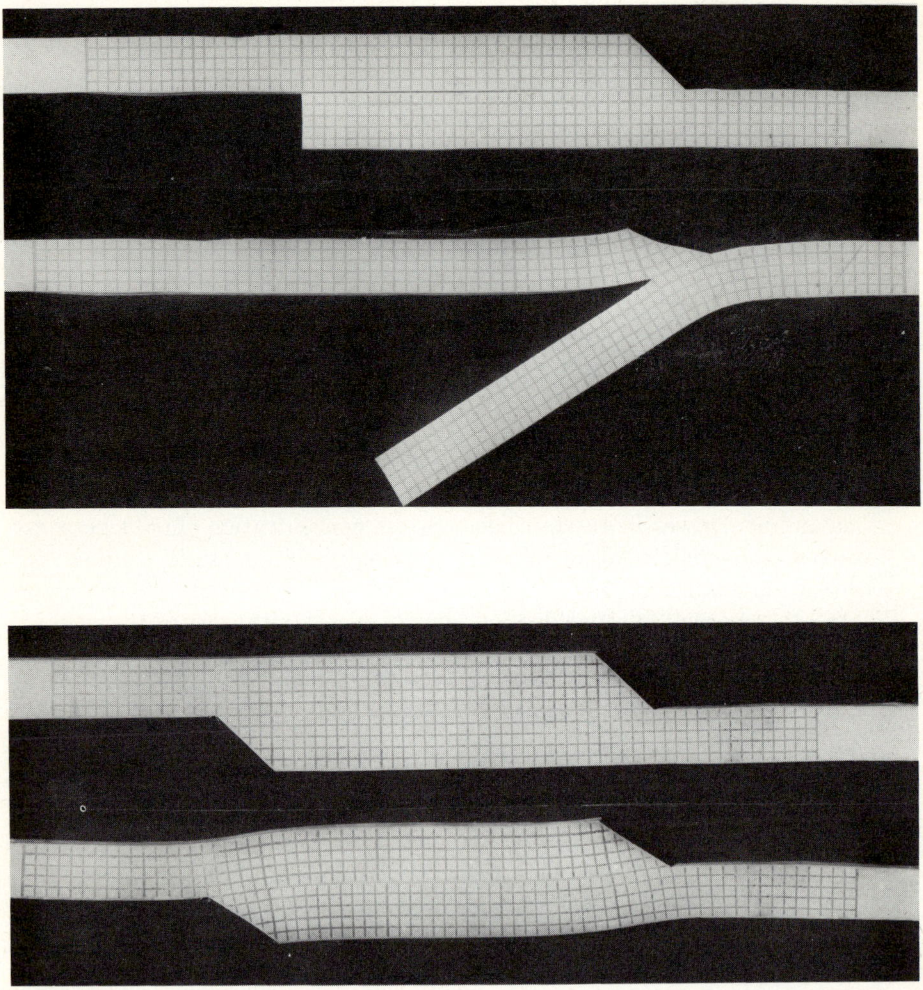

Fig. 8-23 Rubber models of lap joints. (*Courtesy of Lincoln Electric Company.*)

8-11 ECCENTRICALLY LOADED JOINTS

A joint of this type is illustrated in Fig. 8-24. In the design of such a connection, the center of gravity of the weld pattern is located, and at that point, two forces equal and opposite to the external load P are applied. One of these forces produces shearing stresses on the welds which have a direction parallel to P and which may be assumed uniformly distributed over the weld pattern. These are called *primary shear stresses.*

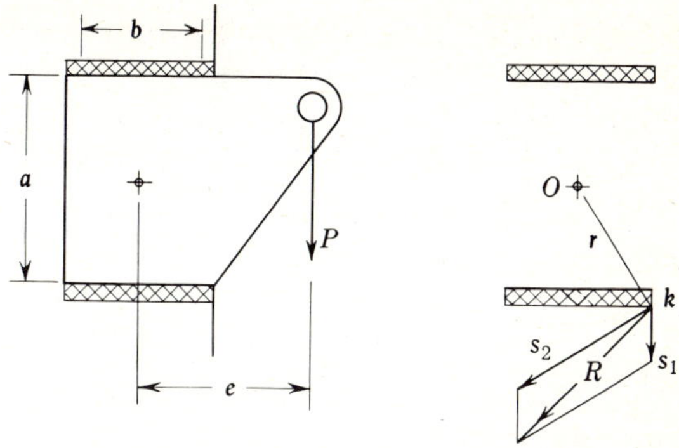

(a) Welded bracket (b) Weld pattern

Fig. 8-24 Bracket attached to support by fillet welds.

The other load P and the externally applied load acting at the eccentricity e produce a twisting moment on the welds which induces additional stresses on the welds, called *secondary shear stresses.* The analysis of loads on this joint is similar to that on eccentrically loaded rivet groups discussed in Art. 8-3.

The polar moment of inertia of the weld pattern may be taken from Table 8-7. Assuming that the plates are relatively stiff and that the deformation under load is concentrated at the welds, the secondary shearing stress at any part of the welds will be proportional to its distance from the center of twist. It follows that the maximum secondary shearing stress will occur at the corners of the weld pattern. The secondary shearing stress may be added vectorially to the primary stress to determine the maximum shearing stress. It can usually be determined by inspection where the resultant stress will be a maximum at any part of the joint. The location is k in Fig. 8-24.

Example 8-3 Assume in Fig. 8-24 that the load $P = 1{,}500$ lb, $e = 5$ in., $a = 3$ in., and $b = 2$ in. The latter dimension is ½ in. shorter than the weld to allow for starting and stopping of the bead. Determine the size of weld required for static loading.

Solution The polar moment of inertia is

$$J = \frac{bt(3a^2 + b^2)}{6} = \frac{2t(3 \times 3^2 + 2^2)}{6} = 10.3t$$

The radius to the most remote area is

$$r = \tfrac{1}{2}\sqrt{a^2 + b^2} = \tfrac{1}{2}\sqrt{3^2 + 2^2} = 1.8 \text{ in.}$$

The primary shear stress is

$$s_1 = \frac{P}{2bt} = \frac{1{,}500}{2 \times 2t} = \frac{375}{t}$$

Table 8-7 Polar moment of inertia for weld patterns

Weld pattern	*Polar moment of inertia J**
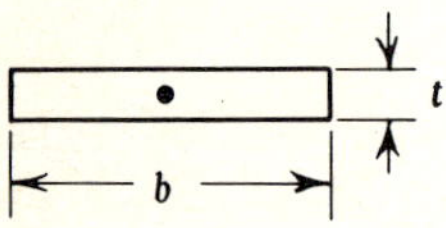	$\frac{tb^3}{12}$
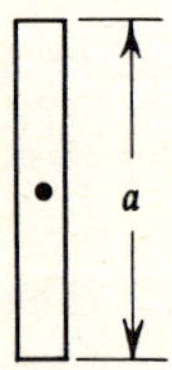	$\frac{ta^3}{12}$
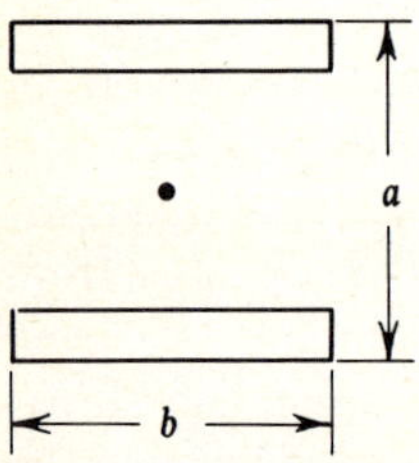	$\frac{bt(3a^2 + b^2)}{6}$
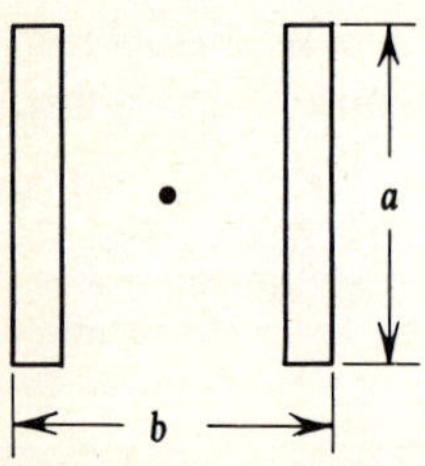	$\frac{at(a^2 + 3b^2)}{6}$
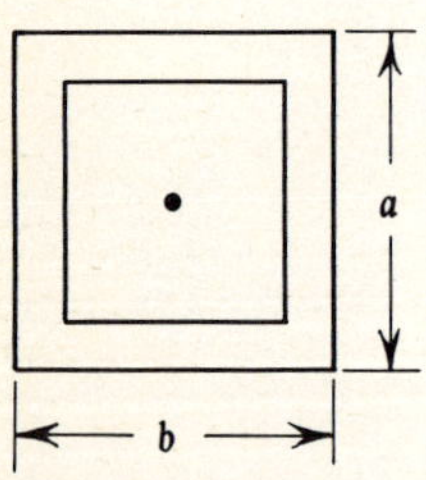	$\frac{t(a + b)^3}{6}$

* Approximate.

The secondary shear stress is

$$s_2 = \frac{Tr}{J} = \frac{(1{,}500 \times 5) \times 1.8}{10.3t} = \frac{1{,}310}{t}$$

The primary shear stress is always directed parallel to P. The secondary shear stress is maximum at the four corners of the weld and is in a direction perpendicular to its radius r drawn from the center of twist to the corner. The resultant stress is the vector sum of the primary and secondary stresses, and it is apparent by inspection that the maximum resultant stress occurs at location k, as indicated in the stress pattern. The resultant stress R may be determined analytically or graphically and is found to be equal to $1{,}550/t$.

This shearing stress would act along the leg of the weld (*cd* in Fig. 8-22). The throat of the weld, however, is smaller than the leg and to be on the side of safety, it is customary to use the throat area, even though the shear stresses as calculated do not lie in that plane. This consideration would require using $0.707t$ instead of t for the section.

The maximum stress then would be equal to

$$\frac{1{,}550}{0.707t} \quad \text{or} \quad \frac{2{,}190}{t}$$

Using an allowable stress from Table 8-5 equal to 14,000 psi, the size of the weld is

$$t = \frac{2{,}190}{14{,}000} = 0.156 \text{ in.}$$

A $\frac{3}{16}$-in. weld would be specified.

8-12 STRESS RELIEVING

As the molten metal solidifies and cools in a welded joint, and as the parts adjacent to the joint cool, stresses are set up in the parts, owing to volumetric contraction. These local stresses may in some cases become very high, and improperly welded joints may crack on cooling, owing to these residual stresses. If the welded parts are rigidly attached to restraining members, additional stresses may be set up because of general cooling of the structure after welding is completed.

If the material of the structure is ductile, the seriousness of the residual stresses may be alleviated, owing to plastic flow. Since the weld metal *itself* is essentially cast, its ductility will generally be lower than that of the parent metal, and it cannot be depended on to relieve residual stresses materially, unless special electrodes are used. Therefore, in order to prevent residual stresses from becoming serious, the parts welded should be made of ductile material and provision should be made, if possible, for flexibility of the parts during cooling of the welds. The principle involved is shown in Fig. 8-25, in which two members joined by fillet welds are not placed in contact, so that contraction of the parts will not

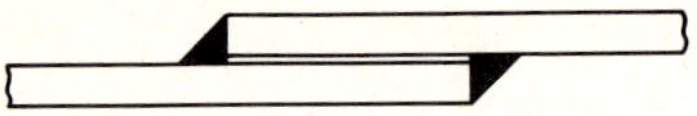

Fig. 8-25 Lap weld with provision for lateral contraction.

be restrained in the vertical direction in the sketch.[1] Intermittent welding may be desirable in long welds in order to prevent concentration of the effects of heating and cooling of the parts welded and the weld itself. Preheating of the parts may also be desirable in some instances. In a structure made of both welded and riveted parts, the welding should precede the riveting so far as possible.

Residual stresses may be relieved after welding by suitable heat treatment. The temperature, rate, and time of heat treatment depend on the size and thickness of the parts and on the plasticizing rate of the material. Highly stressed parts whose failure would involve life hazard or expensive structure damage, such as welded pressure vessels and welded penstocks, should be stress-relieved; this applies also to parts which require close tolerances or those subjected to impact and fatigue loading.

In welded steam-generator drums, stress relieving is effective in reducing caustic embrittlement.

Hand peening of the joint while it is hot is effective in increasing the strength of the joint, particularly in austenitic steel, but overpeening should be avoided. Indications are that peening reduces residual stresses in parts of the weld that may be in the plastic state and that it also introduces some cold working (see Art. 6-3).

It is not generally necessary to stress-relieve welded machine members if they have been properly designed, with special precautions to avoid stress concentrations, and if they have been welded properly. However, in highly stressed parts and in parts subjected to repeated loading, stress relief may be necessary.

8-13 TESTING AND INSPECTION OF WELDS

The mechanical testing of welds in either static or fatigue loading has in general three purposes, as follows:

1. *Comparison of strengths of joints.* This comparison may be for (*a*) welds of different geometry, for example, in comparing the strengths of a single-V butt weld with a double-V butt weld, or of a full-bead with a flush-bead weld, or of a rectangular splice plate in a beam connection with a diamond-shaped plate, as shown in Fig. 8-26. Or (*b*) the comparison may be for the purpose of determining the effect on strength of

[1] S. C. Hollister, Stress Relieving, "Welding Handbook," American Welding Society, 1938.

using, for example, welding rods of different composition, rods of different sizes in electric-arc welding, or bare- versus coated-electrode welds.

2. *Testing of filler metal.* In this type of testing of welds, specimens are prepared by machining from a welded joint a part of the deposited metal and the adjacent parent metal. The specimens are used to determine yield point, ductility, nature of bond between deposited metal and parent metal, and chemical composition.

3. *Qualification tests of welders.* Standard joints are made and they are tested to determine the quality of the welding. These procedure tests are made periodically and aid in maintaining consistent high-quality welding.

Inspection of welds, sometimes called *nondestructive testing,* is specified in some instances. The most commonly used of these methods, each having its own fields of application, are as follows:

Visual inspection A method of inspection available to skilled operators is known as "visual inspection," which means that the operator observes as the bead is "laid down" characteristics of its formation that are a good indication of the quality of the weld. These characteristics are rate of burning off of the electrode, fusion and penetration, forming of the bead in the joint, and the sound of the arc. The evaluation of these dynamic characteristics involves a certain amount of the "personal element" in the operator, but, taken together, they are one of the best indications of the quality of a weld as they permit the operator, as it might be said, to see inside the joint.

X-ray inspection The x-ray equipment is expensive and is not portable; hence such inspection is available only where large numbers of parts must be inspected and where the parts are not too large to be conveniently transported. The limitation of the method is the thickness of material suitable for penetration in a reasonable time. Two inches is generally considered to be the upper limit of metal thickness for ordinary work although up to 5-in. sections have been satisfactorily defined in photographs.

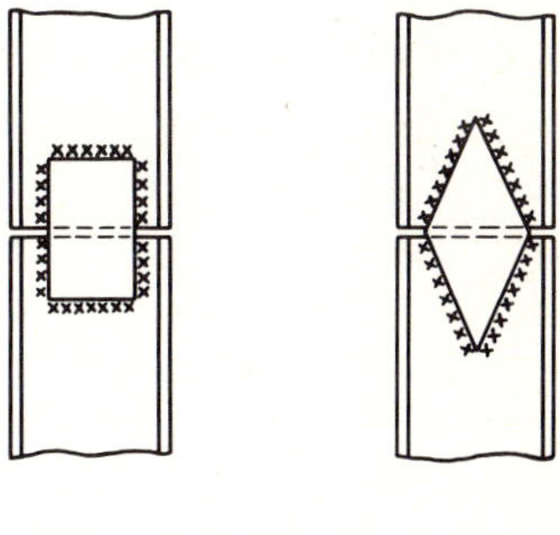

Fig. 8-26 Splice plates.

Gamma-ray inspection The gamma-ray equipment may be portable and hence is adaptable to field work. The cost of the equipment is lower than the x-ray equipment; however, the radium required makes its use expensive. The gamma rays penetrate the work more rapidly than x-rays; hence this method is suitable for inspecting thick plates.

Magnaflux method This method uses magnetic dust, applied either dry or in an oil bath. The welded part, which of course must be magnetic, is magnetized and the magnetic dust assumes a pattern in the vicinity of the joint which gives valuable indications of surface defects and some subsurface defects, such as cracks and other discontinuities and inclusions. If residual magnetism is objectionable, as in aircraft parts, it should be removed by demagnetization.

Stethoscope method In this method, the soundness of the weld may be determined by means of the characteristics of the sound or ring produced by tapping the weld with a light hammer. The sound is analyzed by the operator using a stethoscope as an aid. The condition of the weld can be quite accurately determined by skilled operators.

8-14 ROLLING, SPINNING, AND SEAMING

These are analogous operations in which (usually) circular or cylindrical thin metal parts are rotated on a spindle while forming rolls shape the part. Figure 8-27 shows a welded V-belt sheave for which the edges of the flange were curled by the forming rolls.

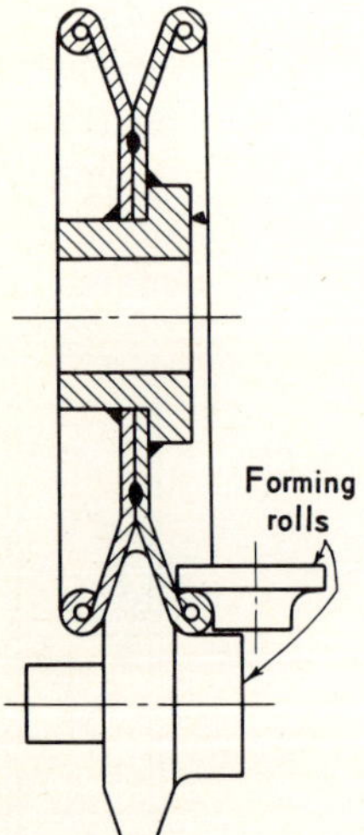

Fig. 8-27 Fabricated V-belt sheave.

8-15 STITCHING

Thin metal sheets may be joined by stitching, in which mild steel clips have their ends punched through the overlapped sheets and then bent over to form the stitch.

8-16 MANUFACTURING PROCESSES[1]

The manufacturing processes may be considered in three groups: (1) *forming* processes, such as casting, forging, rolling, and machining to dimension; (2) *fabricating* processes, such as soldering, brazing, and welding; (3) *finishing* processes, such as finish machining or grinding, honing, lapping, and superfinishing. A brief discussion of these processes follows.

Casting Sand casting requires a separate mold for each piece. Metal-mold casting may be used for relatively low-melting-temperature metals; it employs a metal mold that can be used for producing many castings. Die casting is used in quantity production of small parts made of low-melting-temperature alloys.

Forging *Hand forging*, or *machine forging* using power hammers, is used for parts whose shape requires working of the materials in the semiplastic state, or where it is desired to control the internal structure as regards grain size and grain direction. Examples of forged parts are upset and swaged pieces, crankshafts, connecting rods, valves, and gears. Forging design should strive for as much material as possible in one plane and for the avoidance of undercut surfaces and deep recesses.

Rolling Rolling in a rolling mill is used to form long members of uniform section, such as bars, rails, structural sections, and plates. The material may be hot-rolled for large reduction in section or cold-rolled for accurate sizing or grain control. Rolling may also be used to form gear teeth and screw threads and to form sheet metal to desired shapes.

The cold-metal processes These include shearing; blanking; bending; drawing, which is used primarily for forming rods and shafting and for producing cup-shaped parts from flat blanks; and squeezing, which includes coining and extrusion. Metal spinning is a type of drawing used for forming surfaces of revolution, including beads and seams. The extrusion process is used to form the softer metals, such as aluminum and magnesium alloys, into a variety of shapes.

Sintering This process is used to combine compressed powdered metals or a mixture of powdered metals by heat treatment so that a rigid piece having desirable properties is formed. Nonmetallic substances may

[1] M. L. Begeman and B. H. Amstead, "Manufacturing Processes," 5th ed., John Wiley & Sons, Inc., New York, 1963.

or may not be included. The process may be used to produce accurately sized parts composed of materials that will not mix or that cannot be solidified from the molten state, and it may also be used for parts that can be produced more economically by sintering than by other processes. The strength and porosity of the finished parts can be accurately controlled. Some examples of sintered products are bearing sleeves which are used for bearings inaccessible for lubrication and which are composed of a mixture of bronze and graphite, porous bronze bearing sleeves which are impregnated with oil, and bronze and iron parts which are used as gears for oil pumps and pistons for hydraulic controls; other examples are refractory products and cutting tips for machine tools.

Soldering, brazing, and welding Soldering and brazing are used to join metal parts by the use of a filler metal whose melting temperature is lower than that of the metals to be joined. Brazing uses a harder filler metal than is used in soldering and forms a stronger joint. Brazing in a furnace is very effective and forms strong and neat joints. Welding is used as a fabrication medium to join parts permanently and to form built-up members as substitutes for castings and forgings, as a joining medium to replace fastenings as rivets and bolts, and as a repair medium to replace broken or worn sections of members. Welding is discussed in Art. 8-7.

8-17 DESIGN OF CASTINGS

An important consideration in the design of castings is volumetric contraction or shrinkage, which takes place as the metal cools from the molten state to the solid state. In many cases the contraction is restrained by the mold or cores or by adjacent parts of the casting that have solidified earlier in the cooling process. The restrained contraction gives rise to residual stresses that may increase or decrease the strength of the part in service. See Art. 6-3.

In general, the necessity for considering the effects of contraction and for designing the casting to minimize harmful effects increases with the melting temperature and the volumetric contraction of the metal. Thus castings of low-melting-point metals and alloys generally may be designed without special regard for effects of contraction; iron castings require medium precautions; and steel castings require great care and special considerations.

Some general features that should be considered in the design of castings are discussed in the following paragraphs. They apply specifically to steel castings. However, their incorporation should result in the improvement of any casting.[1]

[1] "The Steel Castings Handbook," Steel Founders' Society of America, Cleveland, 1960.

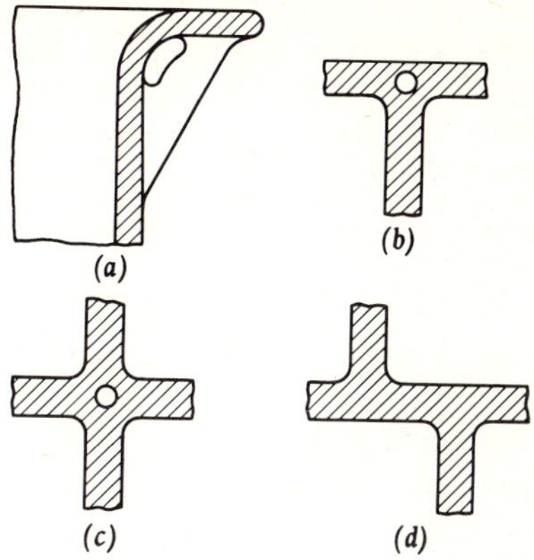

Fig. 8-28 Reduction of concentration of mass.

Large risers are desirable to supply liquid metal to compensate for loss of volume during solidification. To ensure proper flow of molten metal, the position of the mold during pouring should be considered.

Provision should be made for solidification in a progressive manner from the lowest points of the casting to the risers. Progressive solidification requires the avoidance of relatively large isolated masses of metal. Gradual changes in section should be used where possible.

Avoid sharp corners. External corners should be rounded off. Generous fillets should be used to join sections, although it is not advisable to use long fillets with radii greater than the thickness of the sections they adjoin. An oversize fillet may introduce the effect of concentration of mass.

Use properly spaced ribs or brackets. A rib or bracket has great strengthening and stiffening effect in service and in addition, since it is thin, it cools quickly and provides support for the sections it adjoins during cooling, thus lessening distortion of the casting. Webs perform the same functions of lessening distortion of the casting and provide strength and stiffness in service. At the junction of sections and a rib, concentration of metal may be avoided by the use of a cored rib as shown in Fig. 8-28. A T junction may be cored as shown at (*b*) to reduce concentration of masses of metal and possibility of forming shrinkage cavities. An X junction may be similarly cored as shown at (*c*) or offset as shown at (*d*).

When the design of a steel casting becomes complicated, consideration should be given to the use of smaller cast parts assembled by welding or bolting.

The precautions and suggestions discussed above are some of the more important special features that should be incorporated in the design of a casting and are intended to be considered in addition to the usual casting design details, such as providing draft, shrinkage and finish allowances, pads for locating points, avoidance of sections too thin to cast or control, caution in the use of inserts and chaplets, and avoidance of deep recesses and undercuts. It is desirable that the designer have an insight into the effects of size and shape of the casting on the properties of the metal and that he consult with the foundryman and the metallurgist during the design of castings in order that quality and economy be ensured.[1]

8-18 SURFACE FINISH

The surface finish is very important from the standpoint of serviceability, appearance, and production cost of a machine member. The type and quality of surface finish has great influence on friction characteristics, wear, endurance limit, and effects of corrosion of machine members.

Since the fine finishing operations remove surface cracks, a polished member will have a higher endurance limit than one with a rough surface. Highly stressed members subjected to cyclic loading, such as the aircraft connecting rod in Fig. 4-10, may therefore be polished.

Members subjected to a corrosive environment may have a coating or plating to resist the corrosive action.

The appearance of the part, or of the complete machine, may be greatly improved by the use of proper finishes and the judicious use of decorative trims.[2]

8-19 SAMPLING AND INSPECTION

In securing samples for the purpose of determining chemical composition or physical properties of the material of which a machine member is made, it is important to have the sample representative of the metal in the part. For check analyses, samples should be secured in accordance with established standards. For castings, test coupons either attached or separately cast should be provided. The coupons are subjected to the same heat treatment as the castings and used for metallurgical and mechanical testing.

The finished product or samples may be subjected to the usual mechanical tests to determine hardness, tensile strength, yield point, ductility, notched-bar or impact-test values, or for ratings, such as Magnaflux ratings, which indicate seams, cracks, inclusions, and cavities. X-ray or gamma-ray inspection may also be necessary.

[1] The "Cast Metals Handbook," American Foundrymen's Society, and the "Steel Castings Handbook," Steel Founders' Society, are sources of considerable information on the subject of casting design.

[2] Harold Van Doren, "Industrial Design," 2d ed., McGraw-Hill Book Company, New York, 1954.

9
Detachable Fastenings

9-1 GENERAL REMARKS

Some machine parts must be so constructed that they may be readily connected or disconnected without damage to the machine or the fastening. This requirement may be for the purpose of holding or adjustment in assembly or service, inspection, repair, or replacement, or it may be for manufacturing or assembly reasons. The parts may be rigidly connected, or provision may be made for predetermined relative motion.

Detachable fastenings perform the functions mentioned above. They are of many forms and degrees of ease of assembly or disassembly. Some types require special tools for this purpose, some have special locking devices, and others depend solely on friction to hold them in place. They have in common, however, positive connection, or in some cases set limits of motion, and relative ease of connection and disconnection with no damage to parts.[1]

[1] Fasteners Book Issue, *Machine Design,* March 21, 1963.

Some of the more common types of detachable fastenings are described in the following paragraphs.

9-2 SCREW FASTENINGS

A screw fastening is composed of *two* elements, as illustrated by a bolt and nut. Frequently one of the parts to be connected is so constructed that it becomes one of the elements of the fastening. In fact, both of the parts to be connected may serve as the elements of the fastening, as in the case of a handle with a thread turned on one end to engage in a tapped hole in a lever.

The choice of type of fastening is very important; the location or placing also is equally important. Fastenings should be located and placed so that they will be subjected to tensile and/or shear loads, and bending of the fastening should be reduced to a minimum. Bending of the fastening due to misalignment, tightening-up loads, or external loads has been responsible for many failures. To relieve fastenings of bending stresses, the use of clearance spaces, spherical-seat washers, or other devices may be necessary.

Some common types of screw fastenings are described as follows:

Bolts A bolt has a head at one end and a nut fitted to the other, as shown in Fig. 9-1(*a*), where it is used to attach a cylinder head to a cylinder.

Tap bolt Figure 9-1(*b*) shows a tap bolt engaging in a tapped hole in the cylinder wall. If the cylinder head is removed frequently, the thread in the hole may become worn, necessitating costly repairs.

Studs As shown in Fig. 9-1(*c*) studs are threaded at both ends. One end is screwed into the tapped hole in the cylinder wall and the other end receives the nut.

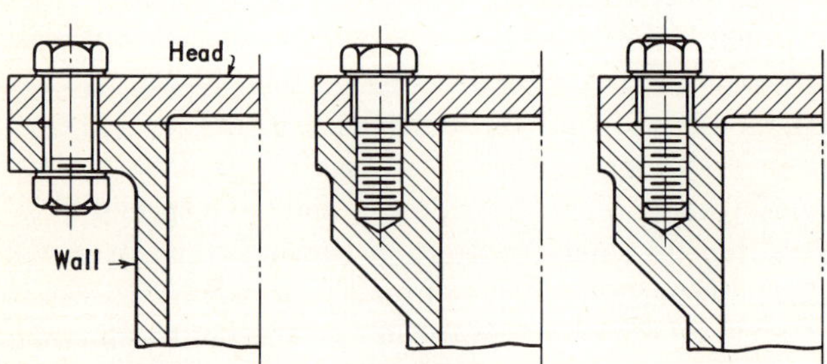

(*a*) Bolt (*b*) Tap bolt (*c*) Stud

Fig. 9-1 Types of screw fastenings.

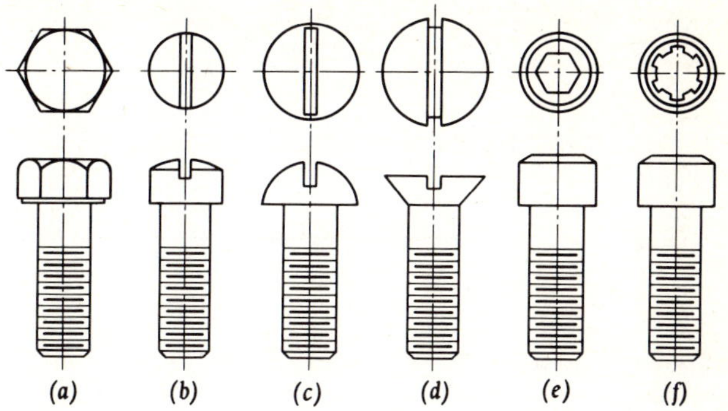

Fig. 9-2 Cap screws with various types of heads: (*a*) hexagonal; (*b*) filister; (*c*) button; (*d*) flat; (*e*) hexagonal socket; (*f*) fluted socket.

Of the three types of fastenings described above, the bolt makes the best connection from the standpoint of construction cost, the tap bolt and stud result in a saving in space and weight since a smaller flange is required, and the stud has a replacement advantage over the tap bolt in the event of wear of the thread.

Cap screws These are similar to small-size tap bolts except that a greater variety of shapes of heads are available, as shown in Fig. 9-2. Shouldered cap screws also are in common use.

Machine screws These are cap screws slotted for a screwdriver and are generally used with a nut.

Setscrews This type of screw is used to prevent relative motion between parts. Setscrews may be used instead of a key to prevent relative motion between a hub and a shaft in light power-transmission members, or they may be used in connection with a key, where they prevent relative axial motion of the shaft, key, and hub assembly. Some types of setscrews having different types of heads and points are shown in Fig. 9-3. Various combinations of heads and points are available.

Locking devices[1] In a tightened screw fastening there are two opposing external forces present: (1) a loosening force due to the action of the axial load which tends to unscrew the fastening and (2) the force of friction which tends to resist the unscrewing motion. Screw fastenings are

[1] Whittemore, Nusbaum, and Seaquist, The Relation of Torque to Tension for Threadlocking Devices, *Natl. Bur. Std.* (*U.S.*), *Res. Paper* 386, 1931.

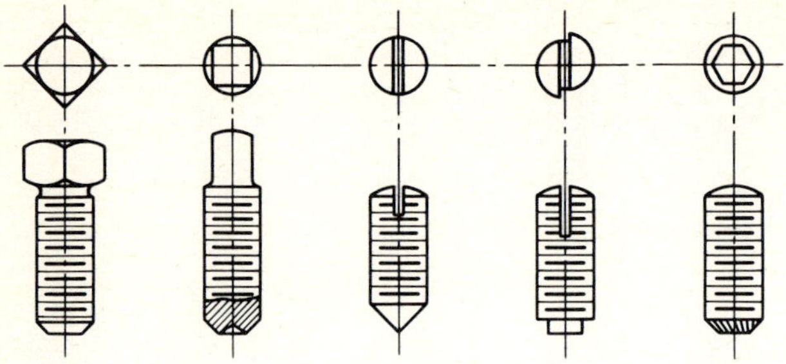

Fig. 9-3 Setscrews.

designed so that the friction force is expected to prevent unscrewing or loosening of the fastening.

Ordinary threaded fastenings generally remain tight under the action of static loads; however, many of these fastenings become loose under the action of cyclic loads. Thus fastenings subjected to pulsating service loads or to vibration require restraint from loosening in addition to that offered by thread and collar friction.[1] A large number of locking devices are available. Some of them are described as follows: The *jam nut*, or *lock nut*, is shown in Fig. 9-4(*a*). The *castle nut* to be used with a cotter pin is shown at (*b*). Two types of *split nuts* are shown at (*c*); the first has the split closed after threading as shown, so that screwing the nut on the bolt will open the split and introduce additional thread friction; the second has the split opened by the small screw after the bolt is tightened. The *Everlock Locknut* is shown at (*d*). The *elastic stop nut* shown at (*e*) employs a hard fiber or nylon collar recessed in the nut that becomes threaded as the nut is screwed on the bolt causing a tight grip. This type of nut is especially effective in parts subjected to vibration. Some types of lock washers are shown at (*f*).

In Fig. 9-4(*a*), the standard thickness nut is placed next to the work with the thin nut (jam nut) on top. The reverse arrangement, i.e., placing the thicker nut on top, is preferred by some designers since, as can be shown,[2] the upper nut carries a greater load than the bottom one and it is concluded that the top nut should be the thick one with the thin one next to the work. However, it has been shown by Goodier and Sweeney that the first few threads next to the loaded face of a nut carry most of the load; hence, it appears that the positioning of the nuts is not of primary importance from the standpoint of loading. Ease of assembly

[1] J. N. Goodier and R. J. Sweeney, Loosening by Vibration of Threaded Fastenings, *Mech. Eng.*, vol. 67, no. 12, p. 798, December, 1945.
[2] Fasteners Book Issue, chap. 13.

is favored by placing the thick nut next to the work where it can be held by a standard wrench during tightening, whereas placing the thin nut on first requires a thin wrench to hold it.

Multiple threads may be used in cases in which a combination of fine threads is required because of space limitations and rapid axial motion, as in a fountain-pen cap or a quick-locking screw. In all standard fastenings, however, a single thread of small pitch is used in order to secure a large holding force and to lessen the tendency to shake loose in service.

Commonly used screw-thread terminology is as follows:

Pitch is the axial distance between corresponding points on two consecutive turns.

Lead is the distance the nut advances during one revolution.

Diameters:

Major, or outside, diameter.

Minor, or root, diameter.

Unless otherwise stated, specifications of screw fastenings refer to the *major* diameter of the fastening. Thus, a ½-in. bolt refers to a bolt ½ in. in outside diameter.

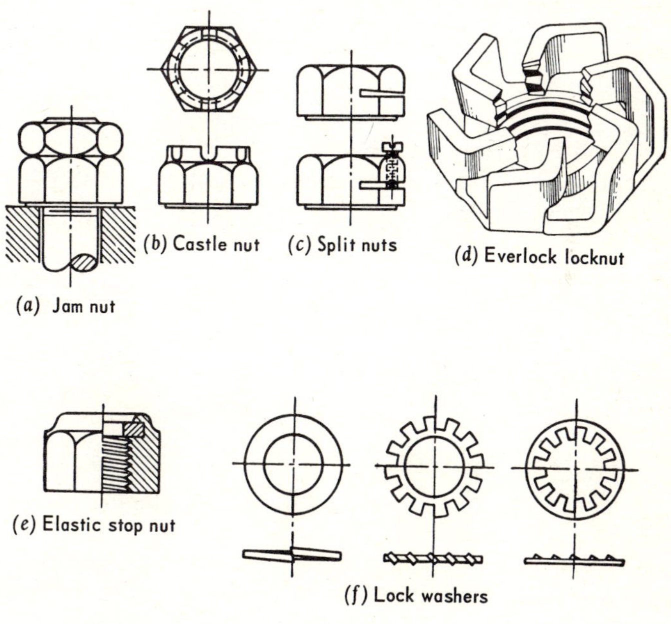

Fig. 9-4 Locking devices.

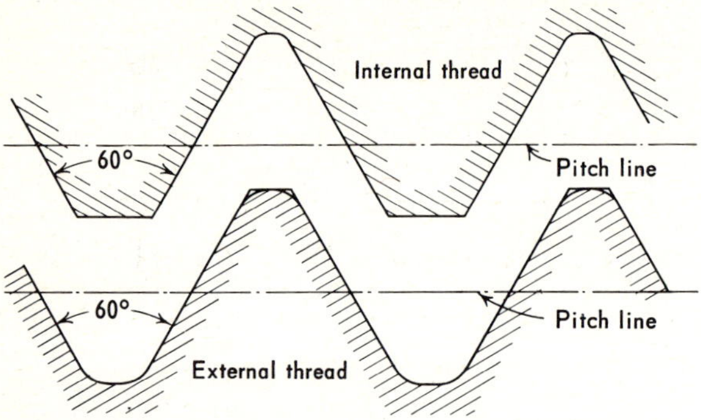

Fig. 9-5 Unified and American screw-thread form. Threads are shown separated for clarity. Crests on external threads may be rounded or flat.

9-3 FORMS OF SCREW THREADS FOR FASTENINGS

The right-hand thread is always used unless there is a special reason for requiring a left-hand thread and, unless otherwise stated, specifications for threads imply right-hand threads.

The *American National thread* was the standard thread form for fastenings in the United States until November, 1948, when the Unified Screw Thread Standard was adopted. The *Unified Standard* now applies to government and industry in Great Britain, Canada, and the United States. The threads now standard in the United States adhere to the *American Standard Unified screw threads.*[1]

A difference is in the tolerance and gauging system which has been reworked to permit interchangeability, and also there are some changes in thread series and in the classes; the old class 1 and class 4 have been dropped and classes 2A and 2B and 3A and 3B have been introduced. In the new classes, A refers to the external thread tolerances and B refers to the internal thread tolerances. Classes 2A and 2B will be used generally for fastenings (the latter for closer fits), and classes 3A and 3B for refined fits. Actually the old classes 2 and 3 have been carried over in the new standards but their use is expected to disappear gradually in favor of the 2A and 2B and 3A and 3B classes.

The thread form for the Unified and American Standard is shown in Fig. 9-5.

For the three series, *coarse* threads are intended for general industrial purposes; *fine* threads are for applications which require fine adjustment or additional tightness and resistance to loosening due to vibration. The *extra-fine* series is used for special applications. Table 9-1 shows thread sizes for the coarse- and fine-series threads.

[1] USAS Bulletin 1–1960, American Society of Mechanical Engineers, New York, 1960.

Table 9-1 Unified and American screw-thread sizes

Size	Coarse threads: Threads per in.	Coarse threads: Root area, in.²	Fine threads: Threads per in.	Fine threads: Root area, in.²
0	....	....	80	0.0015
1 (0.073)	64	0.0022	72	0.0024
2 (0.086)	56	0.0031	64	0.0034
3 (0.099)	48	0.0041	56	0.0045
4 (0.112)	40	0.0050	48	0.0057
5 (0.125)	40	0.0067	44	0.0072
6 (0.138)	32	0.0075	40	0.0087
8 (0.164)	32	0.0120	36	0.0128
10 (0.190)	24	0.0145	32	0.0175
12 (0.216)	24	0.0206	28	0.0226
1/4	20	0.0269	28	0.0326
5/16	18	0.0454	24	0.0524
3/8	16	0.0678	24	0.0809
7/16	14	0.0933	20	0.1090
1/2	13	0.1257	20	0.1486
9/16	12	0.1620	18	0.1888
5/8	11	0.2018	18	0.2400
3/4	10	0.3020	16	0.3513
7/8	9	0.4193	14	0.4805
1	8	0.5510	12	0.6245
1 1/8	7	0.6931	12	0.8118
1 1/4	7	0.8898	12	1.0237
1 3/8	6	1.0541	12	1.2602
1 1/2	6	1.2938	12	1.5212
1 3/4	5	1.7441		
2	4 1/2	2.3001		
2 1/4	4 1/2	3.0212		
2 1/2	4	3.7161		
2 3/4	4	4.6194		
3	4	5.6209		
3 1/4	4	6.7205		
3 1/2	4	7.9183		
3 3/4	4	9.2143		
4	4	10.6084		

In designating screw threads, the nominal size is given first (in inches or numbers), hyphen, number of threads per inch, letters of the thread form and series, hyphen, and class of tolerance. For example: 1/2-20 UNC-2A indicates a Unified bolt having a nominal diameter of $\frac{1}{2}$ in., with 20 threads per inch, National coarse series with allowances and tolerances for class 2A fit for the bolt. The mating nut would carry the same designation except that it would be 2B instead of 2A.

Aero threads The Aero thread employs an insert made of wire of section as shown in Fig. 9-6. The insert is shaped like a coil spring and is screwed into the threaded recess in the nut or threaded member, thus forming a thread for the engagement of the bolt or screw. The thread spaces of the bolt or screw are approximately semicircular in section, and as a result, little stress concentration is introduced. The bottom of the thread in the nut or threaded member is sharp, but stress concentration at this location is generally not serious, since the nominal stress is low. In addition to lowered stress concentration in the screw, the Aero thread has the advantage that it can be used in materials that are difficult to thread, such as cast iron and aluminum. The insert can be screwed into the hole as tapped, thus leaving the insert itself to form the thread for receiving the screw. The insert, usually bronze, will resist wear and may be replaced if necessary. A comparison of endurance limits of standard threads and Aero threads is shown by the diagram in Fig. 9-7. In Fig. 4-11(g) is shown an Aero-thread assembly for an aircraft-engine-cylinder flange fastened to an aluminum crankcase.

Another thread insert known as the "Heli-coil" is similar to the Aero-thread insert except that the cross section of the wire is diamond-shaped so that it will form a thread that will accommodate an American Standard threaded part. Although the Heli-coil thread does not have any stress-concentration advantage over the American Standard thread, it does have

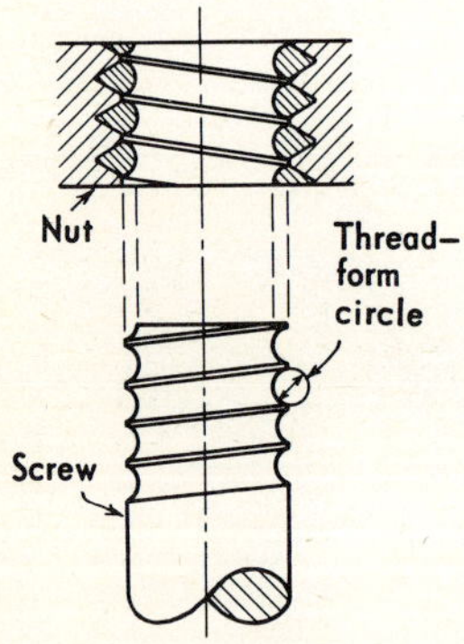

Fig. 9-6 Aero thread.

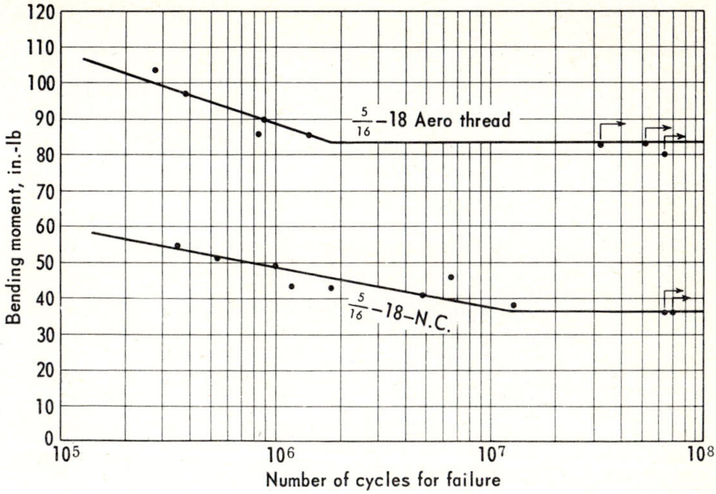

Fig. 9-7 Curves showing results of endurance tests on steel specimens with Aero thread and National coarse thread. (*From H. F. Moore, Strength of Screw Threads, Prod. Eng., November,* 1939.)

advantages for use in materials that are difficult to thread and at the same time maintains interchangeability with American Standard threads. An application of this thread is in the spark plugs in cast-iron or aluminum cylinder heads. This feature is an advantage because worn female threads can be rethreaded to accept the Heli-coil insert to accommodate the male thread, such as on a cylinder-head bolt or a spark plug.

Pipe threads The American Standard taper pipe thread is formed with a taper of ¾ in. per ft. The form of the thread is shown in Fig. 9-8. In addition to their use in pipe fittings, pipe threads are used in attaching oil-feeding devices, gauges, etc.

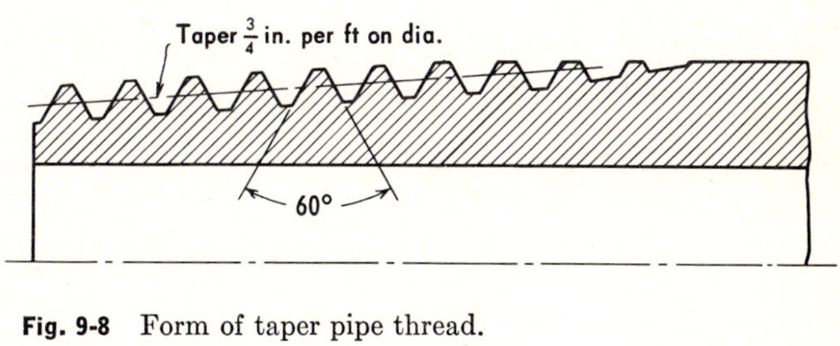

Fig. 9-8 Form of taper pipe thread.

9-4 ROLLED THREADS

The dies for forming rolled threads are of two types: flat dies, used in pairs which have relative motion of translation, and cylindrical dies, which rotate and are used in sets of three. The dies have grooves of the desired thread form, with a slope on the die corresponding to the angle of the helix of the thread. The cylindrical blank is rolled between the dies, thereby forming the threads. The thread is formed by depressing the surface of the blank to form the root of the thread while the displaced material flows outward to form the crest. Thus the original diameter of the blank, or shank of the bolt, lies between the outer diameter and the root diameter of the thread. The process is limited to external threads.

Because of the cold forming of the material, the mechanical properties are increased, which, coupled with favorable flow lines and uniform and smooth surfaces, materially increases the strength of the part, especially in cyclic loading. Close control may be had over dimensions; in fact, excellent micrometer screws are made by the process. Threads may be rolled on all sizes of blanks up to 5 in. in diameter. Although thread rolling has been practiced for over a hundred years, the modern precision-rolled thread is far superior to the rolled carriage bolt of early days, and in production, marked savings may be had in labor and material cost as well as improved performance.[1]

9-5 HEIGHT OF NUT

The approximate thickness of a nut required for equality of strengths of a bolt in tension and of the thread in shear may be determined as follows:

Assumptions: (1) that each turn of the thread of the nut supports an equal share of the load; (2) that stress concentration is neglected; (3) that for standard coarse threads, $d_r = 0.8d_o$ (see Fig. 9-9); (4) that for the

[1] "Engineering Data on Thread Form Rolling," Reed Rolled Thread Die Co., Worcester, Mass.

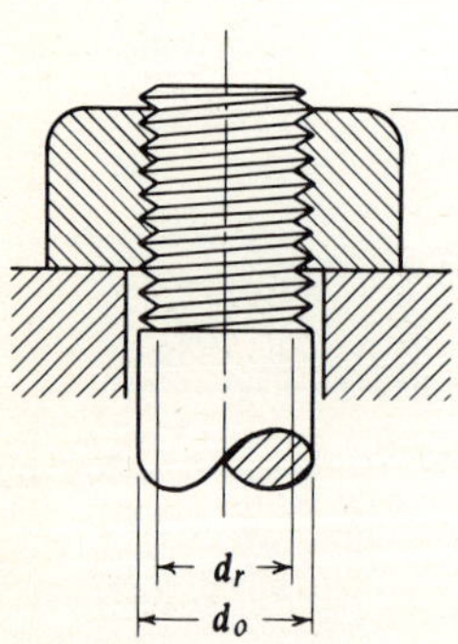

Fig. 9-9 Height of nut.

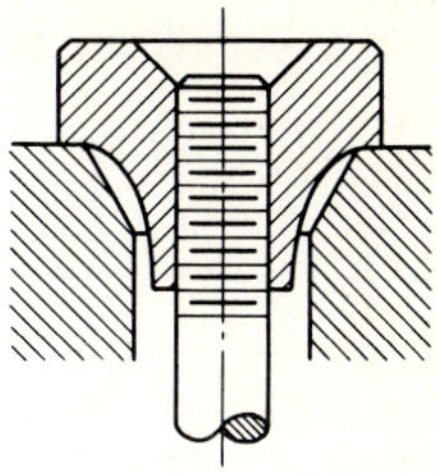

Fig. 9-10 Shape of nut to reduce localization of thread loading.

material of nut and bolt, the yield stress in shear, s_s, is equal to half the yield stress in tension, s_t. The latter is approximate.

The strength of the bolt in tension is equal to

$$F_t = \frac{\pi d_r^2}{4} s_t$$

The strength of the threads in shear is equal to

$$F_s = \pi d_r h s_s$$

By equating F_t and F_s,

$$h = 0.4 d_o$$

Therefore, for standard coarse threads, the threads will be as strong in failure by shear as the bolt in tension if the height of the nut is approximately 0.4 times the nominal diameter of the bolt. American Standard nuts are approximately $\frac{7}{8}d_o$ in height; hence, standard threads will not fail by shear. If such threads do fail, the failure may have been preceded by "thread crossing" or wear. The thickness of a standard nut is made greater than that required for simple strength to provide for sufficient bearing area between the wrench and the nut.

The assumption that each thread supports an equal share of the load is incorrect because of the elongation of the bolt and the compression of the nut under load. It has been shown that the load is concentrated on the part of the thread near the base of the nut.[1] The form of nut shown in Fig. 9-10 allows a more equal distribution of the load and has been used in heavily loaded bolts. In some instances, during assembly the bolts are stretched by a bolt puller and the nuts are screwed on by hand and then loaded as the bolt puller is released. This procedure is followed in order to load the bolts more accurately than is possible by tightening up the nuts and in order to prevent galling of the threads.

[1] J. N. Goodier, The Distribution of Loads on the Threads of Screws, *J. Appl. Mech.*, vol. 7, no. 1, p. A-10, 1940.

9-6 STRESSES IN SCREW FASTENINGS

Tightening-up stresses Tightening-up stresses in bolts are highly indeterminate.[1] The judgment of the mechanic in selecting a wrench and in applying the force cannot be predicted with accuracy.

An equation[2] for estimating the tightening-up load is

$$F_1 = kd$$

where F_1 = initial axial load due to tightening up, lb
d = nominal diameter of bolt, in.
k = 16,000 lb per in., as determined for making steamtight joints (this value may be reduced for fastenings not set up as tight as for a steamtight joint)

Shearing stresses When bolts are subjected to direct shearing loads, they should be located and threaded so that the shearing load comes upon the body of the bolt and not upon the threaded portion (see Fig. 16-2). If several bolts are intended to share a shearing load, high-grade construction requires finished bolts fitted to reamed holes. In some instances, the bolts may be relieved of shear loads by shear pins used for this purpose.

Stress concentration Photoelastic tests indicate that the concentration of stress at the root of a standard coarse thread is large, as shown by a static stress-concentration factor that is equal to 5.62.[3]

This concentration of stress is generally not serious in bolts made of *ductile* material and subject to static loads. However, in dynamic loading, the concentration of stress has been found to reduce the endurance limit of standard coarse threaded bolts by fatigue stress-concentration factors equal to 2.84 for medium-carbon steel and 3.85 for SAE 2320 nickel steel heat-treated.[4] These factors were determined for repeated tension loading of bolts threaded with no special care toward relieving stress concentration at the juncture of the thread and the shank of the bolt. Marked improvement in fatigue strength may be secured by using undercut relieving grooves, as discussed in Art. 4-6.

Effect of rigidity of parts The resultant axial load on a bolt depends on the initial tension and on the external load. It depends also on the relative elastic yielding (springiness) of the bolt and the connected members. If the connected members are very yielding as compared with the

[1] "Torquing of Nuts in Aircraft Engines," *SAE* War Engineering Board, Society of Automotive Engineers, New York, 1943.

[2] J. H. Barr, The Stress on Bolts in Service, *Sibley J. Eng.*, October, 1902.

[3] S. G. Hall, Determination of Stress Concentration in Screw Threads by the Photoelastic Method, *Univ. Illinois Bull.* 245, 1932.

[4] Moore and Henwood, The Strength of Screw Threads under Repeated Tension, *Univ. Illinois Bull.* 264, 1934.

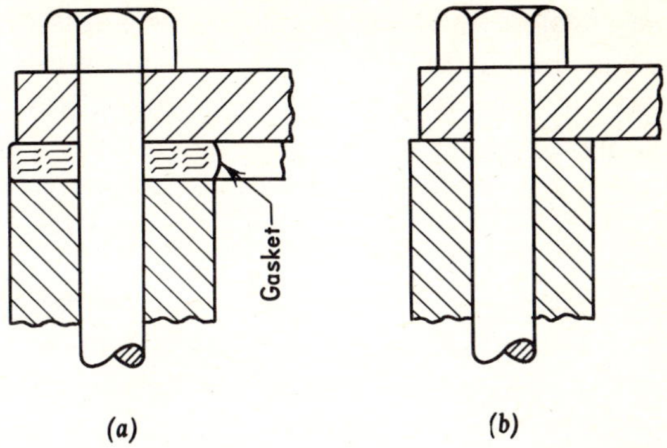

Fig. 9-11 Relative rigidity of bolts and connected parts.

bolt, as in Fig. 9-11(*a*) which shows a soft gasket, the resultant load on the bolt will closely approximate the sum of the initial tension and the external load. If the bolt, however, is very yielding as compared with the connected members, as in Fig. 9-11(*b*), the resultant load will be either the initial tension or the external load, whichever is greater. Actual conditions usually lie between the two extremes.

To estimate the resultant load, the following equation may be used:

$$F = F_1 + \frac{a}{1 + a} F_2$$

where F = resultant axial load, lb
F_1 = initial tension due to tightening up, lb
F_2 = external load, lb
a = ratio of elasticity of connected parts to elasticity of bolt

For soft gaskets and large bolts, the value of a is high and the value of $a/(1 + a)$ approximates unity so that the resultant load is equal to the sum of the initial tension and the external load.

For hard gaskets or metal-to-metal contact surfaces and with small bolts, the value of a is small and the resultant load is due mainly to the initial tension (or to F_2 in the rare case it is greater than F_1).

The value of a may be estimated by the designer to obtain an approximate value for the resultant load.

The designer thus has control over the influence on the resultant load on a bolt by proportioning the sizes of connected parts and bolts and by specifying initial tension in the bolt.

For instance, a connecting-rod bolt may be tightened up with initial tension greater than the external load. The load on the bolt then will be static and the bolt may be designed on the basis of static failure. If the initial tension was not high enough, the resultant load on the bolt would be affected by the external load which is cyclic, and failure of the bolt would be on the basis of cyclic failure and governed by the endurance limit of the material and stress concentration, both of which would be unfavorable.

That is the reason for the practice in the assembly of automotive engines to tighten up connecting-rod bolts so that the stress approaches the yield point.

It is difficult, however, to determine accurately initial stresses even though a torque wrench is used for tightening. This is on account of the variation in surface finish of the threads, fits, and the coefficient of friction at the threads and faces of the boltheads and nuts. For accurate determination of initial tension, it is necessary to measure the elongation of the bolt.[1]

The "turn-of-nut" method is one of the simplest methods used to introduce calculated preload. The angular turn of the nut to introduce the desired preload is calculated according to the bolt length, diameter, material, thread pitch, and joint characteristics. The nut or head of the bolt is tightened by a wrench to seat it well and is then loosened. The nut or head is then turned to an initial tightness referred to as "finger-tight." Finally the nut or head is given the calculated number of turns to induce the desired initial tightness.[2] This method obviates the use of torque wrenches and is virtually independent of the friction between the connected parts and the contacting parts of the bolt.

It can be reasoned that, if a well-designed bolt does not fail in hard tightening up, it will not fail in service. The reasoning is that, when the bolt is being tightened, the stress will be due to the tensile-load stress combined with the torsional stress due to the tightening torque. The latter stress disappears on removal of the tightening torque; hence the remaining stress is less than the maximum stresses during tightening.

In highly stressed dynamically loaded bolts and studs it is important that tightness be maintained. Although properly tightened bolts with high elasticity used with rigid bolted assemblies may be satisfactory, short bolts with low elasticity are apt to fail in service. Consider a short stud of the proportions used to join an automotive cylinder to the crankcase. The thickness of the cylinder flange plus the thickness of the washer may be less than ½ in. When the nut is tightened, the stud is elastically elongated through the length from near the bottom of the nut to near the last thread engaging the crankcase. If this distance is ½ in. and if the nut, at proper tightness, stresses the stud to 20,000 psi, the stud will then

[1] J. O. Almen, Bolt Failure as Affected by Tightening, "Machine Design," p. 133, Penton Publishing Company, Cleveland, 1943.

[2] Fasteners Book Issue, p. 33; T. C. Baumgartner and F. R. Kull, Determining Preload in a Bolted Joint, *Machine Design*, Feb. 13, 1964.

be elastically elongated to 0.002 in. If, during operation, the bolted assembly is reduced in thickness by 0.001 in. through wear, corrosion, embedding, or by displacement of material such as soft plating, the stud will lose one-half of the required tension, and fatigue failure will follow.[1]

9-7 BOLTS SUBJECTED TO SHOCK LOADING

When a bolt is subjected to shock loading, as in the case of a cylinder-head bolt of an internal-combustion engine, the resilience of the bolt should be considered in order to prevent breakage at the thread.

Resilience is defined as the energy that is returned by a member upon release of the load. If a member is not stressed beyond the elastic limit, it is capable of returning all the energy of deformation; if stressed beyond the elastic limit, it will return only a portion of the energy of deformation, the energy not returned being consumed in permanently deforming the member. The term "resilience" applies to a member or specimen; the units of resilience are inch-pounds. See Art. 3-11.

The resilience of a cubic inch of a material when it is stressed to the elastic limit is called the *modulus of resilience.* The modulus of resilience is equal to the area under the stress-strain curve up to the elastic limit. It is a property of a material and is a measure of its shock-resisting ability. Its units are inch-pounds per cubic inch and it is equal to $s_e^2/2E$, in which s_e is the stress at the elastic limit and E is the modulus of elasticity.

If a bolt of the usual form having a full-sized shank and threaded end, as shown in Fig. 9-12(a), is used to support a tensile load, the stress in the threaded part will be higher than that in the shank. If a tensile load is suddenly applied, the energy absorbed by each unit volume of the bolt will be proportional to the square of its stress at the same location; hence, a large part of the energy will be absorbed at the region of the threaded part. If the shank of the bolt is turned down as shown at (b), the shank of the bolt will undergo a higher stress and hence will absorb a greater proportion of the energy, thus relieving the material at the sections near the thread.

[1] J. O. Almen and P. H. Black, "Residual Stresses and Fatigue in Metals," chap. 15, McGraw-Hill Book Company, New York, 1963.

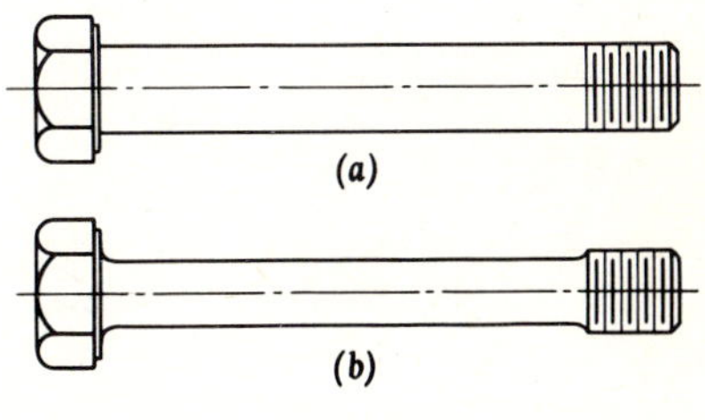

Fig. 9-12 Bolts of same strength but one (b) having greater resilience than the other (a).

If the shank of the bolt is turned down to a diameter corresponding to that at the root of the threads, or even slightly less, the static strength of the bolt will not be decreased.[1]

The resilience of a bolt also may be increased by increasing its length.

9-8 KEYS

The most common function of a key is to prevent relative rotation of a shaft and the member to which it is connected, such as the hub of a gear, pulley, or crank. A large number of types of keys are available, and the choice in any installation depends on several factors, such as power requirements, tightness of fit, stability of connection, and cost. For very light power requirements, a setscrew may be tightened against the round shaft or against a flat spot on the shaft. For most requirements a positive connection, such as that offered by a key, is necessary. A setscrew is frequently used to seat the key firmly in the keyway and to prevent axial motion of the parts.

Square keys as shown in Fig. 9-13(*a*) are common in general industrial machinery.

Flat keys [Fig. 9-13(*b*)] are used where added stability of the connection is desired, as in machine tools. Square or flat keys may be of uniform cross section or they may be tapered. In tapered keys the width is uniform, and the height of the key is tapered by 1/8 in. per ft. Tapered keys may have *gib heads*, as shown in Fig. 9-13(*c*), to facilitate removal.

[1] See "Torquing of Nuts in Aircraft Engines," SAE War Engineering Board, Society of Automotive Engineers, New York, 1943.

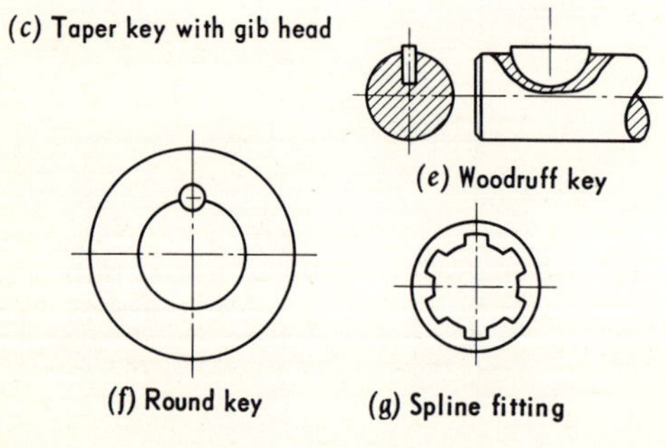

Fig. 9-13 Types of keys.

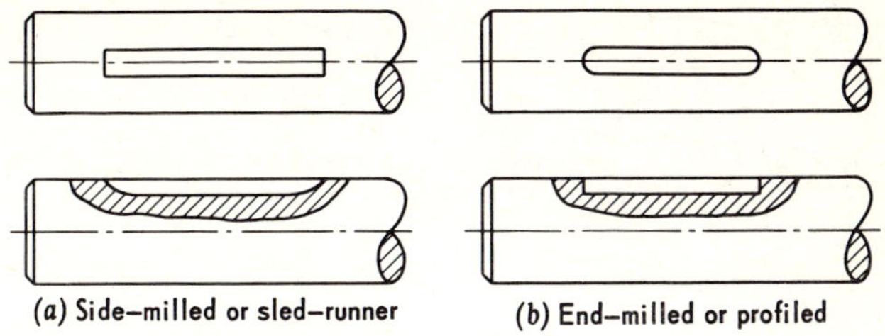

Fig. 9-14 Types of keyways.

The Kennedy key is used for heavy duty and consists of two keys driven as shown in Fig. 9-13(*d*). The hub is bored to fit the shaft and is then rebored off center as shown. The keys force the shaft and hub in concentric position.

The Woodruff key, shown in Fig. 9-13(*e*), requires a key seat formed by a special side-milling cutter. This key will align itself in the key seat. It has the disadvantage of weakening the shaft more than by the use of a straight key.

The round key, or pin [Fig. 9-13(*f*)], introduces less stress concentration at the key seat in the shaft and is satisfactory except for the necessity of drilling the hole to accommodate the pin *after* assembly of the hub and shaft. This may be a disadvantage in production and prevents interchangeability.

A spline fitting, as shown in Fig. 9-13(*g*), is composed of a splined shaft formed by milling and a mating hub with internal splines formed by broaching. The splines in reality are a number of keys integral with the shaft. Splined fittings are adaptable to mass production and are used where radial space must be conserved.

Straight-side splines are being replaced at an increasing rate by stub involute splines. These splines have the advantages of greater strength, a self-centering feature, and production economy. The standards specify fittings with ½ to 48/96 diametral pitches, from 6 to 50 teeth, all with a 30° pressure angle.[1]

Keyways for straight keys are formed either by a side-milling cutter, which forms a *sled-runner* keyway [Fig. 9-14(*a*)] or by an end-milling cutter forming profiled keyways, illustrated at (*b*). The sled-runner keyway requires a longer space between the end of the key and the end of the keyway than does an end-milled keyway. This favors the end-milled keyway in locations near a shoulder. However, the end-milled keyway reduces the endurance limit of a shaft more than does the sled-runner type.

[1] See USAS B5.15–1946, and also "SAE Handbook."

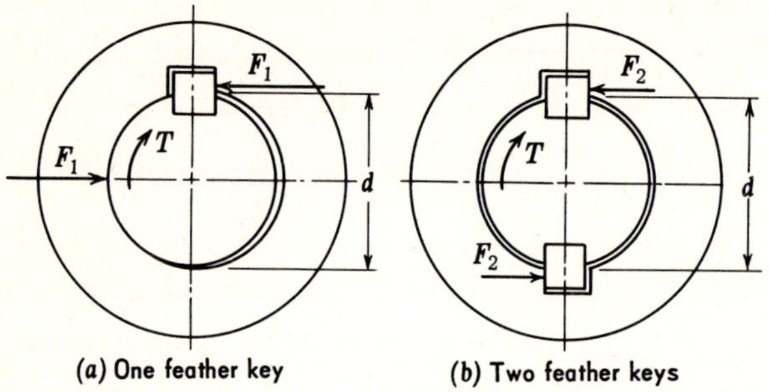

Fig. 9-15 Forces for one and two feather keys. Clearance is exaggerated.

Feather keys are used where it is necessary to slide a keyed gear or pulley along the shaft. The key is generally tight in the shaft and clearance is provided between the key and the hub keyway.

When a gear or pulley must be moved axially along the shaft while power is being transmitted, it is desirable to provide for a minimum of axial force. The use of two feather keys equally spaced requires a smaller axial force than the use of one key. To show this, consider the axial force required to slide the gear hub along the shaft of diameter d when a torque T is being transmitted for the following cases: (1) when one feather key is used and (2) when two feather keys are used.

Because of clearance between the bore and the shaft and between the key and the keyway, the shaft will assume a position approximating that shown in Fig. 9-15(a). For this case,

$$T = F_1 \frac{d}{2} \qquad \text{approximately}$$

or $$F_1 = \frac{2T}{d}$$

The axial force required to move the hub axially is

$$2fF_1 = \frac{4fT}{d}$$

where f is the coefficient of sliding friction.

Where two feather keys are used, the shaft will assume a position as shown in Fig. 9-15(b). For this case

$$T = F_2 d \qquad \text{approximately}$$

or $$F_2 = \frac{T}{d}$$

The axial force necessary to move the hub axially is

$$2fF_2 = \frac{2fT}{d}$$

By comparing the two equations for the axial forces, it is evident that the axial force required for two feather keys is one-half that for one feather key.

Stresses in keys When a key is used in transmitting torque from a shaft to a rotor or hub, the distribution of stresses in the key is three-dimensional and is complicated in nature. These stresses are caused by forces that are of two classes: (1) forces due to the fit of the key in its keyway, as in a tight-fitting straight key or in a tapered key driven in place. These forces produce compressive stresses in the key, and it is impossible generally to predict their magnitude. (2) Forces that are caused by the torque transmitted. These forces produce compressive and shearing stresses in the key. The distribution of the forces along the length of the key is not uniform, since the forces are concentrated near the torque-input end. The nonuniformity of distribution is caused by the twisting of the shaft within the hub. The exponential curve in Fig. 9-16 shows the approximate distribution[1] of the stress along the length of a key.

In order to obtain a relation between stresses and loads that may be used for design purposes, it is customary to neglect the forces due to the fit of the key and to assume that the distribution of forces along the length of the key is uniform.

Proportions of a key A proportion that has given good results in practice makes the *width* of a key one-quarter the diameter of the shaft. The thickness of a key for equal strengths of the key in failure by shearing of the key and compression on the key may be determined for steel by considering the corresponding allowable stresses in shear and compression.

[1] J. P. Den Hartog, *Trans. ASME,* vol. 54, p. APM-54-10, 1932.

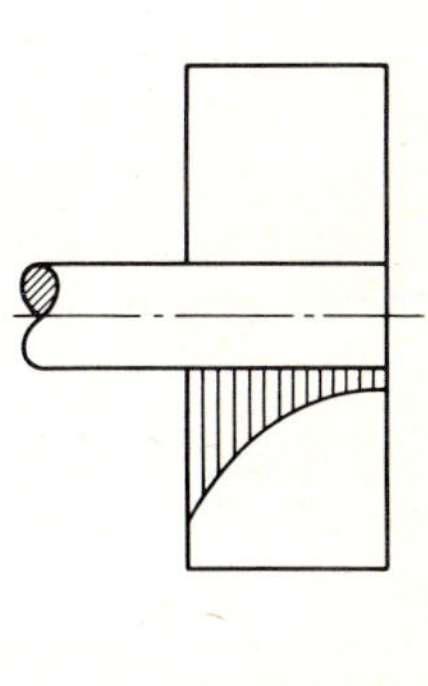

Fig. 9-16 Approximate distribution of tangential load along length of key.

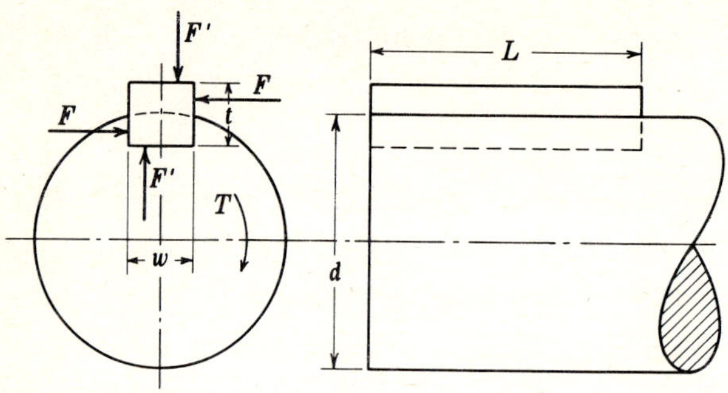

Fig. 9-17 Forces on key due to transmitted torque.

The forces acting on a key for a clockwise torque being transmitted from a shaft to a hub are shown in Fig. 9-17. The forces F' on the top and bottom of the key resist tipping of the key. The force F between the side of the key and the keyway in the hub is due to the resisting torque T and may be determined from the relation

$$T = F\left(\frac{d}{2} + \frac{t}{4}\right) = \frac{Fd}{2} \qquad \text{approximately}$$

On the basis of the assumptions stated above, the strength of the key for failure by shearing of the key is

$$T_s = \frac{Fd}{2} = \frac{wLs_sd}{2}$$

and its strength for failure by compression of the key is

$$T_c = \frac{Fd}{2} = \frac{tLs_cd}{4}$$

By equating the two torques and using the relation that $s_c = 2s_s$, which is approximately true for steel, it is found that $w = t$; hence a square steel key is approximately as strong in failure by shear and crushing.

A square key whose sides are one-quarter the diameter of the shaft may be stronger or weaker than the shaft, depending on the length of the key. The length of the key required for equality of strengths of the key and the shaft may be determined by equating the strength of the key in shear to the strength of the shaft in torsion. Therefore, neglecting all stress concentrations,

$$\frac{wLs_sd}{2} = \frac{\pi d^3}{16} s_s$$

Substituting $d/4$ for w, it is found that

$$L = \frac{\pi d}{2} = 1.5d \qquad \text{approximately}$$

Thus for equal strengths of the key and shaft, the length of the key should be 1.5 times the shaft diameter.

The above discussion is the basis for the proportions of a key that has wide industrial use, i.e., a square key whose sides are one-quarter of the shaft diameter and whose length is at least 1½ times the shaft diameter. For gears, and for rotors of large diameter, stability may require a longer hub, and a key of corresponding length may be used. Also, shafts designed on the basis of stiffness are generally overly strong, and hence short keys may be used.

In order to limit the number of stock sizes of keys, standard key sizes, which may vary somewhat from one-quarter shaft diameter, are approved by the USAS and are given in Table 9-2. Flat-stock key dimensions are also given in the table.

It may be noted that a flat key is weaker in compression than in shear; however, the use of flat keys is generally confined to equipment with the shaft sizes that are required for stiffness, which usually have an excess of strength, so that the standard size flat key has ample strength.

The assumption that stress concentration may be neglected is no doubt justified since, as shown in Fig. 9-16, the key is subjected to a concentration of load and stress; also, the keyway introduces stress concentration in the shaft, so that the stress concentrations will affect

Table 9-2 Standard dimensions of plain parallel keys

Shaft diameter, in. (inclusive)	*Key dimensions, in.*		
		Thickness	
	Width	*Square key*	*Flat key*
½–9/16	⅛	⅛	3/32
⅝–⅞	3/16	3/16	⅛
15/16–1¼	¼	¼	3/16
1 5/16–1⅜	5/16	5/16	¼
1 7/16–1¾	⅜	⅜	¼
1 13/16–2¼	½	½	⅜
2 5/16–2¾	⅝	⅝	7/16
2 13/16–3¼	¾	¾	½
3 5/16–3¾	⅞	⅞	⅝
3 13/16–4½	1	1	¾
4 9/16–5½	1¼	1¼	⅞
5 9/16–6	1½	1½	1

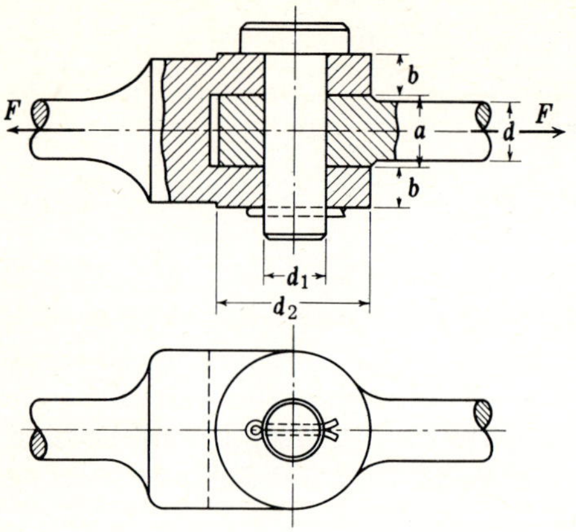

Fig. 9-18 Knuckle joint.

both elements and can be neglected to obtain approximate relations of dimensions. In addition, the effect on seriousness of stress concentration of the ductility of the material should be considered. In a steel shaft and key made of ductile material and subjected to a static load, stress concentration may not be serious. However, with variable loading, as in a shaft with torsional vibration, the stress concentration may be serious.

9-9 KNUCKLE JOINTS AND COTTER JOINTS

Knuckle joints These joints as illustrated in Fig. 9-18 are used to connect two rods that are under the action of tensile loads, although if the joint is guided the rods may support a compressive load. They are used where it is desired to have a joint that can be readily disconnected for adjustments or repairs. The ends of each of the rods are formed by upsetting the rod and machining to form the *eye,* as shown at the right-hand part of the joint in Fig. 9-18, and the yoke or fork, shown at the left in the figure. The knuckle pin forms the connecting element and is held in place by a small cotter pin or other fastening.

The methods of failure of the joint are indicated below, and the *strength* of the joint for each method of failure is indicated. Stress concentration and nonuniform distribution of loads are neglected; therefore the strengths given below will be approximate; however, they serve to indicate a well-proportioned joint. The safe loads for the rod would require an appropriate factor of safety.

1. Failure of the solid rod:

$$F = \frac{\pi d^2}{4} s_t$$

2. Failure of the knuckle pin in shear:

$$F = 2\frac{\pi d_1^2}{4} s_s$$

3. Failure of the rod end by shearing:

$$F = (d_2 - d_1)as_s \qquad \text{approximately}$$

4. Failure by shearing of the fork ends:

$$F = (d_2 - d_1)2bs_s$$

5. Failure of the eye in tension:

$$F = (d_2 - d_1)as_t$$

6. Failure of the fork ends in tension:

$$F = (d_2 - d_1)2bs_t$$

7. Failure of the rod end in compression by bearing against the pin:

$$F = ad_1s_b$$

8. Failure of the fork ends in compression by bearing against the pin:

$$F = 2bd_1s_b$$

9. Failure of the pin in bending. Assuming that the pin is loaded as shown by Fig. 9-19, the maximum bending moment on the pin occurs at the midsection of the pin and is equal to

$$\frac{F}{2}\left(\frac{a}{4} + \frac{b}{3}\right) \qquad \text{approximately}$$

Therefore $F = \dfrac{3\pi d_1^3 s_t}{4(3a + 4b)}$

A well-designed joint is one that has equal strengths in failure for all the above methods of failure. For a particular material, such as steel,

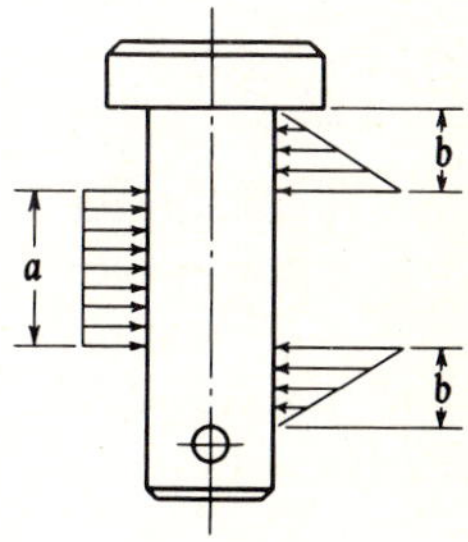

Fig. 9-19 Loading of pin in knuckle joint.

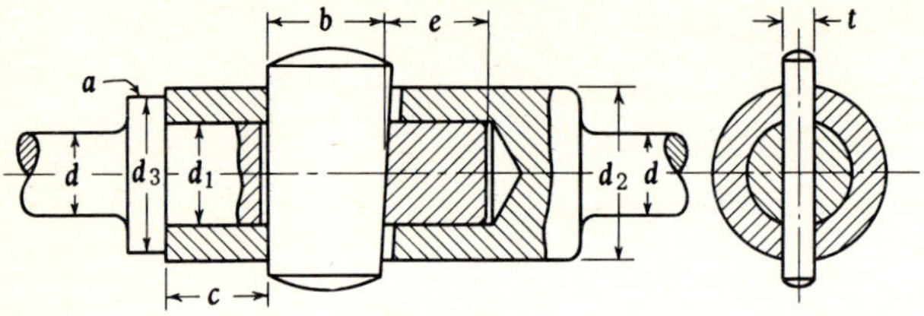

Fig. 9-20 Cotter joint.

the relation between the stresses would be fixed, which would be used to determine the dimensions of the joint. From the above equations, it is obvious that dimension b should be equal to one-half a. However, b is generally made somewhat greater than one-half of a in order to prevent deflection or spreading of the forks, which would introduce excessive bending of the pin.

Cotter joints A cotter joint is used to connect two rods rigidly. The joined rod may carry tensile or compressive loads. As shown in Fig. 9-20, there are three parts to the joint, the *rod end* (left-hand end), *the socket end* (right-hand end), and the *cotter* that fits into the tapered slot. The taper of the slot as well as the cotter is usually on one side, the right as shown in the figure. Clearance between the cotter and the slots in the rod end and socket allow the driven cotter to draw together the two parts of the joint until the socket end comes in contact with the shoulder on the rod end at a. Further driving of the cotter will bend it in the plane of the side view, and this bending will aid in keeping the joint tight under the action of a variable load. The dimensions that may be used in investigating the relative strengths of the rod in various methods of failure are indicated by symbols on the sketch.

Tightening of the cotter introduces initial stresses that are difficult to estimate.

Failure of the joint may occur in several ways, the most probable of which are as follows:

1. Tension failure of the rods at the diameter d
2. Tension failure of the rod or socket across the slot
3. Shear failure of the rod or socket between the slot and the end of the rod or socket
4. Shear of the cotter
5. Bearing between the cotter and the slot in the rod end or in the socket
6. Failure of the cotter in bending

7. Failure of the collar of the rod end in shear due to tightening of the cotter or to any compressive load to which the joint may be subjected.

9-10 PINS AND SNAP RINGS

Pins and snap rings are two examples of a large number of fastenings of a general type that is finding increasingly popular application in machinery, especially in light machinery parts.

Pins may be used in straight, tapered, or grooved forms. In function, they may be classified as *locating pins,* called *dowel pins,* which fix the relative position of two parts; *shear pins* which transmit service loads; and *safety pins* which have the function of a shear pin but which are designed to fail at a predetermined load and thus protect expensive parts from damage. *Safety links,* designed to fail in tension under the action of an overload, perform the same function as a safety pin.

Pins should be accurately fitted into reamed holes and located so that bending of the pin will be eliminated or reduced to a minimum. Provision should be made for pins to be removed. This may require a through hole so that the pin may be punched through, or it may require a plain or a threaded head for drawing.

The grooved pin has a rolled groove, generally tapered, as shown in Fig. 9-21. The rolling of the groove deforms the pin, as shown by the enlarged sectional view, so that when the pin is driven or pressed into a drilled hole, a tight fit is produced between the pin and the hole. The pin deforms elastically, and it is effective under the action of variable loading without loosening of the pin. Some applications of pins are shown in Fig. 9-22.

Snap rings are of two general types, external and internal, as shown in Fig. 9-23(*a*) and (*b*). The relations between the dimensions of the ring and of the recesses that accommodate them should be considered so as to provide for sufficient strength of the ring against permanent set during insertion or under load and to provide for ease of insertion or removal.[1]

[1] Fasteners Book Issue, p. 258.

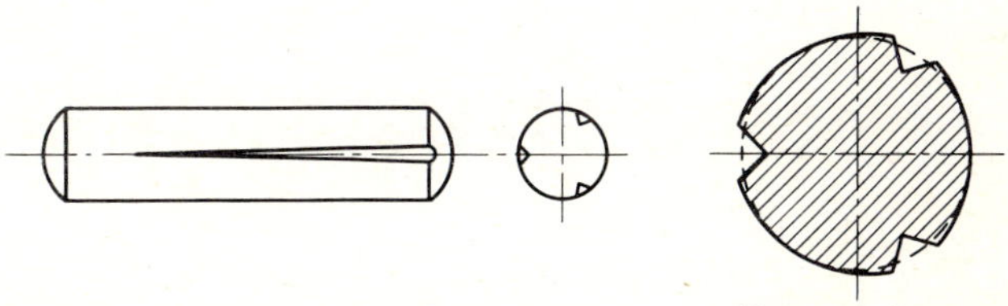

Fig. 9-21 Grooved pin.

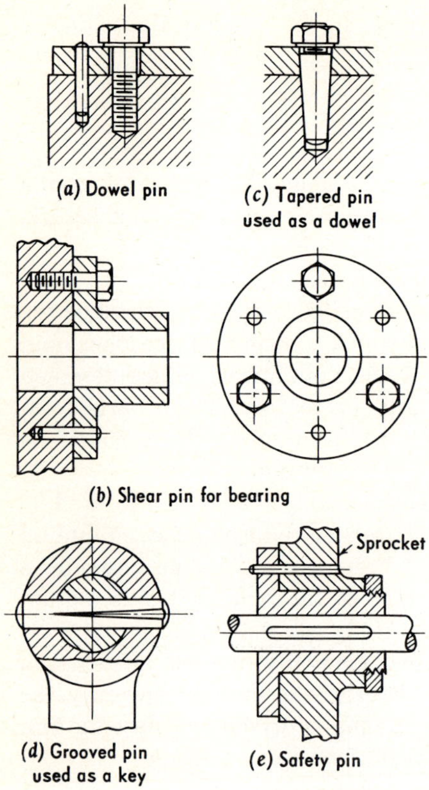

Fig. 9-22 Application of pins.

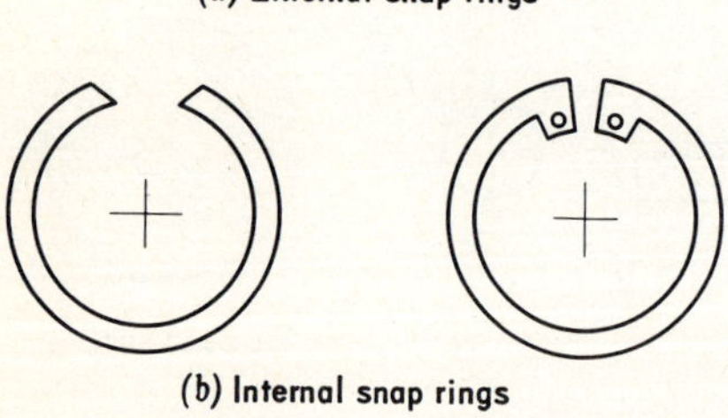

Fig. 9-23 Snap rings.

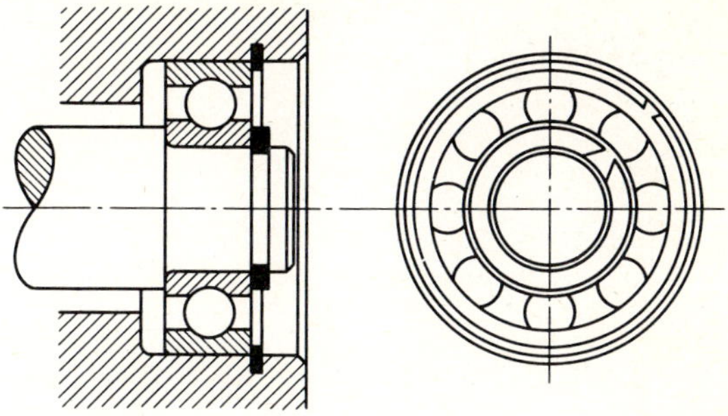

Fig. 9-24 Ball-bearing installation.

Snap rings, and also wire clips that perform the same function, are available in various forms and sizes. Figure 9-24 shows snap rings used to retain a ball bearing in place. The larger ring supports an axial load on the shaft to the right in the figure. The smaller ring retains the inner race of the bearing against the shaft shoulder. Since ease of insertion is necessary in production, clearance must be provided as the rings cannot be seated under axial load. If end play is not permitted, screwed locking fastenings should be used instead of snap rings.

10
Springs

10-1 GENERAL DISCUSSION[1]

Springs are used in a wide range of types and sizes. At one extreme is a coil spring, overall length 0.05 in., outside diameter 0.032 in., with nine coils of 0.004 in. wire, used in a miniature ball-bearing assembly. One thousand of these springs weigh ¼ oz. An example of the other extreme (see Fig. 10-1) is a spring weighing half a ton, which is part of a missile launching unit made for the Defense Command Post in Colorado Springs, Colorado.

Types of springs include helical compression, tension, and torsion; leaf; conical; disk (Belleville); ring, Neg'ator (can be constant force); spiral; torsion bar; air; liquid; and rubber.

Springs have a variety of applications in machinery, as follows:

1. Springs are used to control forces due to impact or shock loading and to control vibration. The energy of impact loading involves the

[1] For an extensive discussion of springs, see A. M. Wahl, "Mechanical Springs," 2d ed., McGraw-Hill Book Company, New York, 1963.

Fig. 10-1 Red-hot, half-ton steel spring rolls off coiling machine at American Steel Foundries' Hammond Division plant. Such springs, the largest ever turned out at this plant, are used to cushion Titan and Atlas ICBMs in underground launching silos. This new missile market has boosted sales of the nation's steel springs. (*Courtesy of American Steel Foundries.*)

product of a force, and a distance that corresponds to the deflection of the resisting member. Thus, to absorb a given amount of energy, increasing the deflection decreases the transmitted force. A spring allows considerable deflection for a moderate load and therefore may be used to reduce the magnitude of the transmitted force in shock loading. Examples of springs for this use are buffer springs for elevators and for car-track terminals, springs for railway cars and automobiles, and springs in cushioned gears and sprockets.

2. Springs are employed to control motions and to apply forces to members. Examples of such uses are valve springs in internal-combustion engines, springs to produce pressure in brakes and clutches, and flexible supports for machinery or equipment.

3. Springs are used for storing energy, for example, in clocks, switch-throwing devices, and starters.

4. Springs are employed to measure forces, as in scales.

The helical compression spring is the most common type used in machinery and will be the only one considered here.

10-2 HELICAL SPRINGS

Figure 10-2(*a*) shows a helical coil spring with ends adapted to support a compressive load. At (*b*) the spring has been deflected by the axial load P, which may be assumed to compress the spring as between two parallel plates. The following notation applies:

Let P = axial load, lb
D = mean diameter of coils, in.
d = diameter of wire, in.
p = pitch of coils, in.
δ = deflection of spring, in.
n = number of active coils
C = spring index = D/d
G = torsional modulus of elasticity, psi
s_s = shearing stress, psi

The number of active coils, n, represents the coils in the spring with the exception of those which lie flat against the compression plates. The inactive coils do not contribute to the deflection of the spring.

Stresses in a helical spring made of round wire Figure 10-3 shows a part of a compression spring that supports a compressive load P and a section of the wire cut by axial plane. The part of the spring shown in

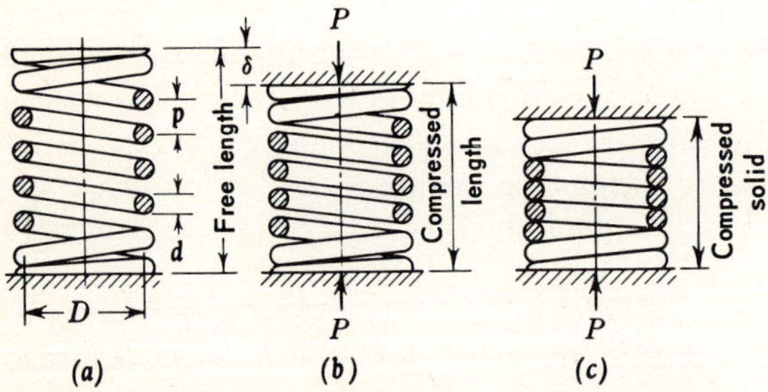

Fig. 10-2 Helical compression spring.

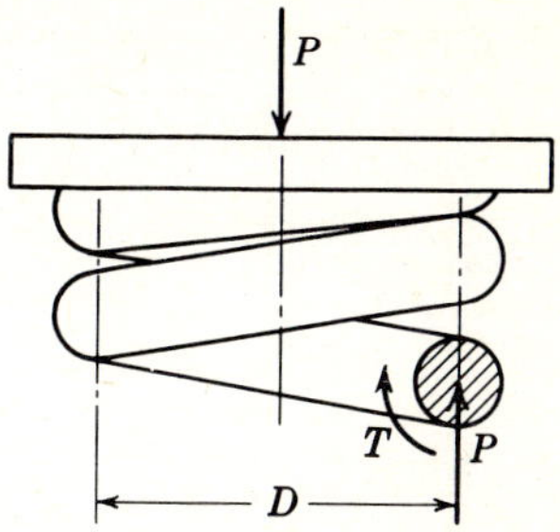

Fig. 10-3 Loading on helical spring.

the figure is in equilibrium under the action of the two forces P and the resisting torsional moment T. The latter equals

$$T = \frac{PD}{2}$$

The shearing stress due to the torque T is

$$s_s = \frac{8PD}{\pi d^3} \tag{10-1}$$

In addition to the stress given by Eq. (10-1), the direct shearing stress due to P should be included. By adding the direct shearing stress, which equals $4P/\pi d^2$, to the torsional stress $8PD/\pi d^3$, the maximum shearing stress, which is located at the inner side of the curved wire, may be determined by

$$\frac{8PD}{\pi d^3} + \frac{4P}{\pi d^2} = \frac{8PD}{\pi d^3}\left(1 + \frac{1}{2C}\right) \tag{10-2}$$

It is evident from Eq. (10-2) that for springs of small index C the effect of direct shear $(1/2C)$ is appreciable.

The effect of curvature of the wire as it forms the coil should be considered also.

In order to include the effects of both direct shear and wire curvature, a stress factor has been determined by the use of approximate analytical methods by A. M. Wahl[1] which may be used with Eq. (10-1) to determine the maximum shearing stress in the wire as follows:

$$s_s = K\frac{8PD}{\pi d^3} = K\frac{8PC}{\pi d^2} \tag{10-3}$$

where $K = \dfrac{4C - 1}{4C - 4} + \dfrac{0.615}{C}$

Values for K may be read from Fig. 10-4. The values of K as indicated by the Wahl equation have been verified by experimental

[1] A. M. Wahl, Stresses in Heavy Closely Coiled Helical Springs, *Trans. ASME*, vol. 51, p. APM-51-17, 1929, or A. M. Wahl, "Mechanical Springs," p. 231.

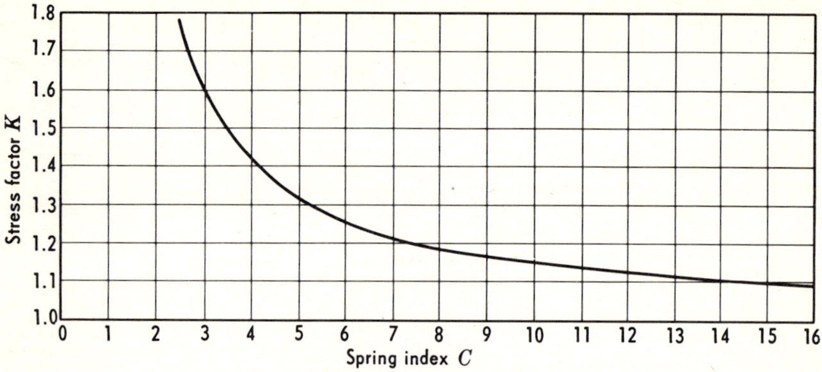

Fig. 10-4 Wahl stress factor for helical springs.

means, using strain measurements secured by sensitive extensometers, and have been verified also by exact calculations based on the theory of elasticity, which indicates that results are correct within 2 percent for spring indexes greater than 3. This includes most springs used in machinery. The Wahl stress factor has been adopted generally for the design of helical springs.

In the design of a helical spring, the designer must use good judgment in assuming the value of the spring index C. The reason for this is the influence of C on the space distribution of the spring for a required load range and deflection. Example 10-1 illustrates this point.

Deflection of helical spring An equation for the axial deflection of a helical spring in terms of the axial load, spring dimensions, and a materials constant may be conveniently determined by equating the work required to deflect the spring to the strain energy in the twisted wire. For close-coiled springs the bending of the wire is small and the strain energy of bending may be neglected.

The axial load as shown in Fig. 10-2 increases linearly from zero to P, and hence the work required to compress the spring is the average force $P/2$ times the deflection, or $P\delta/2$.

The strain energy in a bar twisted by a torsional moment T through a total angle θ is $T\theta/2$. The total angle of twist $\theta = TL/GJ$, where L equals the length of the twisted wire and J equals the polar moment of inertia of the wire section. The active length of wire in the helical spring equals $\pi Dn/\cos\alpha$, where α is the lead angle of the helix, which for close-coiled springs is of the order of 5°; hence $\cos\alpha$ is approximately unity.

For springs of indexes as used in machinery, namely, $2\frac{1}{2}$ or 3 and over, the strain energy in the curved wire forming the coils will be close to the strain energy in a straight wire; hence from the above relations the following equation for δ may be obtained:

$$\delta = \frac{8PD^3n}{Gd^4} = \frac{8PC^3n}{Gd} \tag{10-4}$$

Equation (10-4) gives values that have been accurately verified by test.[1] The value of G for spring steel is approximately 11,500,000 psi.

By solving Eq. (10-4) for P/δ, a useful quantity is obtained, i.e.,

$$\frac{P}{\delta} = \frac{Gd}{8C^3n} \tag{10-5}$$

It may be noted that the right-hand side of this equation is constant for a particular spring; hence P/δ is constant. Its units are pounds per inch, its physical interpretation is the number of pounds required to deflect the spring 1 in., and it is known as the *spring rate*. The spring rate represents the slope of the load-deflection line as shown in Fig. 10-5. Any difference between the ideal line and the actual curve is due principally to the change in number of active coils as the spring is deflected, to mechanical hysteresis of the material, and to friction between the end coils in contact with each other or with the loading plate.

Compression springs in which the free length is more than four times the mean diameter of the coils may fail by sidewise buckling. If space

[1] A. M. Wahl, Further Research on Helical Springs of Round and Square Wire, *Trans. ASME,* 1930.

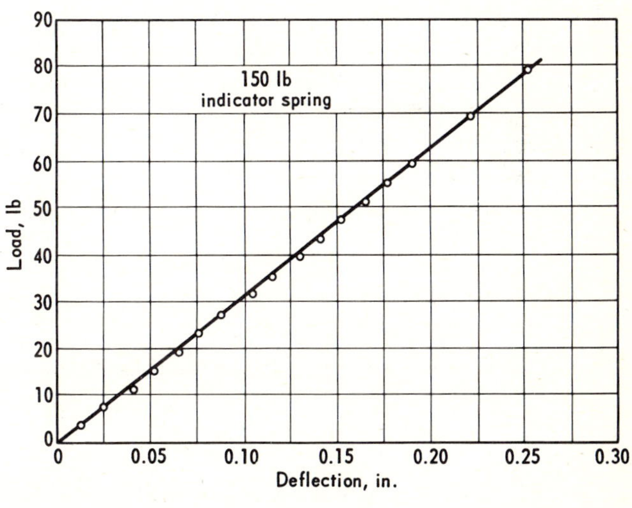

Fig. 10-5 Load-deflection characteristics for indicator spring in compression.

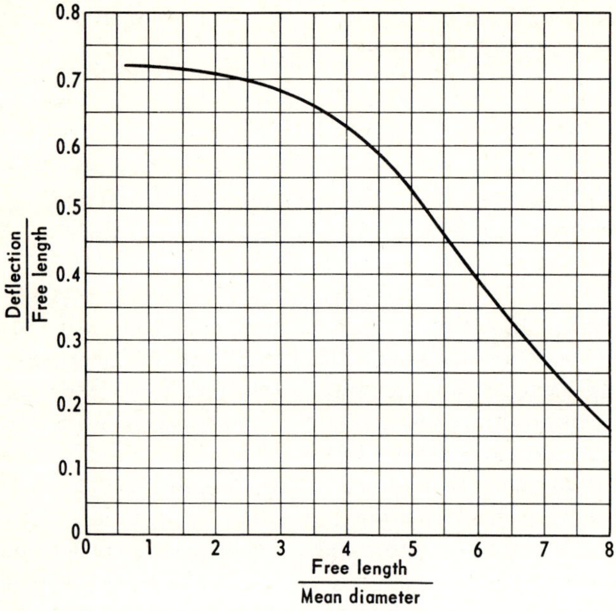

Fig. 10-6 Buckling will occur at points above curve.

limitations make the use of such slender springs necessary, they may be guided by a central rod or mounted in a hole or tube, but the coils are apt to bind and cause wear. Instead, the spring may be resolved into a series of short springs separated by guided bushings. The designer should be aware, however, of the increase in diameter of a spring when it is compressed. This is owing to the changing of slope of the coils and to the unwinding of the coils.[1]

The curve in Fig. 10-6 may be used to determine the possibility of buckling of a helical compression spring with squared and ground ends, mounted so that the ends move only in the axial direction.

10-3 SPRING MATERIAL[2]

To aid the designer in the selection of spring material, the service for which helical springs are used may be broadly classified as severe, average, and light. Although the classification in a particular application depends largely on the designer's judgment, in general, *severe service* includes rapid

[1] See SAE War Engineering Board Manual, "Design and Application of Helical and Spiral Springs for Ordnance," Society of Automotive Engineers, New York, February, 1943.

[2] A. M. Wahl, "Mechanical Springs," p. 231.

continuous loading where the ratio of minimum to maximum stress is one-half or less, as in valve springs. *Average service* includes the same stress range as in severe service but with only intermittent operation, as in engine-governor springs. *Light service* includes springs subjected to loads that are static or very infrequently varied, as in safety-valve springs.

Music wire (ASTM A228-41) is suitable for small springs made of wire up to 0.028 in. in diameter, or if necessary for strength and finish up to 0.125 in. Oil-tempered wire (ASTM A229-41, about SAE 1065) is suitable for low stresses and noncritical uses. Valve-spring wire (ASTM A230-41) is used when long life is important, or alloy-steel wire (SAE 6150) may be used. For high-temperature applications, tungsten high-speed steel (18-4-1) or molybdenum type (6-6-2) is suitable. Nonferrous materials include phosphor bronze, beryllium copper, silicon bronze, nickel-silver alloys (German silver), or nickel alloys, such as Monel, K Monel, Iconel, Iconel X, and Z nickel may be used. As to forming, springs made of wire less than ½ in. in diameter are formed cold; wire over ½ in. is hot-formed.[1]

Glass fiber laminates have been developed for use in internal expanding-sleeve springs and leaf springs and are currently being developed for coil springs. Glass filaments have a modulus of elasticity in tension of the order of 10×10^6 psi and an elastic limit of 300 to 400×10^3 psi. For the internal expanding-sleeve spring, the structure is a continuous, resin-dipped filament which has a modulus of elasticity of 3 to 6×10^6 psi and a maximum stress of 200×10^3 psi.

Recently leaf springs made of glass fiber laminates have been developed with unidirectional and woven-fabric construction; they have superior fatigue characteristics and high damping capacity. There are also being developed coil springs of glass laminates which are light in weight, nonmagnetic, and corrosion-resistant. Applications may be in light airplanes, helicopter parts, and landing gears.

10-4 ALLOWABLE STRESSES

Static loading For helical coil springs subjected to static loading, the elastic limit in torsion may be used as a basis for determining the working stress. For springs subjected to light service, a factor of safety of 1.5 is suggested. The factor may be used to divide into the yield-point stress to obtain the working stress, or to multiply the actual spring load to obtain a limit load to use for design purposes. In case failure of the spring would have serious consequences, the factor of 1.5 should be raised.

[1] See F. P. Zimmerli, Proper Use of Spring Materials, SAE Meeting, French Lick, Ind., June, 1947.

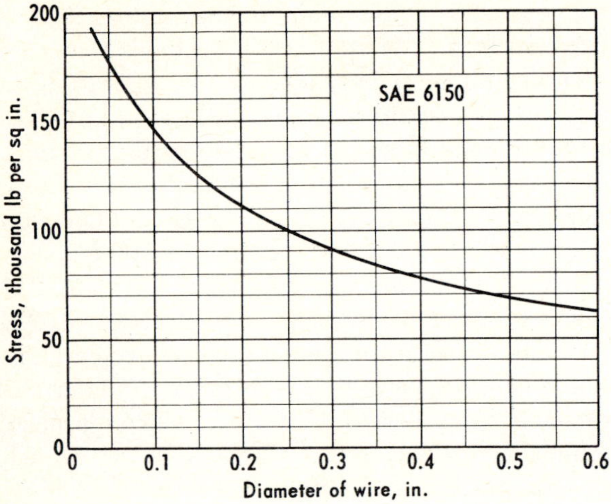

Fig. 10-7 Torsional elastic limit for SAE 6150 spring wire.

In determining the maximum stress induced in the wire, the effect of ductility of the material should be considered. The value of the Wahl stress factor K includes both the effect of curvature of the wire and of direct shearing stress in the wire. The direct shearing stress is uniformly distributed over the wire section. However, the stress augment due to wire curvature is concentrated at the inside of the coil. This latter stress has the nature of a localized stress or a stress concentration and, as discussed in Chap. 4, this type of stress concentration is not serious in *ductile* materials subjected to *static* loads. Therefore the use of the full value of K in Eq. (10-3) will result in conservative design. Equation (10-2) will give satisfactory results for springs subjected to essentially static loads. Figure 10-7 indicates approximate values for elastic limits in torsion for chromium-vanadium spring steel wire, SAE 6150.

Fatigue loading The same general principles that govern working stresses for members subjected to fatigue loading apply to springs; i.e., the working stresses should be based on the endurance limit of the material, and full consideration should be given to stress concentration, surface finish, range of stress, notch sensitivity of the material, and the possibility of corrosion fatigue. The maximum stress induced in the wire can be satisfactorily determined by the use of Eq. (10-3), although for some ranges of stress and for some types of materials this equation gives results that are somewhat high.[1]

[1] See A. M. Wahl, Analysis of Effect of Wire Curvature on Allowable Stresses in Helical Springs, *Trans. ASME*, vol. 61, p. A-25, 1939, and "Handbook of Mechanical Spring Design," American Spring Corporation, Bristol, Conn., 1963.

Failure of round bars in torsion In a straight bar of circular cross section subjected to a torsional moment T, the maximum shearing stresses are as indicated in Fig. 10-8(*a*). If the torsional moment is increased until the shearing stress s_s equals the yield point in shear of the material, failure will occur in shear across the right section shown at (*a*).

Figure 10-8(*b*) shows the principal stresses in the distorted element *ABCD* which was originally rectangular. It is evident that the diagonal *AC* has been lengthened, which induces a tensile stress s_t as indicated at (*b*), and the diagonal *BD* has been shortened, thereby inducing a compressive stress s_c, which acts along the 45° helix *EF*.

In static loading, if the material is weaker in shear than in tension, as in low-carbon steel, fracture will occur in shear along the normal section, as in Fig. 10-8(*a*). However, if the material is weaker in tension, as is common in brittle materials, fracture will occur along a line normal to the maximum tensile stress, that is, along the helical section *EF* at (*b*) in the figure. This is the manner of failure of a piece of chalk (brittle material) if twisted until it breaks.

The same type of failure also occurs in a ductile material subjected to unidirectional repeated torsion, resulting in the typical torsion-bar fracture illustrated in Fig. 10-9. The reason for this is that fatigue cracks can spread only by tensile stress and in directions perpendicular to the tensile stress. Fatigue cracks cannot spread by compressive stress nor by shear stress. See Chap. 6.

In helical coil springs in repeated compression or tension loading, the wire is subjected mainly to shearing stresses due to torsion of the wire, so that fatigue fractures will occur along sections normal to the maximum tensile stress, as shown in Fig. 10-10.

In accordance with the above discussion, helical springs may be designed to be subjected to *static* loading by basing the allowable stress on the yield point in shear and using Eq. (10-3). For springs to be subjected to *fatigue* loading, the allowable stress may be based on the endurance limit in tension, and Eq. (10-3) may still be used because the maximum

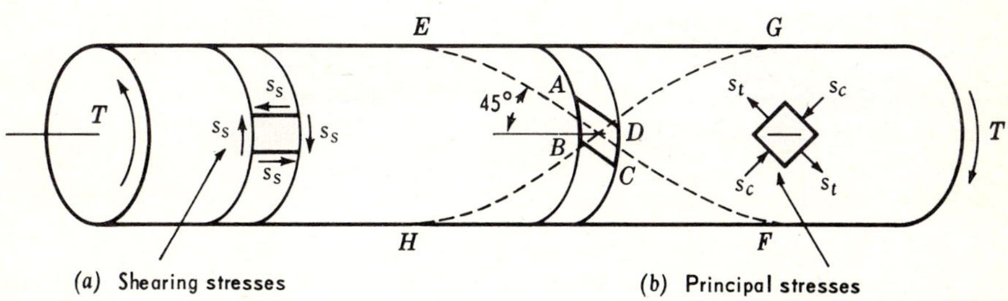

Fig. 10-8 Cylinder subjected to torsion.

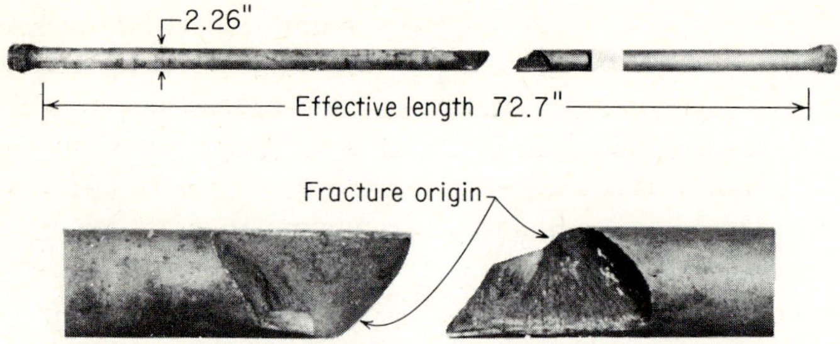

Fig. 10-9 Laboratory fatigue fracture of torsion suspension spring for military vehicle. (*From J. O. Almen and P. H. Black, "Residual Stresses and Fatigue in Metals," McGraw-Hill Book Company, New York,* 1963.)

tensile stress s_t in Fig. 10-8(*b*) is equal, in pure shear, to the shear stress s_s in Fig. 10-8(*a*).[1]

The allowable stresses may be determined from fatigue tests on rotating-bar testing machines, using the same stress range as for the spring being designed and using an appropriate factor of safety, or, preferably, the allowable stresses may be determined from data from laboratory or service tests on coil springs. See Art. 5-3.

There is usually not sufficient time in design and development departments for such orderly procedure; instead, data available in the design office or data supplied by the spring manufacturers must be used, all such data being seasoned by the designer's judgment.

In order to determine allowable shearing stresses, several methods

[1] S. Timoshenko, "Strength of Materials," I, 3d ed., pp. 60 and 283, D. Van Nostrand Company, Inc., Princton, N.J., 1955.

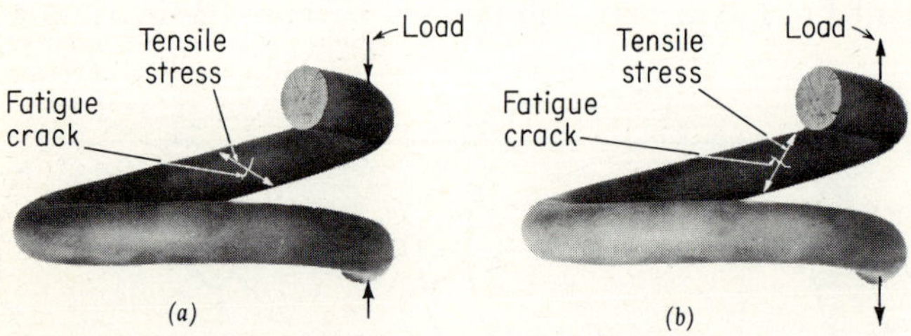

Fig. 10-10 Direction of tensile-stress component and direction of resulting fatigue crack depend upon direction of applied load. (*From J. O. Almen and P. H. Black, "Residual Stresses and Fatigue in Metals," McGraw-Hill Book Company, New York,* 1963.)

have been proposed to allow for the effect of range of stress, i.e., the ratio of minimum stress to maximum stress in the wire. The most promising of these methods[1] is based on the consideration that the actual stress is composed of a variable stress superimposed on a steady stress (see Art. 5-3). For example, an assumed stress range in a spring of $s_{max} = 30{,}000$ psi to $s_{min} = 20{,}000$ psi would be considered as a variable stress $s_v = \pm 5{,}000$ psi superimposed upon a static stress $s_m = 25{,}000$ psi. As a further step, the effects of stress concentration and of the "sensitivity index" are considered separately in their influence on the two components of stress, since one component is static and the other variable.

Since endurance data for spring wire under all representative conditions of service are not available to date, it appears satisfactory at the present time to use Eq. (10-3) with the full value of the Wahl stress factor for determining the induced stress and to use available endurance data, as in Fig. 10-11, to determine allowable stresses. This procedure will result in conservative design.

Figure 10-11 shows allowable torsional-stress ranges for SAE 6150

[1] See A. M. Wahl, Analysis of Effect of Wire Curvature on Allowable Stresses in Helical Springs.

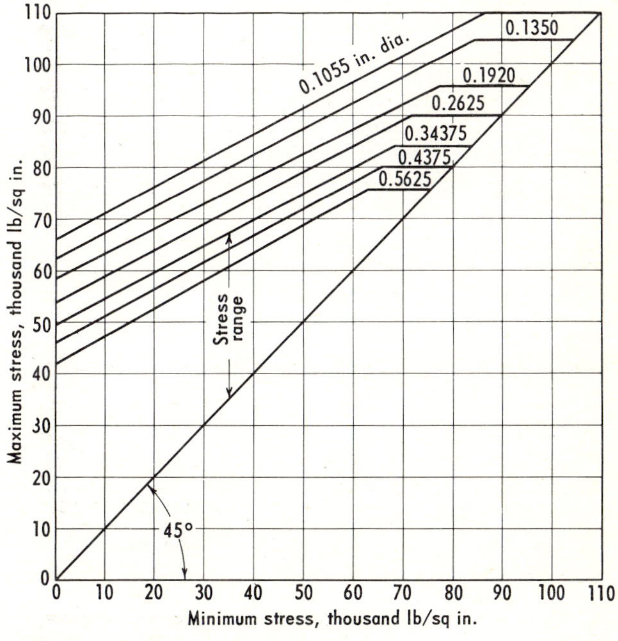

Fig. 10-11 Allowable torsional-stress range for SAE 6150 steel. Load-deflection chart.

Table 10-1 Spring design stresses, psi

Wire diameter, in.	*Severe service*	*Average service*	*Light service*
Up to 0.085	60,000	75,000	93,000
0.085–0.185	55,000	69,000	85,000
0.185–0.320	48,000	60,000	74,000
0.320–0.530	42,000	52,000	65,000
0.530–0.970	36,000	45,000	56,000
0.970–1.5	32,000	40,000	50,000

steel. The ordinate between the 45° line and the line for a particular-size wire indicates the allowable stress range for any value of the minimum stress. For example, if $s_{min} = 35{,}000$ psi, and the wire size is 0.344 in., then $s_{max} = 67{,}000$ psi, or the stress range is 35,000 to 67,000 psi, as indicated in the figure.

For general-purpose helical springs, design stresses used by Westinghouse Electric Corporation are given in Table 10-1. These stresses should be used with Eq. (10-3) and apply to SAE 6150 oil-tempered hot-wound springs heat-treated after forming. The stresses are conservative and may be increased if conditions warrant, particularly for lightly loaded springs.

Example 10-1 A helical valve spring is to be designed for an operating-load range of approximately 20 to 30 lb, that is, 20 lb when the valve is closed and 30 lb when open. The deflection of the spring (valve lift) in the above load range is to be about 0.3 in. Assuming severe service and a spring index of 10, determine the size of wire, size and number of coils, and pitch you would recommend.

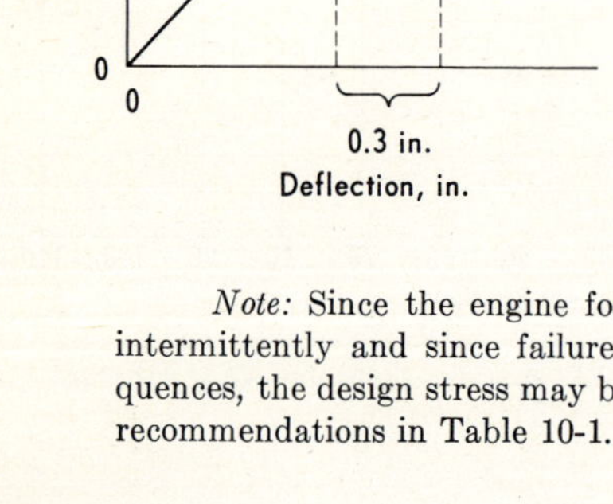

Fig. Example 10-1

Note: Since the engine for which this valve is to be used will operate intermittently and since failure of the spring would not have serious consequences, the design stress may be assumed as about 25 percent in excess of the recommendations in Table 10-1.

Solution From Eq. (10-3),

$$d^2 = \frac{8KPC}{\pi s_s} = \frac{8 \times 1.15 \times 30 \times 10}{\pi s_s} = \frac{880}{s_s}$$

Size of wire: For a trial, assume the size of the wire will be between 0.085 and 0.185 in. in Table 10-1. Then the allowable stress will be 1.25 × 55,000 = 68,700 psi.

$$d^2 = \frac{880}{68{,}700} = 0.0128 \text{ in.}^2$$

or $d = 0.113$ in.

Use No. 11 wire, diameter = 0.120 in., from Appendix XIV. If the calculated size of the wire had not been in the range assumed, further trials would have been necessary.

Number of coils: The figure shows that the spring rate is

$$\frac{P}{\delta} = \frac{10}{0.3} = 33.3 \text{ lb per in.}$$

The spring rate represents the slope of the load-deflection line and is constant over the entire range.

Equation (10-5) may be written as

$$n = \frac{Gd}{8C^3(P/\delta)} = \frac{11{,}500{,}000 \times 0.120}{8 \times 10^3 \times 33.3} = 5.18$$

Use $5\frac{1}{2}$ active coils or $7\frac{1}{2}$ total number of coils to allow for one coil at each end of the spring for squaring and grinding as in Fig. 10-13(*d*).

The actual spring rate is

$$\frac{P}{\delta} = \frac{Gd}{8C^3n} = \frac{11{,}500{,}000 \times 0.120}{8 \times 10^3 \times 5.5} = 31.4 \text{ lb per in.}$$

This differs from the previous value because of rounding out the size of wire and number of coils.

Pitch of coils: A valuable feature of a compression spring is that there is an automatic stop to the deflection when the coils come into contact and the spring becomes solid. If the shear stress in the wire at this condition is below the yield point, the spring will not be damaged owing to overstressing the wire in shear. It is therefore desirable to specify the pitch of the coils so that if the spring is accidentally or carelessly compressed solid the stress will not exceed the yield point in torsion. From Fig. 10-7, the yield point in torsion is 135,000 psi. The corresponding load on the spring is

$$P = \frac{\pi d^2 s_s}{8KC} = \frac{\pi \times (0.120)^2 \times 135{,}000}{8 \times 1.15 \times 10} = 66.4 \text{ lb}$$

The corresponding deflection of the spring is

$$\frac{66.4}{31.4} = 2.11 \text{ in.}$$

From the expressions in Art. 10-5 for spring lengths, the compressed-solid length of the spring is

$$(n + 1)d = (5.5 + 1) \times 0.120 = 0.78 \text{ in.}$$

The free length of the spring is the compressed-solid length plus the deflection required to compress it solid; thus the free length of the spring is

$$0.78 + 2.11 = 2.89 \text{ in.}$$

This is equal to $np + d$, from which the pitch may be found equal to 0.50 in. This is the pitch that would allow a stress at coil closure equal to the yield point. The actual pitch, however, should be less than 0.50 in., to provide a margin of safety as well as to conserve space. By assuming a pitch equal to $\frac{5}{16}$ in., the free length of the spring becomes

$$\begin{aligned} np + d &= 5.5 \times 0.3125 + 0.120 \\ &= 1.84 \text{ in.} \qquad \text{Use } 1\tfrac{7}{8} \text{ in.} \end{aligned}$$

The outside diameter of the spring is

$$(C + 1)d = (10 + 1) \times 0.120 = 1.44 \text{ in.} \qquad \text{Use } 1\tfrac{7}{16} \text{ in.}$$

Care must be used in reducing the pitch of the coils, as was done above, so that the spring does not close up before the maximum service load is reached.

The specifications for the spring are: outside diameter, $1\frac{7}{16}$ in.; size of wire, No. 11 W and M gauge; pitch of coils, $\frac{5}{16}$ in.; $7\frac{1}{2}$ coils with squared and ground ends.

Optimum design There is an infinite number of possibilities for the design of helical springs where the load, deflection, and allowable stress are known or assumed.

In one method of design the spring index can be determined for minimum weight, volume, or length.[1] In the helical-spring design problems at the back of the book the spring index C is usually suggested because the designer can exert good judgment in assuming a value for the spring index; i.e., a small value for C leads to a long, slender spring with large value of the stress factor K, whereas a large value of C leads to a short, fat spring with lower value of the stress factor K. Thus, the designer can use his judgment in selecting the spring index C according to space limitations. Problem 10-1 illustrates this point.

10-5 END CONNECTIONS FOR HELICAL SPRINGS

A compression helical spring has a number of advantages over a tension spring as follows:

1. There is less stress concentration at the end coils of a compression spring than at the loop or other attaching device of a tension spring.
2. Breaking of the wire of a compression spring will not necessarily result in failure of the spring, since the coils adjacent to the break will

[1] R. T. Hinkle and I. E. Morse, Jr., Design of Helical Springs for Minimum Weight, Volume, and Length, *ASME., Paper 58-SA-9*, 1958.

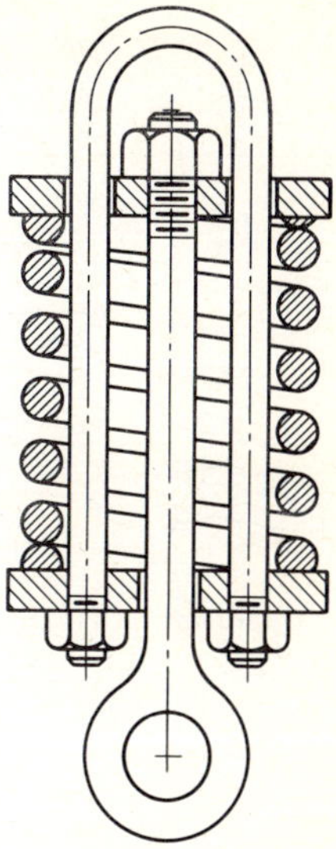

Fig. 10-12 Compression spring arranged for tension rod.

contact and permit the spring, if it is held in line, to function until repairs can be made.

3. Deflection of a compression spring to the solid condition acts as an automatic limit to the stress in the spring.

The free length of a coil spring is approximately $np + d$, and the compressed solid length is approximately $(n + 1)d$ (see Fig. 10-2).

The Washburn and Moen wire gauge is usually standard for steel spring wire (see Appendix XIV); Brown and Sharpe is standard for nonferrous materials. The wire size should be specified in decimals.

A spring used in compression for a tension load is shown in Fig. 10-12. End constructions for helical compression springs are shown in Fig. 10-13,[1] and tension-spring connections are shown in Fig. 10-14. There may be a large stress concentration introduced at the loop of a tension spring, as shown in Fig. 10-14(a). The stress concentration depends on the ratio of r_0/r_1. One spring manufacturer[2] states that a stress factor equal to r_0/r_1 used with Eq. (10-1) gives satisfactory results.

Tension springs[3] are generally wound with initial tension to facilitate handling, so that the spring does not begin to extend until the external load overcomes the initial tension.

[1] See Design Work Sheets, *Product Engineering.*

[2] Barnes, Gibson, and Raymond, "The Art and Science of Spring Making," 4th ed., p. 47, Division of Associated Spring Corp., Detroit and Ann Arbor, Mich., 1941.

[3] "Handbook of Mechanical Spring Design," Division of Associated Spring Corp., Bristol, Conn., 1958.

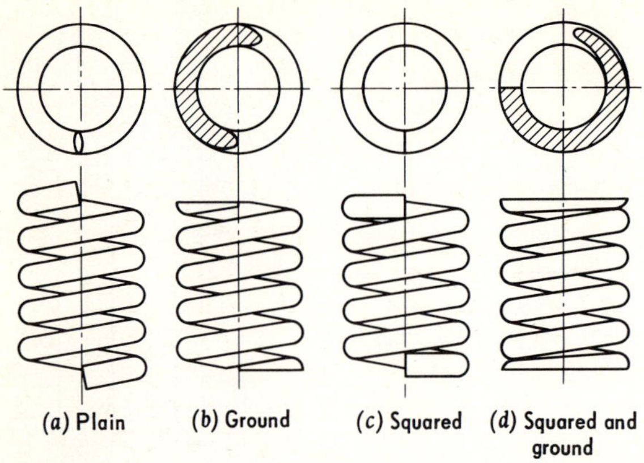

Fig. 10-13 Compression-spring ends.

In all springs there is eccentricity of loading introduced by the end connections. This eccentricity is generally not appreciable in compression springs having squared and ground ends, large spring index, and six or more active coils.[1]

10-6 SURGE IN SPRINGS

When a helical spring rests on a rigid support at one end and the other end is deflected by a suddenly applied load, the coils of the spring will not momentarily share the deflection equally, since time is required for the propagation of stress along the spring wire. An analysis of the

[1] C. T. Edgerton, Stresses in Helical Compression Springs—Present Status of Problem, *Trans. ASME*, vol. 61, 1939.

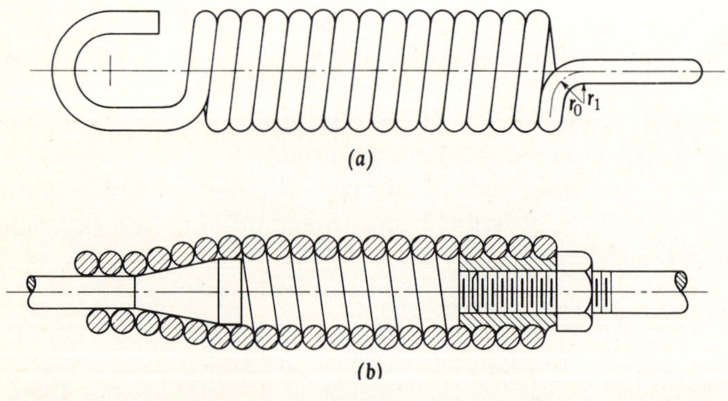

Fig. 10-14 Tension springs.

progression of the deflection is that the end coil in contact with the applied load deflects, and then this coil transmits a large part of its deflection to the adjacent coil. Thus, in progression, a wave of compressed coils travels along the spring to the supported end, where it is "reflected" and travels back to the deflected end. In the absence of damping, this wave travels along the spring indefinitely. The presence of damping, however, causes the wave to die out. If the time required for the wave to travel from one end of the spring to the other and return coincides with the time interval between load applications, a condition of resonance is reached and very large deflections of the coils will be produced with correspondingly high stresses. Under this condition failure of the spring may result. To avoid the possibility of resonance or surge in springs, it is advisable to ensure that the natural frequency of the spring be considerably removed from the frequency of application of the load.[1] The condition of surge has been a problem in valve springs in internal-combustion engines in connection with spring breakage and valve fluttering.

The following equation may be used to determine the natural frequency of a steel coil spring, with both ends fixed, as described above.[1]

$$f_n = \frac{761{,}500d}{nD^2}$$

where f_n = natural frequency, cpm
d = diameter of spring wire, in.
n = number of active coils
D = mean diameter of coils, in.

Stranded-wire springs In some applications of springs, such as in machine-gun mechanisms, the velocity of application of the load is high and the reflection of stress waves in the spring may give rise to resonant stresses. The low damping capacity of steel may permit these stresses to reach large magnitudes which may result in short life of a conventional spring.

Springs made of stranded wire have shown improved performance in these applications. The springs are usually wound with wire made of three strands of preformed wire in which the direction of winding the strands is opposite to the direction of winding the coils to form the spring, so that the deflection of the spring tends to wind the individual wires closer together, which introduces frictional damping and thus decreases the magnitude of the transient stresses. The fatigue life of stranded-wire springs has been increased as much as four times over the life of conventional springs;[2] however, galling of the wires may become serious.

[1] See A. M. Wahl, "Mechanical Springs," p. 231.

[2] H. H. Clark, Stranded Wire Helical Springs for Machine Guns, *Prod. Eng.*, July, 1946, p. 154.

10-7 LEAF SPRINGS

Leaf springs may have the form of a single leaf, as shown in Fig. 10-15(*a*), or they may be laminated, as shown in (*b*). The laminated form is used to secure large resilience within a small space. A common trouble with laminated springs is fatigue failure of the leaves. Some factors contributing to this type of failure are the weakening effect of the hole if the center bolt is used; contact pressures produced by U bolts and rebound clips; stress concentration caused by improperly shaped leaf ends; initial curvature and relative change of curvature of the leaves during loading; vibration of the spring ends during rebound, which may cause breakage of a leaf at an unloaded location; and improper heat treatment, especially surface decarburization. Each of these factors has at least a partial cure, which should be incorporated in leaf-spring design.[1]

Stresses and deflections in leaf springs The equations for maximum stress and for deflection in leaf springs may be closely determined by the method shown in the following example of a cantilever leaf spring. In Fig. 10-16(*b*) is shown a flat cantilever made up by splitting the leaves of the actual spring shown in (*a*) and laying them side by side to form a beam of uniform thickness t, as shown by the dotted line in (*b*) which is a beam of uniform strength. The breadth at the fixed end is b, which is the sum of the widths of the composite leaves.

The equations are for the maximum stress

$$s = \frac{6PL}{bt^2} \tag{a}$$

[1] M. J. Siegel, V. L. Maleev, and J. B. Hartman, "Machine Design," p. 327, International Textbook Company, Scranton, Pa., 1965; SAE War Engineering Board, "Manual on Design and Application of Leaf Springs," Society of Automotive Engineers, New York, 1968.

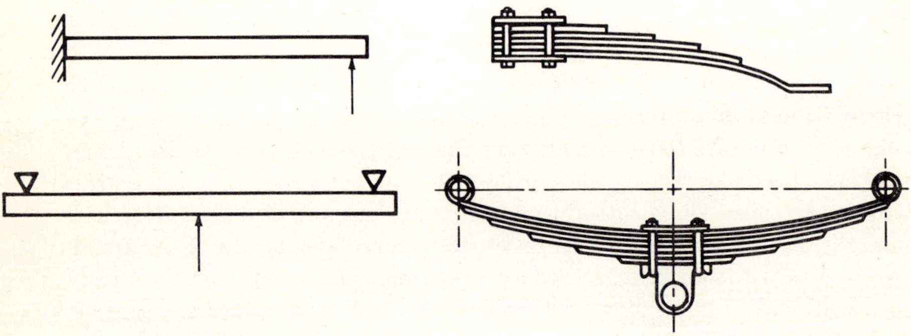

(*a*) Flat springs (*b*) Laminated springs

Fig. 10-15 Forms of leaf springs.

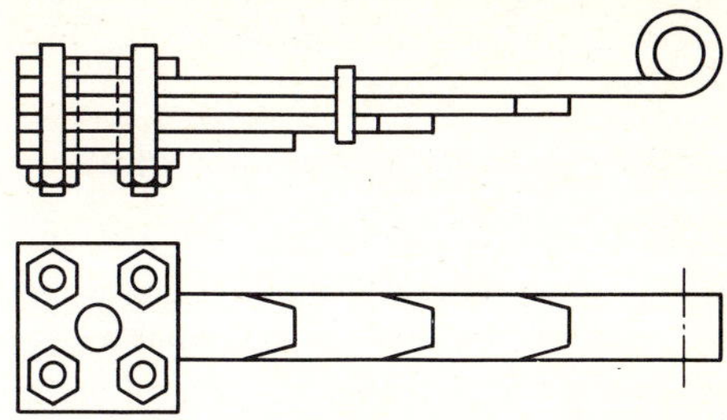

(a) **Leaf spring**

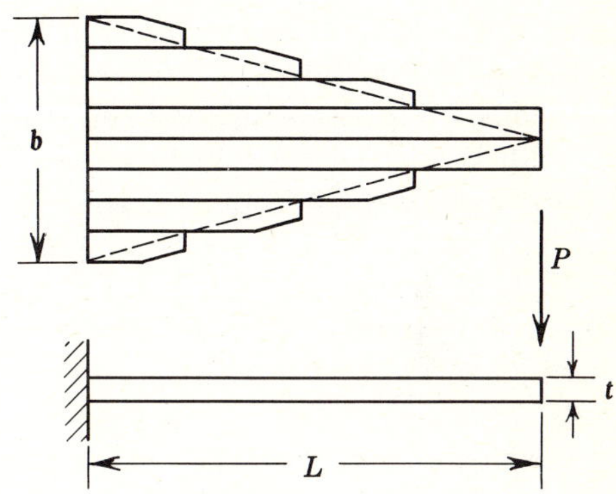

(b) **Equivalent flat cantilever**

Fig. 10-16 Leaf spring and equivalent cantilever.

and for the deflection at the free (loaded) end

$$\delta = \frac{6PL^3}{Ebt^3} \tag{b}$$

where E is the modulus of elasticity in pounds per square inch and the other notation is as shown in Fig. 10-16.

Laminated springs permit a saving in material and a greater deflection than a spring of constant depth; that is, they have greater resilience and shock-absorbing ability. When laminated springs for automotive service are made of leaves of uniform width and thickness, there is sliding between the contacting surfaces of the leaves during deflection; this results in interleaf friction. This friction under varying types of road and speed conditions is not always compatible with the optimum characteristics of the shock absorbers used for rebound control. In tapered single-leaf springs, interleaf friction is eliminated so that the spring lends itself to shot peening, with considerable saving in weight.[1]

10-8 OTHER TYPES OF METALLIC SPRINGS

Concentric helical springs These springs are usually used in compression and provide a large amount of resilience in a small space; they are also used for safety in the event of failure of one of the component springs. The direction of winding of the two helices is opposite to avoid locking of the coils in the event of sideways misalignment or buckling. The design of concentric helical springs should be based on equal maximum stresses in the wires of the two springs.

Conical and volute springs Springs of this type are made of round or rectangular wire with coils of decreasing size to provide for telescoping of the coils. The decrements of coil sizes in relation to the wire size may provide for partial or complete telescoping in springs made of round wire, or for complete telescoping of springs made of rectangular wire wound with the long side of the wire parallel to the axis of the spring.

A comparison of the construction of a conical and a volute spring shows that the conical spring is wound with a constant axial pitch, resulting in a spring with a variable pitch angle and an outline of a truncated cone, as shown in Fig. 10-17(*a*), whereas a volute spring is wound with a constant pitch angle and a variable axial pitch, resulting in a spring with the outline of a truncated paraboloid, as shown in Fig. 10-17(*b*) and (*c*).

Both conical and volute springs have a constant spring rate until the largest active coil starts to "bottom." Thereafter the number of *active* coils decreases as the load is applied and, as a result, the spring rate increases. Thus such springs become "stiffer" as they are loaded.[2] This is a very desiıable characteristic for some applications.

[1] J. O. Almen and P. H. Black, "Residual Stresses and Fatigue in Metals," McGraw-Hill Book Company, New York, 1963.
[2] A. M. Wahl, "Mechanical Springs," p. 231.

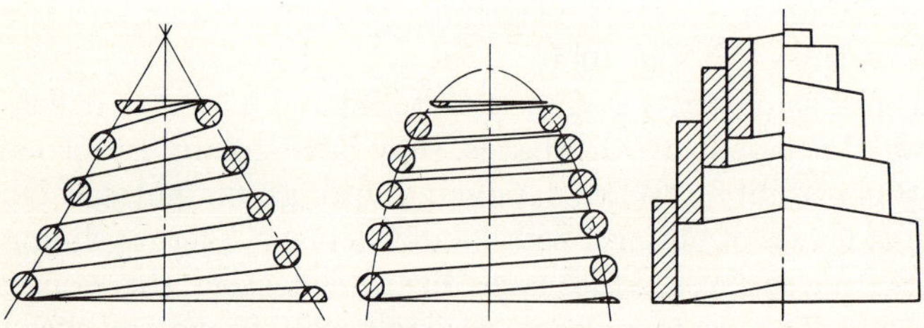

(*a*) Conical (round wire) (*b*) Volute (round wire) (*c*) Volute (rectangular wire)

Fig. 10-17 Conical and volute springs.

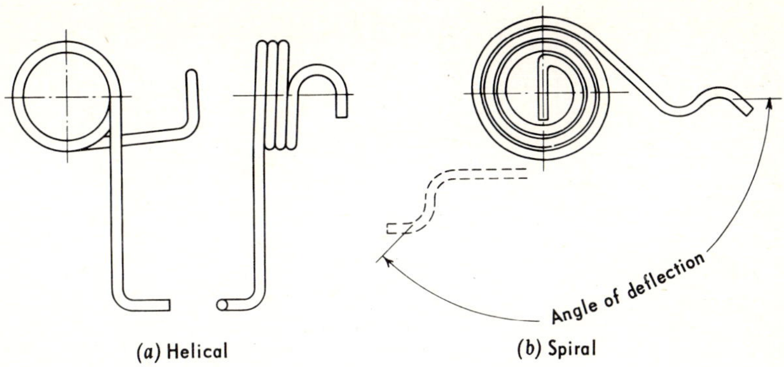

Fig. 10-18 Torsion springs.

The compactness of conical and volute springs and their variable spring rate are desirable in some applications; however, an accompanying disadvantage is an inefficient distribution of stress in the spring. The distribution of stress is complicated in that it varies both *along* the wire and *over* the cross section of the wire. The stress along the wire is a maximum at the largest active coil and it decreases toward the smallest coil. However, as the load on the spring increases, the largest active coil starts to "bottom" and become inactive; hence the remaining active coils support the increasing loads that induce increasing stresses. The maximum stress that is possible occurs when the smallest coil is the only active one, and this condition should be the basis for design if the spring is likely to be compressed solid. The Wahl stress factor for the coil of smallest index, i.e., the smallest coil, should be used.

The influence of the stress factor becomes considerable in springs wound of rectangular wire "on edge," as in Fig. 10-17(*c*). In some volute springs, the wire is tapered toward the small-coil end, giving a somewhat more favorable stress distribution.

To determine the deflection of a conical or volute spring, the equations for the deflection of a helical spring may be used as an approximation in which the mean coil diameter is used as the average of the mean diameters of the active coils.

Torsion springs These springs are of either helical or spiral form, as shown in Fig. 10-18. They may be used in flexible drives, for cushions and return springs in starters, and in various electrical devices and mechanisms, and for spring-closed covers.

Ring springs; plate springs As shown in Fig. 10-19, ring and plate springs are used where high spring rates are desirable, as in flexible support

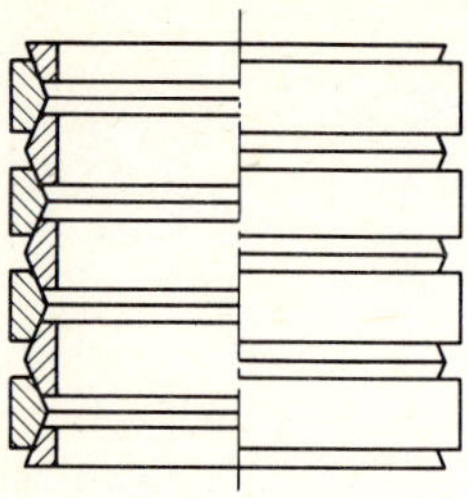

(a) Ring spring

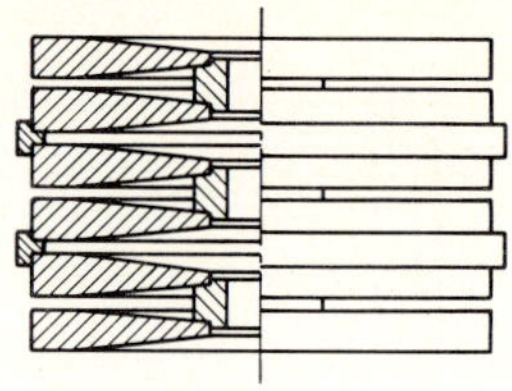

(b) Plate spring

Fig. 10-19 Springs having high rate.

for heavy equipment, in large flexible couplings, and in large valve springs. In Fig. 10-19(*a*), a compressive load causes the external rings to stretch and the internal ones to be compressed. The slope of the contact areas is critical in order to provide for proper ring deformation and allowable compressive forces between the rings. The compressive forces are limited by compressive stresses and by frictional forces to permit self-release of the rings.

Belleville springs A type of plate spring known as the "disk," or Belleville, spring has found many uses in applications requiring high spring rates and compact spring units.[1] A nest of these springs is shown in Fig. 10-20. The unit may be held in alignment by a central bolt or tube.

The Neg'ator spring This spring is called the Neg'ator spring because it can be designed to produce a force-deflection characteristic with a negative slope, as shown in Fig. 10-21(*a*); also it can be designed for a zero slope (constant force) or for a positive slope.

The spring is composed of a flat steel band which is prestressed so that it has a natural curvature and which is wound on a flat-faced reel.

[1] *Ibid.*

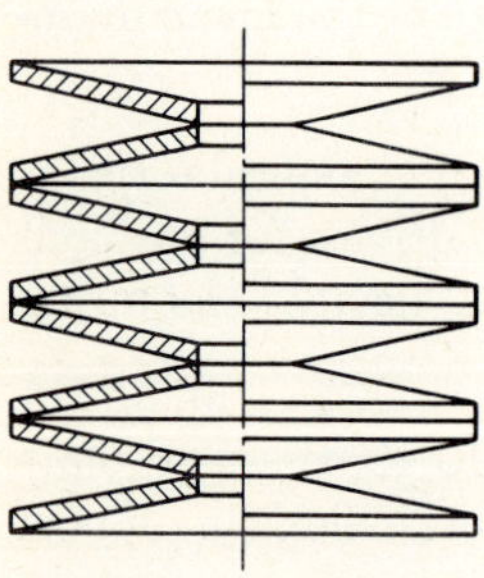

Fig. 10-20 Nest of disk springs.

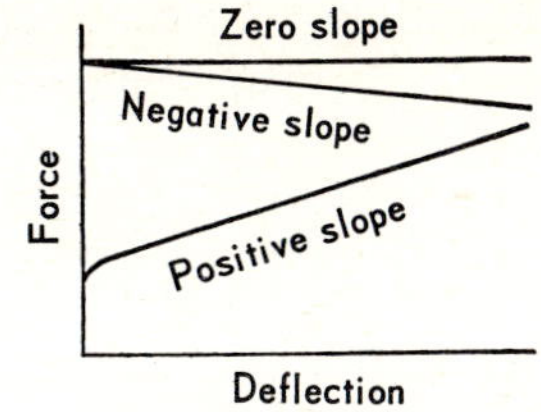

(*a*) Force-deflection charactistics of Neg'ator springs

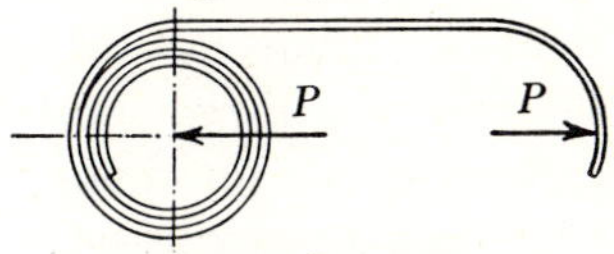

(*b*) Neg'ator spring

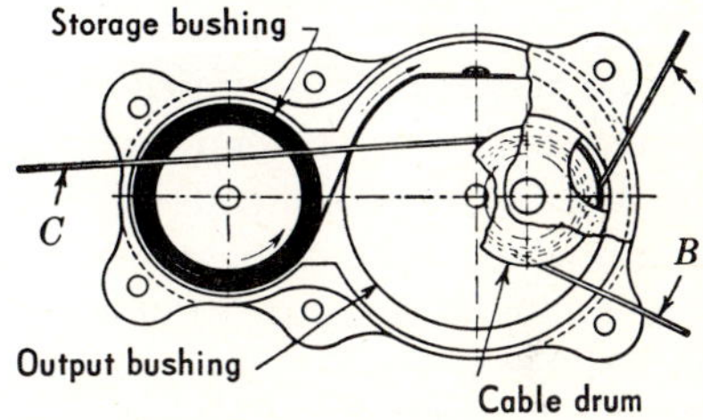

(*c*) Application of Neg'ator spring

Fig. 10-21 Neg'ator springs. Courtesy of Hunter Spring Co. and Republic Aviation Corp.

A force must be exerted on the end of the spring to unwind it from the reel and in turn the spring tends to wind itself back on the reel. Depending on the degree of prestressing, the force on the spring may be constant or it may increase or decrease as the band is unwound from the reel. Figure 10-21(*b*) shows such a spring.

At (*c*) in the figure is an application of a constant-force spring used in an aircraft as a counterbalance to provide for ease in lowering and raising an instrument from the fuselage for inspection so that it will remain in place wherever it is stopped. The Neg'ator spring wound on the storage bushing (or reel) provides a constant tangential force on the output bushing. The output bushing is geared to the cable drum around which are wound cables *A*, *B*, and *C* which run over pulleys and provide three-point suspension for the instrument. In this application, the instrument weighs 36 lb and may be lowered a distance of 18 in. from the latched position. In this case, because of space and weight limitations, dead-weight counterbalances are not possible.

Neg'ator springs have a large variety of applications in the industrial field as well as in mechanisms of all types, such as in business machines, office, household, and building equipment and in servomechanisms. The spring units are light in weight, occupy small space, and have versatile characteristics.[1]

Torsion-bar springs The torsion-bar spring has a range of applications from precision instruments to vehicle suspensions. The bars are usually solid of circular cross section, although hollow tubes and rectangular bars are used. The torsion bar is usually fixed at one end, with a radial bearing and crank arm at the other end. The suspended load at the end of the crank twists the bar. The "spring rate" is usually expressed as torsional moment in inch-pounds per degree of twist (T/θ). The usual equations for torsional stress and angular deflection may be used in the design of torsion-bar springs.

The stresses are limited by "settling" in service or by fatigue failure. Proper selection of material and heat treatment[2] is effective in securing suitable endurance properties and for limiting settling. Shot peening is effective in raising the endurance limit but will not reduce settling.

For fastening the ends of the bar, splines are suitable for heavy service but the rod should be upset at the ends so that the outside diameter of the spline may be 20 to 30 percent greater than the diameter of the bar. The length of the spline is usually about 0.4 times the diameter of the spline. To avoid stress concentration, the transition from the body of the bar to the spline should be gradual.

In many applications the use of torsion bars makes possible a saving in weight and space as well as desirable load-deflection characteristics.

10-9 NONMETALLIC SPRINGS

Air springs This type of spring secures resilience by means of air confined in a cylinder, with the load supported by a plunger. The position of the plunger and the spring rate can be altered by controlling the air pressure. One application is in passenger buses so arranged that, as the bus becomes loaded with passengers, the air pressure is externally and automatically increased in order to maintain the normal level of the bus body.

Liquid springs There are available liquids that can be compressed to 90 percent of their original volume by applying pressures of the order of

[1] See W. J. Cook and P. C. Clarke, The Neg'ator Spring—A Basic New Elastic Member, *Prod. Eng.*, July, 1944; F. A. Votta, The Theory and Design of Long-deflection, Constant-force Spring Elements, *Trans. ASME*, vol. 74, p. 439, 1952. Also see literature from Hunter Spring Co., Lansdale, Pa.

[2] See "SAE Manual of Design and Manufacture of Torsion Bar Springs," SP-26, 1947.

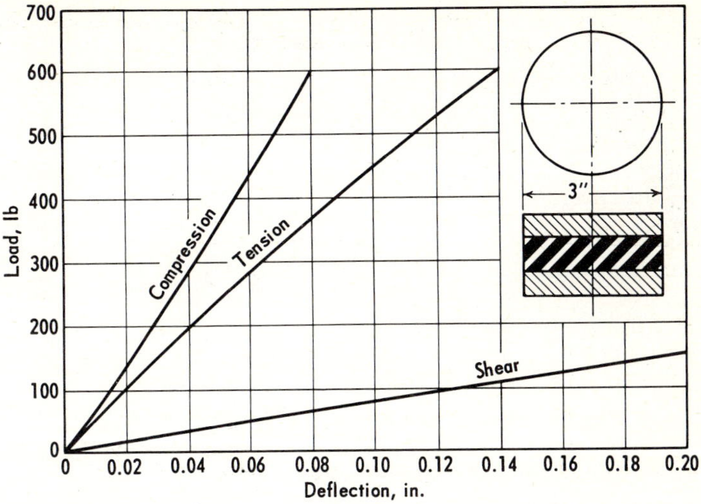

Fig. 10-22 Load-deflection characteristics of rubber mounting.

20,000 psi. Aircraft landing gear is one application of liquid springs. Their advantage for large loads is greatly reduced size as compared with coil springs or conical-disk springs. Disadvantages are sealing, temperature stability, and cost.[1]

Rubber springs A type of flexible support that has advantages in some applications employs rubber or a material with similar properties to provide suitable load-deflection characteristics. The usual support of this type employs rubber in compression and/or shear.

Rubber in shear instead of compression is particularly adaptable to large deformations per unit load. The comparison is shown by Fig. 10-22.

One type of rubber mounting uses a series of shear units which provide a spring rate comparable with that of a steel coil spring. One of these mountings designed for supporting an axial compressive load is shown in Fig. 10-23; the rubber, bonded to the sleeves, is in shear and compression.[2]

As contrasted with metallic springs which have practically no damping, the nonmetallic springs have advantages in their inherent damping properties which are of considerable value in vibration loading. The rubber should be protected from oil and high temperature. If it is not

[1] Lloyd M. Polentz, Saving Space with the Liquid Spring, *Prod. Eng.*, Aug. 17, 1959.
[2] See C. M. Harris and C. E. Crede (eds.), "Shock and Vibration Handbook," chaps. 32 and 35, McGraw-Hill Book Company, New York, 1961; C. E. Crede, "Vibration and Shock Isolation," p. 216, John Wiley & Sons, Inc., New York, 1951.

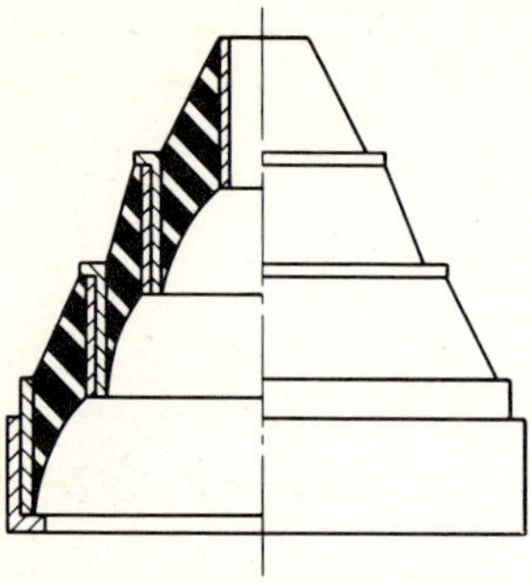

Fig. 10-23 Rubber spring.

practicable to keep oil from the rubber, a material such as neoprene should be used.

10-10 ADDITIONAL CONSIDERATIONS

Presetting.[1] In cold setting, sometimes called "scragging," of helical springs, the spring is coiled to a greater length than the final required free length. The spring is then compressed beyond the elastic limit of the material. Since the wire is stressed in torsion, the maximum *tensile* stress in the wire will exceed the yield point so that, on release of the presetting load, the elastic recovery will induce a residual *compressive* stress in the direction of the tensile stress induced by the prestressing.[2] This operation will decrease the free length of the spring. In service loading, the residual compressive stress due to presetting will reduce the maximum tensile stress due to external loading so that the fatigue strength of the spring will be increased.

Shot peening The residual surface compressive stress induced by shot peening can markedly increase the fatigue durability of springs that are loaded in one direction, such as a helical compression spring, a helical tension spring, leaf springs, and torsion-bar suspension springs. Additional increase can be secured by shot-peening these springs while the springs are held preset in the direction of the service loading. Some leaf springs and torsion-bar springs, such as automobile suspension springs, are loaded in both directions, as in shock loading and in rebound; however, in these applications, the maximum load is in one direction during shock loading in traversing an irregularity in the road, and the rebound load is of smaller magnitude. In one example of an automobile leaf spring, a

[1] See Helical Steel Springs, "Metals Handbook," p. 167, American Society for Metals, Cleveland, 1961. See also "Manual on Design and Application of Helical and Spiral Springs," TR-9, Society of Automotive Engineers, New York, 1958.

[2] See Art. 6-3, *Plastic Deformation.* See also Almen and Black, *op. cit.*, p. 100.

conventional multiple-leaf spring weighed 30 lb, whereas a prestressed, shot-peened single-leaf spring for the same service weighed 10 lb.[1]

Protective coatings For springs subjected to repeated loading in corrosive environment, including intermittent applications of water, the problem of corrosion fatigue appears. The use of nonmetallic protective coatings is, in general, of doubtful advantage, and even metallic coatings may yield disappointing results. Cadmium plating gives comparatively good results, as shown by the following fatigue limits of a specimen of spring steel:[2]

Uncoated, in air	70,000 psi
0.0004 in. cadmium plate, in water	42,000 psi
0.0002 in. cadmium plate, in water	35,000 psi
Uncoated, in water	18,000 psi

Corrosion effects In static loading of springs, the loss of metal may be serious, particularly in springs made of small wire. In fatigue loading, the simultaneous effects of fatigue and corrosion may be particularly serious. Shot-peened springs and springs made of plated wire may have increased fatigue strength, or the spring may be made of corrosive-resistant material such as high-nickel alloy. Design stresses for this type of material are markedly affected by the alloy content, manufacturing processes, and service conditions. The wire manufacturing company should be consulted in such cases.[3]

The same suggestion applies to the design of springs for elevated operating temperatures where relaxation and creep must be considered.[4]

[1] See Almen and Black, *op. cit.*, p. 60.

[2] D. J. McAdam, Jr., Fatigue and Corrosion Fatigue of Spring Material, *Trans. ASME*, vol. 51, p. APM-51-5, 1929.

[3] See Technical Bulletin T-35, Huntington Alloy Products Division, The International Nickel Company, New York, 1960.

[4] See A. M. Wahl, "Mechanical Springs," pp. 45–52.

11
Pressure Cylinders

11-1 INTRODUCTION

Internally pressurized cylinders have a variety of uses in mechanical equipment. They may be classified variously, each group requiring particular equations for their design and the class of material for their construction, as follows:

Dimensions (thin- and thick-walled cylinders) The ratio t/d, which is the ratio of wall thickness t to cylinder diameter d, is the criterion. Thin-walled cylinders are used for pressure cylinders in small presses, tanks, and low-pressure processing equipment. They are generally made of welded or riveted steel or of seamless steel tubing. Thick-walled cylinders are used in high-pressure cylinders, tanks, gun barrels, etc.

End construction (open end or closed). A simple cylinder with a piston, such as a cylinder of a press, is open-ended and has circumferential, or hoop, stresses induced by the fluid pressure. A tank has closed ends and has longitudinal stresses in addition to circumferential stresses. In either open or closed cylinders, the restraint offered by the ends and the relative rigidity of the parts of the cylinder often induce indeterminate stresses at the juncture of the cylinder and the head; these stresses are higher than those indicated by elementary equations and may be of such magnitude in high-pressure installations that accurate calculations based on the theory of elasticity are demanded. Also, stresses due to welding or riveting may be introduced. Such fabrication stresses should be considered when the factor of safety is selected, and reduced to a minimum by appropriate design features, including annealing.

Material The material may be brittle, such as cast iron, or ductile, such as low-carbon steel. The type of material may dictate the method of failure and thereby the appropriate equation for its strength.

Service (conditions of temperature, pressure, and environment) Low-temperature or high-temperature operation may influence the choice of material and the value of the allowable stress, and it may also influence the basis for design, i.e., on elastic state for normal temperatures, or on creep if the temperature for steel cylinders is above 650°F. In addition, thermal stresses may be induced owing to heating or cooling of the cylinder,[1] and in some cases caustic embrittlement is present (see Art. 5-9). High-pressure cylinders may require special considerations because of weight limitations or safety requirements.[2]

Whether the pressure is applied internally or externally will also affect the design. In some pressure vessels which normally have internal pressure, failure because of external pressure should be guarded against in the event of a partial vacuum being formed inside the cylinder, for example, because of condensation in a steam drum or evaporator. Various cryogenic vessels also must be designed to withstand high vacuums.

The case of external pressure being applied on the side or ends of cylinders requires design on the basis of buckling.[3]

In chemical apparatus or refining equipment, stainless steel or a corrosion-resistant lining may be required.

The designs of some of the more common types of cylinders subjected to internal pressure are discussed in the following articles.

[1] S. P. Timoshenko and J. N. Goodier, "Theory of Elasticity," 2d ed., McGraw-Hill Book Company, New York, 1951; C. H. Kent, Thermal Stresses in Thin-walled Cylinders, *Trans. ASME*, vol. 53, p. APM-53-13, 1931.

[2] ASME Boiler Construction Code, sec. VIII, Unfired Pressure Vessels.

[3] S. P. Timoshenko and J. M. Gere, "Theory of Elastic Stability," 2d ed., McGraw-Hill Book Company, New York, 1961; Saunders and Trilling, Collapse by Instability of Thin Cylindrical Shells under External Pressure, *Trans. ASME*, vol. 56, p. 819, 1934; R. G. Sturm, A Study of the Collapsing Pressure of Thin-walled Cylinders, *Univ. Illinois Bull.* 329, 1941; ASME Boiler Construction Code, *loc. cit.*

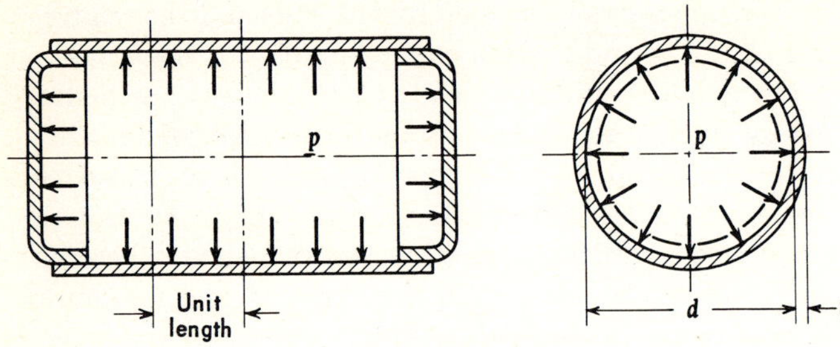

Fig. 11-1 Cylinder with internal pressure.

11-2 THIN-WALLED CYLINDERS

The analysis of stresses induced in a thin-walled cylinder as shown in Fig. 11-1 will be made by neglecting the effects of curvature of the cylinder wall and by assuming that the tensile stresses are uniformly distributed over the section of the walls. This assumption is valid for thin cylinders whose ratio t/d is equal to or less than 0.1. For values of t/d greater than this value, the variation from uniform stress distribution becomes appreciable and thick-cylinder design must be used.

Circumferential tensile stress The fluid force acting on a longitudinal section of unit length is equal to pd, and for equilibrium of forces may be equated to the resisting force equal to $2ts_1$, where s_1 represents the circumferential, or *hoop*, stress. Thus

$$s_1 = \frac{pd}{2t} \tag{11-1}$$

Longitudinal tensile stress The fluid force acting on a ring section is equal to $\frac{1}{4}\pi d^2 p$ and for equilibrium of forces may be equated to the resisting force $\pi d t s_2$, where s_2 represents the longitudinal stress. Thus

$$s_2 = \frac{pd}{4t} \tag{11-2}$$

In Eqs. (11-1) and (11-2), the notation is:

s_1 = circumferential stress, psi
s_2 = longitudinal stress, psi
p = internal pressure, psi
d = internal diameter, in.
t = wall thickness, in.

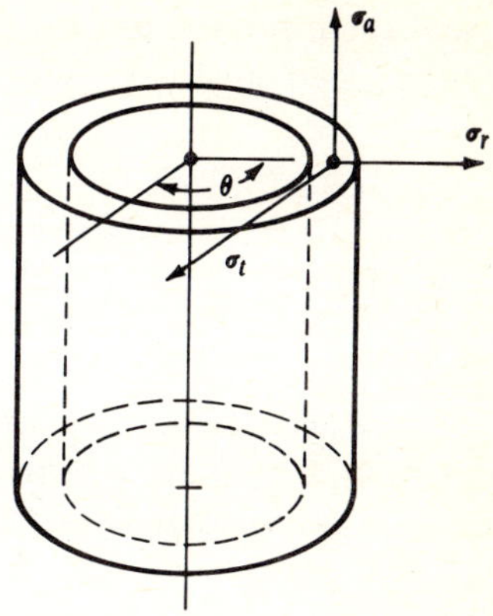

Fig. 11-2 Stresses in a thick-walled cylinder under internal pressure.

11-3 THICK-WALLED CYLINDERS

When the wall thickness of a cylinder is large relative to its diameter, the assumption of uniformly distributed tensile hoop stress no longer applies. For a thick-walled cylinder with closed ends subjected to internal pressure p_i, stresses are induced as shown in Fig. 11-2. Since the pressure loading is independent of the angular direction θ, the axial, radial, and tangential stresses are likewise independent of θ.†‡ These stresses are principal stresses, and shear stresses are not present on planes on which these principal stresses act.

Lamé's solution for principal stresses Consider a cross section of a cylinder perpendicular to its axis, as shown in Fig. 11-3. This figure shows a small element (assuming unit length of element) of the cylinder

† Timoshenko and Goodier, *op. cit.*

‡ F. B. Seely and J. O. Smith, "Advanced Mechanics of Materials," 2d ed., pp. 295–304, John Wiley & Sons, Inc., New York, 1952.

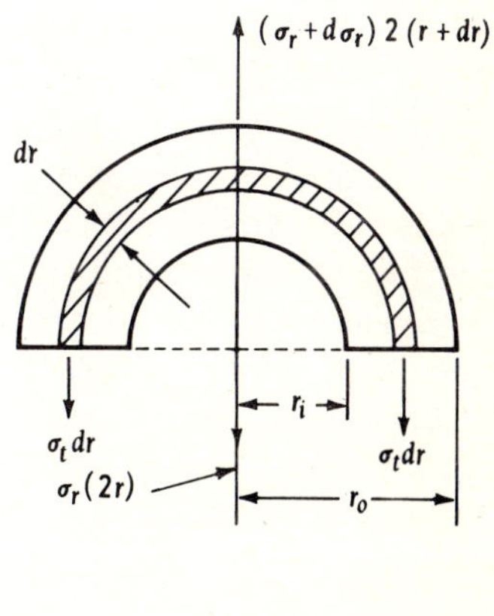

Fig. 11-3 Circumferential element of thick-walled cylinder of unit length.

and the forces under static equilibrium. The notation for the equations to be developed is as follows:

r_o = outer radius of cylinder, in.
r_i = inner radius of cylinder, in.
p_i, p_o = internal and external pressure, respectively, psi
t = wall thickness, $r_o - r_i$
ν = Poisson's ratio
σ_t, s_t = tangential stress, psi
σ_r, s_r = radial stress, psi
σ_a, s_a = axial stress, psi
s_s = maximum shear stress, psi
$s_{t'}$ = allowable tensile stress (limiting value of s_t), psi
$s_{s'}$ = allowable shear stress (limiting value of s_s), psi
ϵ_a = axial strain, in. per in.

By applying the relation for the equilibrium of forces acting on the thin-walled element in the vertical direction, i.e., that the algebraic sum of the vertical components of forces is equal to zero, the following equation is obtained:

$$-2\sigma_t\,dr - \sigma_r \cdot 2r + (\sigma_r + d\sigma_r)2(r + dr) = 0$$

or
$$-\sigma_t\,dr + r\,d\sigma_r + \sigma_r\,dr + d\sigma_r\,dr = 0$$

The term $d\sigma_r\,dr$ is negligible; hence,

$$\sigma_t = \sigma_r + r\frac{d\sigma_r}{dr} \tag{11-3}$$

The three-dimensional stress-strain relationship, with elastic behavior assumed, is as follows:

$$\epsilon_a = \frac{1}{E}[\sigma_a - \nu(\sigma_t + \sigma_r)] \tag{11-4}$$

The longitudinal strains of the axial fibers are assumed to be equal. This means that plane-transverse (parallel) sections before loading remain plane and parallel after loading. This assumption is essentially true for sections of a closed-end cylinder that are well removed from the ends and is true also for a cylinder with open ends. In the case of a closed cylinder, the axial stress is constant, and for a cylinder with open ends the axial stress is zero. Therefore, the quantities ϵ_a, σ_a, E, and ν are constants, or

$$\sigma_t + \sigma_r = \frac{\sigma_a - \epsilon_a E}{\nu} = 2A \tag{11-5}$$

Substituting for σ_t by means of Eq. (11-3) yields

$$2\sigma_r + r\frac{d\sigma_r}{dr} = 2A$$

or

$$\frac{1}{r}\left(\frac{d}{dr}r^2\sigma_r\right) = 2A$$

$$\frac{d}{dr}r^2\sigma_r = 2Ar$$

Integrating yields

$$\sigma_r = A + \frac{B}{r^2} \tag{11-6}$$

By substituting in Eq. (11-3), the tangential stress is

$$\sigma_t = A - \frac{B}{r^2} \tag{11-7}$$

To evaluate the constants A and B, the boundary conditions of the cylinder are applied. These are

At $r = r_o \qquad \sigma_r = -p_o$

$r = r_i \qquad \sigma_r = -p_i$

Substituting in Eq. (11-6) for σ_r at $r = r_o$ and at $r = r_i$, respectively, yields

$$A + \frac{B}{r_o^2} = -p_o$$

$$A + \frac{B}{r_i} = -p_i$$

By solving for B and A,

$$B = \frac{(p_o - p_i)r_o^2r_i^2}{r_o^2 - r_i^2}$$

$$A = \frac{p_ir_i^2 - p_or_o^2}{r_o^2 - r_i^2}$$

By substituting the constants into Eqs. (11-6) and (11-7), the following equations are obtained:

$$\sigma_r = \frac{p_ir_i^2 - p_or_o^2}{r_o^2 - r_i^2} - \frac{(p_i - p_o)r_o^2r_i^2}{r_o^2 - r_i^2}\frac{1}{r^2} \tag{11-8}$$

$$\sigma_t = \frac{p_ir_i^2 - p_or_o^2}{r_o^2 - r_i^2} + \frac{(p_i - p_o)}{r_o^2 - r_i^2}\frac{r_o^2r_i^2}{r^2} \tag{11-9}$$

The distribution of stress in a thick-walled cylinder subjected to internal pressure only is shown in Fig. 11-4. An analysis indicates that the tangential stress is considerably greater at the inside surface than at

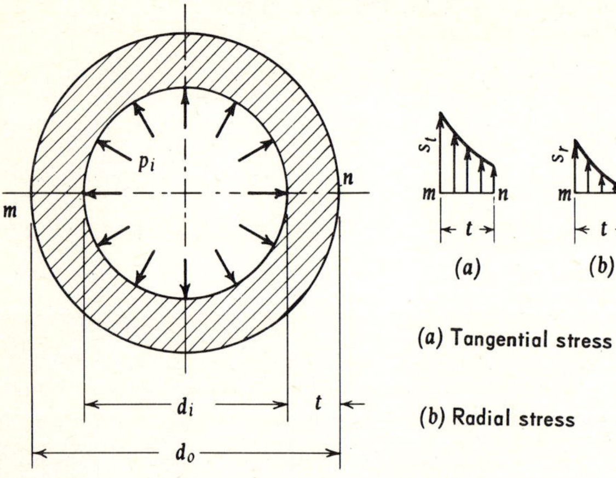

Fig. 11-4 Stress distribution in wall of thick cylinder.

the outside and that the radial stress is of significant magnitude. By substituting a zero value for the external pressure, the following expressions are obtained for the tangential and radial stresses:

$$s_t = \frac{p_i r_i^2}{r_o^2 - r_i^2}\left(1 + \frac{r_o^2}{r^2}\right) \tag{11-10}$$

$$s_r = \frac{p_i r_i^2}{r_o^2 - r_i^2}\left(1 - \frac{r_o^2}{r^2}\right) \tag{11-11}$$

The axial stress for a thick-walled cylinder with closed ends subjected to an internal pressure may be found by writing a static-equilibrium equation for the summation of forces in the axial direction. Thus,

$$s_a = \frac{p_i r_i^2}{r_o^2 - r_i^2} \tag{11-12}$$

The maximum tangential stress at the inside surface of the cylinder (due to internal pressure) becomes

$$s_t(\max) = \frac{p_i(r_o^2 + r_i^2)}{r_o^2 - r_i^2} \tag{11-13}$$

and the maximum radial stress at the same point is

$$s_r(\max) = -p_i \tag{11-14}$$

Brittle material Since the tangential stress is a principal stress and is a maximum for the biaxial system, in designing a cylinder of *brittle material* in accordance with the maximum-stress theory of failure, this stress should not be exceeded, or

$$s_t = s_t(\max) = \frac{p_i(r_o^2 + r_i^2)}{r_o^2 - r_i^2}$$

If $t = r_o - r_i$, the thickness of a cylinder made of brittle material is

$$t_i = r_i\left(\sqrt{\frac{s_{t'} + p_i}{s_{t'} - p_i}} - 1\right) \tag{11-15}$$

Ductile material For ductile materials such as low- and medium-carbon steels, aluminum, brass, etc., failure is assumed to occur at the beginning of yielding.

Maximum-shear-stress theory The maximum shearing stress at any point in a stressed body is equal to one-half the algebraic difference of the maximum and minimum principal stresses at that point. For a thick-walled cylinder subjected to internal pressure, the maximum shear stress occurs at a point on the inner surface ($r = r_i$). The thickness is

$$t = r_i\left(\sqrt{\frac{s_{s'}}{s_{s'} - p_i}} - 1\right) \tag{11-16}$$

Usually $s_{s'} = s_{t'}/2$.

Maximum-strain theory By using the fact that the maximum strain occurs at the inner surface of a thick-walled cylinder, the thickness is

$$t = r_i\left[\sqrt{\frac{s_{t'} + (1 - \nu)p_i}{s_{t'} - (1 + \nu)p_i}} - 1\right] \tag{11-17}$$

This is Birnie's equation for open-end cylinders: alternatively, we have

$$t = r_i\left[\sqrt{\frac{s_{t'} + (1 - 2\nu)p_i}{s_{t'} - (1 + \nu)p_i}} - 1\right] \tag{11-18}$$

known as Clavarino's equation for closed-end cylinders.

Maximum energy of distortion Failure in this case is based upon the energy of distortion per unit volume of the cylinder at the internal surface for closed-end cylinders:[1]

$$t = r_i\left(\sqrt{\frac{\sqrt{q}}{\sqrt{q} - 1}} - 1\right) \tag{11-19}$$

where $\sqrt{q} = \dfrac{1}{\sqrt{3}\,p_i}\,s_{t'}$

[1] Seely and Smith, *op. cit.*, pp. 315–316.

In all the above equations for ductile materials, the allowable tensile stress $s_{t'}$ is equal to either the elastic limit divided by the factor of safety or the yield strength divided by the factor of safety.

The following outline is given for reference to select the equation to be used in the design of thick cylinders:

Brittle material, open ends—use Eq. (11-15)
Brittle material, closed ends—use Eq. (11-15)
Ductile material, open ends—use Eq. (11-16) or (11-17)
Ductile material, closed ends—use Eq. (11-16), (11-18), or (11-19)

For brittle materials such as cast iron, good agreement exists between failure predictions based upon the maximum-principal-stress theory and experimental data. Several of the various failure theories for ductile materials have been found to be in good agreement with experimental results, depending upon the material being tested.[1] However, the wall thickness of cylinders given by the maximum-shearing-stress theory or by the distortion-energy theory is considered to be the most reliable.[2]

For high internal pressures, the required wall thickness of a cylinder becomes very large. Most of the resulting high stresses are located in a region surrounding the inner surface of the cylinder, and the outer region is subjected to lower stresses. There are several techniques of making more effective use of the material near the outer region of the cylinder. All these methods involve the creation of residual compressive stresses (tangential stress) at the inner surface of the cylinder. In these cases, a more favorable stress distribution results since the load stresses induced by the internal pressure must be superimposed upon the residual stresses in order to obtain the resultant favorable stress distribution. The three most common methods of prestressing are the shrinking of a hollow external cylinder over the main cylinder, the construction of laminated cylinders, and "autofrettage" or self-hooping.[3]

Thermal stress As a result of nonuniform temperature distribution, or of different coefficients of thermal expansion, "self-balancing" or thermal stresses are produced. In a solid body these stresses arise during a change of temperature whenever a relatively high volume of material is prevented from assuming the shape and size that it normally should if there were no constraints on the body. During a nonuniform temperature change, each fiber within the body constrains its neighboring fibers from fully expanding (or contracting).

[1] Lloyd E. Brownell and Edwin H. Young, "Process Equipment Design," pp. 274–275, John Wiley & Sons, Inc., New York, 1959.
[2] Seely and Smith, *op. cit.*, p. 319.
[3] Seely and Smith, *op. cit.*, pp. 320–340.

In the case of a thick-walled tube subjected to steady-state heat flow the following thermal stresses are obtained:[1]

$$s_r = \frac{\alpha E(T_i - T_o)}{2(1 - \nu)\log(r_o/r_i)}\left[-\log\frac{r_o}{r} - \frac{r_i^2}{r_o^2 - r_i^2}\left(1 - \frac{r_o^2}{r_i^2}\right)\log\frac{r_o}{r_i}\right] \tag{11-20}$$

$$s_t = \frac{\alpha E(T_i - T_o)}{2(1 - \nu)\log(r_o/r_i)}\left[1 - \log\frac{r_o}{r} - \frac{r_i^2}{r_o^2 - r_i^2}\left(1 + \frac{r_o^2}{r^2}\right)\log\frac{r_o}{r_i}\right] \tag{11-21}$$

$$s_a = \frac{\alpha E(T_i - T_o)}{2(1 - \nu)\log(r_o/r_i)}\left(1 - 2\log\frac{r_o}{r} - \frac{2r_i^2}{r_o^2 - r_i^2}\log\frac{r_o}{r_i}\right) \tag{11-22}$$

for zero axial force, in which α is the linear coefficient of thermal expansion. In the preceding equations, T_i, T_o are the inner and outer temperatures, respectively, in degrees Fahrenheit; s_r, s_t, s_a are the radial, tangential, and axial stress components, respectively; ν is Poisson's ratio; and E is Young's modulus.

In this situation, the maximum magnitudes of stress occur at the inner surface ($r = r_i$), where the tangential and axial stresses are equal. If $T_i - T_o$ is positive, these stresses are compressive, and if negative, the stresses are tensile.

11-4 CYLINDER HEADS

Bending of the shell due to the restraint of the heads is shown exaggerated in Fig. 11-5. It is evident that the stresses in the plate induced by fluid pressure, as determined by the cylinder equations in Arts. 11-2 and 11-3, are increased because of the bending of the plate, and this bending becomes greater in proportion to the relative stiffness of the head and shell. The bending stresses may be allowed for in low-pressure-cylinder designs by the use of an appropriate allowable stress; however, in high-pressure-cylinder designs the bending stresses should be reduced to a minimum by the use of special head construction, as shown in Fig. 11-6(*h*).

[1] J. N. Goodier, Thermal Stresses and Deformation, *J. Appl. Mech.*, September, 1957, pp. 472–473.

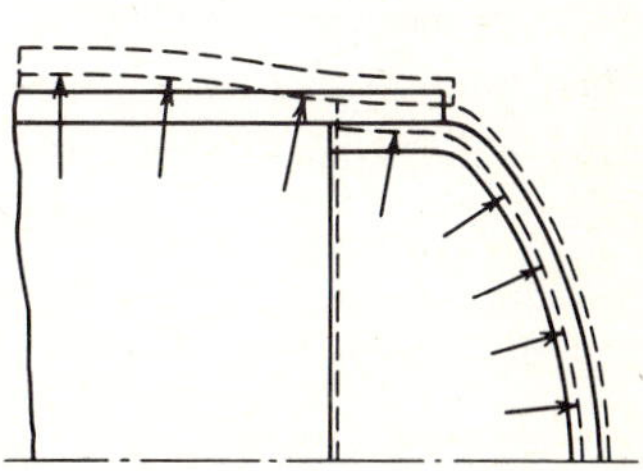

Fig. 11-5 Buckling of shell due to end restraint.

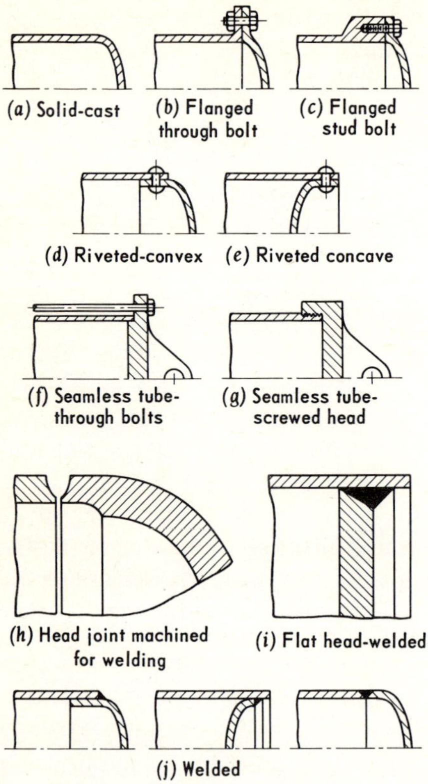

Fig. 11-6 Types of cylinder heads.

Openings for manholes, hand-holes, nozzle attachments, and other connections markedly affect the stresses in pressure vessels and should be carefully considered, especially in high-pressure and high-temperature installations.[1] Examples of head constructions and means of attachment are shown in Fig. 11-6. Recommendations for the design of welded joints and nozzle connections and a procedure for the design of uniformly distributed edge-loaded members may be found in the ASME Boiler and Pressure-vessel Code—Nuclear Vessels.[2]

11-5 ASME CODE DESIGN

The purpose of this article is to convey to the reader that there are rational, generally accepted, and proved procedures, known as "codes,"

[1] Maulbetsch and Hetenyi, Stresses in Pressure Vessels, Design Data, *Trans. ASME*, vol. 58, p. A-107, 1936; Taylor and Waters, The Effect of Openings in Pressure Vessels, *Trans. ASME*, vol. 56, p. APM-56-3, 1934; "Welding Handbook," pp. 1306ff, American Welding Society, New York, 1942.

[2] ASME Boiler and Pressure-vessel Code—Nuclear Vessels, sec. III, 1965.

for the design of pressure vessels. One of the most all-inclusive codes is the ASME Boiler and Pressure-vessel Code for Nuclear Vessels, Section III.

According to this code, stresses are divided into five characteristics, as shown in Fig. 11-7. In order that a design may be acceptable, the calculated stress must not exceed specified allowable limits designated in terms of S_m (Fig. 11-7). *Stress intensity* is defined as twice the maximum

*General membrane**	*Local membrane**	*Bending**	*Secondary membrane plus bending*	*Peak*
Average primary stress across solid section. Excludes discontinuities and concentrations.	Average stress across any solid section. Considers discontinuities but not concentrations.	Component of primary stress proportional to distance from centroid of solid section. Excludes discontinuities and concentrations.	Self-equilibrating stress necessary to satisfy continuity of structure. Occurs at structural discontinuities. Can be caused by mechanical load or by differential thermal expansion. Excludes local stress concentrations.	(1) Increment added to primary or secondary stress by a concentration (notch) (2) Certain thermal stresses which may cause fatigue but not distortion of vessel shape

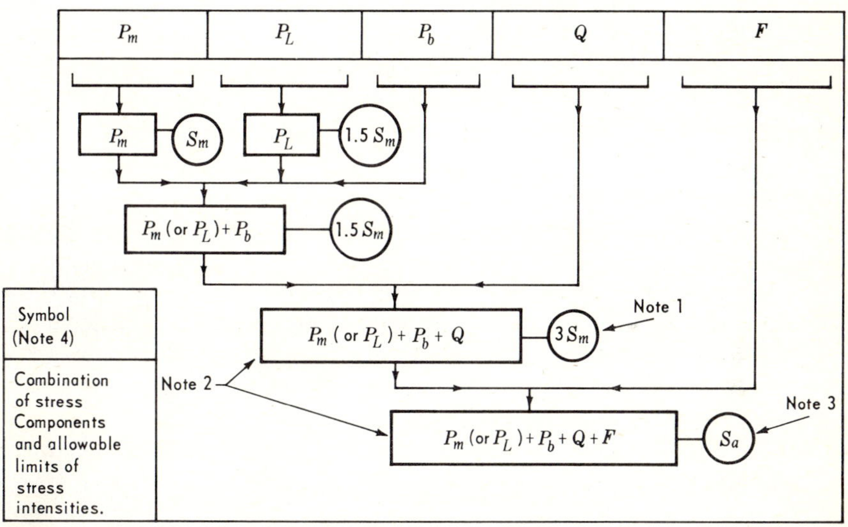

Note 1: Consult code for recommended values of S_m.
Note 2: The stresses in category Q are the parts of the total stress that are produced by thermal gradients, structural discontinuities, etc., and do not include primary stresses which may also exist at the same point. If the stress in category F is produced by a stress concentration, the quantity F is the additional stress produced by the notch, over and above the nominal stress. For example, if a plate has a nominal stress intensity S and has a notch with a stress-concentration factor K, then $P_m = S$, $P_b = 0$, $Q = 0$, $F = P_m(K - 1)$, and the peak stress intensity equals $P_m + P_m(K - 1) = KP_m$.
Note 3: S_a is obtained from the fatigue curves. Consult code.
Note 4: The symbols P_m, P_L, P_b, Q, and F do not represent single quantities but rather sets of six quantities representing the six stress components σ_t, σ_a, σ_r, τ_{ta}, τ_{ar}, and τ_r.
* Primary: produced by mechanical loads only.

Fig. 11-7 Stress categories and limits of stress intensity.

shear stress; in other words, it is the difference between the algebraically *largest* principal stress and the algebraically *smallest* principal stress at a given point. The *membrane stress* is the component of normal stress that is uniformly distributed and is equal to the average value of stress across the section thickness under consideration. *Stress intensity* may be obtained by the following procedure:

I. At a point in the cylinder under consideration, calculate the normal stresses σ_t, σ_r, and σ_a (in the tangential, radial, and axial directions, respectively) and the shearing stresses τ_{ta}, τ_{ar}, and τ_{rt}. The stress components are calculated for each type of loading and assigned to one or a group of the following categories:

A. General primary membrane stress P_m is a normal or shear stress developed by the applied loading that is necessary to satisfy the simple laws of equilibrium of external and internal forces and moments. It is a stress so distributed in the structure that no redistribution of load occurs as a result of yielding. An example is the membrane stress due to internal pressure (or to distributed live load) in a cylindrical or spherical vessel.

B. Local primary membrane stress P_L is the stress produced by pressure or other mechanical loading which results in excessive distortion in the transfer of the load to other regions of the structure. An example is the stress produced in a shell by an external load and a moment at a permanent support or at a nozzle connection.

C. Primary bending stress P_b is a bending stress due to imposed loading.

D. Secondary stress Q is a normal or shearing stress resulting from a constraint of adjacent parts or by self-limiting or, in general, local yielding and minor distortion. General thermal stress (due to a temperature gradient in a shell) and bending stresses at a gross structural discontinuity are examples.

E. Peak stresses F are the increment of stress that may be superimposed upon the primary or secondary stress owing to stress concentration or to a localized thermal stress. They do not cause marked distortion but may be sources of fatigue cracks or brittle fractures. Examples of this stress are the thermal stress in the austenitic steel cladding of a carbon-steel vessel and the thermal stress in the wall of a vessel, due to a rapid change in temperature of the enclosed fluid or to the stress at a localized structural discontinuity.

II. For each rectangular box of Fig. 11-7, calculate the algebraic sum of σ's and τ's that result from the various types of loads. This yields a set of six stress components in each box.

III. From the stress components in the t, a, and r directions, calculate the principal stresses σ_1, σ_2, and σ_3. (In many cases the axes may be such that the shear stresses are zero and σ_1, σ_2, and σ_3 are identical to σ_t, σ_a, and σ_r.)

IV. Calculate the stress differences S_{12}, S_{23}, and S_{31} from the following expressions:

$$S_{12} = \sigma_1 - \sigma_2$$
$$S_{23} = \sigma_2 - \sigma_3$$
$$S_{31} = \sigma_3 - \sigma_1$$

The stress intensity S is the largest absolute value of S_{12}, S_{23}, and S_{31}.

V. The stress intensity calculated as described in step IV from the stresses in each rectangle of Fig. 11-7 must not exceed the allowable values, S_m, shown in the circle adjacent to the rectangle in Fig. 11-7.

The design stress-intensity values S_m, corresponding to various temperatures, are presented in tables contained in the code. In establishing these stress values the following basis was used:

For ferritic steels at a particular temperature the design stress is the lowest of the following:

1. One-third of the specified minimum tensile strength
2. One-third of the tensile strength at temperature
3. Two-thirds of the specified minimum yield strength
4. Two-thirds of the yield strength at temperature

The design stress for austenitic steels is the lowest of these:

1. to 3. Same as the respective items 1 to 3 above
4. 90 percent of the yield strength at temperature but not to exceed two-thirds of the specified minimum yield strength

Example 11-1 Determine the wall thickness for a closed-end cylindrical pressure vessel of 8-in. internal diameter and subjected to a pressure of 1,500 psi and external and internal temperatures of 600 and 500°F, respectively. Assume that bending stresses and thermal stresses due to irradiation may be neglected and that the shell material is carbon steel SA-201, type A.

The general primary membrane stress intensity P_m:

$$S = \frac{p_i r_i}{t} + \frac{p_i}{2}$$

where r_i = internal radius, in.
p_i = internal pressure, psi
t = thickness, in.
S = stress intensity, psi
S_m = design stress, psi
S_m = 14,800 psi, corresponding to maximum metal temperature of area = 600°F

where $p_i r_i / t$ is the hoop (tangential) stress for a thin-walled cylinder, and $-p_i/2$ is the average radial stress.

$$\text{Therefore } t = \frac{p_i r_i}{S_m - 0.5p_i}$$

$$t = \frac{(1{,}500)(4)}{14{,}800 - 750}$$

or $t = 0.427$ in.

The maximum value of the primary plus secondary stress intensity due to pressure occurs at the inside surface and equals

$$S = \frac{2p_i Y^2}{Y^2 - 1} \qquad \text{Same as Eq. (11-16)}$$

in which $Y = r_o/r_i$ and r_o = outer radius, in.

$$Y = \frac{4.427}{4.000} = 1.11$$

$$S = \frac{(2)(1{,}500)(1.23)}{0.23}$$

$$= 16{,}100 \text{ psi} < 1.5S_m$$

The secondary stress intensity:

Calculate the thermal stresses, assuming steady-state heat transfer; maximum value at $r = r_1$.

Using Eqs. (11-21) and (11-22),

$$S_a = s_t = \frac{\alpha E(T_i - T_o)}{2(1 - \nu)\log(r_o/r_i)}\left(1 - \frac{2r_o^2}{r_o^2 - r_i^2}\log\frac{r_o}{r_i}\right)$$

$$= \frac{(8.36)(25.7)(-100)}{2(0.70)\log 1.11}\left[1 - \frac{2(1.11)^2}{(1.11)^2 - 1}\log 1.11\right]$$

$$= 17{,}100 \text{ psi}$$

$$s_r = 0$$

where $\alpha = 9.36 \times 10^{-6}$
$E = 25.7 \times 10^6$ psi
$\nu = 0.30$

Therefore, the total of primary and secondary membrane stress intensity
$= 16{,}100 + (17{,}000 - 0)$
$= 33{,}200$ psi

which is less than

$3S_m = (3)(14{,}800) = 44{,}400$ psi

On this basis, the design is satisfactory.

12
Translation Screws

12-1 INTRODUCTION

Translation screws are used to move machine parts against resisting forces, for instance, in a screw-operated tensile-testing machine, jack, press, or lead screw of a lathe. The usual source of power is a rotating shaft or crank. The design of a translation screw requires considerations of strength of the screw in tension or compression, shearing of the screw itself or of the threads, wearing of the threads, and power requirements. The latter consideration involves frictional resistance, which may account for a large proportion of the power supplied to the screw.

In most translation screws, the nut has axial motion against the resisting axial force while the screw rotates in its bearings. In some screws, however, the screw rotates and moves axially against the resisting force while the nut is stationary, and in others the nut rotates while the screw moves axially with no rotation.

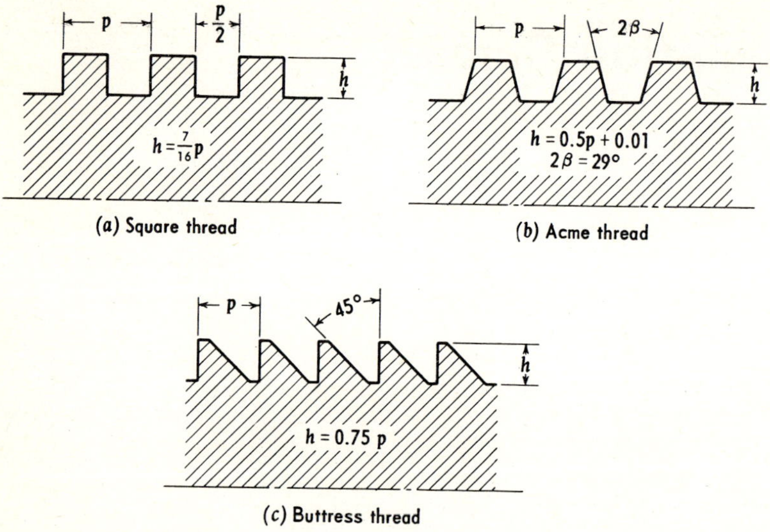

Fig. 12-1 Forms of translation-screw threads.

12-2 FORMS OF THREADS

Square threads This thread has the proportions shown in Fig. 12-1(*a*). The square thread is more efficient than angular types, such as V or acme threads; but it is more difficult to cut, and it is not so adaptable to a split nut; also it has no means of adjustment for wear. A common proportion of square-threaded screws is the Sellers' square thread, as given in Table 12-1.

Acme thread This thread, which is a form of the V or angular thread, has the proportions shown in Fig. 12-1(*b*) and is used where a split nut is required and where provision must be made to take up wear, as in the lead screw of a lathe. Wear may be taken up by means of an adjustable split nut. Proportions of the acme thread are given in Table 12-2. The efficiency of the acme thread is slightly less than that of a square thread.

Buttress thread This thread has the form shown in Fig. 12-1(*c*) and is used for translation under load in one direction only. This thread combines the higher efficiency of the square thread and the ease of cutting and the adaptability to a split nut of the acme thread; furthermore it is stronger than other forms because of the greater thickness at the base of the thread. The angle of the load-carrying face of the thread (0° in

Table 12-1 Sellers' square threads

Diameter, in.	*Threads per in.*	*Root area, in.²*
1/4	10	0.0207
5/16	9	0.0375
3/8	8	0.0555
7/16	7	0.0767
1/2	6 1/2	0.1049
9/16	6	0.1364
5/8	5 1/2	0.1709
11/16	5	0.2063
3/4	5	0.2597
13/16	4 1/2	0.3000
7/8	4 1/2	0.3638
15/16	4	0.4058
1	4	0.4804
1 1/8	3 1/2	0.6013
1 1/4	3 1/2	0.7854
1 3/8	3	0.9201
1 1/2	3	1.1462
1 5/8	2 3/4	1.3414
1 3/4	2 1/2	1.5394
1 7/8	2 1/2	1.8265
2	2 1/4	2.0422
2 1/4	2 1/4	2.7245
2 1/2	2	3.3410
2 3/4	2	4.2000
3	1 3/4	4.9087
3 1/4	1 3/4	5.9396
3 1/2	1 5/8	6.8930
3 3/4	1 1/2	7.8853
4	1 1/2	8.8434

Fig. 12-1) is actually 7° (counterclockwise in Fig. 12-1) as a manufacturing tolerance to avoid concentration of the nut force at the top of the tooth.[1] Although this thread has limited use for power transmission,[2] it is used in some cases as a screw fastening.

Multiple threads Translation screws with multiple threads, such as double, triple, etc., are employed when it is desired to secure a large lead with fine threads or high efficiency. An application is found in high-speed actuators.

[1] See USAS B1.9–1953.
[2] G. H. Atwood, A New Type Screw-luffing Crane for Shipbuilding, *Mech. Eng.*, September, 1944, p. 569.

Table 12-2 Acme screw threads

Threads per in.	*Depth of thread, in.*	*Thickness at root, in.*
1	0.5100	0.6345
1½	0.3850	0.4772
2	0.2600	0.3199
3	0.1767	0.2150
4	0.1350	0.1625
5	0.1100	0.1311
6	0.0933	0.1101
7	0.0814	0.0951
8	0.0725	0.0839
9	0.0655	0.0751
10	0.0600	0.0681

12-3 EFFICIENCY OF SCREWS

Square threads Consider that the nut of a screw is moved against an axial load by the rotation of the screw. The load on the nut will be transferred to the screw as a distributed load on the surface of the threads in contact. For the purpose of analysis, this distributed load may be assumed to be concentrated at a point on the mean circumference of the thread, as at o in Fig. 12-2. The following notation will be used:

Let Q = axial load, lb
d = diameter of mean helix, in.
α = lead angle
ϕ = friction angle
2β = included thread angle [see Fig. 12-1(b)]

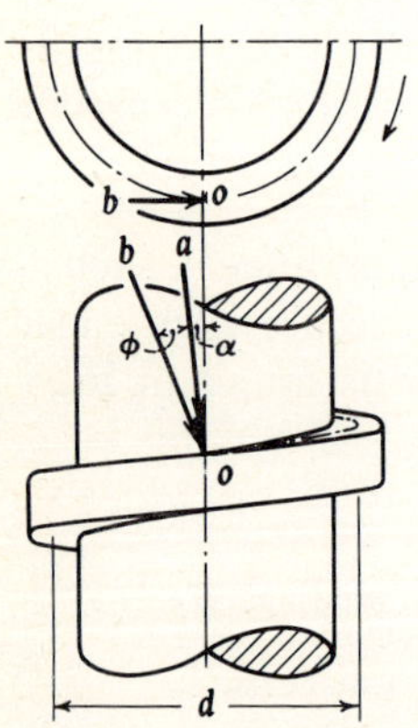

Fig. 12-2 Forces on square-threaded screws.

f = coefficient of thread friction = $\tan \phi$
L = lead of threads, in.
T = torque required to overcome thread friction and to move load, lb-in.
T_o = torque required to move load, neglecting friction, lb-in.
e = efficiency of screw

Under static conditions, the direction of the load on the thread will be normal to the thread surface, as along *ao* in Fig. 12-2. When the screw rotates so that the nut is moved *against* its external load Q, the line of action of *ao* will be rotated through the angle of friction ϕ to *bo*, as shown in the figure.

For equilibrium of forces, the component of *bo* parallel to the axis of the screw is

$$Q = bo \cos(\alpha + \phi)$$

The component of *bo* at right angles to the axis of the screw is

$$F = bo \sin(\alpha + \phi)$$

Hence $F = Q \tan(\alpha + \phi)$

and
$$T = F\frac{d}{2} = \frac{Qd}{2}\tan(\alpha + \phi) \tag{12-1}$$

$$= \frac{Qd}{2}\frac{\pi f d + L}{\pi d - fL} \tag{12-1a}$$

To obtain an expression for the efficiency of the screw, let T_o represent the torque required to raise the load, assuming that friction is not present. Thus from Eq. (12-1)

$$T_o = \frac{Qd}{2}\tan\alpha$$

and the efficiency

$$e = \frac{T_o}{T} = \frac{\tan\alpha}{\tan(\alpha + \phi)} = \frac{QL}{2\pi T} \tag{12-2}$$

Angular, or V, thread Since the normal loads on the thread of an angular, or V, type is oblique to the axial load, the normal load will be greater than that for a square thread. Hence, the frictional force will be correspondingly greater, and the friction terms in Eqs. (12-1) and (12-2) should be divided by $\cos\beta$, or

$$T = \frac{Qd}{2}\frac{\pi f d \sec\beta + L}{\pi d - fL\sec\beta} \tag{12-3}$$

and
$$e = \frac{\tan\alpha[1 - (f\sec\beta\tan\alpha)]}{\tan\alpha + (f\sec\beta)} \tag{12-4}$$

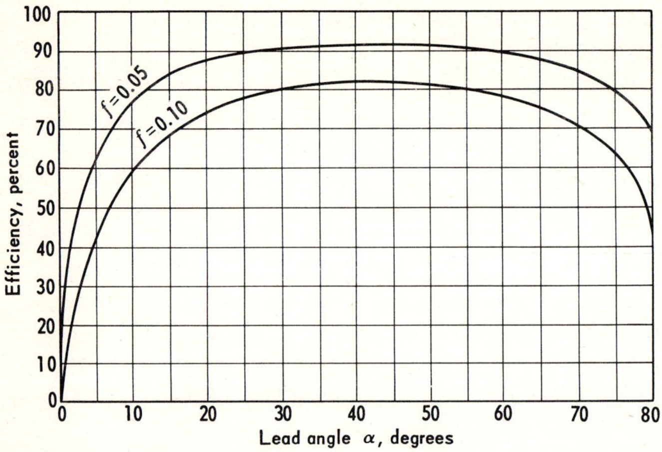

Fig. 12-3 Efficiency of screws.

Reverse motion If the rotation of the screw moves the nut in the *same* direction as the load, as in the case of a screw jack in lowering the load, the vector *bo* in Fig. 12-2 will make an angle with the axis of the screw that is equal to $\alpha - \phi$. The equations for the torque and efficiency as derived for direct motion will be altered. The torque required to lower a load for a square-threaded screw becomes

$$T = -\frac{Qd}{2}\tan(\alpha - \phi) = \frac{Qd}{2}\frac{\pi fd - L}{\pi d + fL}$$

It may be noted from the above equation that, if $\phi < \alpha$, the torque T required to lower the load is negative; i.e., an effort must be applied to the screw to resist the tendency of the load to descend.

If, however, $\phi > \alpha$, the torque T will be positive, indicating that an effort must be applied to lower the load. Such a screw is known as a "self-locking" screw.

Efficiency versus lead angle It is evident that the efficiency of a square-threaded screw depends on the lead angle α and the friction angle ϕ. Figure 12-3 shows the variation of the efficiency of a square-threaded screw for direct motion (raising the load) with the mean lead angle α.

It may be noted that the efficiency reaches a maximum value for α that is equal to 45° approximately.

A qualitative explanation of the increase in efficiency to a maximum value followed by a decrease as α increases may be made by considering that the work of thread friction is equal to the frictional force along the thread multiplied by the distance through which this force acts. For a low angle, the slope of the thread is small, and the distance through which the friction force acts is great compared with the lead of the thread. Hence, the work of thread friction will be great as compared with the useful work, and a low efficiency will result.

For a large lead angle, say 70°, the normal thread force becomes large, and likewise the force of friction and the work of friction become large as compared with the useful work. Between the two extremes that have been discussed there is a value of lead angle that yields a minimum value of work of friction, as compared with the useful work, and hence a maximum efficiency also.

It should be remembered, however, that, if the screw is required to be self-locking, a low lead angle is necessary in order to introduce intentionally a frictional force that is sufficient to prevent reverse motion.

12-4 COLLAR FRICTION

The axial force produced on a screw by the nut must be resisted by a collar or its equivalent. This action and reaction is illustrated in Fig. 12-4.

To determine the total torque required to turn the screw, the frictional torque of the collar must be added to the screw-thread torque, as given by Eq. (12-1) or Eq. (12-3).

The frictional torque of the collar may be determined by assuming

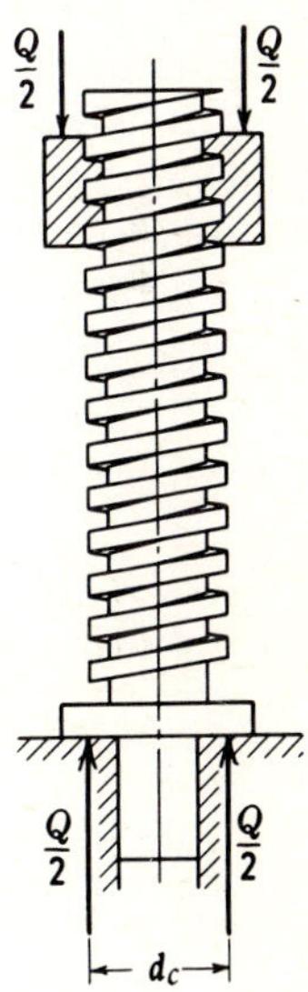

Fig. 12-4 Axial forces on screw. Screw and collar for jack.

that the frictional force on the collar acts at its mean diameter. Therefore

$$T_c = \frac{fQd_c}{2} \tag{12-5}$$

where T_c = torque of collar friction, in.-lb
Q = axial load on screw, lb
f = coefficient of collar friction
d_c = mean diameter of collar, in.

The efficiency of a screw including both thread friction and collar friction may be obtained by dividing the torque required to move the load, neglecting friction, by the torque required to move the load, including thread and collar friction.

12-5 COEFFICIENT OF FRICTION

Values for the coefficient of friction for the threads of translation screws have been determined by investigations[1] and have been found to depend on the quality of materials, workmanship in cutting the threads, degree of "running-in" of the threads, and lubrication. The value of the coefficient of friction does not vary markedly with different combinations of material, load, or rubbing speed, except under starting conditions.

1. For high-grade materials and workmanship and for well run-in and lubricated threads, the coefficient of friction may be taken as 0.10.
2. For average quality material, workmanship, and conditions of operation, the coefficient of friction may be taken as 0.125.
3. For poor-quality material and workmanship and for newly machined surfaces which are indifferently lubricated and which have slow motion, the coefficient of friction may be taken as 0.15.
4. The coefficient of friction for starting conditions may be taken as $1\frac{1}{3}$ times the value for running conditions.
5. The coefficient of collar friction may be taken as the same as for thread friction.

Example 12-1 A 10-ton screw jack with a maximum extension of 4 in. is to have double square threads. (*a*) Using an allowable compressive stress of 5,000 psi, determine the size of screw, size of collar, and length of nut required. (The allowable compressive stress in the screw is a low value in order to give a screw which will accommodate a well-proportioned thread.) (*b*) Determine the torque required to raise the load and also the efficiency of the jack.

[1] Ham and Ryan, An Experimental Investigation of Thread Friction of Screw Threads, *Univ. Illinois Bull.* 247, 1932.

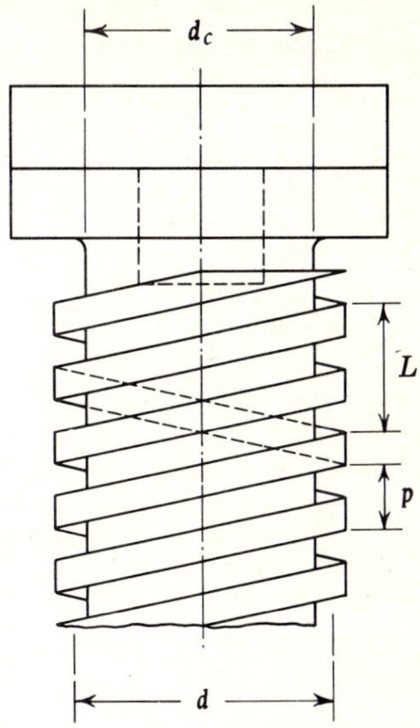

Fig. Example 12-1

Solution (*a*) Dimensions of screw, collar, and nut: The root area of the screw equals

$$\frac{Q}{s_c} = \frac{20{,}000}{5{,}000} = 4 \text{ in.}^2$$

From Table 12-1, a $2\frac{3}{4}$-in. screw with 2 threads per inch is satisfactory. The pitch of the threads is $\frac{1}{2}$ in. and the lead (double thread) is 1 in.

The outside diameter of the collar is usually larger than the outside diameter of the screw to give a reasonable bearing pressure on the collar. Assuming an outer diameter for the collar equal to $3\frac{1}{2}$ in. with a pilot pin 1 in. in diameter, the collar bearing pressure is

$$p_c = \frac{4Q}{\pi(d_o^2 - d_i^2)} = \frac{4 \times 20{,}000}{\pi(3.5^2 - 1^2)} = 2{,}270 \text{ psi}$$

This value is in the safe range as compared with values in Table 12-3 for screw threads. If trouble due to galling is expected, a thin bronze washer may be interposed between the rubbing parts of the collar.

The length of the nut may be determined from the number of threads required to limit the bearing pressure on the threads to a permissible value.

The area of one thread is equal to

$$\pi dh = \pi \times 2.5 \times 0.25 = 1.97 \text{ in.}^2$$

where h is the depth of the thread.

Using an allowable thread bearing pressure $p = 2{,}500$ psi, the number of threads required is

$$\frac{20{,}000}{2{,}500 \times 1.97} = 4.06 \text{ threads}$$

Since there are 2 threads per inch, the length of the nut in inches is one-half of 4.06. On the basis of bearing pressure, a 2-in. nut would be satisfactory, but for stability of the screw it is practice to use its length at least equal to the diameter of the thread. Three inches would be a reasonable length for the nut.

(*b*) Torque to raise the load and the efficiency of the jack: The mean diameter of the thread may be found from the figure for the outside diameter of $2\frac{3}{4}$ in. and pitch of $\frac{1}{2}$ in. Considering the thread as square in section, $d = 2\frac{3}{4} - \frac{1}{4} = 2\frac{1}{2}$ in.

The torque to raise the load and to overcome thread friction, using a coefficient of friction $f = 0.125$ from Art. 12-5, is

$$T = \frac{Qd}{2}\,\frac{\pi fd + L}{\pi d - fL} = \frac{20{,}000 \times 2.5}{2}\,\frac{\pi \times 0.125 \times 2.5 + 1}{\pi \times 2.5 - 0.125 \times 1} = 6{,}400 \text{ in.-lb}$$

The torque for the collar is

$$T_c = \frac{fQd_c}{2} = \frac{0.125 \times 20{,}000 \times 2.25}{2} = 2{,}810 \text{ in.-lb}$$

The total torque is

$$6{,}400 + 2{,}810 = 9{,}210 \text{ in.-lb}$$

The efficiency of the screw and collar is

$$\frac{QL}{2\pi T} = \frac{20{,}000 \times 1}{2\pi \times 9{,}210} = 0.35, \text{ or } 35 \text{ percent}$$

12-6 STRESSES IN SCREWS

Tensile or *compressive stresses* in the body of a screw due to the axial load may be determined by dividing the axial load by the minimum cross-sectional area of the screw, unless the screw is long and slender and subjected to a compressive load in which case column formulas must be used.

The torsional stress in the screw owing to the torsional moment

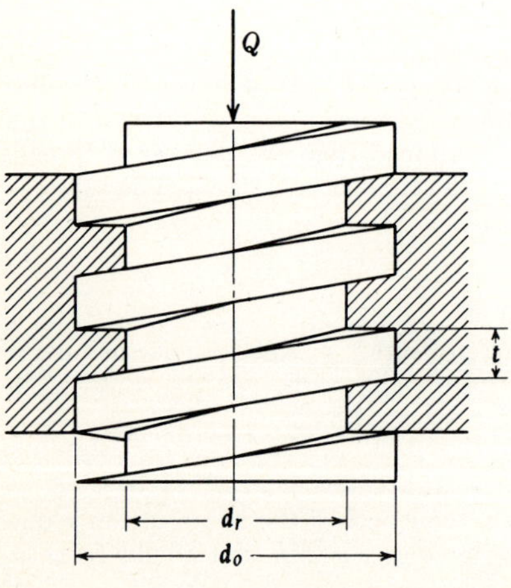

Fig. 12-5 Screw-thread dimensions.

Table 12-3 Safe bearing pressures for screws

Type	*Material*		*Safe bearing pressure, psi*	*Rubbing speed*
	Screw	*Nut*		
Hand press	Steel	Bronze	2,500–3,500	Low speed, well lubricated
Jackscrew	Steel	Cast iron	1,800–2,500	Low speed < 8 fpm
Jackscrew	Steel	Bronze	1,600–2,500	Low speed < 10 fpm
Hoisting screw	Steel	Cast iron	600–1,000	Medium speed 20–40 fpm
Hoisting screw	Steel	Bronze	800–1,400	Medium speed 20–40 fpm
Lead screw	Steel	Bronze	150–240	High speed > 50 fpm

required to rotate the screw may be determined by considering the minimum cross section of the screw.

Shearing stress and *bearing pressure* are generally determined by assuming that the loads are uniformly distributed on the threads in contact.

Let Q = axial load on screw, lb (see Fig. 12-5)
d_o = major diameter of threads, in.
d_r = minor diameter of threads, in.
t = width of thread, in.
n = number of threads in engagement
s_s = shearing stress, psi
s_b = bearing pressure, psi

The shearing stresses for the threads of the screw and of the nut are, respectively, equal to

$$s_s\ (\text{screw}) = \frac{Q}{n\pi d_r t}$$

$$s_s\ (\text{nut}) = \frac{Q}{n\pi d_o t}$$

The bearing pressure on the threads is equal to

$$s_b = \frac{4Q}{n\pi(d_o^2 - d_r^2)}$$

The bearing pressure is limited by lubrication conditions. Table 12-3 gives some limiting values of bearing pressures.

12-7 OTHER TYPES OF SCREWS

Differential screws When slow advance or fine adjustment is necessary and it is not advisable to use fine threads, a differential screw may

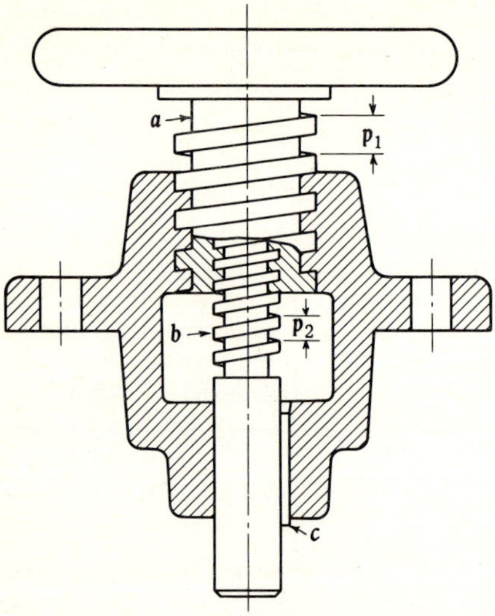

Fig. 12-6 Differential screw.

be used. As shown in Fig. 12-6, this screw is composed of two component screws. One screw a is keyed to the handwheel and has threads of pitch p_1. The other screw b is threaded into the first one but is prevented from rotating by the feather key c. The threads on this screw have a pitch p_2, which is smaller than p_1.

If the handwheel is turned right-handed one turn, the screw a will move downward in the frame a distance p_1. As a rotates, the screw b will be screwed upward into a and will move relative to a a distance p_2. Thus the total movement downward of b relative to the frame will be $p_1 - p_2$. It is apparent that, if the two pitches p_2 and p_1 are equal, the screw b will not move; and if p_2 is made less than p_1, as discussed above, the motion of b will equal their difference, which can be made as small as required.

If the hand of one of the threads is reversed, the motion of b will be equal to $p_1 + p_2$. This type of screw is known as a "compound screw."

Ball-bearing screw and nut A type of screw has been developed which interposes steel balls between properly formed threads on the screw and nut, and thus rolling friction is substituted for sliding friction.[1] Figure 12-7 shows a ball-bearing screw and nut.

The threads in both screw and nut are approximately semicircular in section. When the nut is translated as the screw rotates, the balls run out of the thread on the nut and are returned to the other end of the nut by a recirculating groove.

[1] R. K. Allan, "Rolling Bearings," p. 20, Sir Isaac Pitman & Sons, Ltd., London, 1945; Ball Bearing Screw and Nut, Saginaw Steering Gear Division, General Motors Corp., Saginaw, Mich.

Efficiencies as high as 90 percent are given for this screw, with corresponding increase in life over conventional screws. Applications to date include steering gear for automotive vehicles[1] and power actuators.

Roton bearing screws Instead of using balls for the rolling-contact elements, as in the ball-bearing screw and nut, the Roton bearing screw uses rollers to provide rolling contact between the screw and nut (outer race). As shown in Figs. 12-8 and 12-9, the cage contains three rollers spaced 120° apart. The rollers engage the threads of the triple-threaded screw and roll on the inner surface of the nut. The mechanical efficiency is high as compared with linear actuators of the sliding-contact type.

This unit is used in two arrangements. One has a positive-drive assembly in which the cage is fixed, and as the screw is rotated it is translated axially a distance equal to the lead of the screw threads. The nut serves only to contain the rollers and carries no load as it rotates freely in the direction opposite to that of the screw.

In the second assembly, the nut is fixed, and by rotating the screw there is a planetary action so that the cage rotates in the same direction as the screw but at lesser speed, giving a differential traverse of the screw, a forward one due to the rotation of the screw and a backward one due to the rotation of the cage. Thus, for a certain torque input, the translation speed is less and the axial force is greater by a ratio of 3:2 than in the positive-drive assembly.

In the planetary-drive assembly, there is protection at the end of the

[1] Francis W. Davis, Power Steering for Automotive Vehicles, *S.A.E. Journal*, vol. 53, no. 4, p. 239, April, 1945.

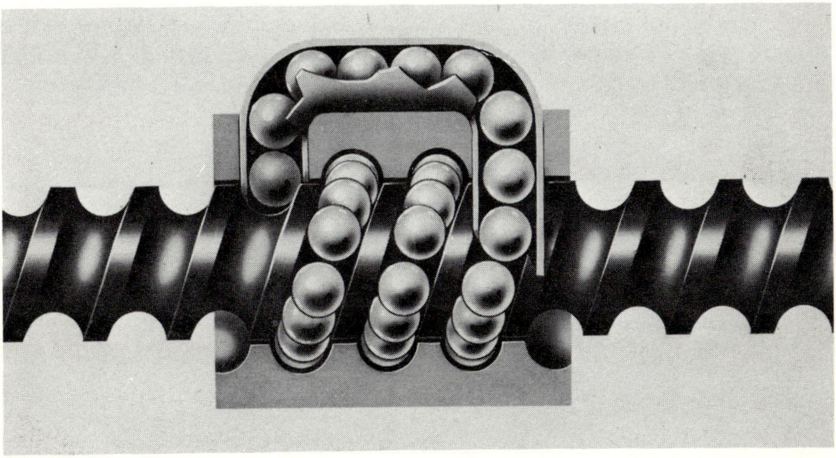

Fig. 12-7 Ball-bearing screw and nut. (*Courtesy of Saginaw Steering Gear Division, General Motors Corporation.*)

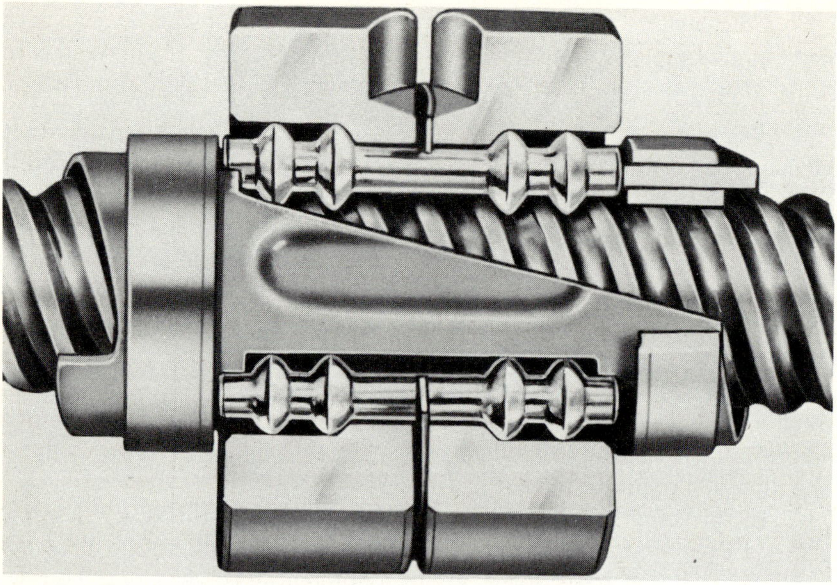

Fig. 12-8 Roton bearing-screw assembly. (*Courtesy of Roton Products Division, The Anderson Company.*)

stroke due to a "free-wheeling" action provided by sliding at the contact of the rollers, thus obviating the use of limit switches. This feature is possible at the end of the stroke for the positive-drive assembly but is not recommended. Both assemblies can be used to transmit rotary to linear or linear to rotary motion.

Some applications of this actuator are automotive brakes, automobile seat and window adjusters, hospital beds and table adjusters, and machine-tool traverse equipment.

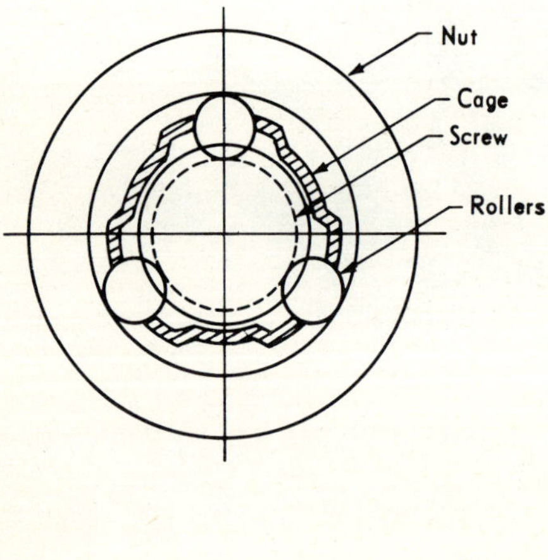

Fig. 12-9 Roton bearing screws.

13
Shafting

13-1 GENERAL CONSIDERATIONS

The design of a shaft may require the interrelated considerations of a number of factors, such as the following: material and heat treatment; strength for power and loading requirements; stiffness as affecting, for instance, bearing performance, gear operation, timing and critical speeds; weight and space limitations; and stress concentration. Some general design methods that involve these considerations are discussed in the following articles.

13-2 TERMINOLOGY

In machinery, the general term "shaft" refers to a member, usually of circular cross section, which supports gears, sprockets, wheels, rotors, etc., and which is subjected to torsion and to transverse or axial loads, acting singly or in combination.

Specifically a *shaft* is a member which rotates and which is subjected

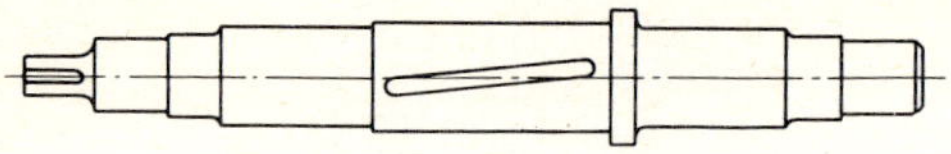

Fig. 13-1 Stepped shaft for induction motor.

to torsion accompanied by transverse or axial loads; an *axle*, with few exceptions, is a member which is subjected to transverse loads but not torsion; a *spindle* is a short shaft.

Most shafts are not of uniform diameter but are *stepped* to provide shoulders for locating gears, pulleys, bearings, or other attached or contacting parts. In Fig. 13-1 is shown a stepped shaft for a motor, with shoulders for locating the rotor, ball bearings, gear, etc.

13-3 MATERIALS

The most common material for shafting is mild steel. For high-strength requirements, an alloy steel, such as nickel, nickel-chromium, or chrome-vanadium steel, is used.

Commercial shafting is made of low-carbon steel similar to SAE 1015 formed by hot rolling and finished to size by cold drawing or by turning and grinding. Cold drawing produces a stronger shaft than hot rolling, but with higher residual stresses. The residual stresses may cause distortion of the shaft when it is machined, especially when slots or keyways are cut. Shafts for special purposes may be forged.

13-4 COMMERCIAL SIZES OF SHAFTING

Formerly shafting was made of bars rolled to nominal dimensions of even quarter-inch intervals and then finished $\frac{1}{16}$ in. under the nominal size. Sleeve bearings are available to accommodate shafts $\frac{1}{16}$ in. under the nominal size. Shafting may be obtained in the sizes given in Tables 13-1 and 13-2. Dimensions are in inches.

Table 13-1 Transmission shafting

Diameter	*Increment*
$1\frac{5}{16}$–$2\frac{7}{16}$	$\frac{1}{4}$
$2\frac{7}{16}$–$5\frac{15}{16}$	$\frac{1}{2}$

Table 13-2 Machinery shafting

Diameter	*Increment*
½–2½	1/16
2½–4	⅛
4–6	¼

13-5 DESIGN CONSIDERATIONS

The size of a shaft for a particular application may be determined on the basis of strength or both strength and rigidity.

In designing a shaft on the basis of *strength*, it is necessary to consider the following: type of loading, i.e., static, shock, or cyclic; the weakening effects at points of stress concentration due to keyways and shoulders; the combination of loading, such as bending and torsion. The size of the shaft must be sufficient to prevent the induced stresses from exceeding the allowable stress for the material.

In most cases the *rigidity* of the shaft is an important design feature. The *twisting* of the shaft may be limited in order to provide accurate prescribed timing or motions, as in the camshaft of an internal-combustion engine. *Transverse deflections* may be limited, for instance, to maintain proper bearing clearances or gear-tooth alignment. Both torsional and transverse rigidity are important in vibration.

13-6 DETERMINATION OF SHAFT SIZES ON THE BASIS OF STRENGTH

The action of the loads on a shaft is generally one of the following: (1) torsion, (2) bending, (3) torsion combined with bending, and (4) torsion combined with axial tension or compression.

Torsion For a simple twisting moment T on a solid circular shaft of diameter d, the maximum shear stress is

$$s_s = \frac{16T}{\pi d^3} \tag{13-1}$$

By assuming that failure is based on the maximum-shear theory, the maximum shear stress must not exceed the allowable shear stress, $s_{s,\text{all}}$. Then

$$\frac{\pi d^3}{16} = \frac{T}{s_{s,\text{all}}} \tag{13-2}$$

Bending For a simple bending moment M on a solid circular shaft of diameter d, the maximum bending stress is

$$s_t = \frac{32M}{\pi d^3} \tag{13-3}$$

By substituting the allowable tensile stress s_{all} for s_t,

$$\frac{\pi d^3}{32} = \frac{M}{s_{all}} \tag{13-4}$$

(See Appendix IV for values of section moduli for standard-size shafts.)

Torsion combined with bending Most rotating shafts carry gears, pulleys, sprockets, or sheaves that cause bending of the shaft in addition to torsion. An effort should be made to mount the gears, pulleys, etc., as near to the bearings as possible in order to reduce the bending moment.

The design of shafting made of ductile material is based on the maximum-shear-stress theory; hence it is necessary to determine the maximum combined shear stress in the shaft due to the applied twisting and bending moments. By using the combined stress equation,

$$s_s\,(\max) = \sqrt{s_s^2 + \frac{s_t^2}{4}}$$

and substituting the values of s_s and s_t from Eqs. (13-1) and (13-3), the maximum shearing stress becomes

$$s_s\,(\max) = \frac{16}{\pi d^3}\sqrt{T^2 + M^2}$$

By limiting the maximum shear stress to the allowable shear stress for the material and solving for the section modulus,

$$\frac{\pi d^3}{16} = \frac{\sqrt{T^2 + M^2}}{s_{s,all}} \tag{13-5}$$

The term $\sqrt{T^2 + M^2}$ is called the *equivalent twisting moment* and is defined as the fictitious twisting moment that will induce the same maximum shearing stress in the shaft as the combined action of the actual twisting moment and the actual bending moment.

Torsion combined with axial tension or compression Some shafts, such as propeller shafts, are subjected to torsion combined with direct axial loads. The shear stress due to torsion for a solid circular shaft is given by Eq. (13-1). The direct stress due to an axial load P is

$$s = \frac{4P}{\pi d^2}$$

These stresses may be combined by the use of the combined stress equation as

$$s_s\,(\max) = \frac{16}{\pi d^3}\sqrt{T^2 + \left(\frac{Pd}{8}\right)^2} \tag{13-6}$$

· By substituting the allowable shear stress for s_s (max), the above equation may be solved by trial. If the axial load is a compressive load, it should not exceed the critical buckling load.

For shafts on which the torque and bending moment are cyclic, stress concentration should be considered. If the fatigue-notch factor for shear stress (K_{tsf}) is applied to Eq. (13-1), and the combined factor for normal stress ($K_{t'f}$) is applied to Eq. (13-3) and both substituted in Eq. (13-5), it becomes

$$\frac{\pi d^3}{16} = \frac{\sqrt{(K_{tsf}T)^2 + (K_{t'f}M)^2}}{s_{s,\text{all}}} \tag{13-5a}$$

In the use of Eq. (13-5*a*), it should be realized that it is a design equation for *strength*. In the design of most shafts in machinery, however, *stiffness* of the shaft usually requires a larger shaft than strength. Since most shafts are not uniform in diameter, it is usually impossible to design them directly on the basis of stiffness. Hence, it is common practice to design such shafts on the basis of strength, using allowable stresses as determined from analyses of similar existing shafts which have performed satisfactorily in service. The allowable stresses so determined usually are low in value and the shafts are "overdesigned" for strength but have the required stiffness characteristics. Allowable stresses in shear equal to 5,000 to 8,000 psi are common in the design of such shafts.

In applications where stiffness is critical, it is usually satisfactory to use Eq. (13-5) with allowable stress s_s in the range 5,000 to 8,000 psi. The allowable stress for shafts carrying the following units of power transmission may be as follows: hypoid, helical, and bevel gears, 5,000 psi; spur gears, 5,500 to 6,000 psi; chains, 7,000 psi; belt drives, 8,000 psi.

After the shaft diameter is determined by Eq. (13-5), the maximum stresses at critical sections in the shaft may be investigated by the use of Eq. (13-5*a*), and the value of the maximum stress, usually at a fillet, compared with the endurance limit of the material. In power-transmission shafts, the torsional moment is usually in one direction and nearly steady so that K_{tsf} can be taken as unity.

If the maximum stress from Eq. (13-5*a*) is too high, the material or its heat treatment may be modified accordingly. However, a change in class of steel or its heat treatment will not affect the deflection characteristics of the shaft. This procedure is illustrated in Example 13-1.

An uncertainty in using Eq. (13-5*a*) is that available values for stress-concentration factors are determined separately for torsion and for bending, and it is not usually known whether the factors so determined apply to combined stresses due to the simultaneous action of torsion and bending. However, if stiffness is not an important requirement, Eq. (13-5*a*) may be used, with an allowable stress determined by dividing the endurance limit in shear by an appropriate value for the factor of safety.

13-7 ALLOWABLE STRESSES

The allowable stresses for the design of shafting should be chosen with consideration for the factors discussed in Chap. 5. Specifically, this bases the allowable stress on the yield point if the loading is static and on the endurance limit if the loading is cyclic.

In many cases of combined loading, one of the loadings may be static and the other repeated. An example of this is a rotating shaft on which are mounted gears transmitting uniform torque in which the bending stresses are completely reversed but the shearing stresses due to the torque may be essentially static. The true allowable stress in this instance should be based on an endurance limit determined from tests simulating the exact combination of stresses that exist in the shaft. Only rarely are such values available. To consider both of the loadings as completely reversed would be on the side of safety. Particular effort should be made to reduce stress concentrations to a minimum in shafting. This is especially necessary in shafts subjected to vibration, either self-induced or parasitic. Such vibration causes fatigue stresses which are superimposed upon those due to power loads.

Rough surfaces should be eliminated. Press and shrink-fit collars and hubs should be carefully considered and their effects minimized, as suggested in Chap. 4.

Example 13-1 The shaft of a speed reducer is mounted on ball bearings A and B, shown opposite. The stub end of the shaft is fitted with a flexible coupling. coupling. The gear is 5 in. in diameter and 5 hp are transmitted at 120 rpm. Assuming an allowable stress s_{sa} = 5500 psi, determine the size of shaft.

Solution An inspection of the sketch shows that the bending moment on the shaft is cyclic and has a maximum value at the midpoint of the gear. The torque on the shaft between that section and the stub end is equal to

$$T = \frac{63{,}030 \text{ hp}}{\text{rpm}} = \frac{63{,}030 \times 5}{120} = 2{,}620 \text{ lb-in.}$$

The tangential force on the gear is

$$F = \frac{2T}{D} = \frac{2 \times 2{,}620}{5} = 1{,}050 \text{ lb}$$

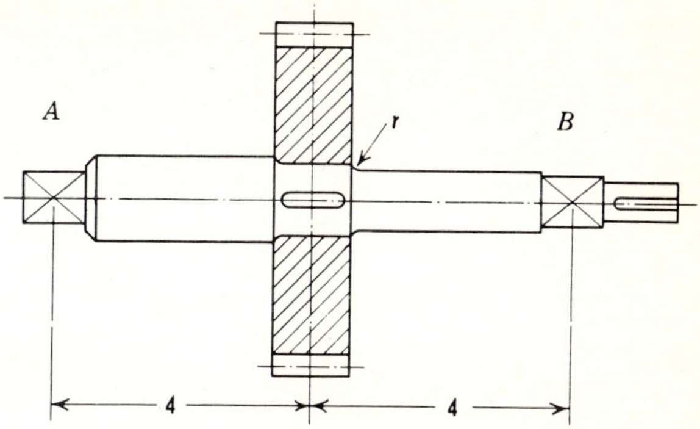

Fig. Example 13-1 Shaft with gear.

The bending moment at the center of the gear may be found equal to 2,100 lb-in. From Eq. (13-5),

$$\frac{\pi d^3}{16} = \frac{\sqrt{T^2 + M^2}}{s_{s,\text{all}}} = \frac{\sqrt{(2{,}620)^2 + (2{,}100)^2}}{5{,}500} = 0.610 \text{ in.}^3$$

From Appendix IV, the required shaft diameter is $1\frac{1}{2}$, for which the value of $\pi d^3/16$ is 0.663 in.^{3}

As an extension to the problem, assume that the diameter of the shaft to the right of the gear is $1\frac{1}{4}$ in. and determine the maximum stress at the fillet ($r = \frac{1}{8}$ in.) by the use of Eq. (13-5*a*); also determine the corresponding factor of safety at this section.

On the assumption that the torsional moment T is steady (not cyclic) we may neglect the effect of stress concentration in torsion at the fillet; that is, K_{tsf} may be taken as unity. This is because stress concentration in ductile material subjected to static loading is not serious; see Art. 4-6.

By making $K_{t'f}$ equal to $K_{t'}$ and solving for the maximum stress, Eq. (13-5*a*) then becomes

$$s_{s(\max)} = \frac{\sqrt{T^2 + (K_{t'}M)^2}}{\pi d^3/16}$$

Using the following values in Fig. X-14, $D/d = 1.2$ and $r/d = 0.1$, the value of $K_{t'}$ is found to be equal to 1.5. Therefore, using $\pi d^3/16$ equal to 0.384 for the $1\frac{1}{4}$-in. shaft at the fillet,

$$s_{s(\max)} = \frac{\sqrt{2{,}620^2 + (1.5 \times 2{,}100)^2}}{0.384} = 10{,}630 \text{ psi}$$

Assuming that the shaft is made of SAE 1020 steel, the following properties may be determined (see Example 7-1).

$s_{\text{ult}} = 62{,}000$ psi (From Appendix V)
$s_e = 25{,}000$ psi (From Appendix VII, with ground finish)
$s_{es} = 0.55 \times 25{,}000$
$= 13{,}750$ psi

The factor of safety, then, is equal to

$$\text{fs} = \frac{s_{es}}{s_{s(\max)}} = \frac{13{,}750}{10{,}630} = 1.29$$

If the factor of safety was found to be less than unity or uncomfortably close to unity, it could be raised by using a steel of higher carbon content or an alloy steel with the appropriate heat treatment. A change in material would not, however, affect the deflection characteristics of the shaft; hence, the nominal diameter of the shaft, 1½ in., would remain unaltered.

13-8 DETERMINATION OF SHAFT SIZES ON THE BASIS OF RIGIDITY

In many machines the sizes of shafts are determined by the limits that are placed on their deflections. Two kinds of rigidity must be considered, *torsional* and *lateral.*

Torsional rigidity The total angle of twist, in radians, for a circular shaft of uniform cross section is given by the equation

$$\theta_r = \frac{TL}{JG}$$

For a solid circular shaft of diameter d, this equation becomes

$$\theta = \frac{584LT}{Gd^4} \qquad (13\text{-}7)$$

where θ = angle of twist, deg
L = length of shaft, in.
T = torque on shaft, lb-in.
G = torsional modulus of elasticity, psi
d = shaft diameter, in.

The permissible amount of twist depends on the particular application. In drive shafts of machine tools, the twist should not exceed 0.08° per ft. In line shafts, 0.75 to 1.0° per ft may be used as limiting values.

Lateral rigidity The diameter of shafts may be determined by permissible lateral deflections as required, for instance, to maintain proper bearing clearances, gear-teeth alignment, and vibration characteristics. The deflections of a shaft of uniform section may be readily found by the use of the appropriate equation from mechanics of materials. If the shaft is of variable cross section, the deflections may be determined from the fundamental equation for the elastic curve of a beam, namely,

$$\frac{d^2y}{dx^2} = \frac{M}{EI}$$

Two integrations of this equation will yield the deflection y. This integration process is not difficult if the moment of inertia of the shaft is uniform and if the distribution of the bending moment along the shaft can be expressed analytically. In most cases, however, neither of these conditions exist. However, a solution may be conveniently reached by use of graphical methods (see Appendix XIII–A).

Alternatively, the deflection y of the shaft at any point x along its centerline may be determined by a digital computer. Equations for the deflections may be written for the four parts of the shaft in terms of the bending moment and x. The bending moments at the junctures of the adjacent parts are equal, and the slopes of the deflection curve dy/dx are equal. The computer program can be written to obtain values for the deflection y at various values of x.

13-9 CODE FOR DESIGN OF TRANSMISSION SHAFTING[1]

The code is based on the failure of ductile ferrous shafting according to the maximum-shear theory. The equation for the outside diameter d is

$$d = \sqrt[3]{\frac{16\sqrt{[K_mM + \frac{1}{8}\alpha Fd(1 + K^2)]^2 + (K_tT)^2}}{\pi s_{s,\,\text{all}}(1 - K^4)}} \tag{13-8}$$

where $s_{s,\,\text{all}}$ = allowable shear stress, psi
K_m = combined shock and endurance factor for bending
K_t = combined shock and endurance factor for torsion
$K = d_i/d$ = ratio of internal to external diameters
α = column-action factor
F = axial tension or compression, lb

[1] A recent committee from ASME indicates that this code is still in use even though it has been withdrawn by the United States of America Standards Institute; however, the code is currently being reevaluated under the sponsorship of ASME.

Table 13-3

	K_m	K_t
Stationary shafts:		
Load gradually applied	1.0	1.0
Load suddenly applied	1.5–2.0	1.5–2.0
Rotating shafts:		
Load gradually applied	1.5	1.0
Load suddenly applied, minor shock	1.5–2.0	1.0–1.5
Load suddenly applied, heavy shock	2.0–3.0	1.5–3.0

This equation considers all the combined actions of loading and is simplified if some of the loads are absent. Hence, if $F = 0$ and $M = 0$, the equation for a shaft in torsion is obtained.

The column-action factor α is given as unity for a tensile load. For a compressive load the following applies:

$$\alpha = \frac{1}{1 - [0.0044(L/\rho)]} \qquad \text{for } \frac{L}{\rho} < 115$$

$$\alpha = \frac{s_y}{\pi^2 nE}\left(\frac{L}{\rho}\right)^2 \qquad \text{for } \frac{L}{\rho} > 115$$

where $n = 1$ for hinged ends and $n = 2.25$ for fixed ends. For ends that are partly restrained, as in bearings, $n = 1.6$. ρ is the radius of gyration.

The allowable shear stresses to be used with the code for commercial-steel shafting are

$$s_{s,\,\text{all}} = 8{,}000 \text{ psi for shafts without keyways}$$
$$s_{s,\,\text{all}} = 6{,}000 \text{ psi for shafts with keyways}$$

13-10 VIBRATION OF SHAFTS

In many rotating shafts, particularly with a mounted disk or rotor, vibrations are set up that may become troublesome or destructive at certain speeds. Frequently violent vibrations are caused by the centrifugal forces on the shaft because of unbalanced gears, pulleys, disks, or other rotating parts (see Chap. 24).

14
Belt Drives and Hoists

14-1 GENERAL CONSIDERATIONS

The part of this chapter that is devoted to belts, Chap. 15 on power-transmission chains, and Chaps. 18 and 19 on gears represent the units of mechanical power transmission. There are also fluid power transmission and electrical power transmission, which are not discussed here; the former is very specialized, and the latter is outside the scope of this book. Other units involved in mechanical power transmission and control are shafts, shaft couplings, clutches, and brakes, which are discussed in other chapters of this book.

Center distance Belts and chains are suitable for long center distances as compared with gears, which may be used for comparatively short center distances. V belts are generally short-center drives.

Velocity ratio Because of the slip and creep of belt drives, the driven pulley will rotate at a speed less than that determined by the use of ratios of pulley diameters. This loss in speed may be compensated for by a

Table 14-1 Comparison of mechanical drives

Type	*Cost*	*Space*	*Limit to power*	*Basis for design*
1. Flat belts	1	Very large	Space	Stretch and slip
2. V belts	3	Large	Space, cost	Stretch and wear
3. Roller chains	5	Medium	Speed, width	Wear of pins, fracture of links
4. Silent chains	7	Medium	Speed, space	Strength
5. Spur gears	15	Small	Space	Strength, wear
6. Helical gears	25	Small	Space	Strength, wear

small change in diameter of one of the pulleys. In chain drives the velocity ratio is not constant during one revolution of the smaller sprocket. In circular gear drives, the velocity ratio is always constant.

Shifting Flat belts with relatively long center distances may be shifted by the use of tight and loose pulleys. Gears may be shifted when provided with feather keys, splines, or clutches.

Maintenance The maintenance of belt drives usually amounts to periodic center-distance adjustment to compensate for stretch of the belt and replacement of the belt when it is no longer serviceable. In chain drives, center-distance adjustment is necessary to compensate for wear of the chain but is generally required at infrequent intervals only. In chain and gear drives, lubrication is an important item in connection with maintenance.

Cost In general, the comparative first cost of the drives under discussion is in the order of flat belts, V belts, roller chains, silent chains, and gears. In the positive drives, i.e., chains and gears, the necessity for greater accuracy in shaft alignment and for provision for lubrication results in further increase in first cost.

A comparison of some aspects of the various mechanical drives is given in Table 14-1. The relative cost figures are approximate and reflect the cost of manufacture, shaft rigidity, and necessity for oiltight cases or housings, with shaft seals for chains and gears.

14-2 MATERIALS

In belt and rope drives, the materials used must be strong, flexible, and durable and must have a high coefficient of friction. The following materials are most commonly used for these drives:

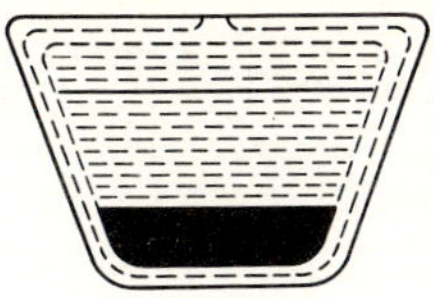

Fig. 14-1 Section of V belt.

Leather Oak-tanned leather is a standard material for flat belts. Chrome leather may be used where a very pliable material is desired.

Fabric and rubber belts Fabric, such as canvas or cotton duck usually impregnated with a filler, is a relatively cheap material and may be used for light power transmission. Such fabric may be vulcanized to form rubber belts for use where exposed to oil or sunlight.

V belts V belts are made of fabric and cords molded in rubber and are generally covered with fabric. As shown in Fig. 14-1, the section that carries the power load is located near the neutral axis of the section where the stresses owing to the bending of the belt around the pulley will not be large. The fabric near the upper surface carries the tension due to bending, and the rubber portion at the bottom is compressed during bending. The belt is vulcanized to form a loop of the exact length required for the particular installation.

In a recent development, the load-carrying element is composed of small endless steel cables, which give high strength and low stretch.

Rope Manila, hemp, and cotton ropes are used occasionally to transmit power. The continuous-rope or the multiple-rope system may be used. The rope sizes vary from $\frac{3}{4}$ to 2 in. in diameter.

Material for pulleys Pulleys are generally made of cast iron, cast steel, or pressed steel. The cast materials have good friction and wear characteristics. Pulleys made of pressed steel are lighter than cast pulleys, but in many instances they have lower friction and may produce excessive belt wear.

14-3 ANGLE OF CONTACT AND LENGTH OF BELTS

The maximum power transmitted by a belt drive is limited either by stretching of the belt or by excessive belt slippage at the pulleys. Stretching may be controlled throughout a reasonable belt life by the use of low allowable stresses in the belt. Since slippage of the belt would occur at the pulley having the smaller angle of contact, the smaller angle of contact should be used in the design of a belt drive.

In the open-belt drive shown in Fig. 14-2, the smaller angle of

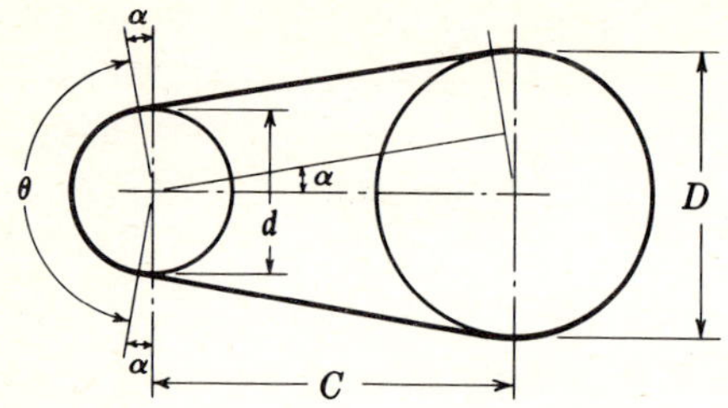

Fig. 14-2 Geometry of open-belt drive.

contact is at the smaller pulley. From the geometry of the figure it may be seen that

$$\theta = \pi - 2\alpha = \pi - \frac{D - d}{C} \qquad \text{approximately} \tag{14-1}$$

In a horizontal flat-belt drive, it is customary to use the top side of the belt as the slack side during the transmission of power.

The length of an open belt may be shown to be approximately equal to

$$\frac{\pi}{2}(D + d) + 2C + \frac{(D - d)^2}{4C} \tag{14-2}$$

For a *crossed belt*, the angle of contact is the same on both pulleys and approximately equal to

$$\theta = \pi + \frac{D + d}{C} \tag{14-3}$$

and the length of the crossed belt is, approximately,

$$\left(\frac{\pi}{2} + \alpha\right)(D + d) + \sqrt{4C^2 - (D + d)^2} \tag{14-4}$$

14-4 FLAT-BELT DRIVES

If the drive shown in Fig. 14-3 is stationary and the belt is tightened by adjusting the center distance, the tensions around the belt will be equal, as shown by the initial tension vector F_0. Now if the smaller pulley is assumed to rotate as a driving pulley in the direction shown to transmit power, the tension in the upper side will be increased and that in the lower side decreased, as shown by F_1 and F_2. In a belt transmitting power, these tensions are known as the "tight tension" F_1 and the "loose tension" F_2. Their difference $(F_1 - F_2)$ is the *net tension*, which may be determined from the horsepower being transmitted and the velocity of the belt.

The ratio of the belt tensions may be determined from the following analysis.

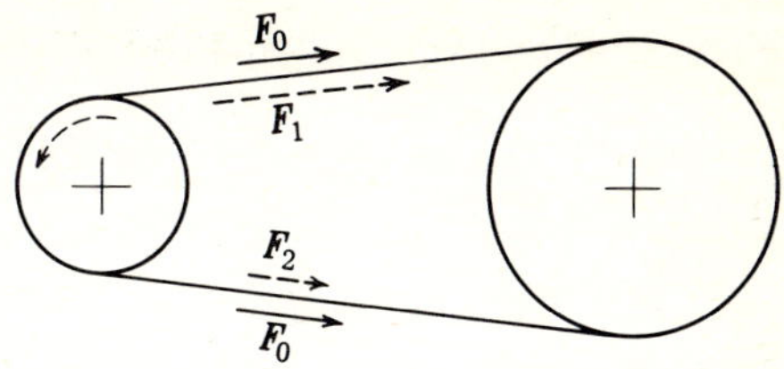

Fig. 14-3 Open-belt drive.

Ratio of belt tensions The forces acting on a belt are shown in Fig. 14-4.

Let F_1 = tight tension, lb
F_2 = loose tension, lb
θ = angle of contact, rad
r = radius of pulley, in.
f = coefficient of friction
b = width of belt, in.
t = thickness of belt, in.
w = weight of belt material, lb per in.^{3}
v = belt speed, fps

The forces acting on an element of the belt shown crosslined in the sketch are as follows:

Let F = force on left-hand end
$F + dF$ = force on right-hand end
P = pulley force
f^P = frictional force
C = centrifugal force

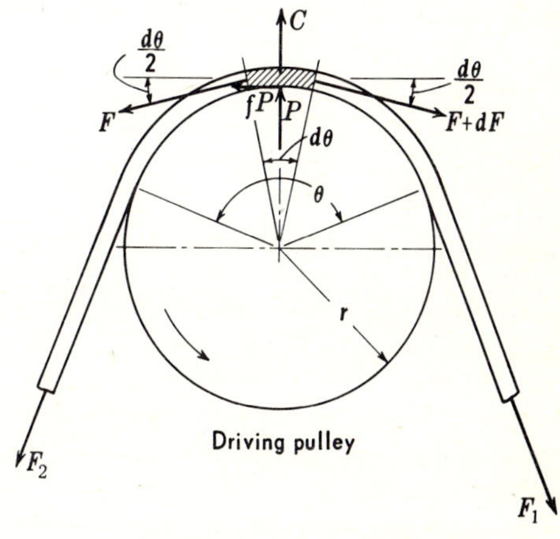

Fig. 14-4 Forces on element of belt.

From the conditions of equilibrium,

$$\Sigma F_v = P + C - F \sin \frac{d\theta}{2} - (F + dF) \sin \frac{d\theta}{2} = 0$$

or $\quad P + C - F\, d\theta = 0 \qquad (a)$

Also $\Sigma F_h = F \cos \frac{d\theta}{2} - (F + dF) \cos \frac{d\theta}{2} + fP = 0$

or $\quad -dF + fP = 0 \qquad (b)$

Now C = mass × acceleration

$$= \frac{r\, d\theta\, btw}{g} \frac{12v^2}{r} = F_c\, d\theta \qquad (c)$$

where $F_c = \dfrac{12wbtv^2}{g}$

By substituting $C = F_c\, d\theta$ into Eq. (a), eliminating P between Eqs. (a) and (b), and assuming f as constant,

$$\int_{F_2}^{F_1} \frac{dF}{F - F_c} = f \int_0^\theta d\theta$$

Therefore $\log_e \dfrac{F_1 - F_c}{F_2 - F_c} = f\theta$

$$\text{or} \quad \frac{F_1 - F_c}{F_2 - F_c} = e^{f\theta} \qquad (14\text{-}5)$$

Power transmitted For convenience in solving belt problems, the belt tensions may be expressed in terms of pounds per square inch of belt cross section. Equation (14-5) then becomes

$$\frac{f_1 - f_c}{f_2 - f_c} = e^{f\theta} \qquad (14\text{-}6)$$

where f_1 = maximum stress in belt, psi
f_2 = minimum stress in belt, psi

$$f_c = \frac{12wv^2}{g}$$

Solving Eq. (14-6) for $f_1 - f_2$, it becomes

$$f_1 - f_2 = (f_1 - f_c) \frac{e^{f\theta} - 1}{e^{f\theta}}$$

The horsepower that may be transmitted per square inch of belt cross section is

$$\text{hp per in.}^2 = \frac{(f_1 - f_2)v}{550}$$
$$= \frac{v(f_1 - f_c)}{550}\,\frac{e^{f\theta} - 1}{e^{f\theta}} \qquad (14\text{-}7)$$

where f_1 = allowable stress in belt, psi
$f_c = 12wv^2/g$
w = weight of belt material, lb per in.^{3}
= 0.035 lb per in.^{3} for leather
v = belt speed, fps
g = acceleration of gravity = 32.2 fps^2
f = coefficient of friction
θ = angle of contact of belt with pulley, rad

The cross-sectional area of the belt required may be found by dividing the horsepower to be transmitted by the horsepower per square inch [Eq. (14-7)]. The width and thickness of belt may then be selected from Table 14-2.

Initial tension The relation between the initial tension F_0 and the tensions F_1 and F_2 depends on the stress-strain characteristics of the belt material. For leather, the stress is not a linear function of strain, and hence the relation between F_1, F_2, and F_0 is not a simple one. The following equation may be used to estimate the effect of initial tension on the power tensions:[1]

$$\sqrt{F_1} + \sqrt{F_2} = 2\sqrt{F_0}$$

A suitable initial tension for leather belts is 200 to 225 psi.

[1] C. Barth, *Trans. ASME,* vol. 31, p. 29, 1909.

Table 14-2 Leather-belt data

	Thickness, in.				
Grade of belting	*Single to 8 in.*	*Double to 12 in.*	*Triple to 24 in.*	*Quadruple*	*Increments of width, in.*
					½ to 1 by ⅛
Light	⅛	¼			1 to 4 by ¼
					4 to 7 by ½
Medium	5/32	5/16	½	11/16	8 to 30 by 1
					32 to 56 by 2
Heavy	3/16	⅜	9/16	¾	60 to 84 by 4

14-5 LEATHER-BELT DATA

Data applying to leather belts are given in Table 14-2. The ultimate strength of oak-tanned leather varies from 3,000 to 4,500 psi. In order to secure reasonable life to a belt, it is necessary to limit the maximum working stress to a value considerably lower than its ultimate strength. On this basis, the value of f_1 in Eq. (14-7) should be limited to from one-tenth to one-eighth of the ultimate strength, say 400 psi for average conditions.

Coefficient of friction The coefficient of friction f depends on the materials of the pulley and the belt, on the slip, and on the belt speed. For leather belts on dry cast-iron or wood pulleys a recommended value of the coefficient of friction is 0.40 to 0.50.

Belt velocities used in practice for maximum power and economy range from 5,000 to 6,000 fpm, but for longer belt life, 3,000 to 4,000 may be used. A considerably lower value may be required, however, on account of low shaft speeds and limited pulley sizes. At high speeds, the pulleys should be dynamically balanced.

Unless the belts are endless, some type of belt joint or fastener must be used. A cemented joint is practically as strong as the material joined. Wire or rawhide lacing decreases the strength of the belt, but if the means of joining the ends does not affect the stretching of the belt or considerably decrease its strength, the belt should operate satisfactorily if low allowable stresses are used in the design of the drive.

14-6 V-BELT DRIVES

Grooves in V-belt sheaves On account of the wedging action of a V belt or rope in the groove of the sheave or pulley, the traction force (force transmitting power) is greater than in a flat belt running on a flat-face pulley. To ensure wedging action in the groove, the belt should make contact with the sides of the groove but not the bottom. It is evident that the wedging action and the traction force are large for small groove angles (see Fig. 14-5); however, the force required to pull the belt out of the

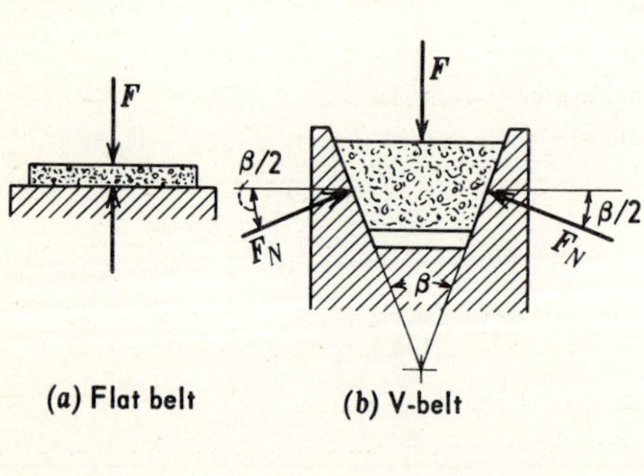

Fig. 14-5 Forces on flat and V belts.

Table 14-3 Selection and dimensions of V belts

Normal horsepower	*Motor speed, rpm* 1,800	1,200	900	720 *and below*
½ to 1½	A	A	A	A
2 to 5	A or B	A or B	A or B	A or B
10	A or B	B	B	B or C
15	B	B or C	B or C	C
20	B or C	C	C	C
25 to 30	C	C	C	C
40	C	C or D	C or D	D
50 to 60	C or D	C or D	C or D	D
75	C or D	C or D	C or D	D or E
100	C	D	D	D or E
125	......	D	D	D or E
150 to 200	......	D	D	E
250	......	E	E	E
300 and over	......	......	E	E

Section	A	B	C	D	E
Width, in.	½	21/32	⅞	1¼	1½
Thickness, in.	5/16	13/32	17/32	¾	29/32

groove as it leaves the sheave is large for small groove angles, resulting in loss of power and excessive wear of the belt. The selected groove angle is therefore a compromise to secure large traction force without unduly large force to pull the belt from the groove. The standard angle between the sides of V belts is 40°. The groove angle in the sheave is less than the belt angle to allow for change in shape or the belt cross section on wrapping around the sheave. Groove angles of 32 to 38° are used.

Standard sizes of belts for power transmission have been adopted and designated by sizes A, B, C, D, and E as indicated in Table 14-3. For low bending stresses as the belt wraps around the sheave, small belts are better for long belt life; however, a large number of small belts requires a wider sheave which increases the load overhang (distance from the belt force to the bearing) which in turn increases the shaft stresses and bearing loads, all of which raises the cost of the drive. Thus optimum design requires compromises in belt selection.

V-belts sheaves The use of large-diameter sheaves and the corresponding high belt speeds decrease the load on the belts required to transmit the power, but the gain with high belt speeds is partly nullified by the

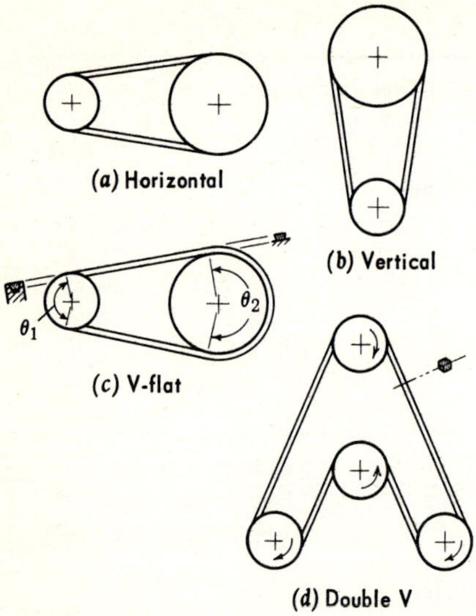

Fig. 14-6 V-belt drives.

increase in centrifugal forces due to the high speeds. Again a compromise must be reached. The accompanying tables will be helpful in selecting the most economical drive.

The sheaves may be made of cast iron, which has excellent friction characteristics on V belts, or of pressed steel, which is lighter and cheaper but may give rise to excessive belt slip, wear, and noise. Sheaves may be purchased with single or multiple grooves and are available also with variable pitch for speed control.

Types of drives In Fig. 14-6 are shown common arrangements of V-belt drives. A disadvantage of the vertical drive at (*b*) is vibration of the slack side of the belt under load, which may cause vibration of the equipment and decrease belt life. In the V-flat drive at (*c*), the small sheave is grooved but the V belt runs on the flat ungrooved rim of the large sheave or flywheel. This drive avoids grooving of the flywheel and is used where the speed ratio is large and the center distance relatively short. Slipping of the belt on the larger sheave is usually not troublesome since the angle of contact is large.

In Fig. 14-6(*d*) is shown a double-V drive used to permit rotation of sheaves in opposite directions. The belt has short life due to high bending stresses, since the belt is double thickness, and to reversed flexure.

Selection of V belts The *belt section* may be selected from Table 14-3, using the rated horsepower of the drive.

The *design horsepower* used in determining the number of belts is the rated horsepower multiplied by the service factor from Table 14-4. The

Table 14-4 V-belt service factors

Agitators, paddle propellers	1.0–1.2
Bakery machinery	1.0–1.2
Flour, feed, and cereal mills	1.0–1.4
Machine tools	1.0–1.4
Conveyors	1.0–1.5
Generators and exciters	1.2
Laundry machines	1.2
Paper machinery	1.2
Printing machinery	1.2
Compressors	1.2–1.4
Screens	1.2–1.4
Brick and clay machinery	1.2–1.6
Oil-field machinery	1.2–1.6
Textile machinery	1.2–1.8
Fans and blowers	1.2–2.0
Pumps	1.2–2.0
Crushing machinery	1.4–1.6
Mills	1.4–1.6
Line shafts	1.4–2.0
Rubber plant machinery	1.4–2.0

service factor takes into account the severity of load transmitted, which depends on the characteristics of the driving and driven units. When shock and peak loads are light, use the lower values from the table; for severe loads, use the higher values.

Where space permits, it is good practice to use a size of *driving sheave* to give belt speeds in the range of 3,500 to 4,500 fpm. Table 14-5 may be used to select a recommended size of driving sheave, and Table 14-6 for available stock sheaves. In these tables, the diameter given is the pitch diameter of the belt in the groove. The AB section is a composite sheave for either A or B section belts. The diameter given is for the A section; for B section, the pitch diameter should be increased by 0.4 in.

Table 14-5 Recommended size of driving sheave

Section	*Minimum pitch diameter, in.*	*Recommended range of small pitch diameters, in.*
A	3.0	3.0–5.0
B	5.4	5.4–7.4
C	9.0	9.0–13.0
D	13.0	13.0–17.0
E	21.0	21.0–28.0

Table 14-6 Stock sheave sizes

*AB section**			*B section*			*C section*			*D section*	
1 *to* 6 *grooves pitch diameter*			2 *to* 10 *grooves pitch diameter*			3 *to* 10 *grooves pitch diameter*			5 *to* 12 *grooves pitch diameter*	
3.0	5.0	9.0	5.4	7.0	16.0	8.5	10.0	20.0	13.0	18.0
3.2	5.2	10.6	5.6	7.4	18.4	9.0	10.2	24.0	13.4	22.0
3.4	5.4	12.0	5.8	8.6	20.0	9.2	10.6	30.0	13.8	27.0
3.6	5.6	15.0	6.0	9.4	25.0	9.4	11.0	36.0	14.2	33.0
3.8	5.8	18.0	6.2	11.0	30.0	9.6	13.0	44.0	14.6	40.0
4.0	6.0	19.6	6.4	12.4	38.0	9.8	16.0	50.0	15.0	48.0
4.2	6.2	24.6	6.6	13.6		.....	.....	.....	15.4	
4.4	6.4	29.6	6.8	15.4						
4.6	7.0	37.6								
4.8	8.2									

* Composite sheave for A or B section. Diameters shown are for A section. For B section, add 0.4 in.

The horsepower ratings of single strand of V belts may be determined from Eqs. (14-8).

Belt section	Horsepower rating per strand	
A	$\text{hp} = V\left(\dfrac{1.95}{V^{0.09}} - \dfrac{3.80}{kd} - 0.0136V^2\right)$	(14-8*a*)
B	$\text{hp} = V\left(\dfrac{3.43}{V^{0.09}} - \dfrac{9.83}{kd} - 0.0234V^2\right)$	(14-8*b*)
C	$\text{hp} = V\left(\dfrac{6.37}{V^{0.09}} - \dfrac{27.0}{kd} - 0.0416V^2\right)$	(14-8*c*)
D	$\text{hp} = V\left(\dfrac{13.6}{V^{0.09}} - \dfrac{93.9}{kd} - 0.0848V^2\right)$	(14-8*d*)
E	$\text{hp} = V\left(\dfrac{19.9}{V^{0.09}} - \dfrac{178}{kd} - 0.122V^2\right)$	(14-8*e*)

where hp = maximum horsepower recommended for one strand of belt of standard quality (multiply by arc-of-contact factor and length-of-belt factor from Fig. 14-7 to obtain corrected horsepower)

V = belt speed, thousands of feet per minute
k = small-diameter factor for speed ratio of drive (Fig. 14-7)
d = pitch diameter of small sheave, in. (the maximum values of kd for belt sections are as follows: A, 5 in.; B, 7 in.; C, 12 in.; D, 17 in.; E, 28 in.)

In the above equations the first term in the parentheses depends on the strength of the belt, the second term is an allowance for bending stress as the belt passes over the sheave, and the third term is an allowance for centrifugal force. In the bending-stress term, the diameter factor k (see Fig. 14-7) allows for the smaller bending stress as the belt passes over the driven sheave if it is larger than the driving sheave. If the speed ratio is 1:1, k is unity, but as the speed ratio increases, the driven sheave becomes larger and the diameter factor increases.

The power transmitted may be limited by slipping of the belt on the small sheave. This is taken into account by the arc-of-contact factor given in Fig. 14-7 in terms of the arc of contact of the belt on the smaller sheave. The factor is unity for 180° contact and decreases with the arc of contact.

A final factor is necessary to allow for the effect of length of belt on the frequency of bending. Length-of-belt correction factors are given in Fig. 14-7.

The horsepower rating from Eqs. (14-8) is corrected by multiplying by the small-diameter factor, the arc-of-contact factor, and the belt-length factor. The corrected rating (horsepower per belt) is then divided into the design horsepower to obtain the number of belts required.

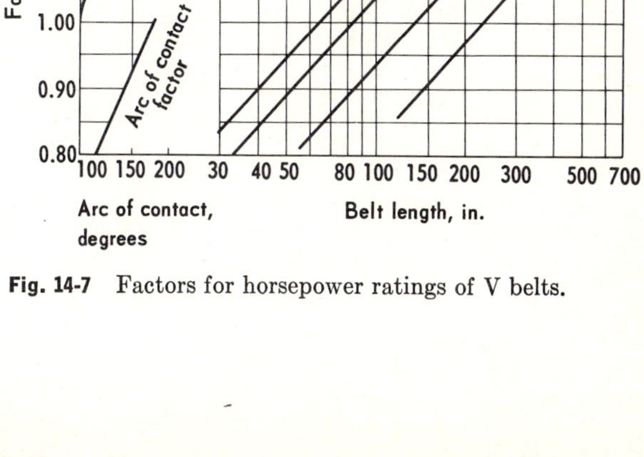

Fig. 14-7 Factors for horsepower ratings of V belts.

If the number of belts is excessive, a larger belt section and/or increased sheave diameters may be considered. A moderate number of belts is desirable so that if one belt breaks or stretches excessively, the remaining belts may carry the load until replacement of the belts can be made. In replacing the belts, a complete set of new belts should be used rather than replacing a single damaged belt, since a new (unstretched) belt would carry more than its share of the load and would have a short life.

In V-belt drives, if assembly or replacement requires means for disconnecting the belt, a metal coupler may be used, or belts made up of removable, laminated leather V-belt sections may be used. In the latter case, tension may be adjusted by adding or removing sections. An application requiring such a belt is a belt-driven generator or compressor mounted under the bed of a passenger railway car and driven by a sheave on the car axle.

Example Select a V-belt drive for a 10-hp 1,160-rpm induction motor to drive a ventilating fan at approximately 400 rpm. The minimum center distance between the sheaves is 40 in.

Solution From Table 14-3, a B-section belt is recommended.

From Table 14-4, assume the service factor is 1.4.

The design horsepower is $1.4 \times 10 = 14$.

Sheave selection: From Table 14-5, the recommended range of size for the small sheave is 5.4 to 7.4, with 5.4 as a minimum. From Table 14-6, assume 7.0 in. as a trial pitch diameter.

The belt speed is

$$\frac{\pi d \text{ rpm}}{12} = \frac{\pi \times 7.0 \times 1{,}160}{12} = 2{,}120 \text{ fpm}$$

The diameter of the large sheave is

$$\frac{1{,}160}{400} \times 7.0 = 20.3 \text{ in.} \qquad \text{Use 20 in.}$$

The speed of the fan will then be 406 rpm.

The speed ratio is approximately 3.1, which gives the small-diameter factor k, from Fig. 14-7, equal to 1.13. Thus $kd = 1.13 \times 7.0 = 7.91$ in. Since the maximum value of kd is 7 in., this value will be used in Eq. (14-8*b*).

$$\text{hp per belt} = V\left(\frac{3.43}{V^{0.09}} - \frac{9.83}{kd} - 0.0234V^2\right)$$

$$= 2.120\left[\frac{3.43}{2.120^{0.09}} - \frac{9.83}{7} - 0.0234 \times (2.120)^2\right]$$

$$= 2.120(3.2 - 1.4 - 0.10) = 3.61$$

The arc of contact, from Eq. (14-1), is

$$\pi - \frac{D - d}{C} = \pi - \frac{20 - 7}{40} = 2.81 \text{ rad, or } 161°$$

From Fig. 14-7, the arc-of-contact factor is 0.95.

The belt length is, from Eq. (14-2),

$$L = \frac{\pi}{2}(D + d) + 2C + \frac{(D - d)^2}{4C}$$

$$= \frac{\pi}{2}(20 + 7) + 2 \times 40 + \frac{(20 - 7)^2}{4 \times 40}$$

$$= 42.4 + 80 + 0.1 = 122.5 \text{ in.}$$

(A standard belt length should be selected from manufacturer's data.)
From Fig. 14-7, the belt-length factor is 1.07.
The corrected rating of one belt is

$$3.61 \times 0.95 \times 1.07 = 3.67 \text{ hp}$$

The number of belts required is

$$\frac{14}{3.67} = 3.82 \qquad \text{Use 4 belts}$$

Thus, four B-section V belts with sheaves 7.0 and 20 in. in diameter would be specified.

As an extension to this problem the shaft loads may be determined. The shaft loads are necessary in order to calculate the required diameter of the shaft. In determining shaft loads, the tight and loose sides may be assumed as parallel because the angle α in Fig. 14-2 is small in this problem ($\alpha = 9°22'$), as it should be in well-designed drives so that the angle of contact for the small pulley will be as large as possible. The assumption that the belt loads F_1 and F_2 are parallel will be on the side of safety in using the shaft loads as the algebraic sum of F_1 and F_2.

To determine the shaft load $F_1 + F_2$, the net tension $F_1 - F_2$ can be found from the torque equation; that is,

$$F_1 - F_2 = \frac{2T}{D}$$

The second equation is Eq. (14-5),

$$\frac{F_1 - F_c}{F_2 - F_c} = e^{f\theta}$$

in which F_c is the centrifugal tension.

$$F_c = A \times \frac{12wv^2}{g}$$

In this equation, A is the cross-sectional area of the belt in square inches,

Table 14-7

Belt section	*Weight, lb per ft*	*Pitch diameter, small sheave*	*Groove angle β, deg*	*Coefficient of friction*
A	0.06	Under 3	30	0.50
		3–5	34	0.45
		Over 5	38	0.40
B	0.106	Under 5.4	30	0.50
		5.4 to 8	34	0.45
		Over 8	38	0.40
C	0.20	Under 9	30	0.50
		9–12	34	0.45
		Over 12	38	0.40
D	0.40	Under 13	30	0.50
		13–16	34	0.45
		Over 16	38	0.40
E	0.60	Under 21.6	30	0.50
		21.6 to 24	34	0.45

and w is the weight of the belt in cubic inches; hence

$$F_c = W\frac{v^2}{g}$$

where W is the weight of a foot length of the belt.

For drives with low belt speeds F_c in Eq. (14-5) can be neglected, but not with high belt speeds.

Thus, for our problem, using W in the above equation as 0.106 lb per ft for the B-section belt from Table 14-7, and f equal to 0.40 for the 7-in. small sheave on the B-section belt,

$$F_c = W\frac{v^2}{g} = 0.106 \times \frac{\overline{2{,}120^2}}{60^2 \times 32.2} = 41.3 \text{ lb}$$

The torque on the driving pulley is

$$T = \frac{63{,}030 \text{ hp}}{\text{rpm}} = \frac{63{,}030 \times 10}{1{,}160} = 544 \text{ in. lb}$$

Therefore $F_1 - F_2 = \dfrac{2T}{d} = \dfrac{2 \times 544}{7} = 155$ lb

For the second equation, the "equivalent" coefficient of friction f depends on the groove angle β in Fig. 14-5, which has values in Table 14-7.

The value of θ from Eq. (14-1) is

$$\theta = \pi - \frac{D - d}{C} = \pi - \frac{20.7}{40} = 2.816 \text{ rad}$$

$$\frac{F_1 - F_c}{F_2 - F_c} = e^{f\theta} = e^{0.40 \times 2.816} = e^{1.125} = 3.08$$

From the three equations,

$$F_c = 41.3 \text{ lb}$$
$$F_1 - F_2 = 155 \text{ lb}$$
$$\frac{F_1 - F_c}{F_2 - F_c} = 3.08$$

the values of F_1 and F_2 can be calculated as

$$F_1 = 310 \text{ lb} \qquad \text{and} \qquad F_2 = 165 \text{ lb}$$
$$\text{or} \quad F_1 + F_2 = 475 \text{ lb}$$

This is the load to be used in determining the bending moments on the shafts for both small and large sheaves. If the centrifugal force were neglected in this problem, the belt loads would be found as 305 lb.

14-7 TOOTH-BELT DRIVE

The sheaves have axial grooves which engage teeth on the belt, as shown in Fig. 14-8. The belt employs a number of small steel cables which carry tension under load, thus permitting a light drive and high speeds. The drive is positive, and the tension to transmit the power does not depend on initial tension. Sheave diameters may be as small as ½ in. to as large as required, and belt speeds of the order of 15,000 fpm are practical. Applications include business machines, sewing machines, timing drives, portable woodworking equipment, and power-transmission

Fig. 14-8 Timing belt drive.

units. Besides being positive, the drive is compact, light, quiet, versatile, and low in maintenance. However, it is more sensitive to misalignment than flat belts or V-belt drives. The first cost is higher on light-duty belt drives but may be more economical on large drives.

14-8 PIVOTED MOTOR MOUNTINGS

When a belt transmitting power passes over its sheaves, the tension in the belt varies from the maximum value F_1 to the minimum value F_2, and it is the difference in tensions $(F_1 - F_2)$ which accounts for the power transmitted. At full power F_1 is large and F_2 small as compared with the values at partial power. It is evident that the initial tension in the belt must exceed the minimum value of F_2 in order to hold the belt against the sheaves. When the drive is at rest, the initial tension must remain in the belt in order to be available for full power requirements; thus, in time, stretching of the belt will shorten its life.

Pivoted motor drives were developed to reduce the initial tension to a minimum and to provide automatically for increasing tension as required by the power being transmitted. Thus when the drive is idle or working at partial load, the belt tensions are comparatively lower than in the fixed-center type of drive and belt life is correspondingly increased.

In Fig. 14-9, the motor pulley is rotating clockwise. The belt tensions F_1 and F_2 tend to overturn the motor in a clockwise direction about the base pivot point. Neglecting centrifugal tension, the overturning

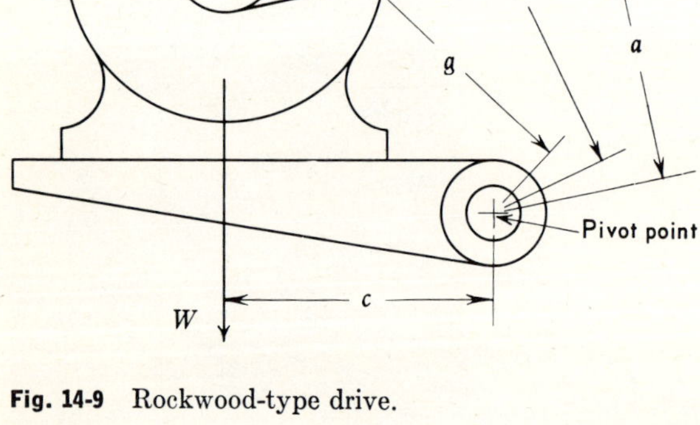

Fig. 14-9 Rockwood-type drive.

moment is $F_1a + F_2b$. The moment that resists the overturning moment is produced in part by the weight of the motor acting at the lever arm c, that is, Wc, and in part by the reaction torque T on the frame of the motor which is counterclockwise and equal to the torque on the rotor. Thus

$$F_1a + F_2b = Wc + T \tag{14-9}$$

The effect of the reaction torque T is to rotate the motor counterclockwise about the pivot point and thus to increase the belt tensions. If the value for the torque T is divided by the distance from the pivot point to the center of the motor, that is, the lever arm g in Fig. 14-9, the result will be a force acting at the center of the motor which in turn is carried by the belt. This action performs the same function as initial belt tension but is induced only as needed by the torque on the motor in transmitting power. The force acting at the center of the motor will depend on the lever arm g; the smaller the lever arm, the greater will be the force, and vice versa.

Two types of pivoted motor mountings are in use. In one, the Rockwood-type drive shown in Fig. 14-9, the distance c and the distance g are both large, so that most of the belt tension is due to the product of W and c and that due to the torque reaction is small. The lever arm c may be altered by shifting the motor along the pivoted support to provide belt tensions as required for the maximum power to be transmitted. When this power is exceeded, the belt will slip on the sheave, which will prevent overloading the motor but may damage the belt.

In the second type of pivoted motor mounting, the American drive shown in Fig. 14-10, the distances c and g are small so that the belt tensions F_1 and F_2 are produced mainly by the reaction torque T. This torque varies with the power transmitted so that there is never more tension in the belt than the load requires. In order to locate the pivot point near the center of the motor, the motor is mounted in a cradle so that the motor will swing about the pivot point shown in the figure. In this drive, belt tensions are built up as required by the load. When the drive is idle or is operating at partial load, the tensions will be low; hence, belt life will be increased. However, if the drive should be overloaded, the belt tensions will be increased accordingly and the belt may be broken or the bearings may be damaged.

14-9 ARRANGEMENTS OF HOISTING TACKLE

A simple hoisting tackle is shown in Fig. 14-11(a), in which a rope passes over a *sheave* so that a load Q may be raised by a downward effort P. To reduce the effort, an arrangement of pulley blocks may be used as shown at (b), in which the lower, or hook, block and the upper block each has two

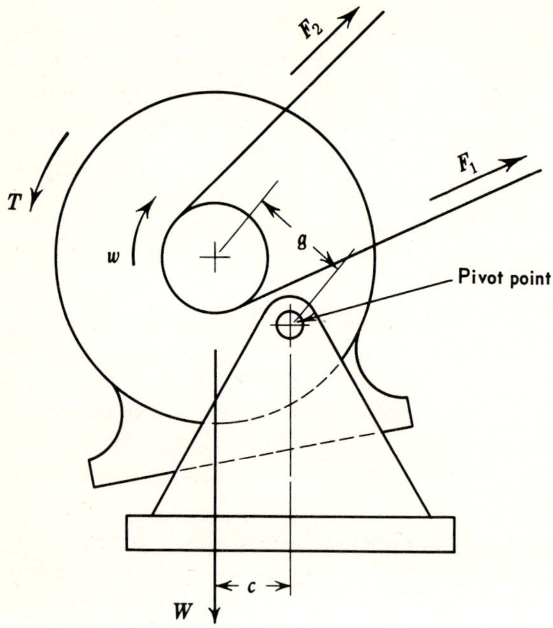

Fig. 14-10 American drive.

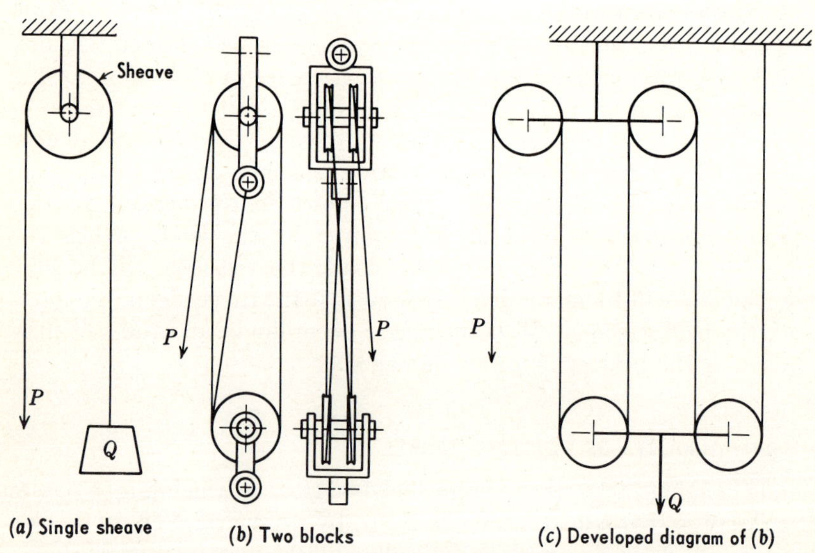

Fig. 14-11 Hoisting tackle.

sheaves. To avoid confusion in representing such a hoist, a *developed* diagram (*c*) may be used. The hoist at (*c*) is known as a "four-part" line hoist, since four lines of rope lead to the hook block.

It is evident that the more lines leading to the hook block, the less will be the effort P required to raise a load Q, but that more rope must be reeled off at P to raise the load a given distance.

14-10 RATIO OF ROPE TENSIONS

When a rope passes over a stationary sheave, as in Fig. 14-12(*a*), the loads P and Q will be equal. Note that the moment of the effort about the sheave center is $PD/2$ and is equal to $QD/2$.

If it is desired to raise Q, it is common experience that it is necessary to apply an effort P which is greater than Q. The increase in P is due to several causes, as indicated at (*b*) in the figure. When the rope runs onto the sheave at f, the resistance of the rope to bending will shift its centerline a distance e to the right of the vertical tangent, as shown in the figure. The moment Qe is that required to bend the rope. As the rope assumes the curve of the sheave in passing from f to g, the wires or fibers of the rope shift on one another and adjust themselves, so that when the rope arrives at g it will be naturally curved and must be straightened as it leaves the sheave. This shifts the rope to the right a distance e', as shown in the figure. Thus the lever arm of the resistance Q has been increased and that of P has been decreased.

In Fig. 14-12(*a*), the support $(P + Q)$ provided by the sheave pin was drawn through the center of the pin; but when the sheave rotates, as in (*b*), the load $(P + Q)$ will be displaced a distance $fd/2$, as shown where

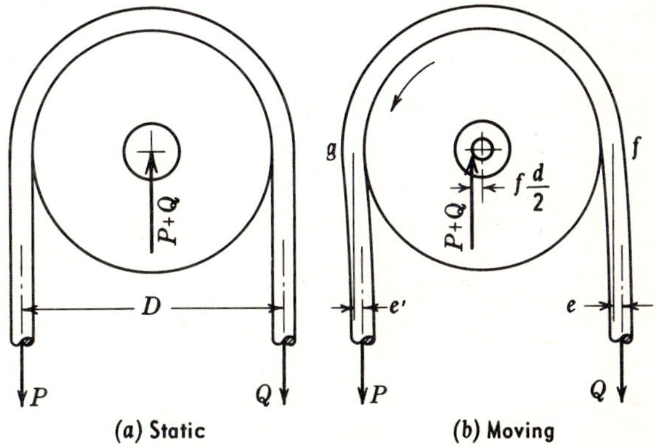

Fig. 14-12 Rope passing over sheave.

f is the ceofficient of bearing friction and d is the diameter of the pin, that is, $fd/2$ is the radius of the "friction circle." Moments may now be taken about the line of action of $P + Q$, for instance, and the following equation obtained for the relation between the effort P and the resistance Q:

$$P = \frac{D + fd + 2e}{D - fd - 2e'} Q = CQ \tag{14-10}$$

where C has a value greater than unity. Values of C depend on the size of rope, on the relative size of sheave and pin, and on the coefficient of friction and have been determined experimentally. For manila rope, an average value for C of 1.14 may be used; for wire rope, see Table 14-11.

If the angle of wrap is different from 180°, the values of C as given may be used without introducing appreciable error.

14-11 LOADS AND EFFICIENCY

As an example of determining the loads on the various ropes of a hoist and its efficiency, the arrangement shown in Fig. 14-13 will be used.

Raising the load Label the ropes T_1, T_2, etc., to represent their tensions. From the analysis in Art. 14-10,

$$\begin{aligned} P &= CT_5 = C^5T_1 \\ T_5 &= CT_4 = C^4T_1 \\ T_4 &= CT_3 = C^3T_1 \\ T_3 &= CT_2 = C^2T_1 \\ T_2 &= CT_1 \end{aligned}$$

From the equilibrium of forces on the lower block

$$\begin{aligned} Q &= T_1 + T_2 + T_3 + T_4 + T_5 \\ &= T_1(1 + C + C^2 + C^3 + C^4) = T_1\frac{C^5 - 1}{C - 1} = P\frac{C^5 - 1}{C^5(C - 1)} \end{aligned}$$

or $$P = \frac{C^5(C - 1)}{C^5 - 1} Q \tag{14-11}$$

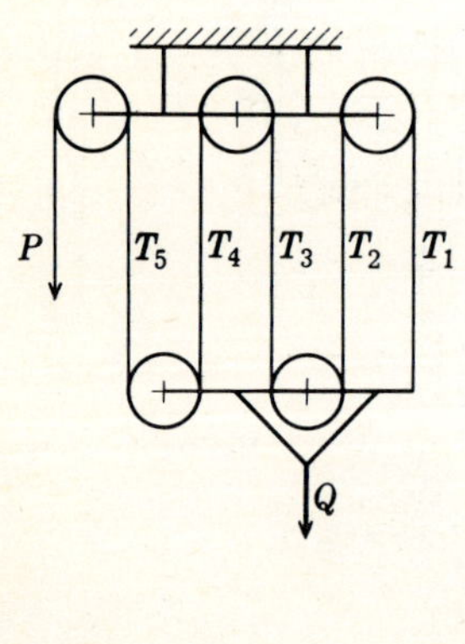

Fig. 14-13 Loads on hoist.

Without friction, $P_0 = Q/5$. Hence, the efficiency of the hoist is

$$\frac{P_0 \times \text{unit distance}}{P \times \text{unit distance}} = \frac{C^5 - 1}{5C^5(C - 1)} \tag{14-12}$$

Lowering the load For this case

$$T_1 = CT_2$$
$$T_2 = CT_3$$
$$T_3 = CT_4$$
$$T_4 = CT_5$$
$$T_5 = CP$$
$$P = \frac{C - 1}{C(C^5 - 1)} Q \tag{14-13}$$

In the above example, note that, in raising the load, the rope P is the maximum loaded one, whereas in lowering the load, the rope T_1 is the maximum loaded one. The capacity of a hoist depends on the most heavily loaded rope, as will be discussed in Art. 14-14.

14-12 MANILA ROPE

Manila rope is used for hoisting only in very small-capacity hoists and where safety is not a prime consideration. Allowable loads for a reasonable life of rope may be found by dividing the ultimate strength of the rope F_{ult} in pounds by a factor of safety equal to 7 for rope speeds up to 100 fpm, increasing to 18 at 300 fpm and to 36 up to 800 fpm (see Table 14-8 for ultimate strengths of manila rope).

14-13 WIRE ROPE

Construction Wire rope is constructed of strands each of which is made of small wires twisted together. This construction permits the rope to be wrapped around a sheave without undue bending stresses in the

Table 14-8 Ultimate strength of manila rope

(d = rope diameter, in.; F_{ult} = ultimate strength, lb)

d	F_{ult}	d	F_{ult}	d	F_{ult}	d	F_{ult}
3/8	1,270	5/8	4,000	1 1/8	10,500	1 5/8	20,000
7/16	1,870	3/4	4,700	1 1/4	12,500	1 3/4	25,000
1/2	2,400	7/8	6,500	1 3/8	15,400	2	30,000
9/16	3,300	1	7,500	1 1/2	17,000	2 1/4	37,000

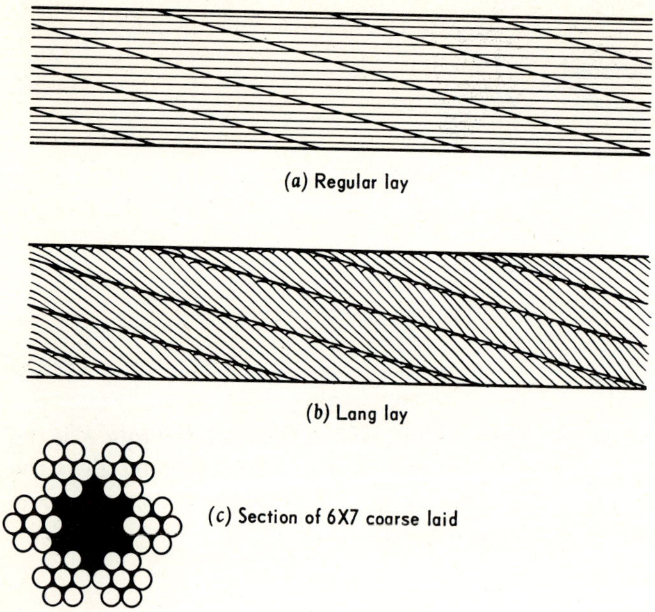

(a) Regular lay

(b) Lang lay

(c) Section of 6X7 coarse laid

Fig. 14-14 Winding of wire rope.

wire. Two types of winding are in use: one type [see Fig. 14-14(*a*)], in which the direction of twist of the wires in the strands is opposite to the direction of twist of the strands, is known as "regular lay." When the directions of twist of the wires and strands are the same, as in Fig. 14-14(*b*), it is known as "lang lay." The regular lay will not untwist or spin under load as much as lang lay, but tests have shown that it has a shorter life.

Wire rope is made in several standard constructions of which the following cover most services:

6 × 19[1]—standard hoisting rope
6 × 37—extra-flexible hoisting rope
8 × 19—extra-flexible hoisting rope
6 × 7—standard coarse laid rope

The hemp core is saturated with lubricant to retard rusting of the wires and to reduce friction between the wires. The rope should be periodically cleaned and relubricated.[2]

Standard sizes of wire ropes may be found from Table 14-9. Ulti-

[1] The first figure represents the number of strands and the second the number of wires in each strand. For example, a ¾-in. 6 × 7 wire rope is ¾ in. in diameter and has 6 strands each with 7 wires [see Fig. 14-14(*c*)]. The core shown in black is hemp.

[2] USAS Safety Code, Elevator Wire Rope Maintenance.

Table 14-9 Increments of standard sizes of wire ropes

Rope	*Diameter, in.*	*Increment, in.*
6 × 19	$\frac{3}{8}$–$\frac{9}{16}$	$\frac{1}{16}$
and	$\frac{5}{8}$–$2\frac{1}{4}$	$\frac{1}{8}$
6 × 37	$2\frac{1}{4}$–$2\frac{3}{4}$	$\frac{1}{4}$
8 × 19		
and	$\frac{1}{4}$–$\frac{9}{16}$	$\frac{1}{16}$
6 × 7	$\frac{5}{8}$–$1\frac{1}{2}$	$\frac{1}{8}$

Table 14-10 Approximate wire rope and sheave data
(d = rope diameter)

Rope	*Ultimate strength, F_{ult}, lb*	*Weight, lb per ft*	d_w	A	*Recommended diameter*	
					Average	*Minimum*
6 × 19	$72{,}000d^2$	$1.60d^2$	$0.063d$	$0.38d^2$	$45d$	$30d$
6 × 37	$68{,}000d^2$	$1.55d^2$	$0.045d$	$0.38d^2$	$27d$	$18d$
8 × 19	$62{,}000d^2$	$1.50d^2$	$0.050d$	$0.35d^2$	$31d$	$21d$
6 × 7	$68{,}000d^2$	$1.45d^2$	$0.106d$	$0.38d^2$	$72d$	$42d$

mate tensile strengths in pounds for plow-steel wire rope and weights in pounds per foot may be approximated from the data in Table 14-10.

14-14 BENDING STRESSES IN WIRE ROPES

As a wire rope is curved to fit its sheave, the *outer* wires are subjected to tension bending. This is in addition to the tension due to the service load. As the rope passes over the sheave, the bending stress is relieved by the readjustment of the wires. When the rope is straightened as it leaves the sheave, the *inner* wires are subjected to tension in bending.

A consideration of the mechanics involved leads to the conclusion that the bending stress is directly proportional to the diameter of the wire and inversely proportional to the diameter of the sheave, or

$$s = k\frac{d_w}{D} \tag{14-14}$$

where k is the constant of proportionality.

Table 14-11 Values of C for wire rope

Rope diameter	C	*Rope diameter*	C
3/8	1.090	5/8	1.064
7/16	1.083	3/4	1.054
1/2	1.076	7/8	1.046
9/16	1.070	1	1.040

If the bending stress is multiplied by the net cross-sectional area A of the wire in the rope, the product sA represents a load that may be referred to as a bending load F_b. Its value from Eq. (14-14) becomes

$$F_b = kA \frac{d_w}{D} \tag{14-15}$$

The sum of this bending load and the external, or service, load on the rope should not exceed the allowable load on the rope, or, expressed as an equation,

$$\frac{F_{\text{ult}}}{\text{fs}} \geqq F_b + F_s \tag{14-16}$$

$$F_b = kA \frac{d_w}{D}$$

Table 14-12 Recommended minimum safety factors for wire rope
(American Steel and Wire Co.)

Track cables	3.2
Guys	3.5
Mine shafts:	
Depths	
To 500 ft	8
1,000–2,000 ft	7
2,000–3,000 ft	6
Over 3,000 ft	5
Miscellaneous hoisting	5
Haulage ropes	6
Overhead and gantry cranes	6
Jib and pillar cranes	6
Derricks	6
Small electric and air hoists	7
Hot ladle cranes	8
Slings	8

where k = 12,000,000 psi
d_w = diameter of wire in rope, in.
D = pitch diameter of sheave, in.
A = net area of wire in rope, in.^{2}
F_{ult} = ultimate strength of rope, lb
fs = factor of safety
F_s = service load on rope, lb

Equation (14-16) is based on the assumption that the significant property of a wire rope is its *strength.* The factor of safety is chosen so that the rope selected on the basis of strength will have a reasonable life. This means that the rope should not fail as a result of fatigue or wear.

An important matter contributing to both fatigue and wear is the compressive stress[1] on the wires caused by bearing against the sheave. This condition, among others, indicates that the analysis given here is approximate and must require the liberal factors of safety suggested in Table 14-12. The values in this table include an allowance for bending stress; hence they are higher than the factors applying to Eq. (14-16).

[1] Drucker and Tachau, A New Design Criterion for Wire Rope, *Trans. ASME,* vol. 67, p. A-33, 1945.

15
Power-transmission Chains

15-1 FEATURES OF CHAIN DRIVES

Important advantages of the chain drive are (1) that it is adapted to long or short center distances and (2) that one chain may be arranged to drive more than one unit. Some arrangements are shown in Fig. 15-1. It is important that the shafts be in good alignment, especially for high-speed chains and for wide chains. Chain drives that are properly selected, installed, and lubricated operate at high efficiency and low maintenance cost.

15-2 CHAIN SPEED

The action of a chain as it runs with a sprocket may be likened to that of a nonslipping belt running with a prism. If the driving prism, or sprocket, rotates uniformly, it is evident that the speed of the chain will vary from a minimum value, as shown by the solid lines in Fig. 15-2(a), to a maximum value, as shown by the dotted lines. Because of this variation in

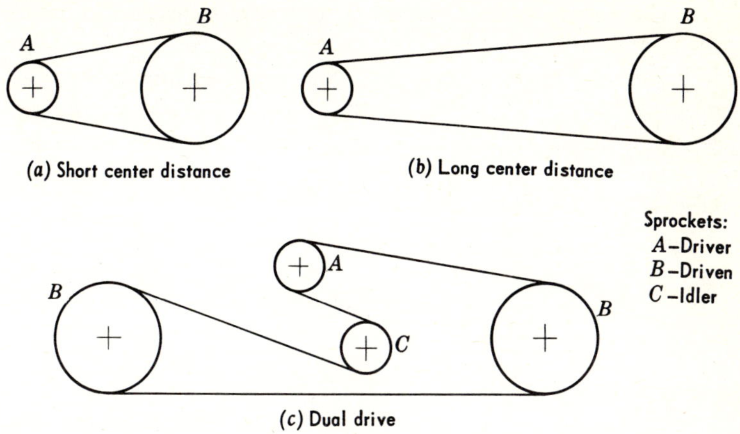

Fig. 15-1 Arrangements of chain drives.

chain speed, the driven sprocket will not rotate uniformly unless it has the same number of teeth as the driving sprocket. By increasing the number of teeth on the sprocket, the variation in chain speed will be reduced.

For a sprocket having 11 teeth, the variation in chain speed amounts to about 4 percent; for 17 teeth, the variation is 1.6 percent; and for 24 teeth, the variation is less than 1 percent. Thus if the minimum number of teeth in the small sprocket is 17 or preferably 24 teeth, the drive should operate smoothly.[1]

[1] R. C. Binder, "Mechanics of the Roller Chain Drive," Prentice-Hall, Inc., Englewood Cliffs, N.J., 1956.

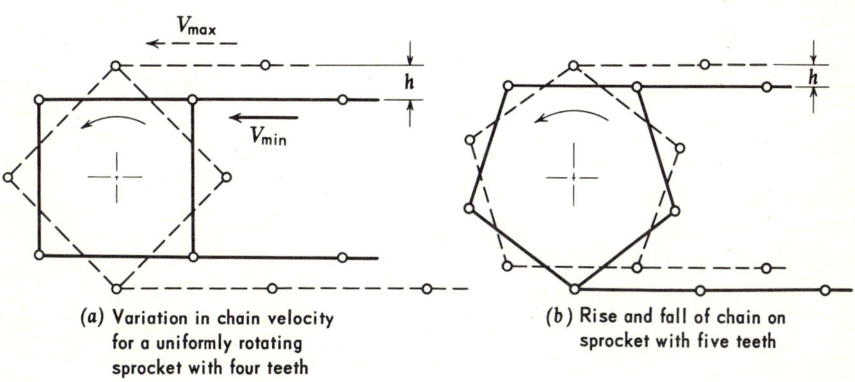

Fig. 15-2 Action of chain engaging sprocket.

The "rise and fall" of the chain on the sprocket is virtually eliminated in the Hy-Vo drive, which employs a rocker joint that shifts the pitch line of the chain automatically as the chain engages with the involute sprocket teeth. This reduces the linear pulsation in the chain, which increases the fatigue strength and the noise level of the drive.[1]

The average chain speed is equal to the length of chain reeled off the sprocket in unit time, or

$$V = \frac{pTN}{12} \tag{15-1}$$

where V = chain speed, fpm
p = pitch of chain, in.
T = number of sprocket teeth
N = sprocket speed, rpm

Although there is a trend toward higher chain speeds for ordinary applications, economical speeds are of the order of 2,500 fpm for roller chains and 4,000 fpm for silent chains. Both types of chains have been operated satisfactorily up to 6,500 fpm.

It is usually desirable to use small pitches in a chain drive to reduce surging of the chain; however, excessively small pitches may increase the first cost unnecessarily. As an aid in selecting a pitch, the following empirical formula, which has given good results in practice, may be used.

$$p \lesseqqgtr \left(\frac{900}{N}\right)^{\frac{2}{3}} \tag{15-2}$$

where p is the pitch in inches and N is the speed of the small sprocket in rpm.

15-3 NUMBER OF TEETH ON SPROCKET

The most desirable number of teeth on a sprocket depends on several considerations. As shown in Fig. 15-3, the angle through which a chain link turns on its pin as it engages with the sprocket is equal to 180° divided by the number of teeth on the sprocket. Thus with a small number of teeth, the angle of rotation will be large and wear of the pin and bushing will be rapid. This consideration requires a minimum of 17 or preferably 24 teeth for steady chain loading. Finally, for a particular pitch of chain, the size of the sprocket—and hence the chain speed—increases with the number of teeth. This means that the load on the chain required to transmit a given horsepower will be less with a large sprocket than with a smaller one; hence a smaller chain may be used.

[1] Morse Chain Company, Detroit, Mich.

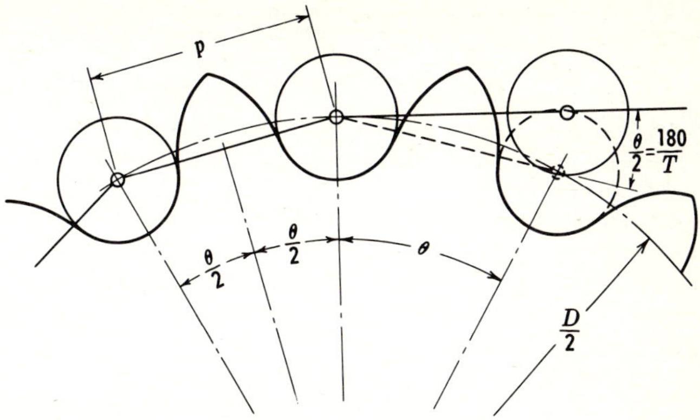

Fig. 15-3 Action of chain engaging sprocket teeth.

There is a limit to the chain speed, however, because of dynamic effects and lubrication difficulties.

The above discussion indicates that a large number of teeth on a sprocket is desirable from an operating standpoint. However, the cost of larger sprockets and cases, space limitations, and lubrication considerations generally favor a small number. Hence, the number of sprocket teeth specified must represent a compromise. Recommended minimum numbers of sprocket teeth are given in the tables for various types of chains.

It is usually preferable to use an odd number of teeth on a sprocket so that wear will be uniformly distributed over the teeth. This is desirable particularly for exposed or partially lubricated drives. An odd number of teeth favors the condition where an old chain is replaced by a new one in that a sprocket with an odd number of teeth would show uniform wear of the teeth, whereas a sprocket having an even number of teeth would show alternate teeth having greater wear than the intermediate ones.

To determine the pitch diameter D of a sprocket, the geometry of Fig. 15-3 may be used, i.e.,

$$\sin\frac{\theta}{2} = \frac{p}{2} \div \frac{D}{2}$$

or $$D = \frac{p}{\sin(\theta/2)} = \frac{p}{\sin(180/T)} \qquad (15\text{-}3)$$

For sprocket cutting, the pitch diameter is calculated to thousandths or tenths of thousandths of an inch, modifying if necessary the calculated value to provide clearance.

Table 15-1 Service factors for power-transmission chains

Type of load	*Roller chain, hr per day* 10	24	*Inverted-tooth chain, hr per day* 10	24
Uniform load, average conditions	1.0	1.2	1.0–1.2	1.3–1.5
Moderate shock	1.2	1.4	1.3–1.5	1.6–1.8
Heavy shock	1.4	1.7	1.6–2.0	2.0–2.5

15-4 DESIGN HORSEPOWER

The useful life of power-transmission chains is usually terminated by troubles arising from wear at the joints. The wear is caused by the rotation of the links as they are seated on the sprocket, i.e., through the angle $\theta/2$ in Fig. 15-3. The wear may be aggravated by shock loads due to speed variations caused by the driving unit or the driven equipment or both. Wear at the joints increases the pitch of the chain so that the links will not seat properly on the sprocket teeth which in turn causes increased dynamic loads on the joints. When this occurs, the rate of wear is increased and the useful life of the chain will be near its end.

To allow for shock loads and hours per day of use, the transmitted horsepower should be multiplied by a service factor to obtain the design horsepower (see Table 15-1).

15-5 BLOCK CHAINS

These chains as shown in Fig. 15-4 are used mainly for conveyor applications. The chain is relatively noisy and wear is rapid because of the impact between the blocks and the sprocket.

15-6 ROLLER CHAINS

The side plates for the roller chain shown in Fig. 15-5 are blanked from cold-rolled steel. The pins, bushings, and rollers are generally made of alloy steel, hardened and ground. The chains are rugged and durable and if properly selected, installed, and lubricated will give excellent service.

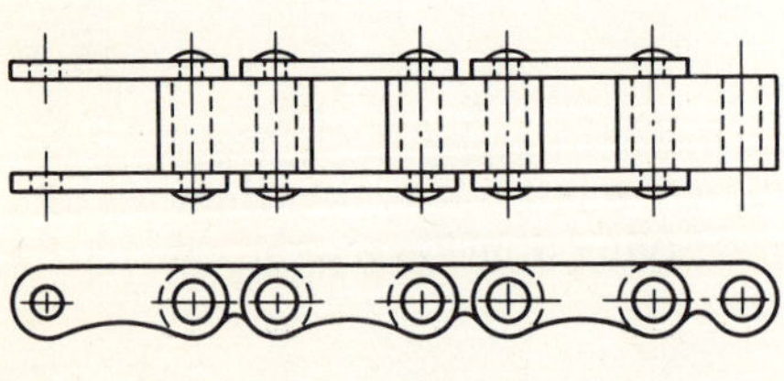

Fig. 15-4 Block chain.

Fig. 15-5 Roller chain.

Roller chains are manufactured in standard pitches as shown in Table 15-2. In the chain number, the right-hand digit 5 indicates a rollerless bushing chain, 1 indicates a lightweight chain, and 0 indicates a chain of usual proportions with roller. The number to the left of the right-hand digit is the number of 1/8 in. in the pitch. The recommended maximum speed of the small sprocket is shown for various numbers of teeth in the small sprocket. Sprockets are available from 5 to 159 teeth. Not fewer than 11 and preferably not fewer than 24 are recommended.

Tables for horsepower ratings for various speeds of small sprocket, chain pitch, and number of teeth on the small sprocket are given in USAS B29.1–1963. Considerable design information is also given in that source, including recommended types of lubrication for different chain speeds.

Figure 15-6 may be used for a trial selection of standard roller chains in terms of speed of small sprocket and horsepower transmitted. Chain sizes in the figure are for RC 25 (1/4-in. pitch) to RC 100 (1 1/4-in. pitch). The upper limit of the horsepower at various sprocket speeds is indicated by the lines in three segments. For instance, for RC 25, segment AB establishes a limit which if exceeded will result in side-bar failure. The segment BC if exceeded will result in roller or bushing failure, and the segment CD if exceeded will result in pin-bushing galling. The life expectancy at rating is 30,000 hrs.

The horsepowers for two- and three-strand chains in Fig. 15-6 were determined by multiplying the single-strand value by 1.7 for two strands and 2.5 for three strands.

Table 15-2 Standard pitches for roller chains

Chain no.	25	35	41	40	50	60	80	100	120	140	160	180	200	240
Pitch	1/4	3/8	1/2	1/2	5/8	3/4	1	1 1/4	1 1/2	1 3/4	2	2 1/4	2 1/2	3

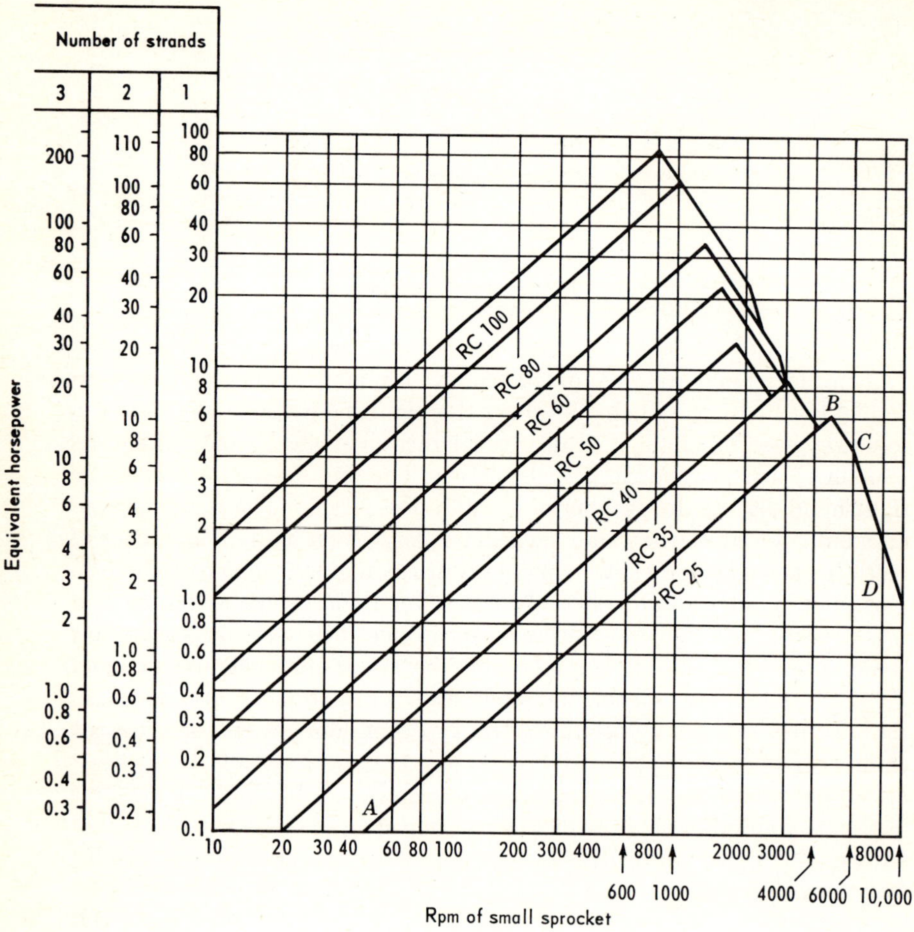

Fig. 15-6 Trial selection of standard roller chains. *(Courtesy of Link-Belt Company.)*

After the trial pitch is determined from Fig. 15-6, the number of teeth for the small sprocket and the type of lubrication can be determined from the horsepower rating tables in Appendix XVI.[1] These tables represent ratings for single-pitch roller chains of pitches equal to ¼ to 1½-in. For chains of 1¾ to 3 in., see the USAS Standard.[1] These standards represent many years of cooperative research by the American Sprocket Manufacturer's Association. For multiple-strand chains, the

[1] United States of America Standards Institute, Transmission Roller Chains and Sprocket Teeth, USAS B29.1–1963.

ratings in Appendix XVI can be multiplied by the "multiple-strand factors" given in the preceding paragraph.

The types of lubrication are as follows:

Type I *Manual lubrication* Oil applied periodically with brush or spout can (100-fpm maximum chain speed)

Type II *Drip lubrication* Oil applied between link plate edges from a drip lubricator (300 fpm maximum)

Type III *Oil bath or oil slinger* Oil level maintained at predetermined height in casing (1,000 fpm maximum)

15-7 INVERTED-TOOTH CHAINS

The inverted-tooth, or silent, chain is made of overlapping links connected by pins in the Reynolds type of chain or by rocker joints in the Morse chain (see Fig. 15-7). Since the loading on the pin or rocker is nearly uniformly distributed over its length, bending of the pins is practically eliminated; thus a small connection may be used. The connection will therefore be so flexible that the chain links will accommodate themselves to small inaccuracies in the sprocket teeth, and pin breakage is uncommon. Thus the width of a silent chain is not so closely limited as in a roller chain. Silent chains over 30 in. wide have been used.

Inverted-tooth chains are manufactured in standard pitches, as shown in Table 15-3. In the chain number, SC refers to silent chain, and the numeral following is the number of $\frac{1}{8}$ in. in the pitch. The table gives the recommended maximum speed of the small sprocket for various pitches and number of teeth on the small sprocket. For best results, the small sprocket should have at least 21 teeth. Sprockets are available with 17 to 150 teeth. As an aid in selecting the pitch of a chain, see Eq. (15-2).

Equation (15-4) gives the recommended ratings for inverted-tooth chains in horsepower per inch of width of the chain. The computed value may be divided into the design horsepower to give the width of chain required. The range of chain widths should be between p and $10p$ but

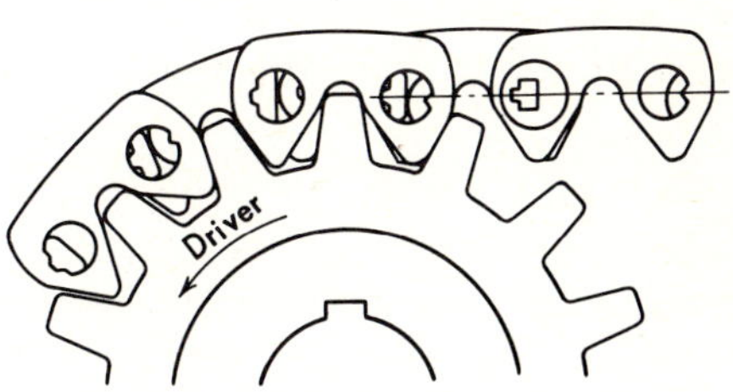

Fig. 15-7 Morse silent chain.

Table 15-3 Maximum rpm of small sprocket for inverted-tooth chains

Chain no.	SC3	SC4	SC5	SC6	SC8	SC10	SC12	SC16
Pitch, in.	3/8	1/2	5/8	3/4	1	1 1/4	1 1/2	2
Max. width, in.	6	8	10	12	16	20	24	30
Number of teeth				*Rpm*				
17	4,000	3,500	2,500	2,000	1,200			
19	5,000	3,500	2,500	2,000	1,500	1,200	1,000	700
21	6,000	3,500	3,000	2,500	1,800	1,200	1,000	700
23	6,000	4,000	3,000	2,500	1,800	1,800	1,200	800
25	6,000	4,000	3,500	2,500	1,800	1,800	1,200	900
27	6,000	4,000	3,500	2,500	2,000	1,800	1,200	900
29	6,000	4,000	3,500	2,500	2,000	1,800	1,200	900
31	6,000	4,000	3,500	2,500	2,000	1,800	1,200	900
33	6,000	4,000	3,500	2,500	2,000	1,800	1,200	900
35	6,000	4,000	3,500	2,500	2,000	1,800	1,200	900
37	5,000	3,500	3,000	2,500	1,800	1,200	1,000	800
40	5,000	3,500	2,500	2,500	1,500	1,200	900	800
45	4,000	3,000	2,000	2,000	1,500	1,000	900	700
50	3,500	2,500	2,000	1,800	1,200	1,000	800	600

preferably between $2p$ and $6p$. The widths vary by $\frac{1}{4}$-in. increments, and maximum standard widths are given in Table 15-3.

$$\text{hp per inch of width} = \frac{pV}{53}\left(1 - \frac{V}{425(T-8)}\right) \tag{15-4}$$

In Eq. (15-4), the first term depends on the allowable bearing pressure on the pins and the last term allows for the centrifugal tension and also for the number of teeth in the small sprocket, which is related to the angle of rotation of the links in seating on the sprocket.

The publication USAS B29.2–1957 and manufacturers' data give tables of horsepower ratings of inverted-tooth chains for various pitches and speeds of sprockets with various numbers of teeth.

15-8 MAXIMUM SPROCKET BORE

In small sprockets, it is necessary to determine whether the number of teeth and the pitch selected will result in a sprocket large enough to be mounted on the shaft with due allowance for a keyway. On this basis, the following relations may be used to determine the minimum number of teeth that may be used with a given pitch for a sprocket bore d.

Roller-chain sprocket:

$$T_{\min} = \frac{4d}{p} + 5 \qquad \text{for pitches to 1 in.} \tag{15-5a}$$

$$= \frac{4d}{p} + 4 \qquad \text{for pitches } 1\tfrac{1}{4} \text{ to } 2\tfrac{1}{2} \text{ in.} \tag{15-5b}$$

Silent-chain sprocket:

$$T_{\min} = \frac{4d}{p} + 6 \qquad \text{for pitches to 2 in.} \tag{15-5c}$$

15-9 LENGTH OF CHAIN

The following equation may be used to determine the approximate length of a chain:

$$L = \frac{T_1 + T_2}{2} + \frac{2C}{p} + \frac{p(T_1 - T_2)^2}{39.5C} \tag{15-6}$$

where L = length of chain in links
T_1, T_2 = numbers of sprocket teeth
C = center distance, in.
p = pitch, in.

The length in links should be used as an *even* number unless it is desired to use an *offset,* or *hunting,* link. The advantage of these links is that adjustment of the center distance may be in units of one link instead of a pair of links. A roller-chain offset link is shown in Fig. 15-5. The center distance for sprockets should not be less than the sum of the diameters of the sprockets.

If an exact chain length is required for a specified center distance or sprocket combination, the following equation may be used:

$$L = \frac{T_1 + T_2}{2} + \frac{2C \cos \alpha}{d} + \frac{d(T_1 + T_2)}{180} \tag{15-7}$$

The notation is the same as for Eq. (15-6); α is the angle between the centerline and the tangent to the sprocket pitch circles (see Fig. 14-2).

$$\sin \alpha = \frac{D_1 - D_2}{2C}$$

15-10 ADJUSTMENT FOR CHAIN TENSION

Means for adjusting the center distance should be provided to secure proper chain tension when the drive is installed and to make adjustment for wear of the chain in service. There should be a small amount of slack in the chain, but excessive slack may permit surging of the chain to take place. In some cases it may be necessary to use idler sprockets or shoes installed on the slack side of the chain to take up slack.

15-11 CHAIN CASES AND LUBRICATION

Chain drives should be enclosed to provide for proper chain lubrication, to keep dirt and grit from the chain and sprockets, and to ensure the safety of the operator. The case should be easy to install and to open for inspection and lubrication. Except at very low speeds the chain should not run through the oil. A high-speed chain running through oil causes an excessive amount of churning which raises the temperature of the oil and which may create high pressure in the case so that leakage may occur. Oil disks or rings may be used to distribute the lubricant.

Example Roller- and silent-chain selection. Select a chain drive to connect an 1,150-rpm induction motor to a centrifugal pump rated at 1,000 gpm against a head of 30 ft at 575 rpm of the pump. The efficiency of the pump is 63 percent and the efficiency of the chain drive may be taken as 98 percent. The fluid pumped is water and the duty is 10 hr per day.

Roller chain:

$$\text{Transmitted hp} = \frac{1{,}000 \times 8.33 \times 30}{33{,}000 \times 0.63 \times 0.98} = 12.2$$

Select 10-hp motor (Art. 25-6). Shaft size = 1⅝ in. From Fig. 15-6, for 1,150 rpm and 12.2 hp: trial chain is RC50, ⅝-in. pitch, single strand. From Fig. XVI-4 in Appendix XVI for 17-tooth sprocket, type III lubrication is satisfactory.

From Eq. (15-5*a*)

$$d_{\max} = \frac{p}{4}(T - 5) = \frac{0.625}{4}\,1{,}705 = 1.875 \text{ in.}$$

The small sprocket is satisfactory for 1⅝-in. bore. Sprocket diameters:

Small sprocket:

$$D_1 = \frac{p}{\sin(180/T_1)} = \frac{0.625}{\sin(180/17)} = 3.40 \text{ in.}$$

$$\text{Velocity ratio} = \frac{1{,}150}{575} = 2{:}1$$

Therefore $T_2 = 17 \times 2 = 34$ teeth

Large sprocket:

$$D_2 = \frac{p}{\sin(180/T_2)} = \frac{0.625}{\sin(180/34)} = 6.76 \text{ in.}$$

(Note that the diameter of the large sprocket is not equal to $2 \times D_1$.)
Center distance:

$$C = D_2 + \frac{D_1}{2} = 6.76 + \frac{3.40}{2} = 8.40 \text{ in.}$$

Chain length:

$$L = \frac{T_1 - T_2}{2} + \frac{2C}{p} + \frac{p(T_2 - T_1)^2}{39.5C}$$

$$= \frac{17 + 34}{2} + \frac{2 \times 8.40}{9.625} + \frac{0.625(34 - 17)^2}{39.5 \times 8.40}$$

$$= 25.5 + 26.8 + 0.545 = 52.8$$

Use 54 links.
Chain speed:

$$V = \frac{pTN}{12} = \frac{0.625 \times 17 \times 1{,}150}{12} = 1{,}020 \text{ fpm}$$

Inverted-tooth chain:

Assume 21 teeth in small sprocket.
Maximum pitch [Eq. (15-2)] is

$$\left(\frac{900}{N}\right)^{\frac{2}{3}} = \left(\frac{900}{1{,}150}\right)^{\frac{2}{3}} = 0.85 \text{ in.}$$

Assume ¾-in. pitch.

$$V = \frac{pT \text{ rpm}}{12} = \frac{0.75 \times 21 \times 1{,}150}{12} = 1{,}510 \text{ fpm}$$

$$\text{hp per in. of width} = \frac{pV}{53}\left[1 - \frac{V}{425(T - 8)}\right]$$

$$= \frac{0.75 \times 1{,}510}{53}\left[1 - \frac{1{,}510}{425(21 - 8)}\right]$$

$$= \frac{14.6}{15.6} = 0.935 \qquad \text{Use 1-in. wide chain.}$$

The numbers of sprocket teeth, sprocket diameters, center distance, and chain length may be calculated as for the roller-chain drive as

$T_1 = 21$ $\quad D_1 = 5.03$ in. $\quad T_2 = 42$ $\quad D_2 = 10.00$ in.
$C = 12.52$ in. $\quad L = 65.55$ $\quad$ Use 66 links.

15-12 TENSION CHAIN LINKAGES

This application is a chain that has lineal motion but not continuous in direction. The chain need not be formed as an endless belt. Figure 15-8(*a*) shows a hydraulic lift with a chain passing over a floating sprocket. In this arrangement the length of the cylinder is one-half the vertical lift of the platform. At (*b*) in the figure is a spring-controlled device to

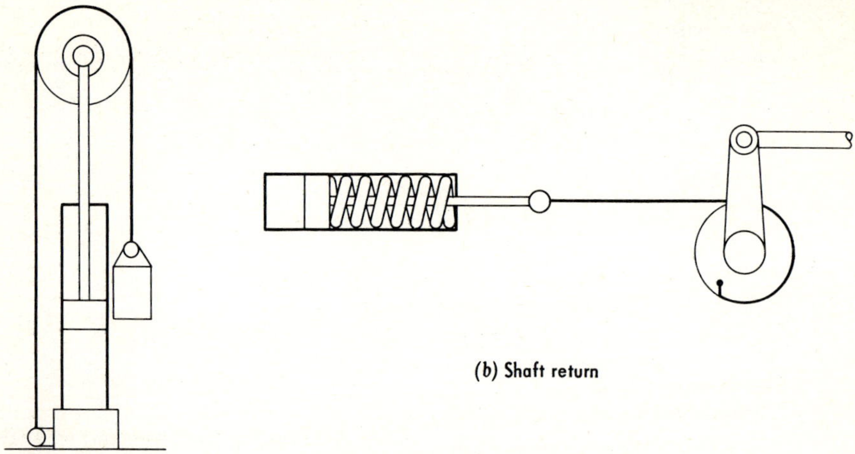

(b) Shaft return

(a) Hydraulic lift

Fig. 15-8 Tension linkages.

reposition a shaft that has been rotated through a limited angle. Other applications of chains in tension linkages are chain hoists, chain earth-drilling rigs, elevators for materials handling, lift trucks, and servomechanisms. For the latter class of applications, a miniature mechanical chain which has a pitch of 0.147 in. is available in roller and inverted-tooth construction, made of stainless steel or of beryllium copper (nonmagnetic) with sprockets as small as 0.3401 in. in pitch diameter.

The advantages of chains versus wire cables in tension linkages are positive connection between chain and sprocket, ease of flexure of chain, simplicity of end connection of chain to attached member, and ease of lubrication.

The allowable load in pounds on a single strand of a standard roller chain used in a tension linkage under steady conditions (no shock) may be determined from the equation

$$\text{Allowable load} = 1{,}900p^2 \tag{15-8}$$

where p is the pitch of the chain in inches. For the lightweight chain, No. 41, use 1,000 instead of 1,900 in the above equation.

Equation (15-8) is based on the ultimate strength of the chain and an average factor of safety of 8. The real factor of safety is lower, however, because the ultimate strength of the chain is not alone the criterion of failure, since bearing pressure, cyclic loading, stress concentration, and endurance limit of the links and pins are also involved. Any effects of shock loading should be allowed for by introducing a shock factor in accordance with the judgment of the designer.

16
Shaft Couplings

16-1 INTRODUCTION

Shaft couplings are used in machinery for several purposes, the most common of which are the following: (1) to provide for the connection of shafts of units that are manufactured separately, such as a motor and a generator, and to provide for disconnection for repairs or alterations; (2) to provide for misalignment of the shafts or to introduce mechanical flexibility; (3) to reduce the transmission of shock loads from one shaft to another; (4) to introduce protection against overloads; (5) to alter the vibration characteristics of rotating units.

Some commonly used types of shaft couplings are described briefly in the following articles.

16-2 RIGID COUPLINGS

This type of coupling has no flexibility or resilience; hence it is necessary for the shafts that are to be connected to be in good alignment, both

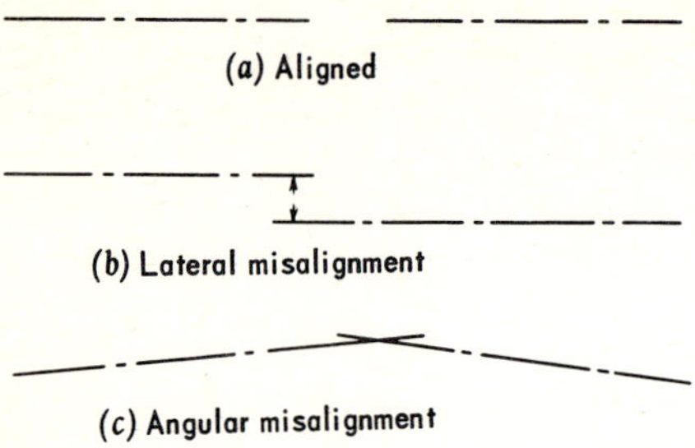

Fig. 16-1 Alignment of shaft centerlines.

laterally and angularly, in order to avoid excessive loads on the coupling, on the shafts, or on the shaft bearings. The two types of shaft misalignment are shown in Fig. 16-1.

Flange coupling This is a common form of rigid coupling; it is composed of two mating halves, which are shown assembled in Fig. 16-2. The coupling bolts should be ground and fitted to holes that are drilled and reamed in the *radial flanges* after the coupling halves are assembled. This procedure is necessary in order that the bolts should share equally the loads imposed on them when the coupling transmits torque. The purpose of the *circumferential flange* is to avoid danger of exposed rotating nuts and boltheads. The coupling should be designed to avoid failure of the key, failure by shearing off of the radial flange at its juncture with the hub, and failure of the bolts.

Compression coupling This coupling utilizes two split cones, as shown at c in Fig. 16-3, which are drawn together by the bolts b in order

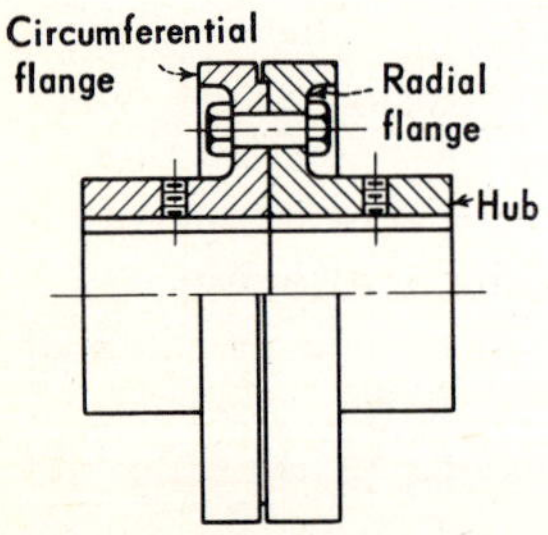

Fig. 16-2 Flanged shaft coupling.

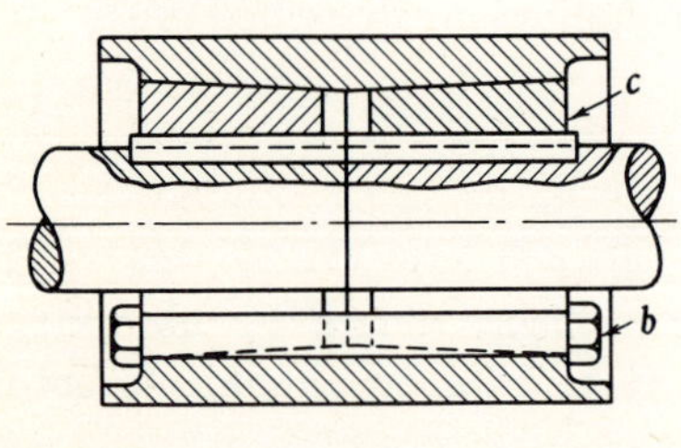

Fig. 16-3 Compression coupling.

to produce a wedging action which tightens the parts of the coupling and the shafts. The coupling in Fig. 16-3 is the Seller's coupling. Some types of compression couplings do not have keys but depend entirely on friction produced by the compression pressure to prevent slipping of the parts in transmitting torque.

16-3 COUPLINGS WITH KINEMATIC FLEXIBILITY

Oldham coupling This coupling may be used to connect shafts that have lateral misalignment. The three parts of the coupling shown in Fig. 16-4 are the two slotted hubs and the central floating part. The floating part has two tongues, one on each face, at right angles to each other. The tongue *a* fits into the slot of the right-hand member and allows for to-and-fro relative motion of the shafts, and the tongue *b* fits into the slot of the left-hand member and allows for vertical relative motion of the parts. The resultant of these two components of motion will accommodate lateral misalignment of the shafts as they rotate.

American flexible coupling This coupling, shown disassembled in Fig. 16-5, utilizes the Oldham principle. By widening the slots and tongues in the Oldham coupling, the central floating member may be formed into a block as shown in the figure so that it may ride between the faces of the jaw flanges of the driving and driven members. The bearing surfaces of the floating block are provided with replaceable nonmetallic strips which are lubricated from the grease reservoir in the hollow block. This coupling may be used for shafts that have lateral misalignment and may be used also with some angular misalignment if the coupling halves are properly mounted with axial clearance.

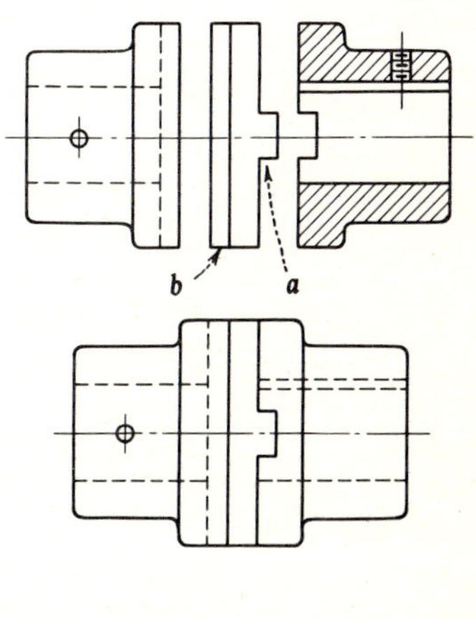

Fig. 16-4 Oldham coupling. Upper view shows parts; lower view shows assembly.

Fig. 16-5 American flexible coupling. (*Courtesy of Zurn Industries, Inc.*)

Amerigear flexible coupling This type of coupling transmits torque from one shaft to the other by means of two hubs and a floating sleeve. The hubs have external gear teeth cut on their peripheries, and the sleeve incorporates matching internal teeth. The two hubs are separated axially, and the gear teeth are of a special design with involute flanks and faces curved in both directions so that they may accommodate angular misalignment by a ball-and-socket action. This coupling has wide application in railway, automotive, air-transportation, and marine equipment as well as extensive industrial use.

Chain coupling This coupling is composed of two sprockets mounted face to face on the ends of the shafts and coupled and connected by a length of chain wrapped around the sprockets. This type of coupling is made by both silent-chain and roller-chain manufacturing companies. Small lateral and angular misalignments may be accommodated. The couplings are enclosed in a case filled with grease.

Universal joint Also known as Hooke's coupling, this is used to join two intersecting shafts. Angular misalignments up to 15 or even 30° may be accommodated. The angular velocity ratio for two shafts connected by one universal joint is not constant, but it may be made constant by the use of an intermediate shaft and two universal joints properly mounted.[1]

A constant-velocity universal joint that utilizes a group of rolling steel balls as the intermediate member has been developed. Two types

[1] Joseph E. Shigley, "Theory of Machines," p. 325, McGraw-Hill Book Company, New York, 1961.

are in use, namely, the Rzeppa[1] and the Bendix-Weiss[2] universal joints. Shafts with angles of from 30 to 40° may be coupled by a single joint of this type. Applications include automotive drives and machine tools.

16-4 FLEXIBLE COUPLINGS WITH RESILIENT PARTS

This type of coupling achieves flexibility by means of resilient parts or members interposed between the coupled parts. The resiliency has several functions, such as allowing for small lateral or angular misalignment, reducing the transmissibility of shock loads from one unit to another, and altering the vibration characteristics of the connected parts.

Ajax coupling This coupling, shown in Fig. 16-6, is composed of two halves with one set of pins rigidly attached to one half of the coupling and the alternate pins attached to the other half. Each steel pin that is attached to one half of the coupling projects into a bronze bushing that is mounted in a rubber sleeve in the other half. The coupling halves are mounted with axial clearance. A small amount of angular and lateral misalignment will be taken up by the rubber sleeves.

Falk coupling This coupling is composed of two slotted members, shown in Fig. 16-7(*a*), which are connected by a continuous steel spring lying in the slots. The elements of the spring provide the flexibility of the coupling.

The sides of the slots in which the spring lies are formed as shown at (*b*) in the figure. It may be noted that the effective length of the elements of the spring changes from a maximum g at no load to a minimum g' at

[1] Michigan Tool Co.—Dana Corporation, Detroit, Mich.

[2] Bendix Products Corp., South Bend, Ind. See H. H. Mabie, Constant Velocity Joints, *Machine Design*, May, 1948.

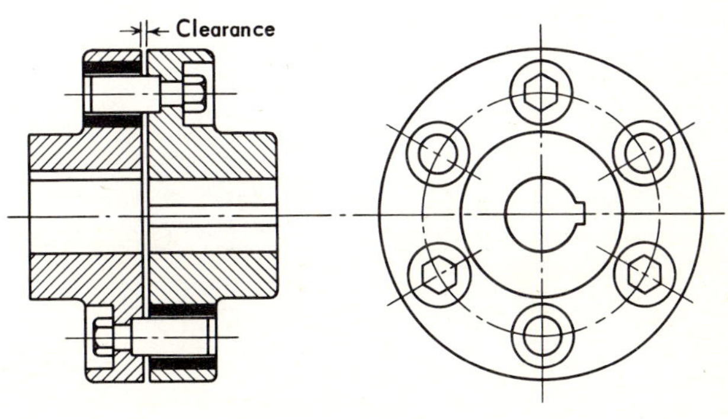

Fig. 16-6 Ajax flexible coupling.

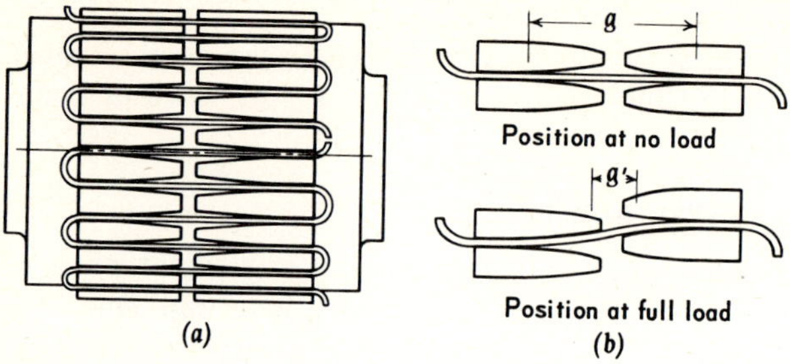

Fig. 16-7 Falk coupling.

full load, so that as the torque on the coupling increases the coupling becomes "stiffer." This is a very desirable characteristic for installations in which torsional vibration of the shaft is an important consideration. The coupling parts may be readily disconnected by removing the steel spring.

Westinghouse-Nuttall coupling This coupling has helical springs as a resilient connection between the two halves, shown in Fig. 16-8. It is used to reduce the transmission of shock loads and to alter the torsional vibration characteristics of rotating members. This type of resilient connection is also frequently interposed between the hub and rim of a gear sprocket or clutch to form a self-contained shock-absorbing member.

Leather-link coupling This coupling employs leather links to connect the driving and driven members, as shown in Fig. 16-9. It should be noted that one set of links are pulling links for one direction of rotation whereas the alternate set are trailing links. In Fig. 16-9 the links a are pulling links for clockwise rotation. The trailing links are necessary to prevent backlash in the event of any tendency to reversal of motion, such as shock loading.

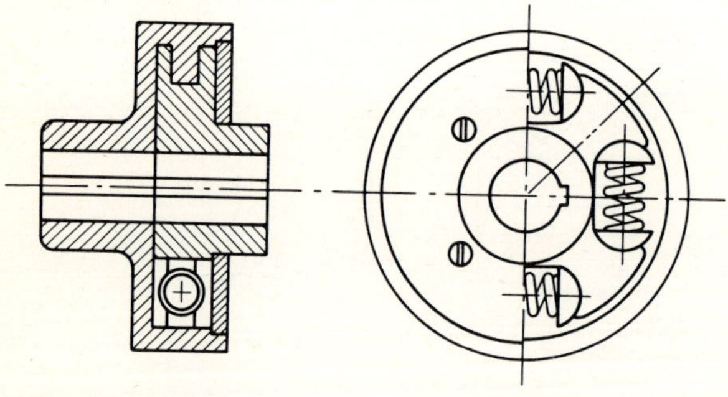

Fig. 16-8 Westinghouse-Nuttall coupling.

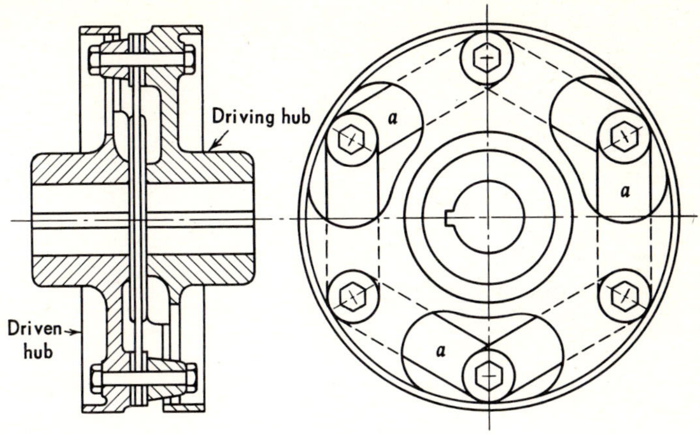

Fig. 16-9 Leather-link coupling.

Flexible disk coupling This coupling is similar in construction to the leather-link coupling, except that a disk made of leather or rubberized canvas is used instead of links. This type of coupling has extensive application in drives transmitting low power.

Lord coupling The two halves of this coupling (Fig. 16-10) are connected by a section made of rubber or neoprene that is bonded to the halves of the coupling.

16-5 SLIP COUPLINGS

This coupling is designed to permit relative rotation, or slip, between the driving and driven parts at a predetermined torque for the purpose of preventing damage to parts because of overloading. There are many variations in form of this type of coupling. A common form for medium and large capacities is illustrated in Fig. 16-11. The flanged hub a is keyed to the driving member. The driven member b is composed of a flanged hub with friction disks attached to both sides of the flange. Several springs, as shown at c, are loaded by tightening the bolts and thus create pressure on the friction surfaces.

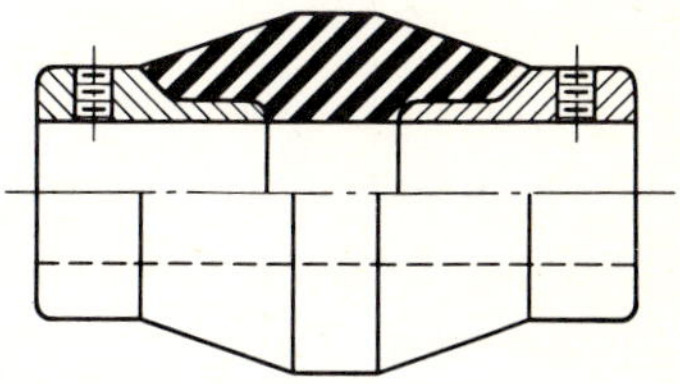

Fig. 16-10 Lord coupling.

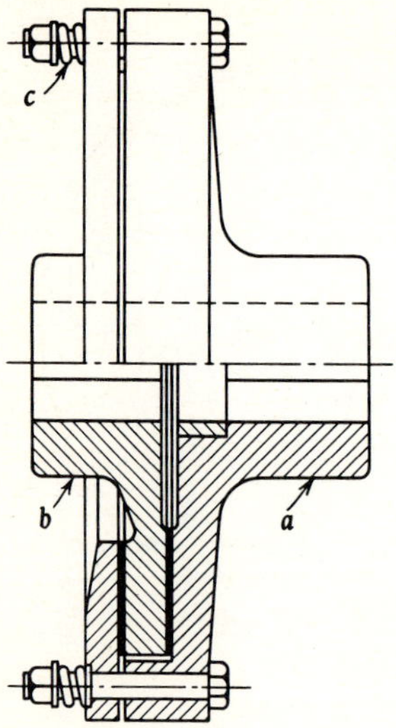

Fig. 16-11 Slip coupling.

16-6 FLUID COUPLINGS

The fluid coupling is composed of a primary, or input, element called the *impeller* and a secondary, or output, element called the *runner*. The enclosed fluid provides the connection between the impeller and runner. The impeller and runner are essentially identical in construction. The principle of operation is as follows: The fluid, usually oil, begins its cycle at the inner radius of a cell that it occupies, as at a in Fig. 16-12(a). When the impeller rotates, the fluid flows outward toward b because of centrifugal force. Since the runner would be initially at rest or *rotating*

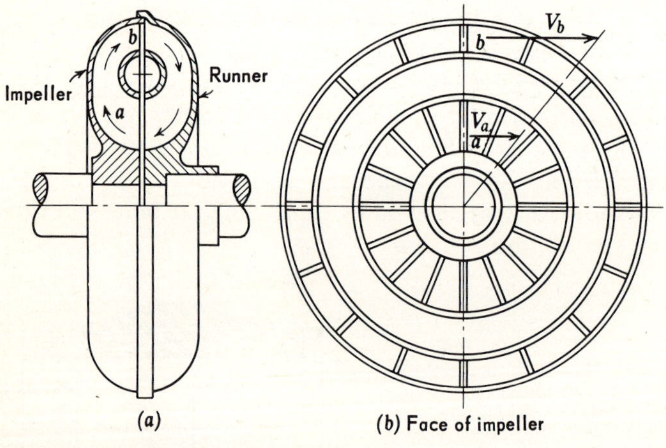

Fig. 16-12 Hydraulic coupling.

at a speed lower than that of the impeller, the centrifugal force of the fluid in the impeller cells which causes fluid flow in a clockwise direction, as shown, will be greater than the centrifugal force of the fluid in the runner cells which tends to cause counterflow. Hence, when the impeller is rotating, there will be fluid flow in a clockwise direction, as shown at (*a*) in the figure.

The above discussion accounts for flow of the fluid from the impeller to the runner and back to the impeller. Now, as shown in Fig. 16-12(*b*), the fluid at position *a* will have a tangential velocity V_a. When the fluid arrives at position *b*, it will have a tangential velocity V_b that is greater than V_a by the ratio of the corresponding radii. A unit mass of fluid will therefore have gained kinetic energy in flowing from *a* to *b*, and this gain represents the input to the coupling.

The next phase in the operation is the transfer of the kinetic energy, which the mass of fluid has gained, to the runner as the fluid passes radially inward through the cells of the runner while its tangential velocity is reduced from V_b to V_a. The energy thus transferred to the runner imparts rotation to the driven member.

The difference between the impeller and runner speeds is known as *slip* and is usually expressed as a percentage of the impeller speed. For well-designed fluid couplings, the slip may be as low as 1 percent at rated capacity.

In a fluid coupling, the fluid connection between the impeller and runner results in a marked decrease in transmission of torsional vibrations.

By controlling the quantity of fluid in the coupling, the speed of the runner may be varied. This feature makes the fluid coupling useful in speed control. Other applications of this coupling are due to its torque-speed characteristics.

In order to avoid self-induced vibrations set up by the coupling on account of the "register," or matching, of the vanes in the impeller and runner, the number of vanes in one half of the coupling is one more or less than the number of vanes in the other half of the coupling.

17
Clutches and Brakes

17-1 INTRODUCTION

In the operation of a *friction clutch* during engagement, the initial condition is that the driving member is moving and the driven member is at rest; in the final condition both members are moving at the same speed, i.e., with no relative motion. In the operation of a *brake* during braking, the initial condition is that one member, such as the brake wheel or drum, is moving and the braking member is stationary; in the final condition both members are at rest and have no relative motion.

It is thus apparent that the principle of operation of both a friction clutch during engagement and of a friction brake during braking is to bring two members having relative motion to the state of no relative motion. The operation of a clutch is therefore essentially the same as that of a brake; however, there are structural differences in the two units because of control requirements and the necessity for providing for heat absorption or dissipation in brakes.

17-2 PLATE CLUTCHES AND BRAKES

Plate clutch In the clutch shown in Fig. 17-1, the part *a* is connected to the driving unit (engine or motor), and the two driving plates *b* are splined to *a*. The driven plates *c* are connected to the driven part by the splines at *d*. When the toggle *efg* is in the position shown, the driven

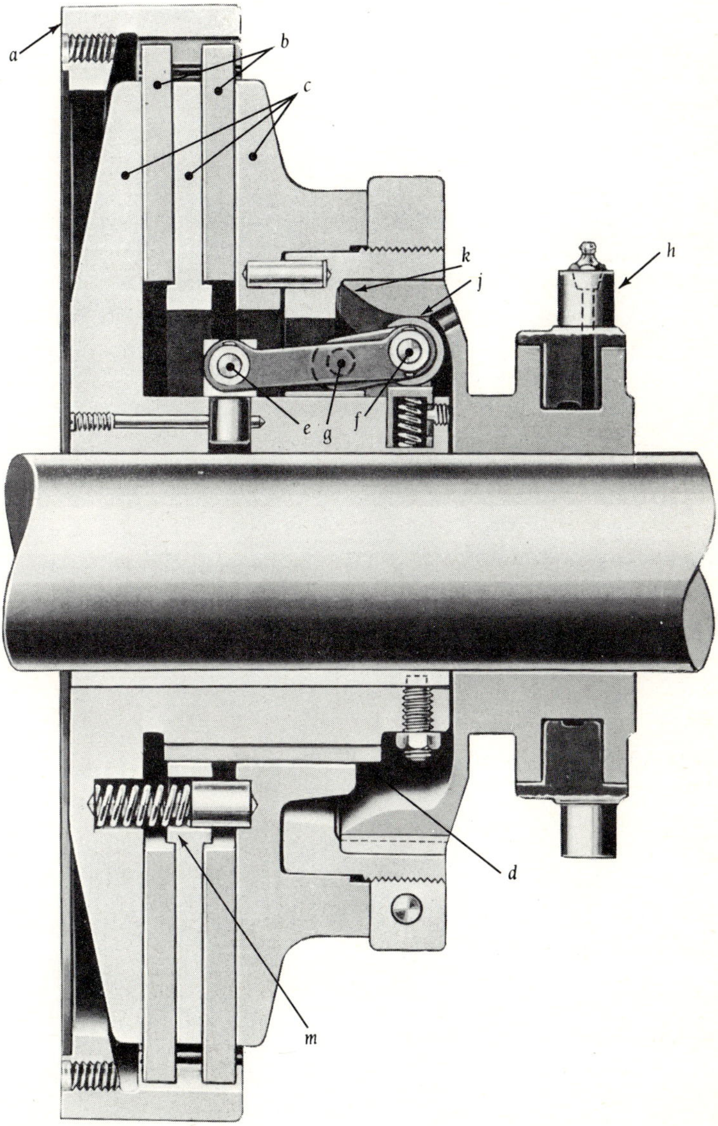

Fig. 17-1 Plate clutch. (*Courtesy of Dodge Manufacturing Corporation.*)

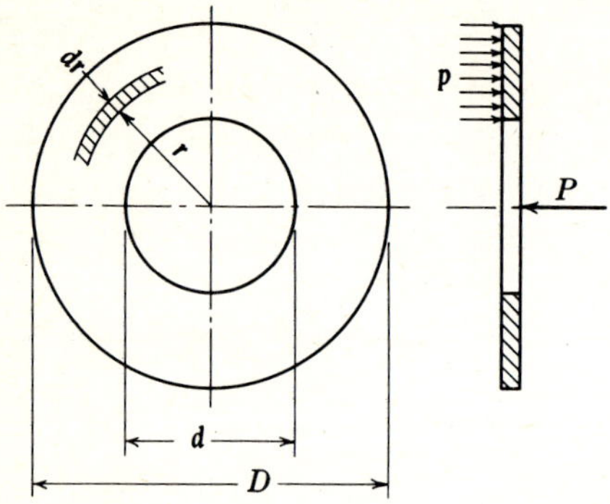

Fig. 17-2 Friction disk.

plates are pulled together to produce rotation of the driven part of the clutch *c*. To release the clutch, a yoke connected to the pins *h* is moved to the right; this allows the rolls at *f* of the toggle mechanism to move radially outward along the internal annular cam *jk*, which relieves the axial force between the disks so that the springs *m* separate the plates. In this clutch there are four pairs of friction surfaces between the driving plates *b* and the driven plates *c*. The output part of the clutch must be "built into" the driven unit to transmit power.

Force analysis In Fig. 17-2 there is shown a friction disk with outer and inner diameters, D and d, respectively. The mating disk, not shown, produces a pressure on the surface because of the axial load P.

$$\begin{aligned} \text{Elementary surface area } dA &= 2\pi r\,dr \\ \text{Normal force on } dA &= p2\pi r\,dr \\ \text{Frictional force on } dA &= fp2\pi r\,dr \end{aligned}$$

where p = surface pressure

f = coefficient of friction, assumed as constant

Therefore

$$P = 2\pi \int_{d/2}^{D/2} pr\,dr \tag{a}$$

and

$$T = 2\pi f \int_{d/2}^{D/2} pr^2\,dr \tag{b}$$

Before Eqs. (*a*) and (*b*) may be integrated, it is necessary to make an assumption regarding the distribution of pressure. Two cases will be considered.

1. *Uniformly distributed pressure.* For new clutches and rigid mountings, the pressure may be assumed to be uniformly distributed over the contact area; hence, p may be regarded as constant.

Therefore, from Eq. (*a*),

$$P = \frac{\pi p}{4}(D^2 - d^2) \tag{c}$$

and, from Eq. (*b*),

$$T = \frac{2\pi pf}{24}(D^3 - d^3) \tag{d}$$

By eliminating p from Eqs. (*c*) and (*d*) and solving for T,

$$T = \frac{fP}{3}\frac{D^3 - d^3}{D^2 - d^2} \tag{17-1}$$

where T is the torque for *one pair* of friction surfaces in contact.

2. *Uniform axial wear.* An inspection of worn clutch plates reveals that the plates are not of uniform thickness. To account for the nonuniformity of wear, it is necessary to make an assumption regarding wear.

The mechanism of wear is the gradual tearing away by interlocking (friction) of particles of the rubbing surfaces. Thus friction and wear are copartners. The rate of wear depends on the friction force at the surfaces in contact, which depends on the pressure, and it depends on the rubbing velocity and on the hardness of the surfaces. Thus the wear on an element of area in one revolution depends on the product of fp and the distance $2\pi r$. If the coefficient of friction is assumed as constant, the wear is proportional to the product pr. This quantity is proportional to the *work of friction* on the unit area; hence, the assumption may be stated *as the normal wear of the surface is proportional to the work of friction.* If n is the normal wear (perpendicular to the surface) and k is the constant of proportionality, then $n = kpr$.

In friction clutches and brakes, it is customary to employ a hard metallic surface, such as cast iron, for one surface, with a replaceable lining or facing made of softer material with a high coefficient of friction for the contacting surface. The latter may be composition fiber, compressed woven asbestos, wood, cork, sintered metal, etc. In the following discussion it is assumed that the wear will be concentrated on the softer material.

In Fig. 17-3(*a*), the pressure is assumed to be uniformly distributed as for new plates; hence the normal wear at the outer radius is equal to kpr_2 and at the inner radius is equal to kpr_1. It follows that wear will be greater at the outer radius; hence the plate will wear until it has assumed a shape such as the one shown at (*b*) in the figure. This changed shape will result in a redistribution of the pressure, so that p_1 at the inner radius will be greater than p_2 at the outer radius. Wear will take place in this

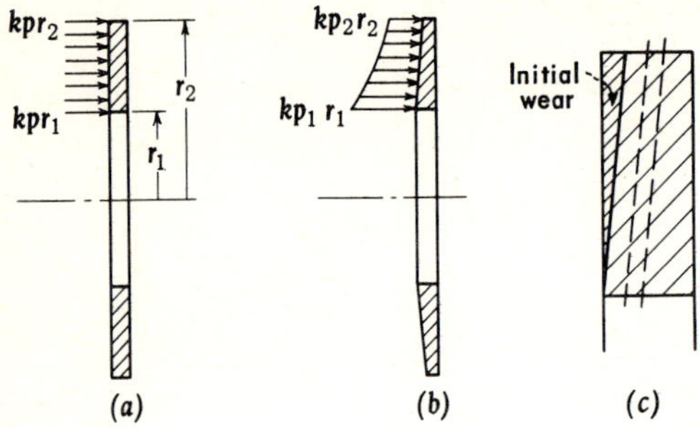

Fig. 17-3 Disk pressure and wear.

manner until the pressure distribution is such that the product p_2r_2 equals p_1r_1. Then the wear will progress uniformly along parallel surfaces.

It is apparent that wear of a friction plate takes place in two stages: (1) the initial wear as the stable pressure distribution becomes established, which is shown in Fig. 17-3(*c*); (2) as the plates wear during the life of the plates, the surfaces follow the dotted lines as shown in the figure.

When the equilibrium condition is reached,

$$n = kpr = \text{const}$$

$$\text{and} \quad p = \frac{C}{r} \tag{e}$$

By substituting the value of p from Eq. (*e*) into Eqs. (*a*) and (*b*), integrating, and eliminating the constant C, the equation for the torque becomes

$$T = f\frac{P}{4}(D + d) \tag{17-2}$$

where T is the torque for *one pair* of friction surfaces in contact.

In Eq. (17-1), which is for *new* clutches and brakes, the term

$$\frac{1}{3}\frac{D^3 - d^3}{D^2 - d^2}$$

may be called the *friction radius*. In Eq. (17-2) the friction radius is $\frac{1}{4}(D + d)$. A comparison of these equations shows that the friction radius for new clutches is slightly larger than for worn-in clutches, and the percentage difference may be expressed in terms of the ratio D/d, as shown in Fig. 17-4. The ratio D/d for industrial clutches and brakes and for automotive clutches is of the order of 1.5. For this proportion the difference in the two equations is very low and is much smaller than the variation in the value for the coefficient of friction; hence on the basis of accuracy the choice between the two equations is unimportant. However, Eq. (17-2) gives values for the axial force which are on the side of safety, it applies to most of the life of the plates, and it can be more

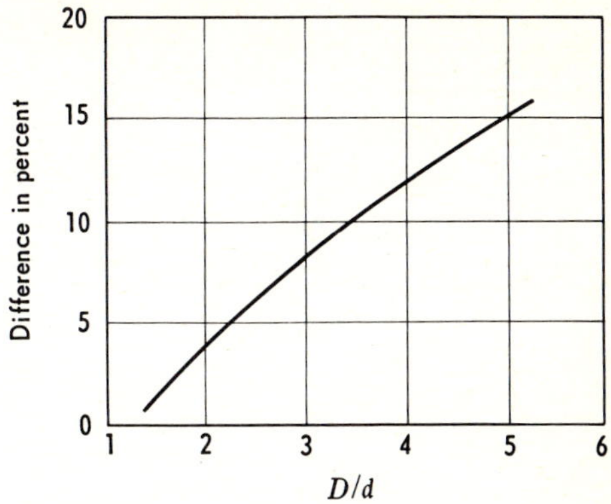

Fig. 17-4 Comparison of Eqs. (17-1) and (17-2).

directly related to space limitations during design and is therefore recommended for use rather than the uniform-pressure equation.

Multidisk clutch Where large torques must be transmitted, a multidisk clutch, as shown in Fig. 17-5, may be used to limit the operating force. The torque transmitted by this clutch may be determined by multiplying the right-hand member of Eq. (17-2) by the number of pairs of friction surfaces in contact.

Disk brakes Disk brakes are used to a limited extent in some installations where speed control is necessary. In Fig. 17-6, there is shown a multidisk brake used with a hoisting drum. The member a is keyed to the shaft b by means of the feather key c, thus causing a to rotate with the shaft but permitting it to slide axially during engagement and disen-

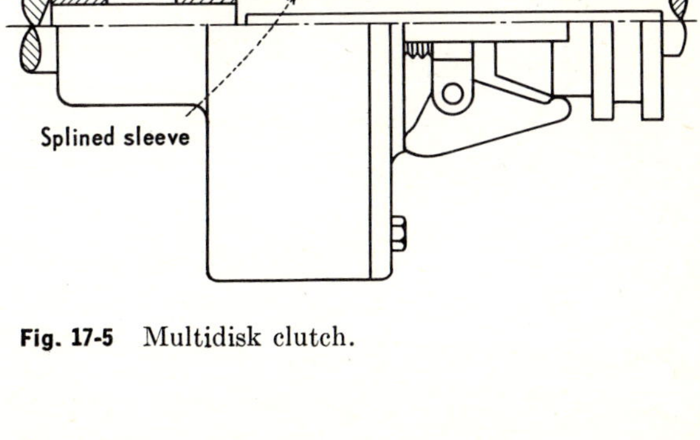

Fig. 17-5 Multidisk clutch.

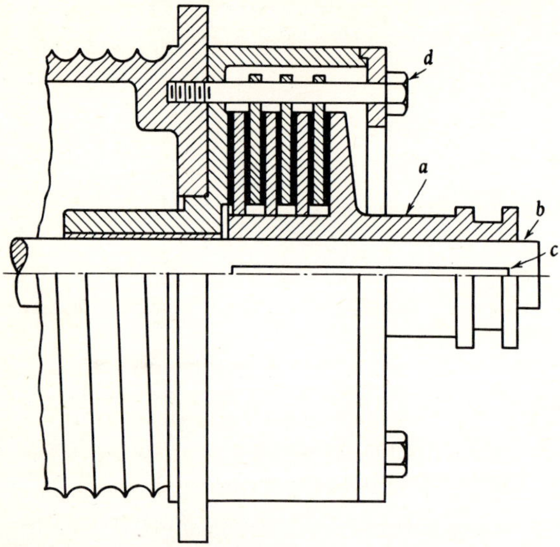

Fig. 17-6 Multidisk brake.

gagement. One set of friction disks is splined to the member *a*, and the other set slides axially on the bolts *d* but rotates with the drum. When it is desired to raise the load, a force on the operating lever slides the member *a* to the left and connects it with the drum, and the unit acts as a clutch. The shaft is prevented from rotating backward by a ratchet. In order to lower the load, the operating force is decreased, which allows the load to be lowered. During this phase the unit functions as a brake.

Single-disk brakes are presently developed for front-wheel automotive service in which the shoe encompasses only an annular segment of the disk. The two parts of the shoe are forced against the disk by a caliper action to provide the pressure. An important advantage of this type of brake is increased cooling of the disk, owing to its exposed areas, which increases its resistance to "fading" because of high temperature of the lining during severe braking.

17-3 CONE CLUTCHES

Cone clutch Figure 17-7 shows a cone clutch in which the outer cone *a* is the driving member and the inner cone *b* is the driven member. The compression spring provides the force to engage the clutch. In order to disengage the clutch, a shifting fork on a collar that runs in the groove *c* may be used to slide the inner cone to the right, thus separating the cones.

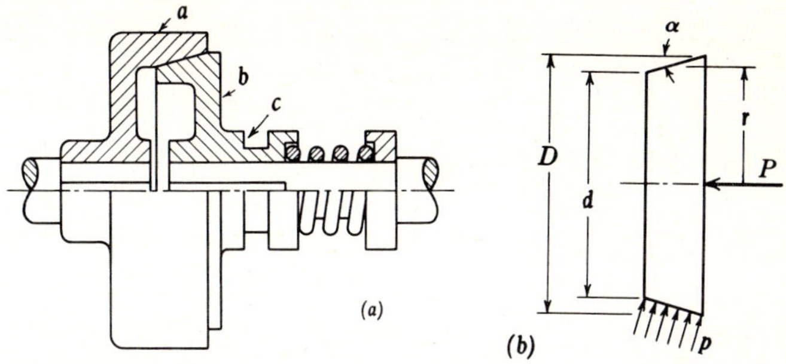

Fig. 17-7 Cone clutch.

Force analysis The relation between the torque transmitted and the axial force may be obtained by following a similar procedure to that for the plate clutch in Art. 17-2. The friction surface of the inner cone of the clutch is shown in Fig. 17-7(*b*). The cone-face angle is α, as shown, and the axial force is P.

$$\text{Elementary surface area } dA = \frac{2\pi r\,dr}{\sin\alpha}$$

$$\text{Normal force on } dA = \frac{p2\pi r\,dr}{\sin\alpha}$$

$$\text{Frictional force on } dA = \frac{fp2\pi r\,dr}{\sin\alpha}$$

$$P = 2\pi \int_{d/2}^{D/2} pr\,dr \tag{a}$$

and $$T = \frac{2\pi f}{\sin\alpha}\int_{d/2}^{D/2} pr^2\,dr \tag{b}$$

From the assumption that "the normal wear is proportional to the work of friction," as discussed in Art. 17-2, it may be deduced that

$$p = \frac{C}{r} \tag{c}$$

By substituting the value of p from Eq. (*c*) into Eqs. (*a*) and (*b*), integrating, and eliminating the constant C, the following equation is determined for the torque transmitted by a cone clutch:

$$T = \frac{fP(D+d)}{4\sin\alpha} \tag{17-3}$$

By making the cone angle α equal to 90°, the cone clutch becomes a plate clutch, and Eq. (17-3) reduces to Eq. (17-2). It may be noted that a given torque may be transmitted by a relatively small axial force if the cone-face angle is decreased. There is a lower limit to the angle α, how-

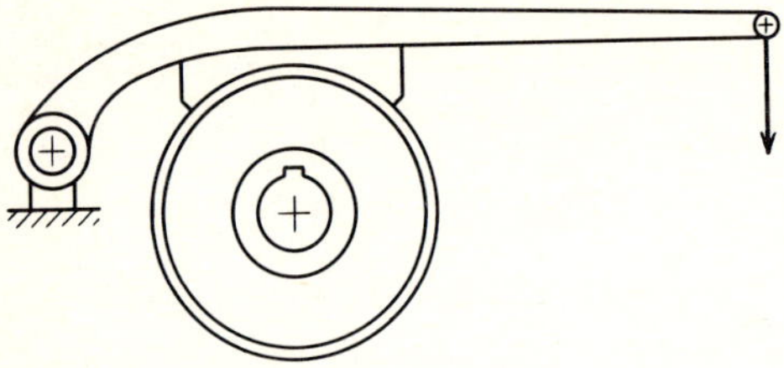

Fig. 17-8 Single-block brake.

ever, since the frictional force that must be overcome in releasing the clutch increases as α decreases. Thus a clutch with a small cone angle requires a relatively small force to engage the clutch but a large force for disengagement. The SAE recommends an angle $\alpha = 12.5°$ for cone clutches faced with leather or asbestos or having cork inserts.

Cone brakes are similar to cone clutches in construction and operation.

17-4 BLOCK BRAKES AND CLUTCHES

In this type of brake or clutch one or more blocks, or shoes, are forced against a wheel that usually has a cylindrical surface.

Single-block brake In Fig. 17-8 is shown a single-block brake in which the block attached to the operating lever is forced against the rotating wheel. The frictional force produced by the block on the wheel will retard the rotation of the wheel.

Force analysis for friction block brake A cylindrical wheel, or drum, that is assumed to rotate as indicated is shown in Fig. 17-9. The block

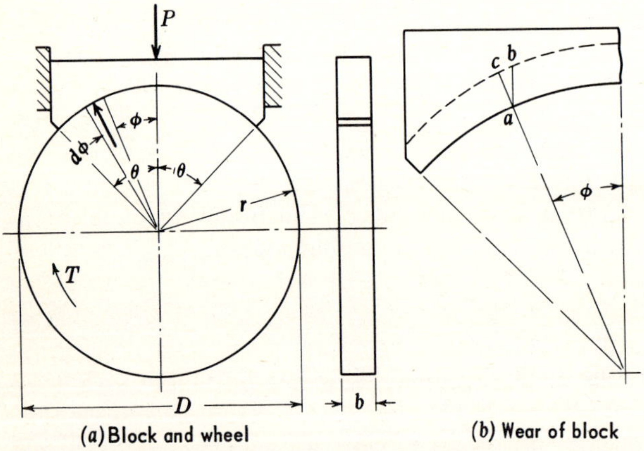

Fig. 17-9 Single-block brake.

is forced against the wheel by a radial force P. The relation between the force P and the torque T may be determined as follows:

Let P = operating force on block in radial direction
D = diameter of wheel
T = torque on wheel
θ = one-half angle of contact surface of block
b = width of wheel
f = coefficient of friction for materials of block and wheel
p = pressure between block and wheel

$$\text{Elementary area of contact } dA = \frac{D}{2} b\, d\phi$$

$$\text{Normal force on } dA = p \frac{D}{2} b\, d\phi$$

The component of this normal force parallel to P is equal to

$$dP = p \frac{D}{2} b \cos \phi \, d\phi$$

or $$P = \int_{-\theta}^{+\theta} p \frac{D}{2} b \cos \phi \, d\phi = \frac{Db}{2} \int_{-\theta}^{+\theta} p \cos \phi \, d\phi \qquad (a)$$

$$\text{Force of friction on elementary area} = fp \frac{D}{2} b\, d\phi$$

$$T = \int_{-\theta}^{+\theta} fp \frac{D^2}{4} b\, d\phi \qquad (b)$$

By assuming that the coefficient of friction is constant, the equation for T becomes

$$T = f \frac{D^2 b}{4} \int_{-\theta}^{+\theta} p\, d\phi \qquad (17\text{-}4)$$

In order to obtain an expression for the pressure in terms of ϕ, it is necessary to make an assumption, a rational one being that the "normal wear is proportional to the work of friction."

In the usual construction of a block brake or clutch, the wear takes place mainly on the block or lining attached to the block. As wear occurs, the block or lining will retain the cylindrical shape of the wheel, as shown in Fig. 17-9(*b*). If it is assumed that the block is constrained to move in a radial direction, the component of wear in the direction of P, that is, $\overline{ab}$, will be constant. Therefore the normal wear $\overline{ac} = \overline{ab} \cos \phi$.

Now the work of friction on an elementary area is proportional to the pressure p; hence from the assumption that the normal wear $\overline{ac}$ is proportional to the work of friction

$$p \text{ is proportional to } \overline{ab} \cos \phi \qquad \text{or} \qquad p = C \cos \phi$$

where C is a constant of proportionality. From Eq. (a)

$$P = \frac{CDb}{2}\int_{-\theta}^{+\theta}\cos^2\phi\,d\phi$$
$$= \frac{CDb}{4}(2\theta + \sin 2\theta)$$

and from Eq. (b)

$$T = \frac{CfD^2b}{4}\int_{-\theta}^{+\theta}\cos\phi\,d\phi = 2Cfb\left(\frac{D}{2}\right)^2\sin\theta$$

By eliminating C from the two preceding equations,

$$T = fP\frac{D}{2}\frac{4\sin\theta}{2\theta + \sin 2\theta} \tag{17-5}$$

The tangential frictional force on the block may be found by dividing the torque by the radius, or $F = 2T/D$; hence,

$$F = fP\frac{4\sin\theta}{2\theta + \sin 2\theta} \tag{17-6}$$

In the above equation, the quantity $f[(4\sin\theta)/(2\theta + \sin 2\theta)]$ may be termed the "equivalent coefficient of friction" and denoted by f'; hence

$$F = f'P \tag{17-7}$$

In Fig. 17-10 is shown the variation of the function $(4\sin\theta)/(2\theta + \sin 2\theta)$ with the angle θ.

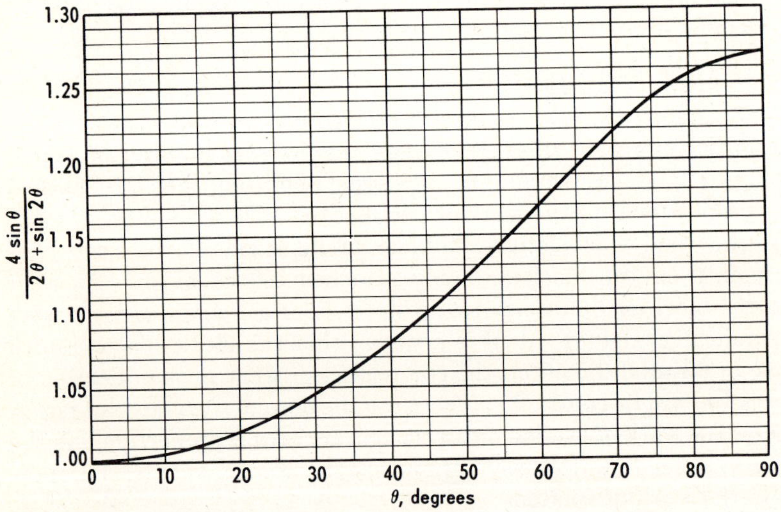

Fig. 17-10 $(4\sin\theta)/(2\theta + \sin 2\theta)$ plotted against semiblock angle θ.

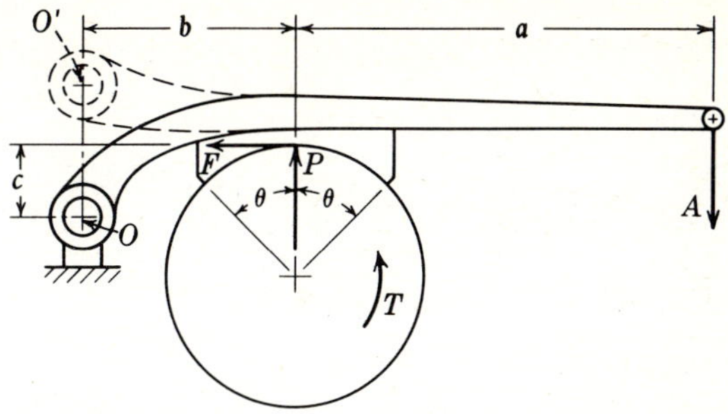

Fig. 17-11 Forces on brake lever.

The forces acting on the operating lever of the single-block brake in Fig. 17-8 are shown in Fig. 17-11 for counterclockwise rotation of the wheel.

For the equilibrium of moments about O of the forces acting on the lever, if the resultant of the frictional forces may be assumed to be a force F acting as shown,

$$A(a + b) - Pb - Fc = O \tag{17-8}$$

From Eqs. (17-7) and (17-8) the equation for the torque in Fig. 17-11 becomes

$$T = \frac{f'AD(a + b)}{2(b + f'c)} \tag{17-9}$$

In Eq. (17-8), the moment Fc is in the opposite direction to the moment due to the operating force A, and hence the self-induced moment Fc retards application of the brake.

For clockwise rotation of the brake wheel in Fig. 17-11, the force F will be reversed and the moment Fc will aid the application of the brake. Hence, such a brake is known as "self-energizing." If the distance c is great enough, the moment Fc will be sufficient to apply the brake without an appreciable operating force A. This may be considered an advantage, but such a brake may "grab" and result in uncontrolled braking.

By locating the pivot point above the line of action of F (at O'), the "self-energizing" feature will be present for counterclockwise rotation.

Figure 17-12 shows one form of a two-shoe (double-block) spring-set brake. The linkage at the top of the brake is arranged so that the compressive force of the spring moves the yokes together and applies, or "sets," the brake. The brake lining is compressed woven asbestos and is fastened to the cast brake shoes by brass (nonscoring) rivets.

The brake is released by means of a General Electric Thrustor mechanism, which is a hydraulic unit operated by the electric motor

Fig. 17-12 Double-block spring-set brake. *(Courtesy of General Electric Corporation.)*

shown.[1] This mechanism, pushing on a floating bell crank, spreads the yokes against the spring force and releases the shoes from the wheel. This control is especially suitable for large brakes because of its smooth operating characteristics. In smaller brakes a solenoid is suitable. Other releasing devices are magnets, torque-motor-driven screws, and hydraulic cylinders.

The brake shoes in Fig. 17-12 are pivoted on the yokes to provide uniform contact between the lining and the wheel. The tangential friction force $f'P$ on the brake shoe introduces a moment on the shoe which makes the distribution of pressure somewhat different from the case in Fig. 17-9; however, if the moment arm is kept small by locating the pivot as near as possible to the contacting surface of the brake lining, i.e., near the line of action of the friction force, the turning moment on the shoe

[1] See General Electric Company literature, or R. T. Hinkle, "Design of Machines," pp. 93 and 98, Prentice-Hall Inc., Englewood Cliffs, N.J., 1957.

will be small and Eq. (17-7) may be used. Thus, the force diagram in Fig. 17-13 may be used without serious error.

An electric device connected in series or shunt with the motor is an ideal arrangement so that the brake is released when the motor is started, and the brake is applied by the spring when the motor switch is opened. In some applications a time-delay relay is used to allow the moving parts to slow down before the brake is applied.

In Fig. 17-13, the relation between the spring force S and the torque on the brake wheel may be determined as follows:

$$\Sigma M_{o_1} = S(a + b) + F_1 c - P_1 b = 0$$
$$\Sigma M_{o_2} = S(a + b) - F_2 c - P_2 b = 0$$

The relation between F and P may be determined by using Eq. (17-7), or

$$\frac{F_1}{P_1} = \frac{F_2}{P_2} = f'$$

From the above equations, F_1 and F_2 may be determined. Then

$$T = (F_1 + F_2)\frac{D}{2} \tag{17-10}$$

Because of the difference in values of F_1 and F_2 the wear of the two brake linings will be unequal if the drum always rotates in one direction. The unequal wear is usually not serious, since the wear is very slight in high-quality linings, and also many brakes operate in both directions of rotation in service, which tends to equalize the wear of the linings of the two shoes. If the pivot points O_1 and O_2 in Fig. 17-13 are located on the lines of action of F_1 and F_2, the shoes will have equal wear. The effect of placing the pivot points as shown in Fig. 17-13 is that a smaller spring force is required to resist a given torque than if the pivot points were located on the lines of actions of the friction forces F_1 and F_2.

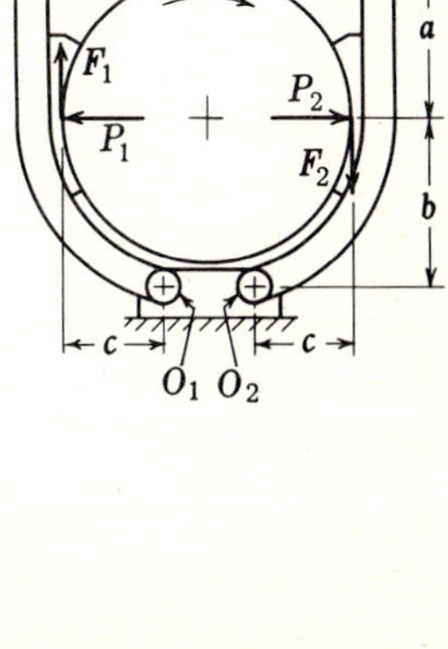

Fig. 17-13 Double-block brake.

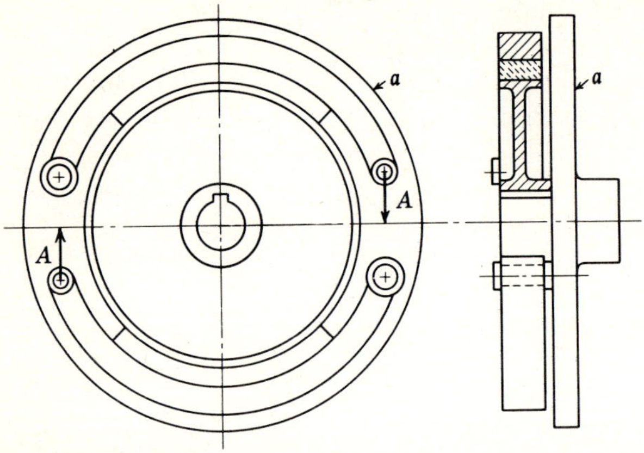

Fig. 17-14 Double-block clutch.

In brakes of the automotive type with internal-expanding shoes,[1] the wear pattern is somewhat different from that discussed earlier in this article; however, the difference in actuating-force–braking-force relations for the internal-expanding shoe brake and the block type in Fig. 17-11 will normally be less than 10 percent; hence the same procedure as used in the example of Fig. 17-13 may be followed. The results will be on the conservative side if that procedure is used for the internal-expanding shoe brake. The analysis of the external-contracting shoe brake can likewise be conservatively made by using the procedure as applied to the example of Fig. 17-13. It is well to remember in all friction brake and clutch applications that the coefficient of friction in service may vary considerably from its value in laboratory tests.

In all types of pivoted-arm or pivoted-shoe brakes, the pin reactions can be readily determined from a free-body diagram of the forces acting on the arm or the shoe.

Double-block clutch The construction of a block clutch is similar to that of a block brake except that provision must be made for rotating the blocks and for controlling the load on the blocks. A double-block clutch is shown schematically in Fig. 17-14. The driving plate a carries the two pivoted levers, to each of which is attached a block. The operating forces A are applied by a suitable linkage.

[1] Joseph E. Shigley, "Mechanical Engineering Design," McGraw-Hill Book Company, New York, 1963.

17-5 BAND BRAKES AND CLUTCHES

Band brakes A band brake is shown in Fig. 17-15 in which the band is tightened by means of the force A on the operating lever. This band may be made of leather, of canvas impregnated with rubber, or of steel faced with wood blocks.

For clockwise rotation of the wheel as shown, the tension F_1 in the band attached to O will be greater than the tension F_2. The sum of the moments of the forces acting on the lever is

$$\Sigma M_o = F_1 O + F_2 a - A(a + b) = 0$$

The relation between F_1 and F_2 may be determined from Eq. (14-5) by deleting the centrifugal force term F_c,

$$\frac{F_1}{F_2} = e^{f\theta}$$

The torque on the brake wheel is given by the following equation:

$$T = (F_1 - F_2)\frac{D}{2}$$

From the three equations above, the relation between the force on the operating lever and the torque on the brake wheel becomes

$$A = \frac{2Ta}{D(a + b)(e^{f\theta} - 1)} \tag{17-11}$$

Band brakes have very extensive use in machinery. They are simple and dependable and are especially suitable for large and rugged installations.

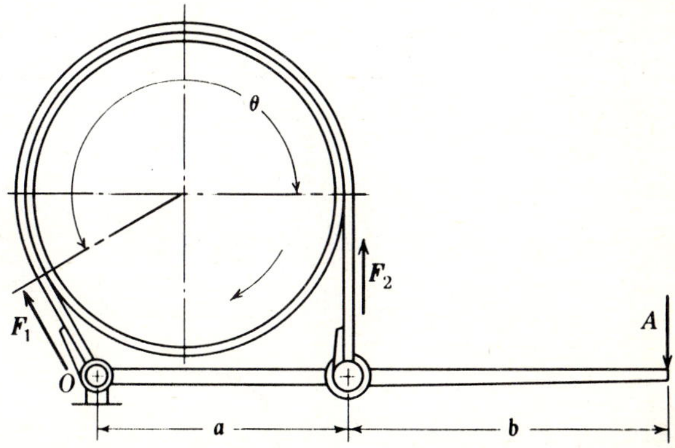

Fig. 17-15 Band brake.

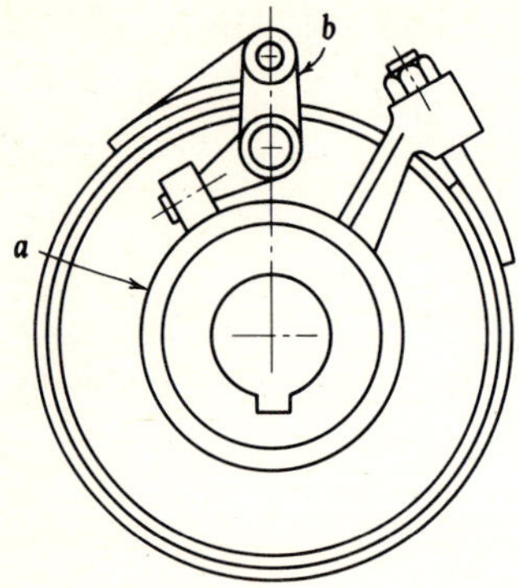

Fig. 17-16 Band clutch.

Band clutches A band clutch in which a conical sleeve *a* in Fig. 17-16 slides on the shaft and rotates the lever *b* which in turn tightens the band is a type used frequently in large, rugged equipment, such as that for road building and field construction.

Differential band brake In this type of band brake, the tension in the band assists in applying the brake, as shown in Fig. 17-17. For this brake,

$$A(a + b) - aF_2 + cF_1 = 0$$

The equation may be reduced to

$$A = \frac{F_1}{a + b} = \frac{a}{e^{f\theta}} - c$$

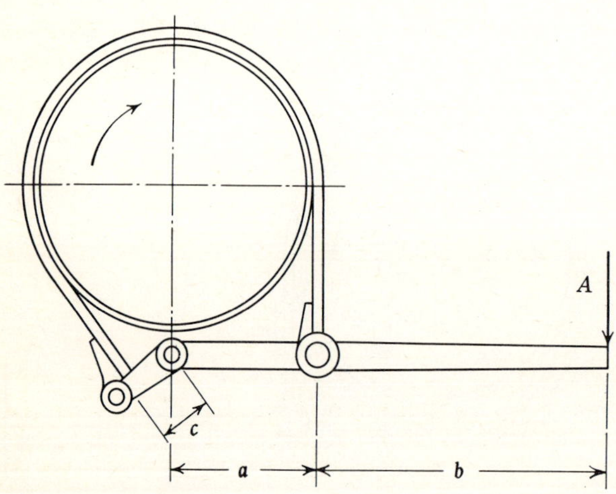

Fig. 17-17 Differential band brake.

In this equation, if the quantity $a/e^{f\theta}$ is less than c, the brake will be self-locking for clockwise rotation of the wheel. This feature is undesirable in a speed-control brake because of "grabbing," but it may be used to advantage in a "backstop," which is a differential band brake used to permit rotation of a shaft in one direction but to prevent backward rotation. An application is on a drive for an inclined conveyor to prevent backward motion in case the power fails while the conveyor is loaded. The brake is placed on the motor shaft where the torque is the least and therefore requires a smaller brake than if it were placed on a lower speed shaft of the drive.

17-6 ENERGY CONSIDERATIONS IN BRAKES

The function of a mechanical brake is to control the speed of a machine or a moving body by transforming the mechanical energy of the moving parts into heat energy and then to dissipate this energy. In the design of a brake it is necessary to determine the amount of energy to be thus transformed and also the time allowed for a braking cycle. The latter is necessary so that the temperature rise of the brake will not be excessive.

Energy equations The energy that a brake must absorb is equal to the difference between the energy given up by the moving parts and the energy losses in bearings, gears, etc. The latter energy is usually relatively small.

The equations given below apply specifically to a motor-driven hoisting drum. A cycle of hoisting and lowering a load, such as that for a traveling crane, is shown in Fig. 17-18. The exact shape of the acceleration curve Oa depends on the torque-speed characteristics of the hoisting motor. The most severe loading on the brake occurs at bc as the weight

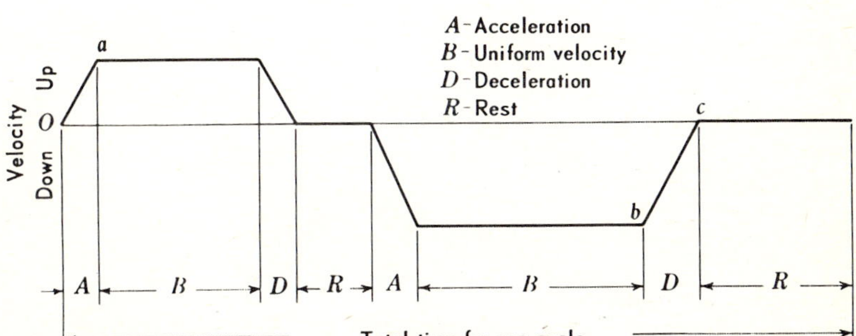

Fig. 17-18 Hoisting and lowering cycle.

is brought to rest. The following equations may be used to determine the required torque capacity for the brake.

Let Q = weight being lowered, lb
v = initial velocity of load, fps
E_t = total energy to be absorbed by brake, lb-ft
T = torque on brake wheel, lb-ft
t = time of application of brake, sec
N = initial speed of brake wheel, rpm
θ = angle through which brake wheel turns during time t, rad

The initial kinetic energy of the load is

$$E_k = \frac{Qv^2}{2g} \tag{a}$$

The change in potential energy of the load is

$$E_p = \tfrac{1}{2}Qvt \tag{b}$$

The initial kinetic energy of rotation of all the rotating parts, such as the drum, gears, motor rotor, etc., is

$$E_r = \frac{WR^2\omega^2}{2g} \tag{c}$$

where W = weight of each rotating part, lb
R = radius of gyration of each rotating part, lb
ω = angular velocity of each rotating part, rad per sec

The sum of Eqs. (*a*), (*b*), and (*c*) equals the total energy to be absorbed by the brake:

$$E_t = E_k + E_p + E_r \tag{d}$$

This energy may be equated to the work done by the brake during the time of its application, or

$$E_t = T\theta \tag{e}$$

From Eq. (*e*),

$$T = \frac{E_t}{\theta} = \frac{2 \times 60E_t}{2\pi Nt}$$

$$= \frac{60E_t}{\pi Nt} \tag{17-12}$$

Heat dissipation The mechanical energy that is transformed into heat energy by a brake will raise the temperature of the brake. The temperature rise depends on the mass of the parts, on the ratio of braking time to the rest time, and on the heat-dissipation capacity of the brake.

The maximum temperature should be limited to prevent rapid deterioration of the materials forming the friction surfaces. The maximum temperature should not exceed the following values:

Leather, fiber, and wood facing	150–160°F
Asbestos	200–220°F
Automotive asbestos block lining	400–500°F

Since the temperature rise of a brake is difficult to predict, it is usually satisfactory in the design of brakes to make use of such design coefficients as foot-pounds of energy absorbed per square inch of surface per minute. For brakes of similar construction with similar materials for the friction surfaces, the pressure on the friction surfaces may be used as a guide for satisfactory temperature conditions. Some values for coefficients of friction and allowable pressures are given in Table 17-1.

The design coefficient mentioned above is usually expressed as the product pV, where p is in pounds per square inch and V equals the rubbing velocity, fpm.

In block brakes, the pressure is usually expressed as pounds per square inch of projected area. Recommended values of pV vary from 30,000 to 80,000, depending on conditions of service and heat dissipation.

The lower value of pV may be used for continuous operation in close surroundings and the higher value for intermittent operation in well-ventilated locations.

In the design of a block brake it is customary to assume a wheel diameter that will be in good proportion to the size of the unit to which it is connected. If the brake is to be attached to a motor shaft, it may be desirable to have the distance between the center of the brake wheel

Table 17-1 Coefficients of friction and allowable pressures

	Coefficient of friction			*Allowable pressure,*
Materials in contact	*Dry*	*Greasy*	*Lubricated*	*psi*
Cast iron on cast iron	0.2–0.15	0.10–0.06	0.10–0.05	150–250
Bronze on cast iron	....	0.10–0.05	0.10–0.05	80–120
Steel on cast iron	0.35–0.25	0.12–0.07	0.10–0.06	120–200
Wood on cast iron	0.35–0.20	0.12–0.08	....	60–90
Fiber on metal	....	0.20–0.10	....	10–40
Cork on metal	0.35	0.30–0.25	0.25–0.22	8–15
Leather on metal	0.5–0.3	0.20–0.15	0.15–0.12	10–40
Wired asbestos on metal	0.5–0.35	0.30–0.25	0.25–0.20	30–80
Wired asbestos on metal, short action	....	....	0.25–0.20	200–300
Steel on cast iron, or cast iron on cast iron, short action	....	....	0.10–0.05	200–300

above the floor the same as the center of the motor shaft above the floor. From this dimension a reasonable diameter of brake wheel may be assumed.

After the arrangement of yokes and the location of the spring have been decided on, the spring force required to set the brake against the rated torque may be determined as well as the remaining forces on the shoes and yokes. If P_L and P_R are then determined, the width of the shoes may be calculated from the pV value. The ratio of the shoe width to the wheel diameter b/D may then be determined. This ratio is usually held between ¼ and ½ in well-proportioned brakes.

If the ratio is less than ¼, it indicates a shoe so narrow that it may not be held in good alignment against the wheel and the lining may wear unequally across the face. If b/D is higher than ½, the lining may also wear unequally across the face because of the difficulty of maintaining a uniform pressure across the face because of its width; also the brake may not be in good proportion to cool properly.

If the tentative assumption of the wheel diameter yields a b/D ratio outside the limits, the wheel diameter should be changed in the second trial solution.

In case P_L and P_R are different, as is usually the case, an average value may be used in connection with the pV value. The justification of this is that the pV value is a heat criterion and, owing to the transfer of heat by the wheel from the shoe that has the larger normal force P to the shoe with the smaller force P, both shoes will operate at approximately the same temperature; hence, an average value of the normal forces may be used.

Example The proposed layout of a brake to be rated at 175 lb-ft torque at 600 rpm is shown in Fig. 17-19. The assumed drum diameter D is 8 in.; the angle of contact for each shoe is 120°. The coefficient of friction f may be assumed as 0.3 and for the conditions of service a pV value of 50,000 ft-lb per in.² of projected area per minute may be assumed. Determine the spring force S required to set the brake and the width of shoes.

Solution Spring force: For the semiangle of contact $\theta = 60°$, the function from Fig. 17-10 is 1.17; therefore

$$f' = 0.3 \times 1.17 = 0.351$$

$$\text{or } \frac{F}{P} = 0.351 \qquad \text{and} \qquad P = 2.85F$$

The equation for the sum of the moments of the forces on the left-hand yoke about its pivot point O_L is

$$\begin{aligned} \Sigma M_{O_L} &= 12S + 2F_L - 6P_L = 0 \\ &= 12S + 2F_L - 6 \times 2.85F_L = 0 \end{aligned}$$

$$\text{or } F_L = 0.795S$$

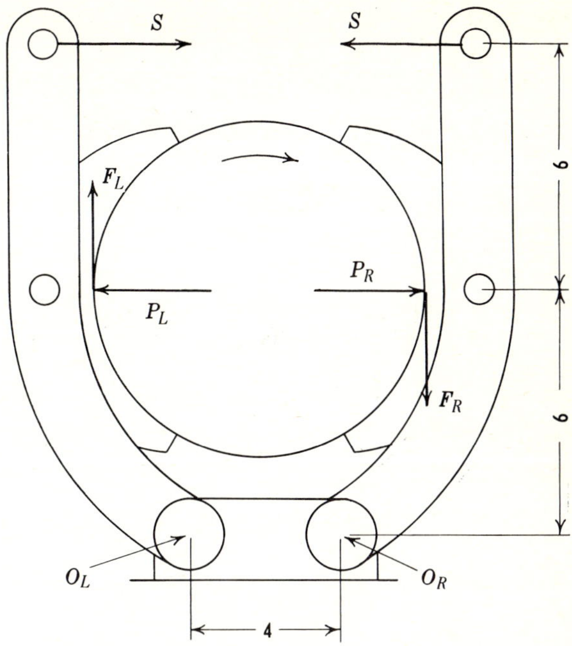

Fig. 17-19 Brake-force diagram.

The corresponding equation for the right-hand yoke is

$$\begin{aligned}\Sigma M_{O_R} &= 12S - 2F_R - 6P_R = 0\\ &= 12S - 2F_R - 6 \times 2.85F_R = 0\end{aligned}$$

or $F_R = 0.628S$

For the rated torque,

$$F_L + F_R = \frac{2T}{D} = \frac{2 \times 12 \times 175}{8} = 525 \text{ lb}$$

$$S(0.795 + 0.628) = 525 \qquad \text{or} \qquad S = 369 \text{ lb}$$

The spring force required to set the brake is 369 lb, but when the spring is compressed further to release the brake the spring force will be greater than 369 lb.

Shoe width:

$$\frac{P_L + P_R}{2} = \frac{2.85(F_L + F_R)}{2} = \frac{2.85 \times 525}{2} = 745 \text{ lb}$$

The projected bearing area for one shoe is

$$b \times 8 \sin 60 = 6.93b$$

where b is the width of the shoe.

The rubbing velocity is

$$V = \frac{\pi D \text{ rpm}}{12} = \frac{\pi \times 8 \times 600}{12} = 1{,}260 \text{ fpm}$$

$$pV = \frac{745 \times 1{,}260}{6.93b} = 50{,}000$$

$$\text{or } b = \frac{745 \times 1{,}260}{50{,}000 \times 6.93} = 2.76 \qquad \text{Use } 2\tfrac{3}{4} \text{ in.}$$

$$\frac{b}{D} = \frac{2.75}{8} = 0.344$$

which is between the limits of $\frac{1}{4}$ and $\frac{1}{2}$; hence the proportions of the shoe width and wheel are satisfactory.

18
Spur and Parallel Helical Gears

18-1 INTRODUCTION

As defined by the American Gear Manufacturers Association (AGMA), gears are machine elements that transmit motion by means of successively engaging teeth. The gear drive is therefore positive, which gives it an advantage in motion-transmission performance over friction drives such as friction wheels and belts.

The first gears had cast teeth, which in their day were satisfactory. However, modern power requirements for greater loads and higher speeds have demanded improvements in tooth form, cutting methods, materials, heat treatment, and lubrication. Progress by the gearing industry is illustrated by the performance of railway-motor gears which operate under severe loading conditions for more than 1½ million miles with negligible wear.

Important items that must be considered in the design of gears for industrial service are strength of teeth under static loading and under dynamic loading, quietness and smoothness of operation, available gear-cutting equipment, proper lubrication, temperature and heat dissipation, and *satisfactory life*. The latter is usually governed by wear, which is the most common method of gear failure.

In this chapter, the general relations of the above considerations to the *design* of gearing are discussed.

18-2 SPUR GEARING

General characteristics (1) The drive is positive and, with circular gears, the angular velocity ratio is constant. (2) The center distance may be relatively short, thus making a compact drive. (3) Provision may be made for shifting gears and in some cases for interchanging them to change the speed of the driven member. (4) The efficiency is high, since the loss of power may be 1 percent or less of the power transmitted. (5) The maintenance of the drive is inexpensive and the life is long.

Spur-gear terminology Spur gears have cylindrical pitch surfaces and operate on parallel axes, and the teeth are straight and parallel to the axis. The terms most frequently used in spur gearing are given here for reference (see Fig. 18-1). For complete discussion, see references on kinematics of gearing.[1]

The *pitch surface* is the surface of the rolling cylinder that the gear may be considered to replace.

The *pitch circle* is a right section of the pitch surface.

The *addendum circle* is the circle bounding the ends of the teeth.

[1] Darle W. Dudley, "Gear Handbook," McGraw-Hill Book Company, New York, 1962; and AGMA Standard, Gear-Nomenclature—Terms, Definitions and Illustrations (USAS B6.10–1954).

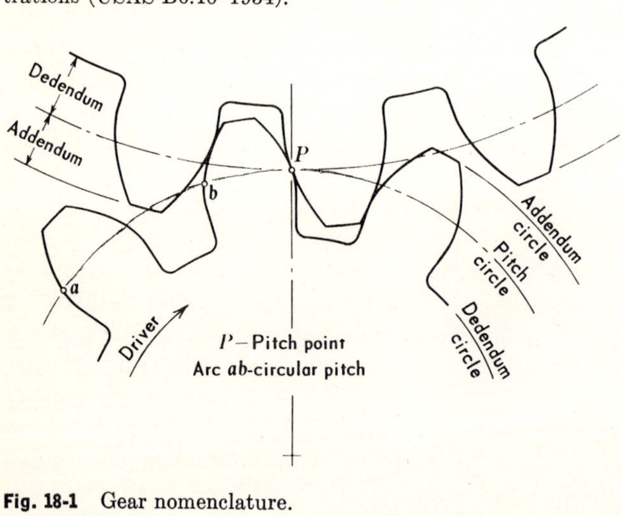

Fig. 18-1 Gear nomenclature.

The *dedendum circle* is the circle bounding the bottom of the spaces between the teeth.

The *addendum* is the radial distance between the addendum circle and the pitch circle.

The *dedendum* is the radial distance between the pitch circle and the dedendum circle.

Clearance is the difference between the dedendum of one gear and the addendum of the mating gear.

Backlash is the difference between the tooth space of one gear and the tooth thickness of the mating gear measured on the pitch circle.

Circular pitch is the distance from a point on one tooth to the corresponding point on the adjacent tooth measured on the pitch circle. Its symbol is p; the units are inches.

Diametral pitch is the number of teeth on a gear per inch of its pitch diameter.[1] Its symbol is P.

Note that the product of the circular pitch and the diametral pitch equals π, that is, $pP = \pi$.

Tooth forms It is important that a satisfactory tooth form be chosen for each application. In general, quietness is favored by the low-pressure-angle forms, but they are weaker than the high-pressure-angle ones. Common tooth forms are given below.

The AGMA has prepared standard tooth proportions for "coarse-pitch" and for "fine-pitch" involute spur gears. Coarse-pitch spur gears are defined as those up to, but not including, 20 diametral pitch; see AGMA Standard, Tooth Proportions for Coarse-pitch Involute Spur Gears (AGMA 201.02–1958). Fine-pitch spur gears are defined as those of 20 or more diametral pitch; see AGMA Standard, Tooth Proportions for Fine-pitch Involute System for Spur and Helical Gears (USAS B6.7–1955). Fine-pitch gears are used primarily for transmitting motion rather than any appreciable power and are not treated in this book.

The two tooth forms for spur-gear teeth now standardized by the AGMA are of full depth ($a = 1/P$ and $b = 1.25/P$) with a fillet radius of the basic rack $0.300/P$ and two pressure angles, 20 and 25°. For the fillet radius of the gear teeth, see AGMA Standard 201.02–1958.

The old standard tooth forms with pressure angles of $14\frac{1}{2}$, 20 ($b = 1.157/P$), and 20° stub are now obsolete. Values of data for the obsolete forms are given in this chapter for the design of replacement gears. For new designs, however, the 20 and 25° full-depth forms should be specified and may be *designed* by using the method and data presented below. These gears then may be *rated* according to the AGMA Standard 220.02, January, 1964.

[1] Frequently the term "pitch" is used. This should be understood to mean "diametral pitch."

Design considerations The specifications for a gear drive generally include (1) the horsepower to be transmitted, (2) the speed of the driving gear, and (3) the speed of the driven gear or the velocity ratio. Frequently the center distance is specified. The usual drive requires a speed *reduction*, for example, 2:1, which means that the speed of the driving gear is twice that of the driven gear. The reason that most drives are reductions rather than increases is that the usual sources of power, i.e., motors, turbines, and high-speed engines, operate at speeds higher than those required by driven units, such as machine tools, pumping or blowing equipment, transportation and propulsion machinery, or conveying equipment. Occasionally, however, speed increases are required.

In the design of a gear drive there are several requirements that must be met, as follows: (1) The gear teeth should have sufficient strength so that they will not fail under static loading, such as that at high starting torques, or under dynamic loading during normal running conditions. (2) The teeth should have good wear characteristics so that their life will be satisfactory. (3) The use of space and material should be economical. (4) The alignment of the gears and deflections of the shafts must be considered because of their effect on the performance of the gears. (5) The lubrication of the gears must be satisfactory. The above general requirements indicate that the final specifications for gears involve many interrelated factors. Compromises are usually necessary in order to secure a desirable balance between good service and minimum cost.

18-3 STRENGTH OF GEAR TEETH—LEWIS EQUATION

The determination of the maximum stresses in a loaded gear tooth is complicated by the variation in magnitude and direction of the load on

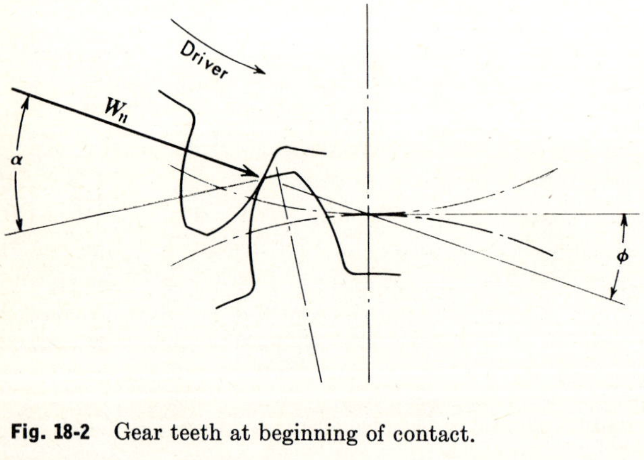

Fig. 18-2 Gear teeth at beginning of contact.

the tooth during contact and by the shape of the tooth, since it has varying width and is joined to the body of the gear by a fillet. At the first point of contact, the load W_n in Fig. 18-2 acts normal to the profile of the tooth in accordance with the fundamental law of gear-tooth action. This normal load may be resolved into tangential and radial components, the tangential component producing a bending moment on the tooth, and the radial component inducing compressive stresses on sections across the tooth. Owing to the radial component being eccentric to the centerline of the tooth, the compressive stress is not uniformly distributed across the tooth.

As the point of contact moves along the profile, the magnitude of W_n changes as well as the moment arm of the bending force. Futhermore, along the line of contact, the number of pairs of teeth that share the transmitted load may vary according to the contact ratio. Finally, because of inaccuracies of tooth spacing and tooth form and because of deflection of the tooth under load, dynamic forces may be induced. These dynamic forces are cyclic and may in some cases be larger than the steady force which transmits power.

As to the gear tooth itself, it is not a simple cantilever because of its stubbiness and nonuniformity of width. Also stress concentration at the fillets enters the scene.

It can readily be appreciated from the above situation that the maximum stresses in an existing gear tooth are difficult to determine. The designer of gears has also to consider that, at the beginning of a design, the sizes of the gears and the teeth are not usually established so that it becomes necessary to approach the design of gears in a somewhat indirect manner.

It should be appreciated in addition that fracture of gear teeth is not the only mode of failure to be considered but that wear of the teeth is also a criterion; in fact, more gears fail by wearing of the teeth than by fracture.

In 1892, Wilfred Lewis made simplifying assumptions[1] regarding the strength of gear teeth which resulted in an equation that has been used extensively by industry in determining the size and proportions of gears.

In this investigation, Lewis assumed that the worst position of loading of the tooth occurred at the first point of contact and that one pair of mating teeth carried the entire load. As shown in Fig. 18-3, the normal load W_n was translated to the centerline of the tooth and then resolved into a radial component W_r and a tangential component W. The tangential component W produces a bending moment on the tooth. The radial component W_r was neglected.

[1] Wilfred Lewis, Investigation of the Strength of Gear Teeth, Engineers' Club of Philadelphia, 1892.

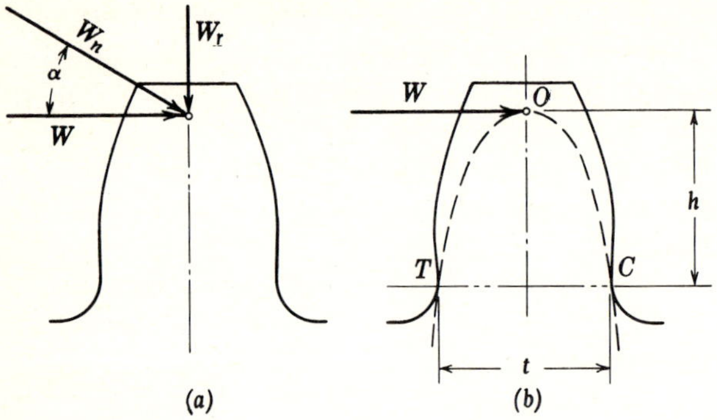

Fig. 18-3 Forces on gear teeth.

Lewis then assumed that the gear tooth could be regarded as a cantilever and made use of the principle that a beam of parabolic outline is one of uniform strength at all sections. He then inscribed within the tooth a parabolic outline drawn through O and tangent to the tooth profile at the fillets. Since the actual tooth is stronger than the parabolic beam at every section except TC, this is the critical section of the tooth, and the flexure formula may be applied to establish a relation between the load W and the stress of this section.

The equation for W will be used here as a basis for determining the proportions of gear teeth. After the proportions are established, the Buckingham equations will be used for determining the hardness for fatigue-strength and wear requirements.

Using the face width of the gear as F in Fig. 18-3(b),

$$Wh = \frac{sFt^2}{6} \tag{18-1}$$

from which $W = \dfrac{sFt^2}{6h}$

The circular pitch may be introduced as follows:

$$W = \frac{sFt^2p}{6hp} = sFpy \tag{18-2}$$

where $y = \dfrac{t^2}{6hp}$

The quantity y is known as the "form factor." To determine its value, t, h, and p may be determined analytically or more conveniently by the graphical method shown in Appendix XII, using a drawing of the tooth of $P = 1$. Note that y is a dimensionless quantity and that, if the gear is enlarged, the distances t, h, and p will each be increased proportionally; hence, the value of y will remain unchanged. The value of y then is independent of the size of the tooth and depends only on the number of teeth on a gear and the system of the teeth. Values for standard tooth forms are given in Table 18-1.

The full tangent radius given in the table is a true radius tangent to both tooth curves and to the root circle; for gears with base circles larger than their root circles, the fillet will be tangent to radial lines through the base points of the involutes.[1] Values for the form factors for the full-depth teeth were calculated by the use of a computer by Thomas R. Kincaid, Ohio University graduate.

The value of W in Eq. (18-2) may be used as the torque on a gear divided by its pitch radius. This gives a value somewhat greater than W in Fig. 18-3, but the difference is relatively small for gears of usual proportions.

In deriving Eq. (18-1), it was assumed that the load was uniformly distributed across the face of the gear. The actual distribution in service depends on the accuracy of cut of the gears, on the initial alignment and on the deflections of the tooth, on the shafting, and on the bearings.[2] For ordinary industrial installations, the face width F is usually made from two to five, or preferably, between three to four times the circular pitch for spur gears.

An extensive investigation of a wide variety of industrial gears that have behaved satisfactorily in service was made by the use of the Lewis equation and values for the allowable stress determined (see Table 18-2). If these allowable stresses are in turn used with the Lewis equation to design similar industrial gears, the resulting gear should have the size and proportion to behave satisfactorily in service.

Note that the values for s_o as calculated by the Lewis equation from service data include allowance for inaccuracies in cutting of teeth, tooth deflection in service, contact ratio, and stress concentration at the fillets of the teeth. The effect of velocity on the tooth stresses is approximated by Eq. (18-3).

As to dynamic stresses and wear, these can be allowed for by alterations of the composition of the material and its heat treatment.

Thus, the sins committed in the assumptions for the derivation of the Lewis equation affect the allowable stress, and in using these allowable stresses in similar designs, the sins are neutralized, and the designer may establish a pair of gears of reasonable size and proportions.

[1] S. J. Johnson, Controlling Tooth Fillet Contours to Increase Finished Gear Strength, Hobbing Notes, vol. 22, no. 1, Barber-Colman Company, Rockford, Ill., January, 1966.

[2] See Poritsky, Sutton, and Pernick, Distribution of Load along a Pinion, *Trans. ASME*, vol. 67, p. A-78, 1945.

Table 18-1 Lewis forms factor y

No. of teeth n	14.5° *composite* $a = 1/P$ $b = 1.157/P$	20° *full-depth* $a = 1/P$ $b = 1.157/P$	20° *stub* $a = 0.8/P$ $b = 1/P$	*Radii from AGMA Standard* 201.02B 20° $a = 1/P$ $b = 1.25/P$	25° $a = 1/P$ $b = 1.25/P$	*Full tangent radius** 20° $a = 1/P$ $b = 1.25/P$	25° $a = 1/P$ $b = 1.25/P$
10	0.055	0.064	0.088				
11	0.062	0.072	0.093				
12	0.067	0.078	0.099		0.077		0.082
13	0.071	0.083	0.103		0.082		0.088
14	0.075	0.088	0.108		0.086		0.092
15	0.078	0.092	0.111		0.091		0.096
16	0.081	0.094	0.115		0.095		0.100
17	0.084	0.096	0.117		0.099		0.104
18	0.086	0.098	0.120	0.081	0.103	0.093	0.108
19	0.088	0.100	0.123	0.083	0.107	0.095	0.112
20	0.090	0.102	0.125	0.085	0.111	0.097	0.116
21	0.092	0.104	0.127	0.088	0.116	0.099	0.119
22	0.093	0.105	0.129	0.090	0.120	0.101	0.123
24	0.095	0.107	0.132	0.094	0.128	0.105	0.130
26	0.098	0.110	0.135	0.098	0.136	0.109	0.136
28	0.100	0.112	0.137	0.010	0.130	0.113	0.131
30	0.101	0.114	0.139	0.105	0.133	0.116	0.134
34	0.104	0.118	0.142	0.113	0.138	0.123	0.139

Table 18-1 Lewis forms factor y (Continued)

No. of teeth	14.5°	20°	20°	*Radii from AGMA Standard* 201.02B		*Full tangent radius**	
	composite	*full-depth*	*stub*	20°	25°	20°	25°
	$a = 1/P$	$a = 1/P$	$a = 0.8/P$	a $1/P$	$a = 1/P$	$a = 1/P$	$a = 1/P$
n	$b = 1.157/P$	$b = 1.157/P$	$b = 1/P$	$b = 1.25/P$	$b = 1.25/P$	$b = 1.25/P$	$b = 1.25/P$
38	0.106	0.122	0.145	0.121	0.143	0.130	0.143
43	0.108	0.126	0.147	0.120	0.147	0.124	0.147
50	0.110	0.130	0.151	0.125	0.152	0.128	0.152
60	0.113	0.134	0.154	0.129	0.156	0.132	0.156
75	0.115	0.138	0.158	0.134	0.161	0.135	0.161
100	0.117	0.142	0.161	0.138	0.166	0.139	0.166
150	0.119	0.146	0.165	0.142	0.169	0.143	0.169
300	0.122	0.150	0.170	0.146	0.175	0.148	0.175
Rack	0.124	0.154	0.175	0.151	0.180	0.154	0.180

* A true radius tangent to both tooth profile and to the root circle.

Table 18-2 Values of basic stress for gears

Material	s_o, *psi*
Cast iron, ordinary	8,000
Cast iron, medium grade	10,000
Cast iron, highest grade	15,000
Cast steel, 0.20% C untreated	20,000
Cast steel, 0.20% C heat-treated	28,000
Forged carbon steel	
SAE 1020 case-hardened	18,000
SAE 1030 untreated	20,000
SAE 1035 untreated	23,000
SAE 1040 untreated	25,000
SAE 1045 untreated	30,000
SAE 1045 heat-treated	30,000
SAE 1050 heat-treated	35,000
Alloy steel	
SAE 2320 case-hardened	50,000
SAE 3245 heat-treated	65,000
SAE 6145 heat-treated	67,500
Bronze SAE 62	10,000
Phosphor bronze SAE 65	12,000
Meehanite metal, grade GA	12,500
Rawhide, Fabroil	6,000
Bakelite, Micarta, Celoron	8,000

The gears may then be investigated for dynamic loads, wear of the teeth, AGMA ratings, and other criteria, as discussed later in this chapter. Although most of these equations are relatively easy to apply to an existing set of gears, they are not of such form that they can readily be used for direct design. The Lewis equation in its original form, or modified for a special class of gears or service, is a most helpful aid in this phase of gear design.

The value of the allowable stress in gear design depends, of course, on the material for which a basic stress s_o may be determined. In order to make allowance for the dynamic effects due to the velocity of tooth action, the Barth equation may be used, i.e.,

$$s_{\text{all}} = s_o C_v \tag{18-3}$$

where s_{all} = allowable stress, psi

s_o = basic stress, psi (see Table 18-2)

C_v = velocity factor = $600/(600 + V)$ for spur gears

V = pitch-line velocity, fpm

The variation of the velocity factor with the pitch-line velocity is shown in Fig. 18-4. The solid line shows the usual range for spur gears.

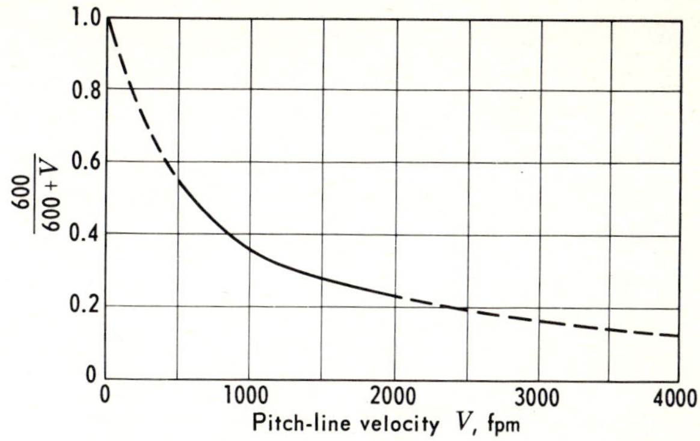

Fig. 18-4 Velocity factor.

There are other velocity factors which are frequently used and which, like the Barth equation, are arbitrary. They are not given here, since the Lewis equation will be used only to determine the proportions of gears, i.e., diametral pitch, pitch diameters, and face, for sufficient *strength*, and Eq. (18-3) is satisfactory for that purpose.

A shortcoming of the velocity factor as given is that it does not take into account the fact that the pinion teeth are loaded a greater number of times than the gear teeth. The number of cycles is important in regard to the endurance limit. A method for considering the relative number of cycles has been proposed by Merritt.[1] This method includes the influence of the number of hours per day of service but does not include the effect of velocity.

In the design of a gear drive, Eq. (18-2) may be used as such, but a solution by trial is necessary. The equation may be placed in more convenient forms for design purposes by introducing the diametral pitch P. Solving Eq. (18-2) for the induced stress gives

$$s_{\text{ind}} = \frac{W}{Fpy} \tag{18-4}$$

Let T = torque to be transmitted, lb-in.
D = pitch diameter, in.
n = number of teeth
P = diametral pitch
$k = F/p = 3$ to 4 for ordinary service

In installations in which space in the axial direction is limited and/or in which the lateral or angular deflection of the shaft is excessive, values of k somewhat lower than 3 may be used.

[1] H. E. Merritt, "Gears," p. 254, Sir Isaac Pitman & Sons, Ltd., London, 1943.

Substituting for F and p in Eq. (18-4) gives

$$s_{\text{ind}} = \frac{WP^2}{k\pi^2 y} \tag{18-5}$$

Equation (18-5) is a convenient form of the Lewis equation for the solution of problems in which the center distance and the velocity ratio are specified.

In many designs, the center distance is not specified, and Eq. (18-5) may be further transformed by substituting for W, as follows:

$$W = \frac{2T}{D} = \frac{2TP}{n}$$

Hence, Eq. (18-5) becomes

$$s_{\text{ind}} = \frac{2TP^3}{k\pi^2 ny} \tag{18-6}$$

Equation (18-6) is a convenient form of the Lewis equation for the solution of problems in which the center distance is not specified. Since there is not a unique solution, the final design will depend on the initial assumptions. The quantities to assume should be those for which judgment is most easily exercised. For this reason the number of teeth on the pinion is a desirable quantity to assume since it affects quietness of operation markedly. By using an average value of k equal to 3.5, the induced stress can be expressed in terms of the diametral pitch. Then by comparison with Eq. (18-3), a probable diametral pitch may be established. This procedure is shown by the example below.

Values for common diametral pitches are given in Table 18-3.

If the gear is to be keyed to a shaft, the maximum permissible bore should be investigated. An equation for this purpose was determined by photoelastic means and has given satisfactory results:[1]

$$d = D(0.50 + 0.0344\sqrt{n - 12}) \tag{18-7}$$

[1] P. H. Black, An Investigation of Relative Stresses in Solid Spur Gears by the Photoelastic Method, *Univ. Illinois Eng. Expt. Sta., Bull.* 288, 1936.

Table 18-3 Common standard diametral pitches

1	2	5	9	16
1¼	2½	6	10	18
1½	3	7	12	20
1¾	4	8	14	24

where d = maximum permissible bore
D = pitch diameter
n = number of teeth

The equation allows for a standard square key with the centerline of the keyway in line with the centerline of a tooth space. This is the best location for the keyway. In addition, it applies to gears having 12 to 24 teeth. For gears having more than 24 teeth, use n equal to 24 in the equation.

The AGMA equation for rating the strength of spur-gear teeth includes the following factors: dynamic loading, bending stress, face width, diametral pitch, geometry, overload, size, and load distribution. This is a rating equation and is not suitable directly for design purposes.[1]

Example 18-1 It is required to determine the proportions of a spur-gear drive to transmit 10 hp from a shaft rotating at 1,170 rpm to a low-speed shaft, with a speed reduction of 3:1. Assume that the teeth are a 20° full-depth system, with 24 teeth on the pinion. The pinion is to be SAE 1045 and the gear SAE 1030 steel. Assume that the torque at starting is 130 percent of torque at rating.

Solution To determine whether the pinion teeth or the gear teeth are weaker, a consideration of Eqs. (18-2) and (18-3) shows that the product $s_o y$ for the pinion and gear determines the relative strengths.

	s_o	n	y	$s_o y$
Pinion	30,000	24	0.107	3,210
Gear	20,000	72	0.136	2,720

The values of $s_o y$ show that the gear is the weaker and hence will be the basis for the design. Using 130 percent rating,

$$T_G = \frac{63{,}030 \times \text{hp}}{\text{rpm}} = \frac{63{,}030 \times 10 \times 1.30}{1{,}170/3} = 2{,}100 \text{ lb-in.}$$

$$s_{\text{ind}} = \frac{2TP^3}{3.5\pi^2 ny} = \frac{2 \times 2{,}100P^3}{3.5\pi^2 \times 72 \times 0.136} = 12.4P^3$$

$$\text{Now } s_{\text{all}} = s_o \frac{600}{600 + V} = 20{,}000 \frac{600}{600 + V}$$

In the above two equations there are two unknowns, the diametral pitch P and the pitch-line velocity V. From the above equation, for the induced stress,

$$\begin{aligned} s_{\text{ind}} &= 4{,}260 \quad \text{for } P = 7 \\ &= 6{,}350 \quad \text{for } P = 8 \\ &= 9{,}040 \quad \text{for } P = 9 \end{aligned}$$

[1] AGMA Standard 220.02, January, 1964.

Note from Fig. 18-4 that the velocity factor varies from 0.5 to 0.2 for the usual range of spur-gear pitch-line velocities. By using tentatively an average value of, say, ⅓ for C_v, the tentative allowable stress will be *approximately* ⅓ × 20,000, or say 7,000 psi. By comparison with the above values for the induced stress, it is seen that the diametral pitch may be 8 or 9. With a trial value of P equal to 9, then $D = 2.67$ in.

$$V = \frac{\pi D \text{ rpm}}{12} = \frac{\pi \times 2.67 \times 1{,}170}{12} = 816 \text{ fpm}$$

$$s_{\text{all}} = s_o \frac{600}{600 + V} = 20{,}000 \frac{600}{600 + 816} = 8{,}480 \text{ psi}$$

Thus $P = 9$ may be satisfactory, since the induced stress is near the allowable stress. The computed face width equals

$$kp = k \frac{\pi}{P} = 3.5 \frac{\pi}{9} = 1.22 \text{ in.}$$

This face width may now be altered in order to reduce the induced stress to the allowable stress, as follows:

$$1.22 \frac{s_{\text{ind}}}{s_{\text{all}}} = 1.22 \frac{9{,}040}{8{,}480} = 1.30 \text{ in.}$$

To use a commercial dimension, $F = 1\frac{1}{4}$ in. would be satisfactory.

Note that the computed face width may be *decreased* by the ratio $s_{\text{ind}}/s_{\text{all}}$ in case the induced stress is less than the allowable stress. Therefore, the face width should not be less than $3p$ or greater than $4p$, or broadly between $2p$ and $5p$.

Results:

$$\begin{aligned} P &= 9 \\ D_P &= 2.67 \text{ in.} \\ D_G &= 8.00 \text{ in.} \\ F &= 1\tfrac{1}{4} \text{ in.} \\ \text{Center distance} &= 5.33 \text{ in.} \end{aligned}$$

18-4 STRENGTH OF GEAR TEETH—BUCKINGHAM EQUATION

In the preceding article the velocity factor was used to make approximate allowance for the effects of dynamic loading. The dynamic loads are due to (1) inaccuracies of tooth spacing, (2) irregularities in tooth profiles, and (3) deflections of the teeth under load.

A closer approximation to actual conditions may be made by the use of equations based on an extensive series of tests,[1] as follows:

$$W_d = W \pm W_i \tag{18-8}$$

[1] See the manual of gear design, "Design Loads on Gear Teeth," by Earle Buckingham, The Industrial Press, New York, 1931.

where W_d = total dynamic load
W = steady load due to transmitted torque
W_i = increment load due to dynamic action

In Fig. 18-5(*a*) there is a representation of Eq. (18-8), where W_i is shown as a cyclic load. In many cases, the value of W_i is greater than W, as shown in Fig. 18-5(*b*), where *e-f* indicates that the teeth separate.

The increment load W_i depends on the pitch-line velocity, the face width, the material of the gears, the accuracy of cut, and the tangential load W. For average conditions of mass of the gears and connected parts, the following equation applies:

$$W_d = W + \frac{0.05V(FC + W)}{0.05V + \sqrt{FC + W}} \tag{18-9}$$

where W_d = maximum dynamic load, lb
W = steady transmitted load, lb
V = pitch-line velocity, fpm
F = width of face of gears, in.
C = deformation factor (Table 18-6)

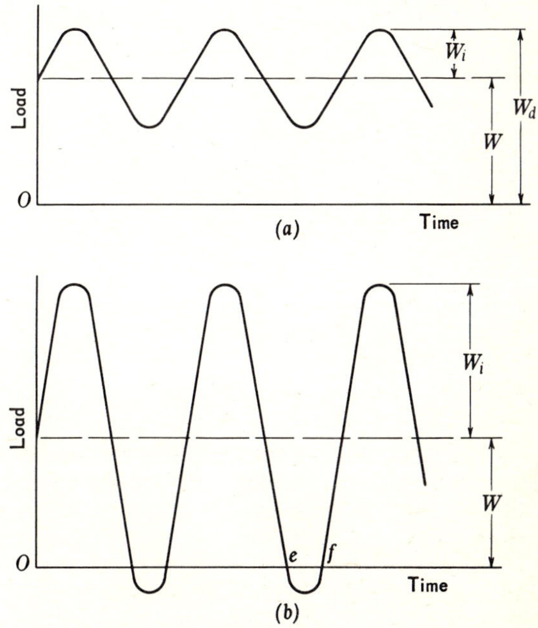

Fig. 18-5 Dynamic loads on gear teeth.

For helical gears:[1]

$$W_d = W + \frac{0.05V(FC\cos^2\psi + W)\cos\psi}{0.05V + \sqrt{FC\cos^2\phi + W}} \qquad W_w = \frac{DFKQ}{\cos^2\psi}$$

ψ = helix angle

The value of the deformation factor C depends on the error in action between teeth, the class of cut of the gears, the tooth form, and the material of the gears.

"Error in action" refers to the maximum deviation of the actual tooth profile from the ideal profile. The deviation is due to inaccuracies in cutting the profiles and in spacing of the teeth and also to the deflections of the teeth under load. Tables 18-4 and 18-5 may be used to estimate the maximum error in action. The values of the deformation factor C may be determined from Table 18-6. These tables are taken from the manual of gear design, as cited, and apply to gears having average conditions of mass, elasticity, accuracy, and pitch-line velocity.

[1] C. D. Albert, "Machine Design Drawing Room Problems," p. 382, John Wiley & Sons, Inc., New York, 1951.

Table 18-4 Maximum allowable error in action between gears

V	*Error*	*V*	*Error*	*V*	*Error*
250	0.0037	1,750	0.0017	3,250	0.0008
500	0.0032	2,000	0.0015	3,500	0.0007
750	0.0028	2,250	0.0013	4,000	0.0006
1,000	0.0024	2,500	0.0012	4,500	0.0006
1,250	0.0021	2,750	0.0010	5,000 and over	0.0005
1,500	0.0019	3,000	0.0009		

Table 18-5 Maximum expected error in action between gears

Diametral pitch P	*Class* 1*	*Class* 2†	*Class* 3‡
1	0.0048	0.0024	0.0012
2	0.0040	0.0020	0.0010
3	0.0032	0.0016	0.0008
4	0.0026	0.0013	0.0007
5	0.0022	0.0011	0.0006
6 and finer	0.0020	0.0010	0.0005

* Class 1: well-cut commercial gears.

† Class 2: gears cut with great care.

‡ Class 3: ground and lapped gears.

Table 18-6 Values of deformation factor C

Materials, pinion and gear	*Tooth form*	*Error in action, in.* 0.0005	0.001	0.002	0.003	0.004	0.005
Cast iron and cast iron		400	800	1,600	2,400	3,200	4,000
Steel and cast iron	14.5°	550	1,100	2,200	3,300	4,400	5,500
Steel and steel		800	1,600	3,200	4,800	6,400	8,000
Cast iron and cast iron	20°	415	830	1,660	2,490	3,320	4,150
Steel and cast iron	full	570	1,140	2,280	3,420	4,560	5,700
Steel and steel	depth	830	1,660	3,320	4,980	6,640	8,300
Cast iron and cast iron	20°	430	860	1,720	2,580	3,440	4,300
Steel and cast iron	stub	590	1,180	2,360	3,540	4,720	5,900
Steel and steel	tooth	860	1,720	3,440	5,160	6,880	8,600
Cast iron and cast iron	25°	455	910	1,820	2,730	3,640	4,550
Steel and cast iron	full	630	1,260	2,520	3,780	5,040	6,300
Steel and steel	depth	900	1,800	3,600	5,400	7,200	9,000

If the dynamic load as determined above is substituted in Eq. (18-2), the calculated stress should not exceed the flexural endurance limit s_{ef}. To provide a margin of safety, the factor f should be introduced.

$$s_{ef} = \frac{fW_d}{Fpy} \tag{18-10}$$

The factor f has the following values.

f = 1.25 for steady loads
 = 1.35 for pulsating loads
 = 1.50 for shock loads

Values of flexural endurance limits are given in Table 18-7.

For high speeds and ultraconservatism, a more sophisticated equation for estimating the dynamic load W_d was developed by Earle Buckingham.[1] This equation considers the impact forces due to accelerations and decelerations of the masses of gears and of connected bodies. For abnormal mass conditions and very high speeds, Eq. (18-9) should be regarded as an approximation.

[1] Earle Buckingham, Analytical Mechanics of Gears, "Gear Handbook," chap. 20, McGraw-Hill Book Company, New York, 1949. This source also contains an interesting account of the history of the analysis of gear teeth in action (chap 18). See also Darle W. Dudley, "Practical Gear Design," art. 14-6 and 14-10, McGraw-Hill Book Company, New York, 1954.

Table 18-7 Fatigue limits of gear materials

Material	*Bhn*	*Flexural endurance limit* s_{ef}	*Surface endurance limit* s_{es}
Gray cast iron	160	12,000	90,000
Semisteel	200	18,000	
Phosphor bronze	100	24,000	
Steel	150	36,000	50,000
	200	50,000	70,000
	240	60,000	88,000
	280	70,000	103,000
	300	75,000	110,000
	320	80,000	118,000
	350	85,000	130,000
	360	90,000	
	400	100,000	
GA Meehanite	...	....	80,000

18-5 WEAR OF GEAR TEETH—BUCKINGHAM EQUATION

The maximum load that gear teeth can carry without premature wear depends on the radii of curvature of the tooth profiles and on the elasticity and the surface-fatigue limits of the materials. The research work of Buckingham led to the following equation for estimating the maximum or limiting load for satisfactory wear of gear teeth.

$$W_w = DFKQ \tag{18-11}$$

where W_w = limiting load for wear, lb

Q = ratio factor = $\dfrac{2r}{r+1}$ (for external gears)

= ratio factor = $\dfrac{2r}{r-1}$ (for internal gears)

r = velocity ratio

K = load-stress factor

D = pitch diameter of pinion, in.

s_{es} = surface endurance limit, psi

ϕ = pressure angle

F = face width of gears, in.

E_1, E_2 = moduli of elasticity of materials, psi

$$K = \frac{s_{es}^2 \sin \phi}{1.4}\left(\frac{1}{E_1} + \frac{1}{E_2}\right)$$

Table 18-8 Values of load-stress factors K

Material in pinion	*Bhn*	*Material in gear*	*Bhn*	*Surface endurance limit* s_{es}, *psi*	K, 14.5° *systems*	K, 20° *systems*	K, 25° *systems*
Steel	150	Steel	150	50,000	30	41	51
Steel	200	Steel	150	60,000	43	58	72
Steel	250	Steel	150	70,000	58	79	98
Steel	200	Steel	200	70,000	58	79	98
Steel	250	Steel	200	80,000	76	103	127
Steel	300	Steel	200	90,000	96	131	162
Steel	250	Steel	250	90,000	86	131	162
Steel	300	Steel	250	100,000	119	162	200
Steel	350	Steel	250	110,000	144	196	242
Steel	300	Steel	300	110,000	144	196	242
Steel	350	Steel	300	120,000	171	233	288
Steel	400	Steel	300	125,000	186	254	314
Steel	350	Steel	350	130,000	201	275	340
Steel	400	Steel	350	140,000	233	318	384
Steel	400	Steel	400	150,000	268	366	453
Steel	150	Cast iron	180	50,000	44	60	74
Steel	200	Cast iron	180	70,000	87	119	147
Steel	250	Cast iron	180	90,000	144	196	242
Steel	150	Phosphor bronze	100	50,000	46	62	77
Steel	200	Phosphor bronze	100	70,000	91	124	153
Steel	250	Phosphor bronze	100	85,000	135	184	228
Cast iron	180	Cast iron	180	90,000	193	264	326
GA Meehanite		GA Meehanite		80,000	105	144	178
Steel		GA Meehanite		80,000	90	123	152

Values of K for various tooth forms and materials are given in Table 18-8. In the wear equation [Eq. (18-11)] the value of W_w should not be less than the dynamic load W_d. In gear design, W_d from Eq. (18-9) may be taken as the minimum allowable value of W_w and when used as such in Eq. (18-11) gives the minimum value of load-stress factor K. The required Brinell hardness numbers may be determined from Table 18-8.

In using the equations for static loads, dynamic loads, and wear loads, it should be realized that the results are not exact but serve for comparison between expected performance of gears being designed and that of the gears of the reported tests.

Example 18-2 Determine the Brinell hardness number (Bhn) for the pinion and gear of Example 18-1 required on the bases of (a) dynamic load and (b) wear load. Assume class 2 gears.

(*a*) Dynamic load:
For 130 percent rating

$$W_d = W + \frac{0.05V(FC + W)}{0.05V + \sqrt{FC + W}}$$
$$= 526 + \frac{40.8 \times 2{,}766}{40.8 + 52.6}$$
$$= 1{,}736 \text{ lb}$$
$$s_{ef} = \frac{fW_d}{Fpy}$$
$$= \frac{1.25 \times 1{,}736}{1.25 \times 0.349 \times y}$$
$$= 52{,}400 \text{ for pinion}$$
$$= 36{,}600 \text{ for gear}$$

From Table 18-7,
Bhn for pinion = 200
Bhn for gear = 150

Auxiliary calculations:

$$W = \frac{33{,}000 \text{ hp}}{V}$$
$$= \frac{33{,}000 \times 10 \times 1.3}{816}$$
$$= 526 \text{ lb}$$
$$0.05V = 0.05 \times 816 = 40.8 \text{ fpm}$$
$$C = 1{,}800 \text{ for class 2 gears}$$
$$FC = 1.25 \times 1{,}800 = 2{,}250 \text{ lb}$$
$$FC + W = 2{,}250 + 526 = 2{,}776 \text{ lb}$$
$$\sqrt{FC + W} = 52.6$$
$$p = \frac{\pi}{P} = \frac{\pi}{9} = 0.349 \text{ in.}$$
$$Q = \frac{2r}{1 + r} = \frac{2 \times 3}{1 + 3} = 1.5$$

(*b*) Wear load:
For 100 percent rating, the dynamic load W_d may be found equal to 1,544 lb.

$$K = \frac{W_w}{DFG} = \frac{1{,}575}{2.67 \times 1.25 \times 1.5} = 309$$

From Table 18-8,
Bhn for pinion = 400
Bhn for gear = 300

Since the second set of Bhn values in Example 18-2 are higher than required on the endurance-limit basis, they should be specified. However, the possibility of heat-treating the pinion and gear to obtain these hardness numbers should be investigated. With the grade of steel originally specified, it would be difficult to harden either the pinion or the gear throughout, but thorough hardening might not be necessary. A form of surface hardening would, in general, be sufficient to obtain the required resistance.

The four papers cited in the footnote are excellent sources of information for specialized gear design and contain additional references.[1]

The American Gear Manufacturers Association has published a

[1] E. J. Wellauer, Applying New Gear Strength and Durability Formulas to Design, *Am. Soc. Mech. Engrs., Paper* 69-*MD*-9, 1964; C. R. Burrell, How Designers Can Short Cut Time in Gear Selection: Coarse Pitch Section, *Am. Soc. Mech. Engrs., Paper* 69-*MD*-9; P. M. Dean, Jr., How Designers Can Cut Time in Gear Selection, Fine Pitch Section, *Am. Soc. Mech. Engrs., Paper* 69-*MD*-20; and Gear Classification Manual AGMA 390.02–1964.

Tentative Standard for Rating the Strength of Spur Gear Teeth (AGMA 220.02, January, 1964). This Standard features a "basic horsepower rating formula" which contains 13 terms, most of which are not known when a gear design is initiated. The Foreword of the Standard states, in part:

> Basic Rating Formula—This section enumerates the factors known to affect strength. Numerical values are presented for those factors which have been evaluated by analytical means, test results or field experience. Suggestions are made for the factors which are not now capable of being expressed accurately. New knowledge and more definite measurement of these parameters will continually necessitate revisions and improvements.

To determine the rated horsepower by the AGMA equation, the following quantities must be considered in addition to the numbers of teeth, diameters of the gears, and the face width. Suggestions for their determination are given in the Standard referred to above. The items in parentheses refer to the considerations for determining each quantity:

K_v = dynamic factor (tooth spacing, velocity, inertia and stiffness, and precision of cutting teeth)
J = geometry factor (shape of tooth, position for most critical stress which depends on contact ratio, stress concentration)
s_{all} = allowable bending stress (material, heat treatment, surface treatment)
K_L = life factor (hardness, number of cycles to failure)
K_o = overload factor (starting and running irregularities)
K_m = load-distribution factor (misalignments and deflections)
K_s = size factor (dimensional and metallurgical)
K_R = factor of safety (reliability)
K_T = temperature factor (lubrication)

In applying the AGMA horsepower rating formula to the gears of Examples 18-1 and 18-2 the various terms in the equation were selected from the various data and charts given in the Standard referred to above. The rated calculated horsepower was found to be 11.05 as compared with the value stated as 10 hp in Example 18-1. The major part of the difference is probably due to the conservative nature of the Buckingham equations. In the AGMA equation, the teeth were assumed to "develop the most critical stress when the load is applied at the highest point of contact where a single pair of teeth is carrying all the load." This condition, rather than a tip-loaded tooth, was assumed because the contact ratio for this pair of gears is equal to 1.71.[1]

18-6 PARALLEL HELICAL GEARS

Because of gradual engagement of the teeth, helical gears run more quietly than spur gears and may be operated at pitch-line velocities up to 10,000 fpm and over. High pitch-line velocities mean low tooth loads, which in turn promote quietness and high efficiency.

Because of end thrust, the helix angle ψ (see Fig. 18-6) is usually limited in single-helical gears to 15°. For smooth operation, one end of a tooth should be advanced a distance over the other end by a distance at least equal to the circular pitch, that is, at least equal to $p/\tan\psi$ in Fig. 18-6. For gradual transfer of the tooth load the minimum face width should be $F \gtreqqless 2p/\tan\psi$.[2]

When helical gears of opposite hand are mounted in pairs on a shaft, or when herringbone gears are used, the axial loads are balanced; hence larger helix angles may be used to secure greater pitch overlap and the accompanying increase in quietness and strength over single-helical gears. For herringbone gears as shown in Fig. 18-7, helix angles from 20 to 30° are used for industrial gears and up to 45° for turbine drives. For both single- and double-helical gears, the maximum width of face should be

[1] See Marks' "Mechanical Engineers Handbook," pp. 8-101, 8-102, McGraw-Hill Book Company, New York, 1958.
[2] Dudley, "Gear Handbook."

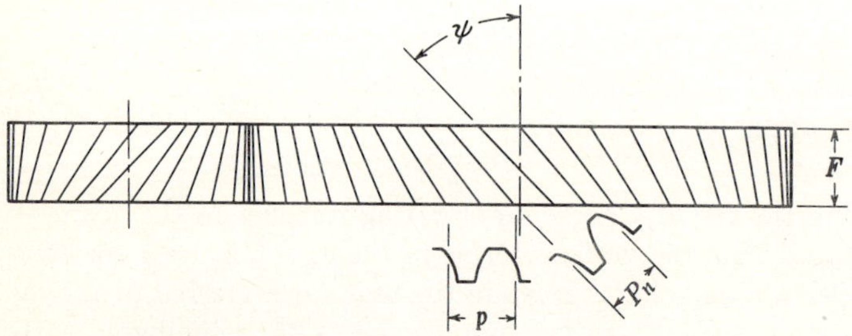

ψ = Helix angle
F = Face
p = Circular pitch in plane of rotation
P_n = Normal circular pitch

Fig. 18-6 Parallel helical gears.

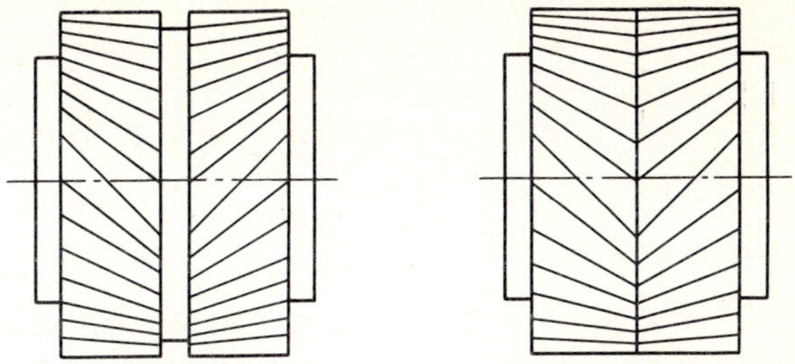

(a) Grooved (b) Continuous tooth

Fig. 18-7 Herringbone gears.

from 1.5 to 2 times the pinion diameter, although 2.5 has been used.[1] In the latter case the mountings were very rigid and accurately aligned.

Strength of parallel helical gears According to the AGMA, a modification of the Lewis equation, Eq. (18-2), may be used for helical gears:

$$W = \frac{s_o F p y C_v}{C} \tag{18-12}$$

where W = tangential tooth load, lb

s_o = basic stress, from Table 18-2

p = circular pitch in plane of rotation, in.

y = form factor; pressure angle is in plane of rotation

V = pitch-line velocity, fpm

F = face width, parallel to axis, in.

C = wear and lubrication factor; for enclosed gears, continuously lubricated with proper oil, $C = 1.15$; for scanty lubrication, regularly inspected, $C = 1.25$; for indifferent lubrication, $C = 1.35$

$$C_v = \frac{78}{78 + \sqrt{V}} \text{ for hardened steel, ground and lapped gears}$$

The relation between the pressure angle ϕ_n in the normal plane and ϕ in the plane of rotation is

$$\tan \phi_n = \tan \phi \cos \psi$$

The *surface durability* of the teeth may be determined by using a modification of the Buckingham equation, Eq. (18-9),[2] or the AGMA standard rating.[3]

[1] W. P. Schmitter, Determining the Capacity of Helical and Herringbone Gearing, *Machine Design*, June and July, 1943.

[2] Albert, *op. cit.*, p. 382.

[3] AGMA Standard Rating, Surface Durability of Helical and Herringbone Gears, AGMA Standard 211.02–1964.

The dynamic load on helical gears may be estimated from an extension of Eq. (18-9) by substituting $F \cos^2 \psi$ for F and by multiplying the last term (increment-load term) by $\cos \psi$. The wear-load equation [Eq. (18-11)] should be divided by $\cos^2 \psi$ for helical gears.[1]

18-7 ADDITIONAL CONSIDERATIONS IN RATING OF GEARS

Profile contact ratio In accurately cut gears of high contact ratio, more than one pair of teeth may be counted on to share the load so that the worst loading on a tooth may not occur at the first point of contact, as assumed in the derivation of the Lewis equation, but instead may take place between that point and the pitch point. For this condition,[2] the inscribed parabola will be shorter than shown in Fig. 18-3 and the form factor will be greater than the value in Table 18-1. In critical gear applications, such as aircraft gears, marine gears, and stationary geared power-plant drives, it may be desirable to consider this more refined procedure, instead of the simple Lewis method, and to include also stress concentration and other influences, as discussed in the remaining sections of this article.

Stress concentration The following equations[3] for stress-concentration factors may be used to determine the magnification of maximum stress at the fillets. In the equations r is the fillet radius and the other notation is the same as in Fig. 18-3.

For 20° involute teeth

$$K = 0.18 + \left(\frac{t}{r}\right)^{0.15} \left(\frac{t}{h}\right)^{0.45}$$

and for 14.5° involute teeth

$$K = 0.22 + \left(\frac{t}{r}\right)^{0.20} \left(\frac{t}{h}\right)^{0.40}$$

Connected loads In high-speed gears in which the rotating parts have large moments of inertia, as in a motor or turbine with large and heavy rotors driving a propeller, flywheel, generator, or any mass with large WR^2, the driving and driven units tend to resist changes in velocity due to irregularities in the gear teeth and to the transfer of load from one

[1] See also AGMA Standard Ratings, Strength of Helical and Herringbone Gear Teeth, AGMA Standard 221.02–1963, and Surface Durability of Helical and Herringbone Gears, AGMA Standard 211.02–1964.

[2] See Dudley, *loc. cit.*

[3] T. J. Dolan and E. L. Broghamer, A Photoelastic Study of Stresses in Gear Tooth Fillets, *Univ. Illinois Eng. Expt. Sta. Bull.* 335, 1942.

pair of teeth to the next. Large dynamic loads may be induced in this case.[1]

Allowable stresses When the refinements of contact ratio, stress concentration, and dynamic loads as affected by tooth errors in action and by connected masses are considered, the allowable stresses in the teeth to prevent failure by fracture may be higher than the stresses used with the simple Lewis equation. It is not possible to state general rules for determining these allowable stresses since the requirements for each class of application of gears are unique, so that the safe procedure in selecting allowable stresses is by experiment and experience. A correct equation for allowable stresses would also include a term to express the degree of necessity of the manufacturer to avoid any failure of the gears.

For instance, the failure of a gear in a jukebox may not cause more than inconvenience until it is replaced; in fact, the result of such a failure may bring pleasure to some listeners.

The failure of a gear in an automobile would probably not endanger life or cause extensive damage; yet if many gears failed in any one make of automobile it would jeopardize the reputation of the manufacturer.

The failure of a gear in an industrial speed reducer would generally not have fatal results and may not be excessively expensive to repair. It might, however, result in expensive shutdown.

The failure of any one of most gears in an aircraft would be likely to cause a serious accident.

A failure in the main propulsion gear of an ocean-going ship may cause a disaster or at best may necessitate an extended layover in drydock, with the necessity, if it were a bull-gear failure, of cutting through several steel decks or the hull to remove the ruined gear and replace it with a new one. To the loss in revenue or service should be added the cost of manufacturing a new gear and drydock charges of the order of $1,000 per hour.

Thus, owing to the widely varying requirements in applications of gears and to the many modes of failure, the choice of allowable stresses requires careful consideration.

Scoring In heavily loaded gears which are operated for a long period of time, pits may develop because of excessive compressive stresses at the region of contact between the teeth. The pits may become large so that a considerable area is affected which may lead to *scoring* of the teeth. Scoring is characterized by radial scratches and may be associated with other types of wear, such as abrasion.

A formula that has been used by automotive-gear designers to evaluate scoring is known as the "PVT formula," in which P is the Hertz contact pressure in pounds per square inch, usually calculated for the

[1] See Buckingham, *loc. cit.*

first and last points of contact on the pinion tooth profile and taking into account the contact ratio. The sliding velocities at the points where P is calculated is represented by V and expressed in feet per minute. The distance in inches along the line of action from the point where P is calculated to the pitch point is represented by T. Values of PVT to prevent scoring are of the order of $1\frac{1}{2}$ million.[1]

Another method for determining limits to avoid scoring and which may be a better criterion than the PVT formula uses the "flash temperature" at the contacting surfaces of the teeth. An equation for the flash-temperature limit to prevent scoring, known as the "Kelley formula," includes terms for the rolling velocities of the tooth surfaces in contact, the coefficient of friction, and the rms value for surface finish after the gears have been run in.

AGMA ratings These ratings expressed in horsepower cover strength of spur, helical, and bevel gears and surface durability of spur, helical, bevel, and worm gears. There are also AGMA thermal-horsepower capacity ratings for enclosed helical-gear speed reducers.

18-8 NONMETALLIC GEARS

In installations where it is desirable to reduce noise and vibration or where the loads are light and long life is a factor, the pinion of a pair of gears may be made of nonmetallic materials, for example, rawhide, Fabroil, molded nylon, or a laminated phenolic, such as Textolite, Micarta, Bakelite, or Celoron. Rawhide and Fabroil gears are usually assembled by compression between two steel plates. Rawhide is especially good for shock loading. To avoid overheating rawhide gears, the pitch-line velocity should normally be limited to 1,750 or 2,000 fpm, with a maximum of 2,500 fpm. Since rawhide is affected by most oils, a paste made of flake graphite and linseed oil forms a good lubricant. High temperature should be avoided.

The laminated phenolic materials are not affected by water, oil, or moderate temperature. Pitch-line velocities up to 3,000 fpm are conservative for industrial installations with commercial standards used in cutting. With precision-cut gears well mounted and balanced, pitch-line velocities up to 5,000 fpm may be used safely.

The nonmetallic pinions are generally run with cast-iron gears, and they may be designed by the use of the Lewis equation, Eq. (18-2), and Eq. (18-3), where

$$C_v = \frac{150}{200 + V} + 0.25 \tag{18-13}$$

[1] J. O. Almen, Surface Deterioration of Gear Teeth, "Mechanical Wear," John T. Burwell Jr. (ed.), American Society for Metals, Cleveland, 1950.

18-9 FAILURE OF GEAR TEETH

Gear teeth may fail by breaking as a result of static load, shock load, or fatigue load, or they may fail by excessive wear. The latter is the usual type of failure, especially in gears that operate continuously. The general types of wear of machine parts, as discussed in Art. 20-7, may be expanded to apply to gear teeth as follows:[1]

Normal wear	Galling
Initial pitting	Burning
Progressive pitting	Rolling
Abrasion	Cracking and checking
Scratching	Chipping
Scoring	Gouging

[1] See AGMA Standard Nomenclature, Gear Tooth Wear and Failure, AGMA 110.03–1962.

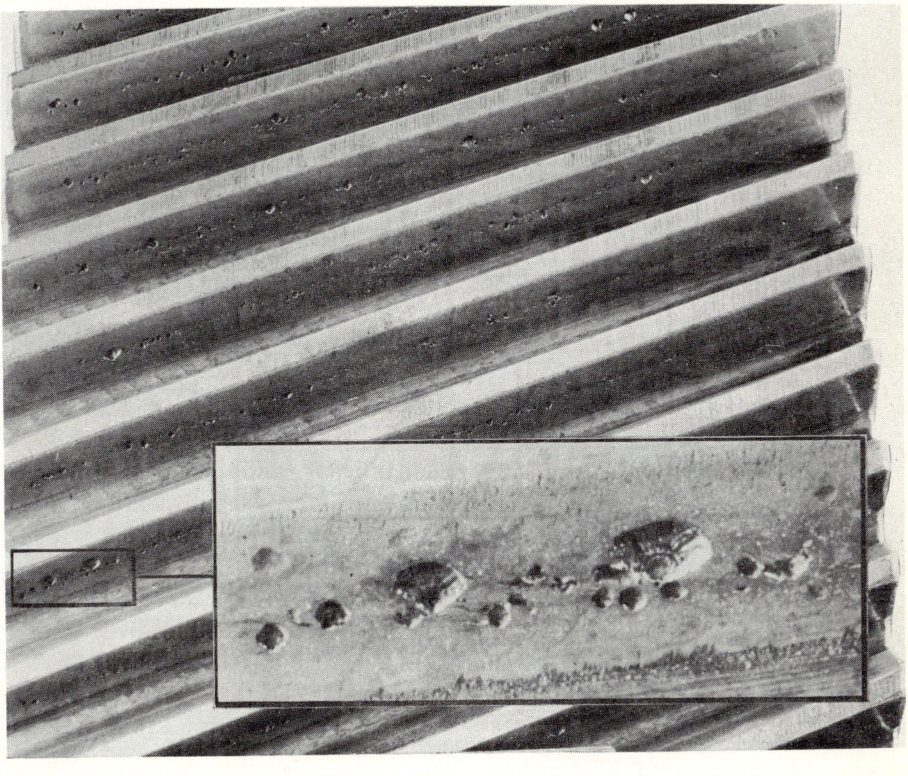

Fig. 18-8 Example of pitted gear teeth. (*Courtesy of The Falk Corporation.*)

Normal wear occurs when the gears are new, but it should cease after the gears are "run in" under proper conditions.

Initial pitting may not be serious but if it leads to *progressive pitting* the gear may fail (Fig. 18-8). This is a common type of gear failure. *Abrasion* is generally due to foreign matter in the lubricant.

The 12 types of failure listed above may occur singly but generally appear in combination. They may be caused by or aggravated by such conditions as the following:

Incorrect center distance	Excessive temperatures
Shaft misalignment or deflection	Excessive speeds
Improper lubrication	Vibration
Excessive loads	Poor surface finish

18-10 BEARING LOADS

The three mutally perpendicular components of the tooth load on a parallel helical gear are F_x, F_y, and F_z, as shown in Fig. 18-9. These components may be determined as follows:

$$F_x = \frac{2T}{D} = \frac{33{,}000 \text{ hp}}{V}$$

$$F_y = F_x \tan \psi$$

$$F_z = F_x \tan \phi$$

where T = torque transmitted, lb-in.
D = pitch diameter of gear, in.
V = pitch-line velocity, fpm
ψ = helix angle
ϕ = pressure angle

To determine the loads on the bearings A and B, the type of support for the shaft must be known. If the shaft is relatively short with long

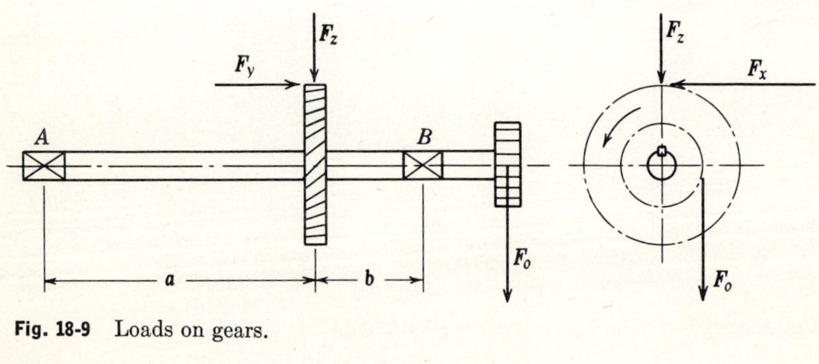

Fig. 18-9 Loads on gears.

sleeve bearings, a distributed load must be assumed. If the shaft is relatively long with short sleeve bearings or ball bearings, the shaft may be assumed to be simply supported.

The loads on the bearings A and B may be determined by considering the components F_x, F_y, and F_z separately. Note that the bearing reactions due to F_y and F_z are in the plane of the left-hand view in Fig. 18-9, whereas the reactions due to F_x are perpendicular to that plane.

If there is an "overhanging load," such as F_o, because of a driven gear, sprocket, or belt, its magnitude may be determined; and, if the direction is known, it can be combined with the bearing reactions caused by F_x, F_y, and F_z. If the direction of F_o is not known, the most unfavorable direction should be assumed, i.e., the same direction as the resultant of the reactions due to F_x, F_y. and F_z. In general, the *maximum* loads that it is possible to impose on bearings A and B should be determined regardless of the direction of rotation of the gears.

Overhanging load The maximum overhung load that a customer may apply to the stub end of the shaft of a speed reducer depends on the type of drive and on the minimum size of sprocket, gear, or sheave which is keyed to the shaft. The following equation may be used to estimate the load:

$$F_o = \frac{126{,}060K \times \text{hp}}{D \times \text{rpm}} \tag{18-14}$$

where F_o = overhung load, lb

K = load factor; for chain drive, $K = 1$, for gear, $K = 1\frac{1}{4}$, for V belt, $K = 1\frac{1}{2}$, and for flat belt, $K = 2\frac{1}{2}$

hp = rated horsepower

rpm = shaft speed

D = diameter of sprocket, gear, or sheave, in.

If D is not known, a value equal to twice the shaft diameter may be used. The location of the line of action of F_0 may be determined on the assumption of its being located one shaft diameter beyond the shaft seal, or at the midpoint of the shaft extension, whichever is the greater distance from the adjacent bearing.

18-11 LUBRICATION OF GEARS

In the discussion of the lubrication of journal bearings in Art. 21-2, it is shown that in order to have fluid-film lubrication exist between two surfaces having relative sliding motion it is necessary for the film to be converging. A pair of gear teeth coming into action provides a converging film so that fluid-film lubrication is possible.[1] Since the surfaces of

[1] E. K. Gatcombe, Lubrication Characteristics of Involute Spur Gears, *Trans. ASME*, vol. 67, p. 177, 1945.

mating gear teeth are both convex, the width of the band of the load-carrying film is small, and the pressures developed will be very high as compared with pressures in a sleeve bearing. In heavily loaded gears, the thickness of the film may become so small that the film is ruptured by the irregularities of the tooth surfaces. In this event, boundary lubrication will exist and special "extreme-pressure" lubricants should be used. Oils of low viscosity should be used for high speeds and light loads, and oils of high viscosity should be used for low speeds and heavy loads.

Grease and oil are the lubricants most frequently used for gears. Oil is generally considered better but it requires an oiltight housing or case with shaft seals. The same oil may be used for the lubrication of the gears and the bearings in many cases. When the oil is circulated by a splash system, the gear should dip into the oil about 1 in. A greater submergence may cause churning of the oil and an excessive temperature rise.

18-12 PARALLEL-SHAFT SPEED REDUCERS

These standard units are manufactured by many companies for general industrial service and whenever possible they should be used rather than a special unit. Frequently, however, gear reducers must be "built in" the machinery of which they are a part.

The standard units may be classified as follows.

1. Number of reductions, such as single, double, triple; single reductions may be had generally from 1:1 to 8 or 10:1; double reductions from 8:1 to 30 or 45:1.*
2. Vertical or horizontal (this refers to the plane in which the shafts are located)
3. Type of gears
4. Type of bearings, such as sleeve, ball, straight roller, or tapered roller
5. External gears, internal gears, or cyclic gear train
6. Geared motor reducer (these units are part of the driving motor)

In addition to the above types there are also worm-gear reducers, bevel-gear reducers, and variable-speed changers.

A single-reduction, single-helical horizontal speed reducer using roller bearings is shown in Fig. 18-10 and a herringbone reducer in Fig. 18-11.

Service factors The AGMA recommends the service factors[1] given in Table 18-9 for application to enclosed reducers. It is necessary that

* AGMA Standard, Ratios for Helical, Herringbone and Combination Spiral Bevel Gear Speed Reducers, 271.03–1963.

[1] See AGMA Standards 420.03–1963 and 423.01–1963.

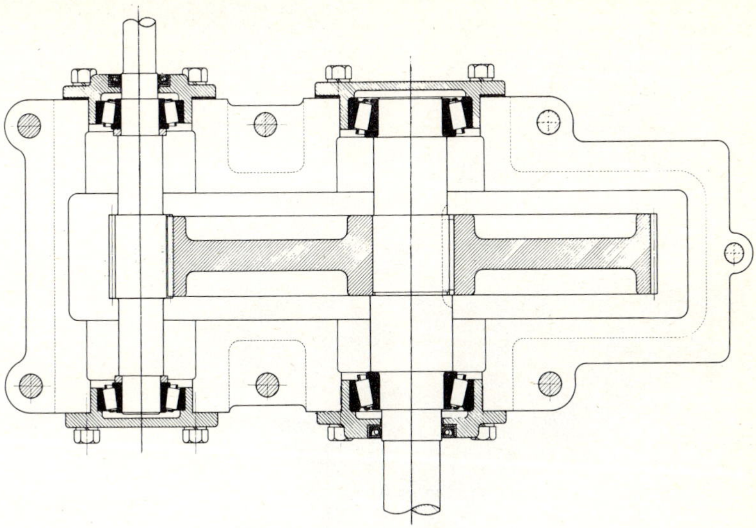

Fig. 18-10 Section of helical-gear speed reducer. (*Courtesy of Timken Roller Bearing Company.*)

the reducer have a capacity at least equal to the actual horsepower multiplied by the service factor.

Thermal horsepower rating When a speed reducer is operated, its temperature will rise until the housing attains a temperature above ambient temperature sufficient to dissipate the heat corresponding to the power

Table 18-9 Service factors*

		Character of load on driven machine		
Prime mover	*Duration of service*	*Uniform*	*Moderate shock*	*Heavy shock*
Electric motor	8 to 10 hr per day	1.00	1.25	1.75
	24 hr per day	1.25	1.50	2.00
	Intermittent, 3 hr per day	0.80	1.00	1.50
	Occasional, ½ hr per day	0.50	0.80	1.25
Multicylinder internal-combustion engine	8 to 10 hr per day	1.25	1.50	2.00
	24 hr per day	1.50	1.75	2.25
	Intermittent, 3 hr per day	1.00	1.25	1.75
	Occasional, ½ hr per day	0.80	1.00	1.50
Single-cylinder internal-combustion engine	8 to 10 hr per day	1.50	1.75	2.25
	24 hr per day	1.75	2.00	2.50
	Intermittent, 3 hr per day	1.25	1.50	2.00
	Occasional, ½ hr per day	1.00	1.25	1.75

* AGMA Standard 420.03–1963.

Fig. 18-11 Single-reduction, herringbone-gear speed reducer; lower view with top half of housing removed. (*Courtesy of DeLaval Steam Turbine Company.*)

loss of the unit. If the temperature of the housing is limited in order to maintain proper lubrication, then the horsepower transmitted is limited and is known as the "thermal capacity" of the unit. The AGMA specifies a thermal rating in terms of the center distance and the face width of the gears. For operation at periods of less than 2 hr with long rest periods, the thermal rating may be ignored, but for operating periods of more than 2 hr, the transmitted horsepower should not exceed the thermal horsepower rating.

For a single-reduction, helical-gear speed reducer, the thermal horsepower depends on the face width of the gears, their center distance, pinion speed, and gear ratio.[1]

[1] AGMA Standard 420.03–1963.

19
Gears for Nonparallel Shafts

19-1 TYPES OF GEARS

Refer to Fig. 19-1. A brief description of some gears is given in the following list.

1. *Straight bevel gears* The axes intersect, generally at right angles. The elements of the teeth are straight lines that converge at the apex of the pitch cone.
2. *Zerol bevel gears* The axes intersect, and the teeth are curved and can be ground.
3. *Spiral bevel gears* The axes intersect, and the teeth are curved and oblique.
4. *Face gears* These gears consist of a spur or helical pinion in combination with a conjugate gear of disk form.
5. *Skew bevel gears* The axes are nonparallel and nonintersecting, and the teeth are straight.
6. *Hypoid gears* The axes are nonparallel and nonintersecting, and the teeth are curved.

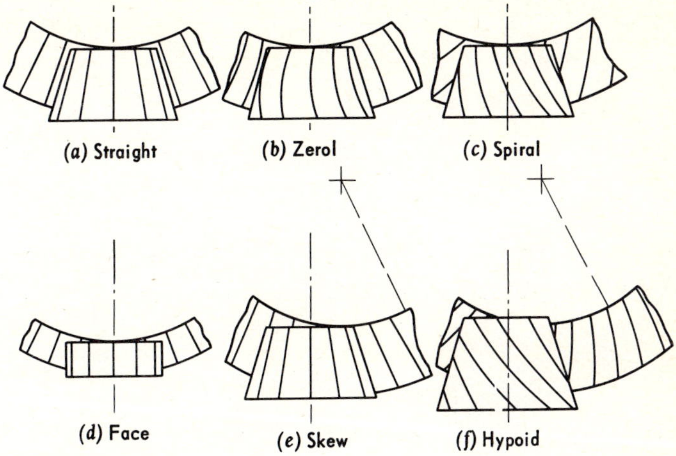

Fig. 19-1 Relation of axes.

7. *Crossed helical gears* The axes are nonintersecting and are at any angle. The teeth have the same or opposite hand (see Fig. 19-8).

8. *Worm gears* These gears include worms and their mating gears. The axes are usually at right angles, as in Fig. 19-12.

In Fig. 19-2 are shown the pitch cones for various relations of axes in the same plane from the external gear to the internal gear.

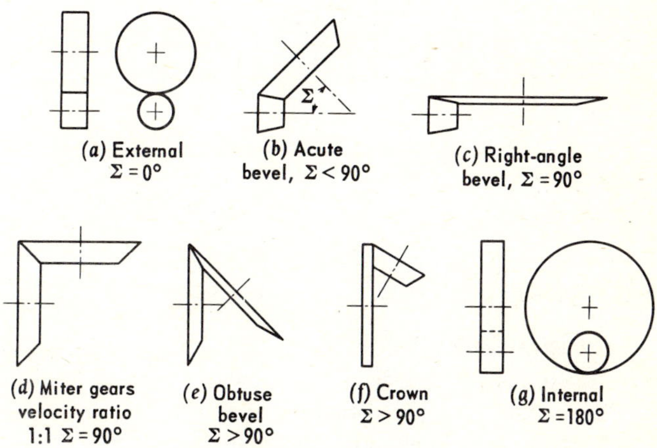

Fig. 19-2 Relation of axes for gears where Σ is angle between shafts.

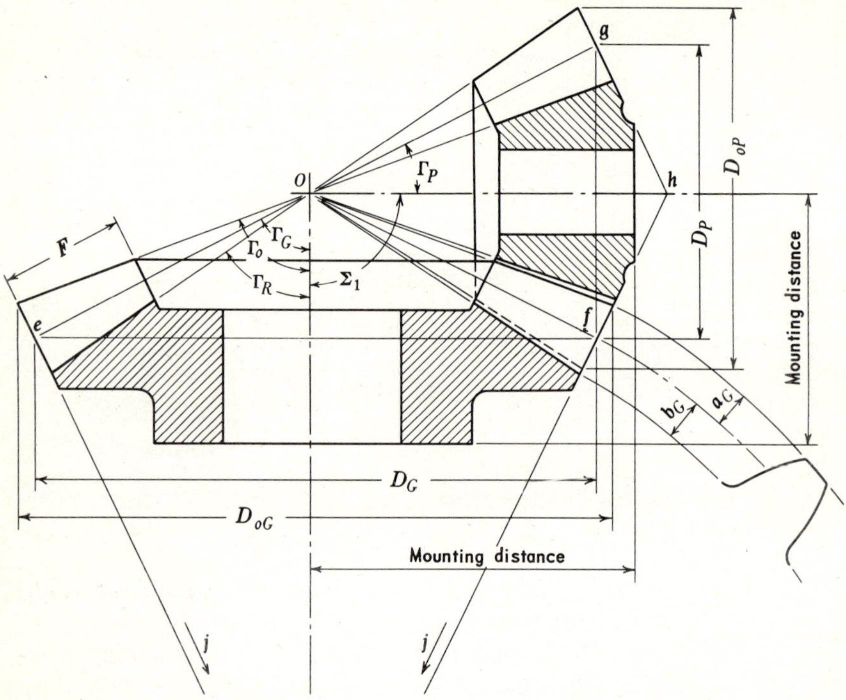

Fig. 19-3 Bevel-gear dimensions.

19-2 BEVEL-GEAR TERMINOLOGY

An axial section of a pair of right-angle bevel gears is shown in Fig. 19-3. The shaft angle Σ equals 90°, Γ is the pitch angle, Γ_o is the face angle, and Γ_R is the root angle. D is the pitch diameter, and D_o is the outside diameter. The addendum is a, and the dedendum b is measured at the large ends of the teeth. The pitch cones are represented by Oef and Ofg and the back cones by jef and hfg. In the figure, the subscripts P and G refer to the pinion and gear, respectively.

19-3 STRENGTH OF BEVEL-GEAR TEETH

Distribution of load along teeth A half section of a bevel gear is shown in Fig. 19-4(a), the large end of a tooth at (b), and a top view of the tooth at (c). Note that the elements of the tooth converge at the apex of the pitch cone, and hence the dimensions of the tooth at any section are proportional to the distance x of the section from the apex of the pitch cone O.

The dotted lines at (*c*) show the deflected tooth under the load. If the tooth does not *distort* under load, the elements of the deflected tooth pass through the apex O. Hence, the deflection of an element dx of the tooth under the load dW will be proportional to x.

The deflection of the element of the tooth as a cantilever is

$$\delta = \frac{dW\,h^3}{3EI} = \frac{12dW\,h^3}{3Et^3\,dx} \tag{19-1}$$

where h and t are the effective height and thickness, respectively, of the tooth at the section at x (see Fig. 18-3). Since the ratio of h and t for every section of the tooth is constant, and since δ is proportional to x, it follows that

$$\frac{dW}{dx} = Cx \tag{19-2}$$

where C is a constant. Equation (19-2) states that the variation of the load W with respect to x is proportional to x; hence the tooth load is greater at the large end of the tooth, as shown in Fig. 19-4(*d*).

The maximum bending stress at the base of the elementary section of the tooth is

$$s = \frac{6h}{t^2}\frac{dW}{dx} = \frac{6h}{t^2}Cx$$

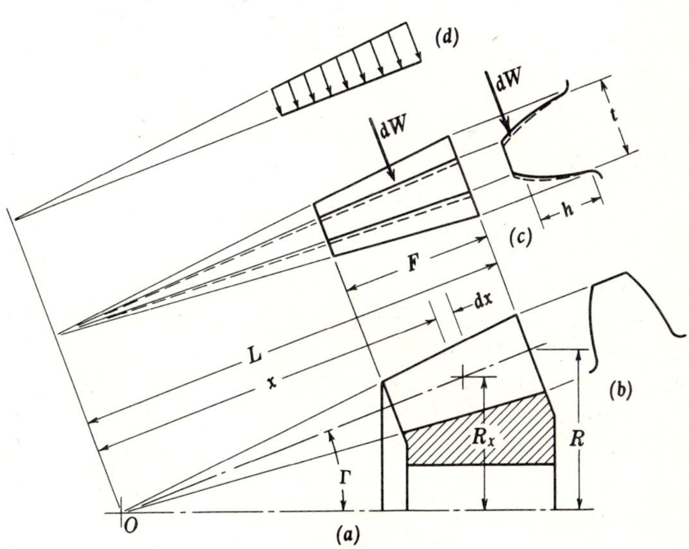

Fig. 19-4 Forces on bevel-gear tooth.

or since h and t are proportional to x, it follows that the stress s is constant along the base of a bevel-gear tooth.

Strength of bevel-gear teeth From the Lewis equation, Eq. (18-2), the load dW on an elementary section of a bevel-gear tooth may be expressed as

$$dW = s\,dx\,p_x y \tag{19-3}$$

where p_x = circular pitch of element of tooth
y = Lewis form factor

The moment of dW about the axis of the gear is equal to $R_x\,dW$ and its integral equals the torque T transmitted by the gear. Hence

$$dT = R_x\,dW = syR_xp_x\,dx \tag{19-4}$$

Since R_x and p_x may be expressed in terms of the length of the pitch-cone element L, and the circular pitch p and the radius R, both measured at the large end of the tooth, Eq. (19-4) becomes

$$T = \frac{sypR}{L^2}\int_{L-F}^{L} x^2\,dx \tag{19-5}$$

$$= sFpyR\left(1 - \frac{F}{L} + \frac{F^2}{3L^2}\right) \tag{19-6}$$

By dividing the torque T by the radius R, the equivalent tangential force W acting at the large end of the tooth may be found, thus

$$W = \frac{T}{R} = sFpy\left(1 - \frac{F}{L} + \frac{F^2}{3L^2}\right) \tag{19-7}$$

In bevel gears, the ratio of width of face to the pitch-cone element, F/L, is usually less than ⅓; hence the last term in the parentheses of the above equation is small compared with the other terms and may be neglected. Then Eq. (19-7) becomes

$$W = sFpy\left(1 - \frac{F}{L}\right) \tag{19-8}$$

Note that a spur gear may be considered as the limiting case of a bevel gear for which L is infinitely long. For a spur gear the ratio F/L is zero and Eq. (19-8) becomes identical with Eq. (18-2).

The tangential force W in Eq. (19-8) may be determined by dividing the torque on the gear by the radius R. The force W is not an actual force but an equivalent force acting at the large end of the tooth for use in designing the tooth and specifying it at its large end. The strength of the tooth at other sections will be equal to that at the large end, as discussed earlier in this article.

In the use of Eq. (19-8) for design purposes, the value of the allowable stress s may be taken from Eq. (18-3) and Table 18-2.

For satisfactory operation of bevel gears, the ratio F/p should be between 2 and 3, and the ratio F/L should not exceed $\frac{1}{3}$. The form factor y should be based on the *equivalent number* of teeth, i.e., the number of teeth on a gear whose radius equals the back-cone radius.

19-4 RESULTANT TOOTH FORCE

For determining the loads on the shafts of bevel gearing, it is necessary to use the actual resultant tooth force that acts at some point along the tooth face rather than the equivalent force at the large end of the tooth, as discussed in Art. 19-3. Since the tooth is loaded heavier at the outer end, the resultant force W_o acts between the midpoint of the tooth and the outer end.

From Eq. (19-2),

$$dW = Cx\,dx$$

or
$$W_o = C\int_{L-F}^{L} x\,dx = \frac{C}{2}[L^2 - (L-F)^2] \tag{19-9}$$

Referring to Fig. 19-4, the moment of dW may be taken about O as

$$dM = x\,dW$$

or, using Eq. (19-2) again,

$$dM = Cx^2\,dx$$

or
$$M = C\int_{L-F}^{L} x^2\,dx = \frac{C}{3}[L^3 - (L-F)^3] \tag{19-10}$$

The distance of the point of application of the resultant load W_o from the apex O is found by dividing M by W_o; hence

$$L_o = \frac{M}{W_o} = \frac{2}{3}\frac{L^3 - (L-F)^3}{L^2 - (L-F)^2} \tag{19-11}$$

The radius R_o corresponding to L_o equals

$$R_o = \frac{L_o}{L}R \tag{19-12}$$

By combining Eqs. (19-11) and (19-12),

$$R_o = R\,\frac{1 - F/L + F^2/3L^2}{1 - F/2L} = ZR \tag{19-13}$$

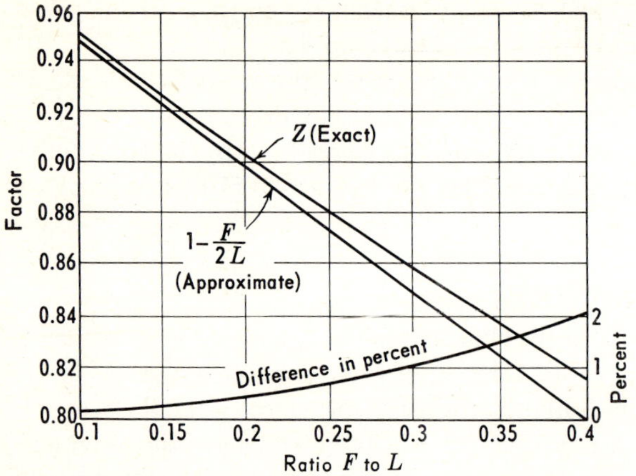

Fig. 19-5 Comparison of exact and approximate radius factors for bevel gears.

The resultant tooth load W_o is equal to the torque on the gear divided by R_o, or

$$W_o = \frac{T}{R_o} \tag{19-14}$$

This is the tangential component of the actual tooth load. The separating component equals W_o multiplied by the tangent of the pressure angle.

The values of Z in Eq. (19-13) for various values of F/L are shown in Fig. 19-5.

From the geometry of Fig. 19-4, it may be shown that the radius corresponding to the midsection of the tooth is

$$R_m = R\left(1 - \frac{F}{2L}\right) \tag{19-15}$$

The values of $1 - F/2L$ are plotted in Fig. 19-5, and also the percentage of difference between Z and $1 - F/2L$. For bevel gears it is evident that the error in considering that the resultant tooth load acts at the midpoint of the face is small, and therefore Eq. (19-15) may be used, except for very precise determinations of shaft loads. In such cases, the effect of tooth inaccuracies; deflection of teeth, shaft, and mounting; friction; and other deviations from the ideal case should be included.

19-5 RATING OF BEVEL GEARS

The following AGMA Standards should be consulted. System for Straight Bevel Gears, 208.02–1964; System for Spiral Bevel Gears, 209.03–1964; Rating the Strength of Straight Bevel and Zerol Bevel Gear Teeth,

222.02–1964. The "Gear Handbook," by Darle W. Dudley,[1] contains good references and information.

19-6 MOUNTING OF BEVEL GEARS

The shafts for bevel gears and the bearing supports should be sufficiently rigid so that the deflections at the teeth under load will not be excessive. Also, the gears should be properly adjusted when they are installed. Only by fulfilling the two requirements of mounting and adjustment and by providing proper lubrication can bevel gears be made to operate satisfactorily in respect to quietness and life.

The bearings should be designed with due consideration of the thrusts along the axes. In spiral bevel gears, the directions of the thrusts depend on the hand of the spiral, the direction of rotation, and whether the gear is the driver or driven member.

A pair of bevel gears mounted in plain bearings are shown in Fig. 19-6(*a*). A ball bearing mounting with the pinion straddle mounted is shown at (*b*). Note that the teeth of the gear are cut on a steel ring that is attached to a separate center.

A typical aircraft bevel-gear mounting is shown in Fig. 19-7. In a mounting of this type, the distance between the pinion bearings should be at least 2.5 times the overhang x.

[1] McGraw-Hill Book Company, New York, 1962.

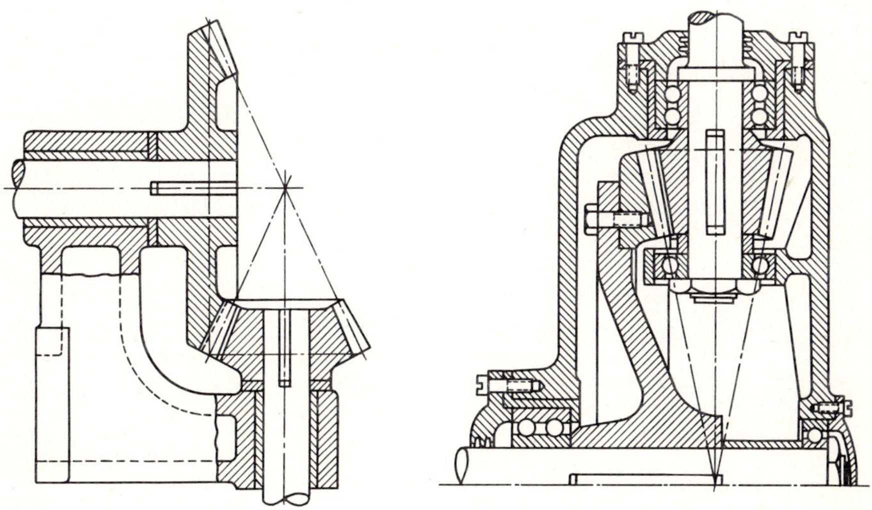

(*a*) (*b*)

Fig. 19-6 Bevel-gear dimensions.

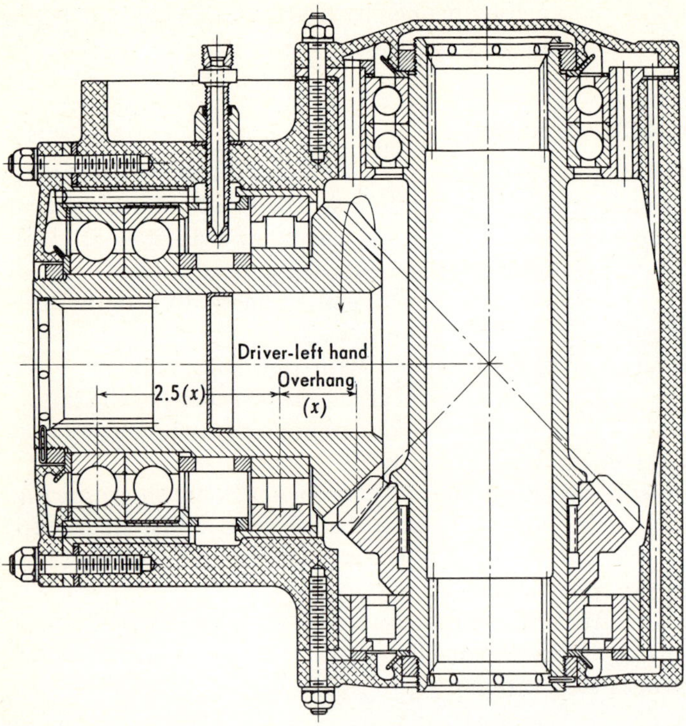

Fig. 19-7 Aircraft bevel-gear mounting. (*Courtesy of Gleason Works.*)

19-7 CROSSED HELICAL GEARS

The most common arrangement of *crossed* helical gears is with the shafts at right angles, as shown in Fig. 19-8. The contact between teeth is, kinematically, point contact. Because of the elasticity of the material the contact point becomes a small area. Because of the high tooth loads,

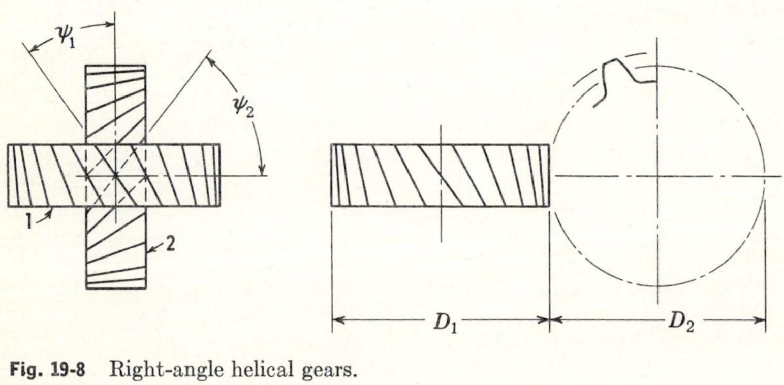

Fig. 19-8 Right-angle helical gears.

high sliding velocity, and the small contact area the contact pressure is very high and wear is comparatively rapid. Hence, these gears are suitable for transmitting light power at moderate speeds. The efficiency is low, seldom more than 85 percent.

The velocity ratio of any type of toothed gearing is equal to the inverse ratio of the numbers of teeth, hence

$$\frac{\omega_2}{\omega_1} = \frac{n_1}{n_2} = \frac{D_1 \cos \psi_1}{D_2 \cos \psi_2} \tag{19-16}$$

where ω = angular velocity
n = number of teeth
D = pitch diameter
ψ = helix angle
(For subscripts, refer to Fig. 19-8.)

Crossed helical gears are used to drive camshafts and auxiliaries on small internal-combustion engines, feed mechanisms on machine tools, and other units that require small power.

19-8 WORM GEARING

This type of gearing is extensively used to transmit power at high velocity ratios between nonintersecting shafts that are generally but not necessarily at right angles. Velocity ratios as high as 300:1 or over are being used in a single reduction. Worm drives may be used as speed *increasers*, but their usual use is as speed reducers, in which the worm is the driver and the worm wheel the driven member. If the driven machinery has large inertia and if there is possibility of the driving power being cut off suddenly, the worm drive must be *reversible*, or a slip coupling or other device must be placed between the driven unit and the worm wheel to prevent damage to the drive. When the lead angle of the worm is greater than the friction angle of the surfaces in contact, the worm is called *reversible*, or *overrunning;* i.e., the worm can be freely driven by the worm wheel. However, under a *fluctuating* load, any worm drive may be reversible in the sense that it may creep backward, as when a worm is used as a brake to hold a load. A similar condition appears in the loosening of bolts under fluctuating axial loads, which was referred to in Art. 9-2.

The speed ratio of a worm drive is equal to the number of teeth on the worm wheel divided by the number of threads on the worm. From one to four threads are commonly used on worms, i.e., single-, double-, triple-, and quadruple-threaded worm.

19-9 TERMINOLOGY AND CONSTRUCTION

Single- and double-threaded worms are shown in Fig. 19-9. The axial pitch p is equal to the circular pitch of the mating worm wheel. Note

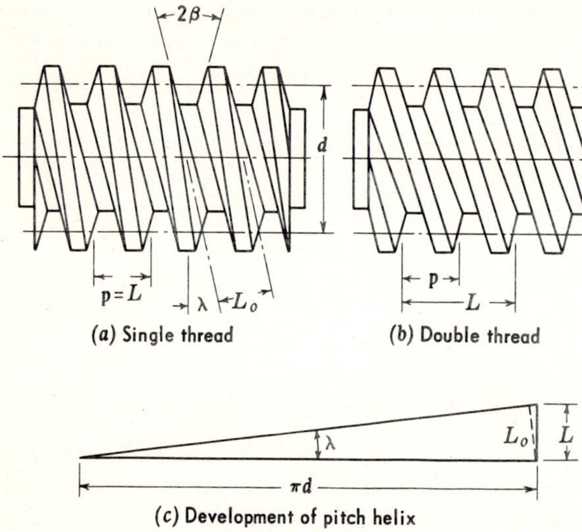

Fig. 19-9 Worm dimensions.

that the lead L of an n-threaded worm is equal to n times the pitch; hence the lead and the axial pitch are equal in a single-threaded worm. The *pitch lead angle* λ is the angle between the tangent to the pitch helix and the plane of rotation. The pitch *helix angle* is the angle between the tangent to the pitch helix and an element of the pitch cylinder and is the complement of the lead angle. The *pressure angle* is β.

The development of the pitch helix is shown at (*c*) in Fig. 19-9; this[1] may be used to determine the relation between λ, L, and the pitch diameter of the worm d, that is, $\tan \lambda = L/\pi d$. The normal lead L_o is equal to $L \cos \lambda$.

The *straight worm* shown in Fig. 19-9 is the type in most common use. The *Hindley*, or double enveloping, *worm* (Fig. 19-10) is used to some extent, but it requires extremely accurate alignment.

[1] See AGMA Standard 341.02–1965 and Dudley, *op. cit.*

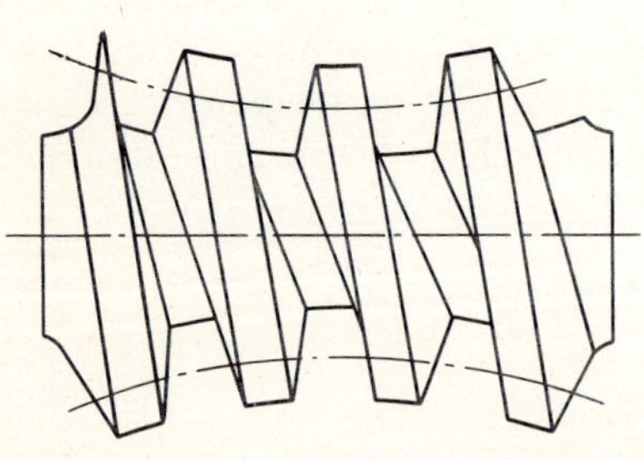

Fig. 19-10 Hindley worm.

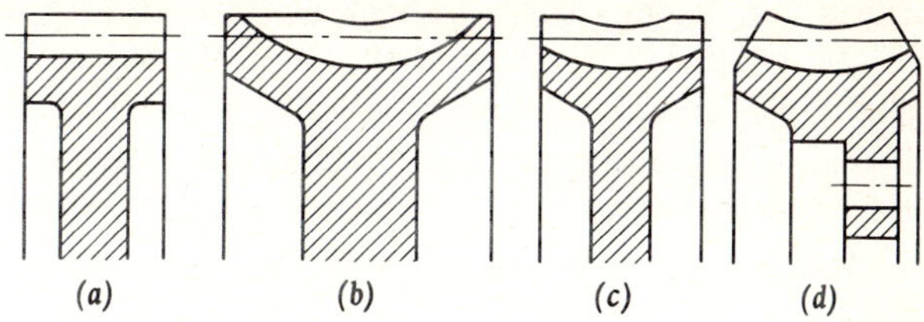

Fig. 19-11 Worm-gear rim construction.

Various types of worm-wheel rim construction are shown in Fig. 19-11. The form at (*a*) is cut with a form cutter and is suitable for light service. The types at (*b*), (*c*), and (*d*) may be hobbed, the latter being a separate ring for attachment to a cast-iron or steel spider.

Materials The worm is generally made of steel, and the gear is made of bronze or of cast iron for light service. Some recommended combinations of materials are given in Table 19-1.

19-10 DESIGN OF WORM GEARING

In the design of a worm drive, the quantities usually specified are the horsepower, speed, velocity ratio, and service. In some cases the exact or the approximate center distance is specified, the former in a replacement drive, and the latter in a drive where the center distance may be limited by space conditions. If the designer is limited in the choice of the center distance, it is usually impossible to achieve the highest efficiency.

In addition to the center distance, the quantities that must be determined are the lead angle, the lead, and the number of threads on the worm. In addition, consideration must be given to the materials of the

Table 19-1 Worm and worm-wheel combinations

Service	*Worm*	*Worm wheel*
Light	SAE 1040	Cast iron or SAE 63
Medium	SAE 2320 or SAE 3120 (case-hardened)	Phosphor bronze
Heavy	Molybdenum or chrome-vanadium steel (hardened)	SAE 65

worm and worm wheel as affecting strength and wear of the teeth, sizes of the shafts as affecting their strength and stiffness, mounting of the shafts, design of a substantial housing of sufficient heat-dissipating capacity, and lubrication. Most of the above items are related, and many of them are not susceptible to exact analysis; hence it is apparent that the design of a worm drive demands the judicious use of theory and experimental and field data.

Proportions of worm drive To determine a satisfactory combination of lead angle, lead, and center distance, the following method may be used.[1]

The center distance C may be expressed in terms of the worm and worm-gear diameters d and D, and it may be further expressed in terms of the axial lead L, the lead angle λ, and the velocity ratio r as follows (see Fig. 19-12):

$$C = \frac{d + D}{2} = \frac{L}{2\pi} (\cot \lambda + r)$$

In terms of L_o, the above equation becomes

$$\frac{C}{L_o} = \frac{1}{2\pi} \left(\frac{1}{\sin \lambda} + \frac{r}{\cos \lambda} \right) \tag{19-17}$$

Since the velocity ratio r is specified, Eq. (19-17) contains three variables, C, L_o, and λ. The right-hand member may be evaluated for various velocity ratios and the results plotted as in Fig. 19-13.

In Fig. 19-13 the low point on each of the curves represents the lead angle which corresponds to the minimum value of C/L_o. This minimum value represents the minimum center distance that can be used with a given lead or, likewise, the maximum lead that can be used with a given center distance.

[1] Courtesy of Thomas P. Colbert, University of Wisconsin, and O. A. Leutwiler, University of Illinois.

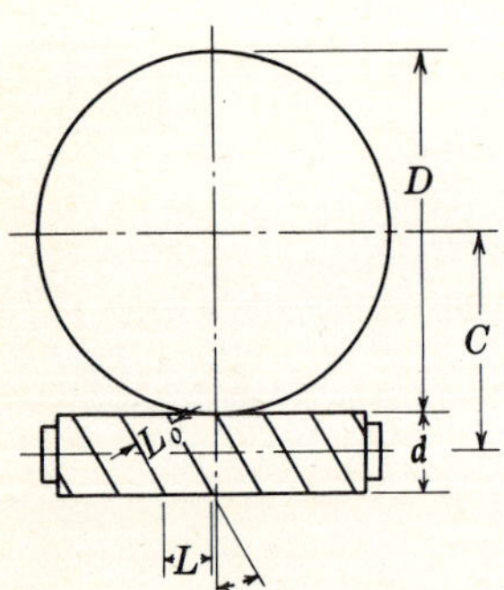

Fig. 19-12 Worm and worm gear.

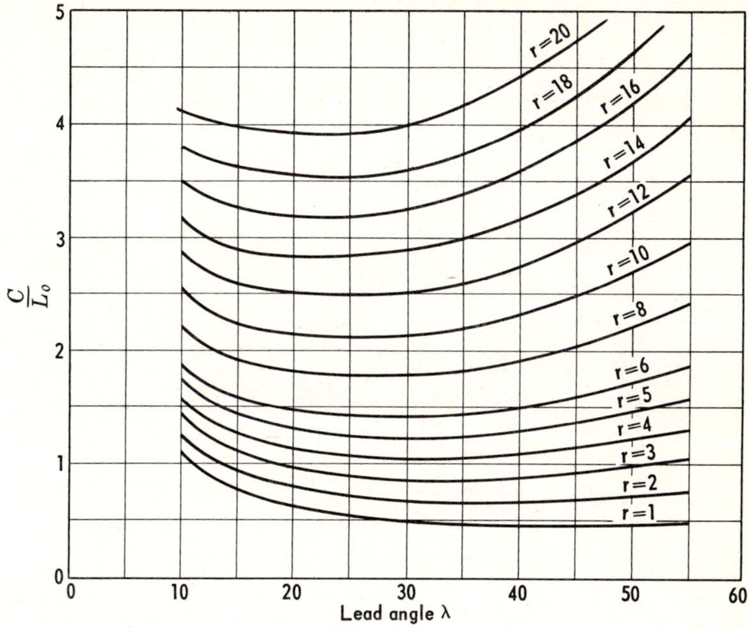

Fig. 19-13 Worm-gear design curves.

Since the numbers of threads that may be used with various velocity ratios have been fairly well established by common use (Table 19-2), and since axial pitches have been standardized (Art. 19-9), it is possible to determine the combination of lead angle, lead, center distance, and diameters to suit specifications. The following example illustrates the procedure.

Example Determine the pitch diameters, number of threads, and axial pitch for a worm drive having a velocity ratio 14:1 and a center distance of approximately 6 in.

Table 19-2 Worm data

Velocity ratio	*Number of threads*
20 and over	Single
12–36	Double
8–12	Triple
6–12	Quadruple
4–10	Sextuple

Solution From Fig. 19-13, the lead angle λ is 22° and the value C/L_o is 2.83. Therefore

$$L_o = \frac{C}{2.83} = \frac{6}{2.83} = 2.12 \text{ in.}$$

$$L = \frac{L_o}{\cos\lambda} = \frac{2.12}{0.927} = 2.29 \text{ in.}$$

From Table 19-2, assume triple threads, that is, $n = 3$. Then

$$p = \frac{L}{n} = \frac{2.29}{3} = 0.763 \text{ in.}$$

The nearest standard axial pitch is ¾ in. Then

$$L = np = 3 \times 0.75 = 2.25 \text{ in.}$$

$$L_o = L\cos\lambda = 2.25 \times 0.927 = 2.09 \text{ in.}$$

$$C = 2.83L_o = 2.83 \times 2.09 = 5.91 \text{ in.}$$

$$d = \frac{L}{\pi\tan\lambda} = \frac{2.25}{\pi \times 0.404} = 1.774 \text{ in.}$$

$$D = 2C - d = 2 \times 5.91 - 1.77 = 10.05 \text{ in.}$$

Strength of teeth Since the teeth of a worm gear are weaker than those of the worm threads, the design for strength may be based on the Lewis equation, Eq. (18-2), applied to the worm-gear teeth. The velocity factor C_v may be taken as $1{,}200/(1{,}200 + V)$, where V is the pitch-line velocity of the worm gear in fpm. Form factors given in Table 18-1 for full-depth teeth may be used.

Since the Lewis equation gives the strength of a single tooth, it may be desirable to consider the number of teeth in contact. The number of teeth in contact may be determined by dividing the angle of action of the teeth by the angle subtended by the circular pitch.

Wear of worm gears Failure at the surfaces in contact is the most prevalent cause of failure of worm gears. The failure may be prevented or satisfactorily delayed by using proper materials and hardness and finish of the surfaces, maintaining bearing pressure within allowable limits and providing correct lubrication. Because of the lack of a standardized method for determining the load limit for wear of worm gears at the present writing, further treatment will not be given here.[1]

Thermal rating of worm gears The amount of heat that must be dissipated in a worm drive is equal to the power loss at the gears, i.e.,

$$H_1 = 33{,}000 \times \text{hp} \times (1 - \text{efficiency}) \tag{19-18}$$

where H_1 = power loss, ft-lb per min.

[1] H. Merritt, Worm Gear Performance, *Proc. Inst. Mech. Engrs.* (*London*), vol. 129, p. 127, 1935; see H. E. Merritt, "Gears," Sir Isaac Pitman & Sons, Ltd., London, 1943.

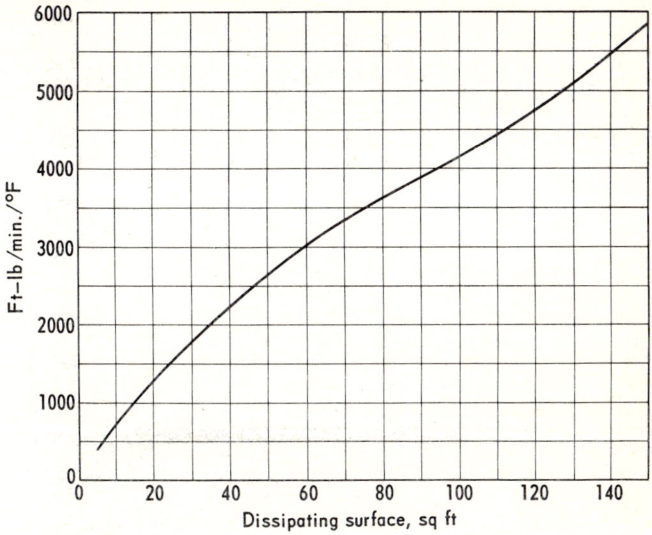

Fig. 19-14 Heat dissipation of worm gearboxes.

This heat must be dissipated by the gearbox or removed by other means, such as artificial cooling in the form of cooling coils in the housing or external oil coolers. The dissipation of heat from the housing depends on the difference in temperature between the surface of the housing and the surrounding air; i.e., heat-dissipation capacity H_2 in foot-pounds per minute is

$$H_2 = KA(t_1 - t_2) \tag{19-19}$$

where A = housing area, ft²
t_1 = temperature of lubricant, °F
t_2 = ambient temperature, °F
K = coefficient of heat dissipation, ft-lb per min per ft² per °F

The maximum temperature of the lubricant should not exceed 180°F. Values for K times square feet may be estimated from Fig. 19-14.[1]

19-11 FORCES ON WORM GEARS

The force components on a worm are similar to those on a power screw, since the axial force on a screw corresponds to the axial thrust on the worm, and the tangential force on the screw and on the worm are similar.

[1] V. M. Faires, "Design of Machine Elements," p. 604, The Macmillan Company, New York, 1965; H. Walker, The Thermal Rating of Worm Gear Boxes, *Proc. Inst. Mech. Engrs. (London)*, vol. 151, p. 326, 1944.

In addition, the worm drive is subjected to separating forces that tend to force the worm and worm wheel out of mesh. The three force components on the worm are shown in Fig. 19-15 and may be determined as follows:

$$\text{Tangential force } F_z = \frac{2 \times \text{torque on worm}}{d}$$

$$\text{Axial force on worm } F_y = \frac{F_z}{\tan\lambda}$$

$$\text{Separating force } F_x = F_y \tan\beta$$

The forces on the worm gear are equal in magnitude but opposite in direction to those shown in Fig. 19-15.

The loads on the worm-shaft bearings may be readily determined by taking moments of the forces F_x, F_y, and F_z and the overhanging load F_o due to a gear, chain, or belt. For determining overhanging shaft loads, the AGMA recommends the following factors: chain, 1; gear, 1¼; V belt, 1½; flat belt, 2. If the direction of rotation of the worm and the direction of the overhanging load are not specified, their directions should be assumed to give the maximum possible loading on the bearings for use in selecting or designing the bearings. The loads on the worm-gear shaft bearings may be similarly determined.

To determine the efficiency of a worm drive, the equation below may be used for approximate results, i.e.,

$$\text{Efficiency} = \frac{\tan\lambda}{\tan(\lambda + \phi_1)} \tag{19-20}$$

where λ = lead angle

ϕ_1 = friction angle (see Fig. 19-16 and Table 19-3 for values of $\tan\phi_1$)

Equation (12-2) was derived for square-threaded screws; hence it does not strictly apply to worms having pressure angles other than zero

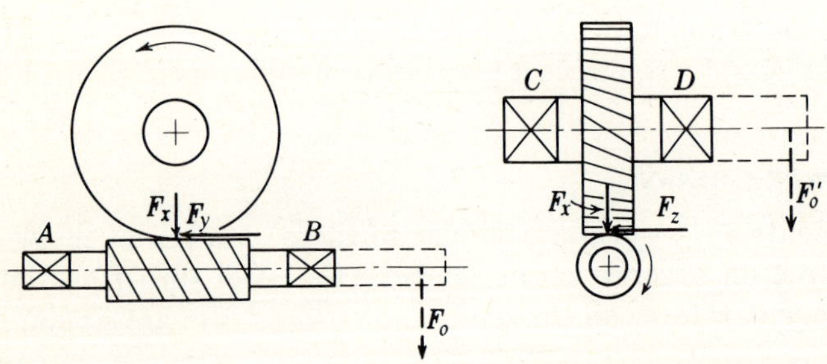

Fig. 19-15 Forces on worm teeth. Forces on worm-gear teeth are in opposite direction.

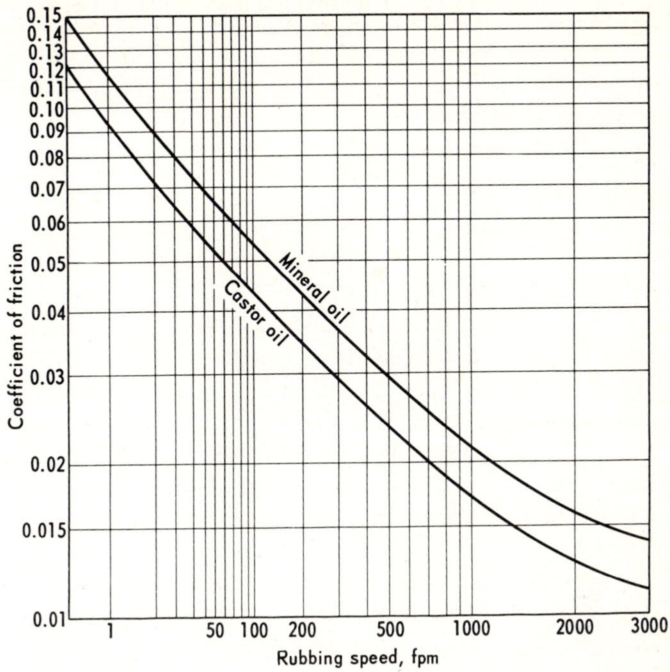

Fig. 19-16 Coefficient of friction for worm gears (see Table 19-3). *(From H. E. Merritt, "Gears," Sir Isaac Pitman & Sons, Ltd., London,* 1943.)

degrees. The error in using the above equation will be slight, however, for worms of usual pressure angles. The values of the coefficient of friction in Fig. 19-16 include allowance for bearing friction and may be used for design purposes. Figure 12-3 shows the variation of efficiency with the lead angle. Note that the maximum efficiency is secured with a lead angle of 45°, but the variation in efficiency is not great for lead angles of 25 to 60°. Care must be exercised in design in this instance since the *loss* in efficiency, i.e., 100 − efficiency, represents the heat that must be dissipated by the gearbox. The difference between 98 and 97 percent in efficiency is about 1 percent, but the difference in loss in efficiency is 50 percent; hence 50 percent more heat must be dissipated by a gearbox for a 97 percent efficient drive than for one of 98 percent efficiency.

19-12 MOUNTING OF WORM GEARS

The requirements for mounting worm gears and bevel gears are the same, in general, as for other types of gears except that they are more exacting.

Table 19-3 Correction factors for coefficient of friction in Fig. 19-16

For	*Working with*	*Multiply values in chart by*
Case-hardened steel	Phosphor bronze	1.0
Cast iron	Bronze	1.15
Cast iron	Cast iron	1.33
Hardened steel	Aluminum	1.33
Steel	Steel	2.0

Since the worm-gear shaft is of relatively large size and its bearings are generally close together, its deflection is usually not excessive. The worm shaft, however, is smaller in diameter because of the low torque it transmits and because the size of the worm shaft should be a minimum because of its great influence on overall size of the drive. Thus the relatively small worm shaft and the necessary large distance between its bearings make its deflection relatively large.

It is necessary to limit the deflection of the shafts to avoid excessive heating at the teeth. For this reason, the shafts should be stiff, the bearings well supported, and the housing rigid. It is important also to mount the shafts so that a rise in temperature will not cause the shaft to buckle, which would in turn cause more heating and result in failure of the teeth. Well-mounted worm and worm gears are illustrated in Figs. 22-12 and 22-24.

19-13 CONCLUDING DISCUSSION

From the considerations of gears in this and the preceding chapter, it is striking that there are so many types of gears, classes of applications, and modes of failure. The development of the many types of gears has been necessitated by the different requirements in the automotive, aircraft, marine, industrial, and other fields. As an illustration of a special application of gears, a differential gear drive for a gravure press is shown in Fig. 19-17. The input shaft is connected to the main way shaft of the press and there may be as many as six or more of these differential gearboxes each coupled to the main way shaft. The output shaft extends through the differential assembly as shown, and the only gear keyed to it is the 35-tooth helical gear. The 50-tooth hypoid gear and the differential spider carrying the adjusting worm gear and the 22-tooth and 23-tooth helical gears all float on the output shaft, which is supported on bearings outside the gears. The adjusting gear and the worm (not shown) are used to justify the rolls to ensure aligned copy.

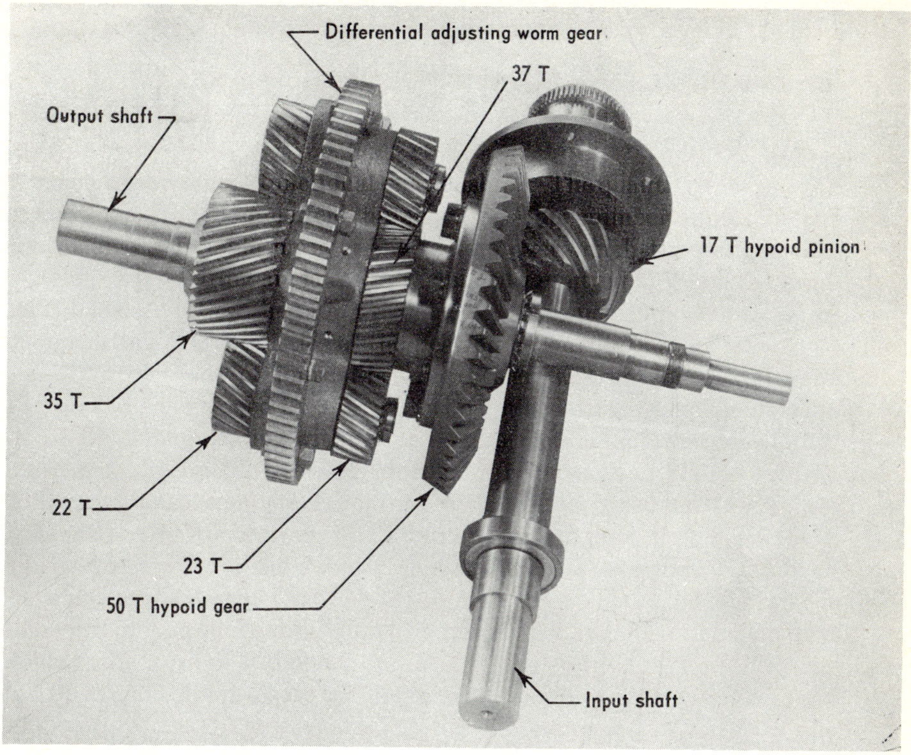

Fig. 19-17 Differential gear drive for gravure press. (*Courtesy of Kidder Press Company.*)

As to modes of failure, there are three common ones, namely, fracture of the teeth, excessive wear of the teeth, and failure of the lubricant.

Fracture of the teeth may be due to large shock loads and/or to exceeding the fatigue life of the material. This type of failure is not common.

Wear is a cumulative process and may be satisfactorily delayed by proper design of the teeth, including limiting the contact pressure and using proper lubrication. A lubricant is coming to be classed as a material of engineering the same as steel, and rightly so. As an example, the wrong grease in the differential housing of an automobile would probably lead to failure of the hypoid gears within a few blocks of travel, and the automobile would be immobilized the same as if the drive shaft broke.

The temperature of the lubricant for gears may be controlled by cooling by convection of heat from the gear housing, using cooling fins, or by the use of forced air or fluid cooling if necessary.

The AGMA Standards are available for strength and durability ratings for spur, helical, and herringbone gears, straight and spiral bevel gears, for the durability of cylindrical and double-enveloping worm gears, and for the thermal ratings of enclosed helical and herringbone, bevel, and worm-gear units. The Standards are used for rating of gear drives. They are precise and, like legal papers, should be read carefully, including the fine print.

20

Surface Finish, Friction, and Wear

20-1 SURFACE IRREGULARITIES

The deviations of the actual surface of a body from its nominal surface may be classified as follows:[1] (1) *Waviness,* as characterized by more or less regular and recurrent deviations of the nature of waves, is shown in Fig. 20-1(*a*). (2) *Roughness,* as shown at (*b*), is characterized by small irregular deviations, as may be felt when the fingernail is drawn over a

[1] See Surface Texture, USAS B46.1–1962.

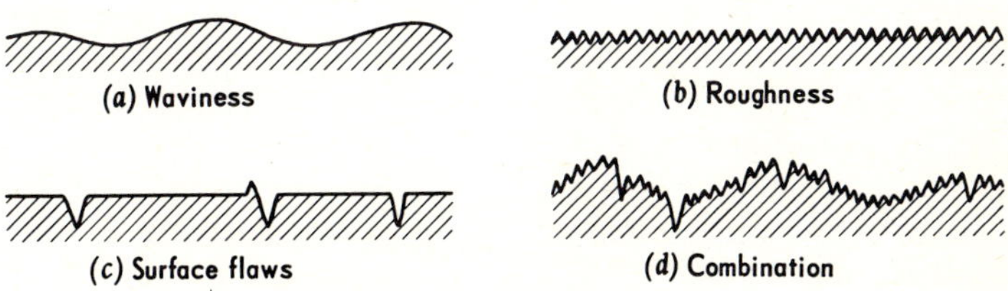

Fig. 20-1 Types of surface irregularities.

ground surface. (3) *Surface flaws,* as shown at (*c*), generally occur at random intervals, for instance, scratches, checks, etc., and are frequently caused by the processing operations. (4) A surface with all three types of deviations is shown at (*d*). (5) *Lay* means the direction of the predominant surface pattern.

20-2 MEASUREMENT OF SURFACE DEVIATIONS

There are a number of methods that have been devised for measuring surface roughness, or the degree to which a surface deviates from a plane. None of these methods is entirely satisfactory, but the commonly used ones, as described below, yield reasonably consistent readings, and on a comparative basis are valuable for judging the surface-finishing methods and for controlling production.

The profilograph is a laboratory instrument in which a stylus with a sharp tracer point with a tip radius of 0.00005 in. is delicately held against and normal to the surface being measured, and then is drawn over the surface at a rate of approximately 0.001 in. per sec. The movement of the stylus is magnified by a reflected light beam and is recorded on a moving sensitized film. Thus a graph is secured that is an indication of the approximate profile of the surface. Since the tracer point has a finite radius, and since it cannot explore beneath overhanging projections, the trueness of the graph is thereby affected. A surface which is described by 5 microinches (abbreviated 5 μin.) irregularity by a profilograph means that the graph indicates a distance from the highest peak to the lowest valley which is equal to 5 millionths of an inch. A series of profilographs is shown in Fig. 20-2.

The profilometer is similar to the profilograph in that a tracer point is used. It differs, however, in that the motion of the stylus normal to the surface is translated into electric voltage which is amplified by a cathode-ray oscillograph arranged to record motions on a moving film. The record on the film is not a graph of the surface but it indicates the "running average of surface deviations," i.e., a root-mean-square value (rms). Thus 5 μin. by a profilometer (recorded as 5 μin., rms) means that the square root of the average of the squares of a number of deviations is 5 millionths of an inch. It is thus apparent that the number of microinches as recorded by a profilograph and by a profilometer cannot be compared directly; however, each is an indication of the irregularities of a surface. The profilometer is a portable instrument.

A recent development is the *Surfagage* (General Motors) which measures the arithmetic mean value of surface deviations rather than the rms value. The Surfagage is a portable tracer-type instrument which converts the motion of the stylus normal to the surface through a thin

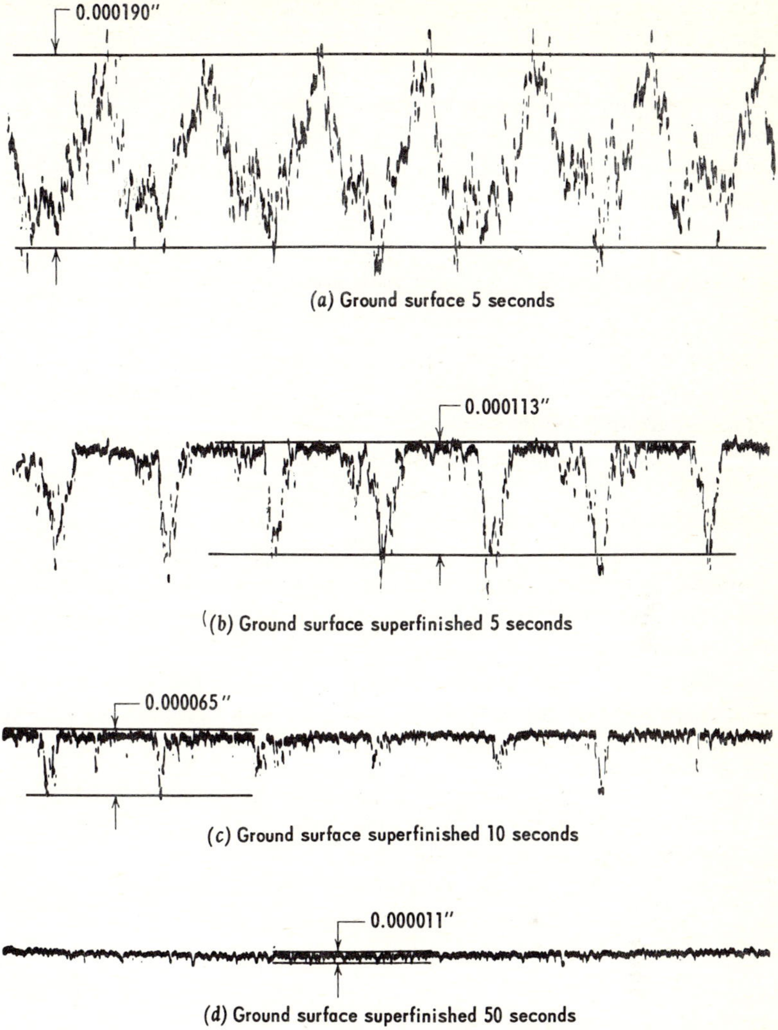

Fig. 20-2 Series of profilographs, each representing about ⅛-in. linear travel of stylus over surface. (*Courtesy of Chrysler Corporation.*)

viscous fluid film in shear to a transducer tube. The amplified impulses yield the arithmetic mean value which may be read directly on the gauge.

The arithmetic mean value is preferred by some manufacturers since it is felt that it is easier for production and inspection personnel to understand readily its significance; also it is closer to what may become the international standard. The British standard now uses the arithmetic mean value. Fortunately, the difference between the two readings is small; for most machine surfaces the arithmetic mean is about 10 percent less than the rms value.

Other methods[1] of determining surface irregularities are the Brush surface analyzer, the electrical profile recorder, piezoelectric pickup, reflection of light from surface, profile photomicrographs (which requires plating of the surface), and the surface dynamometer.

20-3 DAMAGE TO SURFACES DUE TO FINISHING

It is evident from metallographic analyses that the structure of a crystalline metal is changed by the machining process to an appreciable distance below the surface. In some of these processes, such as turning, planing, shaping, milling, and rough grinding, a damaged layer of considerable depth remains on the part as a result of the operations. An ideal finishing process is one that removes the damaged layer and produces a geometrically true surface without introducing a new damaged layer. Some of the finishing processes, such as fine grinding and honing, produce a surface that has only small irregularities, i.e., a good finish, but at the same time generally introduce a new damaged layer from a metallographic standpoint.

The two main types of surface damage will now be discussed, including some features of the surface-finishing methods that avoid these damaging actions.

Metallurgical damage The high pressure required and the high temperature produced by grinding cause a change in structure of an annealed steel specimen, indicated in Fig. 20-3. Fine grinding affects the structure to a distance of the order of 0.001 cm below the surface.

This depth of damage is not surprising in view of the high temperatures, well above 2000°F, regardless of whether or not a coolant is used. The wet grinding affected the surface only about 0.0007 cm below the surface. The last three finishes shown, i.e., lapped, polished, and superfinished, affected the surface about 0.0001 cm, a very thin layer, since the pressures and temperatures were low.

Mechanical damage When a metal-cutting tool, such as a planer tool, makes a cut, the chip that is formed curls over the cutting face of the tool. Since the chip is newly cut metal at high temperature, it removes any film of lubricant or oxide from the face of the tool and produces conditions favorable for particles of the chip to adhere to the tool. Under some

[1] A. M. Swigert, "Story of Superfinish," pp. 91ff, Lynn Publishing Company, Detroit, Mich., 1940; Surface Finish, American Society of Mechanical Engineers, 1942.

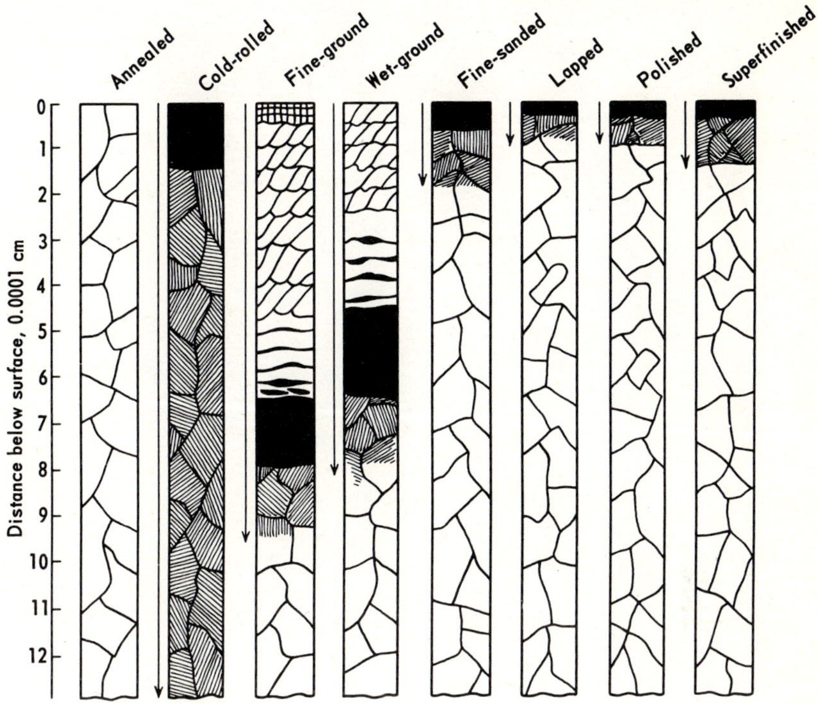

Fig. 20-3 Metallurgical effects of surface-finishing methods. Arrows show depth of damage. (*From J. Wulff, Proceedings of Special Summer Conferences on Friction and Surface Finish, Massachusetts Institute of Technology,* 1940, *p.* 14.)

conditions a sizable accumulation of metal will build up at the cutting edge of the tool and will result in the so-called "built-up edge," as illustrated in Fig. 20-4. This built-up edge will grow until it becomes so large that it breaks off, owing to the cutting forces. Thus the built-up edge periodically forms and breaks off in more or less regular cycles. The built-up edge has the effect of changing the shape of the tool point and likewise the form of the surface being machined and is one cause of the familiar cross marks on machined surfaces, as shown by Fig. 20-5. To reduce such irregularities, finer cuts may be taken, but the built-up edge will always be present even though reduced in size. The built-up edges may appear in all metal-cutting tools, such as planers, shapers, milling cutters, etc.

In the use of a grinding wheel made of a bonded abrasive, cutting is performed by small projections of the abrasive. Small built-up edges or points may appear at the cutting edges or points of such cutters. This is

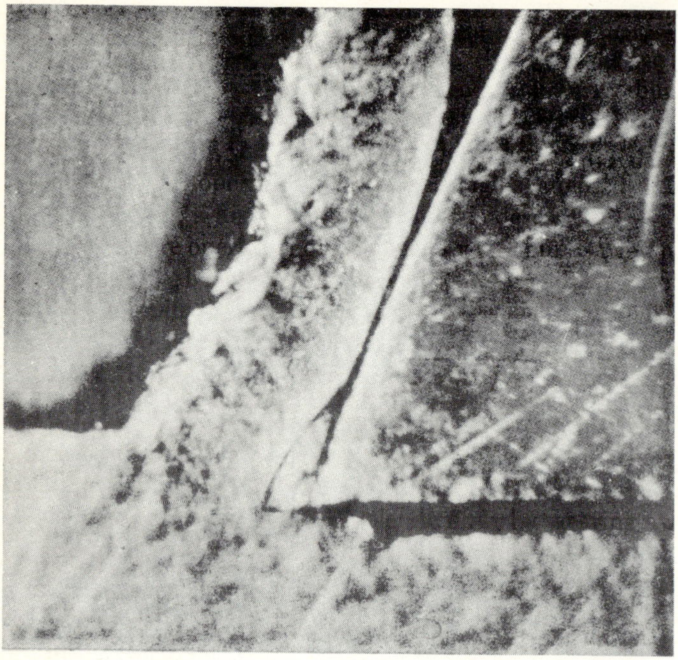

Fig. 20-4 Built-up edge on planer tool. (*Courtesy of Cincinnati Milling Machine Co.*)

one reason for the lines that appear on ground surfaces in the direction of the path of the abrasive over the piece.

In lapping, the motion of the lap should not be in one direction, as in scrubbing, but should periodically change direction, thus removing the built-up points on the abrasive before they have an opportunity to reach

Fig. 20-5 Planed surface showing effects of built-up edge.

a damaging size. One reason for the perfection of the superfinishing process is that the motion of the abrasive stone is random, i.e., continually changing in direction, thus avoiding built-up points on the abrasive.

20-4 PRODUCTION OF SMOOTH SURFACES

The production of a smooth surface is necessarily a step-by-step process. First, rough machining followed by fine machining is necessary to bring the part to approximate dimensions. Then the part may be rough-ground, followed by fine grinding. In the latter operation, the dimensional accuracy of the part is established. For the finishing operation, the part may then be honed, lapped, or superfinished. These operations are similar in that, during the passage of the abrasive element over the surface, the high points of the surface are removed and the surface gradually approaches one without irregularities (see Fig. 20-2).

In honing, if high pressure is used for large stock removal the material beneath the surface may be metallurgically damaged because of pressure and corresponding high temperature; also in the absence of *random motion* the built-up points of the abrasive particles may cause mechanical damage. "Microhoning" uses low pressure and yields a smooth surface with a crosshatched pattern favorable to lubrication and antiscoring.

In lapping, the peaks of the surface irregularities are progressively removed, as in honing; but since the pressure between the lap and the work is comparatively low and the motion slow, a new damaged layer will not be introduced. Also, since the motion is random, scratches due to built-up points are avoided. In hand lapping a considerable time may be necessary to finish a surface.

Superfinishing is similar to lapping except that the relative motion of the abrasive stone and the part is the result of from three to six separate superimposed motions so that the path of a particular abrasive point over the work is very complicated—so-called "multimotion." The pressure is relatively low, being of the order of a few ounces per square inch. A continuous flow of lubricant is used to flush away particles of the abrasive.

When the irregularities are removed, the lubricant forms a continuous fluid film over the surface that prevents further finishing, and the operation will cease. Thus the superfinishing process produces a surface that has only small irregularities, within a few microinches of a true surface, and one that is metallurgically and mechanically undamaged base material to within less than a microinch of the surface. In production, less than a minute is required to superfinish a surface to within 2 μin. rms.

The surface that is described above is useful in reducing wear. This is true since there will not be large surface irregularities to promote failure

Table 20-1

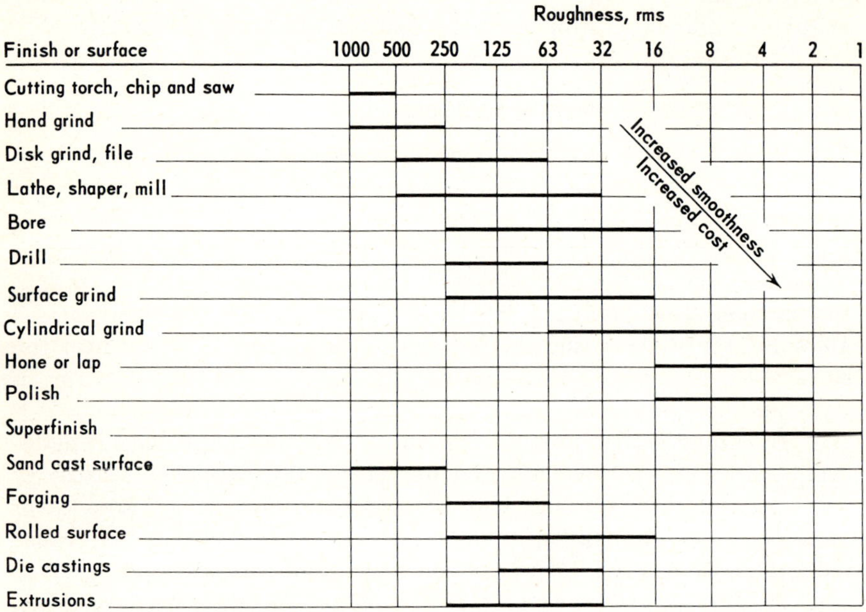

of the oil film under adverse conditions of load and speed. Also, for finely finished surfaces, the wear which takes place during the life of the machine is greatly reduced, so that it is possible to make the parts originally with the clearances which are required for the best operating conditions instead of being forced to use too small clearances when the parts are new and too large clearances after wear has taken place.

In Table 20-1 are shown some surface-finish values which are obtainable in various machined and natural finishes.

20-5 SPECIFICATION OF SURFACE FINISH

Improvement in performance of many machines requires finer finished surfaces, and therefore it has become necessary to establish standard methods for designating surface finishes. USAS Standard B46-1 uses a designation system to grade surface roughness as rms values from 0.00000025 in. ($\frac{1}{4}$ μin.) to 0.063 in. (63,000 μin.) in 13 steps. A designating symbol is used on the drawing, for example, f^4 indicates that the surface is to be finished so that its roughness is limited to a maximum of 4 μin., rms.

It has been shown, however, that surfaces of the same material finished to the same degree of roughness but by different methods sometimes behave differently with respect to fatigue strength, friction, wear, and corrosion. This is true because other characteristics of the surface than its roughness affect the performance. Among these are (1) direction of irregularities in the surface plane, called *lay*, and (2) the pattern of the irregularities. In some instances, therefore, it is desirable to specify the finishing method as well as the degree of roughness.

Sets of machine-finished specimens or their replicas are available on the market. These are used in a comparative manner so that the surface specified by the designer may be produced by the shop mechanic and checked by the inspector.[1] In this system, 10 steps of roughness are designated by the usual finish symbol with an attached number and, when desirable, the finishing method, for example, "f^4 buff," where the 4 represents the degree of smoothness.

Roughness-width cutoff In some applications it is desirable to distinguish between roughness and waviness of the part in specifying the surface finish. As an example, where the contact area between two mating surfaces is important, waviness and roughness both are significant, the former to ensure uniform pressure between the parts, and the latter to aid in maintaining the film of lubricant and in controlling wear. In this case a relatively long distance should be used in obtaining the surface-finish value. As another example, in a part subjected to cyclic loading, surface scratches and roughness are dangerous as stress raisers, and in this case waviness is unimportant. Here, a short distance may be used to inspect the surface irregularities. The distance over which the surface irregularities are to be averaged to obtain the surface-finish value is known as *roughness-width cutoff*. Standard cutoffs are 0.003, 0.010, 0.030, 0.100, and 1.000 in. The value preferred for most applications is 0.030 in.

A complete indication of surface finish is shown by the symbols in Fig. 20-6. In (*a*), one number denotes average roughness. Two numbers in (*b*) show maximum and minimum roughness values. The waviness-height rating in inches is added at (*c*), and at (*d*) the 0.010 value is the roughness-width cutoff value, and the *lay* is indicated as perpendicular to the edge. Other symbols for lay are two parallel lines to indicate parallel to edge of surface indicated, *X* for angular in both directions,

[1] Walter Mikelson, Surface Finishes, *Gen. Elec. Rev.*, vol. 46, no. 3, p. 185, 1943.

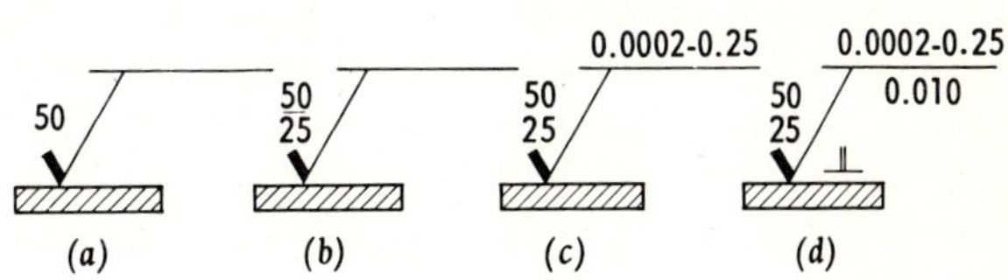

Fig. 20-6 Surface-finish symbols.

C for approximately circular pattern relative to center of piece, *M* for multidirectional and *R* approximately radial relative to center.

20-6 MECHANICAL WEAR

Wear is important to the designer of mechanical equipment because it is the predominant cause of failure of mechanical parts. For instance, it has been stated that the weight of a completely worn-out 5-ton truck may be 5 lb less than its weight when new. Why? The reason is *wear*.

Mechanical wear may be defined as cumulative surface damage or deterioration due to use. Destruction of parts by wear is mechanical, metallurgical, chemical, electrical, and thermal in nature and may be affected by environmental conditions; therefore, it may be and usually is a complex phenomenon. For this reason, general rules for delaying wear cannot be stated; however, in many specific cases of mating parts in sliding or rolling contact, it is possible to identify the contributing factors and on this basis to classify wear into types and to frame various theories of wear. Although agreement on the theories held by researchers is not firm, the following discussion points out the present generally accepted pattern.

20-7 TYPES OF WEAR

(*a*) *Adhesive wear* When two metallic bodies are pressed lightly together, the peaks of the surface irregularities or asperities will be in contact, as shown exaggerated in Fig. 20-7(*a*). When the bodies are pressed together with a large force, the contacting asperities are subjected to very high pressure so that the yield point of the material is reached and the contacting regions are displaced as shown in Fig. 20-8(*b*), where plastic flow is indicated by the shaded area.

The high pressure at the contacting asperities is due to the fact that the *real* area of contact is very small. For instance, if carefully ground surfaces of two steel cubes, 1 in. on their sides, are pressed together with a force of 10 lb, the *real* area of contact is not 1 in.^{2} but is of the order of 1/10,000 in.^{2} This is because only the contacting asperities come into actual metal-to-metal contact. This means that the average contact pressure in this instance is not 10 lb divided by 1 in.^{2}, or 10 psi, but is 10 lb divided by 0.0001 in.^{2}, which equals 100,000 psi.[1]

If the surfaces in contact are clean, there will be strong adhesive forces between the contacting surfaces, and in addition the interface formed by the plastic flow referred to above will be along crystallographic

[1] F. P. Bowden and D. Tabor, "Friction and Lubrication," John Wiley & Sons, Inc., New York, 1957.

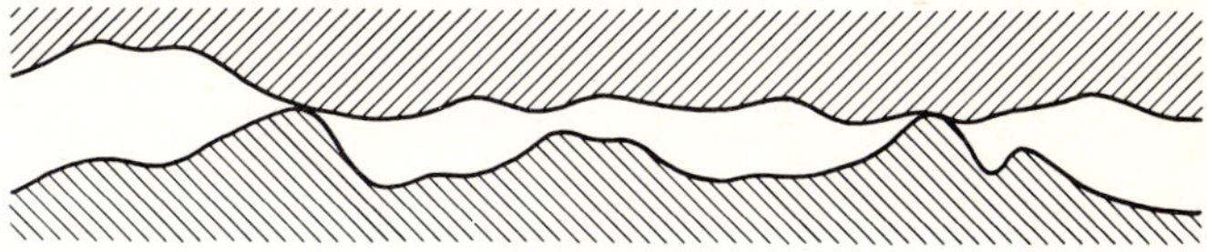

(a) Light pressure

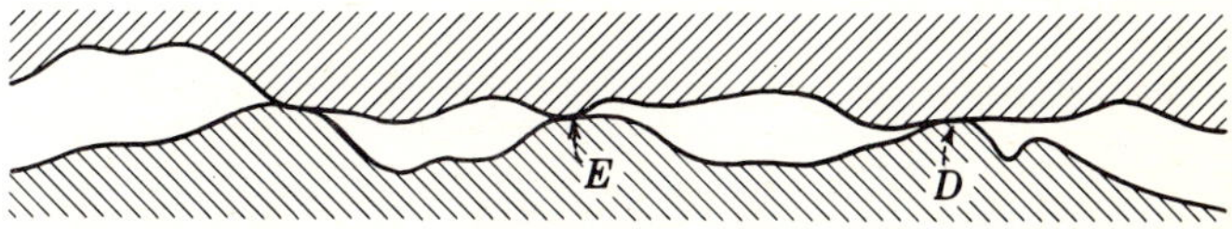

(b) Heavy pressure

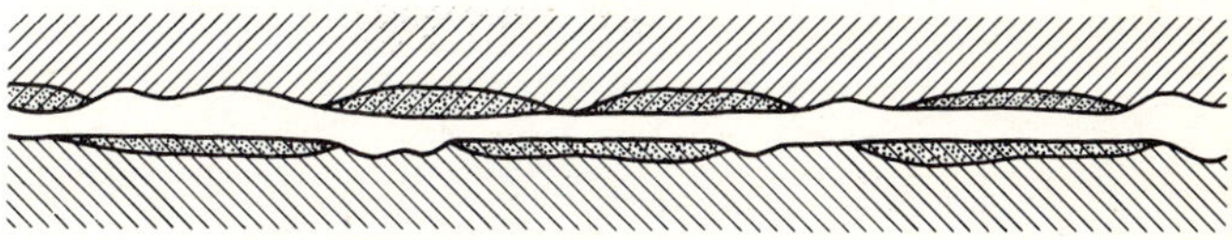

(c) Worn-in surfaces, shown separated

Fig. 20-7 Contact between two bodies.

planes and will probably be interlocked in a stepped manner, as shown in Fig. 20-8(*c*). During the plastic deformation of the asperities, the metal undergoes severe strain hardening so that the hardness of the metal adjacent to the juncture is greater than that of the parent metals.

If the two surfaces now undergo relative sliding motion, the interlocked steps at the interface will be broken and the energy thus liberated will be transformed into heat so that, at the existing temperature, solid-state diffusion may occur to form a metallic junction or bridge. Continued sliding of the two parts will introduce strains in this metallic bridge.

Note that, in the original plastic deformation of the asperity and the strain hardening that accompanied the plastic flow, the hardening effect was more severe near the interface and decreased toward the parent metal. Note also that the cross-sectional area of the joined asperities increases toward the parent metal (away from the interface) so that, at a section some distance from the interface, the "bridge" will be weakest and may fail at this section. A particle of metal will thus be transferred from one surface to the other, as shown in Fig. 20-8(*d*). This type of wear has been called "metal transfer," or adhesive wear.[1]

If this type of wear is severe, it is referred to as galling, scuffing, scoring, or seizing. If it is controlled and induced carefully, particularly

[1] I-Ming Feng, Metal Transfer and Wear, *J. Appl. Phys.*, vol. 23, no. 9, pp. 1011–1019, September, 1952; C. Lipson and L. V. Colwell (eds.), "Handbook of Mechanical Wear," chap. 3, The University of Michigan Press, Ann Arbor, Mich., 1961.

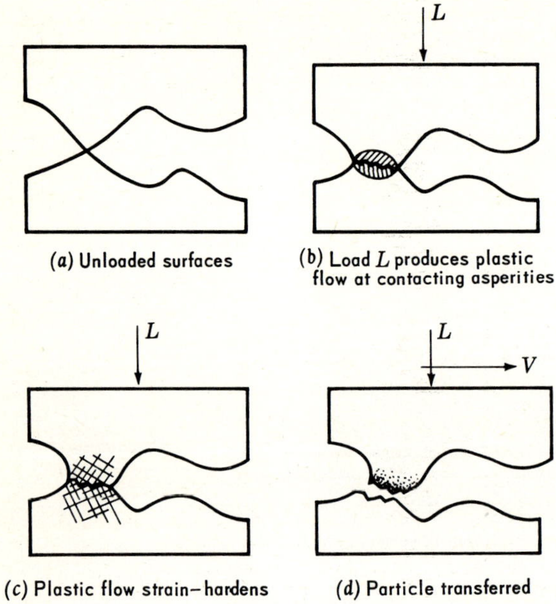

Fig. 20-8 Metal transfer. (*From P. H. Black, "Theory of Metal Cutting," McGraw-Hill Book Company, New York,* 1961.)

with well-lubricated surfaces, it is often of the type experienced in machine parts, as in bearings and cylinders that are "worn in," and amounts to wearing off the prominent irregularities until the surfaces are smooth enough to operate with a continuous thick film of lubricant.

Adhesive wear, or metal transfer, is affected mainly by the hardness of the two surfaces, their surface finish, the normal pressure between the parts, and the reduction of friction by contaminants on the surfaces, such as sulfides, chlorides, certain oxides, and liquid and solid lubricants.[1]

(*b*) *Abrasive wear* This type of wear results from the displacement of the softer of two materials in sliding contact by an asperity of the harder material. It can also be due to a hard particle of foreign material which may become partly embedded in the softer material so that the hard particle behaves as a miniature cutting tool which gouges grooves in the harder of the two surfaces.[2] In mining, earth-moving, minerals benefaction processes, etc., abrasive wear, frequently referred to as "plowing," may occur as a result of dynamic contacts between the

[1] See Lipson and Colwell, *op. cit.*, p. 34.

[2] J. T. Burwell, Jr. (ed.), "Mechanical Wear," American Society for Metals, Cleveland, 1950.

wearing surfaces and the hard particles. Abrasive wear may accompany adhesive wear if asperities of the harder surface gouge the softer surface.

In bearings of machinery, the shaft is usually made of steel for required strength and deflection characteristics. It is desirable to use a softer material for the bushing, or its liner, so that any hard particles of grit, sand, or particles of metal will become *completely* embedded in the bushing, which will prevent scoring of the shaft. Hence, oil filters and, in internal-combustion engines, air filters will reduce wear of the shaft. Abrasive wear is a common type of wear encountered in engineering practice and is the greatest single cause of wear in industrial equipment.

(*c*) *Mechanical pitting*[1] This type of failure is a surface phenomenon which may appear in regions of high contact pressure, such as at the contacting regions of heavy rollers, wheels and rails, gear teeth, cams and followers, and in contacting surfaces of rolling bearings.[2] Experimental evidence indicates that there are three stages in the formation of the pit: (1) A small fatigue crack appears in the surface, owing to the roller pressure; (2) in the presence of a liquid, such as a lubricant that fills the crack, subsequent applications of the roller pressure cause the crack to spread and form a small chip; (3) the chip breaks off and forms a pit. The theory is based on observations which showed that the pits form only in the presence of a liquid and that they are invariably fan-shaped and have the orientation predicted on the hydraulic basis.

(*d*) *Mechanical fretting* This type of damage occurs on metallic surfaces that are finely finished and closely fitted and have slight relative sliding motion. The motion is of very small magnitude and may be due to elastic deformation, such as that between a ball-bearing race and the shaft on which it is pressed, or it may be a slightly oscillatory motion, such as that between the balls or rollers and their races of a variable-pitch propeller-operating mechanism or between the leaves of a spring during flexing. The damage manifests itself as a powder, or muddy substance if oil is present, and a scabby appearing surface. Although the damaged area is usually in the form of isolated shallow pits, it may be serious enough to affect the fit between the parts.

[1] Called here *mechanical pitting* to differentiate this type of pitting from electrical pitting, which is an electrolytic action that affects, for example, bearing surfaces of electrical machinery. Another type of pitting is found in fluid flow in connection with cavitation and affects runners, propellers, blades, and valves.

[2] S. P. Timoshenko and J. N. Goodier, "Theory of Elasticity," 2d ed., p. 84, McGraw-Hill Book Company, New York, 1951; Joseph E. Shigley, "Mechanical Engineering Design," p. 39, McGraw-Hill Book Company, New York, 1963; H. R. Thomas and V. A. Hoersch, Stresses Due to the Pressure of One Elastic Solid upon Another, *Univ. Illinois Engr. Expt. Sta. Bull.* 212, 1930; Charles Lipson, Fretting, Fretting Corrosion, and Pitting, *Machine Design*, Dec. 19, 1963, p. 144.

This type of damage has been believed[1] to be due to corrosion or oxidation of small particles torn from the contacting surfaces and has been termed "fretting corrosion." It is now believed[2] that the corrosive action is secondary and that the primary cause is high pressure at contacting peaks or areas at the interface of the surfaces. This causes plastic deformation at these areas, which in turn forms a mechanical interlocking of the contacting peaks (see Fig. 20-8). When sliding motion takes place, the areas of plastic deformation are strengthened (work-hardened). This makes a bonded interlock stronger than the parent metal, and shearing failure occurs at a distance beneath the surface so that a particle is torn off.[3] Loose particles then accumulate in low areas adjacent to the disrupted peak. These particles may later be drawn into the contacting regions and may produce abrasive wear. To make matters worse, since the particles are very small, they may oxidize, thereby becoming harder and of increased volume so as to accelerate the abrasive action and cause spreading of the affected area. Neighboring areas similarly affected may merge and create a sizable affected region. As the regions increase in size and fill with expanded (oxidized) particles, the pressure at the center of the region increases owing to crowding; this further increases the abrasive action so that a pit may form. These pits are of varying area and depth and may eventually damage the surface so that the fit is ruined.

The cure for fretting is not easy. Finer finishing in order to increase the contacting area and thus decrease the pressure is not a general cure since the contact is more intimate and fretting may be encouraged. Decreasing the loads to reduce the pressure and also decreasing the motion by using larger parts or reducing vibration are effective but not always possible. Lubricating the surfaces is generally ineffective; in fact, it may aggravate the trouble since slipping may be increased.

The introduction of thin films of rubber and of nylon to cushion the contact has met with some success in the laboratory. Badly fretted surfaces have been ground off and rebuilt by plating or metal spraying; however, residual tensile surface stresses in the deposited metal may reduce wear resistance.

(*e*) *Corrosive wear* This type of destruction is due to chemical action and, in general, causes decrease in sizes of members when a protecting film is not formed by the results of corrosion. Corrosion generally accelerates other types of wear, although galling is frequently retarded

[1] Battelle Memorial Institute, "Prevention of Fatigue of Metals," p. 160, John Wiley & Sons, Inc., New York, 1941.

[2] Lipson, *op. cit.*, p. 140.

[3] P. H. Black, "Theory of Metal Cutting," p. 124, McGraw-Hill Book Company, New York, 1961.

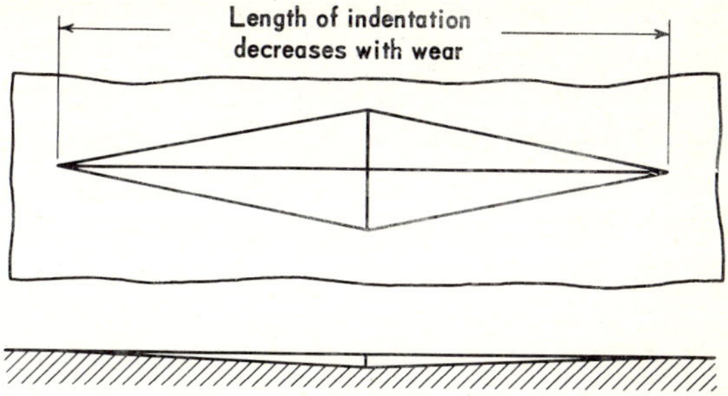

Fig. 20-9 Shape of indentation of McKee wear gauge. Original length slightly less than 1 mm.

by corrosion. Corrosion may be due to active chemicals or gases, water, brine, etc. Special attention should be paid to corrosive agents in lubricant oil.

In addition to the type of corrosion mentioned above, other types are galvanic corrosion, caused by the action of a liquid on the adjacent surfaces of two unlike metals, and "concentration-cell" corrosion, caused by the action of a fluid of nonuniform concentration affecting the surface.

(*f*) *Measurement of wear* Since the actual quantity of material removed by wear is usually very small, measurement of wear may be difficult. Relatively large wear may be determined by comparison of the dimensions of the parts as measured by a micrometer before and after wear. In the laboratory, wear of small parts may be determined by comparing weights before and after wearing. In service, a wear gauge may be used. One type (the McKee wear gauge)[1] makes use of a small indentation in the surface. The indentation is a diamond-shaped pyramidal, shown in Fig. 20-9. The length of the indentation is measured with a microscope. The difference between the lengths before and after wear has taken place is a measure of the wear normal to the surface. The method may be used for flat or curved surfaces. It is stated that the sensitivity of this gauge is 0.000015 in.

The *radioactivity method* for measuring wear has been developed to a high degree of sensitivity. In this method, the part whose wear is to be measured is irradiated; then during test the wear particles pass into the lubricant and the concentration of wear particles in samples of the lubricant can be very accurately determined as a measure of rate of wear. The method requires elaborate laboratory preparations for handling the

[1] C. S. Bruce and J. T. Duck, Cylinder Wear Measured with a Microscope, *S.A.E. Quart. Trans.*, vol. 1, no. 3, July, 1947.

equipment and for safety of the personnel, but it is precise and does not require dismantling of the parts being tested for measurements.

Under conditions of boundary lubrication (see Art. 21-6) in which a *thin film* of lubricant is present at the surfaces in contact, friction and wear do not follow either laws for dry or for lubricated (thick film) conditions. Wear under these circumstances is markedly affected by the composition of the lubricant as well as the material of the surfaces. This field is under extensive investigation since wear of machine parts is largely under these conditions. The references[1] below give results of research work on this subject.

20-8 SOLID FRICTION

When two solids are pressed together, a force parallel to their surfaces in contact is necessary to cause sliding or relative motion. This force is resisted by a force acting along the surfaces that is known as the "force of friction."

If the surfaces are clean and dry, the solid material of the two surfaces will be in contact, and the force of friction will be caused by the overcoming of the interlocking effect of the surface irregularities. These irregularities may be large, as in visibly rough surfaces, or they may be of molecular size.

In connection with Coulomb's laws, it is important to remember that they apply to surfaces that are *clean* and *dry*. In determining solid frictional resistance by experiment, the surfaces should be thoroughly cleaned to remove all contamination, including the adsorbed film. Since the adsorbed film is very difficult to remove completely, many experimental results reported for friction characteristics of clean surfaces were in reality determined for partly clean surfaces. This fact probably accounts in part for the variation in values as quoted by different experimenters.

For solid friction of clean surfaces, the laws of Coulomb apply and are stated as follows:[2]

1. The frictional resistance is approximately proportional to the load on the rubbing surfaces.
2. The frictional resistance is slightly greater for large areas and small pressures than for small areas and large pressures.
3. The frictional resistance, except for low speeds, decreases as the velocity increases.

[1] Charles Lipson, "Wear Considerations in Design," Prentice-Hall, Inc., Englewood Cliffs, N.J., 1967; Ernest Rabinowicz, "Friction and Wear of Materials," John Wiley & Sons, Inc., New York, 1965.

[2] Archbutt and Deeley, "Lubrication and Lubricants," 5th ed., p. 49, Charles Griffin & Co., Ltd., London, 1927.

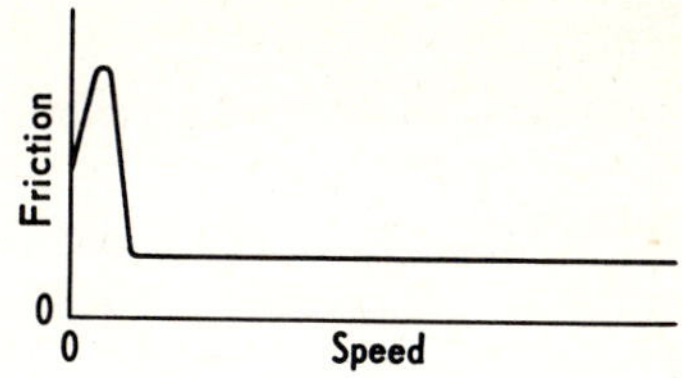

Fig. 20-10 Variation of friction with speed of rubbing.

These laws of friction may be expressed by the formula

$$F = fN$$

where F = force applied tangent to surfaces in contact to overcome resistance due to friction
N = normal force on surfaces in contact
f = coefficient of friction

If F is the force necessary to start sliding of the surfaces, f is the coefficient of *static friction;* if F is the somewhat smaller force necessary to maintain sliding, f is the coefficient of *kinetic friction.* The transition from static friction to kinetic friction is gradual, as indicated by the curve in Fig. 20-10.

Some coefficients of static friction for dry surfaces are given in Table 20-2.

Clinging friction In applications where a lubricated pin or bearing has intermittent operation with periods of rest, it has been found that the coefficient of friction is increased after the rest period. For instance, an electric switchgear may be required to operate quickly and with little resistance after an interval of rest, possibly several months. The friction hindering motion may be greater than that determined from short-time tests. Another example is a plastic molding machine in which the dies

Table 20-2 Coefficients of static friction for dry surfaces*

Wood on wood	0.3–0.5
Metal on metal	0.3 average
Metal on wood	0.2–0.6
Leather on wood	0.3–0.5
Leather on metal	0.3–0.6
Brake lining on steel	0.4–0.5

* Kimball and Barr, "Elements of Machine Design," p. 99, John Wiley & Sons, Inc., New York, 1935.

are forced together by a screw; after the dies are held together under the required load (and usually at a high temperature) it may require more torque on the screw to release the dies than was required to set them.

The friction in these cases is known as "clinging friction." Few data are available for coefficients of friction after rest periods; however, one paper[1] states that in the first 24 hr the increase in coefficient of friction in pivot bearings is of the order of 15 to 40 percent, most of the increase occurring in the first $2\frac{1}{2}$ hr, and a small increase in the subsequent 4 days. The tests covered a range of 60 sec to 90 days.

20-9 ROLLING RESISTANCE

When a roller made of an elastic material is pressed by a force W against a plane surface of a similar material, both the roller and the plane are deformed so that the surfaces in contact appear as shown in Fig. 20-11(*a*), where the dotted lines represent the original outlines and the ABC represents the deformed contacting surface.[2] If the roller rotates, as in moving from left to right, the material in both bodies undergoes increased compression in the forward part of the contact area, AB, and the compression in the rear part BC will be reduced. Thus, the resultant forces on the contacting surface of the roller AB and BC will be as shown by the vectors, where the vector on the area AB is larger than that on BC. Because of the mechanical hysteresis of the materials, the portion of the

[1] M. C. Hunter, Static and Clinging Friction of Pivot Bearings, *Proc. Inst. Mech. Engrs. (London)*, p. 274, 1942.

[2] M. C. Shaw and E. F. Macks, "Analysis and Lubrication of Bearings," p. 423, McGraw-Hill Book Company, New York, 1949; Arvid Palmgren, "Ball and Bearing Engineering," 3d ed., p. 30, S. H. Burbank and Co., Philadelphia, 1959.

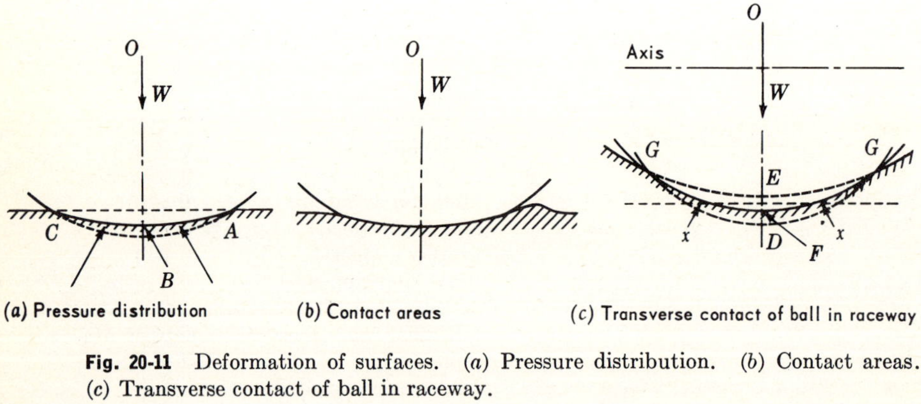

Fig. 20-11 Deformation of surfaces. (*a*) Pressure distribution. (*b*) Contact areas. (*c*) Transverse contact of ball in raceway.

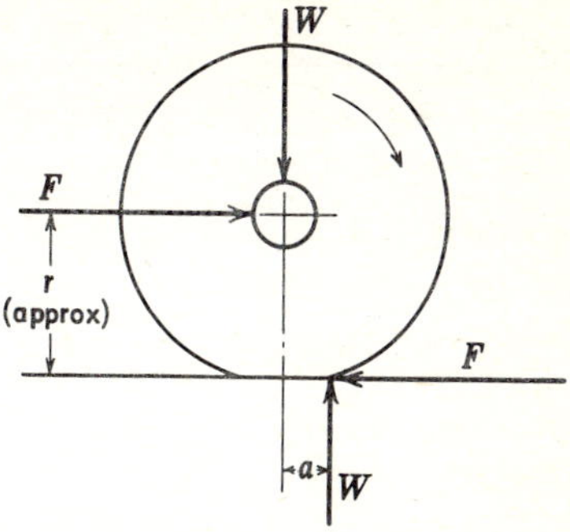

Fig. 20-12 Forces on roller.

stress cycle where the load is increasing will be greater than the portion regained on unloading. This effect constitutes part of the resistance to loading.

If there is an appreciable tangential component due to sliding friction acting at the contact surface, the force on the area AB in Fig. 20-11(a) will be increased and that on area BC will be decreased so that an asymmetrical contact area will be formed, as shown in Fig. 20-11(b). There will always be tangential sliding at the surfaces because all points on the plane are substantially at rest whereas points A and B have different sliding velocities because their radii measured from the center of rotation, O, of the roller are not equal. The tangential sliding will develop elastic deformations in both tangential and perpendicular directions, the latter pushing the material in a hump in front of the roller and stretching it at the rear. This is indicated in Fig. 20-11(b) in an exaggerated manner.

Figure 20-11(c) shows the transverse view of the surface, seen fore and aft, of a spherical roller running in a raceway such as in a ball bearing. Here the surface GEG is the undeformed surface of the raceway, and GDG is the undeformed surface of the ball. GFG is the deformed contacting interface. We see that various points of contact along the line GFG are moving at different velocities about the axis. Thus, only the points of contact, such as at x and x, can be in pure rolling contact. All other points of contact must have relative sliding motion. The points of contact along the surface xFx on the roller will slide backward in relation to the direction of the roller, and the points of contact along the surfaces Gx will slide in a forward direction. This sliding contact is responsible for "spalling" of ball-bearing inner races, as discussed in Art. 22-9.

For the purpose of analyses involving rolling friction, it is convenient to consider that the torque due to frictional resistance is represented by a couple Wa, as shown in Fig. 20-12, where W is the weight on the roller or ball, F is the force required to cause rolling, r is the radius of the roller, and a is the distance (coefficient of rolling friction).

Table 20-3

Annealed steel rolling on annealed steel	0.02
Hardened steel rolling on hardened steel	0.003–0.004
Cast iron or steel rolling on wood	0.10
Ball bearings	0.0008–0.0012
Roller bearings	0.0020–0.0040

For equilibrium of forces for the roller or ball,

$$Fr = Wa$$

or $$a = \frac{Fr}{W} \tag{20-1}$$

The value of a may be determined experimentally, and it has been shown that it depends on the materials of the roller or ball and the surface over which it rolls, but it is generally considered independent of the size of the roller and of the speed and is known as the "coefficient of rolling friction." It has dimensions of length and is given in handbooks in inches. Some values are given in Table 20-3. The resistance to rolling of balls in radial ball bearings is discussed in Chap. 22.

20-10 FLUID FRICTION

When a fluid film separates two surfaces having relative motion, solid friction does not exist, and the only friction present is that within the fluid film. This type of friction is discussed in the following chapter.

21
Sliding-contact Bearings

21-1 INTRODUCTION

In a general sense, the function of machine bearings is to permit constrained relative motion of rigid parts. In *sliding-contact bearings* a lubricant is generally inserted or supplied between the mating surfaces to reduce friction and wear and in some cases to carry away the heat generated. In *rolling-contact bearings*, rolling motion is utilized. Some common types of sliding-contact bearings are the following: (1) journal, or sleeve, bearing; (2) full journal bearing; (3) partial journal bearing; (4) clearance bearing; (5) fitted bearing; (6) thrust bearing; and (7) slipper, or slider, bearing.

Journal bearings are used to furnish *lateral* support to rotating shafts. In a journal bearing, the *journal* is the part of the shaft that runs in the bushing or sleeve, which is usually stationary and supports the journal, as shown in Fig. 21-1. In some applications, the journal is at rest and the bushing rotates; in others, such as connecting-rod bearings, both the journal and the bushing have motion.

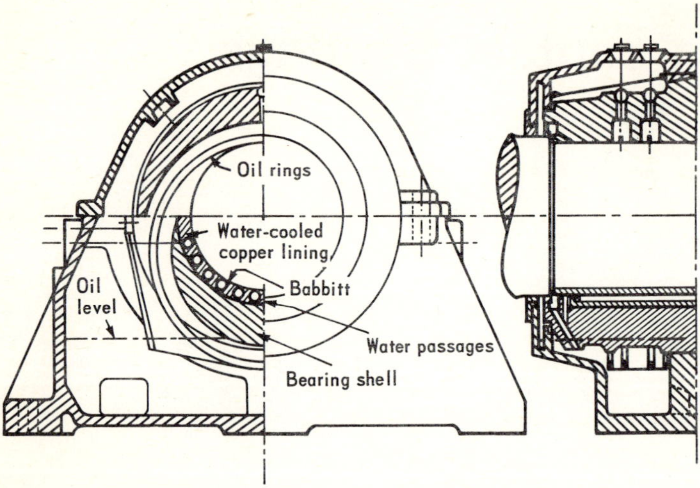

Fig. 21-1 Oil-ring lubricated, water-cooled journal bearing. (*Courtesy of Westinghouse Electric Corporation.*)

In a *full journal bearing*, the angle of contact of the bushing with the journal is 360°. This is the type most commonly used in industrial machinery to accommodate bearing loads in any radial direction.

In a *partial journal bearing*, the angle of contact is 180° or less, 120° being a common value. The partial journal bearing may be used where the direction of the load does not change materially; it is used for structural simplicity, for convenience in applying lubricant, and for the reduction of frictional loss. Railroad-car bearings represent an extensive application of the partial journal bearing.

Clearance in a bearing refers to the thickness of the space allowed for the lubricant that separates the parts having relative motion. A *clearance bearing* is one in which the radius of the journal is less than the radius of the bushing. Most journal bearings are of this type. The shape of the oil film in a clearance bearing is shown in Fig. 21-2(*a*).

A *fitted bearing* is one in which the radius of the journal and bushing

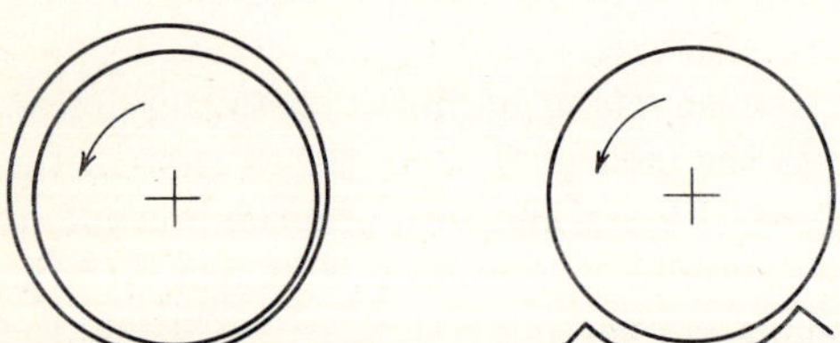

(a) Clearance bearing **(b) Fitted bearing** **Fig. 21-2** Shape of oil film.

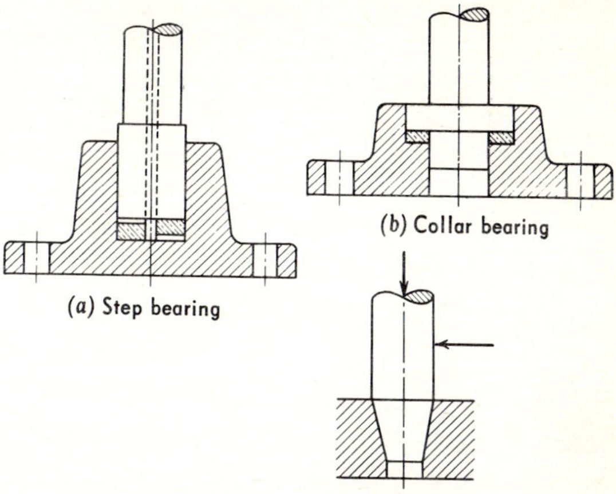

Fig. 21-3 Thrust bearings.

are equal. It is evident that a fitted bearing must be a partial bearing and the journal must run eccentric with the bushing in order to provide space for the lubricant, as shown in Fig. 21-2(*b*). Fitted bearings are more common in England than in the United States.

A bearing designed to support an *axial* load is called a *thrust bearing.* This type of bearing may be broadly divided into *step bearings* [see Fig. 21-3(*a*)], in which the end of the shaft is in contact with a bearing surface; *collar bearings,* at (*b*), in which a collar is attached to or formed integral with the shaft; and *pivoted-segment bearings.* The latter type is discussed in Art. 21-3.

Conical bearings as shown at (*c*) will support a transverse load as well as an axial load.

A *slipper,* or *slider, bearing* is one in which the two surfaces are flat and nearly parallel and the relative motion is translation.

Figure 21-4 is a schematic diagram of a slider bearing that shows a

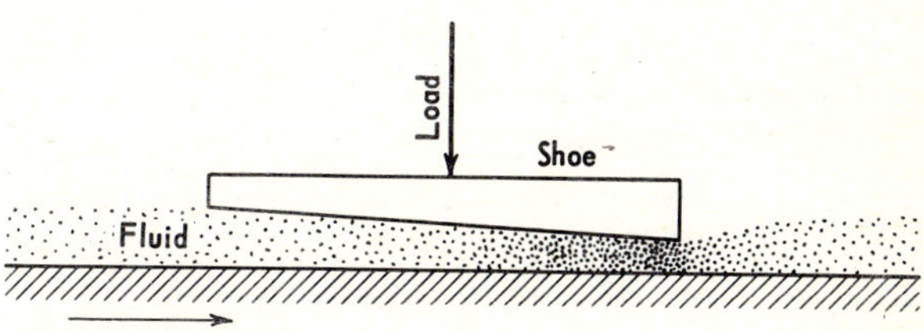

Fig. 21-4 Diagram of slider bearing.

wedge-shaped film of lubricant. It will be shown in Art. 21-2 that a wedge-shaped, or converging, film is necessary in order that a load may be supported by a fluid film.

A *lubricant* in sliding bearings is fed or inserted between the surfaces to reduce friction and wear. Oils and greases are the most common lubricants but other substances may be used, such as water in hydraulic equipment, air in high-speed spindles, and milk in cream separators. In addition to reducing friction and wear, the lubricant may carry away heat generated in the bearing.

21-2 MECHANISM OF FLUID LUBRICATION[1]

By fluid lubrication is meant a condition of operation in which the film of lubricant is so thick that metal-to-metal contact between the parts is prevented, and the only friction is that which occurs within the fluid film. Fluid friction is considerably less than metal-to-metal friction. Since fluid lubrication is essential if friction and wear are to be minimal, the conditions that are necessary to establish and maintain fluid lubrication are important and will be developed in this article. The slider bearing will be used for this purpose, and later it will be shown how the results arrived at for the slider bearing may be applied to other types of bearings, for example, the journal bearing.

A slider bearing is shown in Fig. 21-5. The following assumptions are made:

1. The fluid completely fills the space between the two surfaces.
2. The fluid "wets" or adheres to the surface so that the velocity of the fluid at each surface is the same as that of the surface.
3. Flow in the direction perpendicular to the velocity U is negligible.
4. The fluid is incompressible.
5. The shear stress is proportional to the rate of shear (newtonian liquid).
6. The motion of the fluid in the film is laminar.
7. The viscosity of the fluid is constant throughout the film.
8. The inertia effects of the forces in the oil film are negligible.

In Fig. 21-5(a), the surface of the shoe is parallel to that of the moving surface. When the moving surface has a horizontal velocity U, the horizontal layers, or laminae, of fluid move with velocities that are directly proportional to their distances from the fixed surface as indicated.

[1] M. D. Hersey, "Theory and Research in Lubrication," John Wiley & Sons, Inc., New York, 1966.

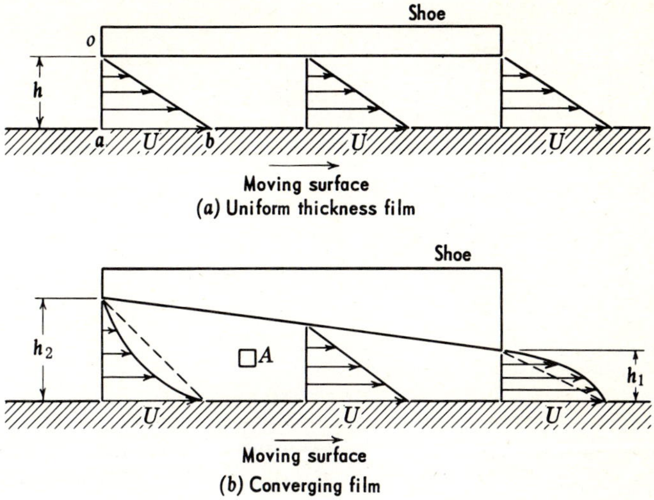

Fig. 21-5 Variation of velocity across film.

It should be noted that the area of the triangle *oab* in the figure is proportional to the volume of fluid passing a unit width of the film in a unit time. Since the volume passing all cross sections of the film must be equal, it follows that the areas of the velocity-clearance diagrams must be equal, as indicated in the figure.

The slope of the line *ob* at (*a*) is known as the "velocity gradient" and may be expressed as du/dy, where y is the distance measured along the film thickness *oa*.

From Newton's law of viscous flow, which states that the shearing stress F is proportional to the rate of shear (velocity gradient), we may write

$$F = \mu \frac{du}{dy} \tag{21-1}$$

where μ is the viscosity of the fluid.

In Fig. 21-5(*a*), the velocity gradient is constant; hence the shearing stress between the horizontal laminae is constant at all points of the film.

In Fig. 21-5(*b*), the surface of the shoe is inclined to the moving surface and the thickness of the film varies from h_2, where the fluid enters the film, to h_1 where it leaves. This is known as a "converging film." It is evident from the figure that the shape of the velocity-clearance diagrams cannot be triangular always, as shown by the dotted lines, since their areas could not be equal as required. The variation of the velocities

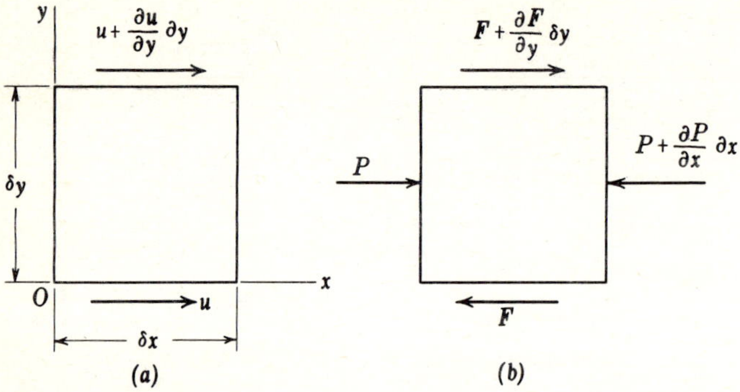

Fig. 21-6 Element of fluid.

as indicated by the curved lines would result in figures of equal areas, and it will be shown that this variation is compatible with velocity-pressure relations. Note that these curved lines represent variable velocity gradients as compared with the uniform gradient for the constant-thickness film.

If an elementary volume of unit width, as shown at A in Fig. 21-5(b), is drawn as in Fig. 21-6(a), it may be referred to axes Ox and Oy, as shown, where δx and δy are dimensions of the element parallel to the Ox and Oy axes, respectively. As originally assumed flow in the Ox direction only will be considered. As shown in Fig. 21-6(a), the lower face of the element will be assumed to move with the velocity u; hence the upper face will move with the velocity $u + (\partial u/\partial y)\ \delta y$. The forces acting in the x direction are shown at (b). The left face is subjected to the fluid pressure P and the right face to $P + (\partial P/\partial x)\ \delta x$. The shear stress on the lower face is represented by F and on the upper face by $F + (\partial F/\partial y)\ \delta y$.

For equilibrium of the horizontal forces on the element,

$$\frac{\partial F}{\partial y}\ \delta y\ \delta x = \frac{\partial P}{\partial x}\ \delta x\ \delta y$$

or

$$\frac{\partial F}{\partial y} = \frac{\partial P}{\partial x} \tag{21-2}$$

This equation states that the change in shear stress in the y direction is equal to the change of pressure in the x direction.

From Eq. (21-1),

$$F = \mu \frac{\partial u}{\partial y}$$

Therefore $\dfrac{\partial F}{\partial y} = \mu \dfrac{\partial^2 u}{\partial y^2}$

From the above equation and Eq. (21-2),

$$\frac{\partial P}{\partial x} = \mu \frac{\partial^2 u}{\partial y^2} \tag{21-3}$$

This equation states that the change of pressure in the x direction is equal to the coefficient of viscosity multiplied by the second derivative of u with respect to y. It is now apparent that when the velocity gradient $\partial u/\partial y$ is constant, as in Fig. 21-5(a), $\partial^2 u/\partial y^2$ will be zero, and therefore $\partial P/\partial x$ will be zero, and there can be no change in pressure along the film in the x direction; hence no pressure can be built up in a parallel film.

It can be shown by referring to Fig. 21-7 that, for the velocity gradient at the entering end of the film, $\partial^2 u/\partial y^2$ is positive; hence $\partial P/\partial x$ is positive, and the pressure in the film will increase in the ox direction. The increase will be positive until the section is reached where the velocity gradient is constant when $\partial P/\partial x$ is zero, indicating a maximum pressure. Further along ox, $\partial^2 u/\partial y^2$ will be negative; hence $\partial P/\partial x$ will be negative, and the pressure will drop. The variation of pressure is shown by the curve in Fig. 21-7.

From the foregoing, it follows that a positive pressure can be built up and a load supported by a fluid *only* by the use of a converging film. All bearings supporting a load by means of a fluid film must make use of the converging film, as, for example, the simple slider bearing shown diagrammatically in Fig. 21-4, the pivoted-segment thrust bearing, and the journal

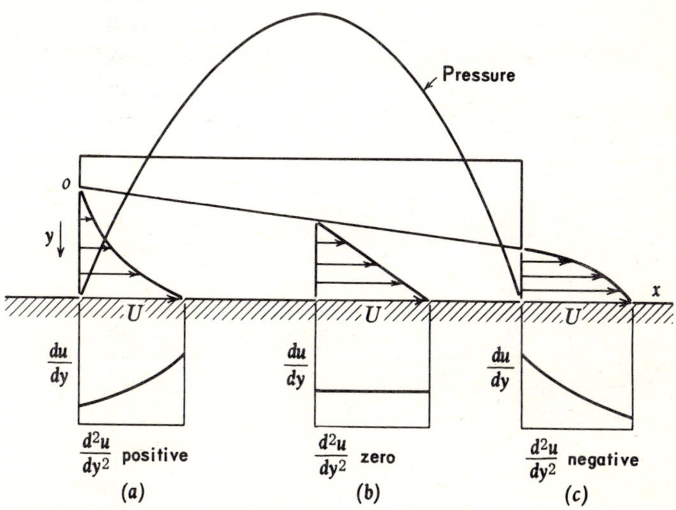

Fig. 21-7 Pressure distribution in converging film.

bearing. The latter types of bearings are discussed in the following articles.

21-3 PIVOTED-SEGMENT BEARING

According to the discussion in the preceding article, a converging film is necessary in a bearing if the advantages of fluid friction are to be realized. It is apparent that the thrust bearings shown in Fig. 21-3 cannot provide converging films, and hence the axial loads will not be supported by a fluid film. Instead there will exist at best partial metal-to-metal contact and the friction will be comparatively high. If the pressure between the rubbing parts is high, the lubricant may be forced from the surfaces in contact, and rapid wear or seizure of the parts may result. Thus the load-carrying capacity of this type of bearing is limited. The use of multiple collars on the collar thrust bearing is an attempt to increase the load-carrying capacity; however, such a bearing occupies considerable space, and it is difficult to secure equal division of the load among the collars.

The pivoted-segment bearing was developed in order to provide a converging film in a thrust bearing. This bearing was developed independently by Kingsbury in the United States and Michell of Australia in about 1905. As shown in Fig. 21-8, the segments are separately pivoted so that they can tilt to form the converging film. Because of leakage of the oil from the sides of the segments, the flow of oil will be as represented on the face view of a segment. The distribution of pressure is also shown.

The segments may be pivoted on a single point pivot by a rocker, or the entire segment may be supported by a number of springs. Loads in

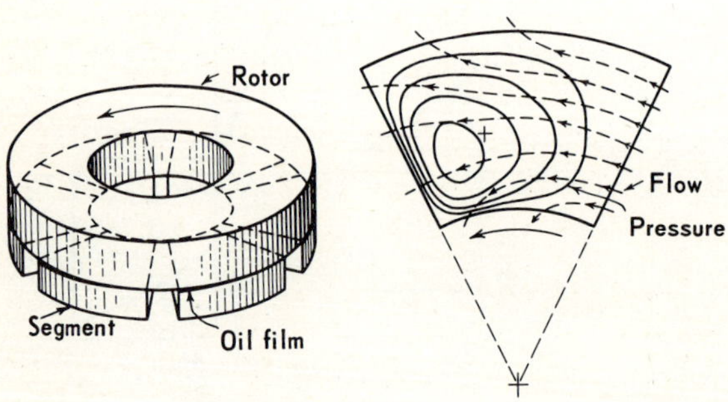

Fig. 21-8 Pivoted-segment thrust bearing, and pressure distribution on a segment.

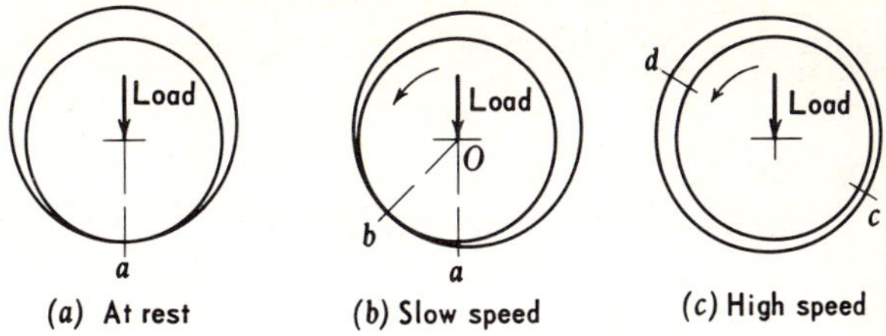

Fig. 21-9 Formation of continuous oil film.

excess of a million pounds are supported in this manner by a single thrust bearing.

21-4 JOURNAL BEARINGS

An oil film of a journal bearing with the clearance exaggerated is shown in Fig. 21-9. At (*a*) in the figure the journal is at rest with metal-to-metal contact at *a* on the line of action of the supported load. When the journal rotates slowly in the direction indicated at (*b*), the point of contact will move to *b*, so that the angle *aOb* will be the angle of sliding friction of the surfaces in contact at *b*. In the absence of a lubricant, there will be dry metal-to-metal friction. If a lubricant is present in the clearance space, a thin adsorbed film of the lubricant may partly separate the surface, but a continuous fluid film completely separating the surfaces will not exist because of the slow speed.

As shown at (*c*), the speed of the journal has been increased so that a continuous fluid film is established, and the center of the journal has moved so that the minimum film thickness is at *c*. It may be noted that, from *d* to *c* in the direction of motion, the film is continually narrowing and hence is a converging film, as was found necessary to support a load in the discussion in Art. 21-2. The curved converging film may be considered as a wedge-shaped film of a slipper bearing wrapped around the journal. It may be noted also in the figure that, from *c* to *d* in the direction of rotation, the film is diverging and cannot give rise to a positive pressure or a supporting action. In fact, a negative pressure will be developed in this region that may serve to draw lubricant from the source of supply if conditions are favorable, or the negative pressure will be prevented from forming if a lubricant is supplied at the proper point by forced feed.

Two views of the bearing of Fig. 21-9(*c*) with the ideal variation of pressure in the converging film are shown in Fig. 21-10(*a*). Actually, because of side leakage, the angle of contact on which pressure acts is

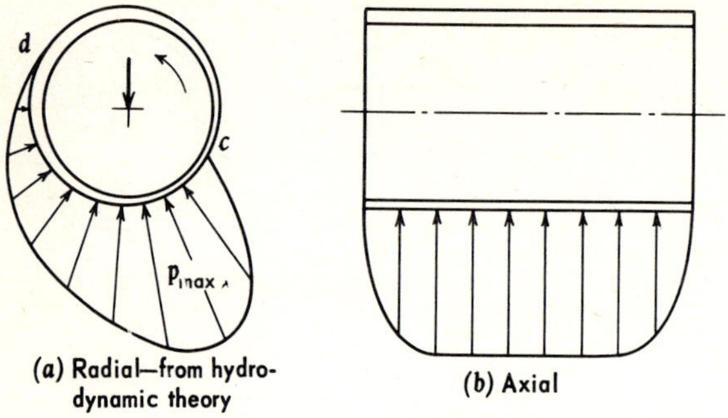

Fig. 21-10 Pressure in journal bearing.

somewhat less than 180°. The distribution of pressure in the axial direction is shown at (*b*) in the figure.

21-5 THE HYDRODYNAMIC THEORY

In the discussion of the mechanism of film lubrication in Art. 21-2, it was shown that a converging film is necessary in order that a positive pressure can be built up in the film to enable it to support a load. For a quantitative determination of the pressures in a bearing corresponding to given operating conditions, it is necessary to resort to the hydrodynamic theory, or to a simplified expression of that theory. A brief résumé of the development of the hydrodynamic theory will be given here, followed by a simplification of the theory leading to results that may be used in bearing design.

In 1883, Petroff published results of an experimental investigation showing a connection between viscosity and friction in a journal bearing. In 1883 and later years, Tower conducted experiments which were subsequently interpreted by Reynolds in his classical paper[1] in which he developed his general equation of the hydrodynamic theory, which can be written as

$$\frac{\partial}{\partial x}\left(\frac{h^3}{12\mu}\frac{\partial p}{\partial x}\right) + \frac{\partial}{\partial z}\left(\frac{h^3}{12\mu}\frac{\partial p}{\partial z}\right) = \frac{U}{2}\frac{\partial h}{\partial x} \tag{21-4}$$

where U is the surface speed, p is the pressure at any point in the film whose coordinates are x and z, μ is the viscosity of the lubricant, and h is the film thickness at any section. U is in the x direction, and h is in the y direction. The derivation of Reynolds' equation is given in Art. 21-14.

[1] O. Reynolds, On the Theory of Lubrication and Its Application to Mr. Beauchamp Tower's Experiments, *Phil. Trans. Roy. Soc.* (*London*), vol. 177, 1886.

A solution to Reynolds' equation, with side leakage assumed to be equal to zero, was made by Sommerfeld and may be expressed in the form

$$\frac{r}{c}f = \phi\left[\left(\frac{r}{c}\right)^2 \frac{\mu n}{p}\right]$$

in which ϕ represents a functional relationship. The quantity $(\mu n/p)(r/c)^2$ is known as the Sommerfeld number S. This number unites in one dimensionless variable most of the factors over which the designer has control. The bearing arc is the only quantity omitted. When the number S has been established for a particular set of operating conditions, all the operating characteristics of the bearing become fixed.

Two methods of journal-bearing design will be considered here. The first was developed by Hersey,[1] who applied dimensional analysis to an infinitely long bearing with a side-flow correction factor as determined experimentally by McKee and McKee.[2] The resulting values for frictional resistance compare favorably with results based on the hydrodynamic theory for bearings operating under high Sommerfeld numbers. This method will be the subject of Art. 21-6. The second method of journal-bearing design (Art. 21-14) is a more sophisticated one and is based directly on the hydrodynamic equation of Reynolds. This method employs the work of Raimondi and Boyd[3] for determining film thickness and its extent, film pressure and its distribution, flow characteristics, and temperature rise. These characteristics are expressed in terms of the Sommerfeld numbers and for various ratios of bearing length to diameter.

21-6 FRICTION IN JOURNAL BEARINGS

The coefficient of friction is important in bearing design, since it affords a means for determining the loss of power due to bearing friction and for estimating conditions of thermal equilibrium. The coefficient of friction in journal bearings is defined as

$$f = \frac{F}{P} = \frac{T}{Pr}$$

[1] M. D. Hersey, "Theory of Lubrication," pp. 70ff, John Wiley & Sons, Inc., New York, 1938.

[2] S. A. McKee and T. R. McKee, Friction of Journal Bearings as Influenced by Clearance and Length, *Trans. ASME*, vol. 51, p. APM-51-15, 1924.

[3] A. A. Raimondi and John Boyd, A Solution for the Finite Journal Bearing and Its Application to Analysis and Design, I, II, and III, *Trans. ASLE*, vol. 1, no. 1, pp. 159–209; also in "Lubrication Science and Technology," Pergamon Press, New York, 1958.

where f = coefficient of journal friction
P = load supported by bearing, lb
F = summation over bearing area of shear stress exerted by lubricant, lb
T = torque required to overcome journal friction, lb-in.
r = radius of journal, in.

By the use of the dimensional analysis referred to in Art. 21-5, it is found that the coefficient of friction for a journal bearing can be expressed as follows:

$$f = \phi\left(\frac{ZN}{p}, \frac{d}{c}, \frac{L}{d}\right)$$

where f = coefficient of friction
ϕ = a functional relationship
Z = absolute viscosity of lubricant, centipoises
N = speed of journal, rpm
p = bearing pressure on projected bearing area, psi
d = diameter of journal, in.
c = difference between diameter of bushing and diameter of journal, in.
L = length of bearing, in.

The quantity ZN/p is termed the bearing characteristic number and is dimensionless if the quantities are expressed in terms of a consistent system of units. The variation of the coefficient of friction with operating values of ZN/p is shown in Fig. 21-11. The part of the curve *ab* represents the region of fluid-film lubrication, which is a straight line only for the full journal bearing. Between *b* and *c* the viscosity Z or the speed

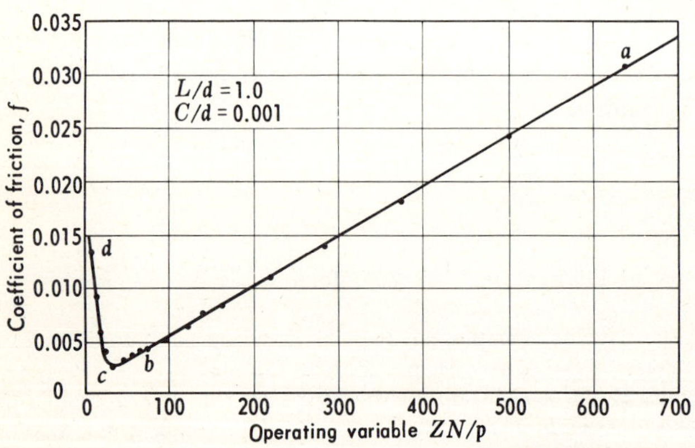

Fig. 21-11 Journal-bearing friction as a function of ZN/p.

N is so low, or the pressure p is so great, that their combination ZN/p will reduce the film thickness so that partial metal-to-metal contact will result. Between c and d on the curve, *boundary lubrication*, or *imperfect lubrication*, exists. This is the region where the viscosity of the lubricant ceases to be a measure of friction characteristics, but the so-called "oiliness" of the lubricant is effective in preventing complete metal-to-metal contact and seizure of the parts.

It may be noted that the part ab of the curve represents stable operating conditions, since from any point of stability a decrease in viscosity Z, for instance, will reduce ZN/p. This will result in a decrease in f, following by a lowering of bearing temperature that will raise the viscosity Z.

The minimum friction for the bearing occurs near point b on the curve, but it is necessary to operate the bearing in service sufficiently far to the right of b to avoid any possibility of breaking down the film of lubricant because of any unpredicted change in operating conditions. It is customary to design bearings for a value of ZN/p at least five times that corresponding to point b on the curve.

The following equation may be used for estimating values of the coefficient of friction f for well-lubricated full journal bearings:

$$f = \frac{473}{10^{10}} \frac{ZN}{p} \frac{d}{c} + K \tag{21-5}$$

where Z = viscosity of lubricant, centipoises
N = speed of journal, rpm
p = bearing pressure, in psi, of projected bearing area = load on journal $\div Ld$
L = bearing length, in.
d = journal diameter, in.
c = difference between bushing diameter and journal diameter, in.
K = factor to correct for end leakage = 0.002 for L/d ratios of 0.75 to 2.8†

Operating values for ZN/p should be compared with values given in Table 21-1 to ensure a safe margin between operating conditions and the point of film breakdown.

The McKee equation (21-5) yields values for the coefficient of friction for lightly loaded bearings that compare favorably with values found by the use of the Raimondi and Boyd data in Art. 21-15. A comparison of the coefficients of friction for a 360° journal bearing found

† McKee and McKee, *loc. cit.;* see also M. C. Shaw and E. F. Macks, "Analysis and Lubrication of Bearings," McGraw-Hill Book Company, New York, 1949; E. I. Radzimovsky, "Lubrication of Bearings," p. 169, The Ronald Press Company, New York, 1959.

Table 21-1 Bearing design practices*

No.	Machinery	Bearing	Maximum p, psi	Operating Z	Operating Zn/p	c/d	L/d
1	Automobile and aircraft engines	Main	800–1,700	7	15	......	0.8–1.8
2		Crankpin	1,500–3,500	to	10	......	0.7–1.4
3		Wrist pin	2,300–5,000	8	8	......	1.5–2.2
4	Gas and oil engines	Main	500–1,200	20	20	0.001	0.6–2.0
5		Crankpin	1,000–1,800	to	10	0.001	0.6–1.5
6		Wrist pin	1,200–2,000	65	5	0.001	1.5–2.0
7	Marine steam engines	Main	500	30	20	0.001	0.7–1.5
8		Crankpin	600	40	15	0.001	0.7–1.2
9		Wrist pin	1,500	30	10	0.001	1.2–1.7
10	Stationary steam engines	Main	200–400	60	20	0.001	1.0–2.0
11		Crankpin	600–1,500	80	6	0.001	0.9–1.3
12		Wrist pin	1,800	60	5	0.001	1.2–1.5
13	Reciprocating pumps and compressors	Main	250	30	30	0.001	1.0–2.2
14		Crankpin	600	to	20	0.001	0.9–1.7
15		Wrist pin	1,000	80	10	0.001	1.5–2.0
16	Steam turbines	Main	100–275	2–16	100	0.001	1.0–2.0
17	Generators, motors, centrifugal pumps	Rotor	100–200	25	200	0.0013	1.0–2.0

* This table omits ambient and inlet pressures and elastic modulus.

by the McKee equation and the Raimondi and Boyd data shows that for high Sommerfeld numbers (i.e., lightly loaded bearings) the two equations are nearly equal.[1]

21-7 BEARING LOADS

An important consideration in the design of a journal bearing to operate under conditions of film lubrication is the conditions affecting the breaking down of the film, which occurs at operating conditions between b and c in Fig. 21-11. In addition to the characteristic number ZN/p, other factors that affect the film breakdown are as follows:

Clearance The clearance in a bearing should be small enough to provide the necessary velocity gradient, so that the pressure built up will

[1] Allen S. Hall, Alfred R. Holowenko, and Herman G. Laughlin, "Theory and Problems of Machine Design," p. 283, Schaum Publishing Company, New York, 1961.

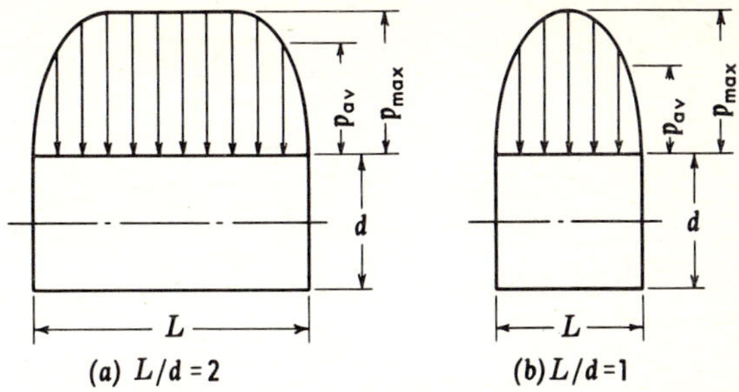

Fig. 21-12 Effect of length-to-diameter ratio L/d on average bearing pressure.

support the load. This consideration requires a small clearance, which has the additional advantage of decreasing side leakage. However, allowance must be made for manufacturing tolerances in the journal and the bushing, for deflection of the shaft, and for space to permit foreign particles, such as grit and flakes of metal, to pass through the bearing. These latter considerations require large clearance. Thus both considerations require a compromise. A commonly used clearance in industrial machines is 0.001 in. per in. of journal diameter. The minimum thickness of the oil film may be of the order of 0.0002 in. per in. of diameter.

Length-to-diameter ratio Because of side leakage of the lubricant from the bearing, the pressure in the film is atmospheric at the ends of the bearing. As shown in Fig. 21-12, the average pressure p_{av} will be higher for a long bearing as shown at (*a*) than for a short bearing as shown at (*b*). Thus from the standpoint of side leakage, a bearing with a large L/d is preferable. However, space requirements, manufacturing tolerances, and shaft deflections are better met with a bearing of short length. Thus a compromise in the L/d ratio is necessary, resulting in a value of from 1 to 2 for general industrial machinery. In crankshaft bearings, the L/d ratio is frequently less than unity, since short bearings are required from space considerations, and since the shaft diameter must be large, which is required by the strength, stiffness, and vibration characteristics. (See Table 21-1 for values of L/d for various types of service.)

Oilholes and grooves Oilholes and grooves in a bearing limit the pressure at their locations to that of the supply pressure, which is generally considerably less than the hydrodynamic pressures required to support the bearing load. Thus a bearing with a central circumferential groove will in effect be equivalent to two narrow bearings, as shown in Fig. 21-13, with considerable reduction in load-carrying capacity. It is evident that an oilhole in the region of high pressure reduces the load-carrying capacity

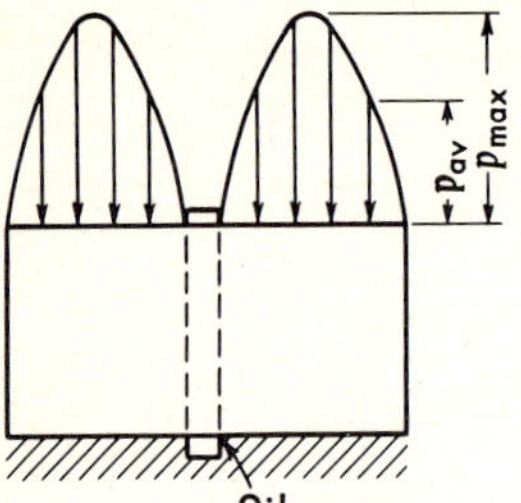

Fig. 21-13 Effect of circumferential oil groove on average bearing pressure. Compare with Fig. 21-12(*a*).

of the bearing. Thus the location of oilholes and grooves is an important consideration. From the standpoint of load-carrying capacity the following items represent good practice:

1. Locate oilholes and necessary axial grooves in the region of low pressure.
2. Avoid diagonal grooving.
3. If circumferential grooves are necessary, locate them near the ends of the bearing.

Surface finish Although the degree of smoothness of the surfaces of the journal and the bushing does not have a marked effect on a bearing operating with fluid-film lubrication, it does have an effect on the point of breaking down of the film. This will be evident from a consideration of the effect of the high points of a rough surface breaking through a film, as shown in Fig. 21-14, at a higher value of ZN/p than for a bearing with smooth surfaces.

The diagram in three dimensions in Fig. 21-15 shows that a bearing operates in the region of fluid friction at a lower value of ZN/p and with a lower value of minimum friction with a smooth surface than with a rough one. For this reason, smooth surfaces are desirable to decrease the wear of a bearing that must operate continually or pass through a low value of ZN/p frequently.

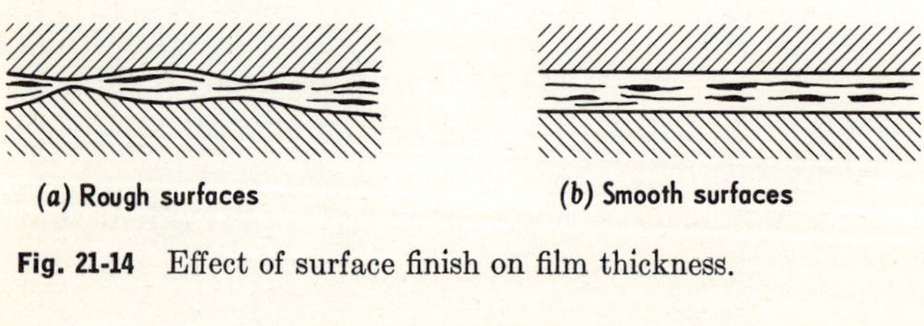

(*a*) Rough surfaces (*b*) Smooth surfaces

Fig. 21-14 Effect of surface finish on film thickness.

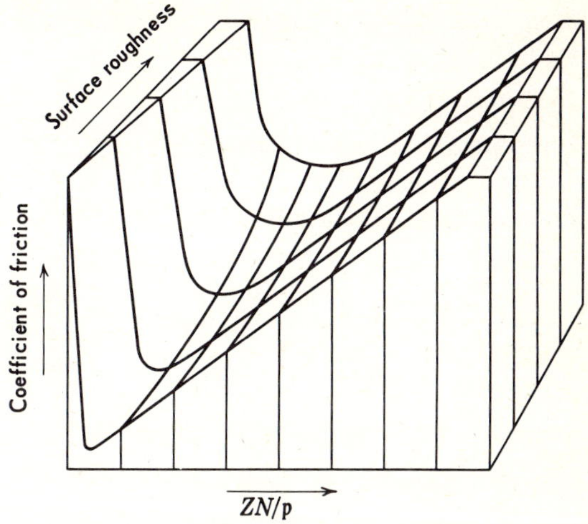

Fig. 21-15 Three-dimensional chart showing effect of surface finish on point of fluid-film rupture.

21-8 BEARING MATERIALS

From the preceding discussion of the mechanism of film lubrication, it is apparent that, when the surfaces of a bearing are separated by a fluid film, the materials of the bearing do not have any effect on its operation so long as the parts have sufficient strength to withstand the imposed loads and sufficient rigidity to maintain alignment.

On account of strength and rigidity requirements, journals are generally made of steel. For the bushing, however, there are other requirements, as discussed below, that must be considered in selecting the best material.

Low-friction properties When a journal is starting, stopping, or running at low speeds, lubrication will be of the boundary type and metal-to-metal contact will be prevented only by the adsorbed film. Metals and alloys that favor the formation of the adsorbed film are therefore desirable for bearings in order to limit wear. The oiliness of the lubricant should be considered in this connection. The low friction and wear characteristics of babbitt metals are partly responsible for their widespread use in bearings.

Compressive strength The bearing material should have high compressive strength to prevent extrusion of the material from the bearing. Since the maximum bearing pressure may be considerably greater than the pressure expressed in pounds per square inch of projected bearing area, yielding may occur in localized regions of the bearing to the extent that the clearance is materially changed.

Fatigue strength Under severe conditions of loading and temperature, surface fatigue cracks may develop. These cracks may cause checking and pitting of the surface, thus leading to its deterioration to the extent that the operation of the bearing is affected. Thus a bearing material having a high fatigue strength is desirable.

For severe operating conditions, a very thin bearing metal bonded to a steel back has been found to prevent the spreading of the fatigue cracks. Such a combination also prevents extrusion of the bearing material because of the restraining effect of the hard backing. In some automotive engines a tin-base babbitt lining 0.003 in. in thickness is used.

Conformability Because of small inaccuracies in the form of the journal and its deflection under the imposed loads, the material of the bearing should adapt its shape to that of the journal. This change of shape may be accomplished by plastic flow, by wearing away, or by local melting. Plastic flow is most desirable to achieve conformability, since wearing away and local melting are accompanied by excessive heat, which may burn out the bearing.

Embeddability All bearings should be designed so that foreign particles will be excluded, but this is difficult to accomplish in plain bearings lubricated with oil, since grit, sand, and metal particles may be introduced with the lubricant or ventilating air. If the bearing materials are hard, the particles may score the surfaces and produce undue wear. But if the bearing lining is soft so that the particles are completely embedded, this trouble may not be serious.

Bonding A bearing material that will bond readily and permanently with the steel or bronze back is necessary for long bearing life. For the thin linings mentioned above, an intermediate layer of sintered alloy is used to secure a good bond.

Corrosion resistance Bearing materials have various degrees of resistance to corrosion; this should be considered where corrosive oils must be used.

The preceding discussion indicates the desirable characteristics of bearing materials. All these have not been incorporated to a high degree in any particular bearing material, and it is evident that the choice of a material for any application must represent a compromise. This is the reason for the large number of bearing materials that have been developed, each of which is most suitable for particular applications.[1] The

[1] See "SAE Handbook."

tin-base and lead-base babbitts are in widespread use since they satisfy most requirements for general applications. However, where the loads are very high, bronze and brass bushings may be used. Where high compressive and fatigue strengths are necessary, copper, lead, and tin alloys may be used. Where the bearing is inaccessible or lubrication is infrequent, "oilless" bearings, composed of porous sintered bronze or iron with graphite or oil filling the porous structure, may be used.

21-9 VISCOSITY AND OILINESS

Newton's law states that at any point in a fluid the shearing stress is proportional to the rate of shear, or

$$F = \mu R \tag{21-6}$$

where F = shearing stress, or tangential force per unit area
R = rate of shear, or velocity gradient

In Eq. (21-6), the proportionality constant μ is known as the "viscosity" of the fluid. In cgs units, it is the *absolute viscosity* and is equal to dynes per square centimeter with a velocity of 1 cm per sec at a distance of 1 cm. The unit of absolute viscosity is the *poise* but is usually used as one-hundredth of a poise, termed a "centipoise."

One standard method of measuring viscosity in this country is by the use of the Saybolt University viscometer by which is determined the time required for a standard volume of oil at a certain temperature to flow under a certain head through a tube of standard diameter and length. The time so determined in seconds is the Saybolt Universal viscosity. Machine oils at room temperatures have a range of viscosity of about 100 to 300 centipoises.

To convert the Saybolt Universal viscosity to the absolute viscosity the following formula may be used:

$$Z = \rho_t\left(0.22S - \frac{180}{S}\right) \tag{21-7}$$

where Z = absolute viscosity at temperature t, centipoises
S = Saybolt Universal viscosity, sec
ρ_t = specific gravity at temperature t
$= \rho_{60} - 0.000365(t - 60)$
ρ_{60} = specific gravity at 60°F

The viscosities of some lubricating oils are given in Fig. 21-16.

Oiliness is a joint property of a lubricant and the bearing surfaces in contact, and it is a measure of the lubricating qualities under boundary

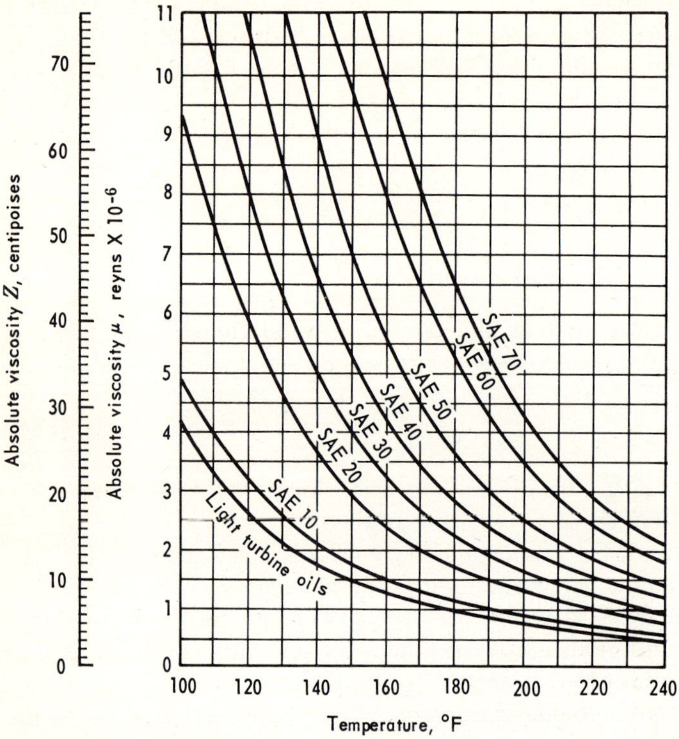

Fig. 21-16 Chart representing viscosity-temperature relationships for a number of representative SAE-classified oils.

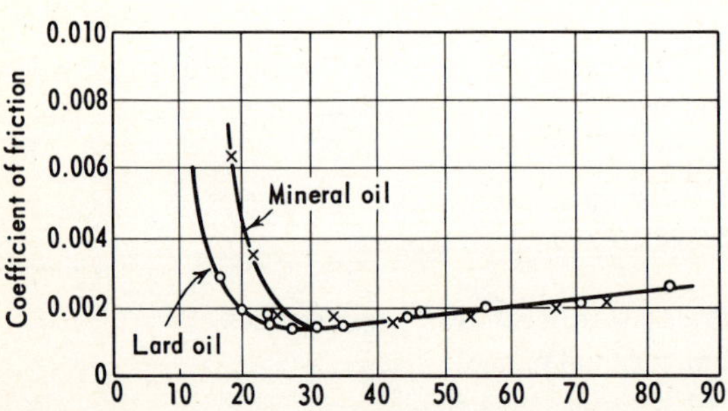

Fig. 21-17 Effect of oiliness on point of fluid-film rupture.

conditions where base metal-to-metal contact is prevented only by the adsorbed film. There is no absolute measure of oiliness. The four-ball testing apparatus[1] gives a comparative measure of this property. Babbitt metals favor the establishment of the adsorbed film, which prevents or retards actual base metal-to-metal contact. Lard oil has better oiliness than mineral oils and has the effect on the coefficient of friction as a function of ZN/p that is shown in Fig. 21-17.

21-10 IMPERFECT LUBRICATION

In bearings in which the operating value of ZN/p is too low to ensure fluid-film lubrication or where there is an insufficient supply of lubricant in the bearing, the friction will be due to metal surfaces in contact. This condition is represented by *cd* in Fig. 21-11. The following formula[2] may

[1] D. Clayton, The Use of the Four-ball Extreme-pressure Lubricant Testing Apparatus for Ordinary Lubricants, in "General Discussion on Lubrication," vol. 2, p. 274, Institution of Mechanical Engineers, London, 1938.
[2] Louis Illmer, High Pressure Lubrication, *Trans. ASME,* 1924, p. 833.

Table 21-2 Factors for Eq. (21-8)

Values of factor C_1

Method of lubrication	*Workmanship*	*Attendance*	*Location*	C_1
Bath, flooded or oil ring	High grade	First class	Clean and protected	1
Oil, free drop, constant feed	Good	Fairly good	Favorable (ordinary conditions)	2
Oil cup or grease, intermittent feed	Fair	Poor	Exposed to dirt or other unfavorable conditions	4

Values of factor C_2

Type and example of bearing	C_2
Rotating journals, such as rigid bearings and crankpins	1
Oscillating journals, such as rigid wrist pins and pintle blocks	1
Rotating journals lacking ample rigidity, such as eccentrics	2
Rotating flat surfaces lubricated from the center, such as annular step or pivot bearings	2
Sliding flat surfaces wiping over ends of guides, such as reciprocating crosshead shoes. (Use two for long and three for short guides.)	2–3
Sliding or wiping surfaces such as marine thrust bearings and worm gears	3–4
Long nuts for power screws and similar wiping parts	4–6

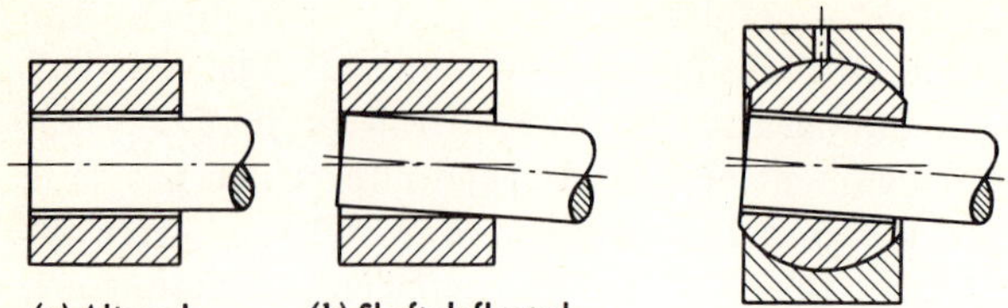

Fig. 21-18 Effect of shaft deflection on bearing alignment.

be used to estimate the coefficient of friction f under this condition:

$$f = \frac{C_1 C_2}{250} \sqrt[4]{\frac{p}{V}} \tag{21-8}$$

where C_1, C_2 = values given in Table 21-2
p = bearing pressure on projected bearing area, psi
V = rubbing velocity, fpm

21-11 STRENGTH AND STIFFNESS OF JOURNAL BEARINGS

The diameter of the journal is usually determined by making it equal to or somewhat less than the adjacent diameter of the shaft. In most cases the shaft diameter is determined from a consideration of the strength and stiffness and possibly of the vibration requirements. However, the deflection of the journal in the bushing is an important consideration in order to prevent undue wear and failure of the oil film. In Fig. 21-18 there is shown diagrammatically a perfectly aligned bearing at (*a*), the effect of shaft deflection at (*b*), and a self-aligning bearing at (*c*), which may be necessary in some instances.

The strength and stiffness of the bearing, bearing cap, and housing should also be considered so that they will support the imposed loads and will not deflect or distort beyond permissible limits. It is especially important to give careful consideration to the deflection of bearings and shafts that are required to support gears whose teeth must be kept accurately in mesh. This is especially true in bevel-gear mountings.

21-12 THERMAL EQUILIBRIUM

The heat generated in a bearing is due to fluid friction or to friction of the parts having relative motion. This heat can be readily determined when the coefficient of friction is known:

$$H_1 = fPV \tag{21-9}$$

where H_1 = heat generated, ft-lb per min
f = coefficient of friction
P = radial load on bearing, lb
V = rubbing velocity, fpm

After thermal equilibrium has been reached, heat will be dissipated at the outer surface of the bearing at the same rate that it is generated in the oil film. The heat dissipated can be expressed in terms of the product of the dissipating area and the temperature drop between the surface and the surrounding air.

For convenience in bearing design, the actual heat-dissipating area may be expressed in terms of the projected area of the journal. Thus, the equation becomes

$$H_2 = CA(t_b - t_a) \tag{21-10}$$

where H_2 = heat dissipated by bearing, ft-lb per min
A = projected bearing area, Ld, in.^{2}
t_b = temperature of bearing surface, °F
t_a = temperature of surrounding air, °F
t_o = temperature of oil film, °F
C = heat-dissipation coefficient, ft-lb per min per in.^{2} of projected bearing area per °F

Values of C have been determined by O. Lasche. The values depend on the type of bearing, its ventilation, and the temperature difference. Figure 21-19 gives values of $C(t_b - t_a)$.

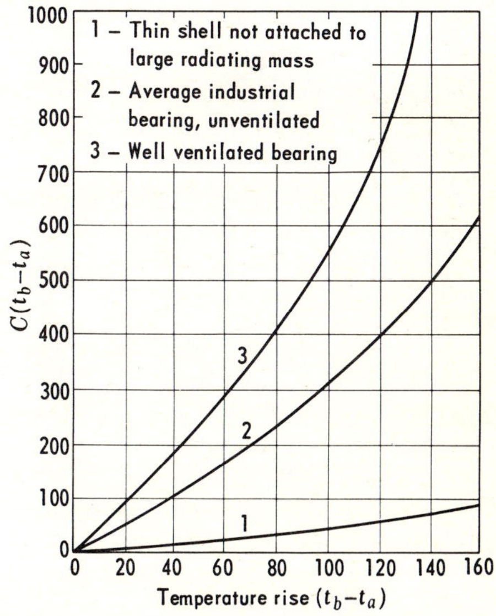

Fig. 21-19 Heat dissipation for plain bearings.

Experiments[1] show that the temperature at the outer surface of industrial bearings is approximately midway between the temperature of the oil film and the ambient temperature, i.e.,

$$t_b - t_a = \tfrac{1}{2}(t_o - t_a)$$

In some bearings, as in cold-water pumps, there may be conduction of heat *from* the bearing, whereas in steam turbines there may be conduction *to* the bearing. Because of the low heat conductivity of steel, however, conduction by the shaft is generally relatively small.

21-13 BEARING DESIGN BASED ON MCKEE AND MCKEE EQUATION

Because of the large number of variables involved, the design of a journal bearing usually requires making reasonable assumptions and then applying available equations to establish their validity. The following procedure is suggested when the bearing load and the diameter and speed of the shaft are known:

1. Determine the bearing length by choosing a ratio of L to d from Table 21-1.
2. Check the bearing pressure $p = P/Ld$ from Table 21-1 for probable satisfactory value.
3. Assume a clearance ratio c/d from Table 21-1.
4. Assume a lubricant from Fig. 21-16, and an operating temperature t_o. This temperature should, in general, be between 80 and 140°F, with 180°F as a maximum for high-temperature installations such as steam turbines. The lower temperatures result in high operating viscosity and resulting high power loss. Higher temperatures may cause film breakdown, rapid vaporization and oxidation of the lubricant, varnishing of the bearing, excessive side leakage, as well as the possibility of burning the hands of the attendant.
5. Determine the operating value of ZN/p for the assumed bearing temperature, and check this value with corresponding values in Table 21-1 to determine the possibility of maintaining fluid-film operation.
6. Determine the coefficient of friction by Eq. (21-5).
7. Determine the heat generated by Eq. (21-9).
8. Determine the heat dissipated by Eq. (21-10).
9. If thermal equilibrium is indicated by comparing the preceding two steps, then the assumed bearing temperature is established, and the operating conditions may be regarded as satisfactory.
10. If approximate equilibrium is not indicated, the designer must assume a different operating temperature, a different oil, or changed

[1] Hersey, "Theory of Lubrication," p. 102.

values of L and d, or artificial cooling if it is impossible to otherwise obtain equilibrium.

The above analysis assumes that the bearing load is essentially constant in magnitude and direction and that the speed is constant. If these conditions do not exist, the designer must consider the variations carefully and arrive at a prediction of probable conditions of service in the light of his own experience and that of others.[1]

Properly designed, installed, and operated journal bearings function very satisfactorily; however, their specifications, including that of the lubricant, require a careful and mature consideration of the numerous general and special considerations involved in any service.

Example 21-1 A journal bearing is proposed for a steam turbine. The data are: Load on journal, $P = 585$ lb, diameter $= 2\frac{1}{4}$ in., length $= 3\frac{3}{8}$ in., speed $=$ 1,800 rpm, diametral clearance ratio $= 0.001$, ambient temperature $= 60°$F. Assume SAE oil No. 10 and an average temperature of the oil film, $t_o = 140°$F. Determine the heat generated and the heat dissipated.

From Fig. 21-16,

$$Z = 14cp$$

$$\text{Pressure } p = \frac{585}{2.25 \times 3.375} = 77 \text{ psi}$$

$$\text{and } \frac{ZN}{p} = \frac{14 \times 1{,}800}{77} = 327$$

Item 16 in Table 21-1 shows that the pressure of 77 psi is well within usual limits and the operating value of ZN/p equal to 327 is high enough to assure that the operating conditions will be in the hydrodynamic range. Hence,

$$f = \frac{473}{10^{10}}\frac{ZN}{p}\frac{d}{c} + K = \frac{473}{10^{10}}(327) \times 1{,}000 + 0.002$$

$$= 0.0156 + 0.002 = 0.0176$$

$$V = \frac{\pi D \text{ rpm}}{12} = \frac{\pi \times 2.25 \times 1{,}800}{12} = 1{,}060 \text{ fpm}$$

The heat generated [Eq. (21-9)] is

$$H_1 = fPV = 0.0176 \times 585 \times 1{,}060 = 10{,}900 \text{ ft-lb per min}$$

The heat dissipated is

$$t_b - t_a = \frac{t_o - t_a}{2} = \frac{140 - 60}{2} = 40°\text{F}$$

From Fig. 21-19, with well-ventilated bearings,

$$C(t_b - t_a) = 190$$

$$\text{Therefore } H_2 = C(t_b - t_a) \times A = 190 \times 2.25 \times 3.38 = 1{,}450 \text{ ft-lb per min}$$

[1] J. T. Burwell, The Calculated Performance of Dynamically Loaded Sleeve Bearings, *J. Appl. Mech.*, vol. 14, no. 3, p. A-231, September, 1947.

Since the heat generated is greater than the heat dissipated, the bearing must be artificially cooled to remove 10,900 − 1,450 or 9,450 ft-lb per min.

21-14 REYNOLDS' EQUATION APPLIED TO BEARING DESIGN

The equation for the change in pressure along a converging fluid film in terms of the velocity gradient across the film was derived in Art. 21-2 as

$$\frac{dp}{dx} = \mu \frac{d^2u}{dy^2} \tag{21-3}$$

or

$$\frac{d^2u}{dy^2} = \frac{1}{\mu}\frac{dp}{dx} \tag{21-11}$$

where p represents the pressure at any point in the film located by x and y in Fig. 21-20.

Integrating Eq. (21-11) twice with respect to y gives

$$\frac{du}{dy} = \int \frac{1}{\mu}\frac{dp}{dx}\,dy = \frac{1}{\mu}\frac{dp}{dx}\,dy + C_1$$

and

$$u = \int \frac{1}{\mu}\frac{dp}{dx}\,y\,dy + C_1\,dy = \frac{1}{2\mu}\frac{dp}{dx}\,y^2 + C_1y + C_2 \tag{21-12}$$

The constants of integration C_1 and C_2 may be evaluated from the boundary conditions (1) $u = U$ when $y = 0$ and (2) $u = 0$ when $y = h$.

From condition 1 it follows that $C_2 = U$ and from condition 2

$$C_1 = -\frac{1}{2\mu}\frac{dp}{dx}h - \frac{U}{h}$$

By introducing the values of these constants into Eq. (21-12), it becomes

$$u = \frac{1}{2\mu}\frac{dp}{dx}y^2 + \left(-\frac{1}{2\mu}\frac{dp}{dx}h - \frac{U}{h}\right)y + U$$

or

$$u = \frac{1}{2\mu}\frac{dp}{dx}(y^2 - hy) + U\frac{h-y}{h} \tag{21-13}$$

In the right-hand member of Eq. (21-13), the first term indicates parabolic distribution of velocity due to flow in the film induced by the pressure. This is shown in Fig. 21-21(*a*). At some section between the entrance of the fluid and its exit the pressure gradient dp/dx is equal to zero, as shown at m in the figure. This is the point of maximum film pressure as shown in Fig. 21-7(*b*).

The second term in the right-hand member of Eq. (21-13) shows a linear distribution of velocity due to the relative motion between the two surfaces. The flow of fluid is shown in Fig. 21-21(*b*), and (*c*) shows the resulting distribution. This is the same as shown in Fig. 21-7.

We now need an expression for the volume of fluid flowing through

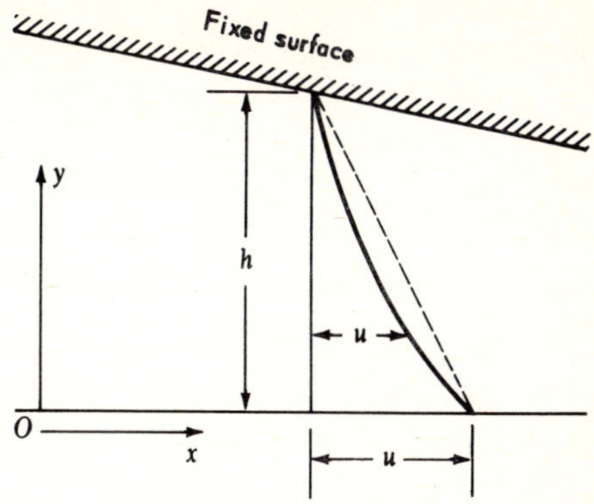

Fig. 21-20 Velocity distribution across oil film.

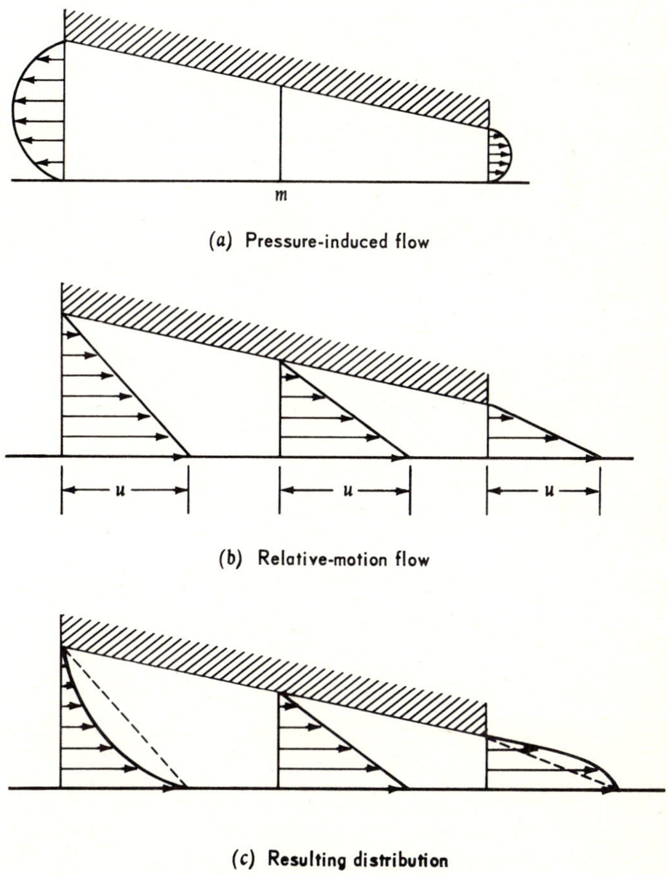

Fig. 21-21 Flow in converging film. (*a*) Pressure-induced flow. (*b*) Relative-motion flow. (*c*) Resulting distribution.

the bearing. Using the assumptions stated in Art. 21-2 and defining Q as the volume of fluid flowing in unit time and through a unit width in the z direction (perpendicular to the x and y axes), then

$$Q = \int_0^h u\,dy$$

By substituting the value of u from Eq. (21-13) and integrating, the equation for Q becomes

$$Q = \frac{Uh}{2} + \frac{h^2}{12\mu}\frac{dp}{dx}$$

For an incompressible fluid, the flow is the same for all cross sections of the film, or

$$\frac{dQ}{dx} = 0$$

Therefore $\dfrac{dQ}{dx} = \dfrac{U}{2}\dfrac{dh}{dx} + \dfrac{d}{dx}\left(\dfrac{h^3}{12\mu}\dfrac{dp}{dx}\right) = 0$

or $$\frac{d}{dx}\left(\frac{h^3}{12\mu}\frac{dp}{dx}\right) = \frac{U}{2}\frac{dh}{dx} \tag{21-14}$$

This is Reynolds' differential equation in two dimensions for the pressure gradient in a converging fluid film, neglecting end leakage (flow in the z direction).

By similar development including end leakage, Eq. (21-14) becomes

$$\frac{d}{dx}\left(\frac{h^3}{12\mu}\frac{dp}{dx}\right) + \frac{d}{dz}\left(\frac{h^3}{12\mu}\frac{dp}{dz}\right) = \frac{U}{2}\frac{dh}{dx} \tag{21-15}$$

which is the same as Eq. (21-4).

Note that, if the coordinate system is as indicated in Fig. 21-20, for a converging film in the direction of motion, dh/dx must be negative; hence the term $U\,dh/dx$ must be negative.

21-15 BEARING DESIGN USING THE SOMMERFELD NUMBER

Since a general solution to Reynolds' equation (21-15) does not exist, it is necessary to consider solutions for particular cases, such as full (360°) bearings and partial bearings and with assumed L/d ratios and also with the assumption that the viscosity of the lubricant is constant in the bearing. Actually, the viscosity varies because of the increase in temperature in the converging film from the lubricant's entrance to its exit, and the viscosity is affected by the pressure in the film and by the rate of shear.

With the above assumptions, Raimondi and Boyd of the Westinghouse Research and Development Center used the computer technique to translate the results of the hydrodynamic equation to the various operating characteristics of journal bearings for L/d ratios equal to $\frac{1}{4}$, $\frac{1}{2}$, 1, and ∞ and for various arcs (β) equal to 60, 120, 180, and 360°.[1] The charts presented in Art. 21-17 are for full journal bearing ($\beta = 360°$) and for L/d ratios $\frac{1}{4}$, $\frac{1}{2}$, 1, and ∞, along with curves for an L/d ratio equal to 2, which were determined by an interpolation formula given in the Raimondi and Boyd paper, using charts prepared by A. A. Raimondi.[2] All these charts are based on the Sommerfeld variable S, and the use of the charts is illustrated by Example 21-2 in Art. 21-17.

The Sommerfeld number, or bearing characteristic number, was proposed by A. Sommerfeld[3] in 1904 as a dimensionless quantity and is defined by the equation

$$S = \left(\frac{R}{C}\right)^2 \frac{\mu n}{p}$$

where S = bearing characteristic number, dimensionless
R = journal radius, in.
C = radial clearance (bearing radius − journal radius), in.
μ = absolute viscosity, reyns
n = speed of journal
p = pressure on projected bearing area (load on journal P ÷ bearing length × diameter), psi

The bearing characteristic number S unites in one dimensionless variable most of the factors over which the designer has control. When this number has been established for the operating conditions, all the bearings operating characteristics become fixed.

21-16 JOURNAL BEARING NOMENCLATURE

P = radial load, lb
n = journal speed, rps
R = journal radius, in.
C = radial clearance = bearing radius − journal radius, in.
L = axial length of bearing, in.
h_0 = minimum film thickness, in.
e = journal displacement or eccentricity, in.
ϵ = e/C, eccentricity ratio
β = angular extent of bearing arc, deg (360° for full bearing)

[1] Raimondi and Boyd, *loc. cit.*

[2] A. A. Raimondi, Journal Bearing Design Curves (with L/D as a Plotting Parameter), Westinghouse Research Laboratories.

[3] A. Sommerfeld, Zur Hydrodynamischen Theorie der Schmiermittelreibung, *Z. Math. Phys.*, vol. 50, pp. 97–155, 1904.

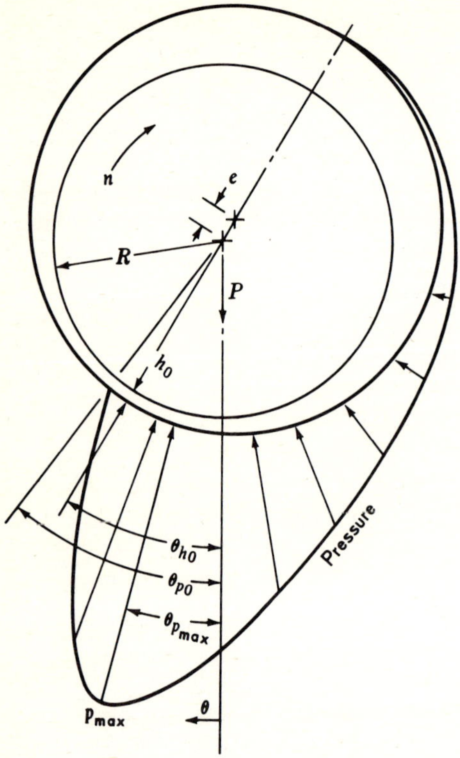

Fig. 21-22 Nomenclature and pressure distribution. (*Courtesy of Westinghouse Research Laboratories.*)

θ = angular coordinate measured in direction of journal rotation from line of centers, deg
p = film pressure = $P/2RL$, psig
p_{max} = maximum pressure, psig
p_a = ambient pressure, psi
θ_{h_0} = position of minimum film thickness, deg
$\theta_{p_{max}}$ = position of maximum film pressure, deg
θ_{p_0} = position of termination of positive film pressure, deg
Q = flow of lubricant drawn into clearance space, in.^{3} per sec
Q_s = flow of lubricant out both ends of bearing, in.^{3} per sec
t_1 = inlet temperature, °F
t_2 = outlet temperature, °F
Δt = temperature rise, $t_2 - t_1$, °F

21-17 BEARING DESIGN USING THE RAIMONDI AND BOYD CHARTS

The 10 charts in this article for full journal bearings (β = 360°) and for L/d ratios equal to ¼, ½, 1, and ∞ are reproduced with the kind permission of Raimondi and Boyd, with curves for L/d equal to 2 added by the authors of this book. The use of these charts is illustrated by Example 21-2.

Example 21-2 The data are the same as for Example 21-1: Load $P = 585$ lb, journal diameter $= 2\frac{1}{4}$ in., length $= 3\frac{3}{8}$ in., speed $= 1{,}800$ rpm, radial clearance ratio $(C/R) = 0.001$, which is the same as the diametral clearance ratio. Assume SAE No. 10 oil and average temperature of the oil film, $t_o = 140°F$. Determine the Sommerfeld number and the bearing operating characteristics. From Fig. 21-16,

$$\mu = 2 \times 10^{-6} \text{ reyn}$$

$$n = \frac{1{,}800}{60} = 30 \text{ rps}$$

$$p = \frac{P}{L \times D} = \frac{585}{3.375 \times 2.25} = 77 \text{ psi}$$

$$\text{and } \frac{L}{D} = \frac{3.375}{2.25} = 1.5$$

$$\text{Therefore } S = \left(\frac{R}{C}\right)^2 \frac{\mu n}{p} = 10^6 \times \frac{2 \times 10^{-6} \times 30}{77} = 0.78$$

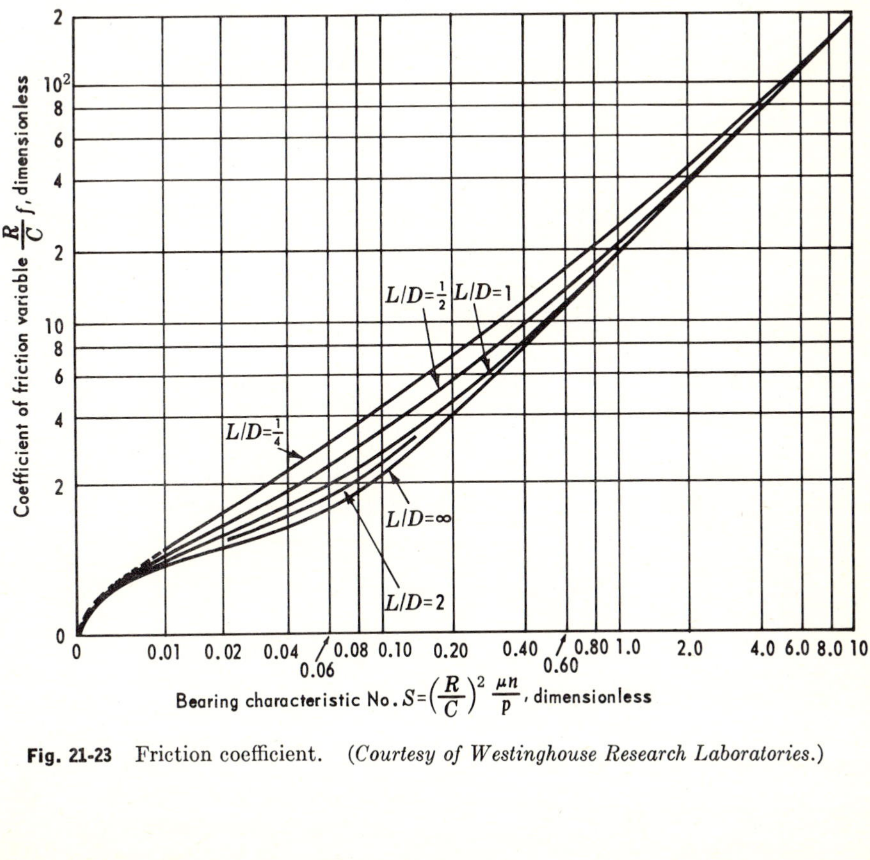

Fig. 21-23 Friction coefficient. (*Courtesy of Westinghouse Research Laboratories.*)

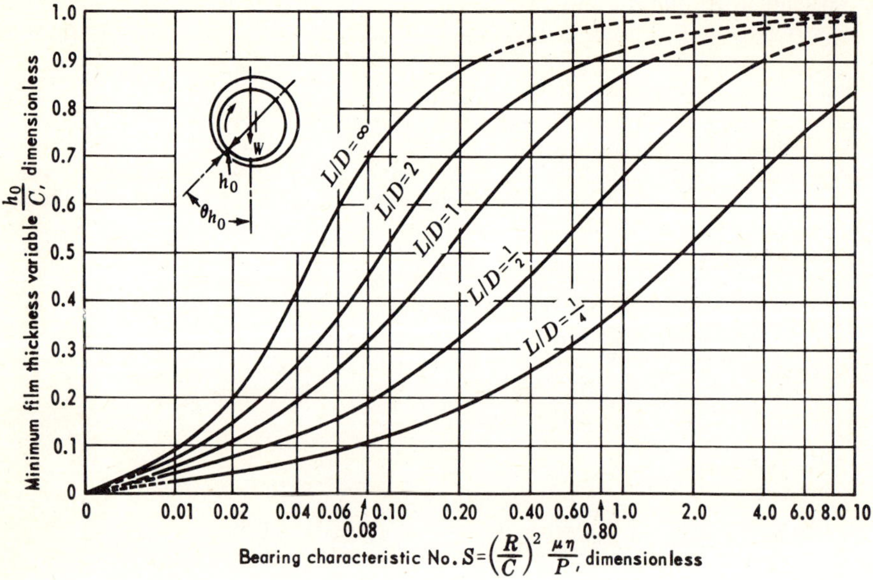

Fig. 21-24 Minimum film thickness. (*Courtesy of Westinghouse Research Laboratories.*)

Coefficient of friction, f

From Fig. 21-23,

$$\frac{R}{C} f = 16$$

$$\text{or } f = \frac{C}{R} \times 16 = 10^{-3} \times 16 = 0.016$$

This value compares well with the value of f in Example 21-1. It can be shown that values of the coefficient of friction by the McKee and McKee equation (21-5) compare well with those by the hydrodynamic theory for bearings operating at high values of the Sommerfeld numbers, that is, about 0.60 and above. This corresponds to values of ZN/p equal to approximately 250 and above. For these values the coefficient of friction by the McKee and McKee equation is around 10 percent lower than that by the hydrodynamic equation. At Sommerfeld number equal to 4.00 (ZN/p = 1,650) the difference in coefficients of friction is negligible.

The friction torque becomes

$$T = f\frac{PD}{2 \times 12} = \frac{0.016 \times 585 \times 2.25}{2 \times 12} = 0.876 \text{ ft-lb}$$

The corresponding power loss is

$$\frac{2\pi \cdot T \text{ rpm}}{33{,}000} = \frac{2\pi \times 0.876 \times 1{,}800}{33{,}000} = 0.3 \text{ hp}$$

Minimum film thickness h_0

From Fig. 21-24,

$$\frac{h_0}{C} = 0.87$$

$C = 0.001 \times R = 0.001 \times 1.125 = 0.001125$
Therefore $h_0 = 0.87 \times C = 0.87 \times 0.001125 = 0.000979$ in.

According to the hydrodynamic theory, the load-carrying capacity of the fluid film increases indefinitely as its thickness decreases. Experiments show, however, that the minimum film thickness consistent with fluid-film lubrication is affected by deviations from ideal conditions as follows:

1. The surface roughness of the rubbing surfaces
2. The deviations in the geometry of the surfaces due to misalignment and to elastic and thermal distortions
3. Contamination of the lubricant by particles of grit, metal, etc.

The minimum allowable film thickness for safe operation is stated as 0.00005 to 0.0002 in. per inch of journal diameter.

The value of h_0 equal to 0.000979 in., given above, appears to be reasonable.

Position of minimum film thickness θ_{h_0}

From Fig. 21-25, this is equal to 75°.

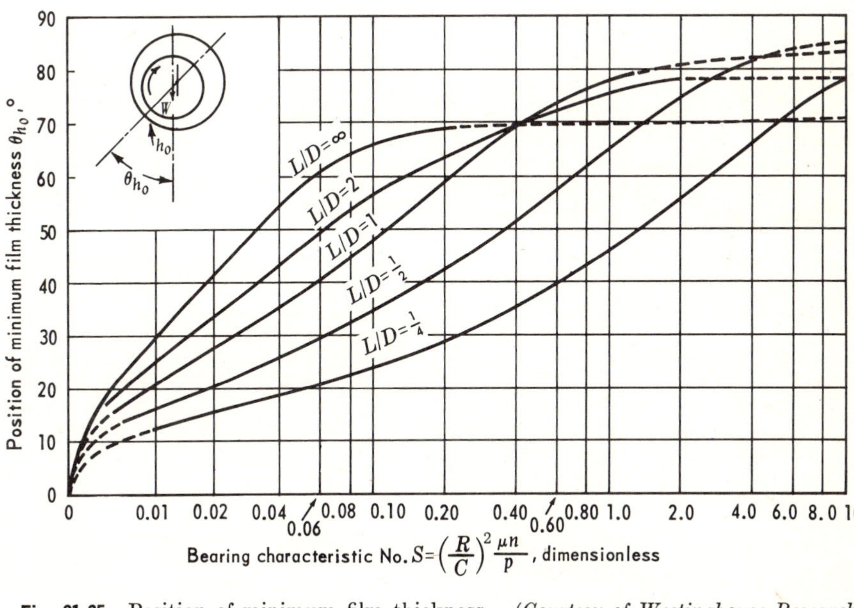

Fig. 21-25 Position of minimum film thickness. (*Courtesy of Westinghouse Research Laboratories.*)

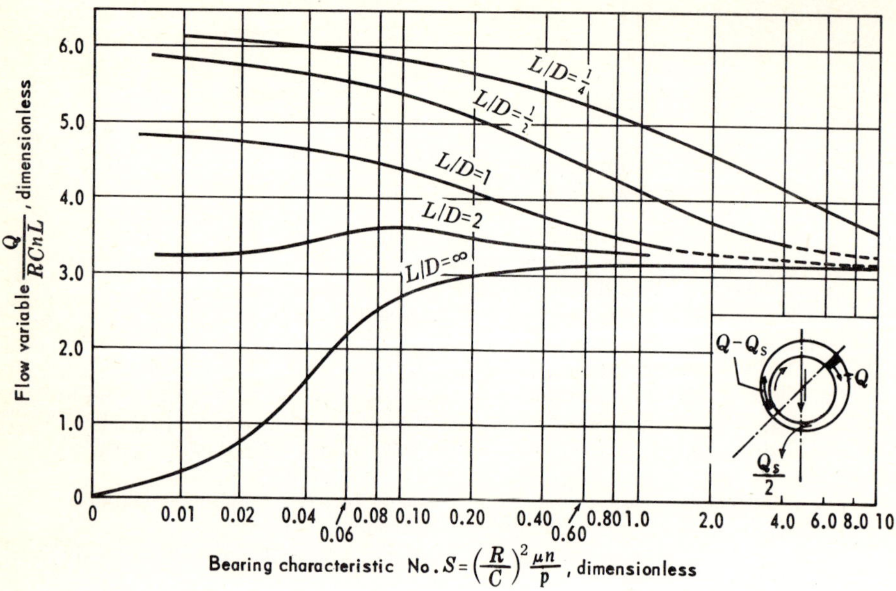

Fig. 21-26 Flow variable. (*Courtesy of Westinghouse Research Laboratories.*)

Flow variable

From Fig. 21-26, this is equal to

$$\frac{Q}{RCnL} = 3.23$$

$$\text{or } Q = RCnL = 3.23 \times 1.125 \times 0.001125 \times 30 \times 3.375$$
$$= 0.415 \text{ in.}^3 \text{ per sec}$$

From Fig. 21-27,

$$\frac{Q_s}{Q} = 0.16$$

$$\text{or } Q_s = 0.16Q = 0.16 \times 0.415 = 0.0664 \text{ in.}^3 \text{ per sec}$$

For the above two quantities, Q is the amount of lubricant that is drawn into the converging space by the hydrodynamic action. Of this, Q_s flows out of the ends of the bearing and is termed side leakage; in this case, it amounts to 16 percent of the total flow and must be replaced through an oil port by an oil ring, splash system, or by pressure feed. If the latter is used, the film pressure will be greater than that given by the hydrodynamic theory.[1]

Temperature rise

On the assumption that all the heat generated in a bearing by fluid friction is effective in raising the temperature of the lubricant and if none of

[1] R. R. Slaymaker, "Lubrication Analysis," John Wiley & Sons, Inc., New York, 1955.

the lubricant leaving the trailing edge of the film is carried over to the leading edge, Fig. 21-28 may be used to determine the temperature rise of the lubricant as it passes through the bearing. Actually, most of the lubricant is carried from the trailing edge to the leading edge, especially in a full journal bearing, which would increase the temperature of the lubricant; however, heat is transferred from the bearing by conduction, convection, and radiation so that precise calculation of the temperature rise in the lubricant is virtually impossible. The use of Fig. 21-28 will yield conservative results.

The temperature-rise variable is

$$\frac{J\gamma C_H\,\Delta t}{p}$$

where J = mechanical equivalent of heat, 9,336 lb-in. per Btu

γ = weight per unit volume of lubricant; may be used as 0.0311 lb per in.^{3}

C_H = specific heat of lubricant; may be used as 0.42 Btu per lb per °F

Δt = temperature rise of lubricant $= t_2 - t_1$, °F (average film temperature $= t_1 + \Delta t/2$)

p = film pressure, lb per in.^{2} of projected bearing area

Therefore, for $S = 0.78$

$$\frac{J\gamma C_H\,\Delta t}{p} = 65$$

$$\text{or } \Delta t = \frac{65p}{J\gamma C_H} = \frac{65 \times 77}{9336 \times 0.0311 \times 0.042} = \frac{65 \times 77}{122} = 41°\text{F}$$

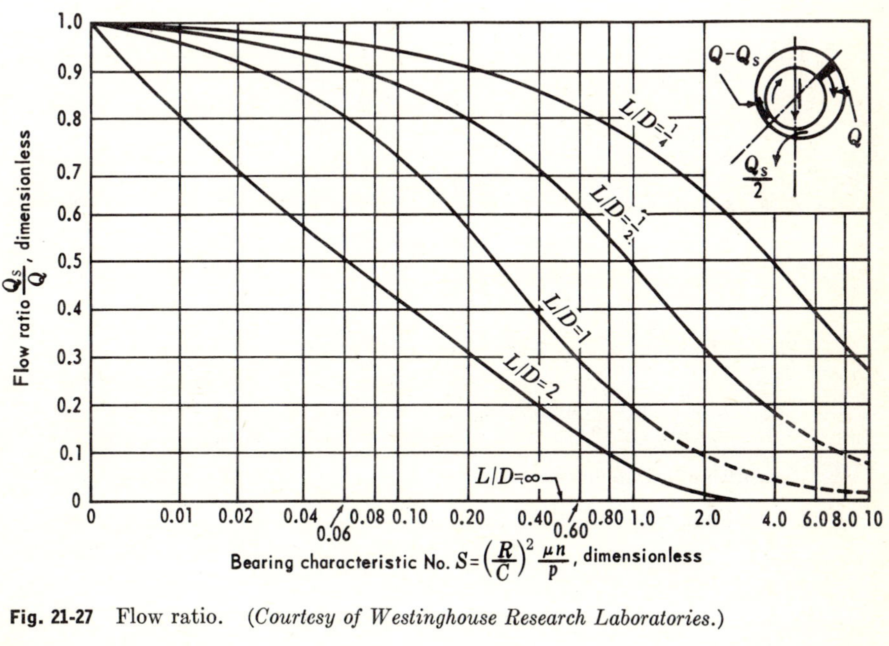

Fig. 21-27 Flow ratio. (*Courtesy of Westinghouse Research Laboratories.*)

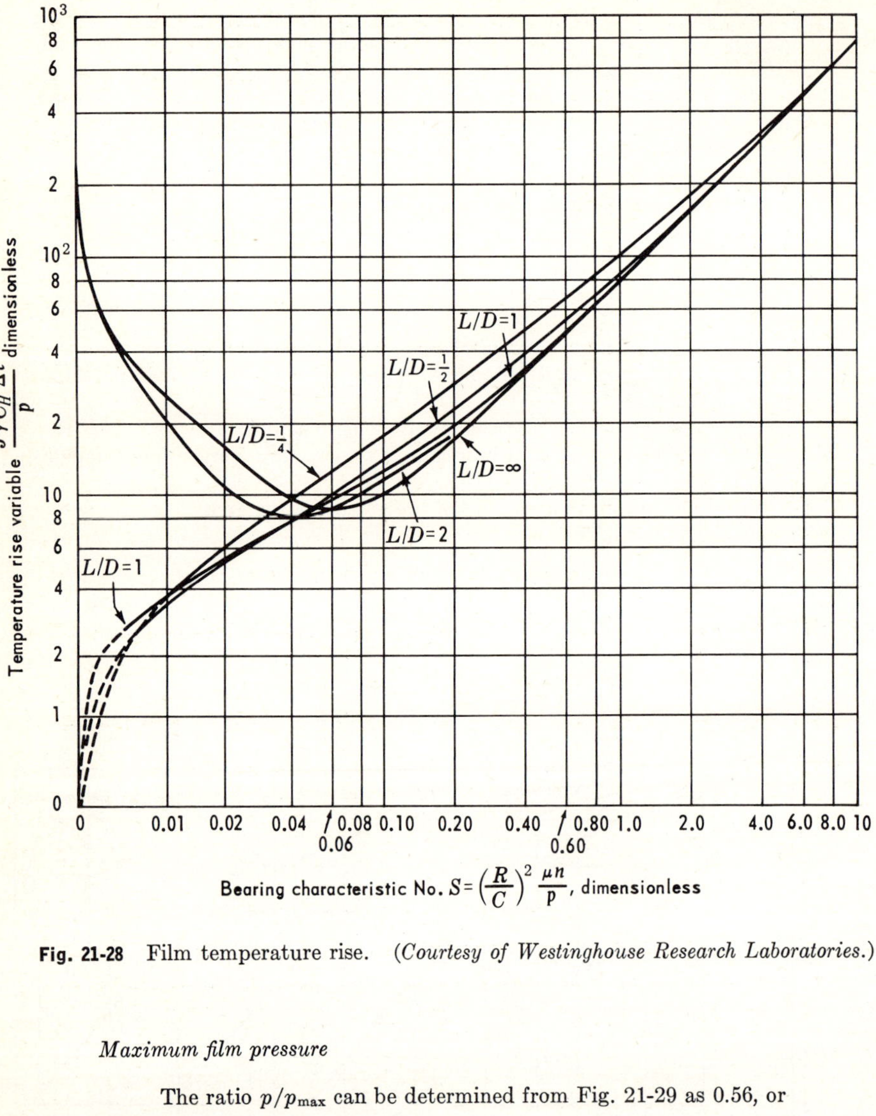

Fig. 21-28 Film temperature rise. (*Courtesy of Westinghouse Research Laboratories.*)

Maximum film pressure

The ratio $p/p_{\max}$ can be determined from Fig. 21-29 as 0.56, or

$$p_{\max} = \frac{p}{0.56} = \frac{77}{0.56} = 138 \text{ psig}$$

Position of maximum film pressure, $\theta_{p_{\max}}$

From Fig. 21-30, the value is 5.0°.

Position of termination of positive film pressure θ_{p_0}

From Fig. 21-31, the value can be approximated as 122°.

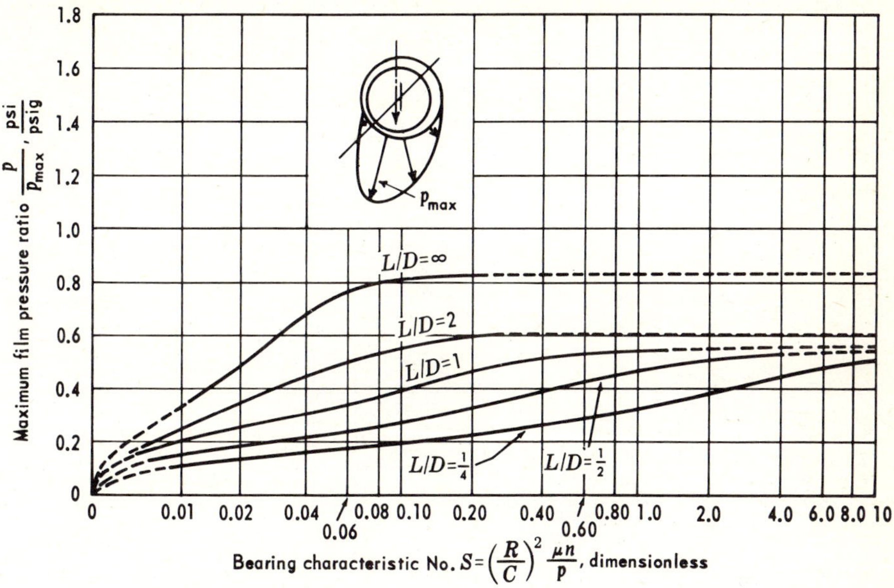

Fig. 21-29 Maximum film pressure. (*Courtesy of Westinghouse Research Laboratories.*)

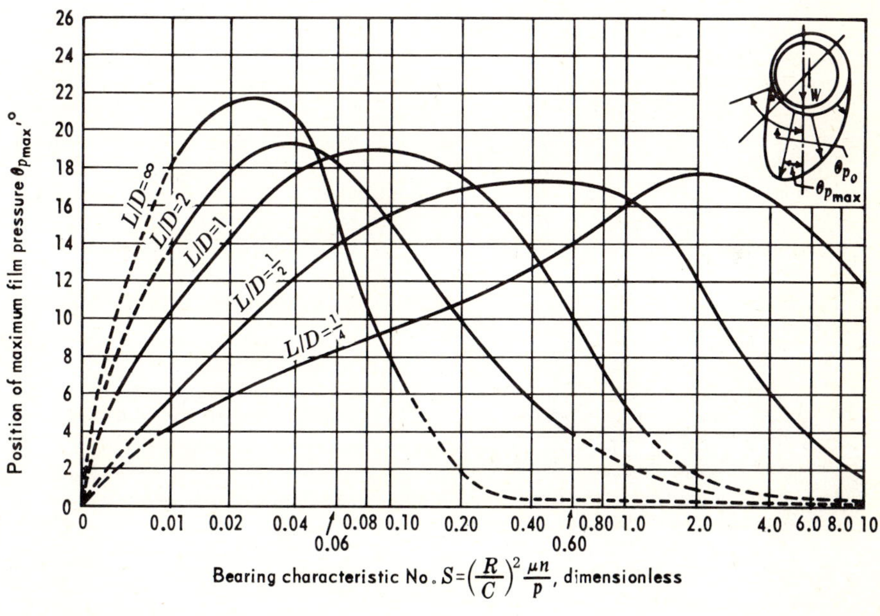

Fig. 21-30 Position of maximum film pressure. (*Courtesy of Westinghouse Research Laboratories.*)

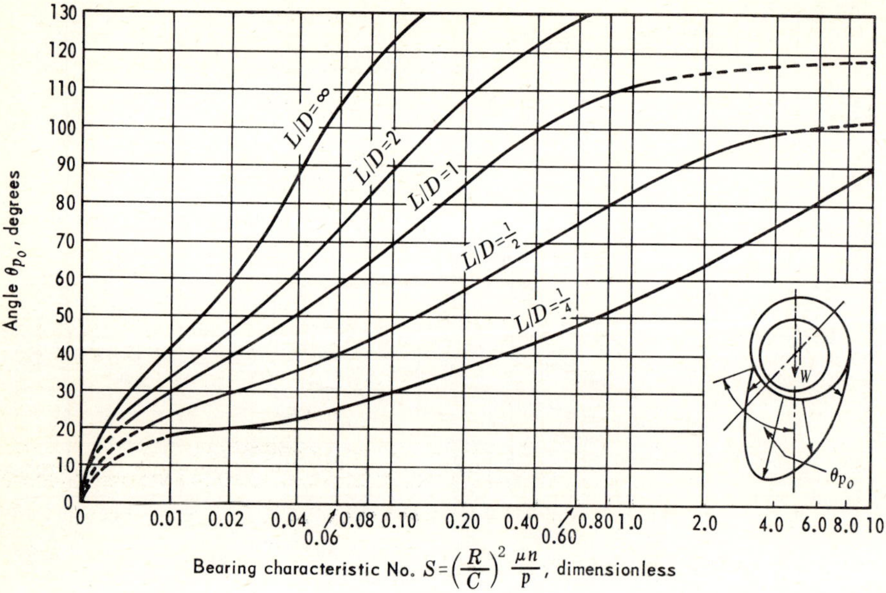

Fig. 21-31 Position of film termination. (*Courtesy of Westinghouse Research Laboratories.*)

Position of beginning of positive film pressure θ_A

From Fig. 21-32, θ_A is equal to $180 - \theta_{h_0}$. This depends on the amount of flow of lubricant supplied to the bearing so that it is quite possible that θ_A will be less than $180 - \theta_{h_0}$ if the supply of lubricant is not adequate.[1] Assuming classical flow rate,

$$\theta_A = 180 - \theta_{h_0} = 180 - 75 = 105°$$

In this example, the position of the minimum film thickness h_0 occurs at a large value of θ_{h_0} (75°), showing that the shaft is "riding high" in the clearance space and is almost concentric ($\epsilon = 0$) with the bearing. At concentric operation, $\theta_{h_0} = 90°$, whereas at contact between the shaft and bushing $\theta_{h_0} = 0°$, and the journal center locus diagram is almost a semicircle between $\epsilon = 0$ and $\epsilon = 1.0$, as shown in Fig. 21-32. If the load on this bearing was increased or if the L/d ratio was decreased, the pressure would be increased, the values of h_0 and θ_{h_0} would be smaller, and the shaft would run "lower" in the clearance space, as shown qualitatively in Fig. 21-32.

[1] H. J. Connors, An Analysis of the Effect of Lubricant Supply Rate on the Performance of the 360-deg Journal Bearing, *Am. Soc. Lubr. Engr. Trans.*, vol. 5, no. 2, November, 1962.

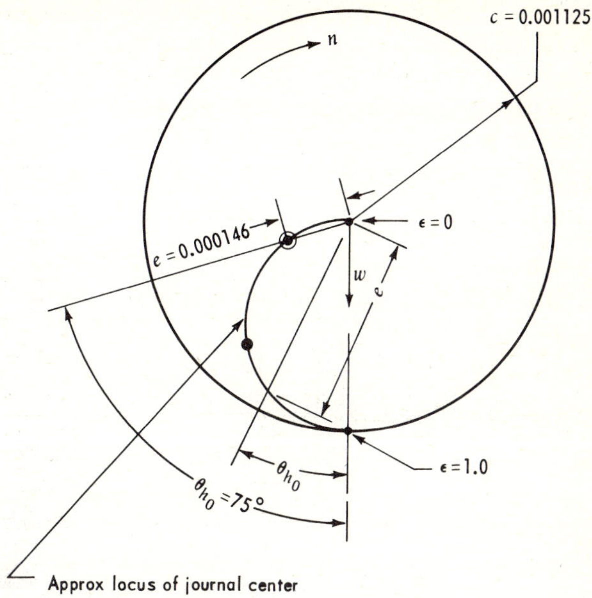

◉ Denotes operating position for design example, p = 77 psi, journal rides "high"

● Denotes operating position for heavier loads, journal rides "low"

Fig. 21-32 Journal center for Example 21-2.

21-18 SYSTEMS OF LUBRICATION

Various means are employed to supply lubricant to bearings and to gears. It is important that the proper method be selected. A brief description of some systems is given here:

1. *Hand oiling.* Suitable for slow speed and lightly loaded bearings.
2. *Drop-feed* or *wick-feed oilers.* Better than hand oiling, since the supply of lubricant is more nearly uniform.
3. *Oil rings* or *chains.* Suitable for horizontal bearings. The ring or chain is slipped over the shaft at the midpoint of the bearing, runs with the shaft, and carries oil to the bearing from the reservoir into which it dips. This type of lubrication has been very widely used for motors, generators, and line shafts.
4. *Splash system.* In this system a rotating part, such as a gear, or a rotating disk, called an *oil slinger,* runs in an oil reservoir and causes a shower of oil in the enclosed casing. The oil runs into channels and ducts

and is led to the bearings. A continuous supply of lubricant may be secured by this method.

5. *Pressure feeding.* Bearings may be positively lubricated by introducing the lubricant under pressure at the proper points of the bearing. This method ensures an adequate supply of lubricant at a pressure sufficient to prevent suction of air into the region of low pressure in bearings operating under conditions of fluid-film lubrication. The flow of lubricant may be in such quantities that most of the heat generated in the bearing will be carried away by the lubricant itself, so that the viscosity of the lubricant may thus be controlled. This may be necessary in heavily loaded bearings.

6. *Grease lubrication.* In parts loaded heavily and with low speeds so that fluid-film lubrication is not possible, grease may be used. Grease may be used also where dripping or spattering of oil is not permissible, as in food- or chemical-processing equipment.

7. *Shaft seals.* In order to prevent leakage of the lubricant from a bearing and to prevent foreign matter from entering, shaft seals may be necessary. Seals are available in many forms, such as felt or leather inserted in grooves in the bearing housings, or they may be purchased separately as a self-contained seal to mount in the housing.

22
Rolling-contact Bearings

22-1 GENERAL CONSIDERATIONS

Ball and roller bearings are frequently referred to as "antifriction bearings." This is a misnomer, however, since friction is always present in such bearings, owing to rolling resistance between the balls or rollers and the races and also to friction between the retainers and contacting parts and to churning of any excessive lubricant. For starting conditions and at moderate speeds, the friction of a ball or roller bearing is lower than that of an equivalent journal bearing, but at high speeds a properly designed and lubricated journal bearing has less friction than an equivalent ball or roller bearing. The friction-speed relations are shown qualitatively by Fig. 22-1.

Because of the kinematic *line* contact of rollers, the roller bearing is frequently used for large bearing loads in preference to ball bearings, which have kinematic point contact; however, an inherent disadvantage of the roller bearing is the variation of pressure along the band of contact, due to deflections of shafts and mountings.

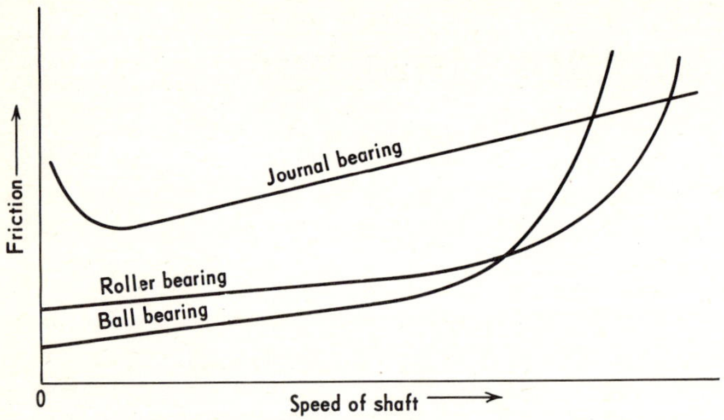

Fig. 22-1 Comparison of friction of bearings.

Some advantages of ball and roller bearings as compared with journal bearings are:

1. Friction is low except at high speeds.
2. Relatively accurate shaft alignment can be maintained.
3. Heavy momentary overloads can be carried.
4. Lubrication is simple.
5. Both radial and axial loads can be carried by some types.
6. Replacement is easy.
7. Selection of bearing from manufacturers' information is relatively simple.

Some disadvantages are:

1. The expense is generally greater because of the cost of the bearing and the necessary provision for mounting.
2. Failure of bearing can occur without warning and cause damage to machinery.

22-2 LOAD RATINGS OF BALL AND ROLLER BEARINGS[1]

Because of the elastic deformations of the surfaces at the regions of contact between a ball or roller and the races, the contact pressure is distributed over the areas in contact. The extent of the areas depends on the elastic properties of the materials, on the size of the balls or rollers, and

[1] New Departure Engineering Data, Analysis of Stresses and Deflections, 2 vols., New Departure Division of General Motors Corp., Bristol, Conn., 1946; New Departure Ball Bearings, General Catalog.

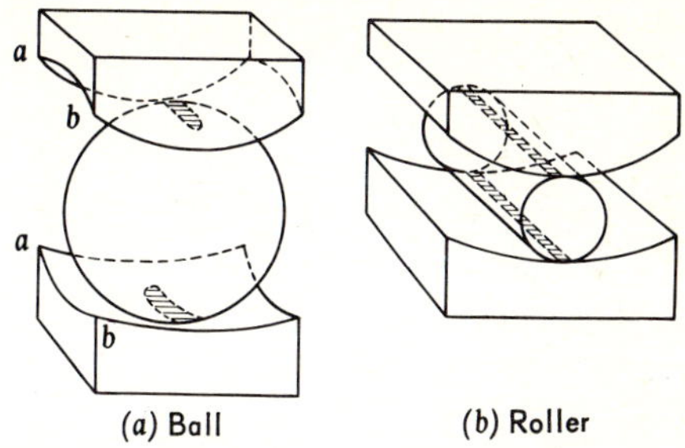

Fig. 22-2 Contact areas in ball and roller bearings.

on the form of the raceways. In order to provide a large area and in turn to reduce the pressure in a ball bearing, and to permit the bearing to carry axial loads in addition to radial loads, the raceways are formed as shown schematically in Fig. 22-2(*a*). The radii of curvatures for the curves *ab* are made slightly greater than the radius of the ball. The contact area in a roller bearing is shown in Fig. 22-2(*b*).

In a ball or roller bearing, the load-carrying capacity is based mainly on the surface endurance limit of the material. The *rated load* of a bearing is the load it will carry for a specific life of the bearing when operating at a definite speed. For other expected periods of life and for other speeds, the load that the bearing should be required to carry must be altered accordingly.

The theoretical basis for the design of ball and roller bearings is found in the works of Hertz and Stribeck, but the commercial ratings and the factors for speed, life, and shock are based mainly on performance data. For satisfactory bearings, it is necessary to make the parts of high-grade material properly heat-treated, accurately formed, and with finely finished surfaces. Failure of bearings generally occurs because of surface deterioration of the balls or rollers and the raceways, or from internal microresidual tensile stresses.

High-speed bearings, however, usually fail at the cage or retainer used to space the rolling elements. The friction at this point is of a sliding nature, and failure is caused by relatively high loads at high speeds and by inadequate lubrication. High-speed bearings in aircraft units may operate at speeds over 20,000 rpm.

22-3 TYPES OF RADIAL BALL BEARINGS

As shown in Fig. 22-3(*a*), the parts of a radial ball bearing are the outer race *c*, the inner race *d*, the balls *e*, and the retainer *f*. The retainer is usually in two parts which are assembled after the balls have been properly spaced. The grooves which form the path for the rolling elements are

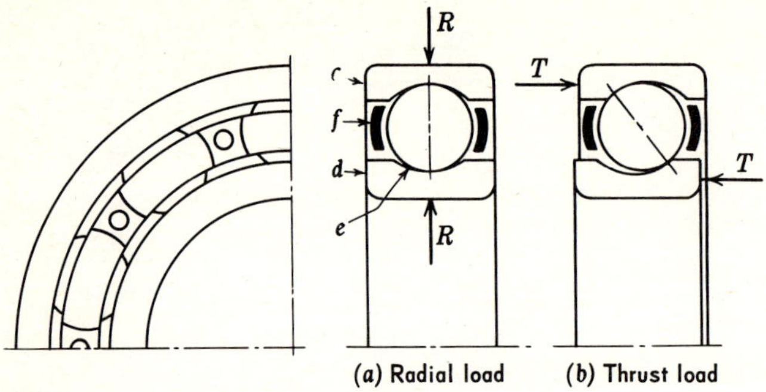

Fig. 22-3 Radial and thrust loads on ball bearings. Section lines are omitted in conformity with general practice.

known as raceways. When a ball bearing supports only a radial load R, as shown in Fig. 22-3(a), the plane of rotation of the ball is normal to the centerline of the bearing. As shown at (b) in the figure, the action of a thrust load T is to shift the plane of rotation of the balls. Both radial and thrust loads may be carried simultaneously.

Many types of radial ball bearings have been developed. The general features of some commonly used industrial types are discussed here.

Single-row deep-groove bearings In the assembly of this bearing, the inner race is placed eccentric with the outer race, and then the balls are inserted in the crescent-shaped space, as shown in Fig. 22-4. The balls are then evenly spaced as the races are brought into concentric relation. The halves of the retainer are then assembled so that the balls will remain evenly spaced but can rotate freely. Even spacing of the balls is desirable from the standpoints of distribution of the load and balance of the bearing.

Radial ball bearing with filling notches By employing a filling notch in each of the races, as shown in Fig. 22-5, additional balls may be inserted

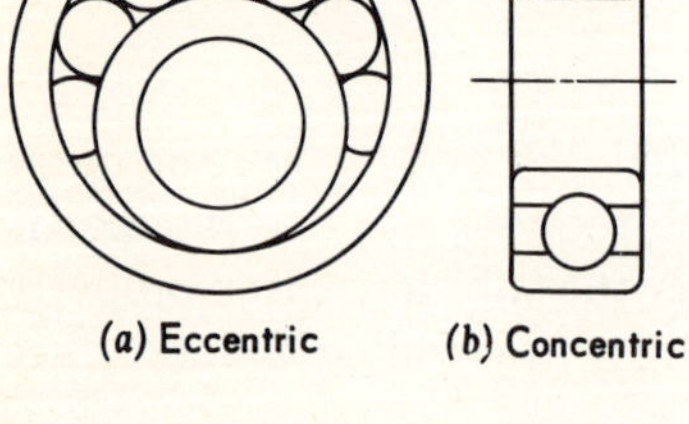

Fig. 22-4 Assembly of deep-groove ball bearing.

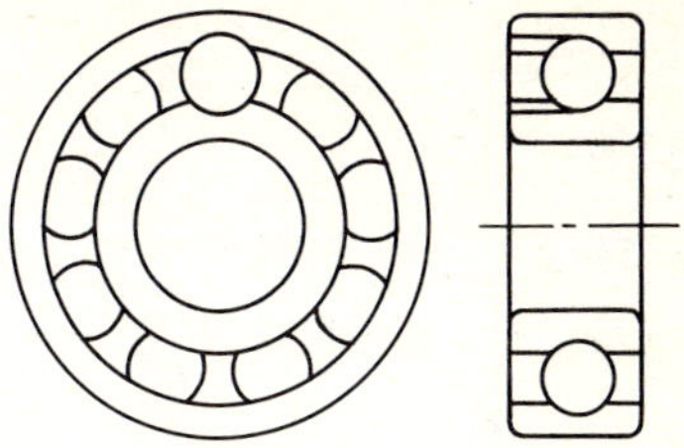

Fig. 22-5 Filling-notch type of radial ball bearing.

after the races are brought into concentric relation. The bottom of each filling notch does not reach to the bottom of its raceway and therefore the balls inserted through the notches must be forced in position. Some bearings of this type, particularly small ones, have as many balls as the annular space will contain and thus a retainer is not necessary to space the balls. Generally less than the maximum number of balls are used and therefore a retainer is necessary.

Since a bearing of this type contains a larger number of balls than a corresponding unnotched one, it has a larger rated *radial* load capacity but the filling notches reduce the allowable *thrust* load.

Proportions of radial ball bearings Ball bearings are available in four series, as shown in Fig. 22-6, for either standard bore dimensions or standard outside diameter. The extra-extra-light series (*LL*00) is used where the loads are light, shaft sizes are comparatively large, and available space is limited. At the other extreme, the medium series (300) has greater load-carrying capacity.

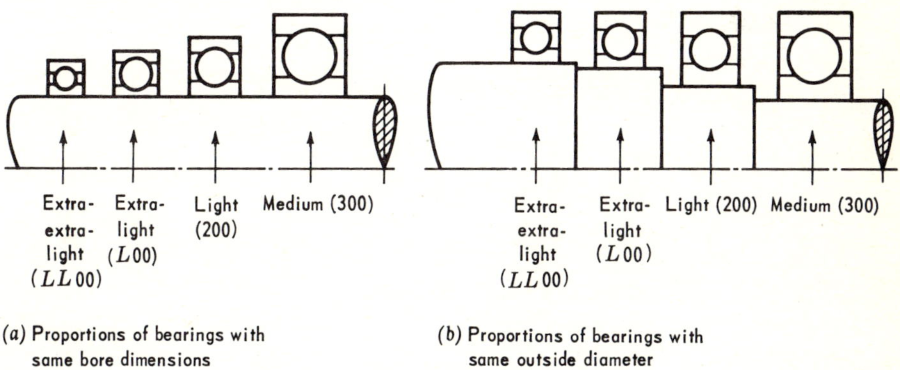

Fig. 22-6 Ball-bearing series. (*a*) Proportions of bearings with same bore dimensions. (*b*) Proportions of bearings with same outside diameter.

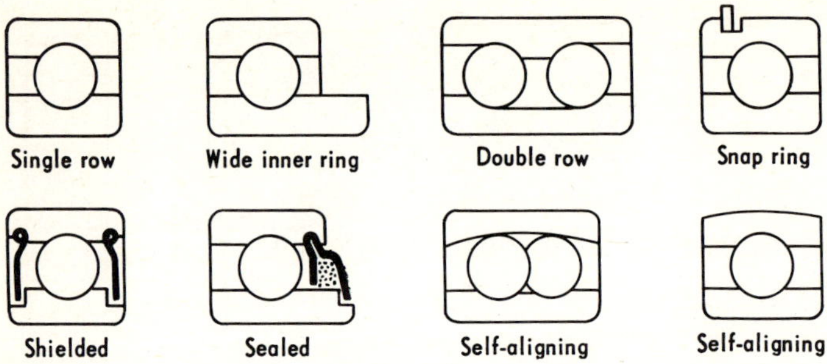

Fig. 22-7 Types of radial ball bearings.

Other types of radial ball bearings Of the many types of radial ball bearings made by the various manufacturers, the single-row type that has been described is in most common industrial use. Other types are shown in section in Fig. 22-7.

22-4 SELECTION OF RADIAL BALL BEARINGS[1]

In determining the type and size of radial ball bearings for a particular installation, the data and procedure used should be that recommended by the company from whom the bearings are to be purchased. The selection procedures used by the various companies are similar in general but vary in detail. For the purpose of illustration, a typical procedure is outlined here.

The rated radial-load capacities for light (200) and medium (300) series of the deep-groove type of bearing for various speeds are given in Table 22-1. The table applies to bearings in which the inner race rotates.

The data necessary for the selection of a bearing are as follows:

Radial load, lb
Thrust load, lb
Speed of shaft, rpm
Desired life of bearing, hr
Conditions of loading

The *rated* radial load of a bearing to comply with the operating conditions may be determined from the following equation; then a bearing

[1] The terminology and data in this chapter are taken, as an example, from New Departure ball-bearing practice: New Departure Engineering Data, *op. cit.;* New Departure Ball Bearings, General Catalog, 1964 printing.

Table 22-1* Rated radial loads for single-row, deep-groove (Type 3000 New Departure) NDur bearings adjusted to speed of the inner race at 1,000 rpm

Basic no.	*Radial load*	*Basic no.*	*Radial load*	*Basic no.*	*Radial load*	*Basic no.*	*Radial load*
3200†	270	3205	690	3210	1,820	3215	3,180
3300	360	3305	1,110	3310†	3,000	3315	4,810
3201	270	3206	1,020	3211	2,250	3216	3,430
3301	465	3306	1,470	3311	3,390	3316	5,260
3202	290	3207	1,590	3212	2,550	3217	4,190
3302	580	3307	1,820	3312	3,780	3218	4,670
3203	510	3208	1,590	3213	2,990	3219	5,180
3303	710	3308	2,200	3313	4,190	3220	5,710
3204	635	3209	1,710	3214	3,180	3221	5,950
3304	1,010	3309	2,780	3314	4,620	3222	6,510

* Courtesy of New Departure Division of General Motors Corporation, Bristol, Conn.

† Bearing numbers 3200 through 3210 are NDur 300; numbers 3310 through 3222 are NDur 600.

may be selected from Table 22-1:

$$C = \frac{RF}{F_s}\sqrt[4]{\frac{L}{Q}} \tag{22-1}$$

where C = rated radial load adjusted to tabular value (Table 22-1) at 1,000 rpm, lb
R = radial load on bearing, lb
T = thrust load on bearing, lb
F = thrust factor (Table 22-2)
F_s = speed factor for operating speed (Table 22-3)
$(L/Q)^{\frac{1}{4}}$ = life factor; see following paragraph
L = B-10 life of bearing, hr
Q = 2,280 for New Departure NDur 300 bearings
= 4,560 for New Departure NDur 600 bearings

Table 22-2 Thrust factor F

T/R	F	T/R	F	T/R	F	T/R	F	T/R	F
0.05	1.01	0.30	1.12	0.60	1.37	1.25	2.02	4.00	4.76
0.10	1.02	0.35	1.16	0.70	1.46	1.50	2.27	5.00	5.77
0.15	1.04	0.40	1.20	0.80	1.56	1.75	2.52	7.50	8.27
0.20	1.06	0.45	1.24	0.90	1.67	2.00	2.77	10.00	10.77
0.25	1.09	0.50	1.28	1.00	1.77	3.00	3.77		

Table 22-3 Speed factors F_S

Rpm	F_S	*Rpm*	F_S	*Rpm*	F_S	*Rpm*	F_S	*Rpm*	F_S	*Rpm*	F_S	*Rpm*	F_S
		270	1.387	825	1.049	1,725	0.8726	3,250	0.7448	5,100	0.6654	8,700	0.5823
10	3.162	280	1.375	850	1.041	1,750	0.8695	3,300	0.7419	5,200	0.6622	8,800	0.5806
15	2.858	290	1.363	875	1.034	1,775	0.8664	3,350	0.7392	5,300	0.6591	8,900	0.5790
20	2.659	300	1.351	900	1.027	1,800	0.8633	3,400	0.7364	5,400	0.6560	9,000	0.5774
25	2.515	310	1.340	925	1.020	1,825	0.8604	3,450	0.7337	5,500	0.6530	9,100	0.5758
30	2.403	320	1.330	950	1.013	1,850	0.8575	3,500	0.7311	5,600	0.6501	9,200	0.5742
35	2.312	330	1.320	975	1.006	1,875	0.8546	3,550	0.7285	5,700	0.6472	9,300	0.5726
40	2.236	340	1.310	1,000	1.000	1,900	0.8518	3,600	0.7260	5,800	0.6444	9,400	0.5711
45	2.171	350	1.300	1,025	0.9938	1,925	0.8490	3,650	0.7235	5,900	0.6416	9,500	0.5696
50	2.115	360	1.291	1,050	0.9878	1,950	0.8462	3,700	0.7210	6,000	0.6389	9,600	0.5681
55	2.065	370	1.282	1,075	0.9821	1,975	0.8436	3,750	0.7186	6,100	0.6363	9,700	0.5666
60	2.021	380	1.274	1,100	0.9765	2,000	0.8409	3,800	0.7162	6,200	0.6337	9,800	0.5652
65	1.981	390	1.265	1,125	0.9710	2,050	0.8357	3,850	0.7139	6,300	0.6312	9,900	0.5637
70	1.944	400	1.257	1,150	0.9657	2,100	0.8307	3,900	0.7116	6,400	0.6287	10,000	0.5624
75	1.911	410	1.250	1,175	0.9605	2,150	0.8258	3,950	0.7093	6,500	0.6263		
80	1.880	420	1.242	1,200	0.9554	2,200	0.8211	4,000	0.7071	6,600	0.6239		
85	1.852	430	1.235	1,225	0.9506	2,250	0.8165	4,050	0.7049	6,700	0.6215		
90	1.826	440	1.228	1,250	0.9457	2,300	0.8120	4,100	0.7027	6,800	0.6193		
95	1.801	450	1.221	1,275	0.9411	2,350	0.8077	4,150	0.7006	6,900	0.6170		
100	1.778	460	1.214	1,300	0.9365	2,400	0.8034	4,200	0.6985	7,000	0.6148		
110	1.736	470	1.208	1,325	0.9321	2,450	0.7993	4,250	0.6965	7,100	0.6126		
120	1.699	480	1.201	1,350	0.9277	2,500	0.7953	4,300	0.6944	7,200	0.6105		
130	1.665	490	1.195	1,375	0.9235	2,550	0.7914	4,350	0.6924	7,300	0.6084		
140	1.635	500	1.189	1,400	0.9193	2,600	0.7875	4,400	0.6905	7,400	0.6063		
150	1.607	525	1.175	1,425	0.9153	2,650	0.7838	4,450	0.6885	7,500	0.6043		
160	1.581	550	1.161	1,450	0.9113	2,700	0.7801	4,500	0.6866	7,600	0.6023		
170	1.557	575	1.149	1,475	0.9074	2,750	0.7765	4,550	0.6847	7,700	0.6003		
180	1.535	600	1.136	1,500	0.9036	2,800	0.7731	4,600	0.6828	7,800	0.5984		
190	1.515	625	1.125	1,525	0.8999	2,850	0.7696	4,650	0.6810	7,900	0.5965		
200	1.495	650	1.114	1,550	0.8962	2,900	0.7663	4,700	0.6792	8,000	0.5946		
210	1.477	675	1.103	1,575	0.8926	2,950	0.7630	4,750	0.6774	8,100	0.5928		
220	1.460	700	1.093	1,600	0.8891	3,000	0.7598	4,800	0.6756	8,200	0.5910		
230	1.444	725	1.084	1,625	0.8857	3,050	0.7567	4,850	0.6738	8,300	0.5892		
240	1.429	750	1.075	1,650	0.8823	3,100	0.7536	4,900	0.6721	5,400	0.5874		
250	1.414	775	1.066	1,675	0.8790	3,150	0.7506	4,950	0.6704	8,500	0.5856		
260	1.400	800	1.057	1,700	0.8758	3,200	0.7477	5,000	0.6687	8,600	0.5840		

The "life factor" $(L/Q)^{\frac{1}{4}}$ is the basis for rating NDur 300 and NDur 600 radial ball bearings made by the New Departure Division of General Motors Corporation. The NDur bearings incorporate two unique features: The first is the use of "vacuum-processed" (carbon vacuum deoxidized) high-carbon–chrome steel for the races and balls, which reduces inclusions; the second is a patented heat-treating process (Marstressing) which gives residual compressive stress in the surface layer of through-hardened material. The nature of both these improvements is discussed in the references below.[1] The NDur 300 bearings use races machined from steel bar or tubing whereas the NDur 600 races are forged.

The life of these bearings was determined by an extensive series of fatigue tests, and the distribution is shown graphically in the "Weibull"[2] plot in Fig. 22-8. The life of the previous standard bearing was plotted as a straight line on the chart, as shown. The other plots represent a conservative summary of the laboratory tests for the NDur 300 and NDur

[1] C. E. Norton and P. J. Baker, The Effect of New Material and Process Developments on Ball Bearing Fatigue Life, 779-B, Society of Automotive Engineers, New York, Jan. 13–17, 1964; C. L. Kobrin, Bearings Get New Odds on Survival, *Iron Age,* Nov. 28, 1963; and J. O. Almen and P. H. Black, "Residual Stresses and Fatigue in Metals," art. 14-6, McGraw-Hill Book Company, New York, 1963; United States Patent 3,117,041, Jan. 7, 1964.

[2] W. Weibull, "Fatigue Testing and Analysis of Results," Pergamon Press, New York, 1961.

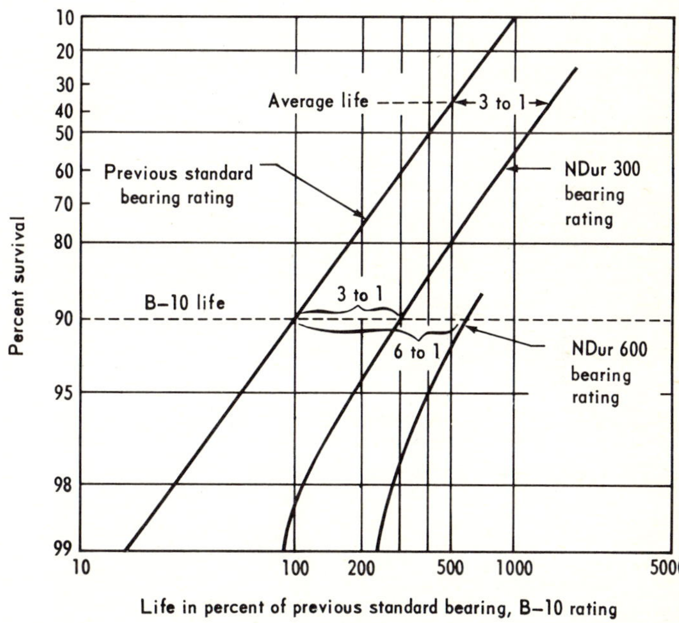

Fig. 22-8 Life ratings of NDur bearings in percent of life rating of previous standard bearings as graphically shown by Weibull diagram. Note that B-10 life (90 percent survival) of NDur 300 bearing has 3:1 life advantage over previous standard bearing. (*Courtesy of New Departure Division of General Motors Corporation.*)

600 series bearings. Note that the NDur 300 bearings have an expected B-10 life (90 percent survival) equal to 300 percent of the ratings of the previous standard-bearing ratings, and the NDur 600 bearings have an expected B-10 life equal to 600 percent of the previous standard-bearing ratings.

The B-10 life in hours can be expressed as $L = Q(F_sC/FR)^4$. This equation is the same as Eq. (22-1).

Example Select a single-row, radial ball bearing (Type 3000) for a radial load $R = 250$ lb, a thrust load of 50 lb, and an operating speed of 3,600 rpm for a B-10 life equal to 16,000 hr. Assume a steady load with no shock, and NDur 300 series bearings.

$$\frac{T}{R} = \frac{50}{250} = 0.20$$

$$F = 1.06 \qquad \text{(Table 22-2)}$$

$$F_s = 0.726 \qquad \text{(Table 22-3)}$$

$$L = 16{,}000 \text{ hr}$$

$$Q = 2{,}280 \text{ for NDur 300 series bearing}$$

$$\text{Therefore } C = \frac{RF}{F_s}\sqrt[4]{\frac{L}{Q}} \qquad \text{[Eq. (22-1)]}$$

$$= \frac{250 \times 1.06 \times 1}{0.726}\sqrt[4]{\frac{16{,}000}{2{,}280}} = 590 \text{ lb}$$

An inspection of Table 22-1 shows that bearing No. 3204, with rated radial load equal to 635 lb, is the nearest conservative value. Its B-10 life would be equal to

$$16{,}000\left(\frac{635}{590}\right)^4 = 21{,}500 \text{ hr}$$

If bearing No. 3302 were considered with its rated load equal to 580 lb, its B-10 life would be equal to

$$16{,}000\left(\frac{580}{590}\right)^4 = 14{,}940 \text{ hr}$$

The designer would decide, on the basis of conservative life versus cost of bearing, whether to specify the 3204 or 3302 bearing. Another consideration affecting cost would be the relation of the bores of the two bearings (0.7874 for No. 3204 versus 0.5906 for No. 3302) to the nominal size of the shaft on which the bearing is to be mounted.

In many cases it is found that a bearing selected on the basis of load-carrying capacity has a bore too small to allow it to be mounted on its shaft, so that it is necessary to select a bearing on the basis of dimensions, even though it has excessive load-carrying capacity. The dimensions of bearings referred to in Table 22-1 are given in Table 22-4. Fortunately, the dimensions of rolling-contact bearings are standardized. Interchangeability tables are available for various makes of bearings.

Table 22-4 Principal dimensions for radial ball bearings

Bear-ing no. plain	*Bore*		*Diameter*		*Width*		*Balls*		*Radius r*
	mm	*in.*	*mm*	*in.*	*mm*	*in.*	*Diam.*	*No.*	
3200	10	0.3937	30	1.1811	9	0.3543	$\frac{7}{32}$	7	0.025
3300			35	1.3780	11	0.4331	$\frac{1}{4}$	7	
3201	12	0.4724	32	1.2598	10	0.3937	0.210	8	0.025
3301			37	1.4567	12	0.4724	$\frac{9}{32}$	7	0.04
3202	15	0.5906	35	1.3780	11	0.4331	0.210	9	0.025
3302			42	1.6535	13	0.5118	$\frac{5}{16}$	7	0.04
3203	17	0.6693	40	1.5748	12	0.4724	$\frac{9}{32}$	8	0.025
3303			47	1.8504	14	0.5512	$\frac{11}{32}$	7	0.04
3204	20	0.7874	47	1.8504	14	0.5512	$\frac{5}{16}$	8	0.04
3304			52	2.0472	15	0.5906	$\frac{13}{32}$	7	
3205	25	0.9843	52	2.0472	15	0.5906	$\frac{5}{16}$	9	0.04
3305			62	2.4409	17	0.6693	$\frac{13}{32}$	8	
3206	30	1.1811	62	2.4409	16	0.6299	$\frac{3}{8}$	9	0.04
3306			72	2.8346	19	0.7480	$\frac{15}{32}$	8	
3207	35	1.3780	72	2.8346	17	0.6693	$\frac{7}{16}$	9	0.04
3307			80	3.1496	21	0.8268	$\frac{17}{32}$	8	0.06
3208	40	1.5748	80	3.1496	18	0.7087	$\frac{15}{32}$	9	0.04
3308			90	3.5433	23	0.9055	$\frac{19}{32}$	8	0.06
3209	45	1.7717	85	3.3465	19	0.7480	$\frac{15}{32}$	10	0.04
3309			100	3.9370	25	0.9843	$\frac{21}{32}$	8	0.06
3210	50	1.9685	90	3.5433	20	0.7874	$\frac{15}{32}$	11	0.04
3310			110	4.3307	27	1.0630	$\frac{23}{32}$	8	0.08
3211	55	2.1654	100	3.9370	21	0.8268	$\frac{17}{32}$	11	0.06
3311			120	4.7244	29	1.1417	$\frac{25}{32}$	8	0.08
3212	60	2.3622	110	4.3307	22	0.8661	$\frac{19}{32}$	10	0.06
3312			130	5.1181	31	1.2205	$\frac{27}{32}$	8	0.08
3213	65	2.5591	120	4.7244	23	0.9055	$\frac{21}{32}$	10	0.06
3313			140	5.5118	33	1.2992	$\frac{29}{32}$	8	0.08
3214	70	2.7559	125	4.9213	24	0.9449	$\frac{21}{32}$	11	0.06
3314			150	5.9055	35	1.3780	$\frac{31}{32}$	8	0.08
3215	75	2.9528	130	5.1181	25	0.9843	$\frac{21}{32}$	11	0.06
3315			160	6.2992	37	1.4567	1	8	0.08
3216	80	3.1496	140	5.5118	26	1.0236	$\frac{11}{16}$	11	0.08
3316			170	6.6929	39	1.5354	$1\frac{1}{16}$	8	
3217	85	3.3465	150	5.9055	28	1.1024	$\frac{25}{32}$	11	0.08
3317			180	7.0866	41	1.6142	$1\frac{1}{8}$	8	0.10
3218	90	3.5433	160	6.2992	30	1.1811	$\frac{27}{32}$	11	0.08
3318			190	7.4803	43	1.6929	$1\frac{3}{16}$	8	0.10

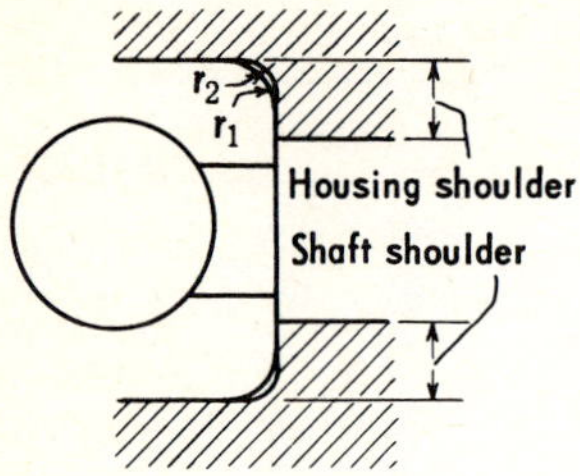

Fig. 22-9 Shoulder heights and fillet radii.

22-5 INSTALLATION OF RADIAL BALL BEARINGS

A common type of ball-bearing installation is that in which the bearing is used to support a rotating shaft. The inner race is fitted to the shaft with a light press fit and may be held firmly on the shaft by a retaining nut or other positive means. All fits and clearances should follow standards. The inner-race fit on the shaft should be sufficient to give a firm mounting but not too tight a fit because this may deform the inner race and destroy the clearance between the rolling elements and the raceways. The outer race should be mounted with a firm fit in the housing. If it is not tight enough, the outer race may tend to "creep." This is undesirable in that it may cause "fretting." Failure records of ball bearings with restrained outer races[1] show that 60 percent are inner-race failures; 20 percent are ball failures; 10 percent are retainer and outer-race breakdowns.

One reason that inner-race failures are more prevalent than failures of the outer race is that the contact pressure is greater at the inner raceway where it curves away from the ball than at the outer raceway where it curves with the ball.

In mounting the bearing, the housing shoulder and the shaft shoulder (Fig. 22-9) should follow standards.

A typical mounting with cap and shaft seal is shown in Fig. 22-10.

[1] Irving Kalikow, Creep of Ball Bearing Races, *Prod. Eng.*, vol. 20, no. 4, April, 1949.

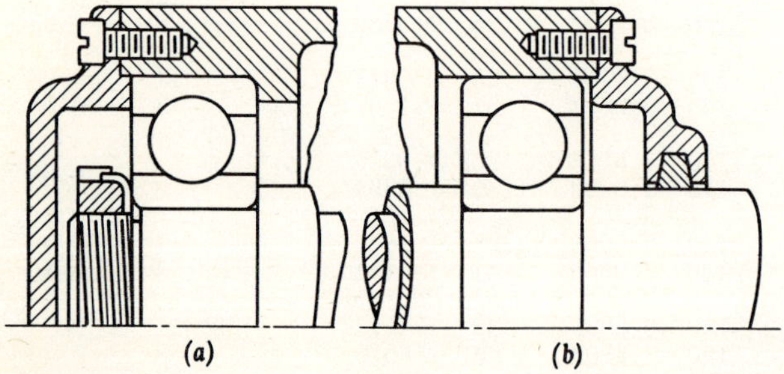

Fig. 22-10 Ball-bearing mounting.

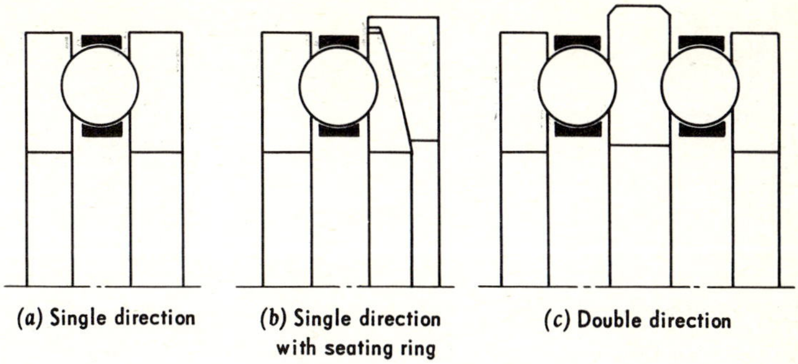

Fig. 22-11 Ball thrust bearings.

The bearing at the left-hand end of the shaft at (*a*) will take thrust in either direction in addition to a radial load. The shaft should have end play of 0.002 to 0.01 in., depending on the size of the bearing.

The outer race of the bearing on the right-hand end of the shaft at (*b*) should be free to move axially to avoid cramping of the bearings because of axial deformation of the shaft produced by thrust loads or expansion caused by temperature changes.

22-6 BALL THRUST BEARINGS

Where heavy thrust loads are to be carried, for example, in worm drives and vertical shafts, ball thrust bearings are applicable. Sections of three types of ball thrust bearings are shown in Fig. 22-11. The mounting of the worm shaft shown in Fig. 22-12 is an application of a double-direction ball thrust bearing to carry the axial load and of radial roller bearings to carry the radial loads.

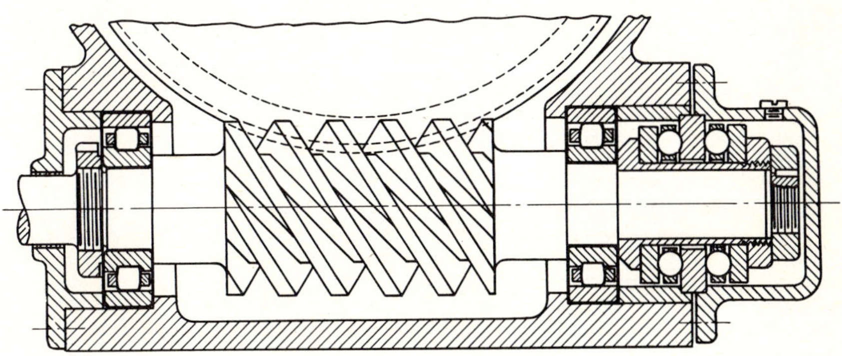

Fig. 22-12 Mounting for worm shaft.

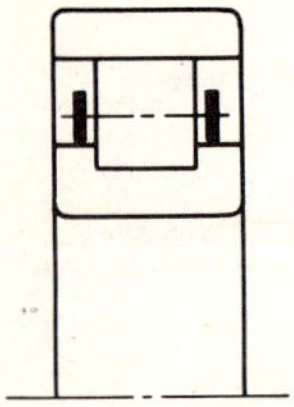

Fig. 22-13 Radial roller bearing.

22-7 RADIAL ROLLER BEARINGS

These bearings have cylindrical rollers that run in cylindrical raceways. In a common construction (see Fig. 22-13) the inner race is provided with retaining shoulders, but the outer raceway is straight. This construction permits axial movement of the shaft. The bearing is interchangeable with ball bearings of corresponding size and has a radial load capacity higher than the ball bearing; but it will not carry a thrust load.

A common type of shaft mounting where a load is carried by a shaft extension is shown in Fig. 22-14. The roller bearing carries the heavy radial load, and the ball bearing carries the lighter radial load and serves to locate the shaft axially.

The *Hyatt* roller bearing (Fig. 22-15) has hollow cylindrical rollers, each made by winding a flat strip of steel in the form of a helix. The flexibility provided by this construction allows the rollers to adjust themselves to slight irregularities in the raceways. It is thus possible to run

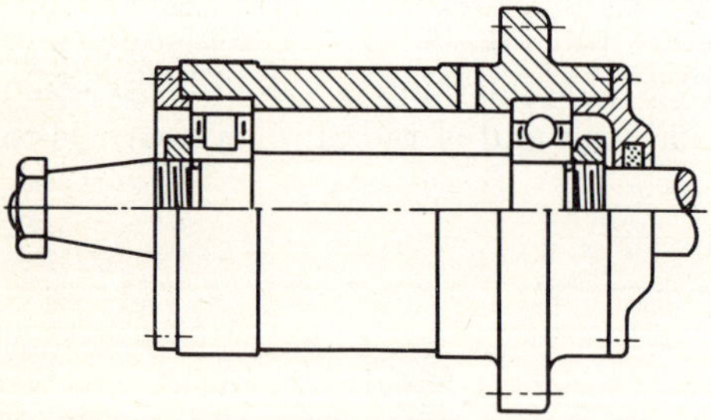

Fig. 22-14 Mounting for shaft with extension.

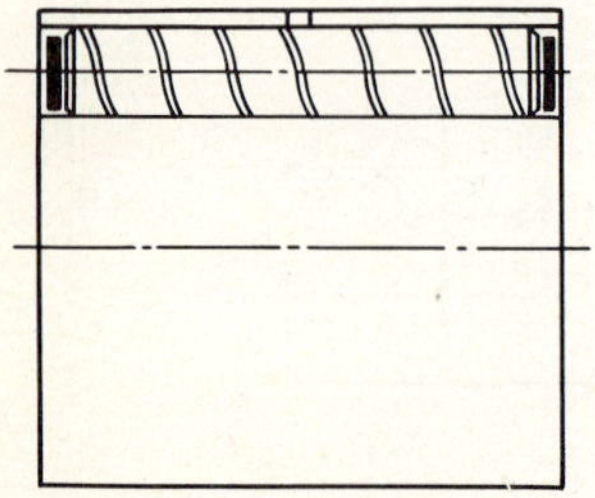

Fig. 22-15 Hyatt roller bearing.

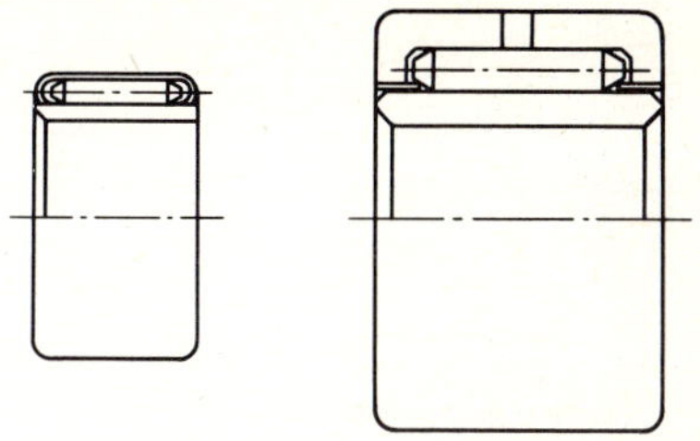

Fig. 22-16 Needle, or quill, bearings. **(*a*) Torrington** **(*b*) Bantam**

the rollers directly on the hardened and ground shaft, saving radial space. The rollers are wound alternately right- and left-hand in order to aid in the distribution of the lubricant along the rollers.

Needle bearings These are roller bearings in which the rollers are small in diameter and relatively long. They run without retainers, as shown in Fig. 22-16. The rollers, or "needles," may run directly on the shaft. This type of bearing is used where space in a radial direction is limited. Two types are available; one has a drawn shell in which the rollers usually run directly on the hardened shaft, and the other, for greater capacities, has a heavy outer race and usually an inner race [see Fig. 22-16(*a*) and (*b*)].

22-8 ANGULAR ROLLER BEARINGS

The angular, or tapered, roller bearing combines the high radial load-carrying capacity of a roller bearing with a thrust capacity in one direction. Since at least two bearings are necessary to support most shafts, angular bearings can be so mounted that one will support the thrust load in one direction and the other in the opposite direction.

As shown in Fig. 22-17, this bearing has four parts: (*a*) the inner race or cone, (*b*) the outer race or cup, (*c*) the rollers, and (*d*) the retainer. The principle of operation of the bearing, as shown in Fig. 22-18, is that the elements of the conical surfaces of the rollers and of the raceways intersect at a common apex on the centerline of the bearing, so that pure rolling

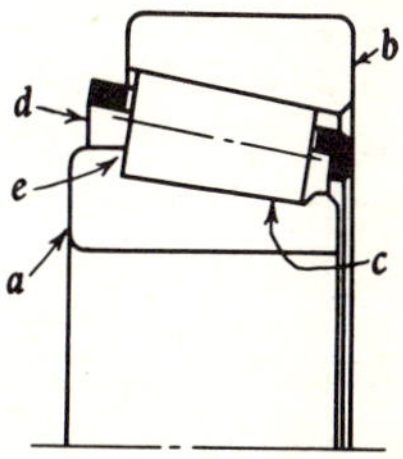

Fig. 22-17 Angular roller bearing.

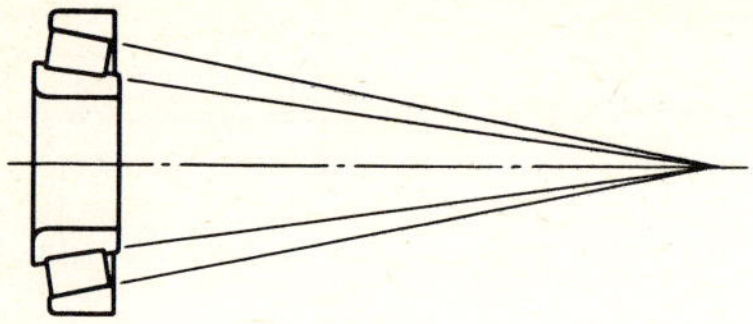

Fig. 22-18 Elements of conical surfaces intersect axis bearing.

motion will exist between the rollers and the raceways. The taper of the rollers is only a few degrees.

Since the conical surface of each roller is subjected to pressures acting normal to the surfaces, as shown in Fig. 22-19, there will be a resultant force tending to force the roller from its place between the races. To retain the roller in place it is necessary to provide a rib on the cone as shown at *e* in Fig. 22-17. The contact between the large end of the roller and the rib introduces sliding friction in the bearing, which in well-designed bearings is usually small. A typical mounting using tapered roller bearings is shown in Fig. 22-24.

22-9 METHODS OF FAILURE

As discussed in Art. 20-9, the contact between "rolling elements" is complicated by surface roughness and by deformation of the elements. The high pressure and the relative sliding between the contact regions may lead to direct mechanical wear, to plastic smearing, or to fatigue fractures. Macroresidual stresses induced by heat treating may reduce or contribute to failure; also, microresidual stresses caused by internal flaws and voids may accelerate fracture. The most common methods of failure in rolling-contact bearings are discussed below. These methods of failure also apply generally to load-translation rollers, gear teeth, cams, railroad wheels, etc. In many of these cases temperature effects are a contributing factor.[1]

[1] J. B. Bidwell, "Rolling Contact Phenomena," American Elsevier Publishing Company, New York, 1962; Almen and Black, *op. cit.*, chaps. 10 and 14.

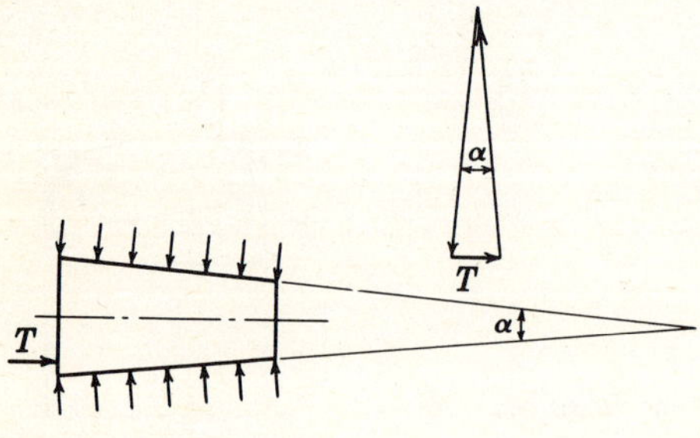

Fig. 22-19 Forces on tapered roller.

(*a*) *Spalling* As shown in Fig. 20-11(*c*), the contact between a bearing ball and the bottom of the inner-race groove subtends a considerable arc, and true rolling can occur only at a greater radius than the undeformed bottom of the inner-race groove, that is, along the line xx in the figure. The direction of sliding at the bottom of the inner-race groove is *opposite* to that of the rolling ball (this is called negative sliding) whereas, at greater radii than the race "pitch circle," sliding is in the *same* direction as that of the rolling ball. The force of friction from negative sliding of the balls against the race caused, in Fig. 22-20(*a*), a succession of cracks in the upper half of the spalled area in a ball-bearing inner race. Each crack began near the upper edge of a pit (see *Pitting* below) and grew in length and depth until the wall of the cavity broke away, thus increasing the tangential length of the spall. Figure 22-20(*b*) shows the cross section through a bearing race near the center of the spall.

(*b*) *Pitting* This type of surface damage was discussed in Art. 20-7(*c*) and is sometimes found in the rolls of roller bearings where the rolling of the roller relative to the inner race is in one direction (tangential) on the rollers rather than random as in ball bearings. The lower part of the spall presumably grew from a subsurface defect and later its area was extended upward, as shown in Fig. 22-20 by successive cracks.

(*c*) *Fretting corrosion* This type of damage was discussed in Art. 20-7(*d*) and appears in bearings that have slight oscillatory motion and

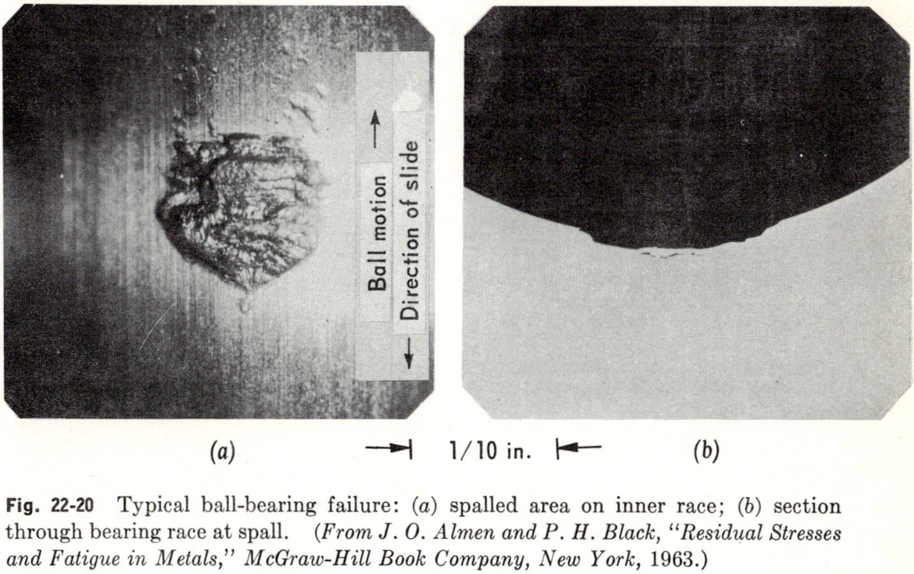

Fig. 22-20 Typical ball-bearing failure: (*a*) spalled area on inner race; (*b*) section through bearing race at spall. (*From J. O. Almen and P. H. Black, "Residual Stresses and Fatigue in Metals," McGraw-Hill Book Company, New York,* 1963.)

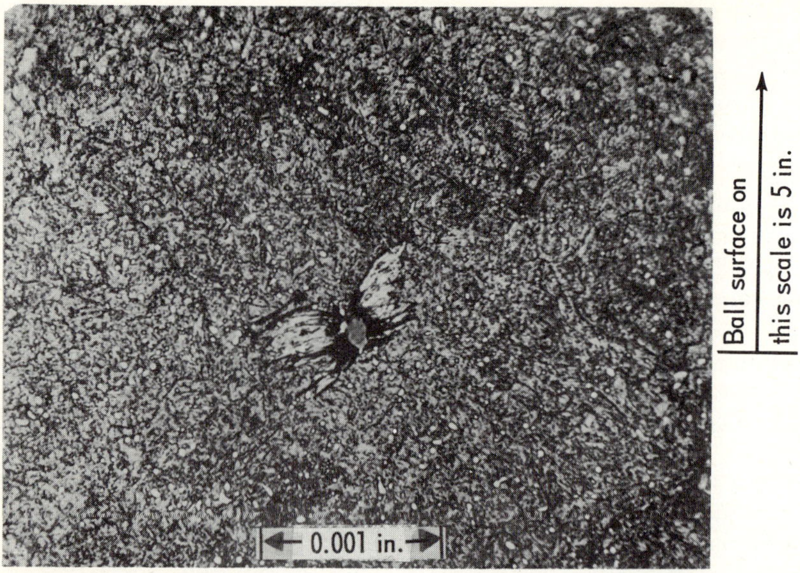

Fig. 22-21 Origin of internal fatigue crack due to tensile stress caused by compressive stress adjacent to defect. (*From J. O. Almen and P. H. Black, "Residual Stresses and Fatigue in Metals," McGraw-Hill Book Company, New York,* 1963.)

usually, but not always, in the absence of lubricant. Lubricant performs two functions: separating the surfaces and excluding oxygen. An example of this type of failure is found in needle bearings of universal joints in which the relative motion is low-amplitude oscillation.

(*d*) *Subsurface defects* Ball-bearing manufacturers agree that pits and cavities in bearings begin as fatigue cracks extending from subsurface defects such as inclusions and voids. These cracks grow outward at acute angles, thus detaching metal to form pits and cavities.

An explanation of this type of failure originating at an internal defect is as follows: In Art. 6-3 on the sources of residual stresses the principle was stated that a "residual stress caused by plastic flow is of the opposite sign to the stress producing the plastic flow." When the ball of a ball bearing is loaded, the compressive stress due to contact pressure around a subsurface defect, such as a void, may exceed the yield point of the material so that plastic flow in compression may occur at opposite sides of the void. This is especially likely to happen because the dynamic yield point is less than the static yield point, as stated by Bauschinger.[1]

When the ball rolls so that the void passes beyond the influence of

[1] Almen and Black, *op. cit.*, p. 12.

the compressive load, the region that suffered compression beyond the yield point will be left with a microresidual *tensile* stress, as indicated by the principle stated above. Because of the repeated loading in the region of the void, the tensile stress will increase repeatedly and a fatigue crack may originate and spread. Figure 22-21 shows such strain lines in a photomicrograph, indicating that microresidual tensile stresses of great magnitude were developed in the steel. As discussed in Art. 6-3 under carburizing and nitriding, the heat treatment induces residual stresses in the case, which, for equilibrium, results in residual tensile stresses in the interior of the ball. The sum of the residual tensile stress in the subsurface material due to the heat treatment and the tensile stress developed by the plastic yielding as discussed above may be a sufficiently high tensile stress to propagate fatigue cracks. Such cracks usually are propagated from the interior of the ball toward the case, expanding as they go, and when they reach the residual compressively stressed case, they spread along the juncture of the compressively stressed case and the tensile stressed core so that a large unsupported area may be eventually involved. The forces on such an area are shown in the free-body diagram of Fig. 22-22(*a*), and at (*b*) the flake has been detached and the ball has started to "spall."

Figure 22-23 shows a bearing in which several successive surface areas or layers have been removed; the end result resembles an onion peel.[1]

Spalling is not as apt to occur on the outer races of ball bearings as on the balls because the forces on the flake of the outer race tend to hold the flake in place rather than to force it outward, as shown for the ball in Fig. 22-22(*a*). The inner races sometimes show spalls because the inner race is formed with a concave surface profile in the axial plane and a convex surface profile in the transverse plane. The surface profile for the outer race is concave in both planes.

[1] See Almen and Black, *op. cit.*, art. 14-4.

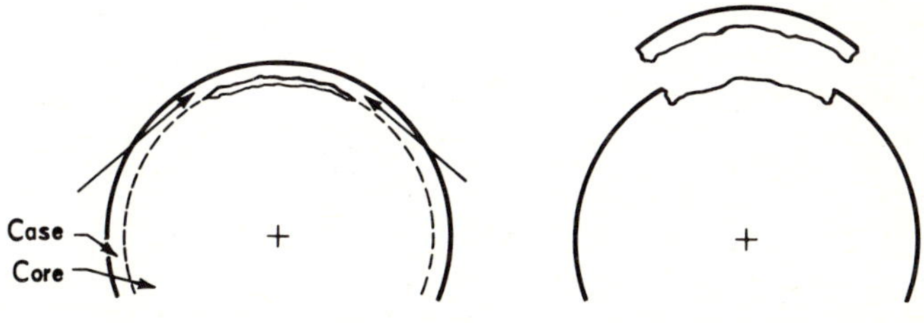

(*a*) Forces on flake (*b*) Flake detached

Fig. 22-22 Spalling of bearing ball. (*a*) Forces on flake. (*b*) Flake detached.

Fig. 22-23 Onion-peel spalling of ball. (*From J. O. Almen and P. H. Black, "Residual Stresses and Fatigue in Metals, McGraw-Hill Book Company, New York,* 1963.)

The cure for spalling is the elimination of internal voids and flaws. Vacuum melting of steel to remove voids is an important step in this direction and, owing to recent developments in this process, this steel is generally available for bearing manufacture.

22-10 LUBRICATION OF BALL AND ROLLER BEARINGS

The pressures existing between the balls or rollers and the raceways are so high that a fluid film of lubricant will not generally exist in loaded bearings. In order to assist in the maintenance of an adsorbed film between the load-carrying parts and thus prevent undue friction and wear, the surfaces of the balls or rollers and the raceways must be finely finished, and the finish must be maintained throughout the life of the bearing.

The functions of the lubricant in ball and roller bearings are to (1) prevent rust and corrosion; (2) aid in preventing entrance of foreign matter, such as water, dust, etc.; (3) reduce friction and wear between rubbing parts of the bearing; (4) dissipate heat.

Oils and greases are the most common lubricants for ball and roller bearings. Oil may be used where the lubricating service is dependable and the loads are relatively light, and also where the unit operates con-

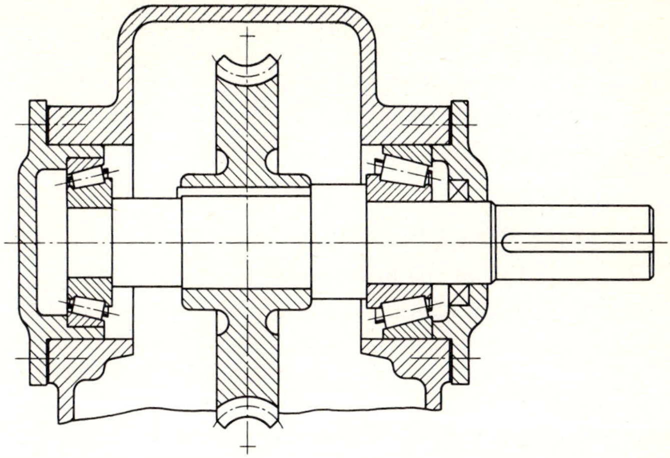

Fig. 22-24 Worm-gear shaft mounted on Timken tapered roller bearing.

tinuously and the loads and speeds are such that it is necessary to make use of a circulating oil-feed system to carry away the heat generated. In order to prevent excessive churning of the oil, especially at high speeds, the level of the oil for a bearing of a horizontal shaft should be near the top of the lowest ball.

Greases are used where the loads are high or where it is desired to extend the lubrication periods because of service conditions or because of

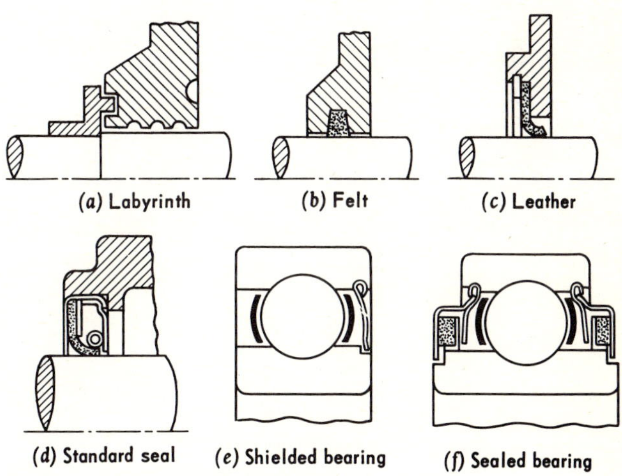

Fig. 22-25 Shaft seals and bearing shields.

the inaccessibility of bearings. In many cases regreasing of bearings is required every 6 to 24 months, depending on the service and the type of grease. To allow space for grease expansion due to temperature change and to prevent churning, it is recommended that the bearing be packed half full of grease.

22-11 SHAFT SEALS AND BEARING SHIELDS

To prevent loss of the lubricant from a ball or roller bearing or from the housing in which it is mounted and to aid in preventing the entrance of foreign matter such as water and dust, a seal or shield should be used. The seal or shield may be an integral part of the bearing, it may be incorporated in the bearing cover, or it may be a separate self-contained seal. Examples of seals and shields are shown in Fig. 22-25. Many other types and arrangements of shaft and bearing seals are available or may be devised to suit the individual installation.

23
Metal Fits and Tolerances

23-1 DIMENSIONAL CONTROL

Two important facts in the control of dimensions of parts that come in contact or near contact are that (1) the more exact the dimensions of the parts, the more satisfactory will be the characteristics of the machine, and (2) the more exact the dimensions of the parts, the higher will be the production cost of the machine.

The exactness of dimensions may be fixed by such requirements as space, leakage, lubrication, assembly or interchangeability, limits to impact loads, expansion or contraction, deflection, and pressure required in force or shrink fits.

In addition to establishing correct dimensions for the parts of a machine being designed, it is necessary for the mechanical-design engineer to select tolerances for the dimensions of the various parts. It has been aptly said that tolerances make or break a machine. Too large tolerances may affect performance and life expectancy. Too close tolerances may mean excessive cost (see Fig. 23-1).

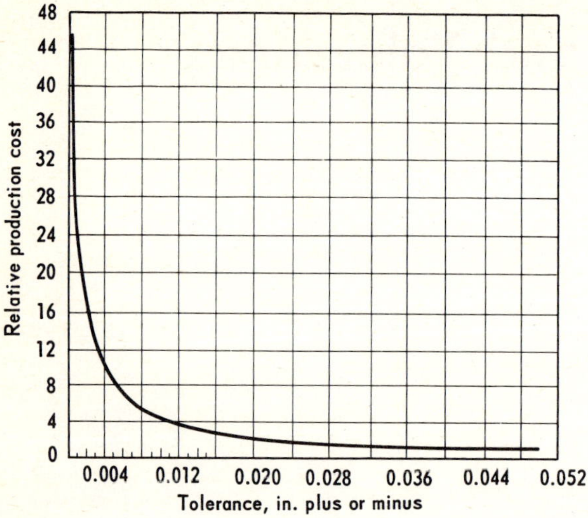

Fig. 23-1 Relative cost of accuracy. *(From F. W. Bolz, Design of Considerations for Manufacturing Economy, Trans. ASME, Paper* 49-*A*-53, 1949.)

Imagine the difficulty in assemblying the parts of an automobile engine if each part were made in a separate department which used for measurement the sixteenth-century rod, which was defined as "the total length of the left feet of the first sixteen men who came out of church on Sunday morning." Imagine the performance of such an engine if it were assembled. Tolerances today in critical engine elements are less than 0.0001 in. Some idea of the relative sizes involved is indicated in the comparison that one ten-thousandth of an inch is to an inch as the diameter of human hair is to the diameter of an oil drum.

23-2 METAL FITS

If the diameter of a member is specified as $1\frac{1}{2}$ in. by the designer, there may be some uncertainty as to the meaning of the size. The degree to which the diameter of the member should conform to exactly $1\frac{1}{2}$ in. depends on the use of the member. If it is to be a rod for holding up a shelf, then, obviously, a high degree of accuracy is not required. If, however, the member is a journal that must run smoothly in a bearing, then it is necessary to hold the dimension within close limits.

In addition to cost, many physical factors must be considered in the selection of fits, such as length of engagement of parts, materials, bearing pressures, lubrication, operating speed, and also ambient conditions such as temperature and humidity.

The *nominal size* of a part is the designation used for the purpose of general identification. *Tolerance* is the total permissible variation in size. The tolerance is the difference between the limits of size. For instance, for a shaft of nominal size $1\frac{1}{2}$ in. in diameter, the size may be designated as 1.500 ± 0.010, which indicates a tolerance of 0.020 in. The tolerance may be *bilateral,* in which case variation in size is permitted so that the part may be either larger or smaller than the nominal size. An example of *unilateral* tolerance is $1.500 \begin{smallmatrix} +\ 0.000 \\ -\ 0.010 \end{smallmatrix}$. Unilateral tolerances are usually used in specifying fits for interchangeable parts. In addition to the definitions of *nominal size* and *tolerance,* other definitions are as follows[1] and as illustrated in Fig. 23-2.

An *allowance* is an intentional difference between the maximum material limits of mating parts. It is a minimum clearance (positive allowance) or maximum interference (negative allowance) between mating parts.

Fit is the general term used to signify the range of tightness that may result from the application of a specific combination of allowances and tolerances in the design of mating parts. Fits are of three general types: clearance, transition, and interference.

A *clearance fit* is one having limits of size so prescribed that a clearance always results when mating parts are assembled.

An *interference fit* is one having limits of size so prescribed that an interference always results when mating parts are assembled.

[1] Extracted from USA Standard Preferred Limits and Fits for Cylindrical Parts (USAS B4.1–1955) with the permission of the publisher, The American Society of Mechanical Engineers, New York.

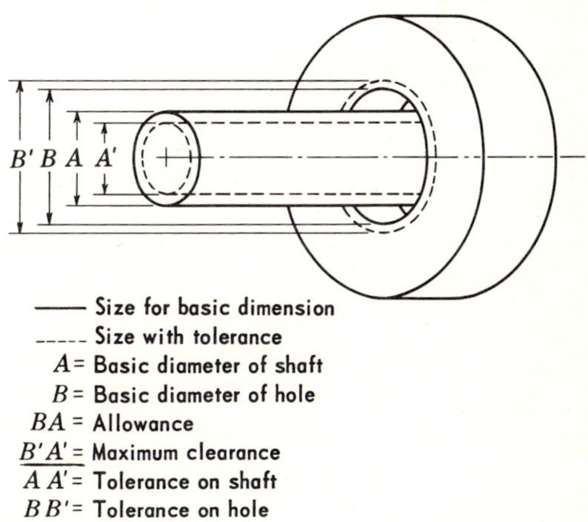

Fig. 23-2 Shaft and hole dimensions.

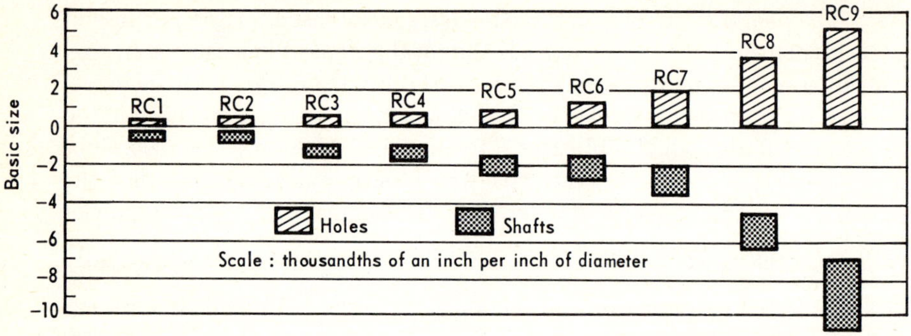

Fig. 23-3 Graphical representation of standard running or sliding fits.

A *transition fit* is one having limits of size so prescribed that either a clearance or an interference may result when mating parts are assembled.

Classification of fits:

1. *Running and sliding fits*
   RC1: close sliding fits
   RC2: sliding fits
   RC3: precision running fits
   RC4: close running fits
   RC5 and RC6: medium running fits
   RC7: free running fits
   RC8 and RC9: loose running fits

The bar diagram in Fig. 23-3 is a graphical representation of these fits, which gives an approximate comparison of the allowances and tolerances.

2. *Locational fits* determine only the location of the moving parts.
   LC locational clearance fits are specified in nine classes as shown in Fig. 23-4.
3. *Transition fits* are a compromise between clearance and interference fits and are specified in seven classes as shown in Fig. 23-5.
4. *Locational interference fits* are used where accuracy of location is of prime importance; they are specified in two classes as shown in Fig. 23-6.
5. *Force fits* or shrink fits, shown in Fig. 23-7, are interference fits classified as follows:
   FN1: light drive fits

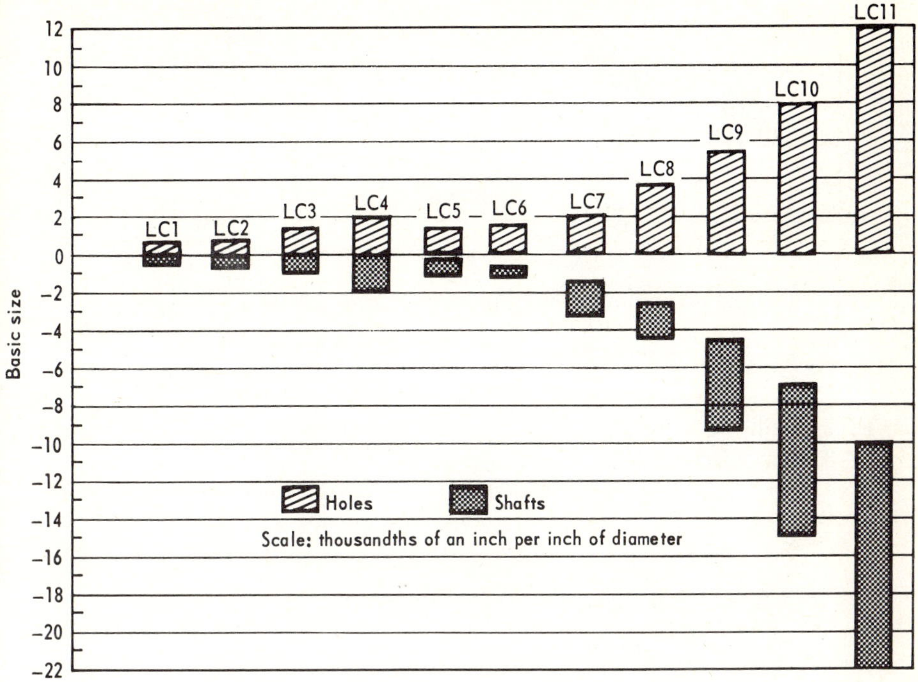

Fig. 23-4 Graphical representation of standard clearance locational fits.

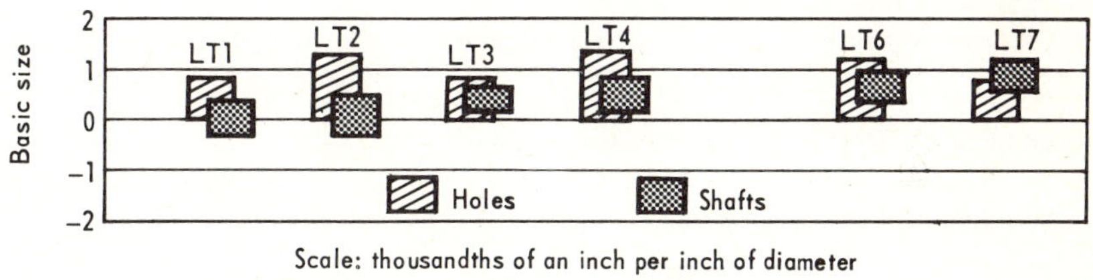

Fig. 23-5 Graphical representation of standard transition fits.

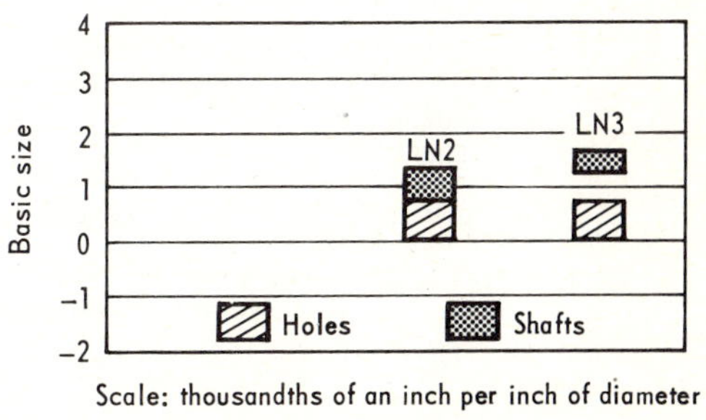

Fig. 23-6 Graphical representation of standard locational interference fits.

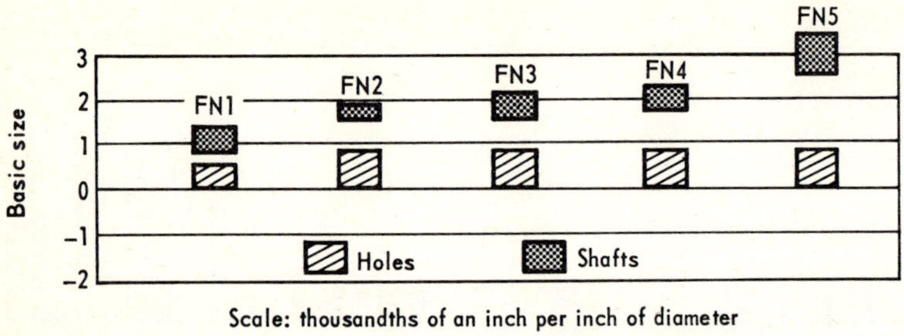

Fig. 23-7 Graphical representation of standard force or shrink fits.

FN2: medium drive fits
FN3: heavy drive fits
FN4 and FN5: force fits or shrink fits in parts that can be highly stressed

The United States of America Standards Institute Standard USAS B4.1–1965 gives limits for dimensions of hole and shaft for all fits from nominal sizes 0.04 to 200 in., except that the transition locational fit, the maximum nominal size, is limited to 19.69 in. It includes also a table of preferred basic sizes of mating parts for applying allowances and tolerances.

The bar diagrams, Figs. 23-3 to 23-7, are not intended to be read for values but are shown in order to give a picture of the various fits.

Example 23-1 Determine the sizes of a $1\frac{1}{2}$-in. shaft and hole for a sliding fit (RC2).

The nominal size of the assembly is 1.500 in. From USAS B4.1–1965, Table 1:

Limits of clearance for hole = 0.0006 to 0 in.
Limits of clearance for shaft = −0.0004 to −0.0008 in.
Therefore Maximum diameter of hole = 1.500 + 0.0006 = 1.5006 in.
Minimum diameter of hole = 1.500 + 0 = 1.500 in.
Minimum diameter of shaft = 1.500 − 0.0004 = 1.4996 in.
Maximum diameter of shaft = 1.500 − 0.0008 = 1.4992 in.

Using unilateral tolerance, the diameter of the hole is specified as $1.500 \begin{matrix} -\ 0.0006 \\ -\ 0.0000 \end{matrix}$ or between 1.500 and 1.5006.

The diameter of the shaft is specified as $1.500 \begin{matrix} -\ 0.0004 \\ -\ 0.0008 \end{matrix}$ or between 1.4996 and 1.4992 in.

The allowance (minimum clearance) is the difference between the smallest hole diameter and the maximum shaft diameter, or 1.5000 − 1.4992 = 0.0008 in.

23-3 FORCE FITS AND SHRINK FITS

A transition fit (class LT; see Fig. 23-5) may or may not be an interference fit (negative allowance). Interference locational fits (class LN; see Fig. 23-6) and force and shrink fits (class FN; see Fig. 23-7) are always interference fits. In these cases the diameter of the hole is less than the diameter of the shaft.[1] Thus there is *interference* of the metal so that assembly must be accomplished by forcing one part over the other, or by shrinking the external member on the internal member. These fits are usually used to prevent slipping of the two members relative to each other. For large interference, the stresses induced in the members may be very high; in some cases the stresses may be sufficiently high to cause failure of the external member.

The maximum stresses induced in a thick cylinder by an internal pressure are given by the Lamé equations (11-15) and (11-17). In a force-fit assembly, the pressure between the parts depends on the amount of interference. The maximum tangential stress and radial stress at the inside surface of the external member are

$$s_t \text{ (max)} = p \frac{1 + (d_i^2/d_o^2)}{1 - (d_i^2/d_o^2)} \tag{23-1}$$

$$s_r = -p \tag{23-2}$$

where d_i = nominal internal diameter of hub, in.
d_o = external diameter of hub, in.
p = contact pressure, psi

The contact pressure between the external member and a solid shaft of the same material is shown to be[2]

$$p = \frac{E\delta}{2d_i}\left(1 - \frac{d_i^2}{d_o^2}\right) \tag{23-3}$$

where δ = interference, in.
E = modulus of elasticity, psi

By substituting the value of p given by Eq. (23-3) into Eqs. (23-1) and (23-2), the principal stresses s_t and s_r become equal to

$$s_t = \frac{E\delta}{2d_i}\left(1 + \frac{d_i^2}{d_o^2}\right) \tag{23-4}$$

$$s_r = -\frac{E\delta}{2d_i}\left(1 - \frac{d_i^2}{d_o^2}\right) \tag{23-5}$$

[1] See M. F. Spotts, "Mechanical Engineering Design," Prentice-Hall, Inc., Englewood Cliffs, N.J., 1964.

[2] S. Timoshenko, "Strength of Materials," pt. II, D. Van Nostrand Company, Inc., Princeton, N.J., 1953; J. H. Faupel, Designing for Shrink Fits, *Machine Design*, January, 1954.

The maximum shear stress is obtained by combining the foregoing principal stresses

$$s_s\,(\max) = \frac{E\delta}{2d_i} \tag{23-6}$$

The allowable stress depends on the material used. As before, for brittle material, the maximum stress should not exceed the ultimate tensile strength of the material. Equating the allowable stress for brittle material to the maximum induced tensile stress in the hub, Eq. (23-4), gives

$$\frac{s_{\text{ult}}}{\text{fs}} = \frac{E\delta}{2d_i}\left(1 + \frac{d_i^2}{d_o^2}\right) \tag{23-7}$$

where fs = factor of safety

s_{ult} = ultimate tensile strength of material, psi

The limiting value of the interference for any fit can therefore be determined.

The allowable shear stress for ductile material based on the maximum-shear theory of failure is given by Eq. (5-1). By equating this stress to the maximum shear stress given in Eq. (23-6), the conditions for yielding of a ductile hub are indicated:

$$\frac{s_y}{\text{fs}} = \frac{E\delta}{d_i} \tag{23-8}$$

where s_y = yield point, psi.

Example 23-2 Determine the maximum allowable interference between a steel shaft and hub if yielding is to be prevented. Assume steel with a yield point of 30,000 psi. Using Eq. (23-8) with a factor of safety of 1, since the yield conditions are to be found, then

$$\delta = \frac{s_y d_i}{E} = \frac{30{,}000 d_i}{30{,}000{,}000} = 0.001 d_i$$

23-4 FORCE FITS—STEEL SHAFT WITH CAST-IRON HUB

The use of unlike materials with different moduli of rigidity will yield a different pressure relation than indicated by the previous article. A safe equation for the contact pressure can be obtained if the ratio of the moduli of elasticity of steel and cast iron is assumed to be 3. A reduction of the Timoshenko[1] equation for a fit of two materials on this assumption gives

$$p = \frac{E_c\delta[1 - (d_i^2/d_o^2)]}{d_i[1.53 + 0.47(d_i^2/d_o^2)]} \tag{23-9}$$

[1] Timoshenko, *op. cit.*

where E_c is the modulus of elasticity of cast iron. The maximum stress is found by substituting this value for p in Eq. (23-1). Equating the results to the allowable stress for brittle material gives

$$\frac{s_{\text{ult}}}{\text{fs}} = \frac{E_c\delta[1 + (d_i^2/d_o^2)]}{d_i[1.53 + 0.47(d_i^2/d_o^2)]} \tag{23-10}$$

Example 23-3 Determine the maximum average interference between a steel shaft and a cast-iron hub if the outer hub diameter is twice the inner diameter. Assume a grade of cast iron with an ultimate tensile strength of 30,000 psi and a modulus of elasticity of 14,000,000 psi.

From Eq. (23-10), using the factor of safety as unity,

$$\begin{aligned}\delta &= \frac{s_{\text{ult}}d_i[1.53 + 0.47(d_i^2/d_o^2)]}{E_c[1 + (d_i^2/d_o^2)]} \\ &= \frac{30{,}000 \times d_i[1.53 + 0.47 \times (0.5)^2]}{14{,}000{,}000 \times [1 + (0.5)^2]} \\ &= 0.000283d_i\end{aligned}$$

23-5 HOLDING ABILITY OF FORCE AND SHRINK FITS

The torque that can be transmitted by a shrink or force fit without slipping between the hub and shaft can be estimated. Assuming that the contact pressure p is uniformly distributed, the total radial force between the surfaces of length L in contact is $p\pi d_i L$. The tangential force due to friction is the coefficient of friction times the radial force. The torque is the tangential force times the radius, or

$$T = \frac{fp\pi d_i^2 L}{2}$$

where T = torque transmitted, lb-in.
p = contact pressure, psi
d_i = diameter of shaft, in.
L = length of hub, in.
f = coefficient of friction, usually from 0.1 to 0.05

23-6 ASSEMBLY OF SHRINK FITS

Whenever fits are used with large interference or large shaft sizes, the force required to press the hub on the shaft is too large for practical assembly. To obviate this difficulty, assembly is frequently accomplished by shrinking the hub on the shaft. The hub is heated until it can be slipped on the shaft, whereupon it shrinks as it cools until a fit is obtained according to the interference specified on the cold parts. The inside diameter of the hub must be expanded by an amount at least as much as the interference before it is slipped on the shaft. Since a uniform heating of the hub increases all its dimensions, the increase in inside diameter for a temperature change ΔT is equal to $\alpha d_i \, \Delta T$, where α is the coefficient of thermal expansion. Therefore the minimum change in temperature for assembly is

$$\Delta T = \frac{\delta}{\alpha d_i} \tag{23-11}$$

where δ = diametral interference, in.

α = coefficient of expansion, in. per in. per °F

ΔT = change in temperature, °F

In assembly, the inside part may be cooled cither as an adjunct to or a substitute for heating of the external part. "Dry ice" (solid carbon dioxide) is widely used for this purpose, and temperature differences of 100°F are easily obtainable by cooling alone.

23-7 STATISTICAL QUALITY CONTROL IN PRODUCTION

In the dimensioning of parts to be made in production quantities, the designer is interested in the causes of variations in dimensions of the finished parts. These causes are considered in two categories: assignable causes and chance causes.

Assignable causes are differences in production or inspection processes from piece to piece or from lot to lot, as, for example, variations in mechanical properties of the material for the parts, variations in mounting the parts in the machine tool, inexperience of machinists, variations due to tool changes, wear of gauges, etc. These variations are not random and can usually be located and eliminated.

When assignable causes of variations are eliminated, the second type, *chance* causes, exist; although they cannot be predetermined and they occur at random, their distribution can be predicted by statistical methods as discussed briefly as follows:

Tolerance is the permissible variation in the size of a dimension, and it is important to keep it small enough to ensure satisfactory conditions of assembly of parts and performance of the unit, but large enough to keep production cost at a near minimum. By statistical methods, the quality-control engineer can determine how the *natural spread* in the dimensions of a manufactured part varies and the magnitude of its probable limits. The diagram in Fig. 23-8 was drawn from the data shown in columns (1), (2), and (3) of Table 23-1. The thickness of 600 brass washers was measured. It was noted that six of the washers were in the range of 0.0180 to 0.01839 in thickness, with 0.0182 as the midpoint (MP) of that group. Thirty washers had 0.0186 as midpoint of the range, etc. The largest group of washers, 120, had thicknesses whose midpoints were 0.0202 in.

Figure 23-8, called a *histogram*, shows the frequency of the thickness of the 11 groups, and the solid line drawn through the midpoints shows their variation. Note that the solid line is not a smooth curve nor is the area below the solid line symmetrical. If the manufacturer had measured a considerably larger number of samples, if he had plotted each value as measured instead of grouping measurements and plotting their midpoints, and if assignable causes of variations had been removed, the irregular line in Fig. 23-8 would have been a smooth and symmetrical curve.

Such a curve, known as the normal or gaussian curve, can be determined from a finite number of readings, and it is a very useful source of information for the quality-control engineer. The normal curve is superimposed by the dashed line in Fig. 23-8, and its determination and interpretation are the subject of the following article.

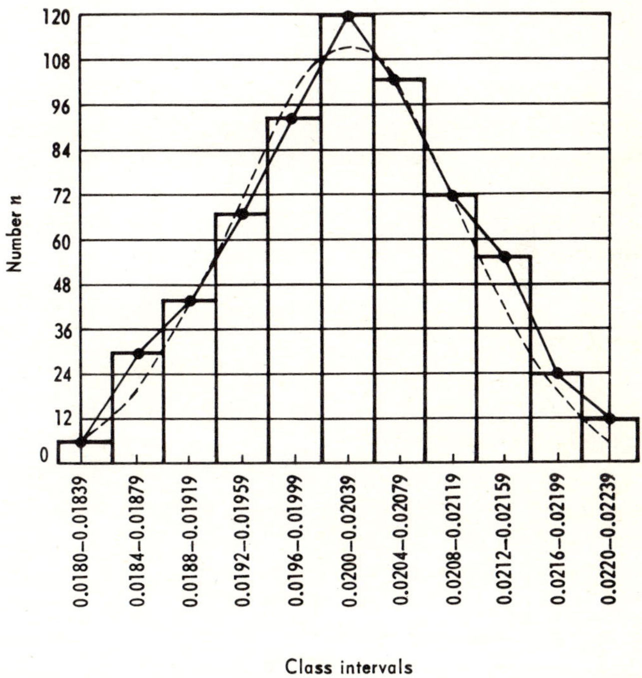

Fig. 23-8 Frequency distribution.

Table 23-1 Method for fitting of normal-curve area with variation of thickness in 600 brass washers manufactured by ABC Company*

(1) Thickness, in.	(2) Midpoints (MP)	(3) No. of washers, f	(3a) Frequency midpoint, $f \times$ (MP)	(4) Deviation of class limit from mean, x	(5) Col. (4) in terms of std. dev., x/σ	(6) % of area between class limit and mean	(7) % of area in class interval	(8) Theoretical frequency, (7) $\times\ \Sigma(3)$
0.0180–0.01839	0.0182	6	0.1092	−0.0022	−2.59	49.55	1.22	7.3
0.0184–0.01879	0.0186	30	0.5580	−0.0018	−2.12	48.30	3.25	19.5
0.0188–0.01919	0.0190	42	0.7980	−0.0014	−1.65	45.05	6.95	41.7
0.0192–0.01959	0.0194	66	1.2804	−0.0010	−1.18	38.10	11.98	71.9
0.0196–0.01999	0.0196	94	1.8424	−0.0006	−0.71	26.12	16.64	99.9
				−0.0002	−0.24	9.48		
0.0200–0.02039	0.0202	120	2.4240	0.0002	−0.24	9.48	18.96	113.8
0.0204–0.02079	0.0206	102	2.1012	0.0006	0.71	26.12	16.64	99.8
0.0208–0.02119	0.0210	60	1.2600	0.0010	1.18	38.10	11.98	71.9
0.0212–0.02159	0.0214	54	1.1556	0.0014	1.65	45.05	6.98	41.7
0.0216–0.02199	0.0218	14	0.3052	0.0018	2.12	48.30	3.25	19.5
0.0220–0.02239	0.0222	12	0.2664	0.0022	2.59	49.52	1.22	7.3
		n = 600	12.1004				99.10	594.2

* Hypothetical data based on smaller distribution given by W. A. Shewhart, "Economic Control of Quality of Manufactured Product," D. Van Nostrand Company, Inc., Princeton, N.J., 1931.

23-8 THE NORMAL CURVE[1]

If there are a very large number of data, such as measurements of a class of parts whose deviations are due to chance causes, and if the deviations are plotted in terms of their frequency, the resulting curve will be bell-shaped, symmetrical, and will extend to plus and minus infinity along the measurement axis, as shown by the dash-line curve in Fig. 23-8 and reproduced in Fig. 23-9. The name of the curve, *normal*, means regular or natural.

This curve, which can be fitted to a histogram as in Fig. 23-8, is of use to the quality-control engineer because, with a comparatively small number of readings or measurements, he can predict a very close estimate of the distribution of errors for a large number of samples. The normal curve can be expressed in terms of the number of parts measured and a quantity termed "standard deviation" σ.

The normal curve actually extends from minus infinity to plus infinity. However, practically all the curve extends from $\pm 3\sigma$, as shown in Fig. 23-9, in which 99.73 percent of the total area under the curve lies between $\pm 3\sigma$; 95.46 percent lies between $\pm 2\sigma$; and 68.26 percent lies between $\pm 1\sigma$.†

The arithmetic mean of the data in Table 23-1 may be found by multiplying the MP of each group by the frequency in that group and

[1] For a discussion of the mathematics of the normal curve, see M. F. Spotts, "Mechanical Design Analysis," Prentice-Hall, Inc., Englewood Cliffs, N.J., 1964.

† Herbert Arkin and Raymond R. Colton, "Statistical Methods," Barnes & Noble, Inc., New York, 1965.

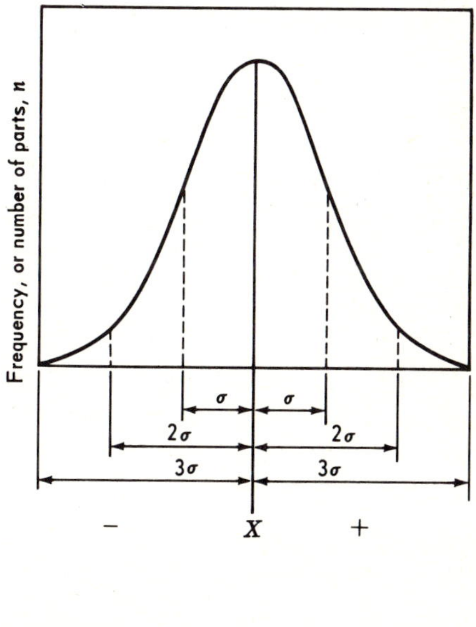

Fig. 23-9 Normal curve.

Table 23-2 Areas under normal curve

$\frac{x}{\sigma}$	Second decimal place in x/σ .00	.01	.02	.03	.04	.05	.06	.07	.08	.09
.0	.0000	.0040	.0080	.0120	.0160	.0199	.0239	.0279	.0319	.0359
.1	.0398	.0438	.0478	.0517	.0557	.0596	.0636	.0675	.0714	.0753
.2	.0793	.0832	.0871	.0910	.0948	.0987	.1026	.1064	.1103	.1141
.3	.1179	.1217	.1255	.1293	.1331	.1368	.1406	.1443	.1480	.1517
.4	.1554	.1591	.1628	.1664	.1700	.1736	.1772	.1808	.1844	.1879
.5	.1915	.1950	.1985	.2019	.2054	.2088	.2123	.2157	.2190	.2224
.6	.2257	.2291	.2324	.2357	.2389	.2422	.2454	.2486	.2517	.2549
.7	.2580	.2611	.2642	.2673	.2704	.2734	.2764	.2794	.2823	.2852
.8	.2881	.2910	.2939	.2967	.2995	.3023	.3051	.3078	.3106	.3133
.9	.3159	.3186	.3212	.3238	.3264	.3289	.3315	.3340	.3365	.3389
1.0	.3413	.3438	.3461	.3485	.3508	.3531	.3554	.3577	.3599	.3621
1.1	.3643	.3665	.3686	.3708	.3729	.3749	.3770	.3790	.3810	.3830
1.2	.3849	.3869	.3888	.3907	.3925	.3944	.3962	.3980	.3997	.4015
1.3	.4032	.4049	.4066	.4082	.4099	.4115	.4131	.4147	.4162	.4177
1.4	.4192	.4207	.4222	.4236	.4251	.4265	.4279	.4292	.4306	.4319
1.5	.4332	.4345	.4357	.4370	.4382	.4394	.4406	.4418	.4429	.4441
1.6	.4452	.4463	.4474	.4484	.4495	.4505	.4515	.4525	.4535	.4545
1.7	.4554	.4564	.4573	.4582	.4591	.4599	.4608	.4616	.4625	.4633
1.8	.4641	.4649	.4656	.4664	.4671	.4678	.4686	.4693	.4699	.4706
1.9	.4713	.4719	.4726	.4732	.4738	.4744	.4750	.4756	.4761	.4767
2.0	.4772	.4778	.4783	.4788	.4793	.4798	.4803	.4808	.4812	.4817
2.1	.4821	.4826	.4830	.4834	.4838	.4842	.4846	.4850	.4854	.4857
2.2	.4861	.4864	.4868	.4871	.4875	.4878	.4881	.4884	.4887	.4890
2.3	.4893	.4896	.4898	.4901	.4904	.4906	.4909	.4911	.4913	.4916
2.4	.4918	.4920	.4922	.4925	.4927	.4929	.4931	.4932	.4934	.4936
2.5	.4938	.4940	.4941	.4943	.4945	.4946	.4948	.4949	.4951	.4952
2.6	.4953	.4955	.4956	.4957	.4959	.4960	.4961	.4962	.4963	.4964
2.7	.4965	.4966	.4967	.4968	.4969	.4970	.4971	.4972	.4973	.4974
2.8	.4974	.4975	.4976	.4977	.4977	.4978	.4979	.4979	.4980	.4981
2.9	.4981	.4982	.4982	.4983	.4984	.4984	.4985	.4985	.4986	.4986
3.0	.4987	.4987	.4987	.4988	.4988	.4989	.4989	.4990	.4990	.4990

dividing their sum by the total number of readings, or

$$\bar{X} = \frac{\Sigma(f \times \text{MP})}{n} = \frac{12.1004}{600} = 0.0202 \tag{23-12}$$

The *deviation* from the mean for any particular value is $X - \bar{X}$, which is called x. Then, the standard deviation is calculated as

$$\sigma = \sqrt{\frac{x_1^2 + x_2^2 + \cdots + x_n^2}{n}} = \sqrt{\frac{\Sigma x^2}{n}} \tag{23-13}$$

The method for determining values to superimpose the normal curve on a histogram will be illustrated by using the histogram of Fig. 23-8 and a continuation of Table 23-1.[1] The method used for columns (4) to (8) for the variation in thickness of 600 brass washers applies to any statistical study of data. The principle in arriving at the normal curve is to express the deviations x of the mean values in terms of the standard deviation σ and to make use of Table 23-2 showing the areas under the normal curve for the intercepts x/σ. The results of this process are shown in Table 23-1 as follows, with the fourth line of Table 23-1 as an example:

Column (4): lowest limit, $0.0192 - \bar{X}$, or $0.0192 - 0.020 = -0.0010$, that is, x

Column (5): $\dfrac{x}{\sigma} = \dfrac{-0.0010}{0.00085^*} = -1.18$

Column (6): using Table 23-2, for $x/\sigma = -1.18$, the area under the normal curve = 38.10 percent.

Column (7): subtract line 5 from line 4, column (6); that is, $38.10 - 26.12 = 11.98$ percent, which is the percentage of area in the class interval.

Column (8): theoretical frequency in the class is 11.98 times the total number of samples, or $11.98 \times 600 = 71.9$. This represents the number of samples whose range of measurements lie between 0.0192 to 0.01959. The total for values in column (8) equals 594.2, or approximately 600 samples.

* See Table 23-3, for computation, $\sigma = 0.00085$, from Frank M. Gyrna, Jr.'s method, J. M. Juran (ed.), "Quality Control Handbook," 2d ed., McGraw-Hill Book Company, New York, 1962.

Since the normal curve is symmetrical about the ± 0 axis, values for frequencies need to be calculated for only one-half of the curve.

23-9 NATURAL SPREAD OF DIMENSIONS, TOLERANCES, AND SCRAP

As defined in Art. 23-7, tolerance is the total permissible variation in size of a manufactured part. The designer wants precise fits between mating

[1] Herbert Arkin and Raymond R. Colton, "Statistical Methods," Barnes & Noble, Inc., New York, 1965.

Table 23-3 Calculation table:* thickness of brass washers by ABC Company

Thickness X (midpoints)	*Frequency f*	d'	fd'	$f(d')^2$
0.0182	6	−5	−30	150
0.0186	30	−4	−120	480
0.0190	42	−3	−126	378
0.0194	66	−2	−132	264
0.0196	94	−1	− 94	94
0.0202	120	0	0	0
0.0206	102	+1	+102	102
0.0210	60	+1	+120	240
0.0214	54	+3	+162	486
0.0218	14	+4	+ 56	224
0.0222	12	+5	+ 60	300
	$n = 600$		$\Sigma fd' = -2$	$2{,}718 = \Sigma f(d')^2$

When $\bar{X}$ = arithmetic mean
X = midpoint
d' arbitrarily called 0
i = interval between X's = 0.0004

$$\bar{X} = X + \frac{\Sigma fd'}{n} i = 0.0202 + \frac{-2}{600} 0.0004 = 0.0202$$

$$\sigma = i \sqrt{\frac{\Sigma f(d')^2}{600} - \left(\frac{\Sigma fd'}{600}\right)^2}$$

$$= 0.0004 \sqrt{\frac{2{,}718}{600} - \left(\frac{-2}{600}\right)^2} = 0.000848$$

* Method of computation from J. M. Juran (ed.), "Quality Control Handbook," 2d ed., McGraw-Hill Book Company, New York, 1962.

parts but the shop likes as much latitude as possible in the production of parts. The quality-control engineer, whose objective it is to control the costs as well as quality of production, is interested in limiting the manufacture of scrap parts that will not work. If the natural spread equals the tolerance limits of the manufacturing process, then it is reliable or has the quality desired or has achieved reliability.[1]

As illustrated in Fig. 23-10(*a*), the natural spread of the dimensions is less than the tolerance so that the parts will work together but the cost

[1] A. V. Feigenbaum, "Total Quality Control," pp. 406–411, McGraw-Hill Book Company, New York, 1961.

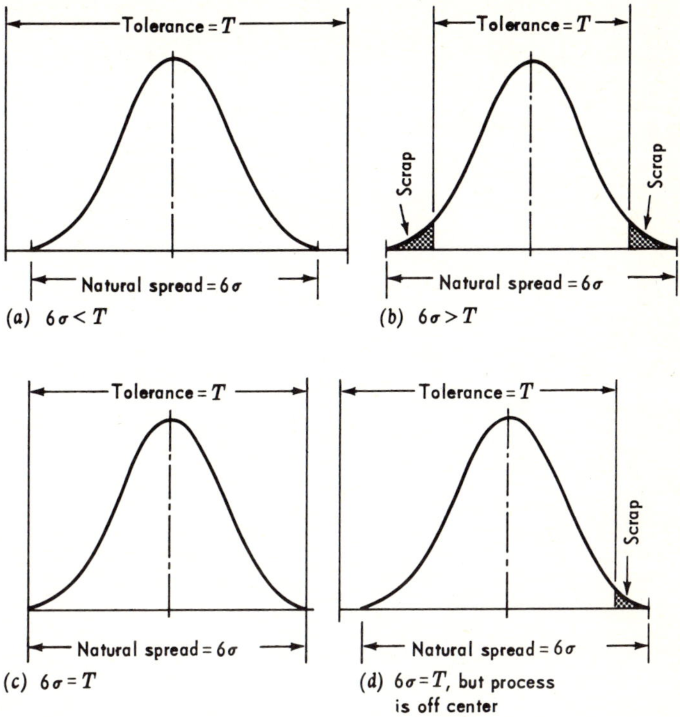

Fig. 23-10 Natural spread, tolerance, and scrap.

is high. As shown at (*b*) in the figure, the natural spread is greater than the tolerance. The cost of production is lowered but scrap is manufactured. An economic compromise may be the point where a limited amount of scrap is produced for the overall minimum cost. Figures 23-10(*a*), (*b*), and (*c*) illustrate "precisely centered" processes whereas (*d*) is off center, with a large amount of scrap.

24

Vibration and Vibration Control

24-1 INTRODUCTION

The major cause of vibration in machinery is the disturbing forces of reciprocating masses, unbalanced rotating masses, and variable fluid pressure. Frequently the disturbing forces can be removed or balanced; if not, their harmful effects can be reduced by proper design of the parts and their mountings.

The simplest vibrating system has one degree of freedom; i.e., the system's position can be specified by one number. This is illustrated by a weight suspended by a spring so that it can move only vertically; the weight can be located at any instant by its distance (one number) from a reference point. If the weight could rotate about a vertical axis at the same time that it moves up and down, it would require an additional number to specify its angular position at any instant and would represent two degrees of freedom. In general, if it requires n numbers to specify the position of a vibrating system, the system has n degrees of freedom.

The single-degree-of-freedom system will be considered here. Many mechanical systems are or can be reduced to this system, and the principles for the simple system can be extended to apply to more complicated ones.[1]

In our discussion, two frequencies are involved; one is the frequency of the disturbing force. This frequency is equal to or related to the operating speed of the machine, which is generally specified and which cannot be altered by the designer.

Another frequency, independent of the operating speed, is the natural frequency of vibration of the system. This frequency depends on the mass of the system and the stiffness of its support.

When the frequency of the disturbing force on a system coincides with its natural frequency, resonance occurs, and the system may vibrate with large amplitudes and develop forces larger than the disturbing forces. These forces may cause annoyance to personnel and failure of parts.

Vibration can often be controlled by removing or balancing the disturbing forces, as mentioned, or by changing the natural frequency of vibration so that it is sufficiently remote from the operating frequency to avoid unsatisfactory operating conditions.

24-2 EQUATION OF MOTION FOR A SYSTEM HAVING A SINGLE DEGREE OF FREEDOM

Consider a spring-supported weight arranged so that the weight can move only in the vertical direction, as shown in Fig. 24-1. The disturbing force is assumed to be $P_0 \sin \omega t$, which is impressed vertically at a time t by an eccentric mass rotating[2] with an angular velocity ω. The spring constant k is equal to the force required to extend the spring a unit distance. The motion of the weight is restrained by *damping*, which is assumed to vary directly with the velocity of the weight so that the

[1] J. P. Den Hartog, "Mechanical Vibrations," 4th ed., McGraw-Hill Book Company, New York, 1956.

[2] A convenient way of producing such a force is to rotate two eccentric masses at the same speed but in opposite directions, as shown in the sketch. The masses may be mounted on mating gears. The vertical components of the centrifugal forces equal $P_0 \sin \omega t$. The horizontal components are balanced. This system is used in an experimental vibrator.

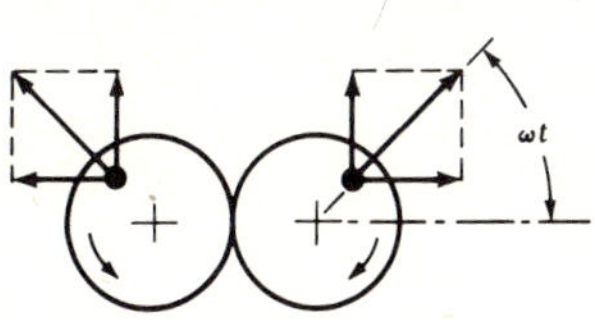

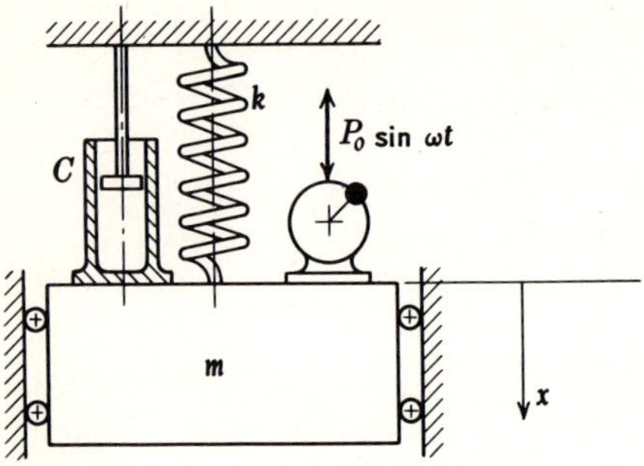

Fig. 24-1 Weight supported by spring.

damping force is the damping constant c multiplied by the velocity of the weight. This is known as "viscous" damping.

Let m = mass of weight
x = displacement of weight measured from its position of rest, positive downward
$\frac{dx}{dt}$ = velocity, represented by $\dot{x}$
$\frac{d^2x}{dt^2}$ = acceleration, represented by $\ddot{x}$

The forces on the weight are

$$\begin{aligned} \text{Spring force} &= -kx \\ \text{Damping force} &= -c\dot{x} \\ \text{Impressed force} &= P_0 \sin \omega t \end{aligned}$$

If the weight is displaced, the force on the weight equals $-kx - c\dot{x} + P_0 \sin \omega t$. This force must equal the accelerating force. Therefore

$$m\ddot{x} = -kx - c\dot{x} + P_0 \sin \omega t$$

or $$m\ddot{x} + c\dot{x} + kx = P_0 \sin \omega t \tag{24-1}$$

Equation (24-1) is the differential equation of motion for a system having a single degree of freedom.

24-3 FREE VIBRATION WITHOUT DAMPING

For this case, Eq. (24-1) becomes

$$m\ddot{x} + kx = 0$$

or $$\ddot{x} = -\frac{k}{m}x \tag{24-2}$$

The general solution of this equation is

$$x = C_1 \sin t\sqrt{\frac{k}{m}} + C_2 \cos t\sqrt{\frac{k}{m}} \tag{24-3}$$

To evaluate the constants C_1 and C_2, let

$$x = x_0 \qquad \text{when } t = 0 \tag{a}$$
$$\text{and} \quad \dot{x} = 0 \qquad \text{when } t = 0 \tag{b}$$

By substituting (*a*) in Eq. (24-3),

$$x_0 = C_1 \times 0 + C_2 \times 1$$
$$\text{or} \quad C_2 = x_0$$

Differentiating the general solution, Eq. (24-3),

$$\dot{x} = \sqrt{\frac{k}{m}}\, C_1 \cos t \sqrt{\frac{k}{m}} - \sqrt{\frac{k}{m}}\, C_2 \sin t \sqrt{\frac{k}{m}}$$

By substituting condition (*b*) in the last equation,

$$0 = \sqrt{\frac{k}{m}}\, C_1 \times 1 - \sqrt{\frac{k}{m}}\, C_2 \times 0$$
$$\text{or} \quad C_1 = 0$$

Therefore Eq. (24-3) becomes

$$x = x_0 \cos t \sqrt{\frac{k}{m}} \tag{24-4}$$

and is the equation of motion for a system having a single degree of freedom vibrating freely without damping. Figure 24-2 is a graphical representation of the motion.

In Eq. (24-4), one cycle of vibration occurs as $t\sqrt{k/m}$ goes through 2π rad. If T represents the period of one cycle,

$$T \times \sqrt{\frac{k}{m}} = 2\pi$$
$$\text{or} \quad T = 2\pi \sqrt{\frac{m}{k}}$$

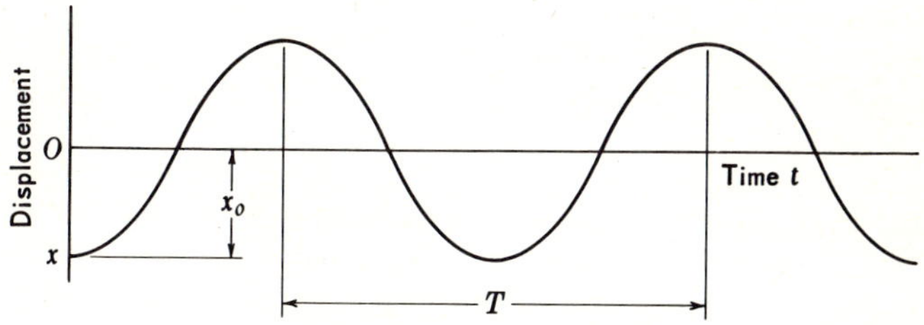

Fig. 24-2 Free undamped vibration.

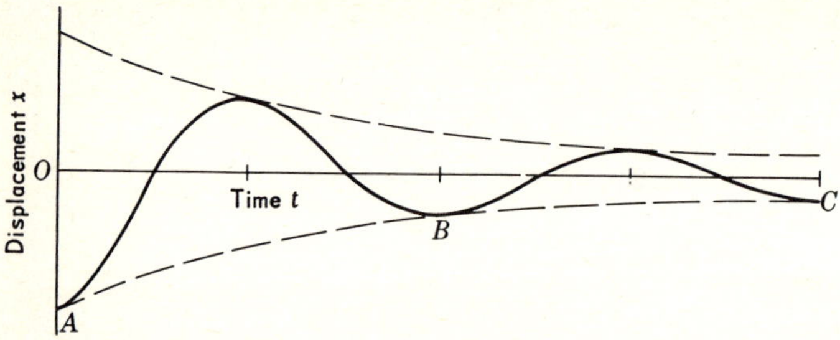

Fig. 24-3 Free vibration with small damping.

The natural frequency of the system is

$$f_n = \frac{1}{T} = \frac{1}{2\pi}\sqrt{\frac{k}{m}} \qquad \text{in cycles per sec} \tag{24-5a}$$

or $$\omega_n = \sqrt{\frac{k}{m}} \qquad \text{in radians per sec} \tag{24-5b}$$

24-4 FREE VIBRATION WITH DAMPING

For this case, Eq. (24-1) becomes

$$m\ddot{x} + c\dot{x} + kx = 0$$

When this equation is integrated, its solution for small damping, as found in machines, may be plotted as shown in Fig. 24-3. The figure shows how the amplitude of the vibrations is decreased in the presence of damping. The smaller the damping, the flatter will be the enveloping curve ABC and the longer will be the time required for the vibrations to die down. The natural frequency of vibration for this case is

$$\omega_n = \sqrt{\frac{k}{m} - \frac{c^2}{4m^2}} \tag{24-5c}$$

Values of c so large that $c^2/4m^2 \gtreqless k/m$ eliminates vibration. Such is the case for a pendulum placed in a very viscous fluid.

24-5 FORCED VIBRATION WITHOUT DAMPING

For this case, Eq. (24-1) becomes

$$m\ddot{x} + kx = P_0 \sin \omega t$$

The solution of this equation is

$$x = \frac{P_0/k}{1 - (\omega/\omega_n)^2} \sin \omega t$$

The quantity P_0/k is the deflection of the mass owing to a statically applied force P_0; hence, $x_s = P_0/k$, and the above equation may be written as

$$\frac{x}{x_s} = \frac{1}{1 - (\omega/\omega_n)^2} \sin \omega t \tag{24-6}$$

The maximum value of x occurs when the value of $\sin \omega t$ is a maximum, i.e., unity. Hence, letting x_0 represent the maximum value of x, Eq. (24-6) becomes

$$\frac{x_0}{x_s} = \frac{1}{1 - (\omega/\omega_n)^2} \tag{24-6a}$$

The graphical representation of Eq. (24-6*a*) is shown in Fig. 24-4 and is useful in interpreting characteristics of vibrating systems. In this figure, the ordinate x_0/x_s is the ratio of the maximum deflection of the mass from its position of rest to the deflection due to a statically applied force P_0. The abscissa ω/ω_n is the ratio of the frequency of the applied force

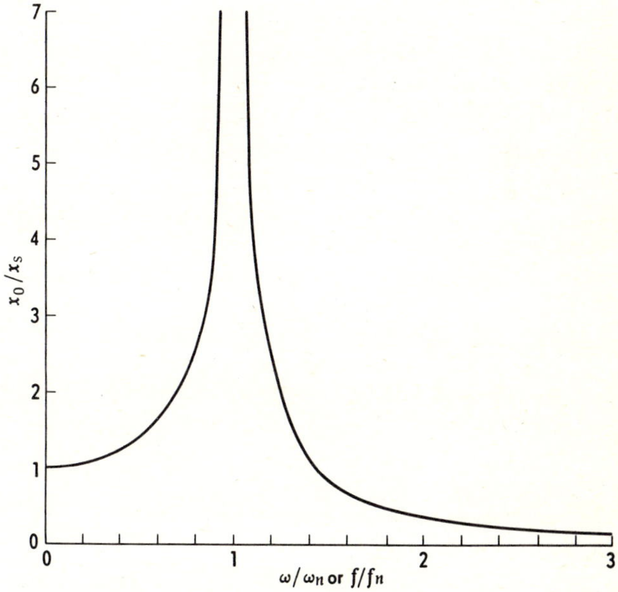

Fig. 24-4 Undamped forced vibration.

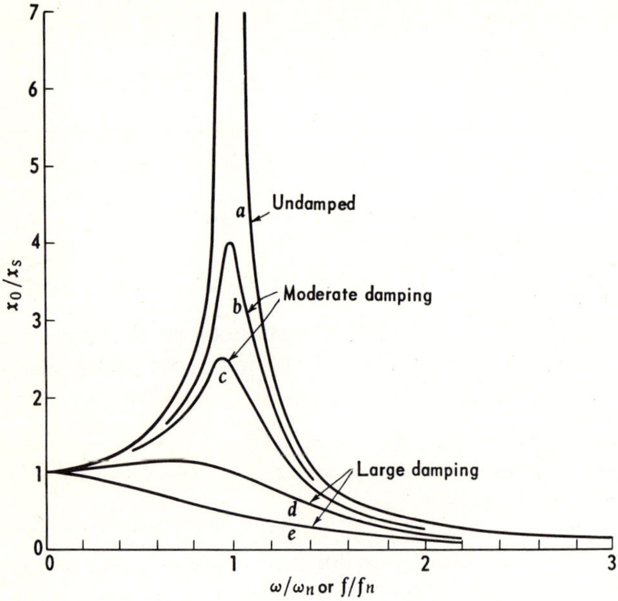

Fig. 24-5 Effect of damping on forced vibration.

P_0 to the natural frequency of the system. Since x_s and ω_n are constants of the system, it may be seen that Fig. 24-4 is a representation of the variation of the maximum deflection x_0 of a vibrating body with the frequency of application ω of the applied force. The figure shows that, when ω is zero, the deflection of the body is equal to the static deflection x_s, and when ω is equal to ω_n, the deflection in the absence of damping is infinite. For operating frequencies greater than $\sqrt{2}\ \omega_n$, the deflection of the body becomes less than x_s. In using Eq. (24-6*a*) for the region where $\omega/\omega_n > 1$, it is necessary to change the sign of the right-hand side of the equation. The physical interpretation of this change is that the deflection and the impressed force are out of phase when ω/ω_n is greater than unity.

24-6 FORCED VIBRATION WITH DAMPING

When damping is present in the system in the form of viscous resistance or of friction, the effect of the damping force is to decrease the amplitude of motion of the vibrating mass. In viscous damping, the resistance to motion of the vibrating mass is proportional to the velocity of the mass. Since high velocities accompany large displacements, viscous damping will affect the curve of Fig. 24-4, as shown in Fig. 24-5.

The equation for the maximum deflection with damping is

$$x_0 = \frac{P_0}{\sqrt{(c\omega)^2 + (k - m\omega^2)^2}}$$

In mechanical systems, there is always some inherent damping present; hence the zero-damping curve is never realized. Curves *d* and *e* represent degrees of damping that are higher than those found in most machinery. The damping in most machinery falls in the intermediate stages, as shown by curves *b* and *c*. These curves show the reduction in amplitude to finite values as the critical speed is passed.

24-7 LATERAL AND TORSIONAL VIBRATIONS

The vibration of a spring-supported weight, as discussed in the preceding article, is an illustration of lateral vibration. Other examples are the vibration of the string of a musical instrument, as shown in Fig. 24-6(*a*), and a beam or shaft vibrating laterally, as shown at (*b*). Another type of vibration in mechanical systems is torsional vibration, illustrated by the motion of a torsional pendulum, as shown at (*c*), in which the mass oscillates through an angle as indicated. In machinery, a shaft may

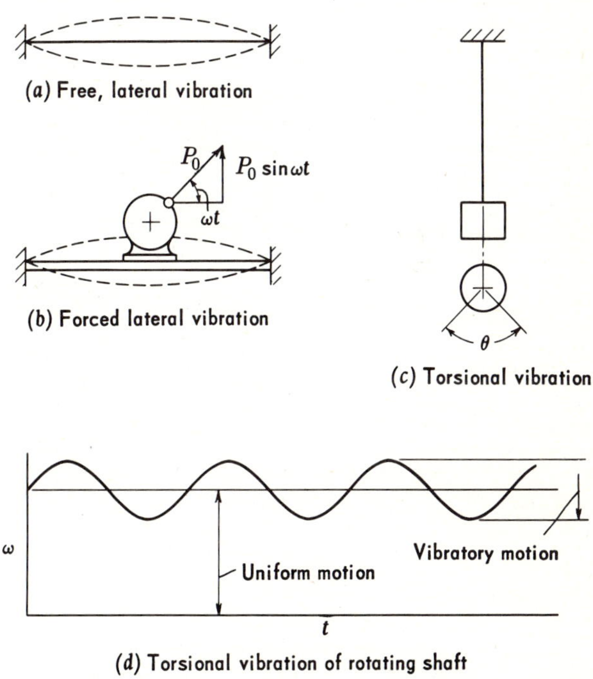

Fig. 24-6 Lateral and torsional vibrations.

rotate uniformly, in which case there is no torsional vibration, or the shaft may have superimposed on its uniform motion one of vibratory character which gives a motion similar to that represented by the curve in Fig. 24-6(*d*) in which ω represents the angular velocity of the shaft. Torsional vibration is quite common and also troublesome in machinery having rotating parts. The frequency of torsional vibration is usually greater than the speed of rotation of the shaft.

The natural frequency in torsional vibration of the system, as shown in Fig. 24-6(*c*), depends on the weight and radius of gyration of the suspended mass, and on the stiffness of the supporting rod. The natural frequency may be determined from the following equation:

$$\text{Natural frequency, cps} = \frac{1}{2\pi}\sqrt{\frac{GJg}{LW\rho^2}}$$

where G = torsional modulus of elasticity of supporting rod, psi
J = polar moment of inertia of rod, $\pi d^4/32$ for circular rod, in.^{4}
g = acceleration due to gravity = 386 in. per sec^2
L = length of supporting rod, in.
W = weight of disk, lb
ρ^2 = radius of gyration of weight, squared
ρ = D^2/d, in.^{2}

Some common examples of forced vibration in machinery are the lateral vibration of a mass, such as an engine resting on its foundation, or the lateral vibration of a shaft subjected to variable loads, or the torsional vibration of a shaft subjected to variable torque. Figure 24-5 was developed by considering a spring-supported weight under forced and damped vibration. However, the quantitative aspects of the vibration phenomena illustrated by the figure apply to all types of vibration, torsional as well as lateral. These vibrations are common in machinery and may be destructive because of the large forces and stresses involved and because vibration induces repeated stresses that may lead to fatigue failure. In addition to the effects of vibration on the machine producing them, the vibrations may have annoying and destructive effects on surrounding equipment.

There is extensive literature dealing with mechanical vibrations of systems with more than one degree of freedom, in addition to the references given in this chapter.[1]

[1] See also F. S. Tse, I. E. Morse, and R. T. Hinkle, "Mechanical Vibrations," Allyn and Bacon, Inc., Boston, 1963; A. H. Church, "Mechanical Vibrations," John Wiley & Sons, Inc., New York, 1963; and an extensive collection of valuable discussions in C. M. Harris and C. E. Crede (eds.), "Shock and Vibration Handbook," 3 vols., McGraw-Hill Book Company, New York, 1961.

24-8 CRITICAL SPEEDS OF SHAFTS

In Eq. (24-5*a*), the natural frequency of vibration of a spring system was given as

$$f_n = \frac{1}{2\pi}\sqrt{\frac{k}{m}} \qquad \text{cps}$$

By neglecting the weight of a shaft to which, for instance, a flywheel is attached, Eq. (24-5*a*) may be used to determine the natural frequency of vibration, or critical speed, where m is the mass of the flywheel and k is the number of pounds required to deflect the weight 1 in.

By substituting in the above equation the static deflection δ in inches of the shaft at the point where the flywheel is located, its equivalent, $12mg/k$, the undamped natural frequency becomes

$$f_n = 187.8\sqrt{\frac{1}{\delta}} \qquad \text{cpm} \tag{24-7}$$

It is evident that a stiff shaft will have a small δ and that the natural frequency will be high, whereas a flexible shaft will have a large δ and the corresponding natural frequency will be low (see Fig. 24-7).

As shown by Eq. (24-5*c*), one effect of damping is to lower the natural frequency of vibration. For the degrees of damping in most practical systems, the difference between the damped and the undamped natural frequencies will be small, and thus Eq. (24-7) may be used for a close approximation to the critical speed of a shaft where the mass can be considered concentrated at one point as a heavy flywheel attached to a light shaft.

For a uniformly loaded shaft in transverse vibration, the mass cannot be considered concentrated at one point, so that it is necessary to consider the deflection of each unit mass of the shaft. For a uniformly loaded shaft in transverse vibration, the critical speed is given by the following equation:[1]

$$f_n = 211.4\sqrt{\frac{1}{\delta}} \qquad \text{cpm} \tag{24-8}$$

where δ is the maximum deflection of the shaft.

The above equation indicates the lowest, or fundamental, frequency of vibration at which it vibrates, as shown in Fig. 24-6(*a*). The vibrating shaft may have nodes at the midpoint, the third points, etc., when it vibrates at frequencies 4 and 9, etc., respectively, times the fundamental frequency.

[1] Thomas Bevin, "Theory of Machines," p. 492, Longmans, Green & Co., Inc., New York, 1939.

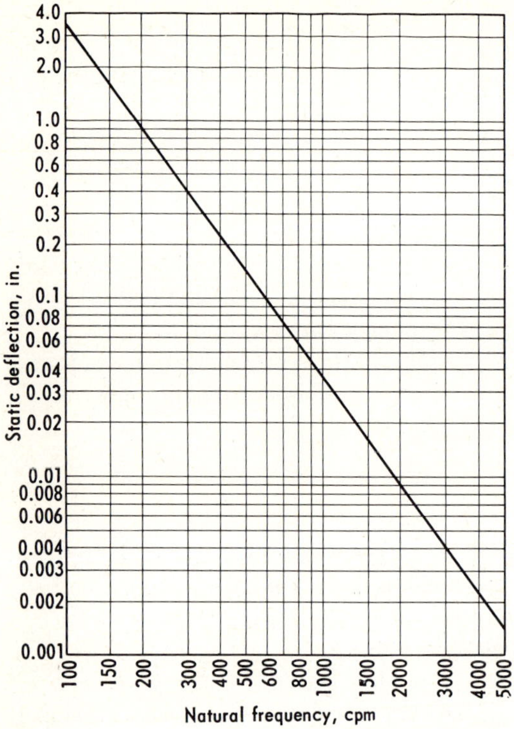

Fig. 24-7 Lowest natural frequency versus static deflection.

24-9 FORCE TRANSMITTED TO SUPPORTS

Since a rigid mounting transmits all the forces impressed on it, an "unsprung" machine subjected to vibration, as represented by Fig. 24-8(*a*), transmits all the impressed force $P_0 \sin \omega t$ directly to the foundation. The introduction of springs between the machine base and foundation alters the transmitted force, and the introduction of damping, such as a dashpot, further alters the transmitted force [see Fig. 24-8(*b*) and (*c*)].

The amplitude of the force transmitted to the foundation and the impressed force is termed *transmissibility*.

$$\text{Transmissibility} = T = \frac{\text{transmitted force}}{\text{impressed force}}$$

$$T = \sqrt{\frac{1 + \left(2 \dfrac{c}{c_c} \dfrac{\omega}{\omega_n}\right)^2}{\left(1 - \dfrac{\omega^2}{\omega_n^2}\right)^2 + \left(2 \dfrac{c}{c_c} \dfrac{\omega}{\omega_n}\right)^2}} \tag{24-9}$$

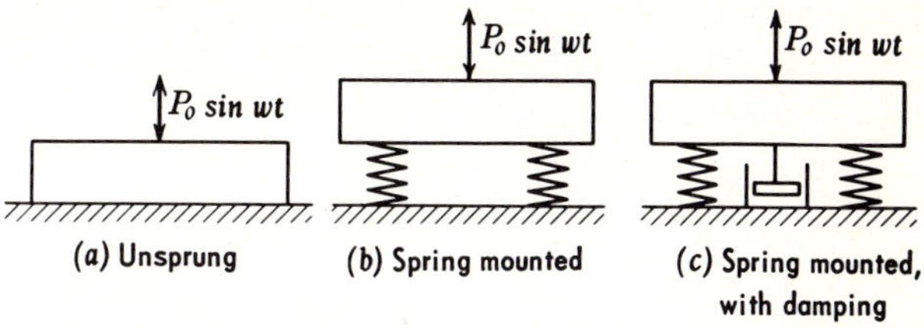

Fig. 24-8 Vibration mountings.

In Fig. 24-9, transmissibility is plotted for various degrees of damping as a function of ω/ω_n. In the figure, it may be noted that, for operating speeds exceeding $\sqrt{2}\ \omega_n$, damping is somewhat detrimental. A comparison of the magnitude of transmissibility for the two ranges cited shows that, as a machine passes through the critical speed, the beneficial effects of damping are pronounced, whereas for speeds higher than the critical speed, the detrimental effects of damping are not marked. The foregoing illustrates that a vibration-absorbing mounting should have damping properties. For speeds several times the critical speed, the transmitted force is a small fraction of the impressed force, and the machine may be said to "run smoothly."

The effect of damping on transmissibility is usually small for the region $\omega/\omega_n > \sqrt{2}$. Equation (24-9) may be reduced to the following

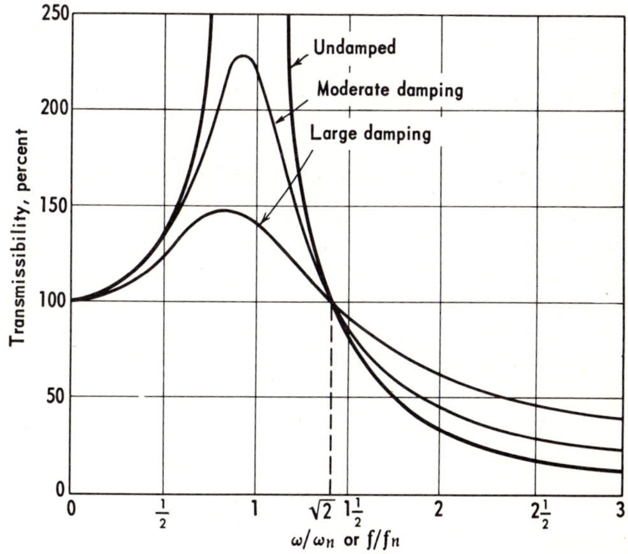

Fig. 24-9 Transmissibility and effect of damping.

form by making the damping coefficient c zero and changing the sign of the right-hand side of the equation for the reason given in Art. 24-5. Therefore

$$T = \frac{1}{(\omega/\omega_n)^2 - 1} = \frac{1}{(f/f_n)^2 - 1} \tag{24-10}$$

24-10 VIBRATION CONTROL

In order to control or to eliminate vibration the following means are available:

1. Remove the disturbing force.
2. Introduce an equal and opposite disturbance.
3. Reduce the effect of the disturbance transmitted to the foundation or support by introducing flexibility and damping.

An example of method 1 is balancing rotors by removing material from the heavy part of a rotor. A simple example of method 2 is a balancing weight placed on the light side of an unbalanced rotor. There are other examples of this method which amount to introducing in a machine or instrument a balancing system that vibrates automatically in such a manner that the effects of the forced vibration are neutralized.

Method 3 is generally used when vibration in machines must be controlled. It may be noted from Fig. 24-9 that, if the normal operating speed f coincides with the natural frequency of the machine f_n, the condition of operation is undesirable. However, by introducing a flexible mounting, the natural frequency may be altered, so that any point along the ω/ω_n axis of Fig. 24-9 can be fixed as the normal operating condition.

In torsional vibration, the natural frequency may be increased by making the shaft more rigid by increasing its diameter, or the natural frequency may be decreased by making the shaft less rigid by decreasing its diameter, by increasing its length, or by introducing a flexible coupling.

Example As an illustration of design procedure, consider a single-cylinder vertical engine that operates at 1,200 rpm and weighs 800 lb. If the maximum unbalanced force in the vertical direction, P_0, equals 30 lb and the engine is mounted directly to the floor, the total force P_0 will be transmitted. If it is desired to reduce the transmitted force to 3 lb (that is, $T = 0.10$), Eq. (24-10) may be used as

$$f_n = \frac{f}{\sqrt{(1+T)/T}} = \frac{1{,}200}{\sqrt{(1+0.1)/0.1}} = \frac{1{,}200}{3.32} = 362 \text{ cpm}$$

Hence, the mounting should be selected so that the 800-lb compressor as a mass vibrates with a natural frequency of 362 cpm.

By solving Eq. (24-7) for the static deflection δ and substituting values,

$$\delta = \left(\frac{187.7}{f_n}\right)^2 = \left(\frac{187.7}{362}\right)^2 = 0.269 \text{ in.}$$

If the engine is mounted on four springs, the spring rate for each spring becomes

$$\frac{800}{4 \times 0.269} = 745 \text{ lb per in.}$$

The maximum amplitude of vibration, from Eq. (24-6*a*), is

$$x_0 = \frac{x_s}{(f/f_n)^2 - 1} = \frac{0.269}{(1{,}200/362)^2 - 1} = 0.116 \text{ in.}$$

The connections for the compressor should allow for this movement.

24-11 VIBRATION-ABSORBING MOUNTINGS

In machines subjected to linear vibrations, such as reciprocating engines, pumps, compressors, etc., it is usually impractical to control vibration by methods 1 and 2 of Art. 24-10. Method 3 is commonly used by employing vibration-absorbing mountings. As previously discussed, the mountings should have flexibility and damping. In addition, the mounting should be so constructed that the machine would not be freed if the resilient part of the mounting should fail. In some mountings, flexibility is secured by mechanical springs, and damping is secured by dashpots or by rubber or cork buffers (see Fig. 24-10).

Other types of mountings employ rubber, cork, or felt for the resilient part. Rubber has a high ratio of deflection to load, especially

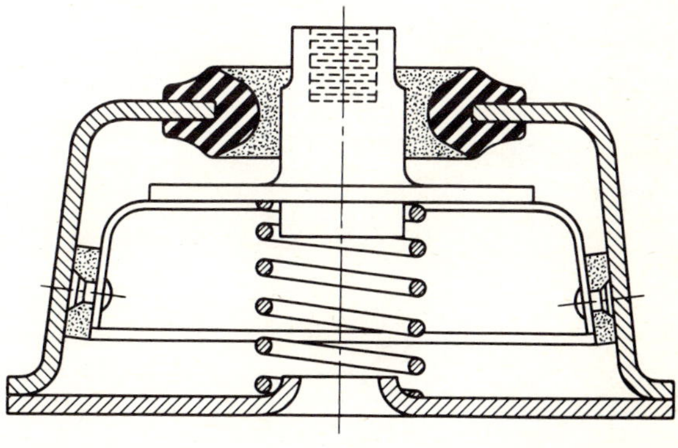

Fig. 24-10 Vibration mounting. (*Courtesy of Robinson Vibrashock Division, Robinson Technical Products, Inc.*)

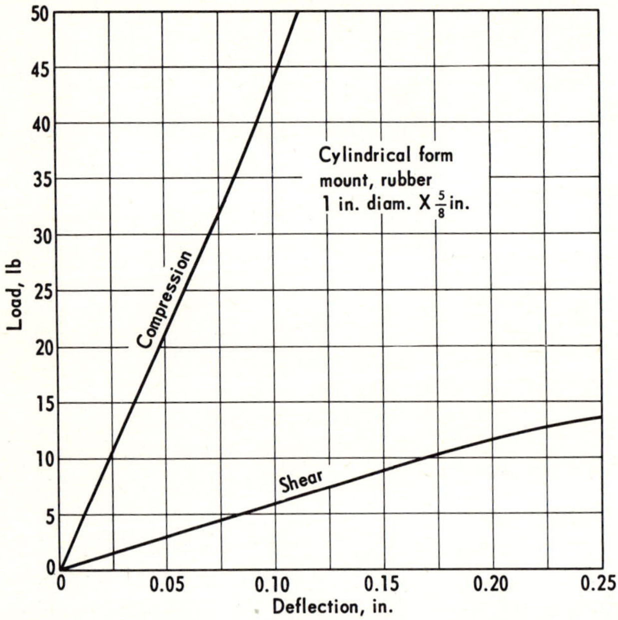

Fig. 24-11 Load-deflection characteristics of cylindrical form of rubber mounting.

when it is used in shear (see Fig. 24-11), and has in addition good damping characteristics. The tube form of mounting shown in Fig. 24-12 has the rubber bonded to the steel tube in the center, which is bolted to the supported member, and to the outer sleeve, which is attached to the supporting member. The plate form in Fig. 24-13 is used particularly for light and medium loads.

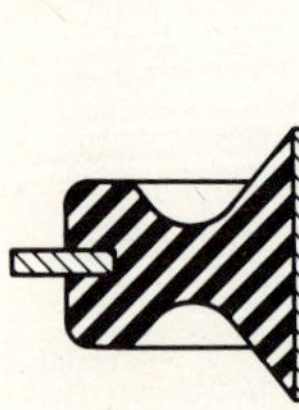

Fig. 24-12 Vibration mounting, tube form.

Fig. 24-13 Vibration mounting, plate form.

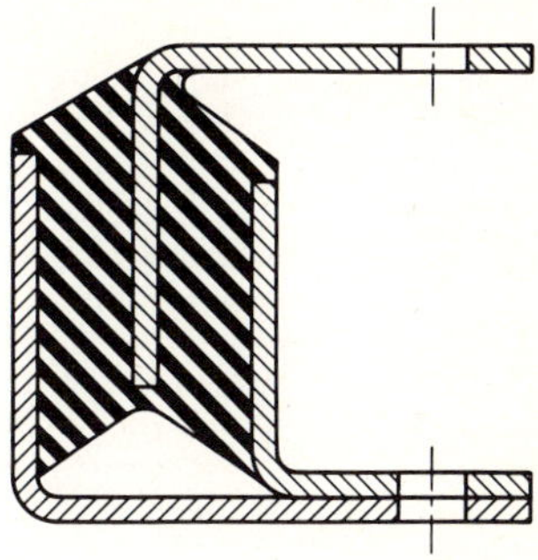

Fig. 24-14 Three-angle type of mounting using rubber in shear.

The three-angle type of rubber mounting, as shown in Fig. 24-14, is suitable for medium and heavy loads.

Many other types of vibration mountings are available in a large variety of forms and sizes and with load capacities that vary from a few ounces, for mounting delicate apparatus and instruments, to 32,000 lb or more for power-plant installations.

24-12 TORSIONAL VIBRATION DAMPERS

It has already been noted that the critical speed in torsion of a shaft can be lowered by making the shaft more flexible in torsion by reducing its diameter, by increasing its length, or by using a flexible coupling, such as a rubber coupling or a Falk coupling. By lowering the critical speed, the normal operating point as shown in Fig. 24-9 is moved to the right into the region of low transmissibility, thus reducing the harmful effects of torsional vibration. To supply damping in addition to that of the material of the rotating parts and their supports, a viscous damper or a Coulomb-friction damper may be used.

There are many other types of torsional-vibration dampers, such as the energy-dissipation type, rotating-pendulum type, and fluid type.[1]

[1] W. K. Wilson, "Practical Solution of Torsional Vibration Problems," vols. I, II, III, Barnes & Noble, Inc., New York, 1956, 1963, 1965.

25 Motor Selection

25-1 MOTOR RATING

The horsepower rating of an electric motor is designated on the basis of its output at rated speed and voltage, with the manufacturer's guarantee that the maximum temperature rise in the motor will not exceed a specified value. The temperature rise for general-purpose motors is 40°C for dripproof motors, 50°C for splashproof motors, and 55°C for totally enclosed motors. It is assumed that the ambient temperature does not exceed 20°C. These are *continuous* ratings.

Intermittent ratings are special ratings for time periods, as for 1 hr, ½ hr, 15 min, etc. This kind of rating is common for motors having intermittent loads, such as crane motors, elevators, and some types of machine tools, so that they may be operated for a short time at relatively high loads followed by a cooling period. The rating and corresponding temperature rise are stamped on the name plate.

25-2 MOTOR TYPES AND APPLICATIONS

The selection of the proper type, horsepower, voltage, and speed, as well as the proper controls, is a most important matter, and the choice may spell the difference between a long life of satisfactory operation or, on the other hand, a continuous round of trouble.

In the design of many machines, the selection of a motor drive and the relation between the motor characteristics and the performance of the machine require careful consideration.[1]

This chapter does not pretend to do more than to bring to the attention of the machine designer some important factors involved in motor selection, in particular for the squirrel-cage induction motor.

In general, direct-current motors are used where special operating characteristics, such as good speed control, are required or where direct current is the only kind available.

By far the bulk of industrial motors are of the alternating-current type. Single-phase current is usually satisfactory up to ½ hp for across-the-line starting and up to 1 hp with resistors or autotransformer starting.

25-3 INDUCTION MOTORS

The most common type of motor for industrial use is the squirrel-cage induction motor. This motor is simple, rugged, and foolproof, and the current is induced in the rotor bars by induction, so that slip rings or commutator is not required. Table 25-1 gives information to assist in selecting National Electrical Manufacturers Association (NEMA) class A, B, C, and D squirrel-cage induction motors for various applications. Figure 25-1 shows curves for approximate speed-torque characteristics for these motors. Class D motor, known as a high-torque high-slip induction motor, is suited for use with a flywheel drive, for example, in a punch press. Since a flywheel cannot transfer kinetic energy to the load without slowing down in the process, it is necessary to use a motor that is capable of speed variation to transfer enough energy to the system to carry the load through the peak in the duty cycle. The motor must also be capable of accelerating the system back to its original speed, thereby restoring the kinetic energy to the flywheel before the next peak in the duty cycle occurs. The high-torque high-slip induction motor is well suited to this purpose.

The *wound-rotor* induction motor is used to drive variable-speed equipment such as hoists, cranes, centrifugal pumps, and fans. The

[1] See C. C. Libby, "Motor Selection and Application," McGraw-Hilll Book Company, New York, 1960. Also see Electric Motors, March 19, 1964, and Electric Controls Book Issue, June 14, 1962, both published in *Machine Design*.

Table 25-1 Characteristics and applications of squirrel-cage induction motors*

NEMA class	*Starting torque*	*Breakdown torque*	*Starting current*	*Slip*	*Application*	*On such machines as:*
A	Normal	High	Normal	Low	For moderately easy-to-start loads; for loads requiring slightly more than full-load starting torque and low slip; relatively high breakdown torque to sustain occasional overloads, where higher starting current can be tolerated	Generators Pumps Lathes Drill presses Grinders Machine tools Conveyors Compressors
B	Normal	Normal	Low	Low	For moderately easy-to-start loads; for loads requiring slightly more than full-load starting torque and low slip; relatively high breakdown torque to sustain occasional overloads	Generators Pumps Lathes Drill presses Grinders Machine tools Conveyors Compressors
C	High	Normal	Low	Low	For hard-to-start loads; for loads requiring high starting torque but not requiring high overload demands after running speed is reached. Not suitable for loads requiring long accelerating time	Electric stairways Pulverizers Compressors Conveyors
D	High	......	Low	High	For hard-to-start intermittent loads and shock loads; no sharply defined breakdown torque point. Surge load causes appreciable decrease in speed with high torque developed to recover speed rapidly. Available in intermittent and continuous ratings	Hoists Elevators Punch presses Machines with large flywheels Centrifuges

* From Westinghouse Electric Corporation.

three-phase windings on the rotor are brought out to three slip rings which are connected to variable external resistors.

25-4 CONTROLS AND MOUNTINGS

Small induction motors may be started across the line, but larger motors require starters employing resistors or autotransformers. Motors may be bolted directly to the bedplate, or adjustable bases may be provided for chain, belt, or gear drives. Many motors are made with special bases or flanges when they are to form an integral part of a machine or appliance.

25-5 STANDARD SIZES AND DIMENSIONS

Motors are built into standard-size frames so that one frame may be used for more than one horsepower rated at different speeds. For instance, frame 145 is used for the following motors:

2 hp at 3,600 rpm
1½ hp at 1,800 rpm
1 hp at 1,200 rpm
¾ hp at 900 rpm

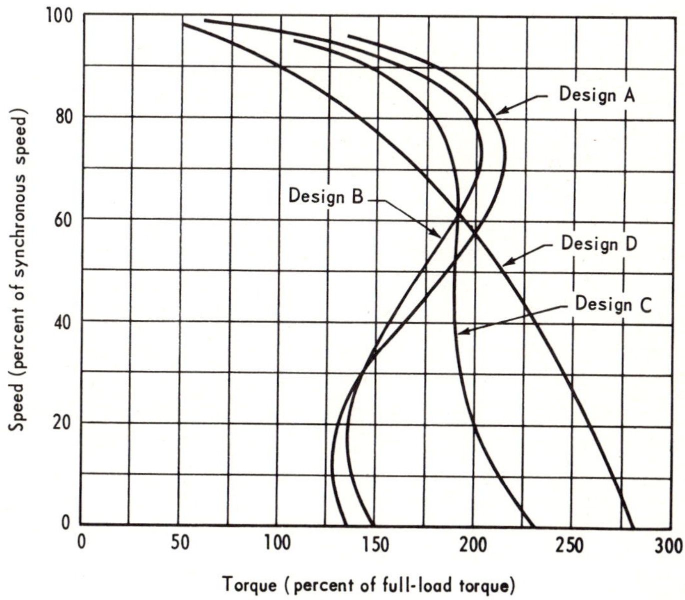

Fig. 25-1 Speed-torque characteristics for induction motors.

Table 25-2 Standard horsepower, speeds, frame numbers, and dimensions for polyphase, squirrel-cage induction motors, Types A and B, NEMA ratings (1965)

	3,600 *rpm*		*O* and E†*				*Dimensions*	
hp	*O**	*E†*	1,800 *rpm*	1,200 *rpm*	900 *rpm*	*Frame size*	*A‡*	*U§*
½	. . .	. . .	. . .	. . .	143	143	7	¾
¾	. . .	. . .	. . .	143	145	145	7	¾
1	. . .	. . .	143	145	182	182	9	⅞
1½	143	143	145	182	184	184	9	⅞
2	145	145	145	184	213	213	10½	1⅛
3	145	182	182	213	215	215	10½	1⅛
5	182	184	184	215	254	254	12½	1⅜
7½	184	213	213	254	256	256	12½	1⅜
10	213	215	215	256	284	284	14	1⅝
15	215	254	254	284	286	286	14	1⅝
20	254	256	256	286	324	324	16	1⅝
25	256	284	284	324	326	326	16	1⅝
30	284	286	286	326	364	364	18	1⅞
40	286	324	324	364	365	365	18	1⅞
50	324	326	326	365	404	404	20	2⅛
60	326	364	364	404	405	405	20	2⅛
75	364	365	365	405	444	444	22	2⅜
100	365	405	404	444	445	445	22	2⅜

* *O*: open motors
† *E*: enclosed motors
‡ *A* = outside diameter of motor, in.
§ *U* = shaft-extension diameter, in. For other designs and motor types, see National Electric Manufacturer's Standards MG1-11.31a.

The price of each of these motors will not vary more than 10 percent from the average of the four.

Therefore, for low motor cost, it is desirable to use high motor speed. However, the driven unit usually operates at comparatively low speed, so that with high-speed motors a greater reduction in speed is usually necessary; this increases the cost of the necessary speed reducer and its maintenance. The choice of motor speed therefore requires a study of the comparative economics.

Table 25-2 gives standard horsepowers, speeds, frame numbers, outside diameters of motors, and diameters of shaft stub ends. The frame numbers are according to NEMA standards.

25-6 SELECTION OF MOTOR SIZE

Considerable care should be used in specifying the size of motor. If it is too small, it will overheat, which will shorten its life; if it is larger than required, first cost will be high, its efficiency will be low, and, for an induction motor, the power factor will be low, thus unduly increasing the cost of operation.

Open 40C integral-horsepower motors are good for *continuous service* greater than the rated horsepower by the following factors: 1 hp, 1.25; 1½ and 2 hp, 1.20; 3 to 200 hp, 1.15.[1]

Power factor On account of the reduction in power factor at part load, which is undesirable in that it increases power loss in the lines and transformers and reduces the kilowatt output of the power-supply generator, it is better to select motors to avoid, if possible, sustained part-load operation. This is particularly desirable in a plant in which a large number of small motors are used. As an example, a 1-hp induction motor, 1,750 rpm, has power factors as follows: full load, 73 percent; ¾ load, 65 percent; and half load, 50 percent.

Efficiency The efficiency of motors increases with the size of motor, speed, and load up to full load. Thus, large, high-speed motors operating at full load are most efficient.

Intermittent rating Since the horsepower rating of a motor is based on temperature rise, advantage may be taken of cooling periods if the duty cycle requires partial load and rest periods in regular cycles. Since most of the heating of the motor is caused by copper losses due to the load current, which (losses) depend on the square of the load current, it is customary to calculate an average horsepower on the basis of the root-mean-square (rms) value. The motor may be selected on the rms horsepower rather than on the maximum-load horsepower.

The rms horsepower for repetitive duty may be determined by multiplying the square of the horsepower for each part of the cycle by its duration in seconds. Divide the sum of these results by the total time in seconds for a cycle. The square root of this last result is the rms value. If the motor is stopped during the cycle, only one-third (for open motors) of the standstill time should be added in the time for one cycle. This is because of the reduction in cooling when the motor is at rest.

In using the rms horsepower to choose a size of motor, the duty cycle should be short as compared with the time required for the motor to reach a steady temperature. Also, if there are extremely high peaks in the duty cycle, such as in drop-forge and punch-press applications, the motor may stall at the peak load. In this case a flywheel should be used.

[1] See NEMA Specification, pt. 4, p. 9.

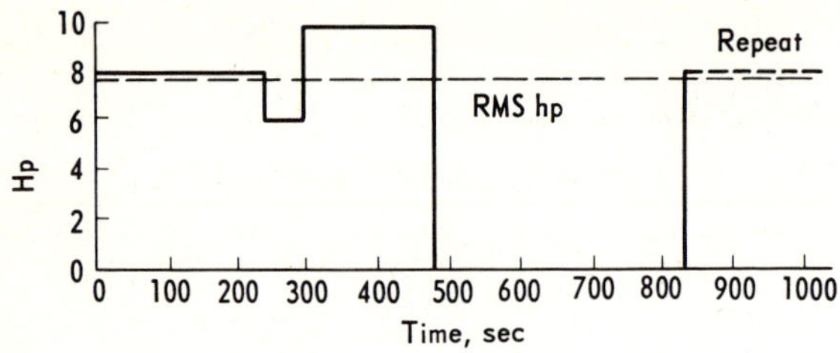

Fig. 25-2 Motor duty cycle and corresponding rms horsepower.

Example Assume a machinery operation in which a motor operates at 8-hp load for 4 min, 6-hp for 50 sec, 10-hp for 3 min, running idle for 30 sec, and standstill for 5½ min. Select a motor size for this repeated cycle of operation (see Fig. 25-2).

$$\text{rms hp} = \sqrt{\frac{(8^2 \times 240) + (6^2 \times 50) + (10^2 \times 180)}{240 + 50 + 180 + 30 + 330/3}}$$
$$= \sqrt{57.5} = 7.6 \qquad \text{Use } 7\tfrac{1}{2}\text{-hp motor}$$

During the 10-hp part of the cycle, the motor will be operating at $(10/7.5) \times 100 = 133.5$ percent of full-load torque. By Fig. 25-1, the pull-out torque for design B motor is 200 percent of full-load torque so that there will be no danger of the motor stalling at the 10-hp part of the cycle.

25-7 CONNECTED ROTATING MASSES

For fast, repeating cycles that require rapid acceleration of rotating masses, it may be necessary to calculate the time required for acceleration. In this case it is usually convenient to make calculations on the basis of torque rather than horsepower.

From the equation for the rate of change of angular momentum, the equation for the time required to accelerate uniformly a rotating body subjected to a constant torque may be found:

$$T = I\alpha$$

where T is the torque, I is the polar moment of inertia of mass, and α is the angular acceleration. By substituting for I its equal $(W/g)R^2$, where W is the weight of the mass and R its radius of gyration, the following equation may be derived for the time required to accelerate uniformly the rotating mass for constant applied torque:

$$t = \frac{WR^2 \times (\text{change in rpm})}{308T} \tag{25-1}$$

where t = time for acceleration, sec
W = weight of rotating mass, lb

R = radius of gyration, ft
T = constant torque, lb-ft

Equations for determining WR^2 for various bodies are given in Appendix XI. Values for WR^2 for standard motor rotors may be found in manufacturers' catalogs.

It is beyond the scope of this book to present methods for the determination of the heating and cooling cycle and the resulting temperature rise in the motor windings. The heat dissipation and temperature rise depend on the thermal characteristics of the motor and its electrical controls and characteristics, on the service (continuous or intermittent), whether starting is frequent, and on the inertia of the load. The calculation of the motor acceleration time may require a "curve matching" technique for the motor-accelerating torque curve and the load torque.[1]

[1] See Libby, *op. cit.*

appendix I

Abbreviations[1]

Abbreviations should be used sparingly in text material and *only* when preceded by numerals. Use lower case for abbreviations except where the word indicated requires a capital. Abbreviations representing compound adjectives should be hyphenated. Periods are not used in abbreviations with few exceptions, as *in.* for inches.

Absolute	abs	Atmosphere	atm
Alternating-current (as compound adjective)	a-c	Average	avg
		Barometer	bar
American Society of Mechanical Engineers	ASME	Boiler pressure	bp
		Brake horsepower	bhp
American Standards Association[2]	ASA	Brinell hardness number	Bhn
		British thermal unit	Btu
Ampere	amp	Calorie	cal

[1] Based largely on USAS list.

[2] Name has been changed to United States of America Standards Institute.

Centimeter	cm
Coefficient	coef
Constant	const
Cubic feet per minute	cfm
Cubic inches	cu in.
Cycles per second	cps
Cylinder	cyl
Degrees	deg
Degree centigrade	C
Degree Fahrenheit	F
Diameter	diam
Direct-current (as compound adjective)	d-c
Dozen	doz
Efficiency	eff
Electromotive force	emf
Equation	Eq.
External	ext
Feet	ft
Feet per minute	fpm
Feet per second	fps
Figure	Fig.
Foot	ft
Foot-pound	ft-lb
Gallon	gal
Gallons per minute	gpm
Gallons per second	gps
Gram	g
High pressure	h-p
Horsepower	hp
Horsepowerhour	hphr
Hour	hr
Inch or inches	in.
Inches per second	ips
Indicated horsepower	ihp
Intermediate-pressure	i-p
Internal	int
Kilogram	kg
Kilovolt	kv
Kilovolt-ampere	kva
Kilowatt	kw

Kilowatthour	kwhr
Liquid	liq
Logarithm (common)	log
Logarithm (natural)	ln
Maximum	max
Mean effective pressure	mep
Meters	m
Miles per hour	mph
Millimeter	mm
Millivolt	mv
Minimum	min
Minute	min
Ounce	oz
Percent	spell out
Pint	pt
Pound or pounds	lb
Pounds per square inch	psi
Quart	qt
Radians	spell out
Revolutions per minute	rpm
Second	sec
Shaft horsepower	shp
Society of Automotive Engineers	SAE
Specific gravity	sp gr
Specific heat	sp ht
Square	sq
Square centimeter	sq cm
Square foot	sq ft
Square inch	sq in.
Standard	std
Temperature	temp
United States of America Standards Institute	USASI
Volt	spell out
Watt	spell out
Watthour	whr
Weight	wt
Yard	yd
Year	yr

appendix II

Elements of sections

Section	Rectangular		Polar	
	Moment of inertia I	$\frac{I}{c}$	Moment of inertia J	$\frac{J}{r}$
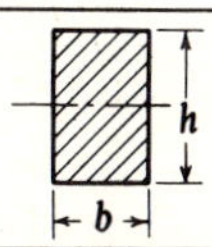	$\frac{bh^3}{12}$	$\frac{bh^2}{6}$		
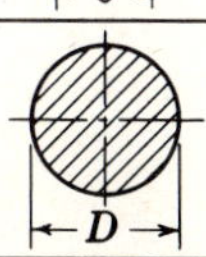	$\frac{\pi D^4}{64}$	$\frac{\pi D^3}{32}$	$\frac{\pi D^4}{32}$	$\frac{\pi D^3}{16}$
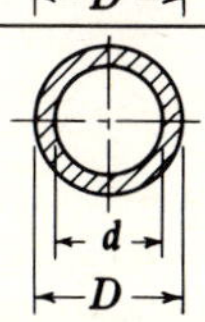	$\frac{\pi(D^4 - d^4)}{64}$	$\frac{\pi(D^4 - d^4)}{32D}$	$\frac{\pi(D^4 - d^4)}{32}$	$\frac{\pi(D^4 - d^4)}{16D}$

appendix III

Bending-moment and deflection formulas

M = maximum bending moment, lb-in.
δ = maximum deflection, in.
P = concentrated load, lb
W = uniformly distributed load, lb
L = length of beam, in.
E = modulus of elasticity, psi
I = moment of inertia of section, in.^{4}

Beam loading and support	M	δ
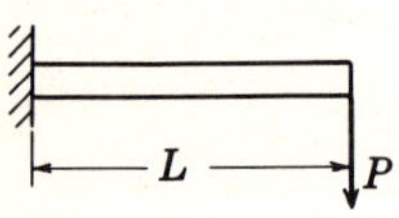	PL	$\frac{PL^3}{3EI}$
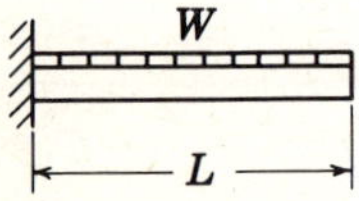	$\frac{WL}{2}$	$\frac{WL^3}{8EI}$
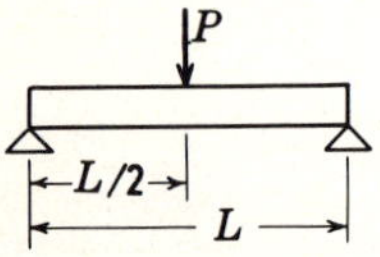	$\frac{PL}{4}$	$\frac{PL^3}{48EI}$
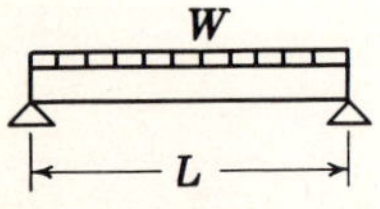	$\frac{WL}{8}$	$\frac{WL^3}{76.8EI}$
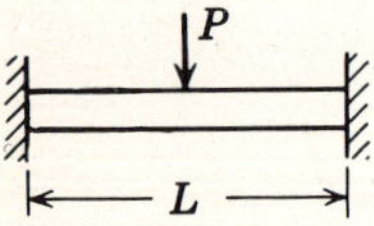	$\frac{PL}{8}$	$\frac{PL^3}{192EI}$
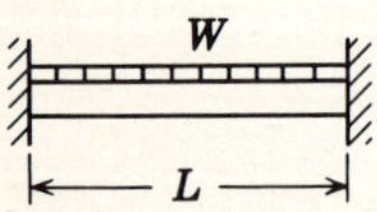	$\frac{WL}{12}$	$\frac{WL^3}{384EI}$

appendix IV Values of $\pi d^3/16$ and $\pi d^3/32$

Values of $\pi d^3/16$ (d = diameter of shaft)

d	0	1/16	1/8	3/16	1/4	5/16	3/8	7/16	1/2	9/16	5/8	11/16	3/4	13/16	7/8	15/16	d
0	0	0.000048	0.00038	0.0013	0.0031	0.006	0.0104	0.0164	0.0245	0.0349	0.0479	0.0638	0.0828	0.1053	0.1315	0.1618	0
1	0.196	0.236	0.280	0.329	0.384	0.444	0.510	0.583	0.663	0.749	0.843	0.944	1.052	1.169	1.294	1.428	1
2	1.571	1.723	1.884	2.055	2.236	2.428	2.630	2.843	3.068	3.304	3.551	3.811	4.083	4.368	4.666	4.977	2
3	5.301	5.639	5.992	6.359	6.740	7.136	7.548	7.975	8.416	8.877	9.352	9.845	10.35	10.88	11.42	11.99	3
4	12.57	13.16	13.78	14.42	15.07	15.75	16.44	17.16	17.89	18.65	19.42	20.22	21.04	21.88	22.75	23.63	4
5	24.54	25.47	26.43	27.41	28.41	29.44	30.49	31.56	32.66	33.79	34.94	36.12	37.33	38.56	39.82	41.10	5
6	42.41	43.75	45.12	46.51	47.93	49.39	50.87	52.38	53.92	55.49	57.09	58.72	60.38	62.08	63.80	65.56	6
7	67.35		71.02		74.82		78.76		82.83		87.04		91.39		95.89		7
8	100.5		105.3		110.3		115.3		120.6		126.0		131.5		137.3		8
9	143.1		149.2		155.4		161.8		168.3		175.1		182.0		189.1		9
10	196.4		203.8		211.4		219.3		227.3		235.5		243.9		252.5		10
11	261.3		270.3		279.6		289.0		298.6		308.5		318.5		328.8		11
12	399.3		350.0		361.0		372.1		383.5		395.1		407.0		419.1		12
13	431.4		443.9		456.8		469.8		483.1		496.6		510.4		524.5		13
14	538.8		553.3		568.2		583.3		598.6		614.2		630.1		646.3		14
15	662.7		679.4		696.4		713.6		731.2		749.0		767.1		785.6		15

Values of $\pi d^3/32$ (d = diameter of shaft)

d	0	1/16	1/8	3/16	1/4	5/16	3/8	7/16	1/2	9/16	5/8	11/16	3/4	13/16	7/8	15/16	d
0	0	0.000024	0.00019	0.00065	0.00154	0.003	0.0052	0.0082	0.0123	0.0175	0.0239	0.0319	0.0414	0.0527	0.0658	0.0809	0
1	0.098	0.118	0.14	0.164	0.192	0.222	0.255	0.292	0.331	0.375	0.421	0.472	0.526	0.585	0.647	0.714	1
2	0.785	0.862	0.942	1.028	1.118	1.214	1.315	1.422	1.534	1.652	1.776	1.906	2.042	2.184	2.333	2.489	2
3	2.651	2.82	2.996	3.18	3.37	3.568	3.774	3.988	4.208	4.439	4.676	4.923	5.176	5.44	5.712	5.995	3
4	6.283	6.58	6.892	7.21	7.535	7.876	8.220	8.580	8.946	9.326	9.712	10.11	10.52	10.94	11.38	11.82	4
5	12.27	12.74	13.22	13.71	14.20	14.72	15.24	15.73	16.33	16.89	17.42	18.06	18.66	19.28	19.91	20.55	5
6	21.21	21.88	22.56	23.26	23.97	24.70	25.44	26.19	26.96	27.75	28.55	29.36	30.19	31.04	31.90	32.78	6
7	33.68		35.51		37.41		39.38		41.42		43.52		45.70		47.95		7
8	50.27		52.66		55.13		57.67		60.29		62.99		65.77		68.63		8
9	71.57		74.59		77.70		80.90		84.17		87.54		90.99		94.53		9
10	98.18		101.9		105.7		109.6		113.6		117.8		122.0		126.3		10
11	130.7		135.2		139.8		144.5		149.3		154.2		159.3		164.4		11
12	169.7		175.0		180.5		186.0		191.8		197.6		203.5		209.5		12
13	215.7		222.0		228.4		234.9		241.6		248.3		255.2		262.2		13
14	269.4		276.7		284.1		291.6		299.3		307.1		315.1		323.1		14
15	331.3		339.7		348.2		356.8		365.6		374.5		383.6		392.8		15

appendix V

Properties of materials

Material	No.	Description	Tension: Ultimate strength*	Tension: Elastic limit*	Endurance limit (bending)*	Modulus of elasticity in tension*	Elongation in 2 in., percent
Cast iron	20 ASTM	$\frac{1}{2}$-in. section	20,000	6,200	6,500	10,000,000	
	20 ASTM	$<\frac{1}{2}$ in.	24,000	7,500	7,500	12,000,000	
	25 ASTM	$\frac{1}{2}$-in. section	25,000	10,000	9,500	13,000,000	
	25 ASTM	$<\frac{1}{2}$ in.	30,000	12,000	11,000	14,000,000	
	50 ASTM	$\frac{1}{2}$-in. section	50,000	30,000	18,000	20,000,000	
Malleable cast iron	....	....	50,000	20,000	24,000	25,000,000	
Nickel cast iron	Grade II	....	35,000	17,000	13,000	16,000,000	
Cast steel	0.20 % C	Annealed	60,000	30,000	24,000	29,000,000	22
	0.30 % C	Annealed	70,000	35,000	28,000	29,000,000	18
	0.40 % C	Annealed	80,000	40,000	32,000	29,000,000	15
Wrought steel	SAE 1010	Hot-rolled	54,000	31,000	See Appendix VII	30,000,000	36
	SAE 1020	Hot-rolled	62,000	35,000		30,000,000	30
	SAE 1030	Annealed	75,000	42,000		30,000,000	26
	SAE 1040	Annealed	90,000	50,000		30,000,000	22
	SAE 1050	Annealed	95,000	52,000		30,000,000	20
	SAE 1095	Annealed	120,000	60,000		30,000,000	20
	SAE 1095	Drawn at 900°F	150,000	80,000		30,000,000	16
Nickel steel	SAE 2320	Annealed	70,000	45,000		30,000,000	29
	SAE 2320	Drawn at 1000°F	120,000	80,000		30,000,000	25
	SAE 2340	Annealed	95,000	55,000		30,000,000	26
	SAE 2340	Drawn at 1000°F	120,000	95,000		30,000,000	22

* Values in psi.

For representative carbon steels, the above values are given by the Society of Automotive Engineers (SAE) numbers. These are the same as the American Iron and Steel Institute (AISI) numbers except that the AISI number carries a letter prefix which indicates the process by which the steel is made. The values in the table are nominal and are altered by heat treatment and size of members.[1]

[1] See "SAE Handbook," Society of Automotive Engineers, New York, 1966; "Metals Handbook," American Society for Metals, Cleveland; and Joeseph H. Faupel, "Engineering Design," John Wiley & Sons, Inc., New York, 1964; also, "Isothermal Transformation Diagrams," U.S. Steel Corporation, Pittsburgh, Pa., 1963.

appendix **VI**

Tensile strength and yield point versus Brinell hardness for steel[1]

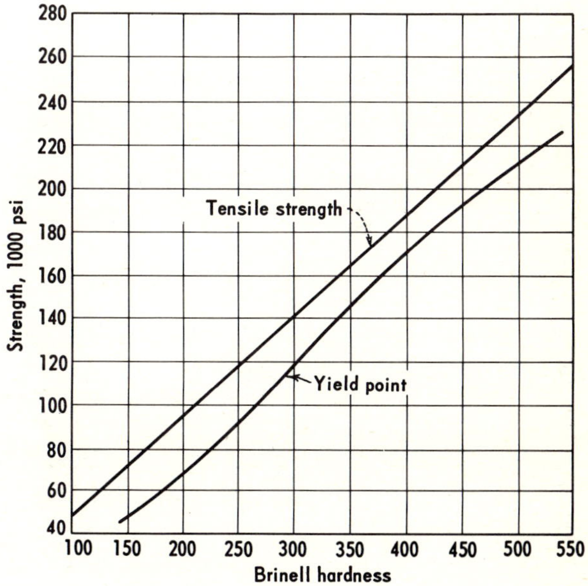

[1] From Noll and Lipson, Allowable Working Stresses, *Proc. Soc. Expt. Stress Anal.*, vol. III, no. II, 1946.

appendix VII

Endurance limit versus tensile strength[1]

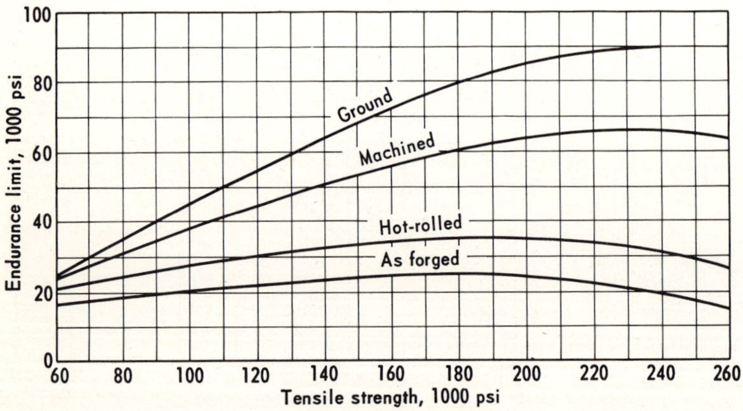

[1] From Noll and Lipson, Allowable Working Stresses, *Proc. Soc. Expt. Stress Anal.*, vol. III, no. II, 1946.

appendix VIII

Yield points for steel

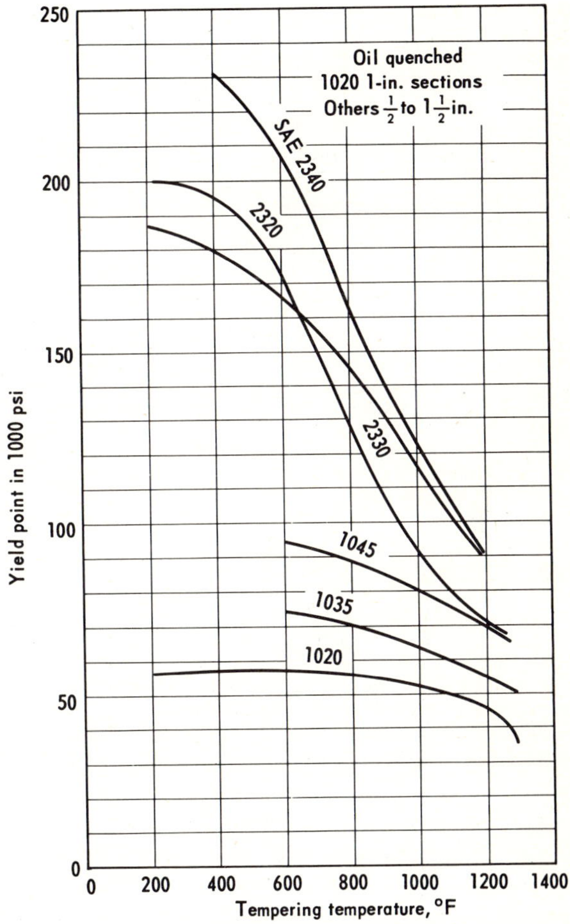

Fig. VIII-1 Yield points for some oil-quenched steels.

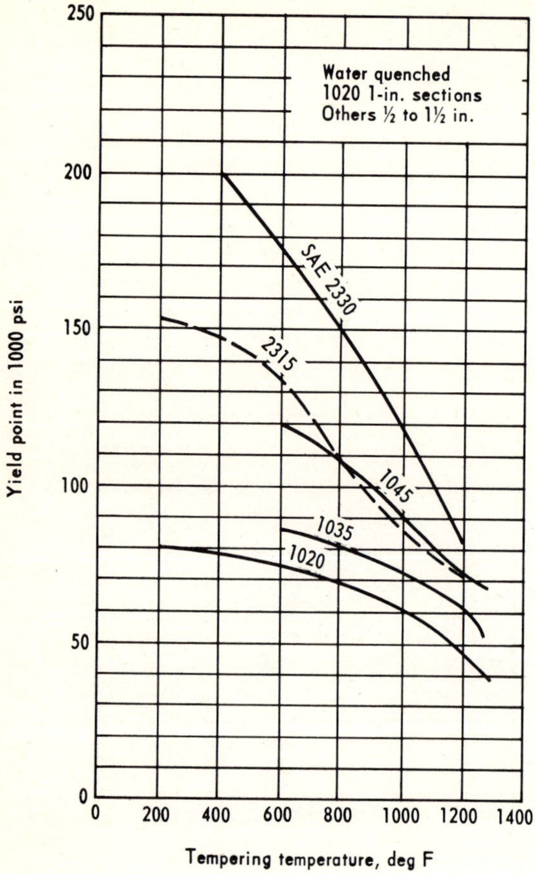

Fig. VIII-2 Yield points for some water-quenched steels.

appendix IX

Hardness numbers conversion chart[1]

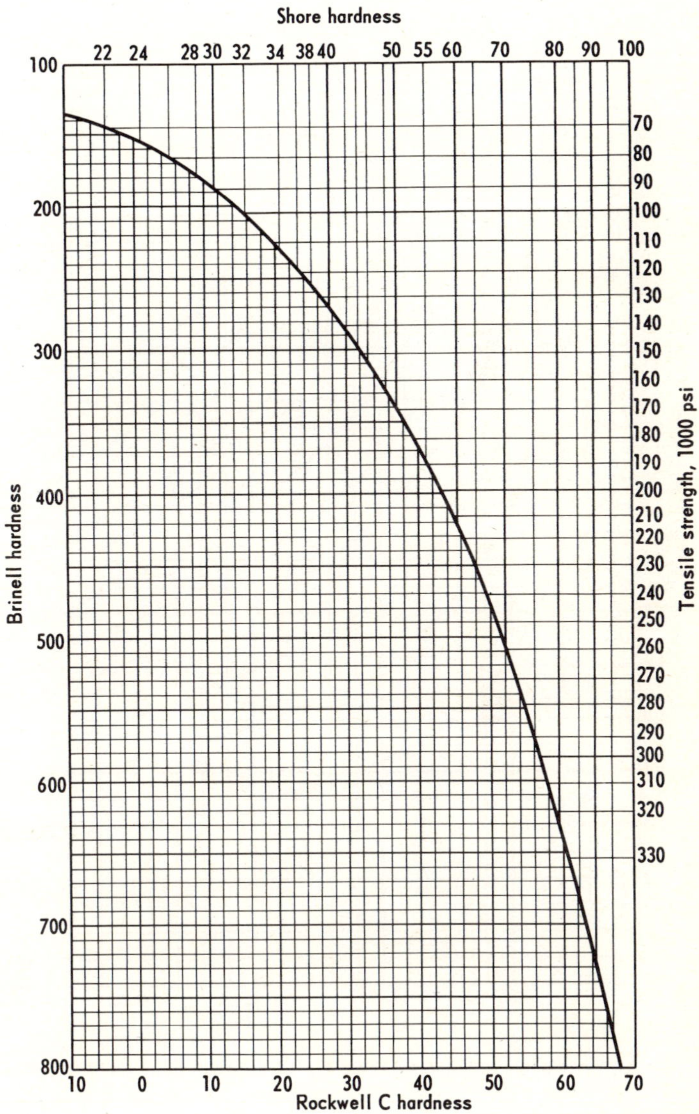

[1] Courtesy of International Nickel Company, Inc.

appendix X

Curves for stress-concentration factors[1]

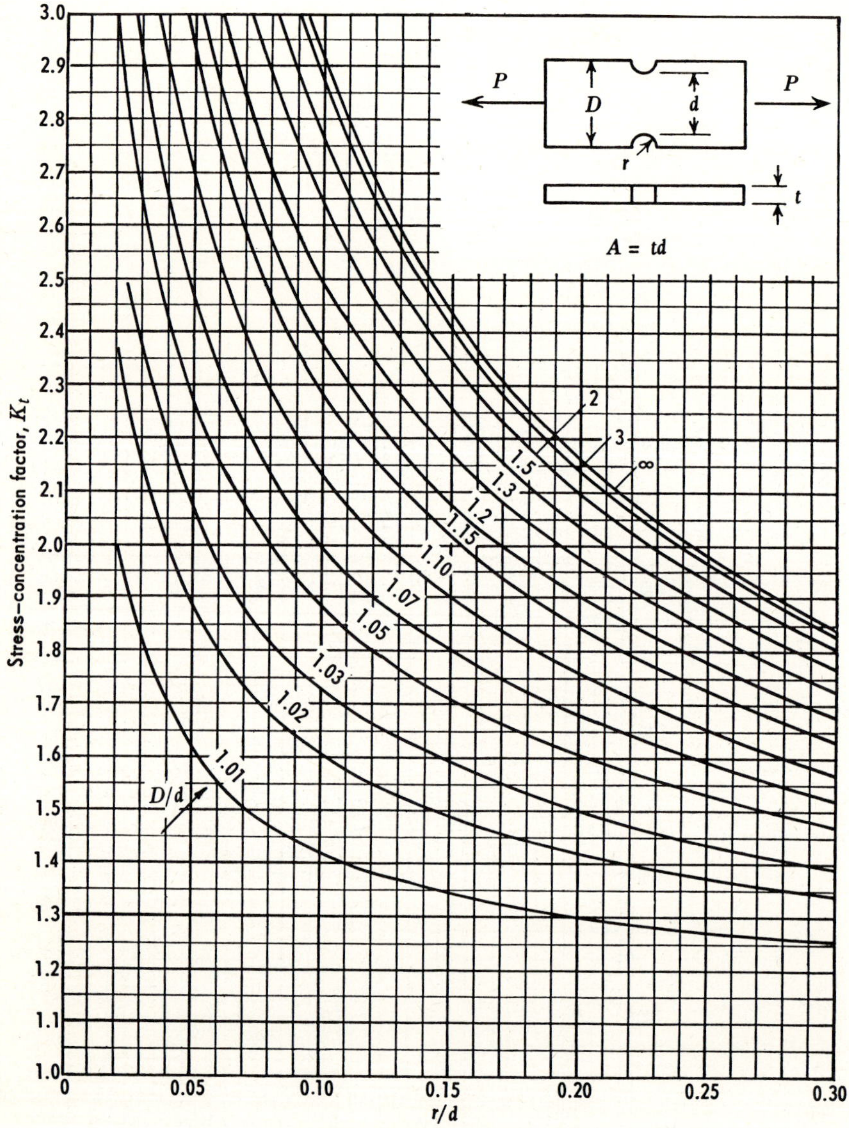

Fig. X-1 Stress-concentration factor K_t for notched flat bar in tension.

[1] Curves taken by permission of the author and publisher of "Stress-concentration Design Factors," by R. E. Peterson, John Wiley & Sons, Inc., New York, 1953.

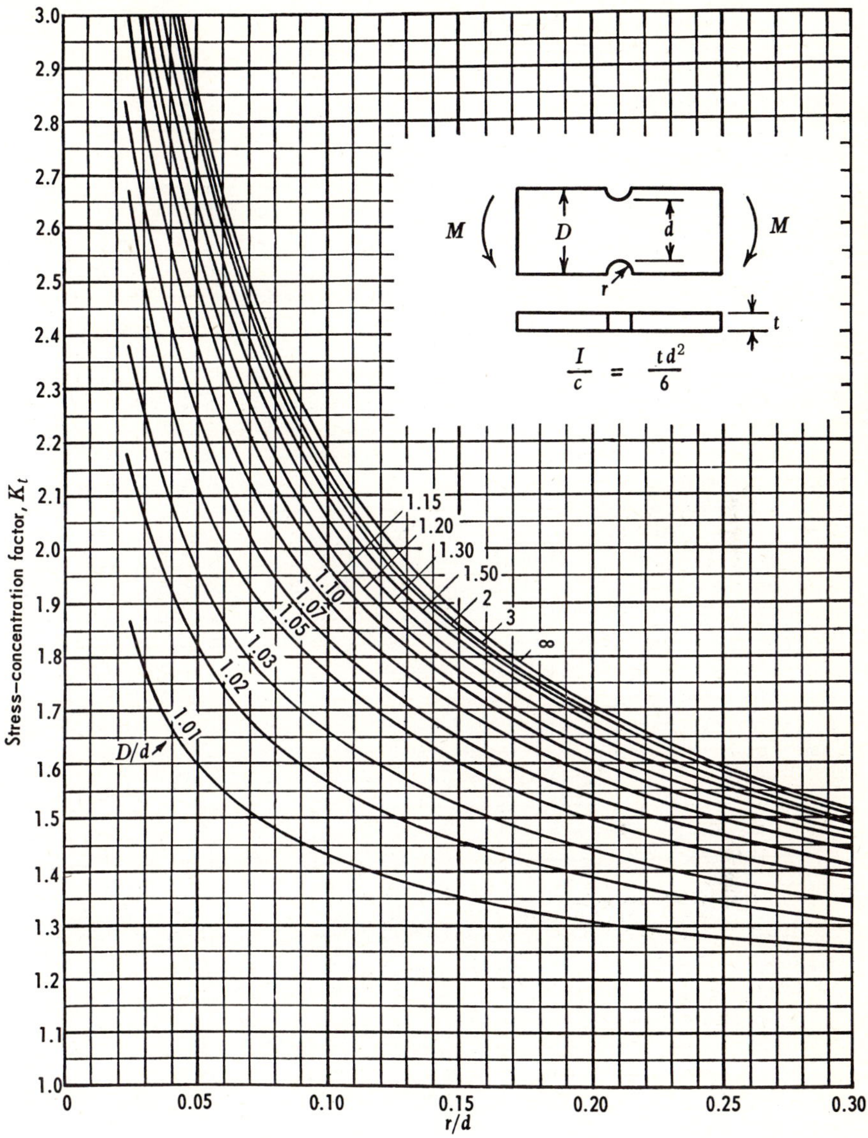

Fig. X-2 Stress-concentration factor K_t for notched flat bar in bending.

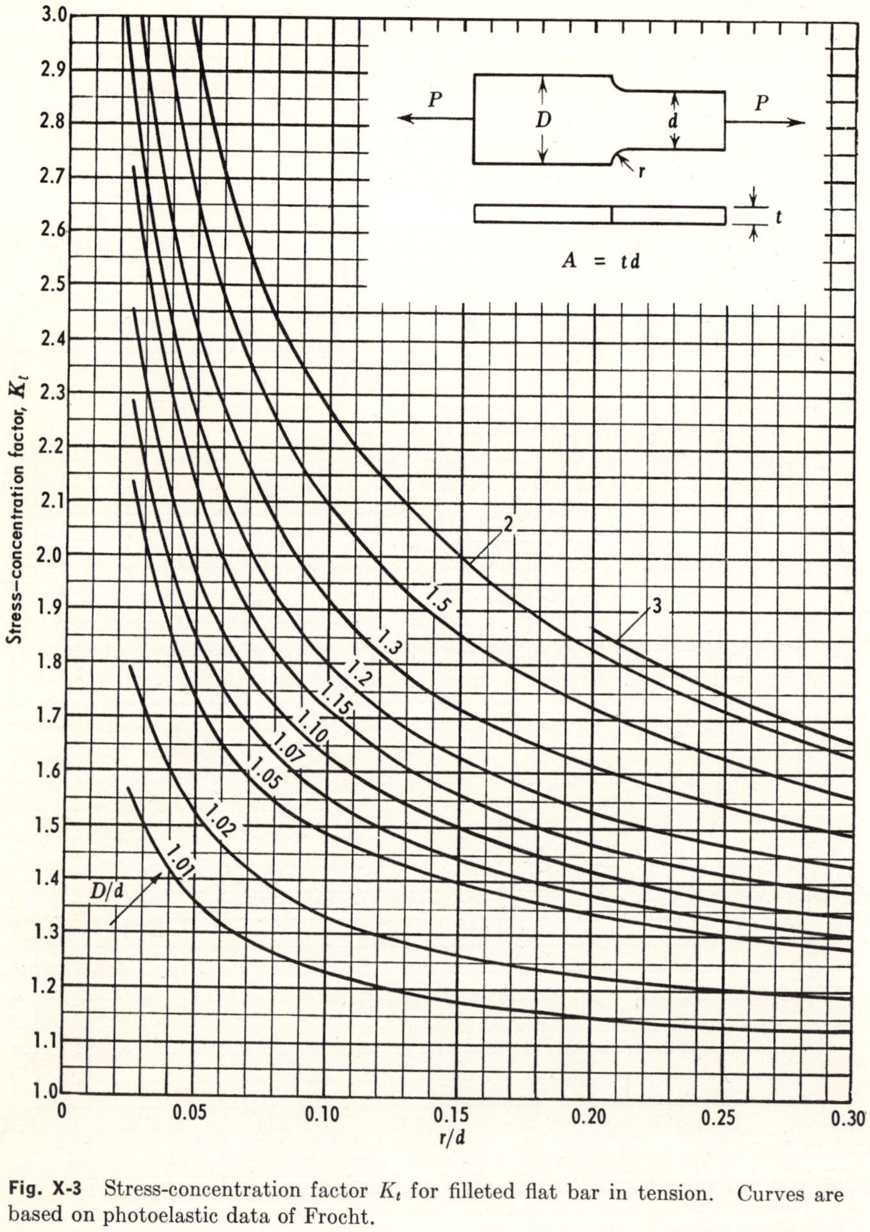

Fig. X-3 Stress-concentration factor K_t for filleted flat bar in tension. Curves are based on photoelastic data of Frocht.

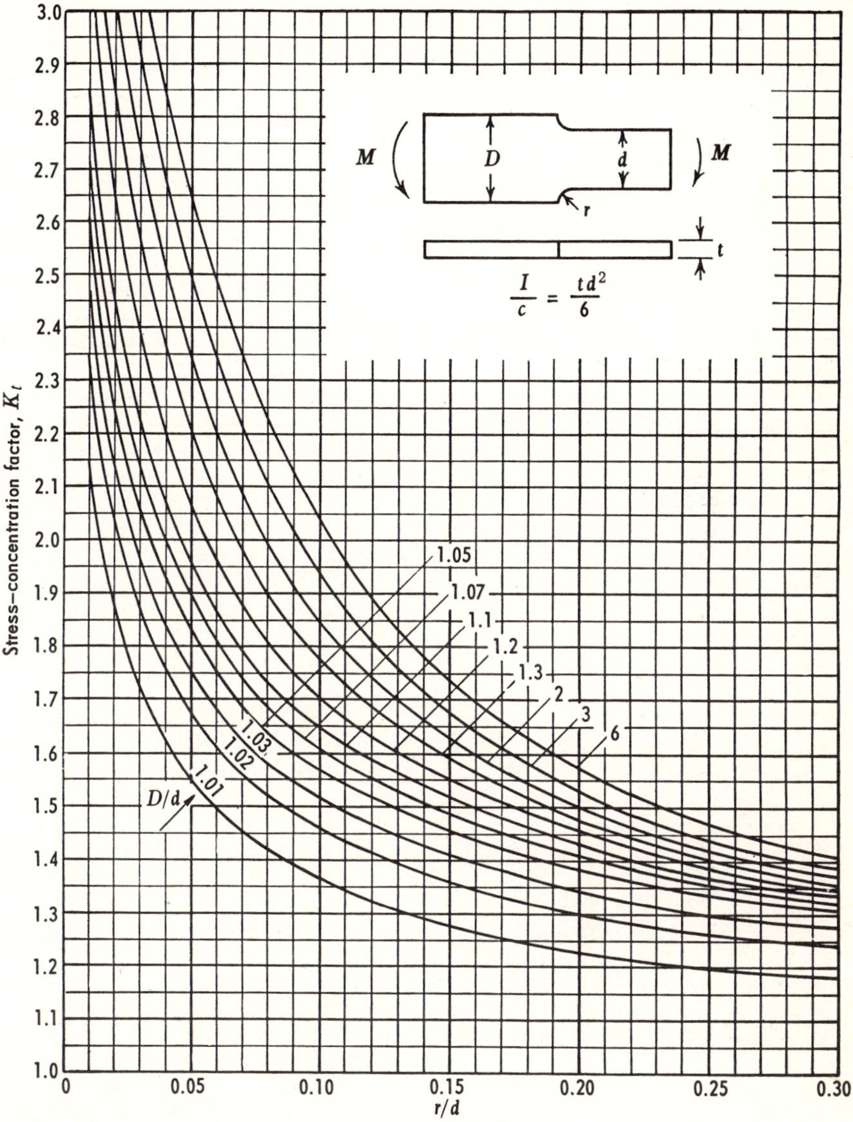

Fig. X-4 Stress-concentration factor K_t for stepped bar in bending.

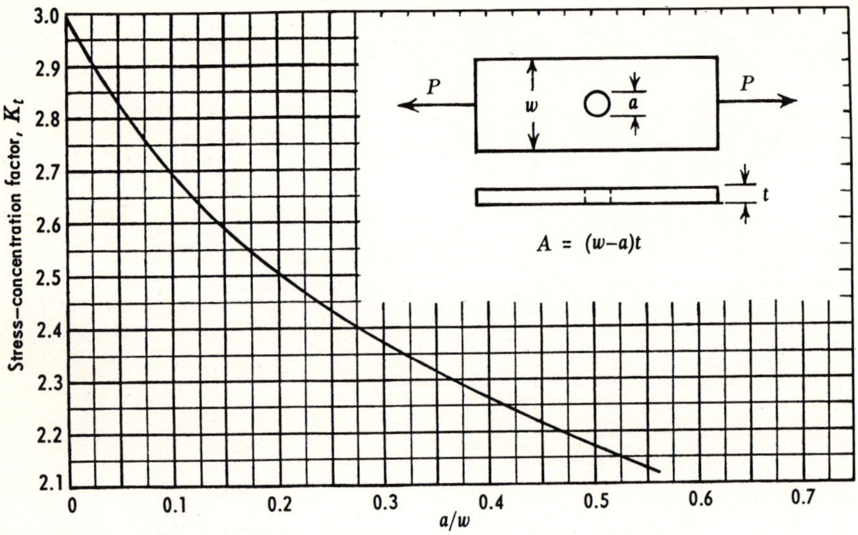

Fig. X-5 Stress-concentration factor K_t for plate with hole, in tension.

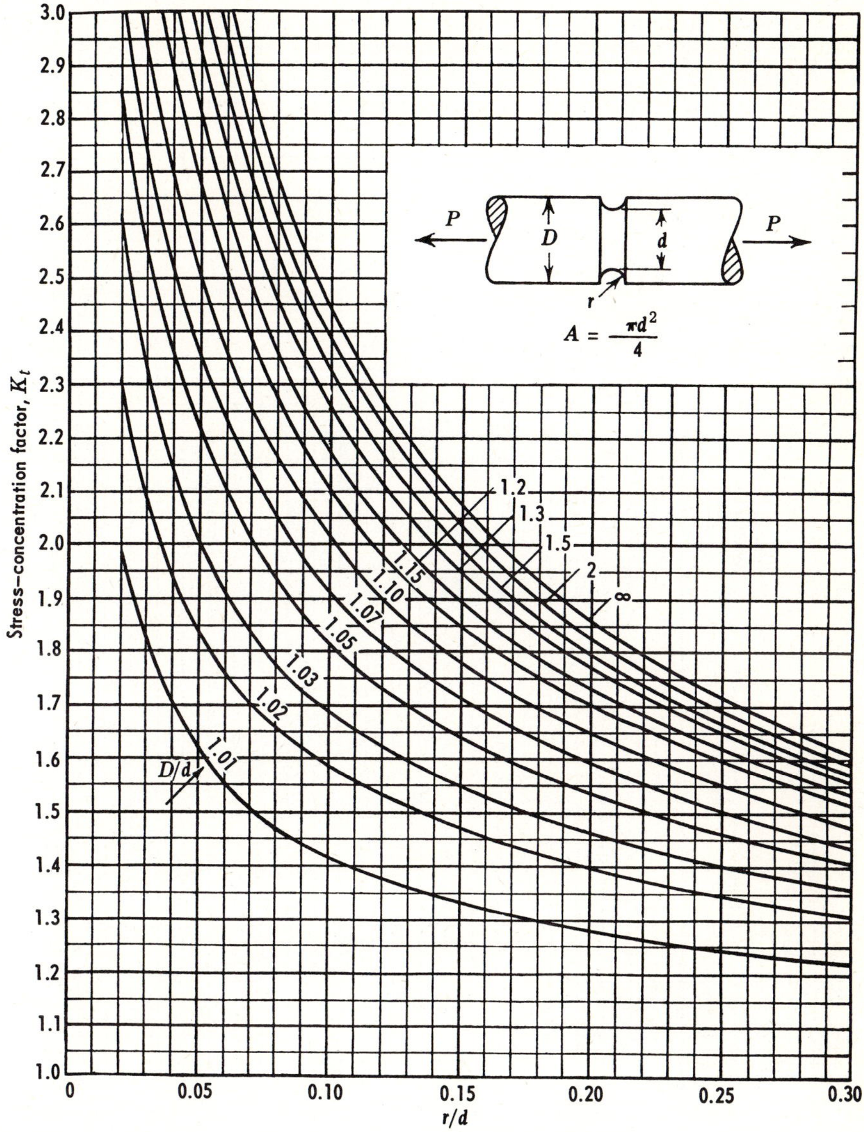

Fig. X-6 Stress-concentration factor K_t for grooved shaft in tension.

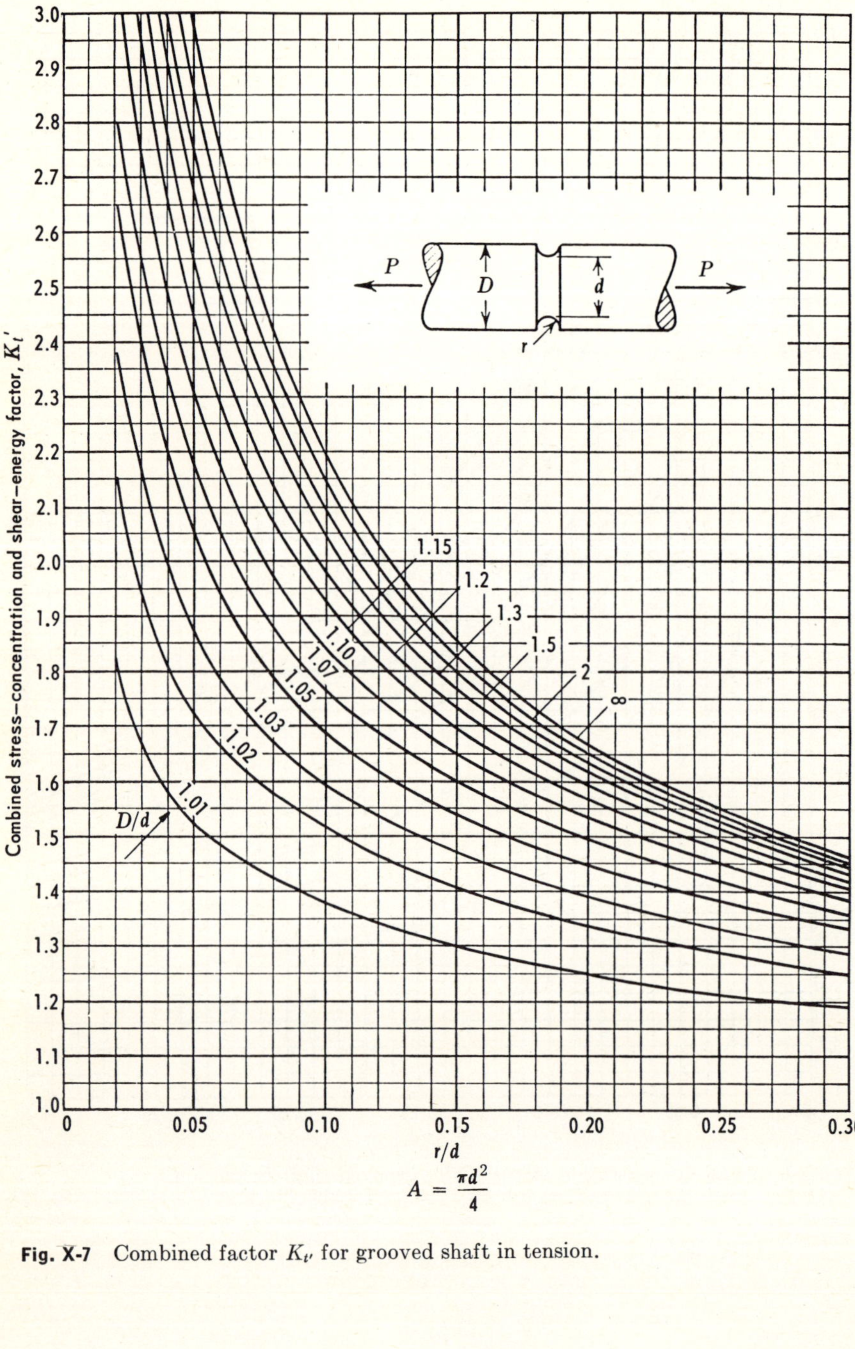

Fig. X-7 Combined factor $K_{t'}$ for grooved shaft in tension.

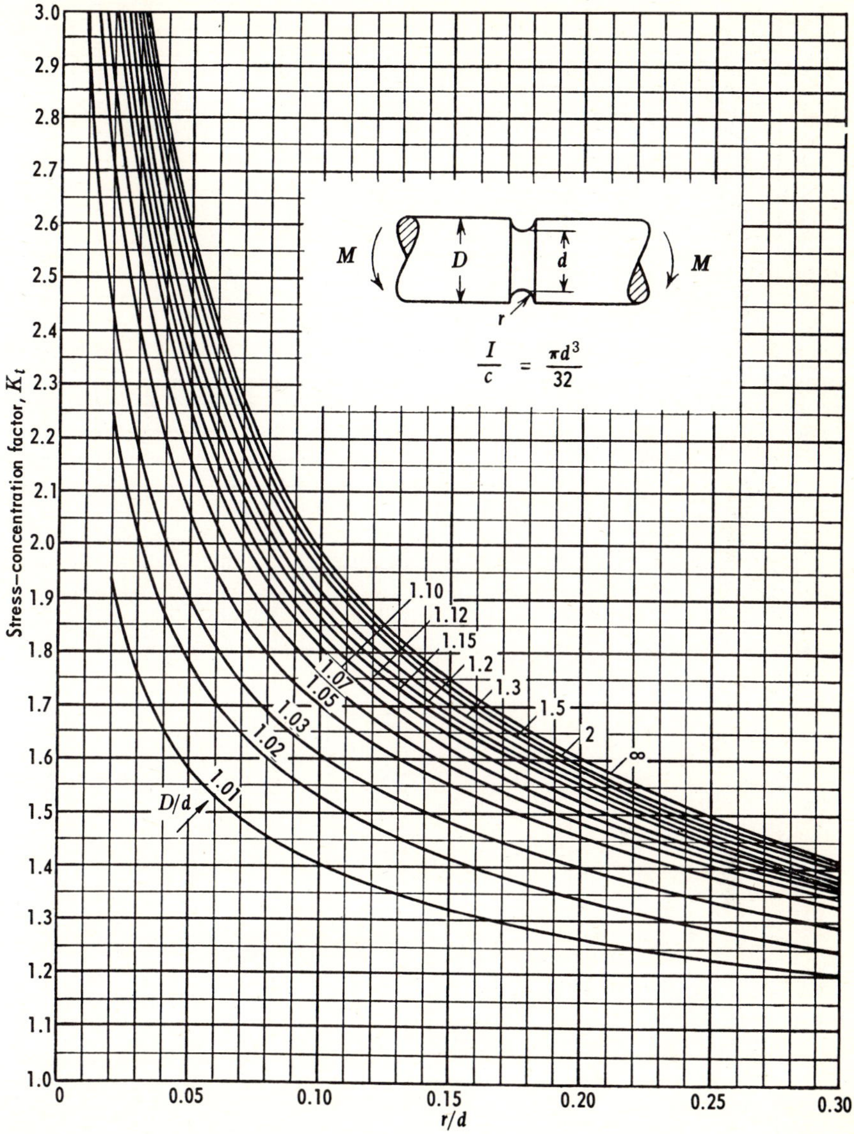

Fig. X-8 Stress-concentration factor K_t for grooved shaft in bending.

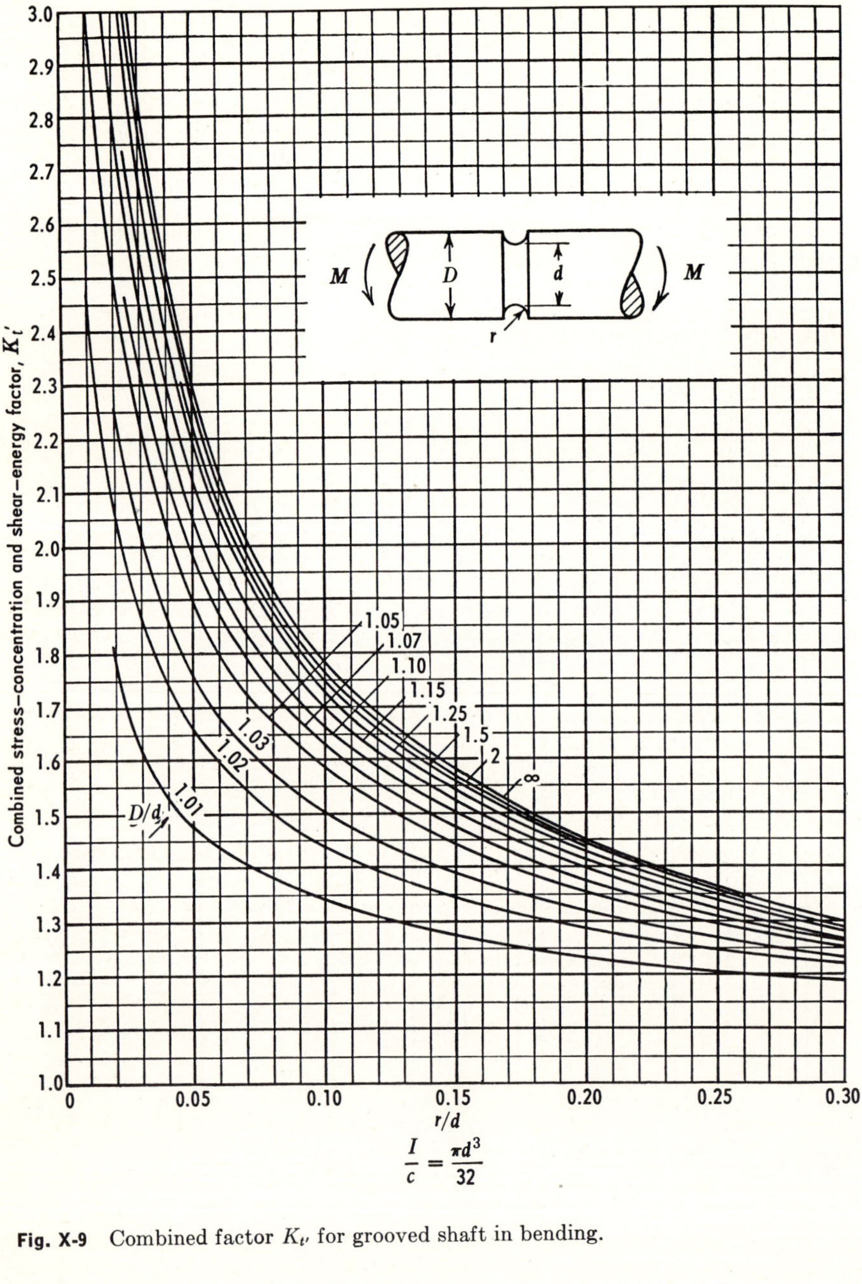

Fig. X-9 Combined factor $K_{t'}$ for grooved shaft in bending.

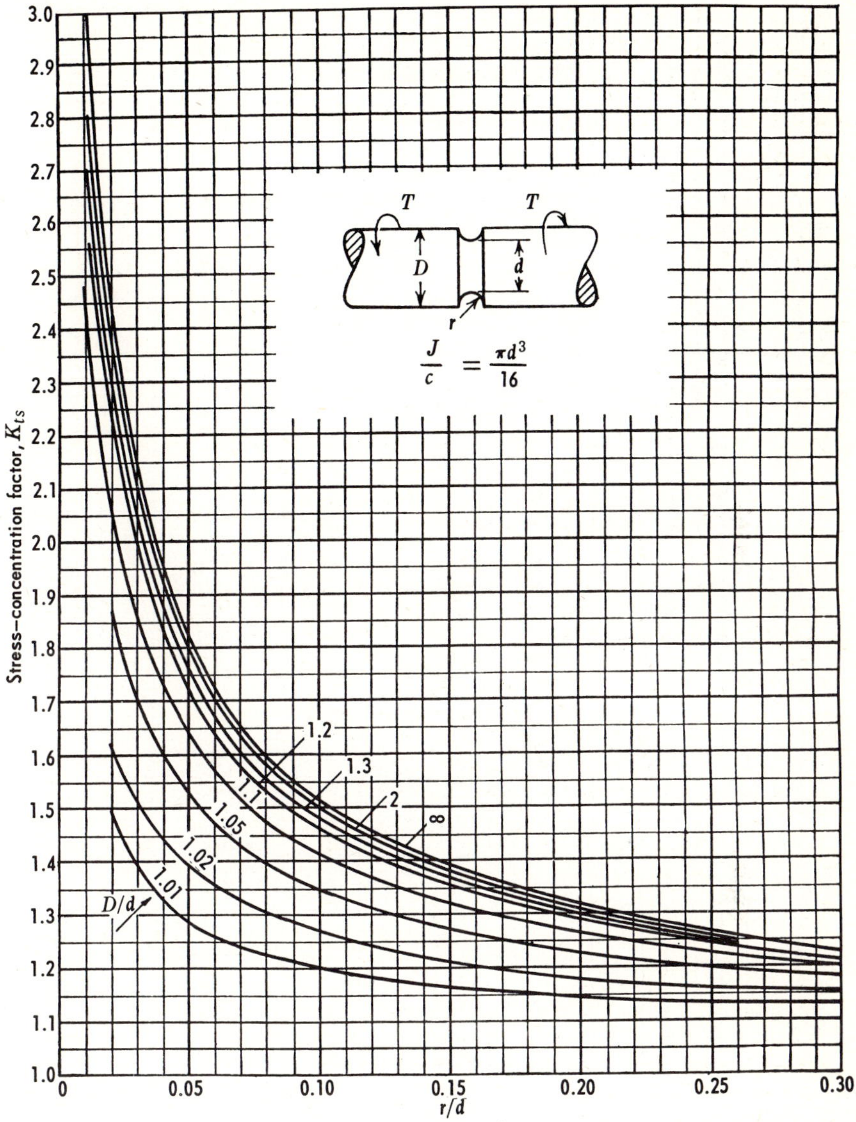

Fig. X-10 Stress-concentration factor K_{ts} for grooved shaft in torsion.

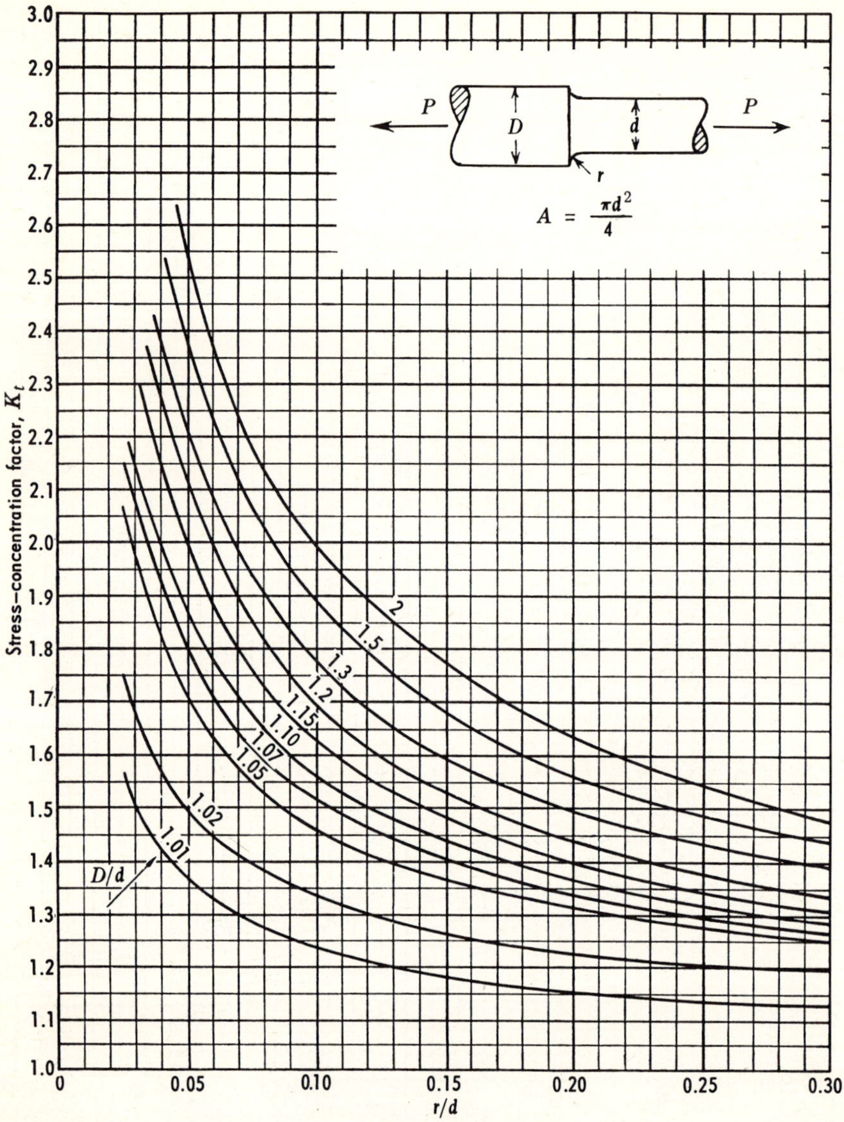

Fig. X-11 Stress-concentration factor K_t for stepped shaft in tension.

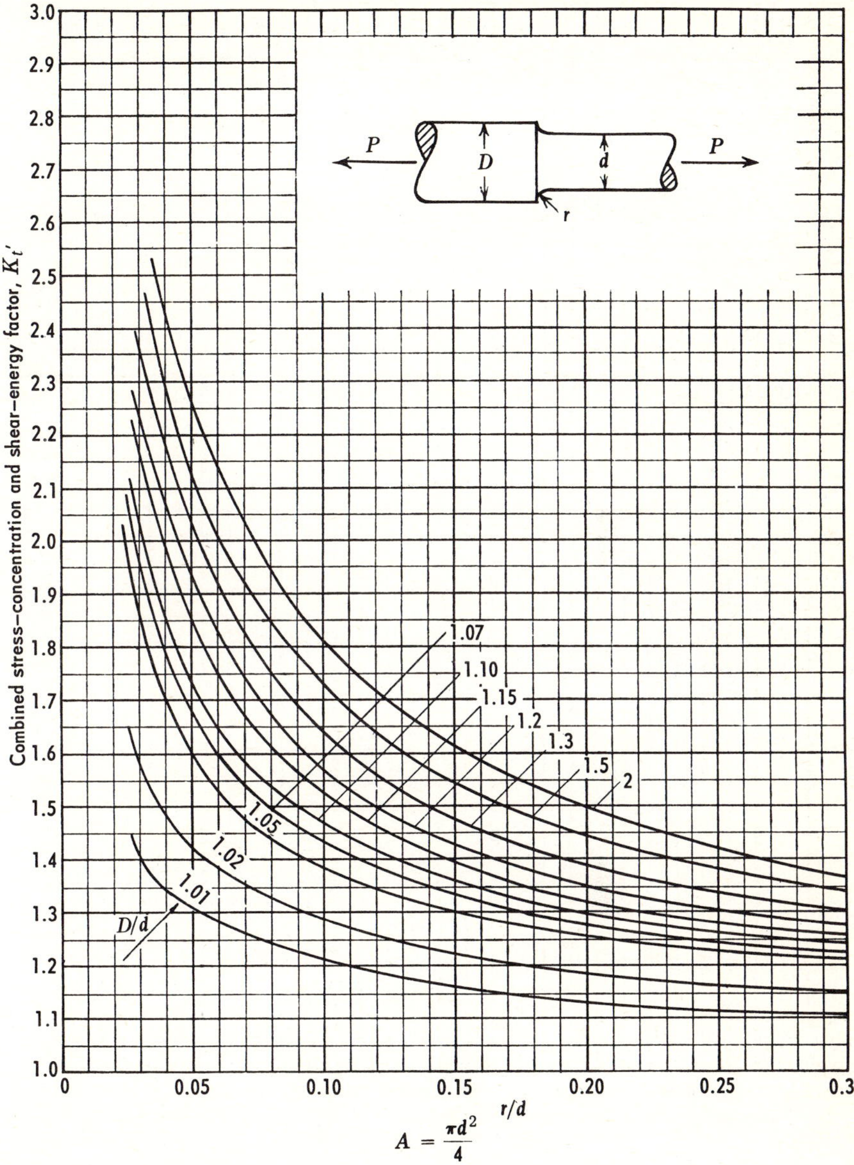

Fig. X-12 Combined factor $K_{t'}$ for stepped shaft in tension.

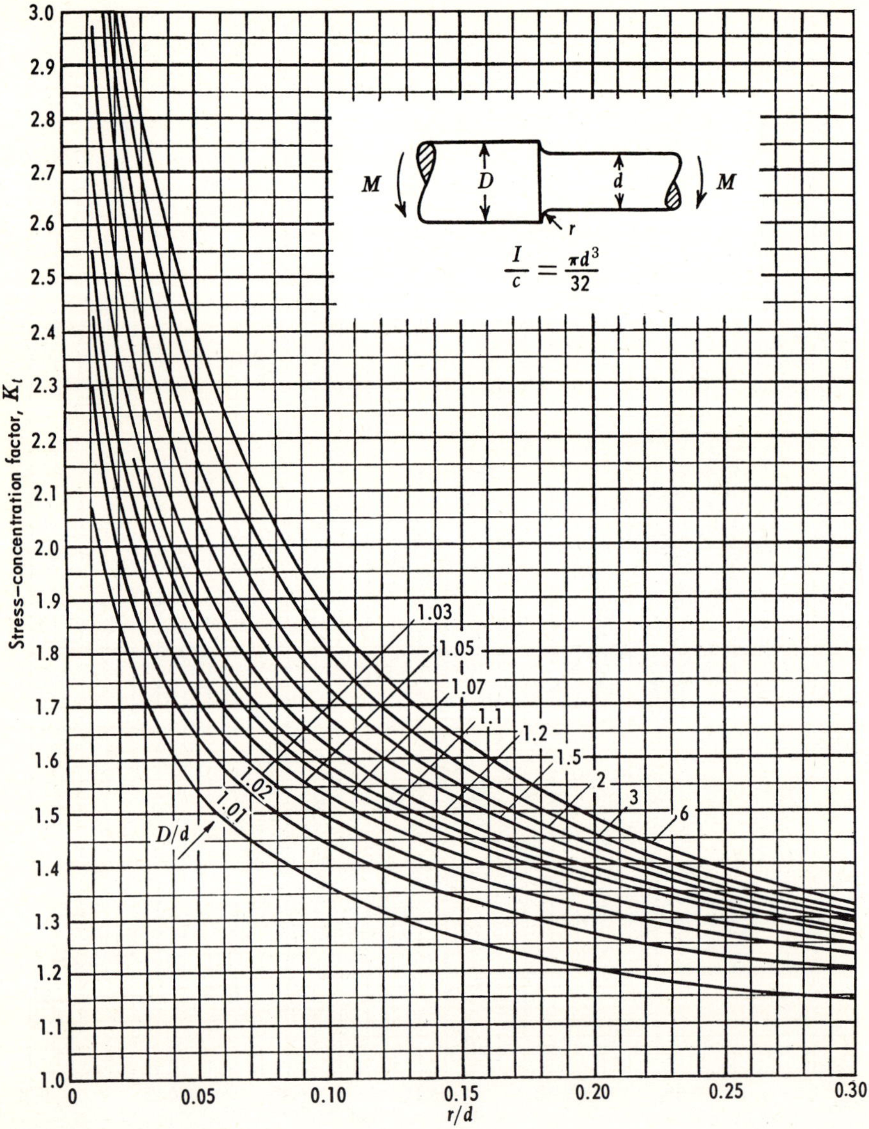

Fig. X-13 Stress-concentration factor K_t for stepped shaft in bending.

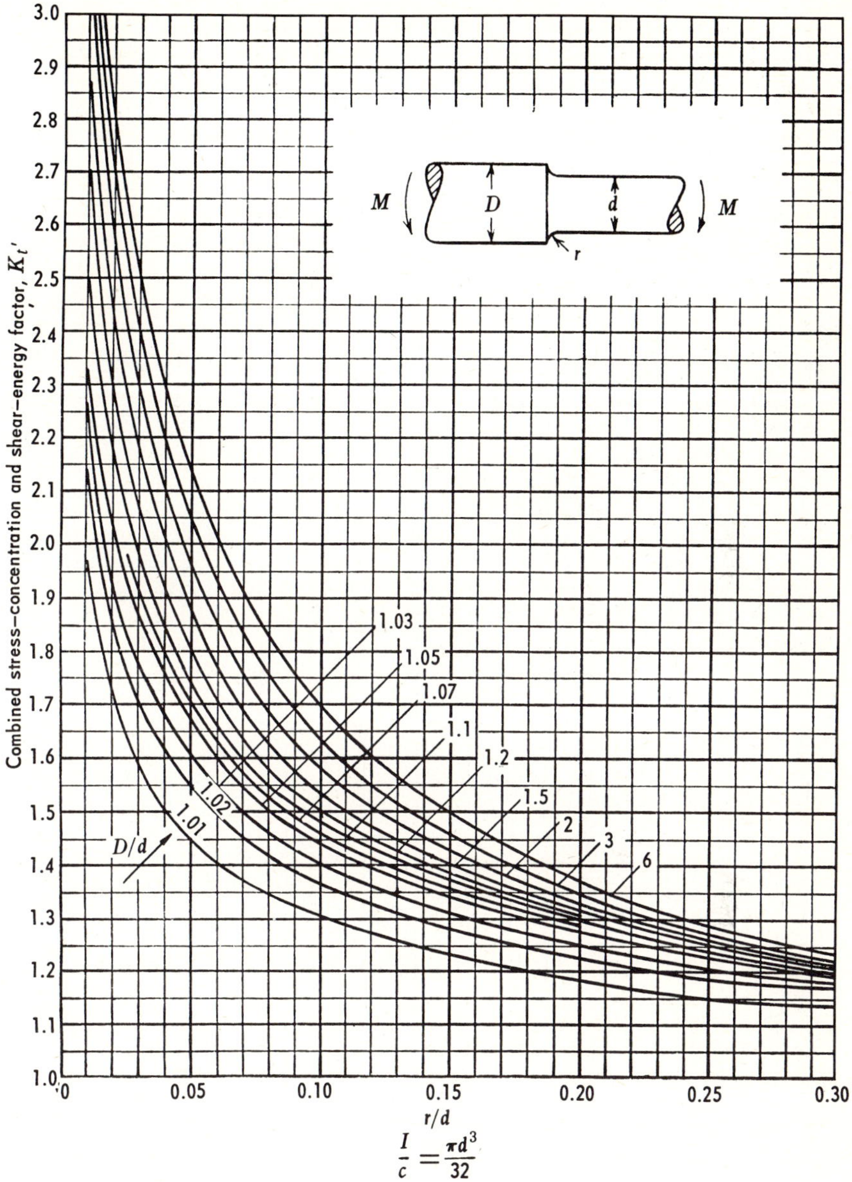

Fig. X-14 Combined factor $K_{t'}$ for stepped shaft in bending.

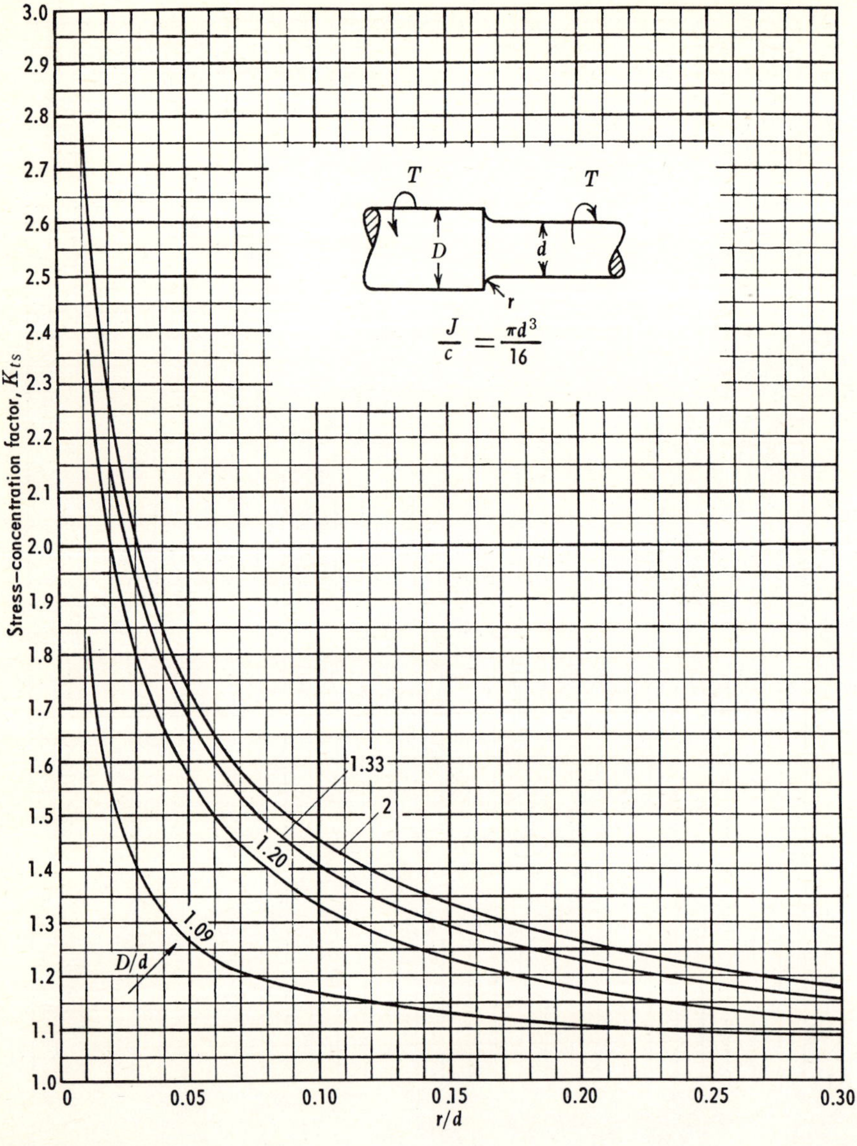

Fig. X-15 Stress-concentration factor K_{ts} for stepped shaft in torsion.

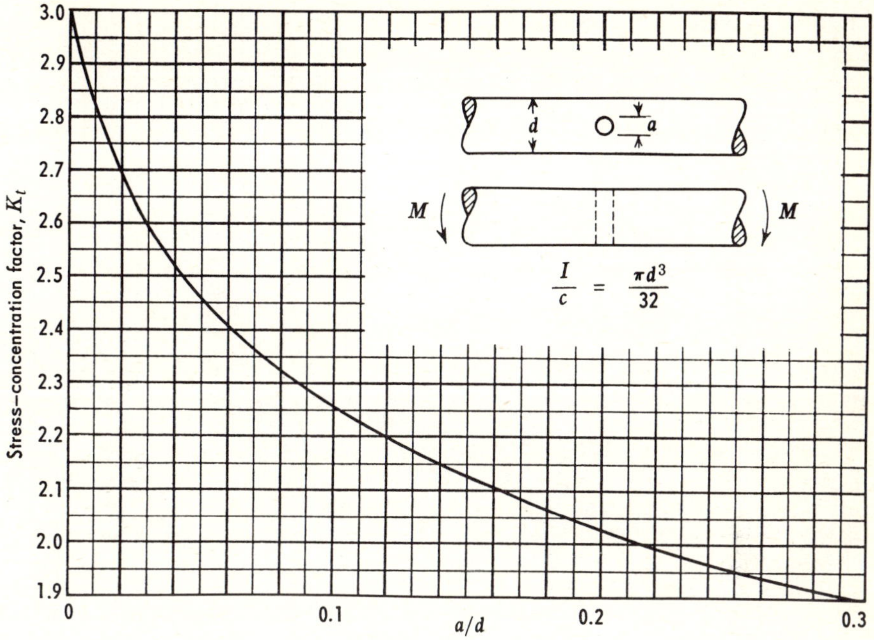

Fig. X-16 Stress-concentration factor K_t and K_{ts} for a shaft with transverse hole.

appendix XI

WR^2 for rotating bodies

W = weight of body, lb
R = radius of gyration, ft
D = outer diameter of cylinder, ft
d = inside diameter of hollow cylinder, ft
Steel weighs 0.283 lb per in.^{3}
Cast iron weighs 0.260 lb per in.^{3}

For rotating solid cylinder,

$$WR^2 = \frac{WD^2}{8}$$

For rotating hollow cylinder,

$$WR^2 = \frac{W(D^2 + d^2)}{8}$$

Equivalent WR^2 referred to shaft at different speed is

$$(WR^2)_2 = (WR^2)_1 \left[\frac{(\text{rpm})_1}{(\text{rpm})_2}\right]^2$$

Equivalent WR^2 for mass in translation at velocity (fpm) is

$$WR^2 = W \frac{\text{fpm}}{2\pi \text{ rpm}}$$

appendix XII

Method for determining Lewis factor

On a drawing of a gear tooth, extend the normal to the tooth profile at the corner A until it intersects the centerline B. Draw BC perpendicular to the centerline of the tooth. By means of a scale, determine by trial a line tangent to the profile at D and such that DC equals CE. It can be shown then that point D locates the weak section of the tooth.[1] Now, draw line BD. Through point D, draw DG perpendicular to BD, intersecting the tooth centerline at G. Through point D, draw DF perpendicular to the tooth centerline. Measure FG labeled k. It can be shown that the Lewis form factor is equal to

$$y = \frac{2k}{3p}$$

where p is the circular pitch of the teeth.[2] The value of y is the same as used in Art. 18-3.

[1] M. A. Durland, *Machinery*, vol. 29, no. 12, p. 958, August, 1923.

[2] Earle Buckingham, "Analytical Mechanics of Gears," p. 475, McGraw-Hill Book Company, New York, 1949; also, AGMA (tentative) Standard 220.02, Rating the Strength of Spur Gear Teeth, 1964.

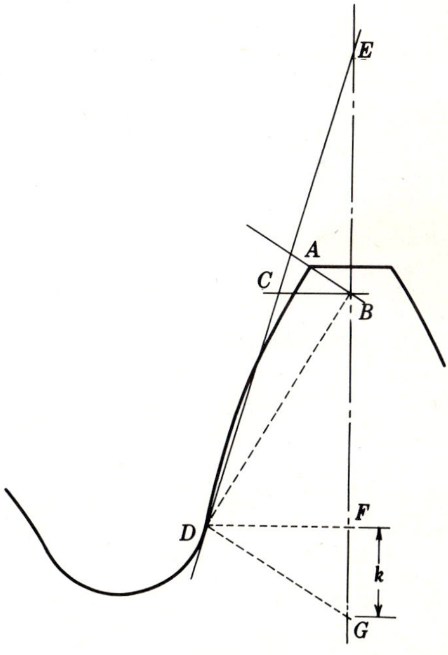

Fig. XII-1

appendix XIII-A
Graphical integration

BENDING MOMENTS

As an example of the graphical determination of bending moments, assume that the shaft or beam in Fig. XIII-1 is supported on two bearings at its ends and loaded as shown. By placing letters A, B, C, and D, the force between A and B may be labeled ab, that between B and C may be labeled bc, the right-hand reaction may be labeled cd, and the left-hand may be labeled reaction da. In Fig. XIII-2, the forces ab and bc were laid off to a scale, and a pole o placed to the right of ac at a convenient location. The "rays" oa, ob, and oc were then drawn. The polygon thus formed is a force polygon. In this polygon, the vectors oa, ab, and bo represent forces in equilibrium since they form the sides of a closed polygon. By laying off in Fig. XIII-1 the lines a and b parallel, respectively, to the rays oa and ob intersecting the line of action, AB, these three lines will then represent concurrent forces. The diagram being formed is known as a "funicular polygon" and the lines a, b, etc., are known as "strings."

Now if the string b is extended until it intersects the line of action of BC, and through that intersection the string c is drawn, the directions of the three concurrent forces, BC, co, and ob will be introduced into the funicular polygon.

So far nothing new has been determined, but if the point where the string a intersects the line of action of DA is connected with the point where c intersects the line of action of CD, then the string d thus drawn will represent the direction of the force which will be in equilibrium with DA and CD. In Fig. XIII-2, the ray od is drawn parallel to the string d and, by so doing, the magnitude of the reactions cd and da is determined graphically.

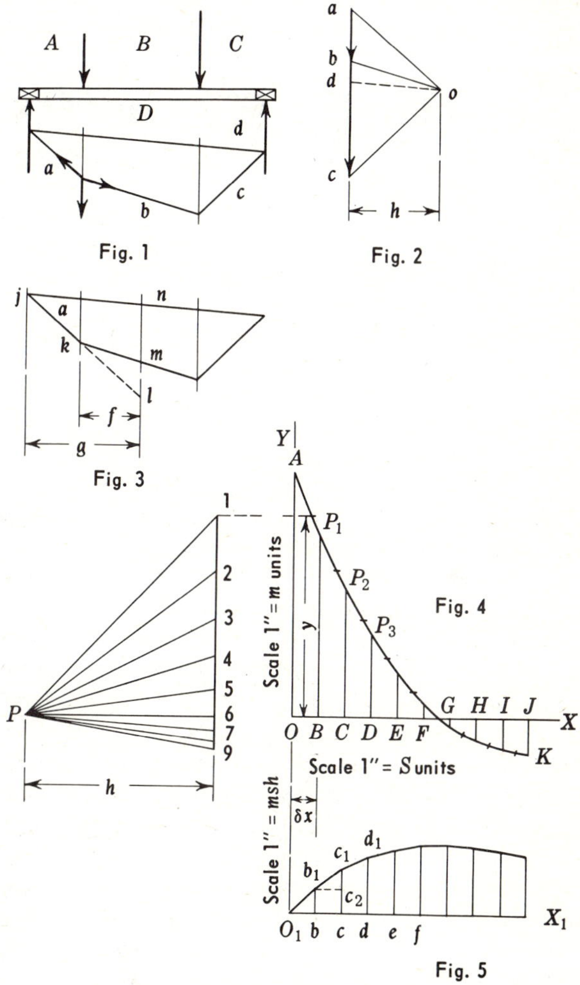

Fig. XIII-1–5 Graphical integration.

In the above procedure, the funicular polygon was drawn solely to determine the direction of the string d but, as will be shown below, the funicular polygon is the bending-moment diagram.

In Fig. XIII-3 the funicular polygon was redrawn. The bending moment at a section such as nm is equal to the sum of the moments of the forces to one side or the other of the section. To the left, the moment

$$M = (da)g - (ab)f$$

The triangle ljn is similar to oda in the force polygon, or

$$\frac{ln}{g} = \frac{da}{h} \qquad \text{or} \qquad (da)g = (ln)h$$

where h is called the "pole distance" and is measured from the line of action of the forces to the pole o in the force polygon.

Now, extend the ray a until it intersects nm at l, and from the similar triangles lkm and oab,

$$\frac{ln}{f} = \frac{ab}{h} \qquad \text{or} \qquad (ab)f = (ln)h$$

Therefore, $M = (da)g - (ab)f = (ln)h - (lm)h$
$= h(ln - lm) = h(mn)$

Thus, the vertical distance mn is a measure of the bending moment at the section of the shaft.

The scale of the bending-moment diagram is sph, where

s = space scale; 1 in. on drawing = s in. on shaft
p = force scale; 1 in. = p lb
h = pole distance, in.

GRAPHICAL INTEGRATION

Assume that the integral of the curve AK in Fig. XIII-4 is to be determined graphically.

Divide the curve into a convenient number of strips OB, BC, etc. Label the points corresponding to the mean ordinates of the strips p_1, p_2, etc., and project these points horizontally to give the points 1, 2, etc., on a line parallel to the Y axis. Choose a pole P on OX extended and draw the rays P_1, P_2, etc. These rays are the chords of the first-integral curve and may be drawn through the points O_1, b_1, etc., as shown in Figs. XIII-5.

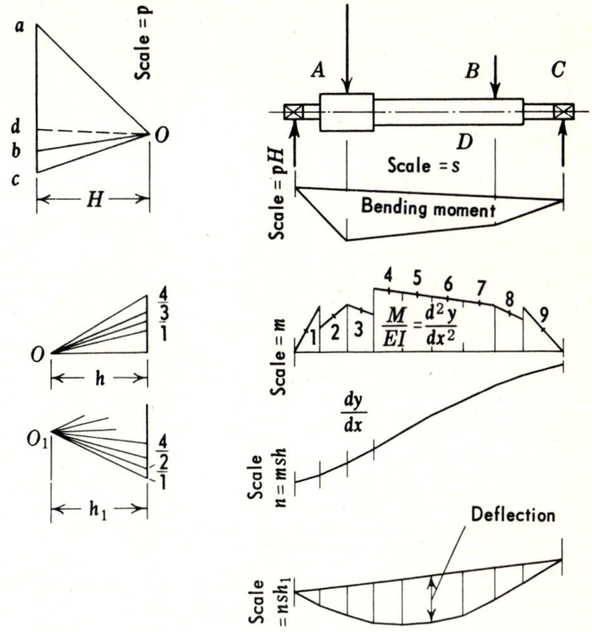

Fig. XIII-6 Deflection of shaft.

In Fig. XIII-4, let y be the mean ordinate of the strip OB and δx its width so that $y\ \delta x$ is closely its area. By construction,

$$\frac{bb_1}{\delta x} = \frac{y}{h} \qquad \text{or} \qquad y\ \delta x = h(bb_1)$$

Hence, the ordinate bb_1 is a measure of the area of the original curve to the point B.

Similarly, it may be shown that the intercept c_1c_2 is a measure of the strip BC, and when c_1c_2 is added to bb_1, the ordinate cc_1 is a measure of the area of the original curve to the point C.

DEFLECTION OF SHAFTS

As shown in Fig. XIII-6, the bending-moment diagram may be determined graphically; then a modified bending-moment diagram may be drawn by dividing the actual bending moment at each section of the beam by the product EI for that section. The modified bending-moment diagram is M/EI which equals d^2y/dx^2, where y is the deflection of the shaft. By integrating this curve twice, the deflection curve y/x is determined.

appendix XIII-B

Shaft deflection by digital computer

The determination of the deflection pattern of a simply supported, stepped shaft is facilitated by the use of a digital computer. By this method, the equations for the slope and deflection of each uniform-diameter section of the shaft accommodate themselves to a general form involving one and two constants of integration, respectively. If there are i lateral loads on the shaft with k steps, then there are $2[(i + k) + 1]$ boundary conditions with the same number of constants of integration. There are two end conditions; that is, the deflection of the shaft is zero at each bearing. In addition, at each step in the shaft and at each point of application of the load, both the deflection and the slope computed for the left and the right of these locations must be equal, respectively. The latter conditions provide for the "continuity of deformation" of the shaft.

The resulting $2[(i + k) + 1]$ algebraic equations may be readily solved by the digital computer by one of the routines for solving linear algebraic equations.

The following example shows a general case for determining the deflections in a stepped shaft (Fig. XIII-7).

As shown in Fig. XIII-7, there are two steps in the shaft, $k = 2$; there are n lateral forces, $Q_1, Q_2, \ldots, Q_n$; and their distances from the right support are $c_1, c_2, \ldots, c_n$, where $c_1 < c_2 < \cdots < c_n$.

The bending moment between the loads Q_m and Q_{m+1} is

$$M = \frac{x}{L} \sum_{i=1}^{m} Q_i c_i + \frac{L-x}{L} \sum_{i=m+1}^{n} Q_i(L - c_i)$$

The equations for the general form of slope and deflection between the loads Q_m and Q_{m+1} may be obtained as follows:

$$\frac{d^2y}{dx^2} = \frac{1}{EI_p}\left[\frac{x}{L} \sum_{i=1}^{m} Q_i c_i + \frac{L-x}{L} \sum_{i=m+1}^{n} Q_i(L - c_i)\right]$$

The slope:

$$\frac{dy}{dx} = \frac{1}{EI_p}\left[\frac{x^2}{2L} \sum_{i=1}^{m} Q_i c_i - \frac{(L-x)^2}{2L} \sum_{i=m+1}^{n} Q_i(L - c_i) + C_r\right] \qquad (a)$$

The deflection:

$$y = \frac{1}{EI_p}\left[\frac{x^3}{6L} \sum_{i=1}^{m} Q_i c_i + \frac{(L-x)^3}{6L} \sum_{i=m+1}^{n} Q_i(L - c_i) + C_r x + C_s\right] \qquad (b)$$

in which I_p is the moment of inertia of a particular cross section.

Equations (a) and (b) are used to meet all the boundary conditions. The number of boundary conditions equals $2[(i + k) + 1]$; therefore, the values of C_r and C_s can be obtained.

As an example, consider the shaft in Fig. XIII-8 which has one step

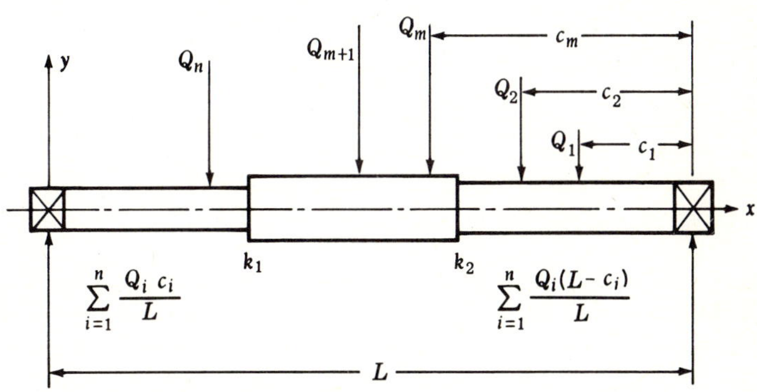

Fig. XIII-7 Shaft layout and nomenclature.

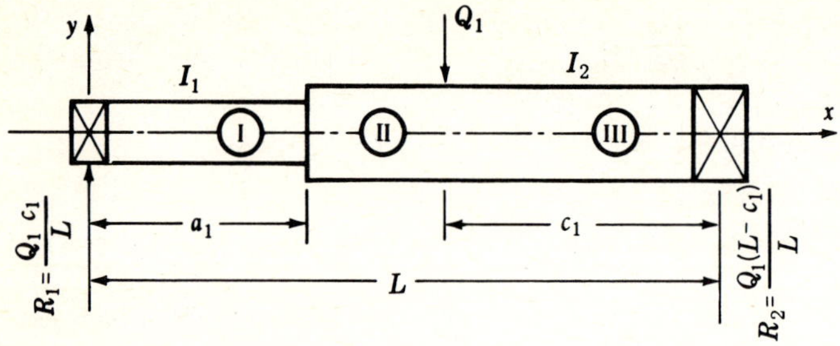

Fig. XIII-8 Shaft layout.

and one load. Then,

$$2[(i + k) + 1] = 2[(2) + 1] = 6$$

so that six constants must be determined and six equations are required. At the left end of the shaft, $x = 0$ and $y = 0$,

$$\frac{1}{EI_1}\left(\frac{x^3}{6L} Q_1c_1 + C_1x + C_2\right) = 0$$

$$C_2 = 0 \tag{a}$$

At the step,

$$x_1 = a_1$$

$$y_{\text{I}} = y_{\text{II}} \qquad \text{and} \qquad \left(\frac{dy}{dx}\right)_{\text{I}} = \left(\frac{dy}{dx}\right)_{\text{II}}$$

$$\frac{1}{EI_1}\left(\frac{a_1^3}{6L} Q_1c_1 + C_1a_1 + C_2\right) = \frac{1}{EI_2}\left(\frac{a_1^3}{6L} Q_1c_1 + C_3a_1 + C_4\right)$$

Therefore $a_1C_1 + C_2 - C_3\left(a_1\frac{I_1}{I_2}\right) - C_4\frac{I_1}{I_2} = \left(\frac{I_1}{I_2} - 1\right)\frac{a_1^3Q_1c_1}{6L}$ (b)

and $\frac{1}{I_1}\left(\frac{a_1^2}{2L} Q_1c_1 + C_1\right) = \frac{1}{I_2}\left(\frac{a_1^2}{2L} Q_1c_1 + C_3\right)$

Therefore $C_1 - \frac{I_1}{I_2}C_3 = \left(\frac{I_1}{I_2} - 1\right)\left(\frac{a_1^2}{2L} Q_1c_1\right)$ (c)

At load Q_1

$$x = L - c_1$$

and $y_{\text{II}} = y_{\text{III}}$

Therefore $\frac{1}{EI_2}\left[\frac{(L - c_1)^3}{6L} Q_1c_1 + C_3(L - c_1) + C_4\right]$

$$= \frac{1}{EI_2}\left[\frac{c_1^3}{6L} Q_1(L - c_1) + C_5(L - c_1) + C_6\right]$$

or $$C_3(L - c_1) + C_4 - C_5(L - c_1) - C_6 = \frac{Q_1c_1(L - c_1)}{6L}[c_1^2 - (L - c_1)^2] \qquad (d)$$

Now $$\left(\frac{dy}{dx}\right)_{\text{II}} = \left(\frac{dy}{dx}\right)_{\text{III}}$$

Therefore $$\frac{(L - c_1)^2}{2L} Q_1c_1 + C_3 = -\frac{c_1^2}{2L} Q_1(L - c_1) + C_5$$

or $$C_3 - C_5 = -\frac{Q_1c_1(L - c_1)}{2} \qquad (e)$$

At the right-hand end

$$x = L \qquad y = 0$$
$$C_5L + C_6 = 0 \qquad (f)$$

or, in matrix form,

$$\begin{bmatrix} 0 & 1 & 0 & 0 & 0 & 0 \\ a_1 & 1 & -\frac{a_1I_1}{I_2} & -\frac{I_1}{I_2} & 0 & 0 \\ 1 & 0 & -\frac{I_1}{I_2} & 0 & 0 & 0 \\ 0 & 0 & L - c_1 & 1 & -(L - c_1) & -1 \\ 0 & 0 & 1 & 0 & -1 & 0 \\ 0 & 0 & 0 & 0 & L & 1 \end{bmatrix} \begin{bmatrix} C_1 \\ C_2 \\ C_3 \\ C_4 \\ C_5 \\ C_6 \end{bmatrix}$$

$$= \begin{bmatrix} 0 \\ \left(\frac{I_1}{I_2} - 1\right)\frac{a_1^3Q_1c_1}{6L} \\ \left(\frac{I_1}{I_2} - 1\right)\left(\frac{a_1^2}{2L}Q_1c_1\right) \\ \frac{Q_1c_1(L - c_1)}{6L}[c_1^2 - (L - c_1)^2] \\ -\frac{Qc_1(L - c_1)}{2} \\ 0 \end{bmatrix}$$

$$\begin{bmatrix} C_1 \\ C_2 \\ C_3 \\ C_4 \\ C_5 \\ C_6 \end{bmatrix} = \begin{matrix} \uparrow \\ \text{Solution} \\ \text{vector} \\ \downarrow \end{matrix}$$

The deflections at the critical locations may then be calculated by substituting the coefficients as determined above into the appropriate deflection equations.

appendix XIII-C

Rupture-stress diagram curve for elevated-temperature systems

Using the data shown in the table, plot rupture-stress versus the Larson-Miller parameter. See Art. 5-7. Use this curve to obtain a rupture-stress versus temperature design curve for a life of 10,000 hr. Assume $C = 20.0$ in the Larson-Miller extrapolation. The data below are for stainless steel.

Rupture data for a stainless steel*

Temperature, °F	*Stress, psi* $\times 10^{-3}$	*Rupture time, hr*	*Temperature, °F*	*Stress, psi* $\times 10^{-3}$	*Rupture time, hr*
1200	40	1.0	1500	15	0.315
	30	6.31		10	5.0
	20	159		9	12.6
	15	1588		8	22.4
	12	3160		6	178
				5	630
1300	30	0.40			
	20	6.30			
	10	1000	1600	10	0.417
	9	2000		8	2.0
	8	5000		6	14.1
				4	200
1400	20	0.50		3	1260
	10	44.6			
	9	126			
	8	251			
	6	3160			

* Abstracted from Ian Finnie and W. R. Heller, "Creep of Engineering Materials," p. 248, fig. 9.9, McGraw-Hill Book Company, New York, 1959.

appendix XIV

Table of wire sizes

Washburn & Moen wire gauge no.	*Fraction of inch*	*Decimal of inch*	*Washburn & Moen wire gauge no.*	*Fraction of inch*	*Decimal of inch*
	1/2*	0.5000		3/32	0.09375
7–0		0.490	13		0.092
	15/32	0.46875	14		0.080
6–0		0.462	15		0.072
	7/16	0.4375	16	1/16	0.0625
5–0		0.431	17		0.054
	13/32	0.40625	18		0.047
4–0		0.394	19		0.041
	3/8	0.3750	20		0.035
3–0		0.3629	21		0.032
	11/32	0.34375		1/32	0.03125
2–0		0.331	22		0.0286
	5/16	0.3125	23		0.0258
0		0.307	24		0.0230
1		0.283	25		0.0204
	9/32	0.28125	26		0.0181
2		0.263	27		0.0173
	1/4	0.250	28		0.0162
3		0.244		1/64	0.0156
4		0.225	29		0.0150
	7/32	0.21875	30		0.0140
5		0.207	31		0.0132
6		0.192	32		0.0128
	3/16	0.1875	33		0.0118
7		0.177	34		0.0104
8		0.162	35		0.0095
	5/32	0.15625	36		0.009
9		0.148	37		0.0085
10		0.135	38		0.008
	1/8	0.125	39		0.0075
11		0.120	40		0.007
12		0.105	41		0.0066
			42		0.0062

* Between 1/2 and 1 in., wire sizes vary by 1/32-in. intervals.

appendix XV-A

Decimal equivalents of fractions

(Advancing by sixty-fourths)

1/64			= 0.015625	33/64			= 0.515625
	1/32		= 0.03125		17/32		= 0.53125
3/64			= 0.046875	35/64			= 0.546875
		1/16	= 0.0625			9/16	= 0.5625
5/64			= 0.078125	37/64			= 0.578125
	3/32		= 0.09375		19/32		= 0.59375
7/64			= 0.109375	39/64			= 0.609375
		1/8	= 0.125			5/8	= 0.625
9/64			= 0.140625	41/64			= 0.640625
	5/32		= 0.15625		21/32		= 0.65625
11/64			= 0.171875	43/64			= 0.671875
		3/16	= 0.1875			11/16	= 0.6875
13/64			= 0.203125	45/64			= 0.703125
	7/32		= 0.21875		23/32		= 0.71875
15/64			= 0.234375	47/64			= 0.734375
		1/4	= 0.25			3/4	= 0.75
17/64			= 0.265625	49/64			= 0.765625
	9/32		= 0.28125		25/32		= 0.78125
19/64			= 0.296875	51/64			= 0.796875
		5/16	= 0.3125			13/16	= 0.8125
21/64			= 0.328125	53/64			= 0.828125
	11/32		= 0.34375		27/32		= 0.84375
23/64			= 0.359375	55/64			= 0.859375
		3/8	= 0.375			7/8	= 0.875
25/64			= 0.390625	57/64			= 0.890625
	13/32		= 0.40625		29/32		= 0.90625
27/64			= 0.421875	59/64			= 0.921875
		7/16	= 0.4375			15/16	= 0.9375
29/64			= 0.453125	61/64			= 0.953125
	15/32		= 0.46875		31/32		= 0.96875
31/64			= 0.484375	63/64			= 0.984375
		1/2	= 0.50				

appendix XV-B

Conversion factors, U.S. and metric units

Each unit in boldface type is followed by its equivalent in other units.

Cubic inch = 16.387162 cubic centimeters
Foot = 0.3048006 meter
Gallon = 0.13368 cubic foot; 0.83268 gallon (British); 231.00 cubic inches; 0.0037854 cubic meter; 3.7853 liters
Inch = 2.540005 centimeter
Kilogram = 2.2046223 pounds (avoirdupois)
Meter = 1.093611 yards; 3.280833 feet; 39.3700 inches
Ounce (avoirdupois) = 28.349527 grams
Pound (avoirdupois) = 0.453592 kilogram; 453.5924 grams
Quart = 0.94633 liters
Radian = 57.29578 degrees
Square inch = 645.16258 square millimeters
Ton (short) = 2,000 pounds; 907.185 kilograms

appendix XVI

Horsepower ratings of USAS Standard roller chain, single strand

No. of teeth small spkt.	100	500	900	1200	1800	2500	3000	3500	4000	4500	5000	5500	6000	6500	7000	7500	8000	8500	9000	10,000
	Revolutions per minute—small sprocket																			
	Type I				Type II Lubrication															
11	0.054	0.23	0.39	0.50	0.73	0.98	1.15	1.32	1.42	1.19	1.01	0.88	0.77	0.68	0.61	0.55	0.50	0.46	0.42	0.36
12	0.059	0.25	0.43	0.55	0.80	1.07	1.26	1.45	1.62	1.36	1.16	1.00	0.88	0.78	0.70	0.63	0.57	0.52	0.48	0.41
13	0.064	0.27	0.47	0.60	0.87	1.17	1.38	1.58	1.78	1.53	1.30	1.13	0.99	0.88	0.79	0.71	0.64	0.59	0.54	0.46
14	0.070	0.30	0.50	0.65	0.94	1.27	1.49	1.71	1.93	1.71	1.46	1.26	1.11	0.98	0.88	0.79	0.72	0.66	0.60	0.51
15	0.075	0.32	0.54	0.70	1.01	1.36	1.61	1.85	2.08	1.89	1.62	1.40	1.23	1.09	0.98	0.88	0.80	0.73	0.67	0.57
16	0.081	0.34	0.58	0.75	1.09	1.46	1.72	1.98	2.23	2.08	1.78	1.54	1.35	1.20	1.07	0.97	0.88	0.80	0.74	0.63
17	0.086	0.37	0.62	0.81	1.16	1.56	1.84	2.11	2.38	2.28	1.95	1.69	1.48	1.31	1.18	1.06	0.96	0.88	0.81	0.69
18	0.092	0.39	0.66	0.86	1.23	1.66	1.95	2.25	2.53	2.49	2.12	1.84	1.62	1.43	1.28	1.16	1.05	0.96	0.88	0.75
19	0.097	0.41	0.70	0.91	1.31	1.76	2.07	2.38	2.69	2.70	2.30	2.00	1.75	1.55	1.39	1.25	1.14	1.04	0.95	0.81
20	0.103	0.44	0.74	0.96	1.38	1.86	2.19	2.52	2.84	2.91	2.49	2.16	1.89	1.68	1.50	1.35	1.23	1.12	1.03	0.88
21	0.108	0.46	0.78	1.01	1.46	1.96	2.31	2.65	2.99	3.13	2.68	2.32	2.04	1.80	1.61	1.46	1.32	1.21	1.11	0.95
22	0.114	0.48	0.82	1.06	1.53	2.06	2.43	2.79	3.15	3.36	2.87	2.49	2.18	1.93	1.73	1 56	1 42	1.29	1.19	1.01
23	0.119	0.51	0.86	1.12	1.61	2.16	2.55	2.93	3.30	3.59	3.07	2.66	2.33	2.07	1.85	1.67	1.51	1.38	1.27	1.08
24	0.125	0.53	0.90	1.17	1.69	2.26	2.67	3.07	3.46	3.83	3.27	2.83	2.48	2.20	1.97	1.78	1.61	1.47	1.35	1.16
25	0.131	0.56	0.94	1.22	1.76	2.37	2.79	3.20	3.61	4.02	3.48	3.01	2.64	2.34	2.10	1.89	1.72	1.57	1.44	1.23
28	0.148	0.63	1.07	1.38	1.99	2.67	3.15	3.62	4.08	4.54	4.12	3.57	3.13	2.78	2.49	2.24	2.04	1.86	1.71	1.46
30	0.159	0.68	1.15	1.49	2.14	2.88	3.39	3.90	4.40	4.89	4.57	3.96	3.47	3.08	2.76	2.49	2.26	2.06	1.89	1.62
32	0.170	0.73	1.23	1.60	2.30	3.09	3.64	4.18	4.71	5.24	5.03	4.36	3.83	3.39	3.04	2.74	2.49	2.27	2.08	1.78
35	0.188	0.80	1.36	1.76	2.53	3.40	4.01	4.61	5.19	5.78	5.76	4.99	4.38	3.88	3.48	3.13	2.85	2.60	2.38	2.04
40	0.217	0.92	1.57	2.03	2.93	3.93	4.63	5.32	6.00	6.67	7.04	6.10	5.35	4.75	4.25	3.83	3.48	3.17	2.91	2.49
45	0.246	1.05	1.78	2.31	3.32	4.46	5.26	6.04	6.81	7.58	8.33	7.28	6.39	5.66	5.07	4.57	4.15	3.79	3.48	2.97
50	0.276	1.18	1.99	2.58	3.72	5.00	5.89	6.77	7.64	8.49	9.33	8.52	7.48	6.63	5.93	5.35	4.86	4.44	4.07	3.48
55	0.306	1.30	2.21	2.86	4.12	5.54	6.53	7.51	8.46	9.41	10.3	9.83	8.63	7.65	6.85	6.17	5.60	5.12	4.70	4.01
60	0.336	1.43	2.43	3.15	4.53	6.09	7.18	8.25	9.30	10.3	11.3	11.2	9.83	8.72	7.80	7.03	6.38	5.83	5.35	4.57
		Type II			Type III						Type IV Lubrication									

Fig. XVI-1 Horsepower ratings of No. 25 (1/4-in.) pitch USAS standard roller chain, single strand.

No. of teeth small spkt.	Revolutions per minute—small sprocket																			
	100	500	900	1200	1800	2500	3000	3500	4000	4500	5000	5500	6000	6500	7000	7500	8000	8500	9000	10,000
	Type I					Type II						Type III						Type IV		
11	0.18	0.78	1.32	1.72	2.47	3.32	2.93	2.32	1.90	1.59	1.36	1.18	1.04	0.92	0.82	0.74	0.67	0.61	0.56	0.48
12	0.20	0.86	1.45	1.89	2.72	3.65	3.35	2.66	2.17	1.82	1.56	1.35	1.18	1.05	0.94	0.85	0.77	0.70	0.64	0.55
13	0.22	0.93	1.59	2.05	2.96	3.98	3.76	2.99	2.45	2.05	1.75	1.52	1.33	1.18	1.06	0.95	0.86	0.79	0.72	0.62
14	0.24	1.01	1.72	2.23	3.21	4.31	4.21	3.34	2.73	2.29	1.96	1.70	1.49	1.32	1.18	1.06	0.97	0.88	0.81	0.69
15	0.26	1.09	1.85	2.40	3.45	4.64	4.66	3.70	3.03	2.54	2.17	1.88	1.65	1.46	1.31	1.18	1.07	0.98	0.90	0.77
16	0.27	1.17	1.98	2.57	3.70	4.98	5.14	4.08	3.34	2.80	2.39	2.07	1.82	1.61	1.44	1.30	1.18	1.08	0.99	0.84
17	0.29	1.25	2.12	2.75	3.95	5.31	5.63	4.47	3.66	3.06	2.62	2.27	1.99	1.77	1.58	1.42	1.29	1.18	1.08	0.93
18	0.31	1.33	2.25	2.92	4.20	5.65	6.13	4.87	3.98	3.34	2.85	2.47	2.17	1.92	1.72	1.55	1.41	1.29	1.18	1.01
19	0.33	1.41	2.39	3.10	4.46	5.99	6.65	5.28	4.32	3.62	3.09	2.68	2.35	2.09	1.87	1.68	1.53	1.39	1.28	1.09
20	0.35	1.49	2.53	3.27	4.71	6.33	7.18	5.70	4.67	3.91	3.34	2.90	2.54	2.25	2.02	1.82	1.65	1.51	1.38	1.18
21	0.37	1.57	2.66	3.45	4.97	6.68	7.73	6.13	5.02	4.21	3.59	3.11	2.73	2.42	2.17	1.96	1.77	1.62	1.49	1.27
22	0.39	1.65	2.80	3.63	5.22	7.02	8.27	6.58	5.38	4.51	3.85	3.34	2.93	2.60	2.33	2.10	1.90	1.74	1.60	1.36
23	0.41	1.73	2.94	3.81	5.48	7.37	8.68	7.03	5.75	4.82	4.12	3.57	3.13	2.78	2.49	2.24	2.03	1.86	1.71	1.46
24	0.43	1.81	3.08	3.98	5.74	7.71	9.09	7.49	6.13	5.14	4.39	3.80	3.34	2.96	2.65	2.39	2.17	1.98	1.82	1.55
25	0.44	1.89	3.21	4.16	6.00	8.06	9.50	7.97	6.52	5.47	4.67	4.05	3.55	3.15	2.82	2.54	2.31	2.11	1.93	1.65
28	0.50	2.14	3.63	4.71	6.78	9.11	10.7	9.44	7.73	6.48	5.53	4.80	4.21	3.73	3.34	3.01	2.73	2.50	2.29	1.96
30	0.54	2.31	3.91	5.07	7.30	9.81	11.6	10.5	8.57	7.18	6.14	5.32	4.67	4.14	3.70	3.34	3.03	2.77	2.54	2.17
32	0.58	2.47	4.20	5.44	7.83	10.5	12.4	11.5	9.44	7.91	6.76	5.86	5.14	4.56	4.08	3.68	3.34	3.05	2.80	2.39
35	0.64	2.72	4.62	5.99	8.63	11.6	13.7	13.2	10.8	9.06	7.73	6.70	5.88	5.22	4.67	4.21	3.82	3.49	3.20	0
40	0.74	3.15	5.34	6.92	9.96	13.4	15.8	16.1	13.2	11.1	9.45	8.19	7.19	6.37	5.70	5.14	4.67	0		
45	0.84	3.57	6.06	7.85	11.3	15.2	17.9	19.2	15.8	13.2	11.3	9.77	8.57	7.60	6.80	6.14	0			
50	0.94	4.00	6.79	8.80	12.7	17.0	20.1	22.5	18.5	15.5	13.2	11.4	10.0	8.91	0	0				
55	1.04	4.44	7.53	9.76	14.1	18.9	22.3	25.6	21.3	17.8	15.2	13.2	11.6	0						
60	1.15	4.88	8.27	10.7	15.4	20.7	24.4	28.1	24.2	20.3	17.4	0	0							

Fig. XVI-2 Horsepower ratings of No. 35 (3/8-in.) pitch USAS standard roller chain, single strand.

No. of teeth small spkt.	50	200	400	600	900	1200	1800	2400	3000	3500	4000	4500	5000	5500	6000	6500	7000	7500	8000	9000
	Revolutions per minute—small sprocket																			
	Type I Lubrication			Type II Lubrication				Type III Lubrication					Type IV Lubrication							
11	0.23	0.80	1.50	2.16	3.11	4.03	4.66	3.03	2.17	1.72	1.41	1.18	1.01	0.87	0.77	0.68	0.61	0.55	0.50	0
12	0.25	0.88	1.65	2.37	3.42	4.43	5.31	3.45	2.47	1.96	1.60	1.34	1.15	1.00	0.87	0.77	0.69	0.62	0.57	0
13	0.28	0.96	1.80	2.59	3.73	4.83	5.99	3.89	2.79	2.21	1.81	1.52	1.29	1.12	0.98	0.87	0.78	0.70	0.64	0
14	0.30	1.04	1.95	2.80	4.04	5.23	6.70	4.35	3.11	2.47	2.02	1.69	1.45	1.25	1.10	0.98	0.87	0.79	0.71	0
15	0.32	1.12	2.10	3.02	4.35	5.64	7.43	4.82	3.45	2.74	2.24	1.88	1.60	1.39	1.22	1.08	0.97	0.87	0	
16	0.35	1.20	2.25	3.24	4.66	6.04	8.18	5.31	3.80	3.02	2.47	2.07	1.77	1.53	1.34	1.19	1.07	0.96	0	
17	0.37	1.29	2.40	3.45	4.98	6.45	8.96	5.82	4.17	3.31	2.71	2.27	1.94	1.68	1.47	1.31	1.17	1.05	0	
18	0.39	1.37	2.55	3.68	5.30	6.86	9.76	6.34	4.54	3.60	2.95	2.47	2.11	1.83	1.60	1.42	1.27	1.15	0	
19	0.42	1.45	2.71	3.90	5.62	7.27	10.5	6.88	4.92	3.91	3.20	2.68	2.29	1.98	1.74	1.54	1.38	1.25	0	
20	0.44	1.53	2.86	4.12	5.94	7.69	11.1	7.43	5.31	4.22	3.45	2.89	2.47	2.14	1.88	1.67	1.49	1.34	0	
21	0 46	1.62	3.02	4.34	6.26	8.11	11.7	7.99	5.72	4.54	3.71	3.11	2.66	2.30	2.02	1.79	1.60	1.45	0	
22	0.49	1.70	3.17	4.57	6.58	8.52	12.3	8.57	6.13	4.87	3.98	3.34	2.85	2.47	2.17	1.92	1.72	0		
23	0.51	1.78	3.33	4.79	6.90	8.94	12.9	9.16	6.55	5.20	4.26	3.57	3.05	2.64	2.32	2.06	1.84	0		
24	0.54	1.87	3.48	5.02	7.23	9.36	13.5	9.76	6.99	5.54	4.54	3.80	3.25	2.81	2.47	2.19	1.96	0		
25	0.56	1.95	3.64	5.24	7.55	9.78	14.1	10.4	7.43	5.89	4.82	4.04	3.45	2.99	2.63	2.33	0			
28	0.63	2.20	4.11	5.93	8.54	11.1	15.9	12.3	8.80	6.99	5.72	4.79	4.09	3.55	3.11	2.76	0			
30	0.68	2.38	4.43	6.38	9.20	11.9	17.2	13.6	9.76	7.75	6.34	5.31	4.54	3.93	3.45	0				
32	0.73	2.55	4.75	6.85	9.86	12.8	18.4	15.0	10.8	8.54	6.99	5.86	5.00	4.33	3.80	0				
35	0.81	2.80	5.24	7.54	10.9	14.1	20.3	17.2	12.3	9.76	7.99	6.70	5.72	4.96	0					
40	0.93	3.24	6.05	8.71	12.5	16.3	23.4	21.0	15.0	11.9	9.76	8.18	6.99	0						
45	1.06	3.68	6.87	9.89	14.2	18.5	26.6	25.1	17.9	14.2	11.7	9.76	0							
50	1.18	4.12	7.70	11.1	16.0	20.7	29.8	29.4	21.0	16.7	13.6	0								
55	1.31	4.57	8.53	12.3	17.7	22.9	33.0	33.9	24.2	19.2	0									
60	1.44	5.02	9.37	13.5	19.4	25.2	36.3	38.6	27.6	0										

Fig. XVI-3 Horsepower ratings of No. 40 (½-in.) pitch USAS standard roller chain, single strand.

No. of teeth small spkt.	50	100	300	500	900	1200	1500	1800	2100	2400	2700	3000	3300	3500	4000	4500	5000	5400	5800	6200
	Revolutions per minute—small sprocket																			
	Type I		Type II				Type III Lubrication							Type IV Lubrication						
11	0.45	0.84	2.25	3.55	6.07	7.86	7.44	5.58	4.42	3.62	3.04	2.59	2.25	2.06	1.68	1.41	1.21	1.07	0.97	0
12	0.50	0.92	2.47	3.91	6.66	8.64	8.50	6.38	5.04	4.13	3.47	2.96	2.57	2.35	1.92	1.61	1.38	1.22	1.10	0
13	0.54	1.01	2.70	4.26	7.26	9.42	9.56	7.17	5.67	4.65	3.90	3.33	2.89	2.65	2.16	1.81	1.55	1.37	1.24	0
14	0.58	1.09	2.92	4.60	7.87	10.2	10.7	8.01	6.34	5.20	4.36	3.72	3.23	2.96	2.41	2.02	1.73	1.53	0	
15	0.63	1.17	3.15	4.97	8.48	11.0	11.9	8.89	7.03	5.76	4.83	4.13	3.58	3.28	2.68	2.24	1.92	1.70	0	
16	0.67	1.26	3.37	5.33	9.09	11.8	13.1	9.79	7.74	6.35	5.32	4.54	3.94	3.62	2.95	2.47	2.11	1.88	0	
17	0.72	1.34	3.60	5.69	9.70	12.6	14.3	10.7	8.48	6.95	5.83	4.98	4.32	3.96	3.23	2.71	2.32	2.05	0	
18	0.77	1.43	3.83	6.05	10.3	13.4	15.6	11.7	9.24	7.58	6.35	5.42	4.70	4.31	3.52	2.95	2.53	0		
19	0.81	1.51	4.06	6.42	10.9	14.2	16.9	12.7	10.0	8.22	6.89	5.88	5.10	4.68	3.82	3.20	2.74	0		
20	0.86	1.60	4.30	6.78	11.6	15.0	18.2	13.7	10.8	8.87	7.44	6.35	5.51	5.05	4.12	3.45	2.96	0		
21	0.90	1.69	4.53	7.15	12.2	15.8	19.3	14.7	11.6	9.55	8.01	6.83	5.93	5.44	4.44	3.71	3.19	0		
22	0.95	1.77	4.76	7.52	12.8	16.6	20.3	15.8	12.5	10.2	8.59	7.33	6.36	5.83	4.76	3.98	3.42	0		
23	1.00	1.86	5.00	7.89	13.4	17.4	21.3	16.9	13.3	10.9	9.18	7.83	6.79	6.23	5.08	4.26	0			
24	1.04	1.95	5.23	8.26	14.1	18.3	22.3	18.0	14.2	11.7	9.78	8.34	7.24	6.64	5.42	4.54	0			
25	1.09	2.04	5.47	8.63	14.7	19.1	23.3	19.1	15.1	12.4	10.4	8.88	7.70	7.06	5.76	4.83	0			
28	1.20	2.30	6.18	9.76	16.6	21.6	26.3	22.7	17.9	14.7	12.3	10.5	9.13	8.37	6.83	0				
30	1.33	2.42	6.66	10.5	17.9	23.2	28.4	25.1	19.9	16.3	13.7	11.7	10.1	9.28	7.57	0				
32	1.42	2.66	7.14	11.3	19.2	24.9	30.4	27.7	21.9	18.0	15.0	12.9	11.1	10.2	8.34	0				
35	1.57	2.93	7.86	12.4	21.2	27.4	33.5	31.7	25.1	20.5	17.2	14.7	12.8	11.7	9.55	0				
40	1.81	3.38	9.08	14.3	24.4	31.1	38.7	38.7	30.6	25.1	21.0	18.0	15.6	14.3	0					
45	2.06	3.84	10.3	16.3	27.8	36.0	43.9	46.2	36.5	29.9	25.1	21.4	18.6	0						
50	2.30	4.30	11.6	18.2	31.1	40.3	49.2	54.1	42.8	35.1	29.4	25.1	0							
55	2.56	4.77	12.8	20.2	34.5	44.7	54.6	62.4	49.3	40.5	33.9	0								
60	2.81	5.25	14.1	22.2	37.9	49.1	59.9	71.1	56.2	46.1	0									

Fig. XVI-4 Horsepower ratings of No. 50 (5/8-in.) pitch USAS standard roller chain, single strand.

No. of teeth small spkt.	50	100	200	500	700	900	1200	1400	1600	1800	2000	2200	2400	2600	2800	3000	3500	3800	4000	4600
	Revolutions per minute—small sprocket																			
	Type I		Type II				Type III Lubrication							Type IV Lubrication						
11	0.78	1.44	2.69	6.14	8.32	10.5	11.9	9.45	7.70	6.49	5.51	4.78	4.20	3.72	3.33	3.00	2.38	2.10	1.95	0
12	0.85	1.58	2.96	6.73	9.14	11.5	13.6	10.8	8.79	7.42	6.29	5.46	4.79	4.25	3.80	3.42	2.72	2.40	2.23	0
13	0.93	1.72	3.22	7.34	9.96	12.5	15.3	12.1	9.89	8.34	7.08	6.14	5.39	4.78	4.28	3.85	3.06	2.70	2.50	0
14	1.00	1.87	3.49	7.95	10.8	13.6	17.1	13.6	11.1	9.32	7.91	6.86	6.02	5.34	4.78	4.30	3.42	3.01	2.80	0
15	1.08	2.01	3.76	8.57	11.6	14.6	18.9	15.0	12.3	10.3	8.77	7.61	6.68	5.93	5.30	4.77	3.79	3.34	3.10	0
16	1.16	2.16	4.03	9.19	12.5	15.7	20.3	16.6	13.5	11.4	9.66	8.38	7.36	6.53	5.84	5.25	4.17	3.68	3.42	0
17	1.24	2.30	4.31	9.81	13.3	16.7	21.7	18.2	14.8	12.5	10.6	9.18	8.06	7.15	6.40	5.75	4.57	4.03	3.74	0
18	1.32	2.45	4.58	10.4	14.1	17.8	23.0	19.8	16.1	13.6	11.5	10.0	8.78	7.79	6.97	6.27	4.98	4.39	4.08	0
19	1.40	2.60	4.86	11.1	15.0	18.8	24.4	21.5	17.5	14.7	12.5	10.9	9.52	8.45	7.56	6.80	5.40	4.76	4.42	0
20	1.48	2.75	5.13	11.7	15.9	19.9	25.8	23.2	18.9	15.9	13.5	11.7	10.3	9.12	8.17	7.34	5.83	5.14	0	
21	1.56	2.89	5.41	12.3	16.7	21.0	27.2	24.9	20.3	17.1	14.5	12.6	11.1	9.82	8.79	7.90	6.27	5.53	0	
22	1.64	3.04	5.69	13.0	17.6	22.1	28.6	26.7	21.8	18.4	15.6	13.5	11.9	10.5	9.42	8.47	6.73	5.93	0	
23	1.72	3.19	5.97	13.6	18.4	23.2	30.0	28.6	23.3	19.6	16.7	14.4	12.7	11.3	10.1	9.06	7.19	0		
24	1.80	3.34	6.25	14.2	19.3	24.3	31.4	30.4	24.8	20.9	17.7	15.4	13.5	12.0	10.7	9.65	7.66	0		
25	1.88	3.49	6.53	14.9	20.2	25.4	32.9	32.4	26.4	22.3	18.9	16.4	14.4	12.8	11.4	10.3	8.15	0		
28	2.12	3.95	7.38	16.8	22.8	28.7	37.1	38.4	31.3	26.4	22.4	19.4	17.0	15.1	13.5	12.2	9.66	0		
30	2.29	4.25	7.95	18.1	24.6	30.9	40.0	42.6	34.7	29.2	24.8	21.5	18.9	16.8	15.0	13.5	0			
32	2.45	4.56	8.53	19.4	26.3	33.1	42.9	46.9	38.2	32.2	27.3	23.7	20.8	18.5	16.5	14.9	0			
35	2.70	5.02	9.40	21.4	29.0	36.5	47.3	53.6	43.7	36.9	31.3	27.1	23.8	21.1	18.9	17.0				
40	3.12	5.80	10.9	24.7	33.5	42.1	54.6	62.7	53.4	45.0	38.2	33.1	29.1	25.8	23.1	0				
45	3.55	6.90	12.3	28.1	38.1	47.8	62.0	71.2	63.7	53.7	45.6	39.5	34.7	30.8	0					
50	3.97	7.38	13.8	31.5	42.7	54.0	69.5	80.0	74.6	62.9	53.4	46.3	40.7	0						
55	4.40	8.18	15.3	34.9	47.3	59.4	77.0	88.4	86.1	72.6	61.6	53.4	0							
60	4.84	8.99	16.8	38.3	51.9	65.3	84.6	97.2	98.1	82.7	70.2	0								

Fig. XVI-5 Horsepower ratings of No. 60 (3/4-in.) pitch USAS standard roller chain, single strand.

No. of teeth small spkt.	25	50	100	200	300	400	500	700	900	1000	1200	1400	1600	1800	2000	2200	2400	2600	2800	3000
	Revolutions per minute—small sprocket																			
	Type I		Type II					Type III					Type IV							
11	0.97	1.80	3.36	6.28	9.04	11.7	14.3	19.4	23.0	19.6	14.9	11.8	9.69	8.12	6.94	6.01	5.27	4.68	4.18	0
12	1.06	1.98	3.70	6.89	9.93	12.9	15.7	21.3	26.2	22.4	17.0	13.5	11.1	9.28	7.92	6.86	6.02	5.34	4.78	0
13	1.16	2.16	4.03	7.51	10.8	14.0	17.1	23.2	29.1	25.2	19.2	15.2	12.5	10.4	8.91	7.72	6.78	6.01	5.38	0
14	1.25	2.34	4.36	8.14	11.7	15.2	18.6	25.1	31.5	28.2	21.4	17.0	13.9	11.7	9.96	8.63	7.57	6.72	6.01	0
15	1.35	2.52	4.70	8.77	12.6	16.4	20.0	27.1	34.0	31.2	23.8	18.9	15.4	12.9	11.0	9.57	8.40	7.45	0.42	0
16	1.45	2.70	5.04	9.40	13.5	17.5	21.4	29.0	36.4	34.4	26.2	20.8	17.0	14.2	12.2	10.5	9.25	8.21	0	
17	1.55	2.88	5.38	10.0	14.5	18.7	22.9	31.0	38.9	37.6	28.6	22.7	18.6	15.6	13.3	11.5	10.1	8.97	0	
18	1.64	3.07	5.72	10.7	15.4	19.9	24.4	33.0	41.3	41.0	31.2	24.8	20.3	17.0	14.5	12.6	11.0	9.79	0	
19	1.74	3.25	6.07	11.3	16.3	21.1	25.8	35.0	43.8	44.5	33.9	26.9	22.0	18.4	15.7	13.6	12.0	10.6	0	
20	1.84	3.44	6.42	12.0	17.2	22.3	27.3	37.0	46.4	48.1	36.6	29.0	23.8	19.9	17.0	14.7	12.9	0.95	0	
21	1.94	3.62	6.76	12.6	18.2	23.6	28.8	39.0	48.9	51.7	39.4	31.2	25.6	21.4	18.3	15.9	13.9	0		
22	2.04	3.81	7.11	13.3	19.1	24.8	30.3	41.0	51.4	55.5	42.2	33.5	27.4	23.0	19.6	17.0	14.9	0		
23	2.14	4.00	7.46	13.9	20.0	26.0	31.7	43.0	53.9	59.2	45.1	35.8	29.3	24.6	21.0	18.2	15.9	0		
24	2.24	4.19	7.81	14.6	21.0	27.2	33.3	45.0	56.4	62.0	48.1	38.1	31.2	26.2	22.3	19.4	17.0	0		
25	2.34	4.38	8.17	15.2	21.9	28.4	34.8	47.0	59.0	64.9	51.1	40.6	33.2	27.8	23.8	20.6	8.34	0		
28	2.65	4.94	9.23	17.2	24.8	32.1	39.3	53.2	66.6	73.3	60.6	48.1	39.4	33.0	28.2	24.4	0			
30	2.85	5.33	9.94	18.5	26.7	34.6	42.3	57.3	71.8	78.9	67.2	53.3	43.6	36.6	31.2	24.5	0			
32	3.06	5.71	10.7	19.9	28.6	37.1	45.3	61.4	77.0	84.7	74.0	58.7	48.1	40.3	34.4	0				
35	3.37	6.29	11.7	21.9	31.6	40.9	50.0	67.7	84.8	93.3	84.7	67.2	55.0	46.1	39.4	0				
40	3.89	7.27	13.6	25.3	36.4	47.2	57.7	78.1	98.0	108	103	82.1	67.2	56.3	0					
45	4.42	8.25	15.4	28.7	41.4	53.6	65.6	88.8	111	122	123	98.0	80.2	4.09	0					
50	4.96	9.25	17.3	32.2	46.4	60.1	73.5	99.4	125	137	145	115	64.6	0						
55	5.49	10.2	19.1	35.7	51.4	66.6	81.4	110	138	152	167	132	0							
60	6.03	11.3	21.0	39.2	56.5	73.2	89.4	121	152	167	190	48.8	0							

Fig. XVI-6 Horsepower ratings of No. 80 (1-in.) pitch USAS standard roller chain, single strand.

No. of teeth small spkt.	Revolutions per minute—small sprocket																			
	10	25	50	100	200	300	400	500	600	700	800	900	1000	1100	1200	1400	1600	1800	2000	2200
		Type I			Type II					Type III							Type IV			
11	0.81	1.85	3.45	6.44	12.0	17.3	22.4	27.4	32.3	37.1	32.8	27.5	23.4	20.3	17.8	14.2	11.6	9.71	6.95	0
12	0.89	2.03	3.79	7.07	13.2	19.0	24.6	30.1	35.5	40.8	37.4	31.4	26.8	23.2	20.4	16.2	13.2	11.1	0	
13	0.97	2.22	4.13	7.72	14.4	20.7	26.9	32.8	38.7	44.5	42.1	35.3	30.1	26.1	22.9	18.2	14.9	12.5	0	
14	1.05	2.40	4.48	8.35	15.6	22.5	29.1	35.6	41.9	48.1	47.0	39.4	33.6	29.2	25.6	20.3	16.6	13.9	0	
15	1.13	2.59	4.82	9.00	16.8	24.2	31.4	38.3	45.2	51.9	52.2	43.7	37.3	32.4	28.4	22.5	18.5	15.5	0	
16	1.22	2.77	5.17	9.65	18.0	25.9	33.6	41.1	48.4	55.6	57.5	48.1	41.1	35.6	31.3	24.8	20.3	17.0	0	
17	1.30	2.96	5.52	10.3	19.2	27.7	35.9	43.9	51.7	59.4	62.9	52.7	45.0	39.0	34.2	27.2	22.2	18.2	0	
18	1.38	3.15	5.87	11.0	20.4	29.5	38.2	46.7	55.0	63.2	68.6	57.5	49.1	42.5	37.3	29.6	24.3	16.6	0	
19	1.46	3.34	6.23	11.6	21.7	31.2	40.5	49.5	58.3	67.0	74.4	62.3	53.2	46.1	40.5	32.1	26.3	6.70	0	
20	1.55	3.53	6.58	12.3	22.9	33.0	42.8	52.3	61.6	70.8	79.8	67.3	57.5	49.8	43.7	34.7	28.4	0		
21	1.63	3.72	6.94	12.9	24.2	34.8	45.1	55.1	64.9	74.6	84.1	72.4	61.8	53.6	47.0	37.3	30.5	0		
22	1.71	3.91	7.30	13.6	25.4	36.6	47.4	58.0	68.3	78.5	88.5	77.7	66.3	57.5	50.4	40.0	32.8	0		
23	1.80	4.10	7.65	14.3	26.6	38.4	49.7	60.8	71.7	82.3	92.8	83.0	70.9	61.4	53.9	42.8	35.0	0		
24	1.88	4.30	8.02	15.0	27.9	40.2	52.1	63.7	75.0	86.2	97.2	88.4	75.5	65.5	57.4	45.6	37.3	0		
25	1.97	4.49	8.37	15.6	29.2	42.0	54.4	66.5	78.4	90.1	102	94.1	80.3	69.6	61.1	48.5	36.3	0		
28	2.22	5.07	9.47	17.7	33.0	47.5	61.5	75.2	88.6	102	115	111	95.2	82.5	72.4	57.5	4.95	0		
30	2.40	5.47	10.2	19.0	35.5	51.2	66.3	81.0	95.5	110	124	124	106	91.5	80.3	63.7	0			
32	2.57	5.86	10.9	20.4	38.1	54.8	71.1	86.9	102	118	133	136	116	101	88.5	70.2	0			
35	2.83	6.46	12.0	22.5	41.9	60.4	78.3	95.7	113	130	146	156	133	115	101	48.4	0			
40	3.27	7.46	13.9	26.0	48.4	69.8	90.4	111	130	150	169	188	163	141	124	0				
45	3.71	8.47	15.8	29.5	55.0	79.3	103	126	148	170	192	213	194	166	76.4	0				
50	4.16	9.49	17.7	33.0	61.6	88.8	115	141	166	190	215	239	209	97.8	0					
55	4.61	10.5	19.6	36.6	68.3	98.4	128	156	184	211	238	265	138	19.1	0					
60	5.07	11.6	21.6	40.2	75.1	108	140	171	202	231	261	181	56.8	0						

Fig. XVI-7 Horsepower ratings of No. 100 (1¼-in.) pitch USAS standard roller chain, single strand.

No. of teeth small spkt.	Revolutions per minute—small sprocket																		
	10	25	50	100	150	200	300	400	500	600	700	800	900	1000	1100	1200	1300	1400	1500	1600
	Type I		Type II Lubrication				Type III Lubrication					Type IV Lubrication								
11	1.37	3.12	5.82	10.9	15.6	20.3	29.2	37.8	46.3	54.5	46.2	37.8	31.7	27.1	23.5	20.6	18.3	16.3	14.7	0
12	1.50	3.43	6.39	11.9	17.2	22.3	32.1	41.5	50.8	59.8	52.8	43.2	36.2	30.9	26.8	23.5	20.9	18.7	16.8	0
13	1.64	3.74	6.97	13.0	18.7	24.3	35.0	45.3	55.4	65.3	59.4	48.6	40.7	34.8	30.1	26.5	23.5	21.0	18.9	0
14	1.77	4.05	7.55	14.1	20.3	26.3	37.9	49.1	60.0	70.7	66.4	54.3	45.5	38.9	33.7	29.6	26.2	23.5	12.6	0
15	1.91	4.36	8.13	15.2	21.9	28.3	40.8	52.9	64.6	76.1	73.6	60.3	50.5	43.1	37.4	32.8	29.1	26.0	4.50	0
16	2.05	4.67	8.72	16.3	23.4	30.4	43.8	56.7	69.3	81.7	81.1	66.4	55.6	47.5	41.2	36.1	32.0	28.7	0	
17	2.19	4.99	9.31	17.4	25.0	32.4	46.7	60.5	74.0	87.2	88.8	72.7	60.9	52.0	45.1	39.6	35.1	31.4	0	
18	2.33	5.31	9.90	18.5	26.6	34.5	49.7	64.3	78.6	92.7	96.7	79.2	66.4	56.7	49.1	43.2	38.3	28.6	0	
19	2.47	5.62	10.5	19.6	28.2	36.5	52.7	68.2	83.4	98.3	105	85.9	72.0	61.4	53.3	46.7	41.5	20.8	0	
20	2.61	5.95	11.1	20.7	29.8	38.6	55.7	72.1	88.2	104	113	92.8	77.7	66.4	57.5	50.5	44.8	11.1	0	
21	2.75	6.27	11.7	21.8	31.4	40.7	58.7	76.0	93.0	110	122	99.8	83.6	71.4	61.9	54.3	48.2	0		
22	2.89	6.59	12.3	23.0	33.1	42.8	61.7	79.9	97.7	115	131	107	89.7	76.6	66.4	58.2	50.2	0		
23	3.03	6.92	12.9	24.1	34.7	45.0	64.8	83.9	103	121	139	114	95.9	81.8	70.9	62.3	42.4	0		
24	3.18	7.25	13.5	25.2	36.4	47.1	67.8	87.9	107	127	145	122	102	87.2	75.5	66.3	32.6	0		
25	3.32	7.57	14.1	26.4	38.0	49.2	70.9	91.8	112	132	152	130	109	92.8	80.4	70.5	0			
28	3.75	8.55	16.0	29.8	42.9	55.6	80.1	104	127	149	172	154	129	110	95.3	0				
30	4.04	9.22	17.2	32.1	46.3	59.9	86.3	112	137	161	185	170	143	122	106	0				
32	4.33	9.88	18.4	34.4	49.5	64.1	92.4	120	146	172	198	188	157	134	102	0				
35	4.77	10.9	20.3	37.9	54.6	70.7	102	132	161	190	218	215	180	154	68.3	0				
40	5.51	12.6	23.5	43.8	63.1	81.7	118	153	186	220	252	255	204	103	0					
45	6.26	14.3	26.6	49.7	71.6	92.8	134	173	212	249	286	230	144	23.9	0					
50	7.01	16.0	29.8	55.7	80.2	104	150	194	237	280	310	220	90.8	0						
55	7.77	17.7	33.1	61.8	89.0	115	166	215	263	310	288	154	0							
60	8.58	19.6	36.5	68.2	98.2	127	183	237	290	342	250	104	0							

Fig. XVI-8 Horsepower ratings of No. 120 (1½-in.) pitch USAS standard roller chain, single strand.

appendix XVII

Case histories

The four case histories presented here are capsule résumés illustrating the engineering design process as discussed in Chap. 1. At the beginning of each is a statement of the "goal," and the résumé includes, particularly, considerations of concept, analysis, and decisions.

CASE 1 LOG SPLITTER

In preparing logs for firewood, cutting the logs to convenient length is relatively easy by hand or power saws, but splitting them is not nice exercise. The age-old splitting with wedges and an axe is effective but hardly efficient by today's standards. Thus, the objective was to develop a log splitter that is inexpensive, portable, and easy to operate.

This case history is a résumé of the development of a log-splitting machine designed, produced, and marketed by the Mackinaw Products Division of the Detriot Tap and Tool Company, Warren, Michigan.

The GOAL (see Fig. 1-1) was to develop a low-cost, wood-splitting device which would split the toughest wood encountered in the United

States and Canada and which would comply with the three objectives stated above: inexpensive, portable, and easy to operate.

It was decided that a gasoline engine would be the most suitable power source because the cost is low, service is readily available, and the engine is a rugged source of power. An electric motor was an alternative power source.

(*a*) *Methods of splitting* Three methods were considered: mechanical (using levers), pneumatic impact, and hydraulic.

The BASIC SPECIFICATION ruled out the first two systems in favor of the hydraulic method because the latter would provide better control of force dependability under "field conditions," and efficiency. After two years of design and testing of prototype machines of the hydraulic type at the factory and in the field, the following design was selected for further development. This design incorporated a sharp wedge (cast steel) against which the log would be forced by the ram, as shown in Fig. XVII-1.

(*b*) *Frame* The initial design of the frame which carries the ram at one end and the wedge at the other was a 4-in. by 4-in. by 6-ft I beam. It was soon learned that this beam did not have sufficient stiffness. Bending caused by the eccentric load produced by the splitting operation interfered with the motion of the ram; also, the beam was unduly heavy. A 5-in. I beam had sufficient rigidity but was too heavy.

Next, a hollow, rectangular beam 6 by 2 in. in section was formed of 11 gauge (0.20) sheet steel and welded along the seam. Weight was now reduced but the bending deflection was only slightly better than that in the original I section.

The hollow rectangular section of 6 by 2 in. now required a 4-in.-wide plate welded to the top of the tubular frame as a guide for the ram and to form a platform to which the wedge could be welded and on which the cylinder could be mounted. A plate 4 in. wide, $\frac{1}{4}$ in. thick, and 6 ft long would provide the guiding and mounting facilities and would improve the strength and deflection characteristics of the frame. Further improvements could be made by bending the beam in the direction opposite to its deflection in service and, while the beam was held in that prestressed condition, welding the $\frac{1}{4}$-in. plate in place. Upon release of the frame from the holding fixture, it would have favorable residual compressive stresses on the top side of the composite frame, which would increase its strength. See Art. 6-3.

The specification that the composite frame be straight and square with tolerances of ± 0.031 in. at any point along the length of the frame were met. Thus, the required strength and stiffness were realized, and the overall weight was satisfactorily reduced.

(c) *Control* To provide oil pressure in the cylinder, a single-stage, constant-volume gear pump was considered. The cylinder pressure during a typical cycle is, qualitatively, as shown in Fig. XVII-2 for pressure to "crack" the log, split it, and return the ram.

It was evident that a constant-oil-volume pump requires too long a time for the cycle. To reduce the time for a cycle, while providing a *maximum* force of 19,000 psi, required a pump that can "automatically" be adjusted or adjust itself to *high-volume, low-force* or *low-volume, high-force* characteristics, as is required for splitting various types of wood.

Fortunately, there was commercially available a pump that could be adapted to meet the specific oil-volume and line-pressure requirements.

Fig. XVII-1 Log splitter in action.

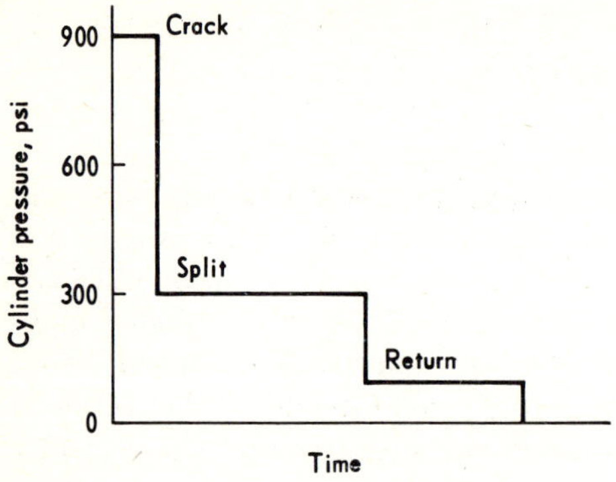

Fig. XVII-2 Splitting cycle.

This was a "duplex" pump containing two sets of gears. With both gear sets in the circuit, high oil volume at low force was provided. When greater force was required, an adjustable, pressure-sensitive, unloading valve would "drop out" one set of gears from the circuit, thus reducing the oil volume and allowing the pressure to increase.

To determine the relation between the quantity of fluid pumped (which determines the speed of the ram) and the pressure at discharge of the pump (which determines the force on the ram), we assume that the fluid is incompressible, that the flow is steady, and that there are no losses in the system.

An equation for the energy E is

$$E = V\,\Delta p$$

where V is the volume of fluid pumped per unit time.[1] Δp is the change in pressure, pump to outlet.

Let us put the above expression in terms of gallons per minute (gpm), pressure p in pounds per square inch, and gauge (psi). We introduce constants so that the units of the equation for the energy E are in terms of foot-pounds per minute and the expression may be divided by the quantity "33,000 ft-lb per min per hp" and also divided by the efficiency of the pump and cylinder e, to obtain the required engine horsepower to drive the system. Therefore,

$$V = \text{gpm} \times 8.33 \times (\text{sp. gr.}) \qquad \text{units in lb per min}$$

$$\Delta p = \frac{\text{psi} \times 144}{62.4 \times (\text{sp. gr.})} \qquad \text{units in ft (head)}$$

or $$V\,\Delta p = \text{gpm} \times \text{psi} \times \frac{8.33 \times (\text{sp. gr.}) \times 144}{62.4 \times (\text{sp. gr.})} \qquad \text{units in ft-lb per min}$$

[1] See any book on thermodynamics.

$$\text{or} \quad \text{hp} = \text{gpm} \times \text{psi} \times \frac{8.33 \times 144}{62.4 \times 33{,}000e}$$

$$\text{hp} = \frac{\text{gpm} \times \text{psi}}{1{,}716e} \qquad \text{(XVII-1)}$$

where gpm = quantity of fluid pumped, gal per min
psi = pressure at pump outlet, psig
e = mechanical efficiency of pump

From the data obtained in early tests, it was noted that the majority of hardwood logs required approximately 900-psi cylinder pressure to "crack" them initially and a sustained pressure of 300 psi to complete the splitting of the log. Exceptions were stringy, twisted types of wood, such as elm, cottonwood, black gum, etc., which required an average sustained line pressure of approximately 1,200 psi.

The power source chosen was a 3-hp, four-cycle gasoline engine. The pump was a "duplex gear pump" with two sets of gears. With one gear set in the circuit, low oil volume at high pressure is provided, and with both sets, high oil volume at low pressure is delivered. Thus, for "cracking" the log, one set of gears is required for high cylinder pressure, and for completing the crack and for return of the ram both sets of gears are required for low pressure. The gears are "shifted" by an adjustable, pressure-sensitive valve which "drops out" one set of gears as required

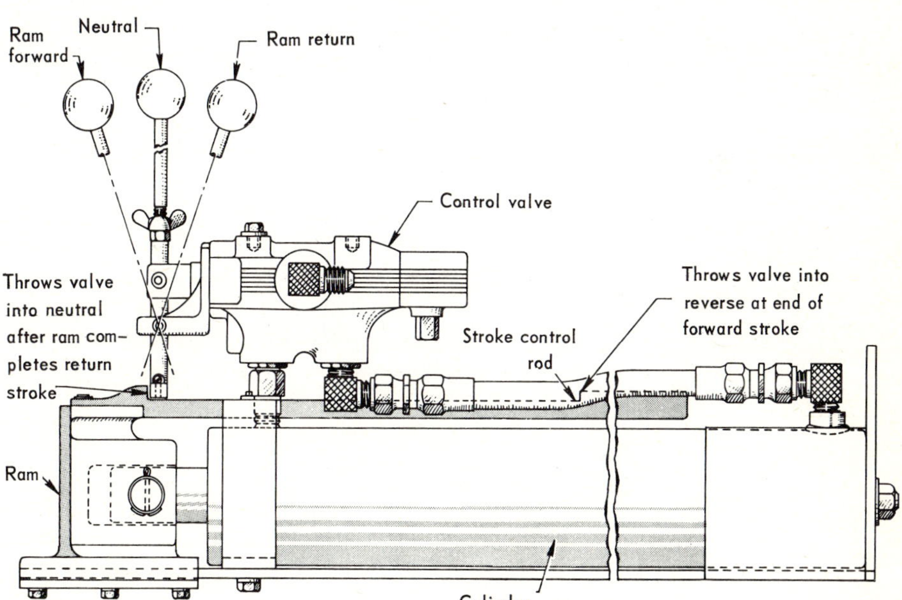

Fig. XVII-3

Fig. XVII-4

for the cracking part of the cycle. By this combination, high pressure is available for cracking the log at low ram speed, whereas the remainder of the cycle is at lower cylinder pressure but at high ram speed.

The characteristics of the pump are that, with one set of gears, the delivery is 1.5 gpm at 2,000 psi, and with both sets of gears, the delivery is 6 gpm at 500 psi. Assuming a pump efficiency of 85 percent, the power required to drive the pump, by Eq. (XVII-1), is

$$\text{hp} = \frac{\text{gpm} \times \text{psi}}{1{,}716 \times e} = \frac{1.5 \times 2{,}000}{1{,}716 \times 0.85} = 2.06$$

or
$$\text{hp} = \frac{6 \times 500}{1{,}716 \times 0.85} = 2.06$$

Thus it is evident that the 3-hp engine will provide ample power to crack the toughest logs, complete the splitting, including the encountering of knots or other tough obstructions, and return the ram to its initial position in a short time.

As shown in Fig. XVII-3, the operator places the log against the wedge and moves the control lever to "ram-forward" position so that the ram moves at high speed to begin the crack; then the control valve increases the ram speed, as explained above for the splitting operation. When the ram approaches the end of its forward stroke, the control lever is moved (by the stroke-control rod) into reverse so that the ram stops and begins its return stroke until the detent on the ram end of the stroke-control rod sets the control lever in neutral. Thus the operator places a log in position, moves the lever to ram-forward position, and "reaches for another log." See Fig. XVII-4.

CASE 2 TWO-CARTRIDGE REVOLVER

This case study was abstracted from a history furnished by courtesy of Smith and Wesson, Inc. The GOAL was a revolver capable of firing both 22-caliber Remington jet cartridges and conventional, low-cost, 22-caliber ammunition (short, long, or long rifle) for practice. The jet cartridge, 2,460-fps muzzle velocity, has a necked-down magnum case for a center-firing pin, and the conventional cartridge has a cylindrical case for a rim-firing pin. Figure XVII-5 shows the outlines of these two cartridges.

The idea of a two-cartridge revolver was to provide in one revolver the economy of the 22-caliber rim-fire cartridge and alternatively the power of the 22-caliber, high-velocity center-fire cartridge. An unwritten requirement in the design was that standard parts should be used where-ever possible, instead of special parts, and that special production processes should be avoided if possible. In this project, few equations were necessary, with the exception of trigonometric relations; instead, *ideas* were required which required analysis and development.

Although some ten ideas were proposed, they were narrowed down

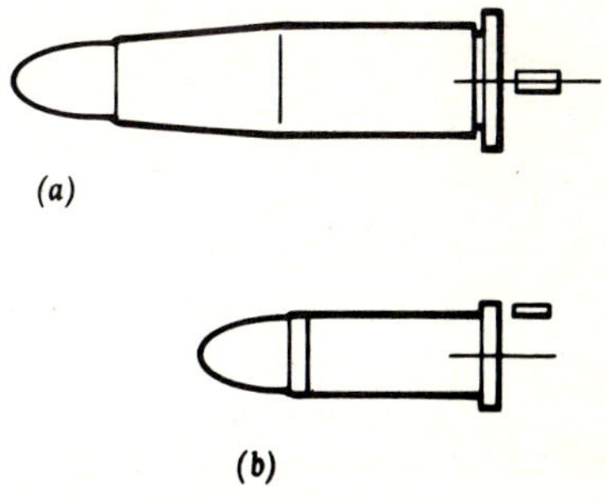

Fig. XVII-5 Cartridges: (*a*) jet center-fire; (*b*) conventional rim-fire.

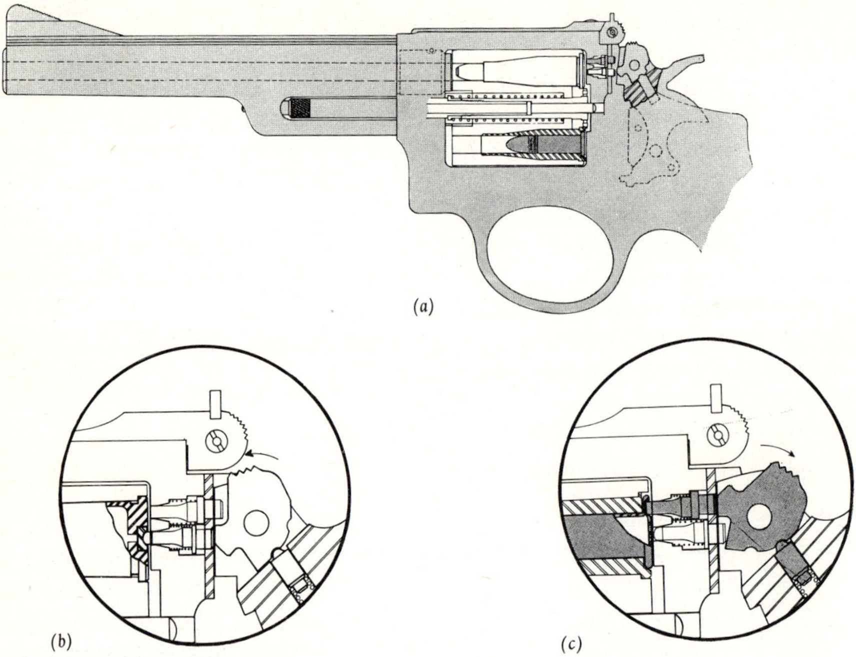

Fig. XVII-6 (*a*) Cylinder arrangement. (*b*) Center fire. (*c*) Rim fire.

to the following three basic approaches:

(*a*) Interchangeable cylinders, one for conventional rim-fire and the other for high-velocity center-fire cartridges
(*b*) "Bent-bore" cylinders
(*c*) Inserted chamber cylinders

The first system, interchangeable cylinders, was discarded because it did not provide for both types of cartridges in the gun at the same time.

In the bent-bore system, the axis of the 22-caliber *rim-fire* chamber was tipped to position the rim for striking by the same firing pin used to fire the 22-caliber *center-fire* cartridge. The gun functioned satisfactorily in test firing but did not have as high accuracy as the straight-through chamber; also, the machining of the rim-fire chamber with its axis tipped at an angle to the bullet bore in the cylinder was difficult and appeared impractical for production.

The idea of chamber inserts was then explored. The cylinder designed for the conventional 22-caliber rim-fire cartridge was "chambered" (bored out) for the necked-down high-velocity cartridge, as shown in the upper chamber in Fig. XVII-6(*a*), and the lower chamber is shown provided with an insert to accommodate a 22-caliber rim-fire cartridge. Previous experience with similar cylinders made for rim-fire cartridges but using high-pressure ammunition indicated adequate cylinder strength; this was confirmed by high-pressure proof tests.

As soon as the decision was reached to develop the inserted-chamber system, a firing-pin and hammer arrangement, quite similar to the final version, was conceived. The novel feature is that the hammer is fitted with twin firing pins and a selective striker on the hammer, so that, as shown in Fig. XVII-6(*b*), the hammer-mounted striker is in position for the center-fire magnum cartridge. At (*c*) the striker has been moved clockwise to bring the rim-firing pin in position for the 22-caliber rim-fire conventional cartridge. The striker is retained in one or the other position by the spring-loaded dog which falls into the selected detent on the back of the striker.

Initial layouts of the arrangement were sketched two-times size on a revolver-frame print and five-times size on a hammer drawing print; from these, sketches were made and the parts were fabricated in the Experimental Department. The revolver was test-fired, using two cylinders. Although some "bugs" were found, it was apparent that the system could be refined to function satisfactorily. At this point, a 10-times-size layout was made. It suggested several alterations to improve the functioning of the dual-firing mechanism and to display the possibility of improvements to the Production Department.

Next, three sample revolvers were made and tested. Their performance indicated the need for some modifications in the shape of the detent and in the heat treatment of parts to reduce distortion.

With some minor alterations suggested by the Production Department in the interest of tooling costs, ease of production, and service replacement, the design was considered complete and tooling started.

Lack of space precludes giving a complete account of the details of this case history which is quite complicated but, in its progress, equally exciting.

In service on the testing range, one final trouble appeared: The 22-caliber jet cases were difficult to extract; this was found to have been caused by excessive oil in the chambers. The trouble was readily eliminated.

CASE 3 AIRCRAFT WING-FLAP DRIVE SYSTEM

This history presents the predominant features of the system proposal and the design approach for the wing-flap drive set of the Grumman Model A-6A Navy Intruder airplane.[1]

[1] Data by courtesy of J. M. Olchawa, Manager, Precision Products Engineering, Foote Bros. Gear Operation, Hewitt-Robins, Inc., Chicago, Ill.

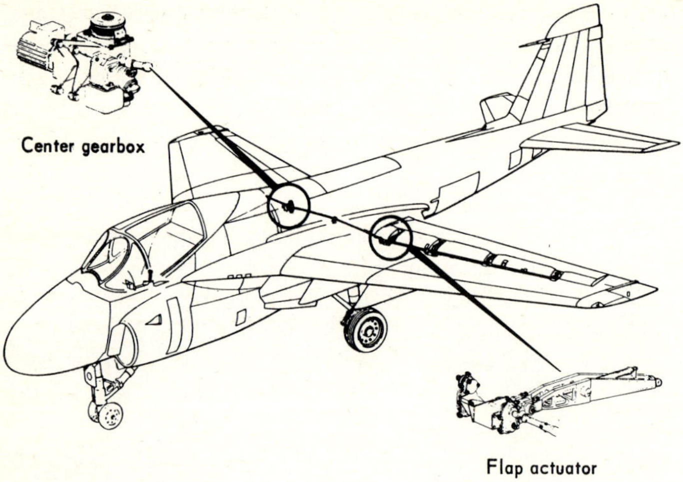

Fig. XVII-7 Flap drive.

The chosen basic system consists of a pilot-controlled, center gearbox, torque tubes, cross-axis helical- and bevel offset-gear boxes, and linear actuators to extend and position the four wing flaps. (See Fig. XVII-7.) The center gearbox distributes power to the eight wing-flap actuators, four on each wing, so that the flaps can be extended and positioned for flight conditions.

The fundamental design philosophy for the system embraces an all-mechanical drive from the center gearbox and also maximum reliability, interchangeability and symmetry of parts (for two wings), minimum weight, maximum efficiency, and maintainability. These objectives are referred to in the last section of this history.

(*a*) *Center gearbox* The center gearbox (see Fig. XVII-8) is connected to two power sources. The normal-operating source is a 12,000-rpm hydraulic motor using the aircraft's 3,000-psi pressure. The auxiliary source is a 400-cycle, a-c electric motor, which can be used in an emergency in the event that the hydraulic power is lost.

Either power source may drive the system through a planetary-gear set. In normal operation, the hydraulic motor, driving through a hydraulically released, spring-actuated brake, turns the ring gear (see Fig. XVII-8), and the sun gear connected to the electric motor is held locked by an integral brake at the hydraulic motor. The time required for full flap extension by the hydraulic motor is approximately 8 sec. When electrical operation is selected, the hydraulic brake holds the ring gear and the electric motor turns the sun gear.

An inertia clutch is provided to absorb the inertia of the relatively high-speed motors and the related gears to limit momentary transmitted torques. The torque-limiting inertia clutch is provided in the gear-train set at about 50 percent above maximum load requirements to prevent high inertia of the prime movers from damaging the gearboxes or torque tubes in the event that the flap jams.

(*b*) *Torque tubes* In order to accommodate wing angularity and structural obstacles between the center gearbox and the four actuators on each wing, the torque tubes were provided with universal joints positioned by offset gearboxes leading to the right-angle gearboxes to drive the actuator screws. In addition, the necessity for wing folding for aircraft storage required a telescoping spline joint at each torque tube.

Critical speeds of the torque tubes presented a problem since the output speed from the center gearbox was approximately 1,080 rpm, decreasing to 550 rpm for the outboard wing actuators. This reduction in speed was necessary because of varying strokes required of the carriages from 16 in. for the inboard actuators to 9½ in. for the outboard actuators. From consideration of the critical speeds dictated by torsional deflection and for minimum weight, hollow tapered aluminum tubes were specified.

(*c*) *Wing-flap drive* Each wing flap is driven through a right-angle, bevel-gear box, as shown in Fig. XVII-9, to transmit motion from the torque tube to the cam-track-guided carriage which positions the wing flap.

A safety problem arose at this location, owing to the possibility that one of the wing-flap drives would "bind"; in this event the full power

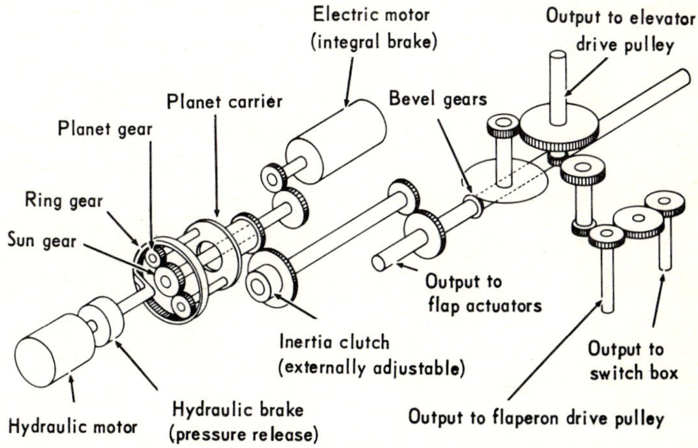

Fig. XVII-8 Center gearbox.

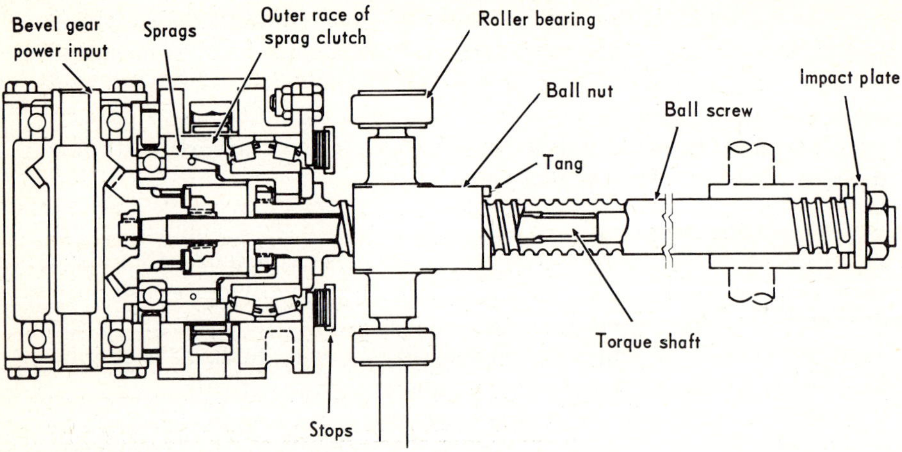

Fig. XVII-9 Flap actuator.

from the center gearbox would be transmitted through the jammed wing-flap drive.

As shown in Fig. XVII-9, a sprag-type load-sensing clutch is provided between the bevel-gear power input and the ball screw in order to limit the transmitted load to 150 percent of normal rating. At this point, the sprag-type clutch locks and stalls the entire system in the event of overload.

In case of an overrun of the carriage during extension, a torsion stop at the after end of the ball screw engages a tang on the nut portion of the assembly, causing a torsion windup of the screw to ease impact.

(*d*) *Flap actuator* The drive from the bevel-gear box to the guided carriage was decided in favor of ball screws (see Art. 12-7) rather than the acme type, because of the high efficiency of the ball screw (over 90 percent) and also to provide a high degree of reversibility. The reversibility is desirable in order to utilize aerodynamic forces for partial retractions of the flaps if "blow-back" load becomes excessive. This occurs at higher than normal landing speed at approximately 170 knots.

In case of overrun of the screw during retraction of the flap, cushioned stops, as shown in Fig. XVII-9, absorb the shock. Both stops are nonjamming.

General design features This is a statement of system concept and design features for the major components of the drive. Some of the design features were referred to earlier but are included here for completeness (see Fig. 18-10).

(*a*) *Operating units* The center-drive gearbox and offset- and bevel-gear boxes are all of similar design and construction, with emphasis on simplicity, reliability, minimum number of parts, and maximum efficiency.

Housings are cast aluminum for maximum strength-weight ratio, without liners or inserts. Sealed rolling-contact bearings are used for minimum friction and weight and are lubricated for 6,000 cycles of operation. Emergency electric-motor operation is designed for fewer cycles of operation. All gearboxes are sealed to exclude contamination and are vented to equalize pressure.

The center-drive gearbox has case-carburized, hardened, and heat-treated gears, whereas the offset- and bevel-gear boxes use only liquid-carburized, case-hardened gears. The offset- and bevel-gear boxes are designed for use on either right- or left-wing installations.

(*b*) *Torque tubes* Thin-walled aluminum-alloy tubing is used for optimum weight-strength ratio and stiffness; it is anodized for corrosion protection. End spline fittings are steel, cadmium-plated for corrosion protection. Universal joints are grease-packed for 6,000 cycles of operation, with sealed joints.

(*c*) *Actuators* Most of the design features have been noted earlier in this account. In addition, the screws are "gun-drilled" with axial holes 0.38 in. in diameter to reduce weight. The column capacity of screws was determined for adequate compressive loads, and the wear life of the screws was checked for 6,000 cycles of operation.

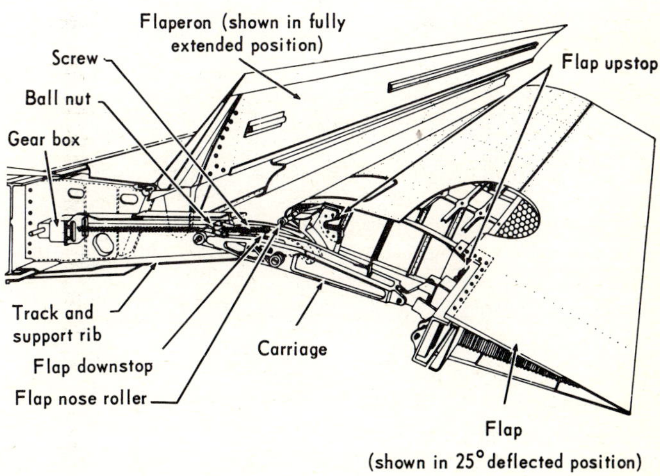

Fig. XVII-10 Flap actuation.

(*d*) *Qualification tests* One set of test units and equipment was provided to include the center gearbox with one complete wing system. The tests included the following:

Endurance	Vibration
Low-temperature operation	Sand and dust
High-temperature operation	Salt spray
Temperature shock	Humidity
Altitude	Static limit and ultimate load

CASE 4 "GORGO," THE MECHANICAL MONSTER

Since this device for horror motion pictures has a "single purpose," it was made as simple as possible consistent with its intended functions. Because of the unique requirements in the design of the beast, i.e., to attain lifelike motions, several means of control were explored: hydraulic, pneumatic, electromagnetic, and one of "biasing" the moving parts to a central position by means of flexible connectors (rubber bands). See Fig. XVII-11.

It was decided early in the design that the movements of the parts of the body—trunk, legs, and arms—could best be accomplished by a man inside, with the coordinated movements of the tail, ears, eyes, and jaw remotely controlled by one of the methods mentioned above.

Hydraulic control was discarded because of the weighty pipework. Electromechanics was discarded because of excessive complexity of controls. The flexible connector was discarded because movements of parts would have to be carried out by a man inside the rubber skin.

It appeared the optimum solution was externally controlled pneumatics as the source of power, mainly because of ease of installation, flexibility of use, and realism of movements.

The external controls were divided into four units: tail, ears, eyes, and jaw. The tail movements were achieved by the use of three air cylinders which were supplied with power on the inlet port adjacent to the rod end; these cylinders maneuvered the tail by means of tension wires running along the "vertebrae" of the tail. The force needed to give the tail a required "whiplash" effect was found by experiment to be around 140 lb. A reasonable cylinder diameter was $\frac{1}{2}$ in. so that, with a cylinder efficiency of 80 percent, an air pressure of 80 psi would produce the required force of 140 lb.

The three cylinders involved were so situated that sideways movements to the left or right would result in the cylinders not being used and the tail would be pulled to a fully extended position.

It was decided that the ears and eyes would operate as pairs, thus enabling one single-acting cylinder to operate both ears and another single-acting cylinder to operate both eyes, the movements being achieved by wires connected to the piston rods.

The jaw requirements were not quite so simple, since the varying speeds of the jaw opening and closing were not necessarily required in a set pattern. By use of variable orifices on the exhaust ports of the control values, suitable lifelike speeds of jaw operation were attained. Care had to be exercised here with regard to the power of the cylinder to avoid too

Fig. XVII-11 Gorgo

hard a shutting action, which would damage the fiber-glass teeth. Before the correct speeds of shutting were obtained, the monster lost at least one set of teeth.

Although the air cylinders were mounted as nearly as possible on the parts of the body requiring the operation, the main valve pack assembly was mounted on a board and fixed to a framework so that it could be clipped over the shoulders of the man inside the skin. This meant that the equipment must be as small and light as possible. It was then learned that the man did not have the time nor room to operate switches or buttons to operate the control valves. It was therefore decided that the control valves would be solenoid, air-pilot-operated, with signals sent to the solenoids from a master control box at some distance from the monster. By this means, the man inside could concentrate on walking and arm movements, without the worry of controlling other movements.

Low-voltage direct current was used in the controls, and it was found that the movements of the tail, ears, and eyes could be arrested part way through their motion, thereby giving a twitching effect. Nylon piping was used extensively through the monster.

The pneumatic system for Gorgo was designed by Baldwin Industrial Controls, Dartford, Kent, England, to whom appreciation is expressed; in particular, to Messrs. Allan and White of Baldwin, who coordinated the developments of controls[1] with Metro-Golden-Mayer engineers who designed the linkages and the monster.[1]

[1] See also *Machine Design,* Aug. 16, 1962, p. 148, for a layout of the controls.

Problems

In many of the following problems, assumptions are stated which would ordinarily be made by the designer. They are given here in order to limit the number of solutions for convenience in checking results. The student should consider whether the assumptions are reasonable, and he may wish to make other solutions with altered assumptions.

The majority of the problems are of the design type (not analysis) with data taken from actual installations. In some cases overall dimensions, capacities, and weights are stated which may not have direct use in the solution of the problem; in these cases they are included to give the student a comprehension of the size of the unit or component he is designing.

CHAPTER 2. MACHINE-DESIGN COMPUTATIONS

2-1. Determine the relative error in percentage due to an absolute error of 1 oz in weighing (*a*) 1 lb, (*b*) 7½ lb, and (*c*) 10 tons.

2-2. A rectangular bulkhead is found by measurement to be 12 ft wide and 8 ft high with errors not exceeding 1 in. Determine the possible relative error in percentage in the area of the bulkhead.

2-3. The seconds pendulum of a clock expands 0.05 percent per degree centigrade. How many seconds would the clock lose per day if the temperature is $1\frac{1}{2}$°C above that for which the pendulum beats true seconds?

Note:

$$t = 2\pi \sqrt{\frac{L}{g}}$$

2-4. A straight cantilever is made of a bar with rectangular section. If the load on the beam is increased 5 percent, and if it is desired that the deflection of the beam remains unchanged, determine the required percentage change in (*a*) the width of the beam or (*b*) the depth of the beam.

2-5. The natural frequency of vibration of a mass on the end of a cantilever is $f_n = 188\sqrt{1/\delta}$, where $\delta = PL^3/3EI$. If the length of the cantilever L is shortened by 1 percent, what will be the percentage of change in the natural frequency?

2-6. The indicated horsepower of an engine is given by the expression

$$\text{Ihp} = \frac{PLAN}{33{,}000}$$

Assume that the approximate values of the quantities are as follows:

P = mean effective pressure = 50 psi
L = length of stroke = 2 ft
A = area of piston, diameter D = 16 in.
N = working strokes per minute = 100

How precisely should P, L, D, and N be determined in order that the computed horsepower may be reliable to 2 percent?

CHAPTER 3. STRESS AND DEFLECTION ANALYSIS

3-1. A machine member is subjected to an impact load P produced by a weight W falling through a height h in. and striking the member so that it is stressed within the elastic limit of the material. Derive an equation for the impact load P in terms of W, h, and the flexibility constant C equal to pounds per inch of deflection (P/δ). The final equation should be reduced to its simplest terms.

3-2. A 12-in. steel I beam resting on supports 15 ft apart is impulsively loaded by a weight of 1,000 lb falling through a height of h in. and striking the beam at its midpoint. The moment of inertia I for the beam section is 246 in.^{4} Assuming that the supports are rigid, determine the value of h for a maximum induced stress in the beam equal to 20,000 psi.

3-3. Determine the maximum bending stress in the cast-iron beam with a hollow section shown in Fig. P 3-3. F_1 = 2,000 lb, and F_2 = 3,000 lb.

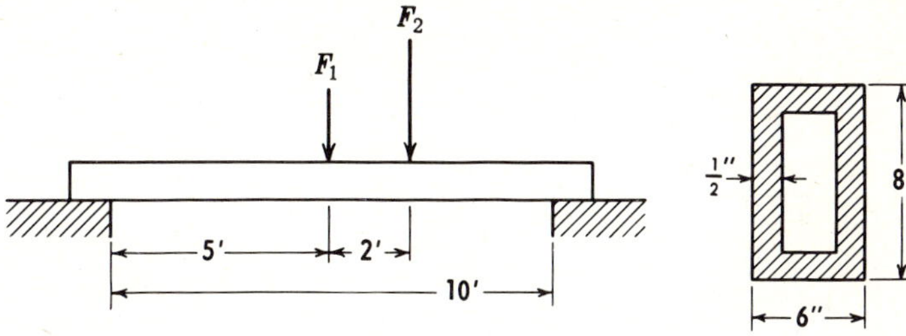

Fig. P 3-3

3-4. The spindle shown in Fig. P 3-4 is used to connect the brake shoe of an industrial spring-set brake to the brake arm. The loads P are each equal to 825 lb. Assuming an allowable stress equal to 17,500 psi, determine the diameter of the spindle that you would recommend.

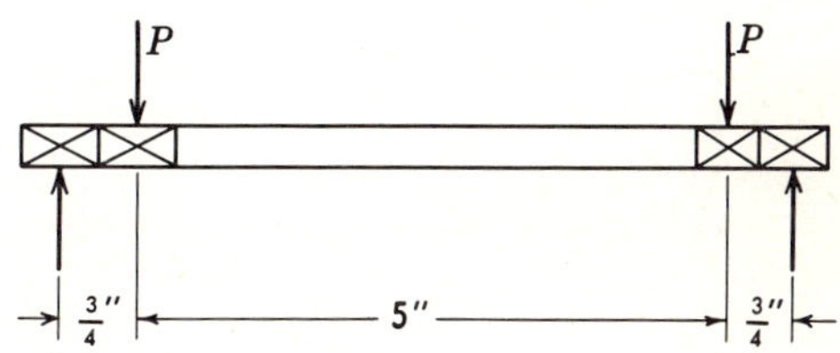

Fig. P 3-4

3-5. A simply supported steel beam, rectangular in section and 8 ft long, is to carry a concentrated load of 4,000 lb at a point 3 ft from one end. The depth of the section is to be twice the width. Using an allowable working stress in tension and compression of 10,000 psi, determine the recommended dimensions of the section.

3-6. A hollow nickel-steel shaft, whose outside diameter is twice the inside diameter, is to transmit 2,000 hp at 220 rpm. The shaft is subjected essentially to torsion. (*a*) Assuming an allowable stress of 32,500 psi, determine the size of hollow shaft required. (*b*) Determine the size of solid shaft for the above conditions, using the allowable stress as 25,000 psi. (*c*) Determine the percentage of saving in weight of the hollow shaft over the solid one.

3-7. A standard railway-car axle carries a total load of 15,000 lb. On account of the rounding of curves and swaying of the car, it may be assumed that one bearing can take two-thirds of the load. Draw the bending-moment diagram, and determine the maximum stress in the axle.

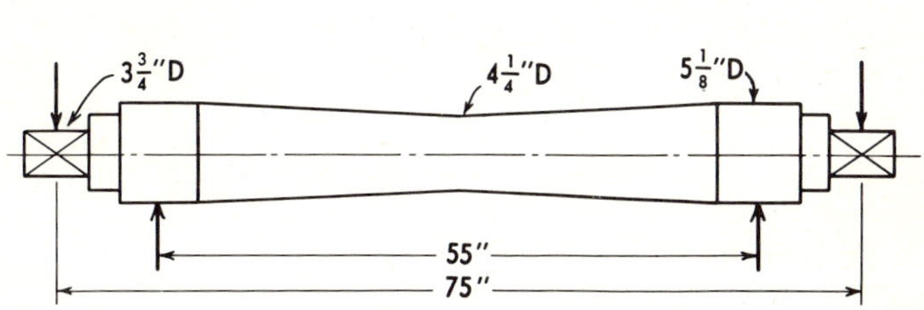

Fig. P 3-7

3-8. The load P on the C clamp shown in Fig. P 3-8 is 5,000 lb. Assuming that the clamp is made of steel casting, $h = 2b$, and $e = 6$ in., and that there is an allowable stress of 14,000 psi, determine the recommended dimensions b and h.

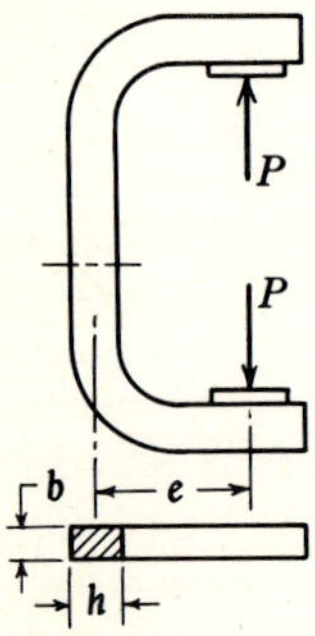

Fig. P 3-8

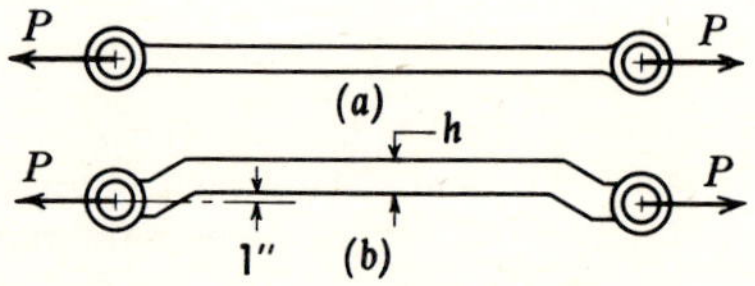

Fig. P 3-9

3-9. A steel link, as shown in Fig. P 3-9(a), has a cross section ½ in. square. It is necessary to replace this link by an offset one to provide clearance, as shown in (b).

Assuming the depth of the replacement link $h = 2$ in., determine the width required so that the strength of the offset link will be equal to that of the straight link.

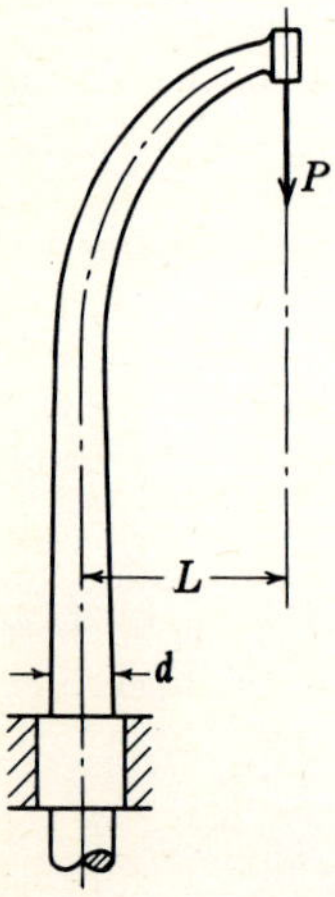

Fig. P 3-10

3-10. A 28-ft lifeboat to carry 50 persons is supported from two davits. The weight of the boat fully equipped is 3,000 lb, and an estimate of the weight of the occupants is 7,000 lb. Assuming that the overhang $L = 76$ in. and that the davits share the load equally, determine the diameter d for an allowable nominal stress of 15,000 psi.

3-11. The piston rod of a steam engine may be considered to be a column. Determine the recommended diameter of the rod, assuming the following:

Diameter of cylinder	18 in.
Maximum net steam pressure on piston	85 psi
Distance from piston to crosshead	64 in.
Material	SAE 1030 steel
Factor of safety	4
End-restraint coefficient n	3.5

3-12. A compression link of a valve mechanism may be considered as a column. The link is to be circular in cross section and 18 in. long. The estimated axial load on the link is 800 lb. Determine the recommended diameter for the link, assuming the following data:

Material	SAE 1020 steel
Factor of safety	2
End-fixity coefficient n	3.0

3-13. Figure P 3-13 shows the end view of a wall rack for storing 18-ft-long steel tubes triangularly stacked as shown. The supports are two cantilevered I beams each 8 ft long. The total weight of the stored tubes is 16,000 lb.

Assuming an allowable bending stress for the beams as 16,000 psi, determine their required section modulus and specify the size of standard steel beams.

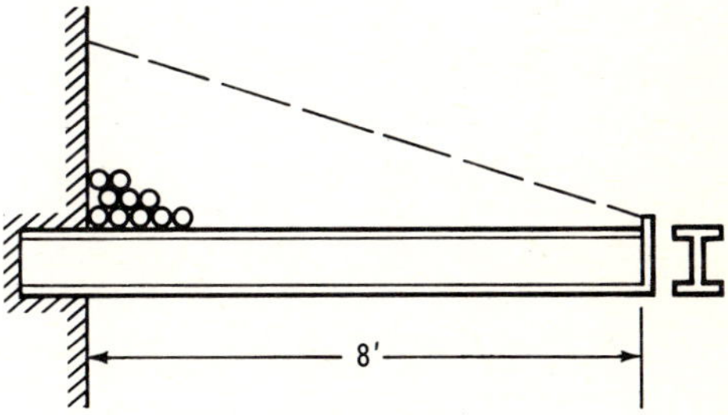

Fig. P 3-13

3-14. A part of a nail-heading machine is to be circular in cross section and 6 in. long, as shown in Fig. P 3-14, and is to carry a compressive load of 5,000 lb. Assuming SAE 1020 steel and a factor of safety of 2, determine the diameter required.

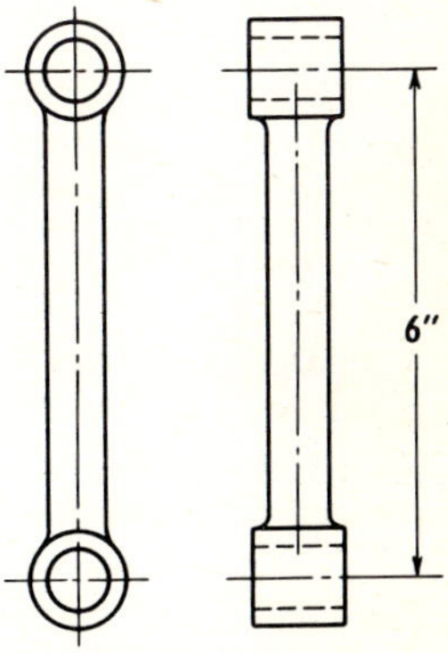

Fig. P 3-14

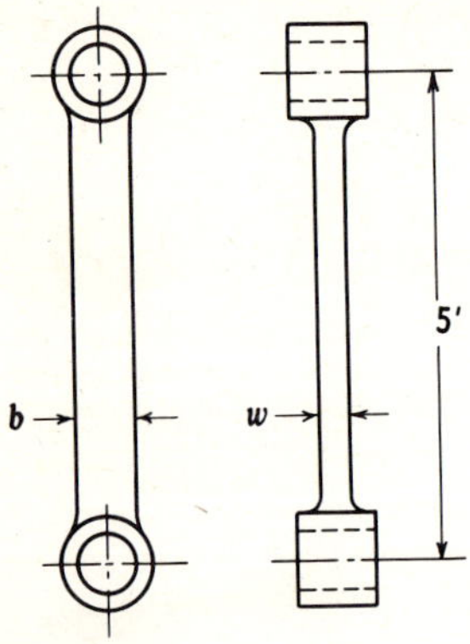

Fig. P 3-15

3-15. A steel strut, as shown in Fig. P 3-15, is rectangular in cross section and is to be part of the operating mechanism of a bascule bridge. The compressive load to be carried is 40,000 lb. Determine the following: (*a*) the relation between *b* and *w* required for equal strengths of the strut in failure by buckling in either plane; (*b*) the required dimensions *b* and *w*, assuming a factor of safety of 4 and SAE 1020 steel.

3-16. Assume a reasonable material, and recommend the diameter of the piston rod for the cylinder in Prob. 11-2.

Note: Assume that the length of the piston rod is 30 in., that its right-hand end is pin-connected to the operating lever, that the piston end of the rod is fixed-ended, and that the factor of safety is 2.5.

3-17. A weight W equals 100 lb, drops freely through a distance $h = 0.1$ in., as shown in Fig. P 3-17, and produces an impact on the stepped steel rod. Determine the maximum stress in the rod.

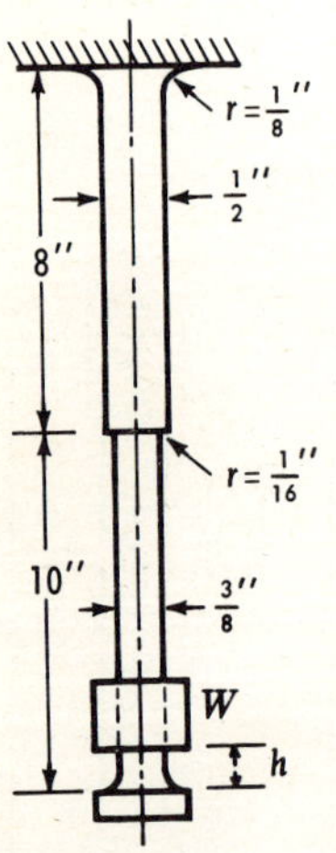

Fig. P 3-17

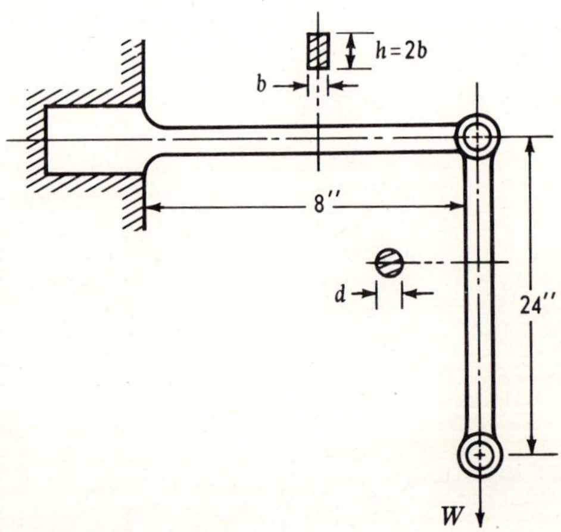

Fig. P 3-18

3-18. A support is composed of a horizontal cantilever of rectangular cross section and a vertical circular rod, as shown in Fig. P 3-18. Assume SAE 1030 steel for both parts and a factor of safety of 1.5. (*a*) Assuming that a weight W equal to 1,000 lb is gradually applied, determine the dimensions to be specified for the cross section of the cantilever and for the diameter of the vertical rod. (*b*) Assuming that the load of 1,000 lb is suddenly applied to the support, as designed, by dropping the weight W through a distance equal to $\frac{1}{32}$ in., determine the maximum stress in the support. Neglect stress concentration.

3-19. A steel circular rod as shown in Fig. P 3-19(*a*) is subjected to repeated impact produced by a weight $W = 150$ lb falling freely and repeatedly a distance $h = 0.05$ in., producing tension in the rod. The rod is rigidly supported by means of a fastener at the top.

(*a*) For the rod in Fig. P 3-19(*a*), assume diameter of rod $= \frac{1}{2}$ in.; radius of groove, $r = \frac{1}{16}$ in.; and $L = 10$ in. Determine the maximum induced stress at the groove and the required ultimate strength of the material.

(*b*) Trouble developed because of fracture in the rod as shown in (*a*), and improvement was suggested by decreasing the diameter of the rod to $\frac{3}{8}$ in. and increasing the length L to 20 in., as shown in Fig. P 3-19(*b*). The radius of the fillet at the top is $r = \frac{1}{16}$ in. Assume all other data the same as in part *a*. Determine the required ultimate strength of the material and recommend the SAE number.

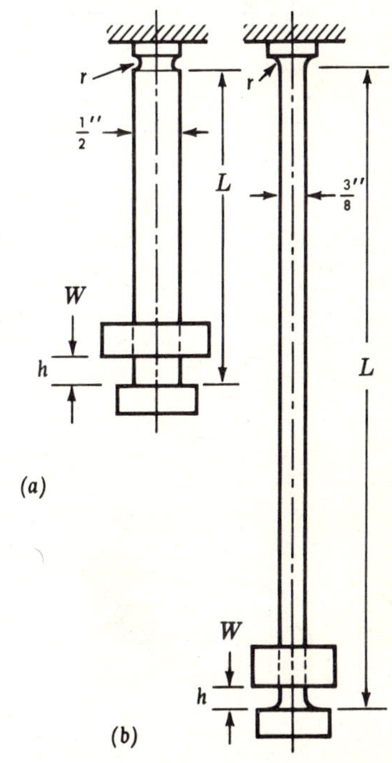

Fig. P 3-19

3-20. A bar shown in Fig. P 3-20(*a*) is loaded in tension by a weight $W = 20$ lb which drops repeatedly through a distance $h = 0.02$ in., thus producing an impact load P on the bar. This design gave trouble because the impact load broke the connection that supported the upper end of the bar.

To introduce resilience in the member and thus reduce the load on the connection at the upper end of the bar, a semicircular bar was proposed as a substitute for the straight bar, as shown in Fig. P 3-20(*b*). The weight W and the distance h are the same in both cases. The radius $R = 3$ in.

Assuming steel bars, and disregarding stress concentration at the eyes, determine the impact load P, and the maximum stress in the body of the bars for the following: (*a*) the straight bar; (*b*) the curved bar, $\delta = \pi P R^3/2EI$.

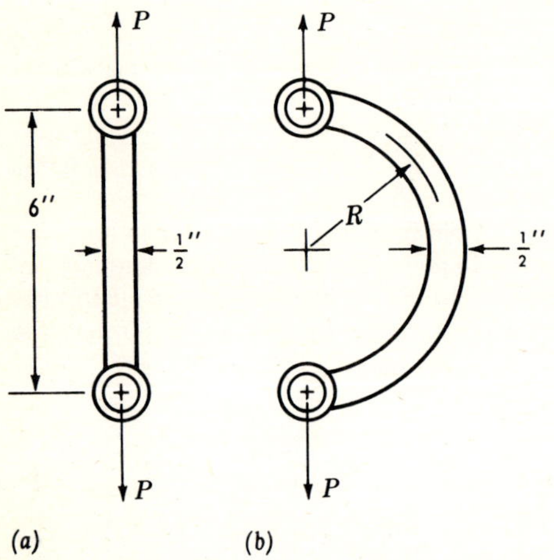

Fig. P 3-20

CHAPTER 4. STRESS CONCENTRATION IN MACHINE MEMBERS

4-1. A bar as shown in Fig. P 4-1 carries an axial load $P = 6{,}500$ lb. Determine the following: (*a*) the maximum nominal stress in the bar; (*b*) the maximum stress in the bar for the following fillet radii r: ½ in., ¼ in., ⅛ in.

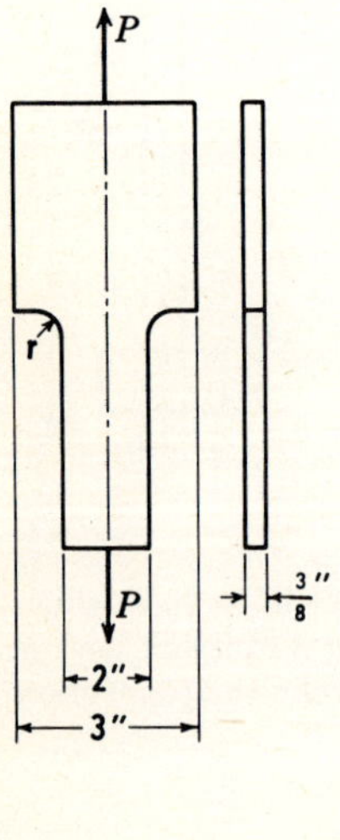

Fig P 4-1

4-2. A flat bar as shown in Fig. P 4-2 carries an axial load $P = 8{,}000$ lb. Determine the minimum radius r required so that the maximum stress in the member will not exceed 16,000 psi.

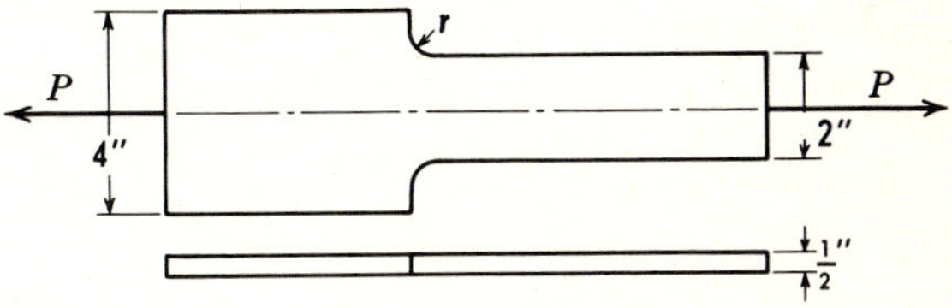

Fig. P 4-2

4-3. A stepped shaft of circular cross section carries a load $P = 2{,}500$ lb and is supported as shown in Fig. P 4-3. Determine (*a*) the maximum nominal stress in the shaft and (*b*) the maximum stress in the shaft.

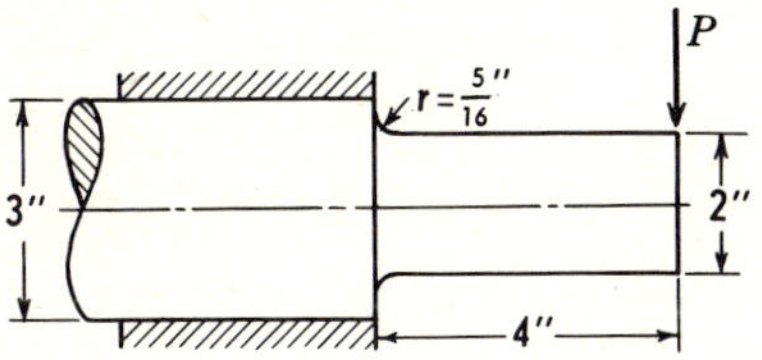

Fig. P 4-3

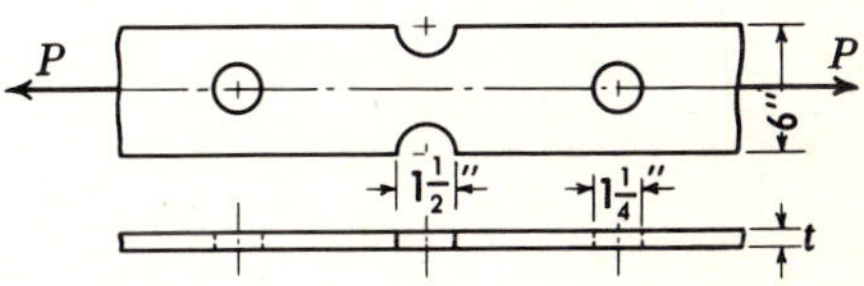

Fig. P 4-4

4-4. A flat bar as shown in Fig. P 4-4 is subjected to an axial tensile load P equal to 50 tons. Assuming that the stress in the bar is limited to 30,000 psi, determine the thickness t required.

4-5. A shaft transmitting 50 hp at 870 rpm rests in a bearing as shown in Fig. P 4-5 and has attached to it a pinion 3 in. in pitch diameter. The shaft fillet radius $r = \frac{1}{8}$ in. Assuming that the shaft is rigidly supported by the bearing and that the tooth load is uniformly distributed over the face of the gear, determine the following: (*a*) the torsional moment in inch-pounds transmitted by the shaft; (*b*) the maximum shearing stress in the shaft due to torsion; (*c*) the maximum bending stress in the shaft due to bending.

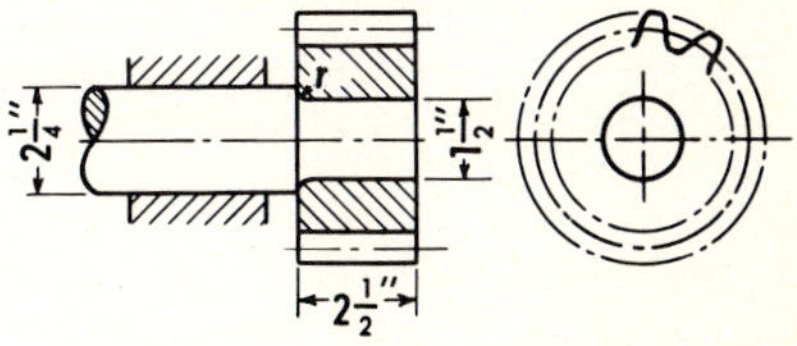

Fig. P 4-5

4-6. The spindle shown in Fig. P 4-6 is part of an industrial brake and is loaded as indicated. Each load $P = 825$ lb and is applied at the midpoint of its bearing area as shown.

Assuming that $d_1 = \frac{7}{8}$ in. and that $d_2 = \frac{5}{8}$ in., determine the radius r of the fillet required in order that the maximum stress in the part of the spindle of length equal to $\frac{3}{8}$ in. will not exceed the maximum stress in the part of diameter d_1.

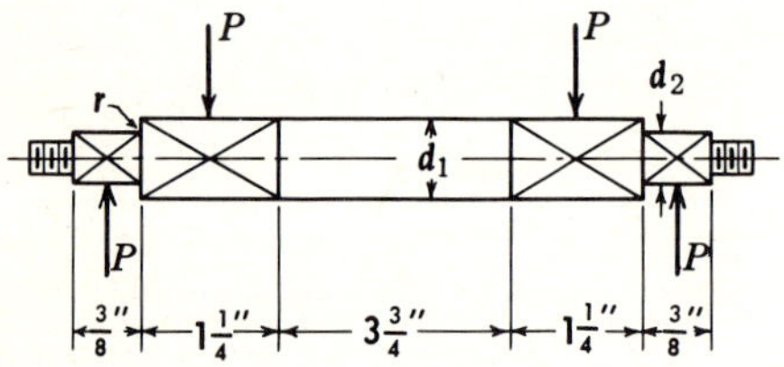

Fig. P 4-6

4-7. A tension member, as shown in Fig. P 4-7(a), supports an axial load equal to P lb. It is necessary to replace this member by one having a $\frac{5}{8}$-in. hole, as shown in (b). Assuming that the width of the replacement member is equal to 2 in., determine the thickness t and the fillet radius so that the maximum stress will not exceed that in (a).

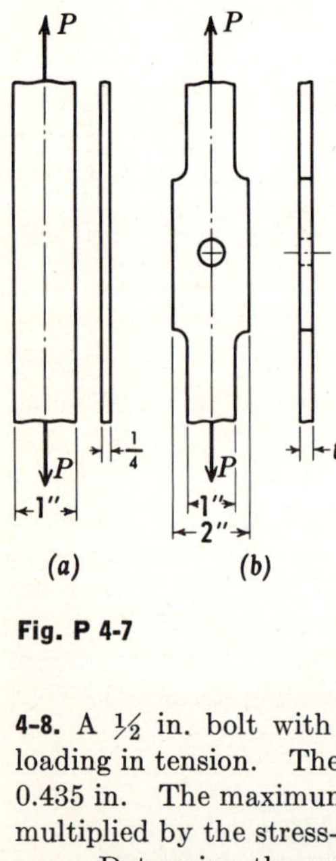

Fig. P 4-7

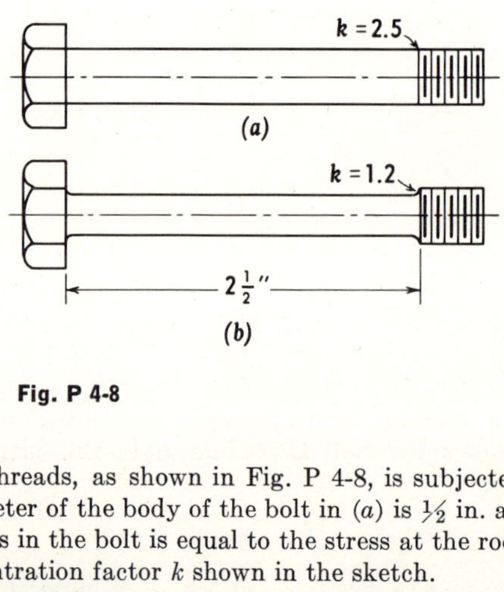

Fig. P 4-8

4-8. A $\frac{1}{2}$ in. bolt with NF threads, as shown in Fig. P 4-8, is subjected to shock loading in tension. The diameter of the body of the bolt in (a) is $\frac{1}{2}$ in. and in (b) is 0.435 in. The maximum stress in the bolt is equal to the stress at the root diameter multiplied by the stress-concentration factor k shown in the sketch.

Determine the percentage of improvement in energy-absorbing capacity for the bolt in (b) over that in (a), assuming that the maximum stress of both bolts is of the same value, that the material of the bolts is steel, and that the energy is absorbed in the $2\frac{1}{2}$-in. portion of the bolts.

4-9. The spindle in Fig. P 4-9 is repeatedly loaded by a weight $W = 600$ lb which falls through a distance equivalent to $h = 0.05$ in. The length $L = 5$ in. The material of the spindle is SAE 2345 nickel steel.

Assuming that the support for the spindle is rigid, determine the maximum stress induced in the spindle.

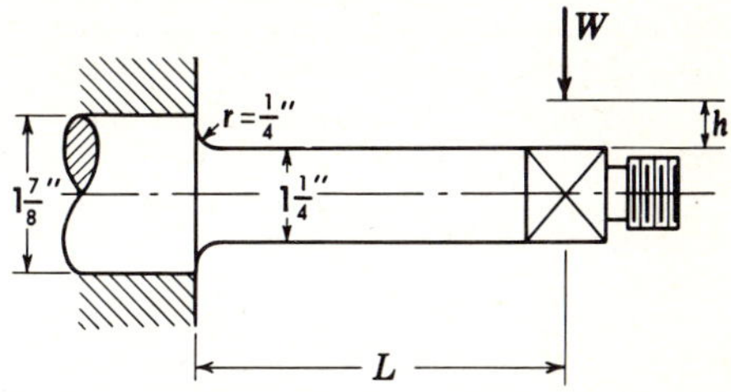

Fig. P 4-9

4-10. The cutter bar of a hand-operated embosser shown in Fig. P 4-10 is operated by pressing at C. The cutting edge is AB. The bar is made of hardened steel and broke along section BD after about 100 operations. What do you think was the cause of the fracture and how could the part be improved? Note that the sharp corner near B was produced by a bevel-edge grinding wheel.

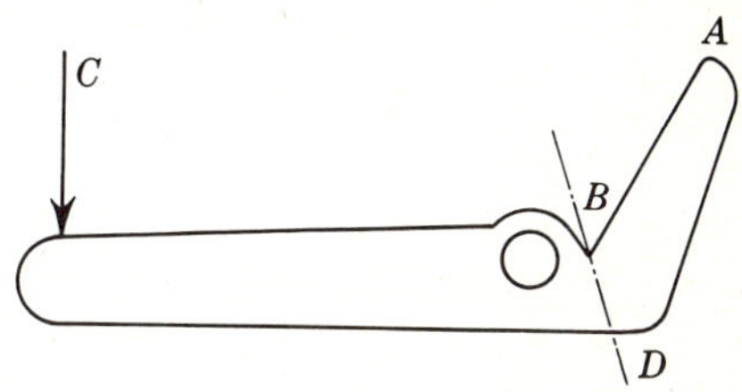

Fig. P 4-10

4-11. A business machine has 9,999 punchings in its mechanism, as the one shown in Fig. P 4-11. A part slides along AB and comes in contact with the cheek BC, thus producing an impact load P. The corner at B was made square to accommodate the sliding part which cannot be rounded, so that the punching has a very small radius at B. Trouble developed when the nose broke off owing to a fatigue crack originating at B.

Assuming that the material (SAE 1025) and the outline dimensions cannot be changed, what is your recommendation for improving the design? What tooling changes would be required?

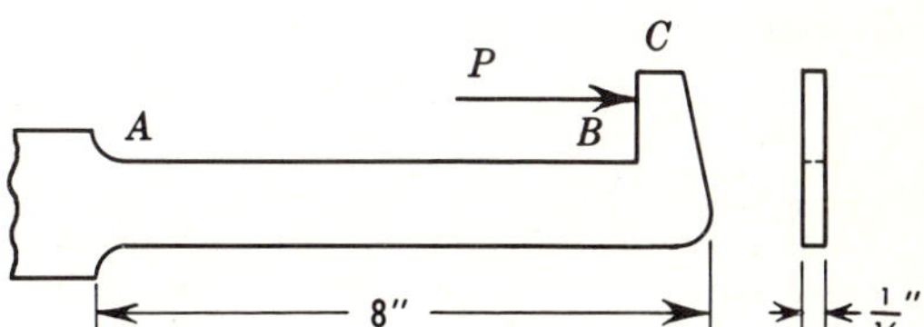

Fig. P 4-11

4-12. In a card-notching machine, the cutter is made by stamping a steel plate, as shown in Fig. P 4-12. The cutter is bent upward (before heat treatment) as shown, the card inserted at A, and the free end of the cutter is pressed down by the ram, thereby cutting the notch in the edge of the card.

Allowing a stress-concentration factor at the fillets at the end of the slots equal to 1.3, determine the minimum length L to limit the bending stress to 20,000 psi. The cards are 0.015 in. thick, and to notch the card the deflection of the cutter at A is 0.020 in. The plate thickness t is $\frac{1}{32}$ in.

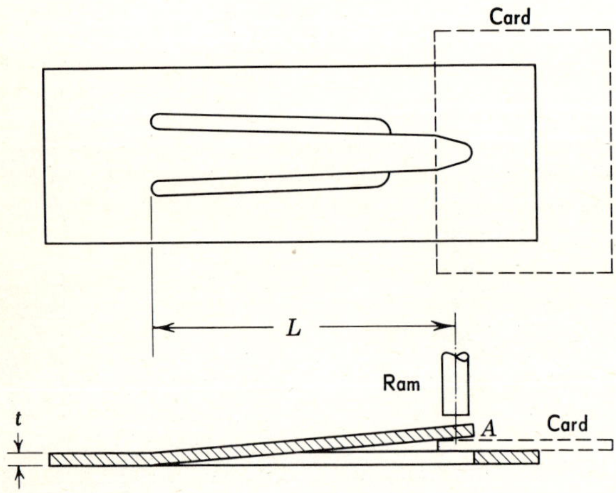

Fig. P 4-12

4-13. A cantilever as shown in Fig. P 4-13 was made of mild carbon steel (SAE 1020) and was designed to support a static load $W = 15$ lb. After the beam was installed, it was discovered that the weight W was not static but vibrated, so that actually the beam was subjected to repeated load. Measurement showed that the weight rose and fell a distance h equal to 0.05 in. Shortly thereafter the beam fractured at the fillet which had a "machined" radius $r = \frac{1}{32}$ in.

(*a*) Make calculations to show and state your conclusion whether or not the beam should have been expected to fail.

(*b*) For the same data, design a new cantilever with the same dimensions except with a fillet radius $r = \frac{1}{8}$ in. Determine whether a carbon steel would be satisfactory and, if so, its SAE number and the method for finishing the beam.

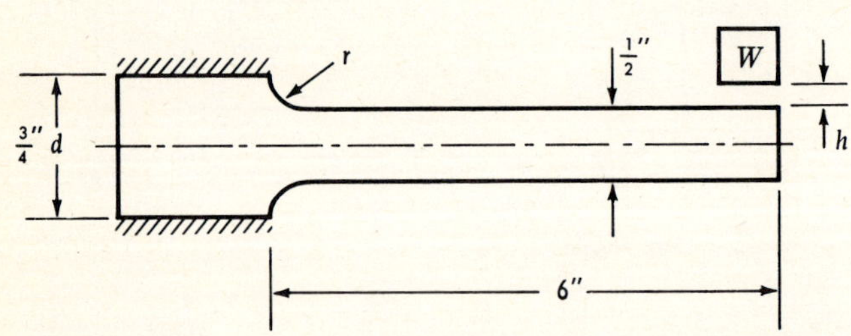

Fig. P 4-13

4-14. In Prob. 4-11, one means of strengthening the punching would be to introduce a three-quarters circle with radius r, as shown in Fig. P 4-14. The dimensions of the critical section are also shown.

(*a*) Develop an equation for the maximum stress in the member, assume the load $P = 1$ lb, and determine the maximum stress for values of r ranging from $\frac{1}{64}$ to $\frac{5}{16}$ in.

(*b*) Plot the values of maximum stress as a function of r.

(*c*) Draw a sketch of the end of the punching, showing to scale the recommended radius.

(*Note:* Figure X-4 in Appendix X may be used for estimating values of stress-concentration factors for the punching.)

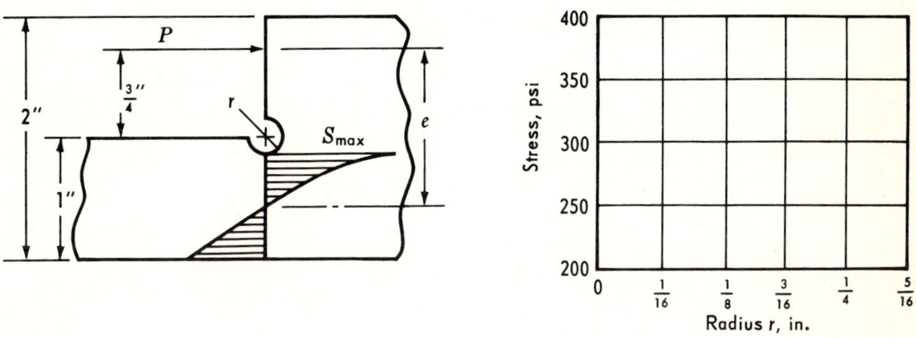

	r	$1-r$	$\frac{D}{d} = \frac{2}{1-r}$	$\frac{r}{d} = \frac{r}{1-r}$	K	$(1-r)^2$	$0.0104\,(1-r)^2$	$1.25 + \frac{r}{2}$	S
$\frac{1}{64}$									
$\frac{1}{32}$									
$\frac{1}{16}$									
$\frac{1}{8}$									
$\frac{3}{16}$									
$\frac{1}{4}$									
$\frac{5}{16}$									

Fig. P 4-14

CHAPTER 5. DESIGN CRITERIA

5-1. A flat bar, as shown in Fig. P 5-1, is to carry an axial load $P = 12{,}000$ lb. The material is to be SAE 1040 steel.

Assuming a factor of safety of 2, determine the thickness t of the bar for the following conditions: (*a*) assuming that the load P is a static load; (*b*) assuming that the load P is completely reversed and applied repeatedly.

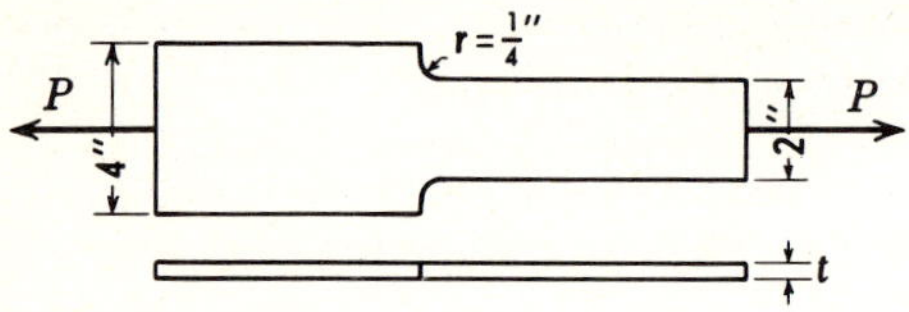

Fig. P 5-1

5-2. A stepped shaft of circular cross section, as shown in Fig. P 5-2, is made of SAE 1045 steel. The load P is repeated and completely reversed with a value of 2,500 lb. Assuming that r/d is approximately $\frac{1}{8}$, determine the diameter d and the fillet radius r so that the maximum stress will be limited to a value corresponding to a factor of safety of 2.

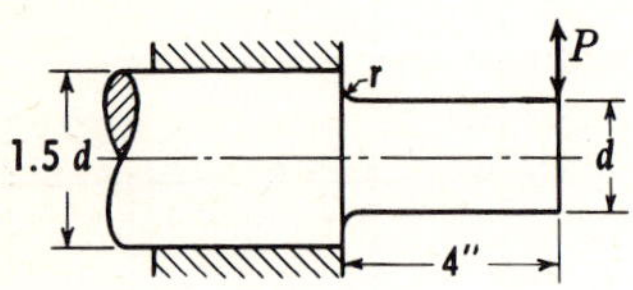

Fig. P 5-2

5-3. The machine part shown in Fig. P 5-3 is to be loaded repeatedly by a force $P = 360$ lb. Determine the required thickness b for the part, assuming SAE 1045 steel and a factor of safety of 2.

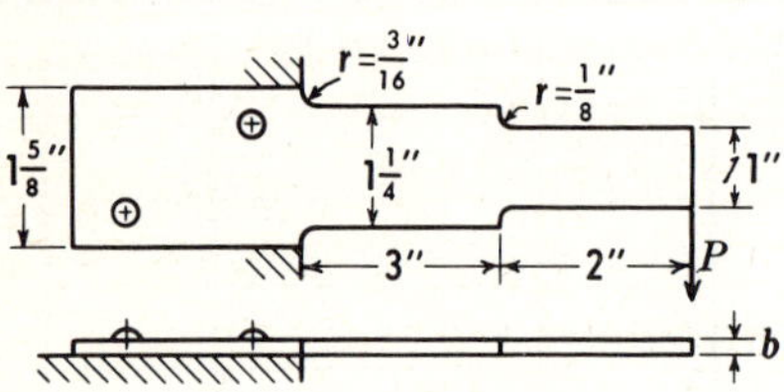

Fig. P 5-3

5-4. A bar, as shown in Fig. P 5-4, is to be subjected to a completely reversed, repeated load $P = 5{,}000$ lb. Assume SAE 1020 steel. Determine the thickness t of bar required, using a factor of safety of 2.

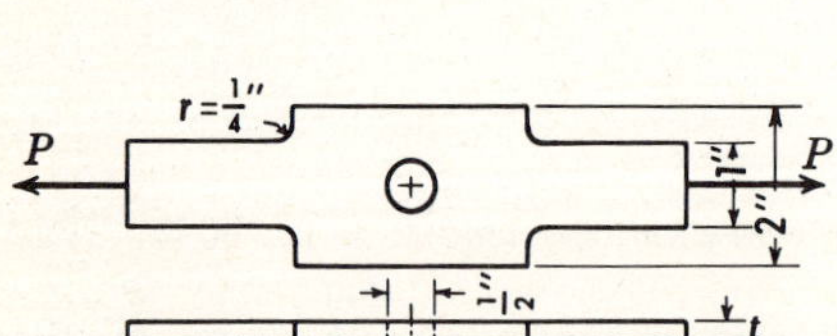

Fig. P 5-4

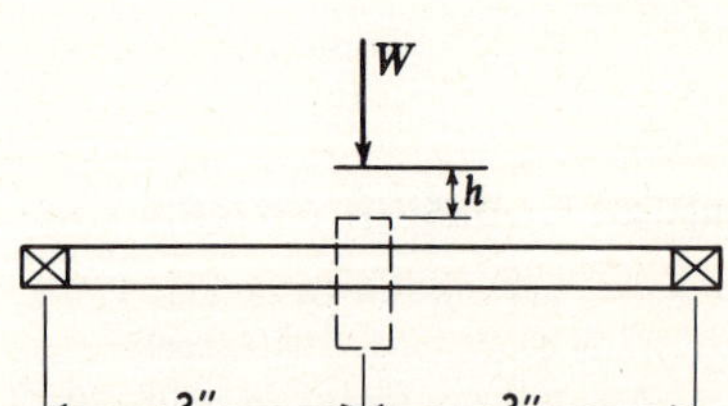

Fig. P 5-5

5-5. A camshaft ¾ in. in diameter is mounted on ball bearings and carries a cam midway between the bearings, as shown by Fig. P 5-5. The follower, which weighs 625 lb, is out of adjustment so that impact is produced when it is contacted by the cam. The height h of fall equals 0.01 in. The shaft is made of SAE 1045 steel. (*a*) Determine the maximum bending stress in the shaft. (*b*) Would the shaft be expected to fail, and if so, under what conditions?

5-6. A machine member as shown in Fig. P5-6 is subjected to a load P which is repeated, completely reversed, and suddenly applied without velocity of approach. The member is rectangular in cross section.

Assume the following:

Load applied in 15-sec cycles for 2 years each of 300 days, 4 hr per day
$P = 1{,}000$ lb
Material: SAE 4340 (see *S-N* curve to the right)
"Machined" finish
Factor of safety = 1.5

Determine the thickness of bar b that you recommend.

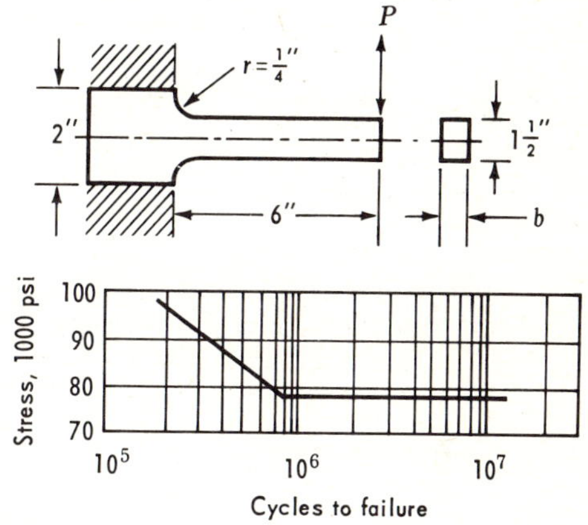

Fig. P 5-6

5-7. Determine the diameter of a solid circular shaft that is subjected to an alternating torque of 16,000 to 20,000 in.-lb. The shaft is made of SAE 1020 hot-rolled steel. Assume a factor of safety of 2, and base the design on the maximum tensile stress.

$s_{yp} = 35{,}000$ psi
$s_e = 22{,}000$ psi
fs = 2.00

CHAPTER 8. MECHANICAL FABRICATION AND PROCESSES

8-1. A static load $P = 70{,}000$ lb is to be supported by steel plates welded as shown in Fig. P 8-1 by shielded-arc welds. (*a*) Assuming that ½-in. plates are to be used and that each weld carries one-half of the load, determine the width of plate to be specified. (*b*) Determine the maximum stress in the plates, and compare it with the allowable value.

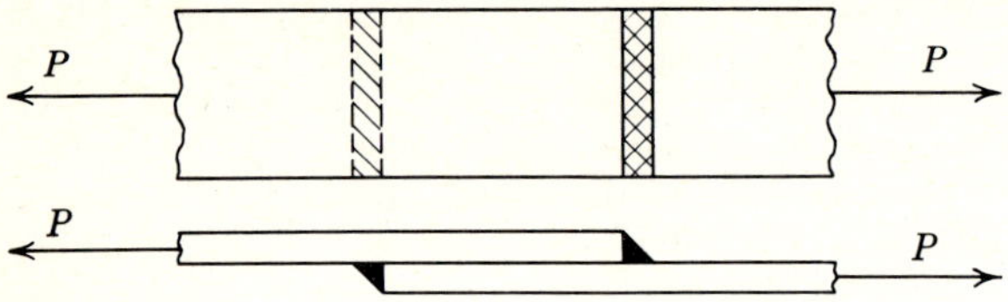

Fig. P 8-1

8-2. A 6- by 4- by ½-in. angle is to be welded to a steel plate by fillet welds along the edges of the 6-in. leg. The angle is to support a tension load of 60,000 lb. Determine the lengths of shielded-arc welds to be specified.

8-3. A gear is to be made by welding a steel hub, ring, and rim as shown in Fig. P 8-3. The torque on the gear is steady at 2,000 lb-ft. Determine the size of welds to be recommended at the hub and the rim.

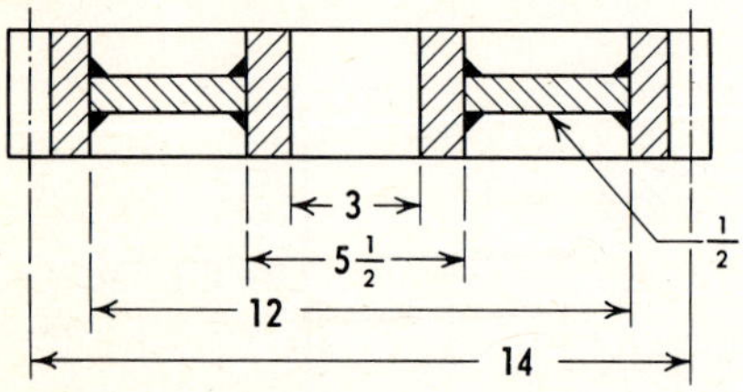

Fig. P 8-3

8-4. Determine the thickness of the plate and size of weld to be specified for the bracket in Fig. P 8-4, where $F = 8{,}000$ lb.

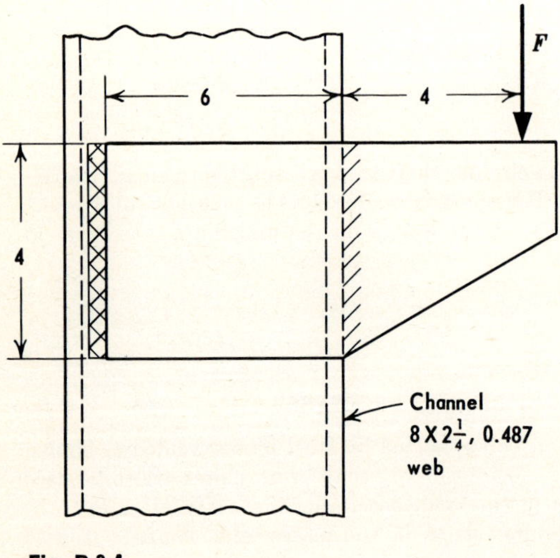

Fig. P 8-4

CHAPTER 9. DETACHABLE FASTENINGS

9-1. For a standard bolt and nut made of steel, determine the relation between the thickness of the nut and the outside diameter of the thread for equal strengths of the bolt in tension and the threads in shear. Assume that the root diameter of the threads is 0.85 × the outside diameter and that the allowable stress in shear, s_s, equals 0.55 × the allowable stress in tension, s_t.

9-2. A pulley bracket is supported by four bolts, two at A and two at B, as shown in Fig. P 9-2. The weight of the pulley and bracket equals W = 175 lb, and the load on the rope P is 6,000 lb. Assuming that the bracket is held against the wall and prevented from tipping about O by the two bolts at A (i.e., neglecting the bolts at B), and using an allowable tensile stress in the bolts as 5,000 psi, determine the size of NC bolts required.

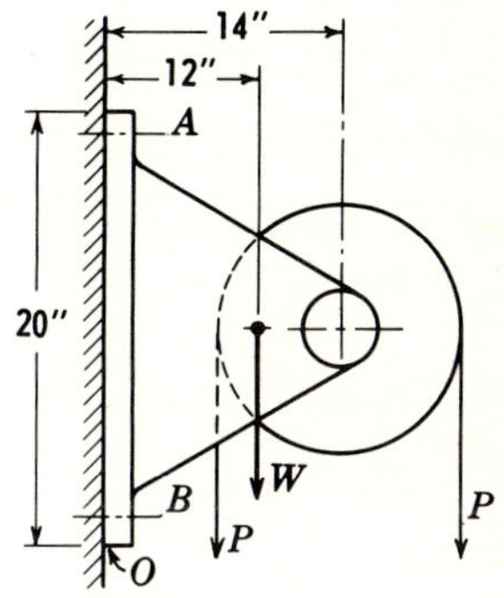

Fig. P 9-2

9-3. A bolt is subjected to axial shock loading so that the maximum applied load equals 2,400 lb. (*a*) Assuming an allowable stress of 8,000 psi, determine the size of bolt required, using NC threads. Neglect stress concentration. (*b*) It is desired to increase the shock-absorbing capacity of the above bolt. Determine the diameter of the hole required to reduce the cross-sectional area of the shank to that at the root of the threads. (*c*) Determine the percentage of improvement in shock-absorbing capacity by using the drilled bolt.

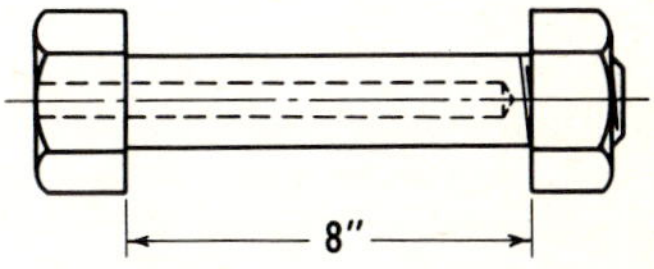

Fig. P 9-3

9-4. Assuming that the load on a bolt due to tightening is equal to 16,000 times the nominal bolt diameter d, derive an equation for the stress due to tightening. The result should be in terms of d. Assume the root diameter is 0.8 × the nominal diameter. What should be the minimum size of the bolts used on the basis of the above?

9-5. A safety stop is composed of steel tension bolts arranged to catch a platform in case of failure of a support. In this case each bolt may be subjected to the impact of

$W = 2{,}500$ lb, dropping through a distance $h = 0.05$ in., as shown schematically in Fig. P 9-5. The bolts are turned down to $\frac{7}{8}$ in. in diameter, which is less than the root diameter of the threads. It is expected that the safety stop will be required to function infrequently. Using a factor of safety equal to 1.25, what mechanical property of the material and what value of that property would you specify?

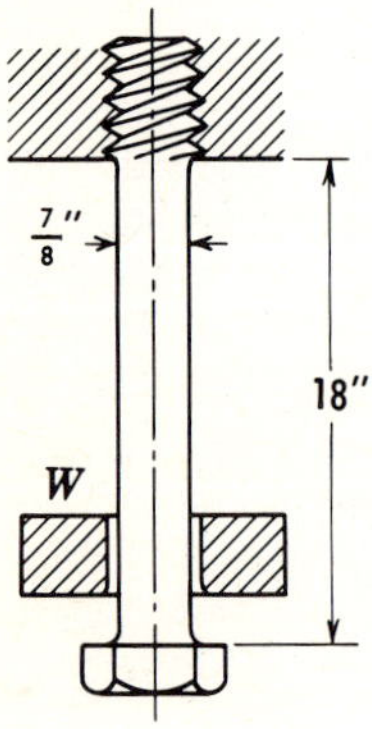

Fig. P 9-5

9-6. For a sliding hub connected to a shaft by means of feather keys, show that the friction of sliding when two keys are used is one-half the friction when one key is used.

9-7. Draw a neat sketch of a shaft fitted with a key, and derive equations for (*a*) the bearing pressure on the key and (*b*) the shearing stress in the key. The results should be in terms of the transmitted torque T, the shaft diameter D, length L of key, width w of key, and depth h. (*c*) Assuming that the key is equally strong in bearing and shear and that the yield point in compression is twice the yield point in shear, determine the proper proportions of the cross section of the key. (*d*) Assuming that there is a square key whose sides are one-quarter the shaft diameter and that the shaft, key, and hub are made of the same material, determine the length of key in terms of the shaft diameter so that the key will be as strong as the solid shaft.

9-8. The low-speed shaft of a 20-hp speed reducer rotates at 384 rpm and is $1\frac{3}{4}$ in. in diameter. Determine the size of key that should be recommended, and compare the shearing stress in the key with the shearing stress in the solid shaft.

9-9. Determine the size and length of the key you would specify for a 5-hp 1,800-rpm motor (see Table 25-2).

9-10. A line shaft $2\frac{7}{16}$ in. in diameter is to be driven by a belt running over a 36-in. pulley. The belt tensions are $F_1 = 1{,}500$ lb and $F_2 = 1{,}000$ lb. The hub of the pulley and its key are both 6 in. long. (*a*) Using a standard square key, determine the shearing stress and the bearing pressure for the key. (*b*) Determine the loads on the bearings A and B.

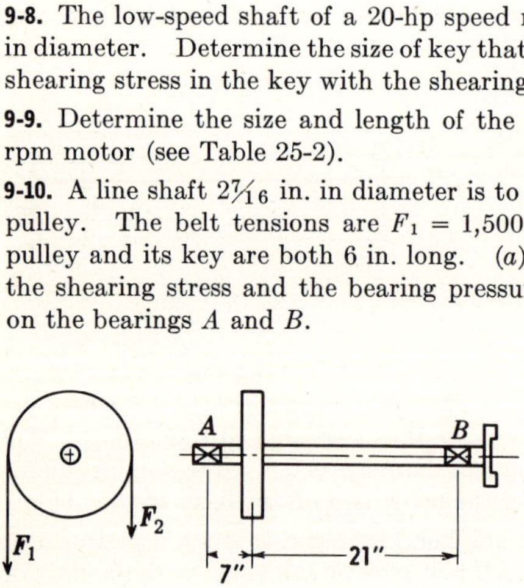

Fig. P 9-10

9-11. Make a neat sketch showing two views of a knuckle joint, and write equations showing the strength of the joint for the most probable methods of failure.

Note: Assign notation to represent the various dimensions of the joint, and express the equations for strength in terms of this notation.

9-12. Make a neat sketch showing two views of a cotter joint with key, and write equations showing the strength of the joint for the most probable methods of failure.

Note: Assign notation to represent the various dimensions of the rod end, socket, and cotter, and express the equations for strength in terms of this notation.

9-13. For the cottered connection shown in Fig. P 9-13, determine the maximum load P that should be applied for the following allowable stresses: s_t = 5,000 psi, s_s = 2,500 psi, and s_b = 12,000 psi. Neglect stress concentration.

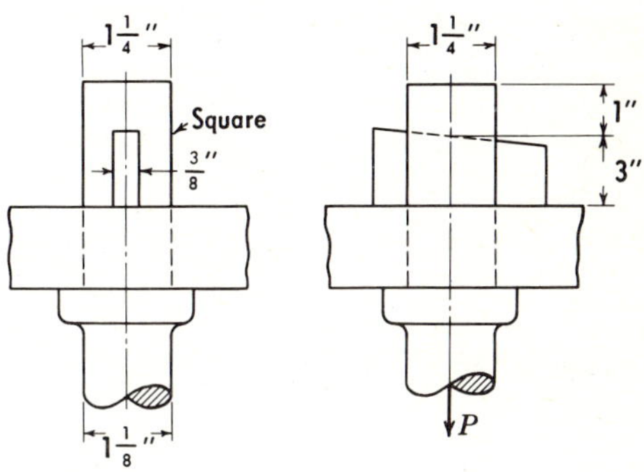

Fig. P 9-13

9-14. The torque capacity of a spline fitting is based on the allowable pressure on the active surfaces of the splines. Assuming a value of this pressure as 1,000 psi, determine for the spline in Fig. P 9-14 the following: (*a*) the length of spline required

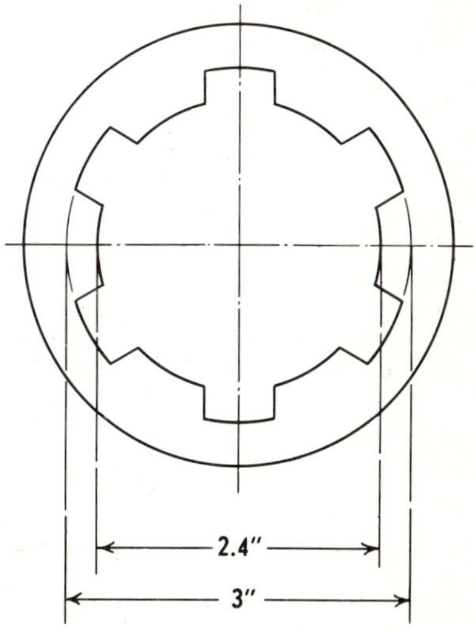

Fig. P 9-14

for a torque capacity of 500 lb-ft; (*b*) the force required to slide the member axially under load, assuming the coefficient of friction as 0.1.

CHAPTER 10. SPRINGS

10-1. A valve for a hydraulic pump is to have a lift of $\frac{3}{8}$ in. and is to be closed by a helical compression spring having a load range of approximately 100 to 175 lb from valve-closed to valve-open conditions. Assume SAE 6150 steel, average service, and spring index $C = 4, 5, 6, 7,$ and 8.

(*a*) Determine the diameter of wire, outside diameter, and number of active coils.

(*b*) Assuming the pitch of the coils is twice the wire diameter, determine the free length of the spring.

(*c*) Make a neat freehand sketch of the springs to half-size scale.

10-2. Figure P 10-2 shows the schematic arrangement of a spring-controlled lever. The spring is to be inserted with an initial compression to produce a force equal to 25 lb between the right-hand end of the lever A and the stop; when the force at A equals 40 lb, its maximum value, the end of the lever A will move downward 1 in.

Determine the following for the spring: (*a*) the load-deflection diagram for the spring; (*b*) the spring rate; (*c*) the size of the wire to be specified, assuming a spring index $C = 8$ and average service; (*d*) the outside diameter to be specified; (*e*) the number of active coils to be specified; (*f*) the actual spring rate; (*g*) the free length, assuming squared and ground ends and $p = 2d$.

Also, for the lever, determine (*h*) dimensions b and h, assuming that $h = 2b$, an allowable stress $s_t = 5{,}000$ psi, and that stress concentration may be neglected.

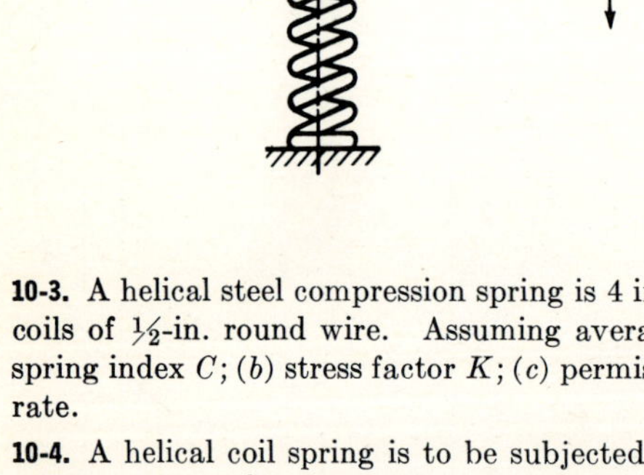

Fig. P 10-2

10-3. A helical steel compression spring is 4 in. in outer diameter and has eight active coils of $\frac{1}{2}$-in. round wire. Assuming average service, determine the following: (*a*) spring index C; (*b*) stress factor K; (*c*) permissible axial load; (*d*) deflection; (*e*) spring rate.

10-4. A helical coil spring is to be subjected in service to loads ranging from 520 to 650 lb. The axial compression of the spring over the above load range is to be approximately $\frac{1}{4}$ in. Assuming a spring index of approximately 5, determine the following: (*a*) the size of wire; (*b*) the outside diameter of the spring; (*c*) the number of active coils.

10-5. A loaded narrow-gauge car weighing 4,000 lb and moving at a velocity of 240 fpm is brought to rest by a bumper consisting of two helical steel springs. In bringing the car to rest the springs are to be compressed 8 in. Assume the following:

Allowable stress	52,000 psi
Spring index	6

Determine the following: (*a*) maximum load on each spring; (*b*) diameter of wire; (*c*) mean diameter of coils; (*d*) number of effective coils.

10-6. A weight of 160 lb located at a position 2½ ft above the center of a platen is to drop on the platen. The platen is to be supported on four helical springs. Assume the following:

Deflection of each spring	3 in.
Spring index	5
Maximum stress	60,000 psi

Determine the diameter of wire, outside diameter of the coils, and the number of active coils.

10-7. A valve for a hydraulic pump is to have a lift of ⅜ in. and is to be closed by a helical compression spring having a load range of approximately 100 to 175 lb from valve-closed to valve-open conditions. Assuming SAE 6150 steel and a spring index of 6, determine the wire diameter, outer diameter of coils, and number of active coils to be recommended.

10-8. A circular cam 8 in. in diameter rotates off center with an eccentricity $e = 1$ in. and operates the roller follower that is carried by the arm as shown in Fig. P 10-8. The roller follower is held against the cam by means of an extension spring. The device is part of a bookbinder. Assuming that the force between the follower and the cam is to be approximately 50 lb at the low position and 80 lb at the high position and that there is a spring index of 7, determine the diameter of wire, outside diameter of spring, and number of active coils you would recommend. The spring may be assumed to be attached at the third point on the arm.

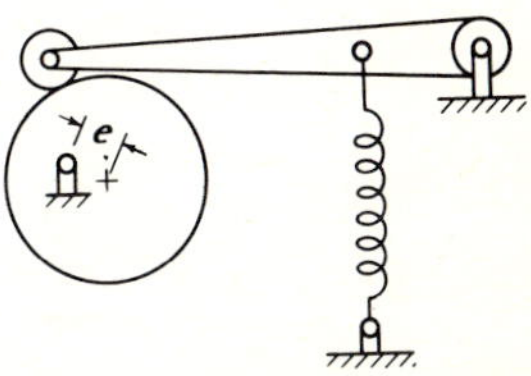

Fig. P 10-8

10-9. In a free-rolling conveyor, a crate loaded with nonfragile material, 60 lb tare, reaches station A at a velocity $v = 10$ fps and is normally unloaded at station B. In case it "gets away" from B, a pair of helical buffer springs at C are expected to arrest the motion of the crate. Assume that the springs are set up with an initial compression of 2 in. and that the compression of the spring due to impact is to be 6 in. Determine the following: (*a*) maximum load on each spring; (*b*) diameter of wire and diameter of coils; (*c*) number of coils.

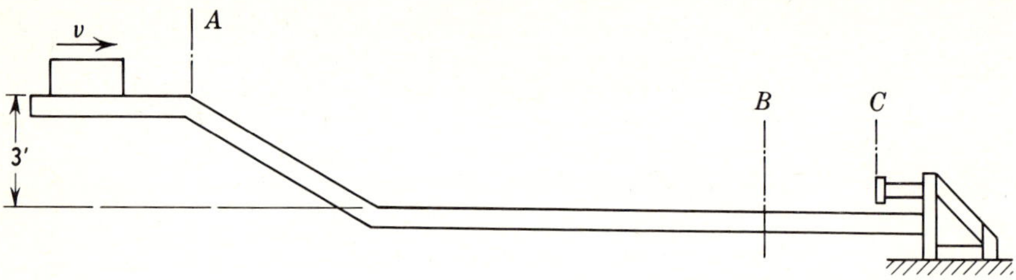

Fig. P 10-9

10-10. A helical coil spring for an industrial spring-set brake is to be made of SAE 6150 wire and is to be loaded in compression by an axial load varying from a minimum of 650 lb to a maximum. The deflection of the spring over the service load range is to be approximately ⅜ in.

(*a*) Assuming average service, a spring index of 5, and that the range of stress during the service loading will be from 80 to 100 percent of the maximum, determine the diameter of wire and the mean diameter of coils.

(*b*) Determine the number of active coils required and the actual spring rate in pounds per inch.

(*c*) Determine the yield point in torsion for the wire, and compute the corresponding spring load and deflection.

(*d*) Assuming that the spring ends are squared and ground, and using the deflection found in (*c*), determine the pitch of the coils and the actual free length of the spring so that the closure stress will not exceed the yield point.

(*e*) Make a copy of the table shown below and enter values.

(*f*) Check the possibility of buckling.

(*g*) Make a neat freehand sketch, approximately to scale, of the spring in its free condition.

Specifications		*Characteristics*				
Outside diameter, in.			Load	Stress	Deflection	Length
Wire diameter, in.		Free				
Total number of coils			650			
Pitch of coils, in.						
Free length, in.		Solid				

10-11. A Westinghouse-Nuttall coupling, as shown in Fig. 16-8, is to have a rated torque of 2,800 lb-in. There are to be four springs located as shown in Fig. P 10-11 in which the radius of the spring centerline $R = 4$ in. (*a*) Assuming that the angular displacement between the coupling halves at twice rated torque is to be 60°, determine

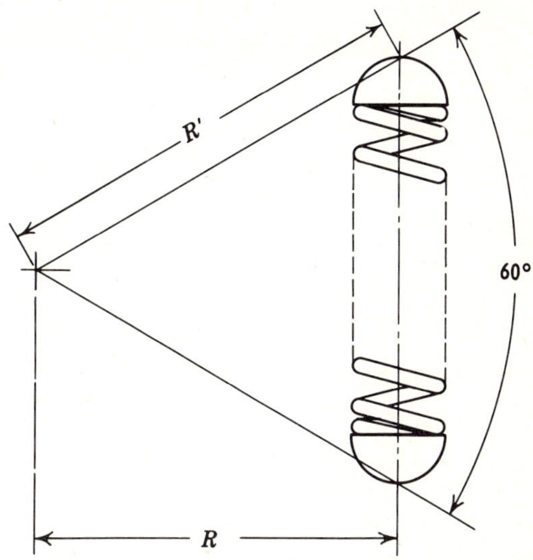

Fig. P 10-11

the spring rate. (*b*) Assuming that the springs are to have an initial compression of 0.1 in. when the coupling is unloaded, determine the load on the springs at twice rated torque.

10-12. A torsion-bar spring is made of a solid, circular steel bar 2.27 in. in diameter with upset ends 2.75 in. in diameter for spline fittings. The splines are composed of 55 serrations 0.037 in. deep with sides having 100° included angle. One end of the bar is anchored to the piece of ordnance equipment of which it is a part, and the other end carries a torque arm 14 in. long. The length of the bar between splines is 72.7 in. The maximum force on the torque arm is 23,800 lb. (*a*) Make a sketch of the bar. (*b*) Determine the maximum torque on the bar in inch-pounds. For a maximum torque, determine the following: (*c*) shearing stress in the body of the bar; (*d*) pressure on the flanks of the splines; (*e*) wind-up of the bar in degrees; (*f*) motion of the free end of the arm in inches. Note that the length of the spline is $1\frac{1}{8}$ in.

10-13. Figure 10-21 shows a Neg'ator spring counterbalance in which the Neg'ator spring in tending to rewind itself on the storage bushing rotates the output bushing and the cable drum to support the weights on the cables A, B, and C. The total weight on the three cables is 36 lb. Determine the constant spring force required to produce the counterbalancing. The diameter of the cable drum is 1.74 in., the diameter of the output bushing is 4 in., and the gear ratio between the output bushing and the cable drum is 4:1.

10-14. A steel torsion bar as shown in Fig. P 10-14 is to carry a repeated reversed load $P = 7{,}500$ lb acting at the end of the lever arm as shown. The length of the bar $L = 48$ in., and the lever arm a is 8 in.

(*a*) Determine the diameter d of the bar to allow it to twist at an angle $\theta = 5°$ on each side of the midposition. Neglect the effect of bending of the arm a.

(*b*) Assuming that the diameter D for the spline fitting at the fixed end of the

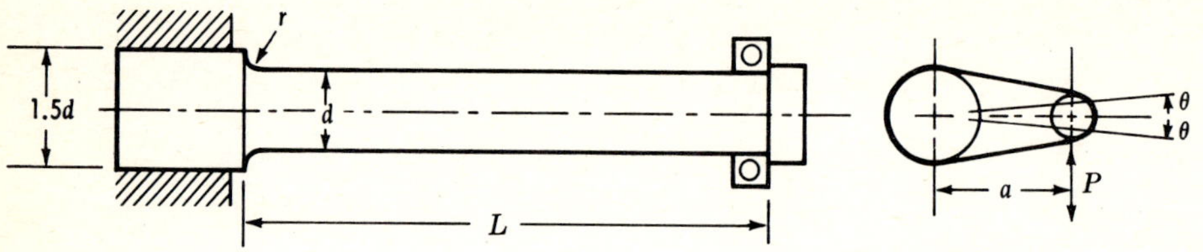

Fig. P 10-14

bar is 1.25d, determine the maximum stress in the bar, and select a material for the bar and its finish, assuming a factor of safety equal to 1.2.

10-15. Assume that there is available a long helical spring whose outside diameter is $1\frac{1}{4}$ in., made of $\frac{3}{16}$-in. wire, with pitch of coils 0.3 in. The material is SAE 6150 spring steel. It is planned to use part of the long spring for a valve spring for the following conditions:

Load when valve is open, 100 lb
Load when valve is closed, 75 lb
Valve lift, $\frac{1}{8}$ in.

Determine the following: (a) maximum operating stress (is this satisfactory?); (b) total number of coils; (c) stress if the spring is compressed solid. Consider that the ends are squared and ground. Compare with the elastic limit in Fig. 10-7.

10-16. A "skip hoist" is an inclined single-bucket conveyor (see Fig. P 10-16) arranged to elevate material such as coal, gravel, iron ore, etc., from ground level to an upper level as required. The transport may be vertical (this requires double rails), but the conveyor is usually installed at a reclining angle so that the weight of the buckets always bears against the supporting rails. When the bucket reaches the top station B, it is stopped, dumped, and then returned empty to the loading station A.

This type of hoist is usually used in *pairs* so that when one loaded bucket is being raised, its companion is falling. This takes advantage of the energy of the falling bucket to assist the rising one.

An experimental investigation shows that the frictional resistance of the bucket rollers corresponds to a tangential rolling friction force F which equals fN, where N is the *normal* force of the rollers on the rails and f is the coefficient of rolling friction.

Assuming the values given below, determine the horsepower required to operate a "paired" hoist in terms of the following:

Q = vertical rise of hoist, ft
R = horizontal "set," ft
W = weight of unloaded bucket, lb
L = payload on bucket, lb
f = coefficient of rolling resistance
V = velocity of conveyor, fpm
e = overall efficiency of drive (other than rollers of buckets)
δ = maximum spring deflection
α = angle of incline

Determine, in terms of the above notation, the maximum force on a buffer spring located at station A to arrest a loaded bucket in case of hoisting-rope breakage at station B.

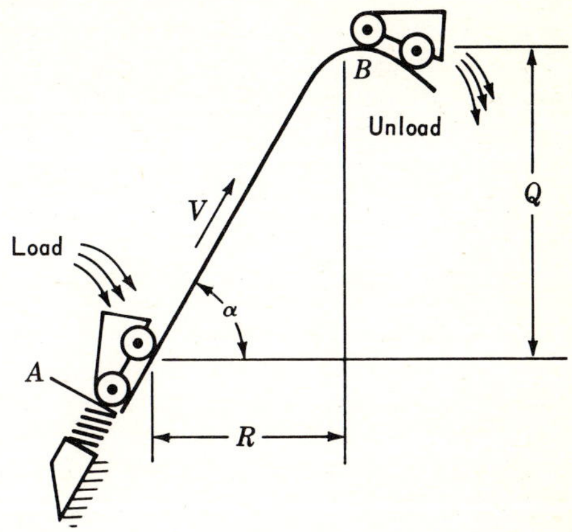

Fig. P 10-16

CHAPTER 11. PRESSURE CYLINDERS

11-1. The drum of a steam-generating unit is 18 ft long and 48 in. in diameter. The steam pressure is 500 psi gauge. Assuming an allowable tensile stress of 10,000 psi and that the efficiencies of the circumferential and longitudinal joints are 65 and 90 percent, respectively, recommend the thickness of shell.

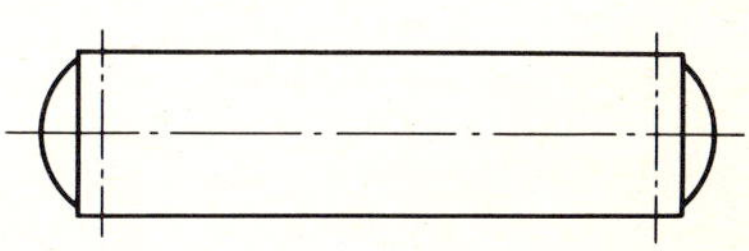

Fig. P 11-1

11-2. The force analysis of a 7½-ton air-operated arbor press shows that the piston rod for the operating cylinder must exert a maximum force $F = 4{,}500$ lb. The air pressure p in the cylinder is 100 psi. (*a*) Determine the diameter of cylinder bore required, assuming that overall friction due to stuffing box and piston packing is equivalent to 10 percent of F. The piston bore should be selected on the basis of ⅛-in. increments. (*b*) Determine the thickness of cylinder, assuming that it is seamless-steel tubing. The allowable tensile stress is 3,000 psi. (*c*) Determine the thickness of the cylinder head, assuming that it is made of cast iron and that the allowable stress is 3,000 psi.

Fig. P 11-2

11-3. A hydraulic control for a straight-line motion utilizes a spherical pressure tank A that is connected to a work cylinder B, as shown in Fig. P 11-3. A pump maintains pressure in the tank equal to 400 psi.

(*a*) Assuming that the tank A is 32 in. in diameter, that it is welded with joints having strength equal to that of the plate, and that the tank is made of steel plates having an allowable tensile strength of 7,500 psi, determine the thickness of the plate required for the tank.

(*b*) Assuming a pressure drop of 30 psi between the tank and the cylinder, determine the diameter of piston required to produce an operating force F of 5,000 lb. Assume an allowance for friction in the cylinder and packing equal to 10 percent of F.

(*c*) Determine the thickness of the cylinder wall, assuming that it is made of cast iron having an allowable tensile stress of 4,000 psi.

(*d*) Determine the horsepower output of the cylinder during a working stroke, assuming the piston stroke is 18 in. and that 5 sec is required for a work stroke.

(*e*) Assuming that the work cycle of the piston rod occurs once every 30 sec, that the overall efficiency of the hydraulic control is 80 percent, and that the pump efficiency is 60 percent, determine the horsepower required for the motor continuously operating the pump.

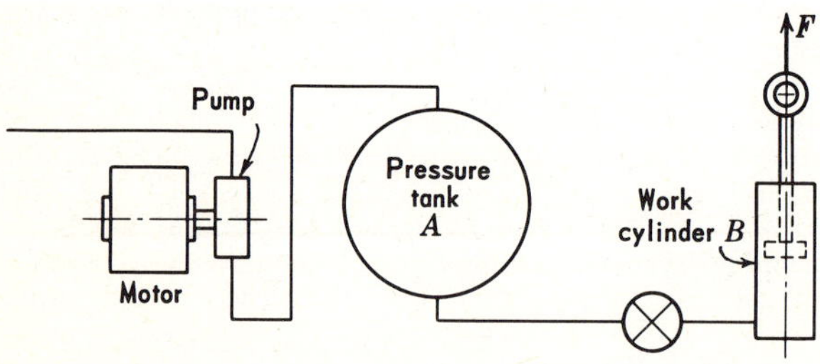

Fig. P 11-3

CHAPTER 12. TRANSLATION SCREWS

12-1. (*a*) Make a neat sketch showing one turn of a right-hand square-threaded screw, and indicate thereon the following notation:

Q = axial load, lb
d = mean diameter of thread, in.
α = lead angle at mean diameter
ϕ = friction angle
L = lead of thread, in.

(*b*) Derive an equation for the torque T required to raise the load and to overcome thread friction in terms of Q, d, f, and L.

(*c*) Write the expression for the torque T_c required to overcome collar friction.

12-2. A sluice gate weighing 60 tons is raised and lowered by means of two $2\frac{3}{4}$-in. Sellers' standard square-threaded screws. The screws are operated by a 570-rpm motor. A ball thrust bearing is used at the collar, reducing the coefficient of collar friction to 0.03 at a diameter of 4 in. The coefficient of thread friction may be assumed to be 0.14.

If the gate is to be raised at the rate of 2 fpm, determine the following: (*a*) the rpm of the screws; (*b*) the horsepower of motor required to raise the gate, assuming a mechanical efficiency of 0.85 for the speed-reducing equipment.

12-3. A screw having a double square thread is to raise a load of 20,000 lb at a speed of 3 fpm. The following data apply:

Outside diameter of thread	$2\frac{7}{8}$ in.
Pitch	1 in.
Mean collar diameter	5 in.
Coefficient of thread friction	0.10
Coefficient of collar friction	0.12

Determine the following: (*a*) the horsepower required to drive the screw; (*b*) the efficiency of the screw.

12-4. The screw of a 30-ton toggle press rotates at 72 rpm and is driven by means of the gear g shown in Fig. P 12-4. The crossheads n move axially along the screw and in opposite directions since one head has left-handed threads and the other right-handed threads. They move against the axial forces S in operating the press. Assuming that $S = 23{,}000$ lb, that the screw is a square-threaded type $2\frac{1}{4}$ in. in outside diameter and has two threads per inch, and that there is coefficient of friction of 0.10, determine the horsepower required to drive the gear g. The friction at the supporting bearings may be neglected.

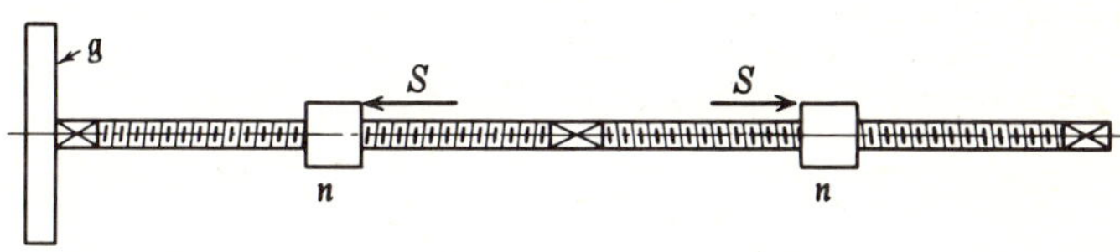

Fig. P 12-4

12-5. The screw of a shaft straightener, as shown in Fig. P 12-5, exerts a load $P = 6{,}000$ lb. The screw is 3 in. in outside diameter and has four square threads per inch.

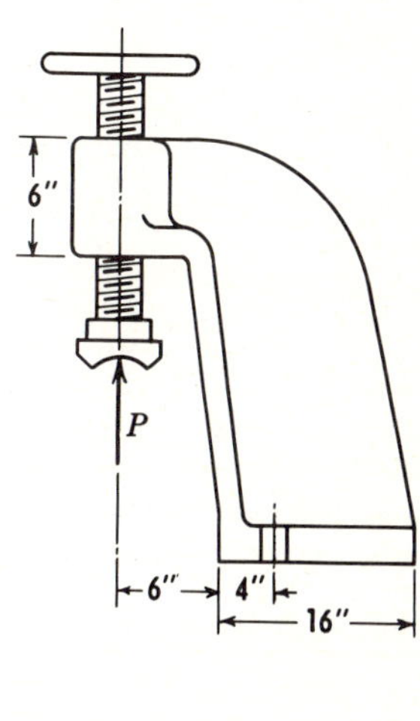

Fig. P 12-5

(*a*) Determine the force required at the rim of the 12-in.-diameter hand wheel, assuming that the coefficients of thread and collar friction are 0.125 and that the mean diameter of the collar is 2½ in. (*b*) Determine the maximum compressive stress in the screw, the bearing pressure on the threads, and the shearing stress in the threads. (*c*) Determine the efficiency of the straightener. (*d*) Determine the size of the two bolts required to fasten the straightener to the base, assuming the allowable stress is 8,000 psi. Neglect tightening-up stresses.

12-6. Figure P 12-6 shows the arrangement for a type of linear actuator for aircraft controls, for example, landing gear and bomb-bay doors. These units weigh from 4 to 15 lb. The screw for such an actuator is to be designed and the motor selected.

Data:

Operating force on actuator	1,000 lb
Maximum static force	2,500 lb
Extension of actuator	5 in.
Time for extension	6 sec
Motor speed	7,200 rpm

(*a*) Determine the root area of the screw for the maximum static force, using an allowable stress in direct tension or compression equal to 10,000 psi. Select a Sellers' standard square thread.

(*b*) Using a coefficient of thread friction of 0.08 and neglecting collar friction since a ball thrust bearing is used, determine the torque on the screw in pound-inches, corresponding to the operating force on the screw.

(*c*) Assuming an overall efficiency for the triple-reduction gear between the motor and screw such as 94 percent, determine the horsepower of motor required.

(*d*) Determine the overall gear ratio from the motor to the screw.

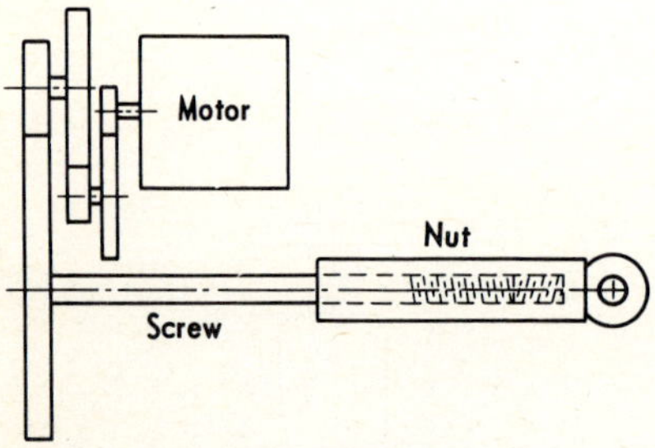

Fig. P 12-6

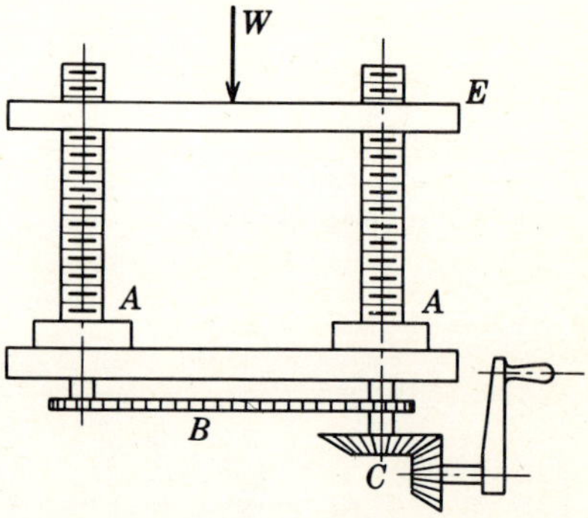

Fig. P 12-7

12-7. A portable conveyer is to be raised and lowered by means of the double-screw arrangement shown in Fig. P 12-7. The two screws are supported by collars A and are connected by the roller chain B and rotated by the bevel gears C, operated by the hand crank as shown. The movable frame E is raised and lowered by the screws against a load W = 1,000 lb, but when the frame is in position the load may be increased to 2,000 lb.

Assume the following data:

Allowable compressive stress in the screw	1,000 psi
Mean diameter of collar	outside diameter of screw
Coefficient of friction	0.15
Chain efficiency	90 percent
Bevel-gear efficiency	90 percent
Bevel-gear ratio	2:1 (pinion on crankshaft as shown)
Maximum tangential force on crank	20 lb

Use Sellers' square threads.

Determine the size of the screws to be used and the length of crank arm required.

12-8. In a shipbuilding crane, as shown in Fig. P 12-8(*a*), the travel of the load is controlled by three operations: (1) raising and lowering the load vertically by reeling in or out of the cable on the main hoisting drum, (2) raising or lowering (luffing) the boom, and (3) rotating the crane on a vertical axis. Each of these operations is controlled by a separate drive.

In one of the largest cranes in the world (see *Mech. Eng.*, September, 1944, p. 569), luffing of the boom is accomplished by rotating two screws located side by side, as the one shown in (*a*) in the figure. Each of the two screws is driven by a motor through a gear keyed to the screw at its midpoint. The screw has left- and

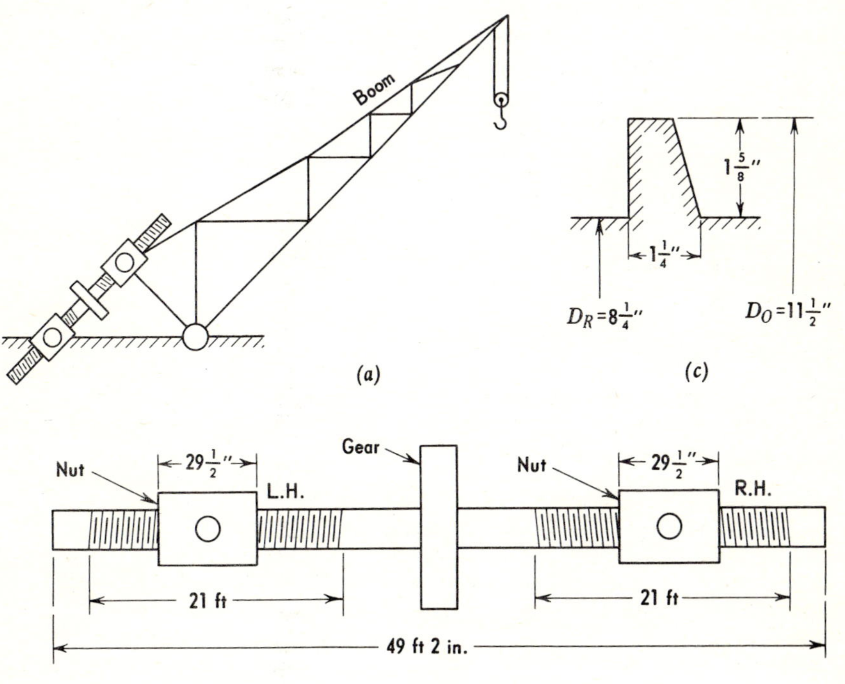

Fig. P 12-8

right-hand threads, each running with a bronze nut. One nut is trunnioned to the base of the crane and the other trunnioned to the boom so that when the screw is rotated the boom will be raised or lowered. The dimensions of the screw are shown in (*b*), and at (*c*) is shown a detail of the buttress thread. Other data are as follows:

Capacity of crane = 75 tons
Maximum tensile load on screw = 327,000 lb
Material of screw is SAE 2340 steel with yield point in tension = 70,000 psi
Material of nut is bronze alloy with yield point in tension = 55,000 psi
Pitch of threads = $2\frac{1}{8}$ in.

Determine the following:

(*a*) Maximum torque in inch-pounds on each screw, assuming a coefficient of friction equal to 0.07. (The threads are very accurately cut, finely finished, and EP grease is used as a lubricant.)

(*b*) The factor of safety for the screw.

(*c*) The factor of safety for strength of the threads.

(*d*) The time required to raise the boom (same as for the nut to traverse the threaded portion of the screw), assuming that the screw rotates at an average speed of 30 rpm.

(*e*) The horsepower required for hoisting at the lowest position of the boom, at which time the screw rotates at 20 rpm. This is the maximum horsepower, and the selection of motor would depend on the duty cycle for the hoist.

12-9. A shaft-positioning device arranged as shown in Fig. P 12-9 has a knob *A* operating the screw to move the lever through a range of 20°. The spring *k* is inserted with an

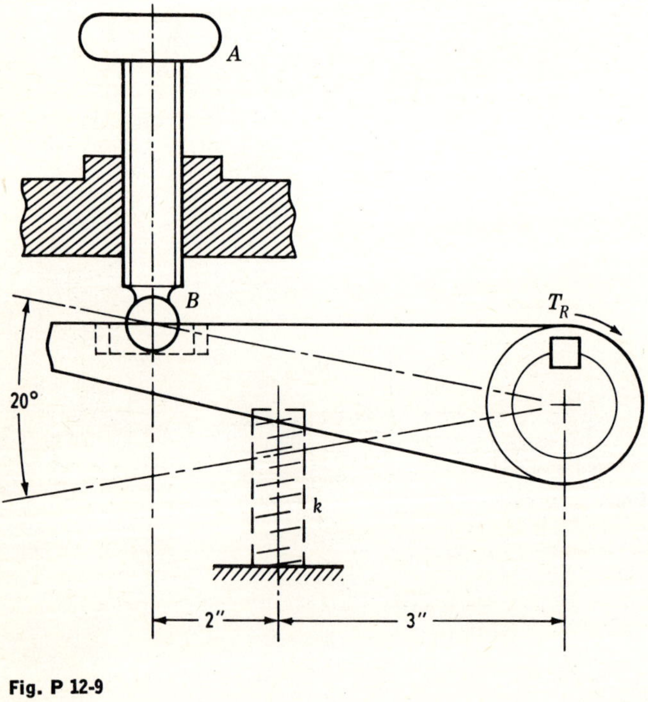

Fig. P 12-9

initial compression load equal to 20 lb when the lever is at the top of the travel. The spring is 0.35 in. in outside diameter and has eight active coils of 0.05-in.-diameter wire. The shaft is rotated against an external resistance torque T_R equal to 40 in.-lb. The screw is acme standard ½ in. in diameter with six threads per inch, and the friction diameter of the ball at B is 0.35 in. Assuming the coefficient of friction for screw threads and ball as 0.15, determine the maximum torque in inch-pounds required to turn the handwheel.

12-10. Before starting a printing press, it is necessary to *adjust* the rolls by rotating a shaft A (shown in Fig. P 12-10) a few degrees. The shaft A is connected to the arm AB, so that by turning the screw E the arm may be moved through a maximum angle of 15°, as shown. In rotating the shaft A, a torque equal to 200 in.-lb is required to overcome the rolling resistance of the rolls and friction of the bearings.

By turning the screw E so that it moves downward, the arm is rotated in a clockwise direction. This compresses the spring H. By "backing off" the screw, the spring must provide the force necessary to move the arm in a counterclockwise direction. Note that in turning the rolls the resistance acts in a direction opposite to their motion.

It is required to design the *spring* and the *screw*. In calculating movements of the arm, the points g and h may be assumed to move vertically on chords instead of arcs.

(*a*) *Spring*. Assume that when the arm is at its upper position the spring is inserted with an initial compressive force required to move the arm upward plus an additional 10 lb for a safe margin, and assume also that the axial compression of the spring when it is inserted equals 1 in. Draw the load-deflection diagram for the spring and determine the maximum compressive force on the spring during operation of the justifier.

(*b*) Assuming average service for the spring and a spring index of 8, determine the size of wire, outside diameter of coils, and number of coils to be specified.

(*c*) *Screw*. Determine the maximum axial load on the screw.

(*d*) For the screw, assume Sellers' standard square threads, coefficient of thread friction $f = 0.125$, and maximum compressive stress in the screw as 2,500 psi, and neglect collar friction. Determine the outside diameter of the screw, number of threads per inch, maximum torque required to turn the screw, and number of turns of screw to move the arm from upper to lower position.

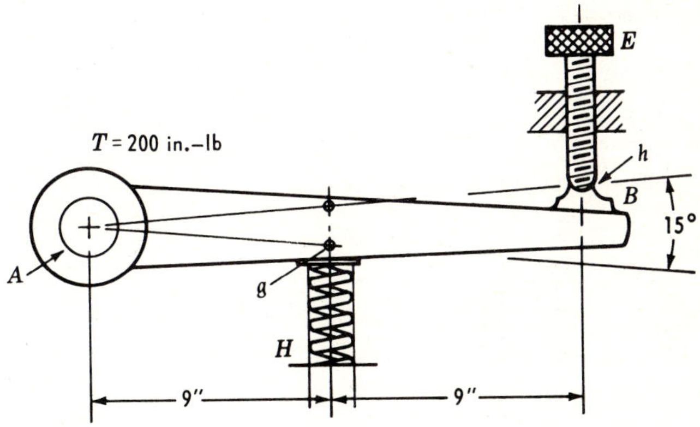

Fig. P 12-10

12-11. An adjusting device is composed of a Scott-Russell mechanism as shown in Fig. P 12-11, in which the slider at B is moved against the spring force at S by turning the knurled screw at A, a slidable connection. Note: $OC = BC = CA$. Assume the following data:

Spring S
Outside diameter = ⅝ in.
Wire diameter = ⅛ in.
Number of active coils = 8
Screw
Mean diameter = 0.365 in.
Pitch (single thread) = ¼ in.
Coefficient of thread friction = 0.1
Neglect collar friction at ball A.

Assume that the spring is set up at the position of the mechanism shown, with a deflection of spring $\delta = 1$ in. Determine (*a*) the spring rate P/δ; (*b*) the spring force P; (*c*) the axial load Q on the screw; (*d*) the torque T required at the thumb screw to compress the spring at the position shown.

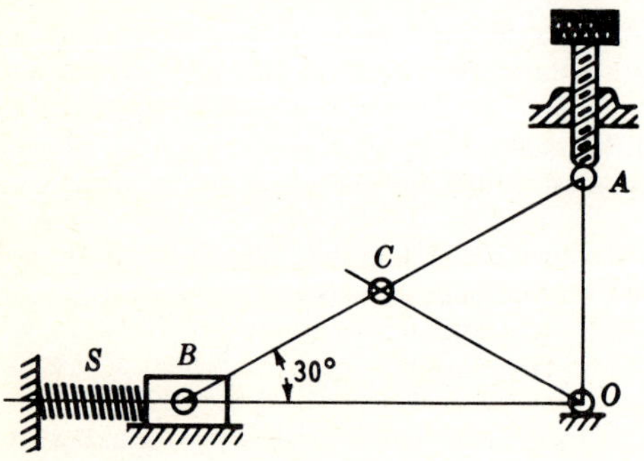

Fig. P 12-11

CHAPTER 13. SHAFTING

13-1. A shaft transmitting 50 hp at 870 rpm is supported in a bearing as shown in Fig. P 13-1 and has attached to it a pinion 3 in. in pitch diameter. Assuming that the shaft is rigidly supported in the bearing and that the tooth load is uniformly distributed over the face of the gear, determine the following: (*a*) the torsional moment in inch-pounds transmitted by the shaft; (*b*) the maximum bending moment on the shaft, assuming 20° involute teeth; (*c*) the shaft diameter required, assuming an allowable stress in shear equal to 6,600 psi.

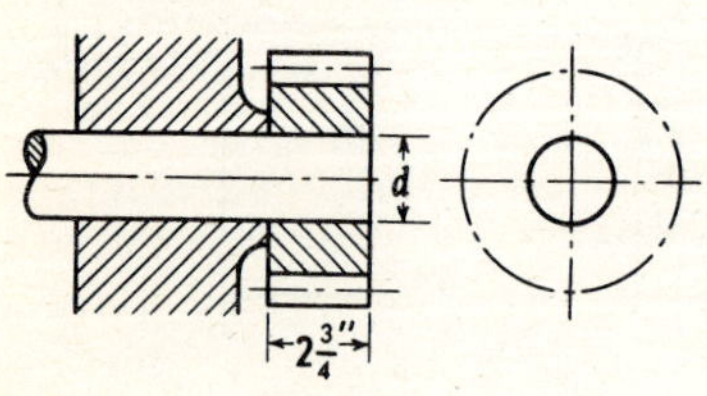

Fig. P 13-1

13-2. An overhung crank, as shown in Fig. P 13-2, supports a load $P = 1{,}000$ lb. Assuming an allowable stress in shear equal to 6,600 psi, determine the diameter of shaft required.

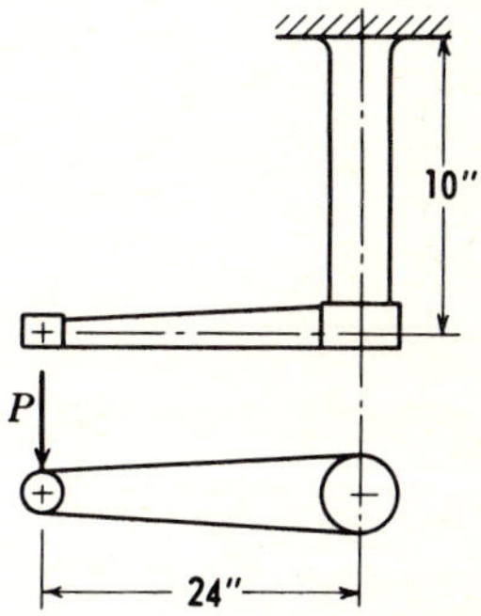

Fig. P 13-2

13-3. A steam engine that has a stroke of 12 in. has an overhung crank as shown in Fig. P 13-3. The maximum tangential force P on the crank may be assumed as 7,500 lb. Assuming an allowable stress in shear as 4,400 psi, determine the recommended diameter d for the crankshaft.

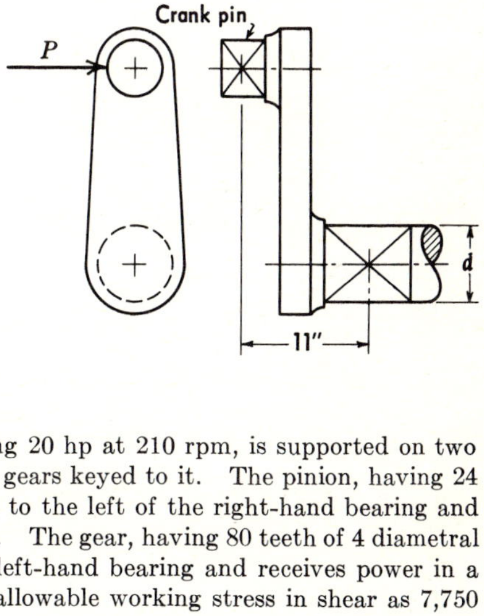

Fig. P 13-3

13-4. A mild-carbon-steel shaft, transmitting 20 hp at 210 rpm, is supported on two bearings that are 30 in. apart and has two gears keyed to it. The pinion, having 24 teeth of 4 diametral pitch, is located 4 in. to the left of the right-hand bearing and delivers the power horizontally to the right. The gear, having 80 teeth of 4 diametral pitch, is located 6 in. to the right of the left-hand bearing and receives power in a vertical direction from below. Using an allowable working stress in shear as 7,750 psi, determine the diameter of the shaft.

13-5. A centrifugal circulating pump for a steam-turbine condenser is to be direct-connected to a motor through a short shaft and a pair of flexible couplings. The pump delivers 5,500 gpm at 865 rpm against a dynamic head of 25 ft. The pump efficiency is 79 percent. (*a*) Determine the horsepower of motor required. (*b*) Assuming hot-rolled SAE 1020 steel, determine the diameter of shaft required, using a factor of safety equal to 4. (*c*) Compare the diameter of the connecting shaft with the diameter of the motor shaft, and state reasons for any difference.

13-6. The solid shaft of a high-speed boat is 32 ft long from sprocket to propeller, as shown in Fig. P 13-6. At maximum torque, the horsepower transmitted to the shaft is 300 at 130 rpm of the shaft. The sprocket is 12 in. in diameter. Assuming the maximum-shear theory for the design, with an allowable stress in shear of 8,000 psi, determine the diameter of the shaft to be specified.

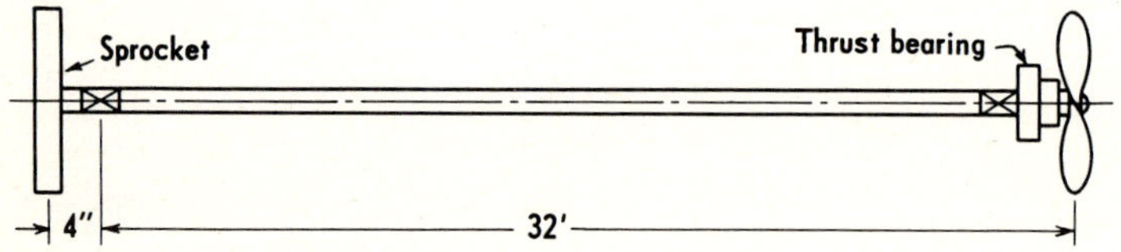

Fig. P 13-6

13-7. The shaft of a hand-operated wire-rope hoist is supported on the bearings *A* and *B* shown in Fig. P 13-7 and has keyed to it the drum, which is 6 in. in diameter. The tangential load on the drum at rated capacity of the hoist is 125 lb. Assuming 100 percent overload capacity and an allowable shear stress of 5,000 psi, determine the shaft diameter to be recommended.

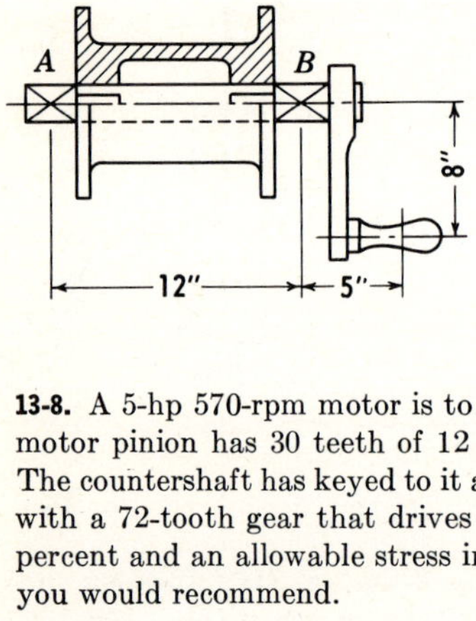

Fig. P 13-7

13-8. A 5-hp 570-rpm motor is to drive a countershaft, as shown in Fig. P 13-8. The motor pinion has 30 teeth of 12 diametral pitch and meshes with a 120-tooth gear. The countershaft has keyed to it a pinion having 24 teeth of 4 diametral pitch meshing with a 72-tooth gear that drives a rotary kiln. Assuming a starting overload of 30 percent and an allowable stress in shear of 5,500 psi, determine the diameter of shaft you would recommend.

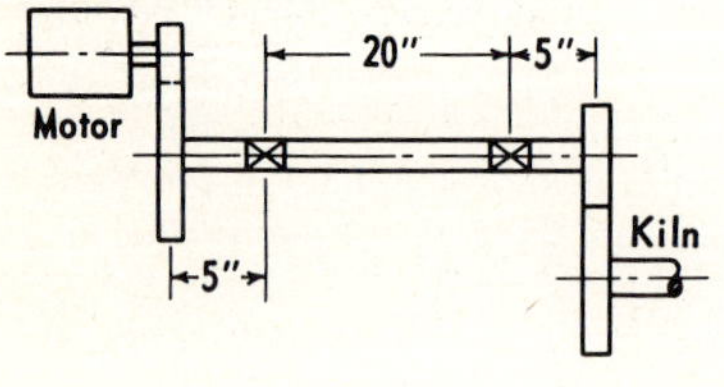

Fig. P 13-8

13-9. The pinion of a water pump is attached, as shown in Fig. P 13-9, to the shaft which is supported on ball bearings A and B. The data are as follows: pinion has 15 teeth; diametral pitch = 5; rpm = 1,200; horsepower = 50. (*a*) Using an allowable stress s_s = 8,000 psi, determine the diameter of shaft required. (*b*) Using a standard square key for the gear, determine the shear stress in the key.

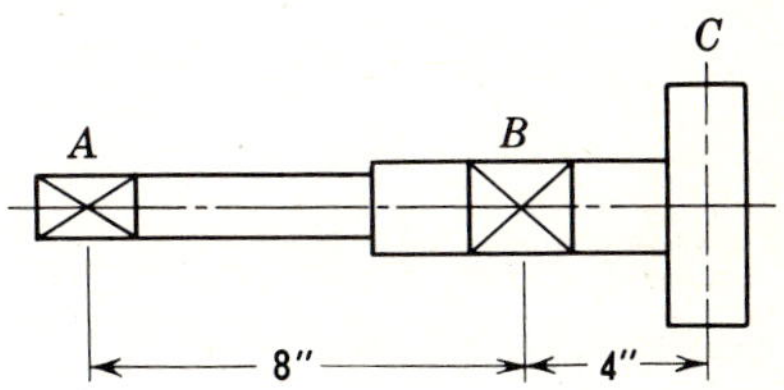

Fig. P 13-9

13-10. Refer to the data and sketch in Prob. 3-7. The detail of the ends of the axle is shown in Fig. P 13-10. (*a*) Determine the maximum bending stress at the section A in the figure. (*b*) Determine the radius r of the fillet so that the maximum stress at the fillet will not exceed the stress at the section A.

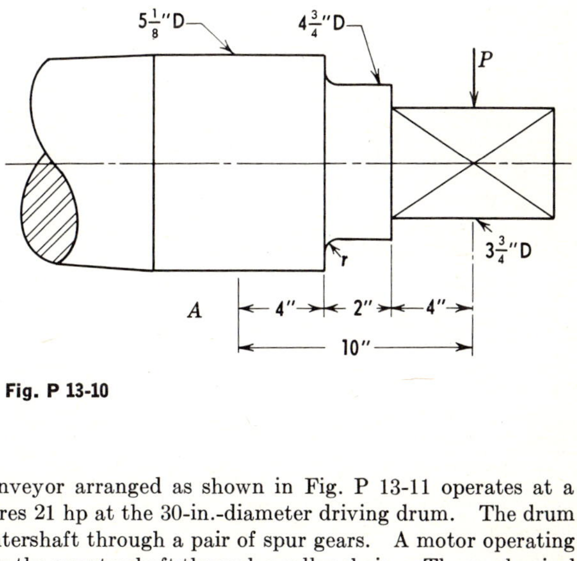

Fig. P 13-10

13-11. A horizontal belt conveyor arranged as shown in Fig. P 13-11 operates at a speed of 270 fpm and requires 21 hp at the 30-in.-diameter driving drum. The drum receives power from a countershaft through a pair of spur gears. A motor operating at 860 rpm delivers power to the countershaft through a roller chain. The mechanical efficiency of the chain drive is 94 percent and of the gears is 96 percent. (*a*) Determine the horsepower of the motor you would recommend. (*b*) Calculate the number of teeth on the large sprocket required to run the conveyor belt at the speed specified. (*c*) Determine the torque in inch-pounds on the countershaft corresponding to the rated horsepower of the motor. (*d*) Assuming an allowable stress in shear equal to 5,500 psi, determine the diameter of countershaft you would recommend.

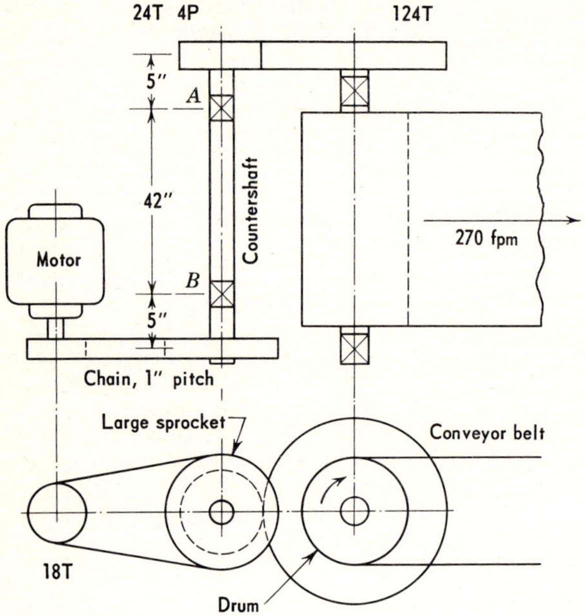

Fig. P 13-11

13-12. The following data apply to motor and gear drive for a centrifugal pump:

Pump output, 500 gpm
Total head, 60 ft water
Pump speed, 65 rpm
Pump efficiency, 64 percent

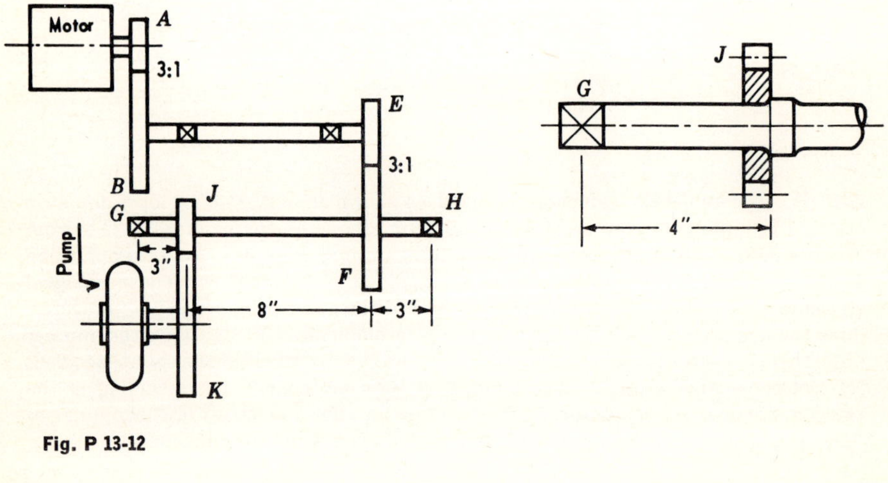

Fig. P 13-12

Motor speed, 1,750 rpm
Power cost, 3 cents per kwhr
Diameter of gear F, 20 in.
Diameter of pinion J, 4 in.

Determine the following:

(a) Maximum horsepower required of the motor. Assume the gears to be 100 percent efficient.

(b) Rated motor horsepower recommended for continuous operation.

(c) Size of shaft GH recommended, assuming an allowable stress $s_s = 5{,}000$ psi.

(d) Factor of safety at the fillet, near J (see Fig. P 13-12), assuming SAE 1025 steel for the shaft and stress-concentration factor at the fillet. Assume $D/d = 1.5$, $r/d = 0.125$.

(e) Power costs, dollars per 24-hr day. Assume a motor efficiency of 87 percent.

13-13. An 1,150-rpm motor drives a processing unit through two pairs of reduction gears AB and EF and a countershaft $BCDE$. Each gear reduction is 3:1, with pinion A having 21 teeth of 7 diametral pitch and pinion E having 16 teeth of 4 diametral pitch. The processing unit requires full-load torque of 9,000 in.-lb at the gear F.

(a) Determine the horsepower of the type B induction motor you would recommend, assuming the following duty cycle: full load, 3 min.; half load, 2 min.; running at no load, 2 min.; rest, 1 min.

(b) Assuming an allowable shear stress $s_s = 6{,}000$ psi, determine the recommended size of countershaft d at gear E.

(c) Assuming that the countershaft is made of SAE 1025 ground finish, the radius r of the fillet is approximately $\frac{1}{8}d$ (see detail at right in Fig. P 13-13), and the torque is steady (static), determine the factor of safety at the critical section at the fillet r.

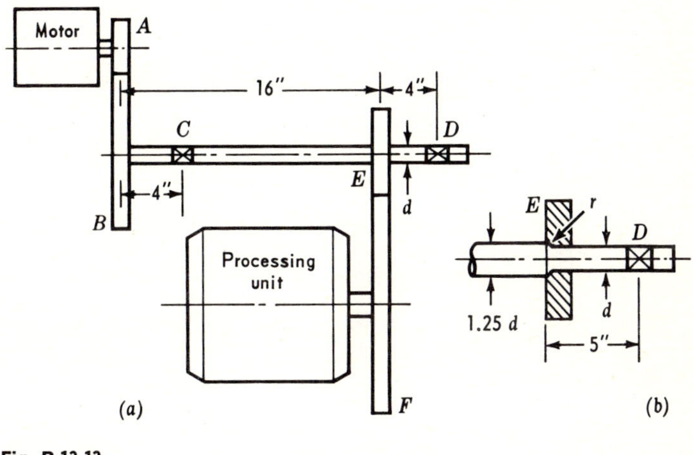

Fig. P 13-13

13-14. The drive for an escalator, as shown in Fig. P 13-14, is composed of a V-belt speed reduction, a helical-gear reduction, and a spur-gear reduction.

(*a*) Explain why the belt drive, helical gears, and spur gears are used in the order shown.

(*b*) In a commercial escalator drive, a brake is placed at *C* so that the escalator may be used as ordinary stairs, or to prevent its running backward in the event of power failure. What factors determine the location of the brake, that is, on motor shaft *A*, on *B*, *C*, or on the head shaft *D*?

(*c*) Assuming the data given below, determine the class of induction motor and its horsepower to be specified. Do not consider a duty cycle because at times the escalator will be operated for protracted periods at full load, the cost of the motor is a small portion of the cost of the escalator, and the burning out of the motor would be serious.

Motor speed: 1,170 rpm
Torque on shaft *D*: 6,100 ft-lb
Speed reductions: V belt, 5:1
Helical gears, 4.2:1
Spur gears, 4.2:1
Mechanical efficiency: motor to shaft *D*, 88 percent.

(*d*) Assuming the escalator driving sprockets shown at the ends of shaft *D* are 30 in. in diameter, determine the speed of the traveling steps in feet per minute.

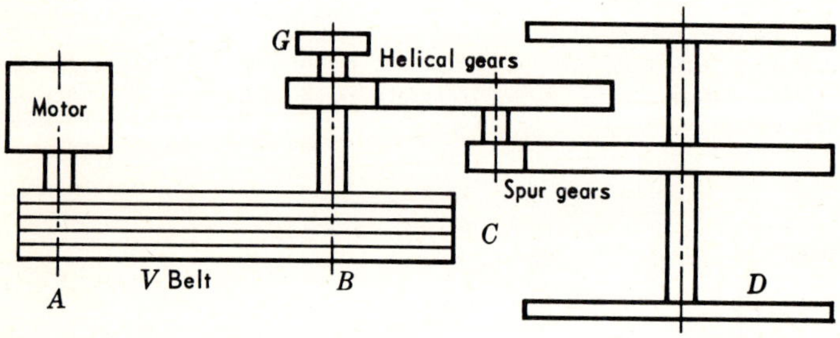

Fig. P 13-14

13-15. A belt-conveyor system is arranged as shown in Fig. P 13-15 to transport cafeteria trays full of dishes, etc., from the serving counter to the kitchen. The conveyor is located under the counter and delivers trays to the platform at *A* where they are unloaded. The top side of the conveyor belt is supported by a steel plate, as shown in the sketch, to prevent sag of the belt. The conveyor belt is driven by the drum *BH* from the motor *M* through two V-belt reductions: *C* to *E* and *G* to *H*. Assume these two speed reductions are equal. The problem is in three parts: (*a*) to determine the required horsepower for the drive; (*b*) to determine the size of countershaft *EG*, and (*c*) to suggest a plan for automatically stopping the drive if the trays are not unloaded at *A*.

Data: Conveyor-belt speed, 100 fpm
Weight of each loaded tray, 8 lb
Maximum spacing of trays on belt, 20 in.
Diameter of conveyor drum *BH*, 10 in.

Diameter of V-belt pulleys C and G, 3 in.
Motor speed, 870 rpm

(a) Determine the maximum horsepower of motor required. Assume the following: coefficient of sliding friction between the tray-loaded belt and the metal supporting plate as 0.3; the efficiency of each V-belt reduction, 95 percent.

(b) Determine the size of countershaft EG. Assume that the ratio of V-belt tensions for each reduction $F_1/F_2 = 3$; pulley G is located 4 in. from adjacent inboard bearing, and an allowable shear stress for the shaft is $s_s = 5{,}000$ psi. All belt tensions may be assumed as horizontal.

(c) Suggest a plan for automatically stopping the conveyor if the trays are not unloaded promptly at station A.

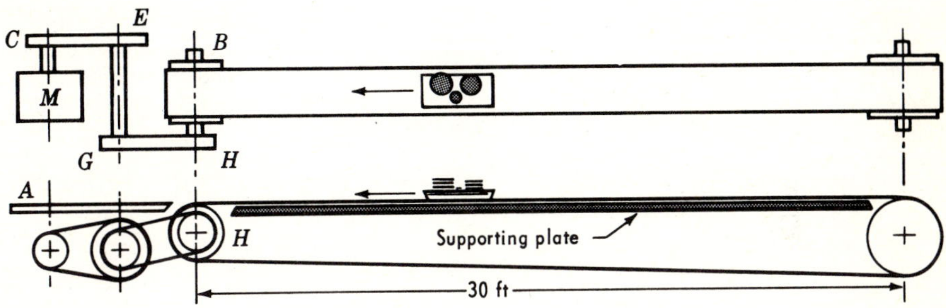

Fig. P 13-15

CHAPTER 14. BELT DRIVES AND HOISTS

14-1. (a) Derive the equation for the ratio of belt tensions in a flat belt transmitting power. The following notation should be used:

F_1 = tension in tight side of belt, lb
F_2 = tension in slack side of belt, lb
f = coefficient of friction between belt and pulley
θ = angle of contact of belt with pulley
b = width of belt, in.
t = thickness of belt, in.
w = weight of belt material, lb per in.^{3}
v = belt speed, fps

(b) Assuming that the ratio of belt tensions is $(F_1 - F_c)/(F_2 - F_c) = e^{f\theta}$, derive the equation for the horsepower transmitted per square inch of belt section. The final equation should be in terms of f_1, f_c and of the belt velocity v, where f_1 is the maximum stress in pounds per square inch and f_c is equal to $12(wv^2/g)$.

14-2. A 25-hp 870-rpm induction motor is used to drive a ventilating fan at 485 rpm. The motor pulley is 10 in. in diameter, and the center distance is 5 ft. Determine the following: (a) the minimum angle of contact; (b) the cross-section area of leather belt required, and the thickness, width, and length of the belt.

14-3. Two pulleys which are 24 in. in diameter and which run at 370 rpm are connected by a leather belt ⅜ in. thick. If the belt transmits 30 hp and the distance between the shafts is 10 ft, determine the width and length of belt that is required.

14-4. A 40-hp 1,200-rpm motor is to drive an air compressor through the medium of V belting. The mean diameters of the motor pulley and the compressor pulley are 9 in. and 30 in., respectively. The shaft center distance is 55+ in. Determine the size and number of V belts to be recommended.

14-5. Select a V-belt drive for connecting a 2-hp 1,170-rpm motor to a 300-rpm blower which is part of an air-conditioning unit. The center distance may be taken as approximately 1½ times the diameter of the large sprocket.

14-6. A block and tackle, having two sheaves at the top block and two at the hook, with the rope anchored at the top block is reefed with wire rope. Determine the following: (*a*) an equation for the effort P required to raise a load Q, in terms of Q and the coefficient C; (*b*) the force P required to raise a load $Q = 2$ tons, assuming $C = 1.076$; (*c*) the efficiency of the hoist.

14-7. A hoist is arranged as shown in Fig. P 14-7, in which a hydraulic cylinder applies a force P in order to raise a load Q. The ratio of rope tensions is equal to C, where C is greater than unity. Derive an equation for P in terms of Q and C.

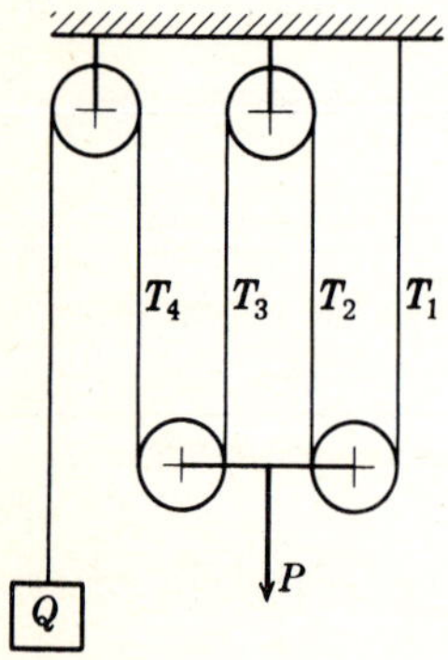

Fig. P 14-7

14-8. A hoist is arranged as shown in Fig. P 14-8, in which a hydraulic cylinder applies a force P in order to raise the load $W = 7{,}500$ lb. Determine the diameter d of the rope required, assuming 6 × 19 steel rope, a factor of safety of 4, sheaves $45d$ in diameter, and a coefficient C of 1.10. Neglect weight of rope.

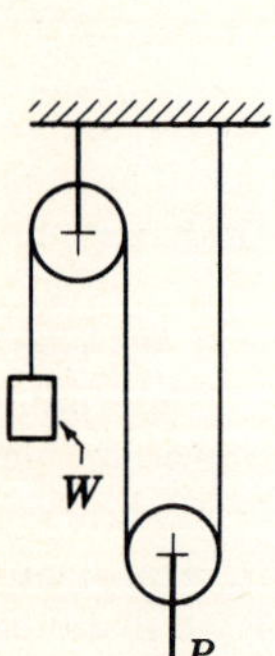

Fig. P 14-8

14-9. The pulley system shown in Fig. P 14-9 is used to raise a load of 10,000 lb with a maximum acceleration of 10 fps². Determine the diameter d of rope required, assuming 6 × 19 steel rope with a factor of safety of 5, sheaves $45d$ in diameter, and a coefficient C of 1.10. Neglect weight of rope.

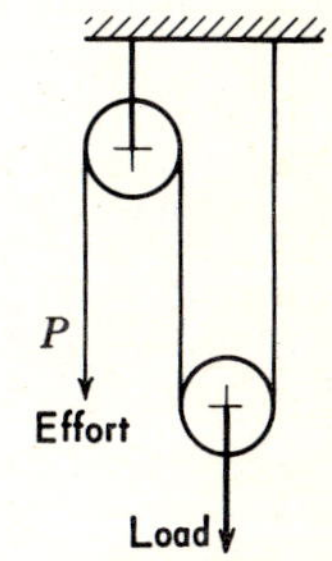

Fig. P 14-9

14-10. The pulley system shown in Fig. P 14-10 is to be used to raise a weight of 10,000 lb with a maximum acceleration of 10 fps². Determine the diameter d of the wire rope, assuming a factor of safety equal to 5, a coefficient C of 1.10, and sheaves of the following diameters: $20d$, $30d$, $40d$, and $50d$.

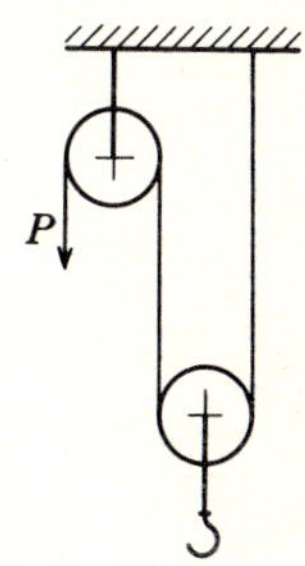

Fig. P 14-10

14-11. A hoisting winch for a Maritime Commission vessel is arranged as shown in Fig. P 14-11. The following data apply:

Rated load on drum cable	3,800 lb
Maximum load on cable	10,000 lb
Drum diameter for first layer of 6 × 19, ¾-in. rope	16¾ in.
Motor speed (at rating)	520 rpm
Mechanical efficiency, motor to drum	85 percent

Determine the following: (*a*) speed of drum cable at rating; (*b*) horsepower of motor for rated load and speed; (*c*) factor of safety for cable at maximum load; (*d*) the load that may be hoisted and the speed of hoisting, using a three-part line with 16¾-in. sheaves (assume that $C = 1.076$); (*e*) the required rating of the brake in pound-feet for rated load on the winch.

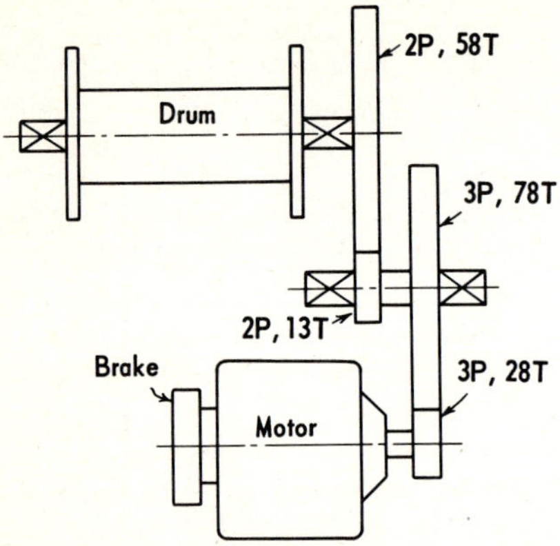

Fig. P 14-11

14-12. A hand-operated wire-rope hoist is to be arranged as shown in Fig. P 14-12 to raise a load $Q = 250$ lb. The force F on the operating lever is limited to 35 lb. The lever arm $L = 12$ in., the diameter of the drum A is 8 in., and the diameter of the sheaves is 10 in. Determine the following: (*a*) the number of strands of rope leading to the hook block; (*b*) the expression for the tension in the rope T in terms of Q and C; (*c*) the factor of safety of the hoist, assuming that it is reefed with ⅜-in. 6 × 19 wire rope.

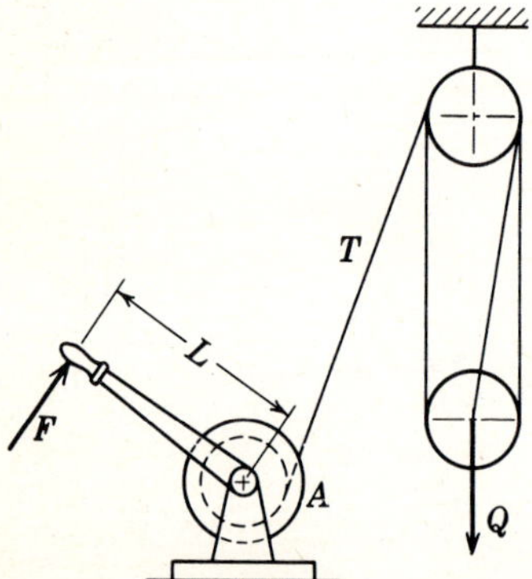

Fig. P 14-12

CHAPTER 15. POWER-TRANSMISSION CHAINS

15-1. A 25-hp squirrel-cage induction motor running at 1,150 rpm is to be connected to a centrifugal pump by means of a roller chain. The pump operates at a speed of approximately 400 rpm and is located 30+ in., center to center, from the motor. A service factor of 1.2 may be assumed. Determine the pitch, width, and length of the chain you would recommend, and also the pitch diameters of the sprockets.

15-2. A 10-hp motor running at 1,150 rpm is to drive a compressor through the medium of a roller chain. The compressor runs at approximately 300 rpm and is to be located 20+ in. from the motor shaft. Determine the following: (*a*) pitch of chain; (*b*) number of strands; (*c*) number of links; (*d*) pitch diameters of sprockets.

15-3. It is desired to transmit 40 hp by means of a roller chain at a motor-sprocket speed of 720 rpm. The pitch diameter of the motor sprocket must not exceed 6¾ in. Assuming uniform service 24 hr per day, determine the pitch and number of strands required for the chain.

15-4. A 10-hp motor running at 1,750 rpm is to drive a blower through the medium of a silent chain. The blower runs at approximately 600 rpm and is to be located 28+ in. from the motor shaft. Determine the following: (*a*) pitch of chain; (*b*) width of chain; (*c*) number of links; (*d*) pitch diameters of sprockets.

15-5. A ventilating fan having a delivery of 18,000 cfm when operating at 298 rpm requires 2.84 hp under these rated conditions. The fan is to be installed in a hospital where quietness of operation is essential.

The fan is to be connected to a squirrel-cage induction motor by means of a chain drive.

Assumptions:

Motor speed	1,160 rpm
Minimum number of teeth on small sprocket	19
Efficiency of chain drive	95 percent

Determine the following: (*a*) horsepower of motor required; (*b*) type of chain; (*c*) pitch and width of chain; (*d*) number of teeth and diameters for sprockets; (*e*) length of chain.

15-6. A silent chain drive, as specified below, is to be replaced by a V-belt drive. Assuming that 6-in. sheaves are to be used, determine the number of B-section V belts required for the replacement drive.

Pitch	¾ in.
Width	1¼ in.
Sprockets	21 teeth
Speed	1,120 rpm
Velocity ratio	1:1

15-7. An 1,160-rpm motor is to be used to drive an oil-field pumping unit that requires 280 lb-ft torque at approximately 320 rpm. Assuming that a roller chain drive is to be used and that there is an efficiency of 95 percent and a service factor of 1.5, determine the following: size of motor, number of sprocket teeth, and pitch and number of strands of chain you would recommend.

15-8. In a roller-chain drive arranged as shown in Fig. 15-1(*c*), a chemical processing unit is connected to each of the two sprockets *B*. Each of the units requires 250 in.-lb torque. The driving sprocket at *A* rotates at 100 rpm and has 11 teeth, and the sprockets at *B* have 33 teeth each. The chain is one-strand No. 41 chain. Determine the service factor for the drive.

15-9. The layout of the drive for an engine-driven snow thrower is shown in Fig. P 15-9. The engine drives a small V-belt sheave *A* which drives a large sheave *B*. On the same shaft as *B* is mounted a small chain sprocket *C* which drives the large sprocket *E*. The "auger" that throws the snow is connected to the shaft *L*. The engine is rated at 7.30 hp at 2,680 rpm of the drive shaft *J* and will clear a 52-in. path of snow and throw it 40 ft.

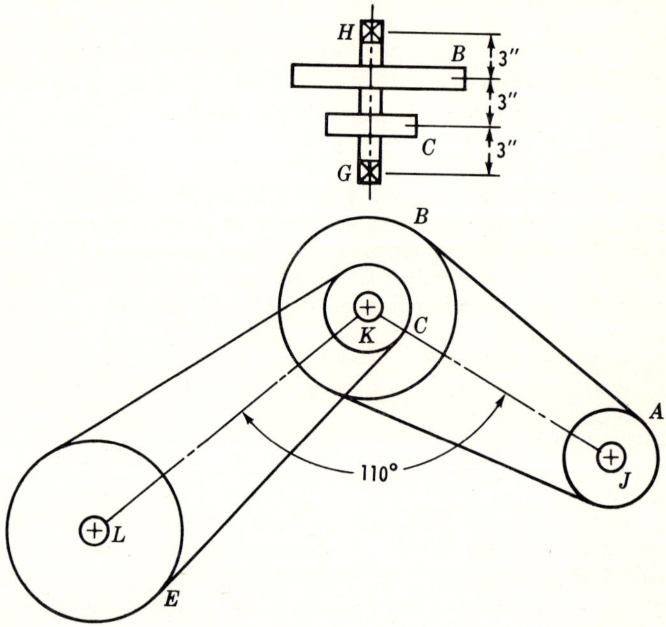

Fig. P 15-9

Additional data are: Sheave A, 4.7 in. in diameter; sheave B, 7.0 in. in diameter; sprocket C, 20 teeth; sprocket E, 45 teeth; pitch of links = ½ in.; G and H represent ball bearings. Shaft dimensions are as indicated in the sketch. Angle JKL = 110°.

Using a value of $e^{f\theta}$ for the belt as 3.5, and allowable stress in shear for the shaft as 8,000 psi, determine the nominal diameter of the shaft GH.

CHAPTER 16. SHAFT COUPLINGS

16-1. (*a*) Make a neat sketch of a flanged shaft coupling showing a face view and a cross-sectional view. (*b*) Derive expressions in terms of the torque T in inch-pounds for the following: the shearing stress in the shaft; the shearing stress in the coupling bolts; the bearing pressure on the bolts.

Use the following notation:

d = shaft diameter
R = radius of bolt circle
n = number of bolts
d_1 = bolt diameter
t = thickness of radial flange

16-2. In a flanged shaft coupling having a 1½-in. bore, it is desired that the torsional stress in the shaft will not exceed 3,500 psi. The outside diameter of the coupling is 7½ in. There are three ⅝-in. bolts on a bolt circle 2⅝ in. in radius. The radial flange thickness is $\frac{3}{4}$ in. Determine the following: (*a*) the horsepower that may be transmitted at 600 rpm; (*b*) the shearing stress in the bolts; (*c*) the bearing pressure on the bolts.

CHAPTER 17. CLUTCHES AND BRAKES

17-1. Assuming that the pressure is uniformly distributed over the friction surfaces in contact, derive an expression for the torque T transmitted by a plate friction clutch in terms of the coefficient of friction f, the axial force P, and the outer and inner diameters of the plates, D and d, respectively.

17-2. Assuming that the normal wear of the surfaces in contact is proportional to the work of friction, derive an expression for the torque T transmitted by a plate friction clutch in terms of the coefficient of friction f, the axial force P, and the outer and inner diameters of the plates, D and d, respectively.

17-3. Assuming that the normal wear of the surfaces in contact is proportional to the work of friction, derive an expression for the torque T transmitted by a cone clutch in terms of the coefficient of friction f, the axial force P, outer and inner diameters, D and d, and the cone angle α.

17-4. A multiple-disk clutch with 17 disks, 8 disks turning with one shaft and 9 with the other shaft, is rated at 25 hp at 500 rpm. The outside and inside diameters of the disks are $5\frac{1}{2}$ and $2\frac{1}{2}$ in., respectively. The coefficient of friction is 0.2. (*a*) Determine the axial force required to transmit the rated torque; (*b*) determine the pressure between the disks.

17-5. An automotive engine develops maximum torque at 1,000 rpm, at which speed the horsepower equals 34.6. The engine is to be equipped with a single plate clutch having two pairs of friction surfaces. (*a*) Using the equation for torque derived on the basis of "normal wear is proportional to the work of friction," and assuming that the coefficient of friction is equal to 0.4 and that the mean diameter of the clutch disks is $7\frac{5}{8}$ in., determine the axial force required to apply the clutch. (*b*) Assuming that the allowable average normal pressure on the clutch faces is 18 psi, determine the outer and inner diameters required for the disks.

17-6. The large diameter and face of the disks of a multiple-disk clutch are 10 in. and 1 in., respectively. The helical compression spring used to engage the clutch has 9.5 effective coils of $\frac{3}{8}$-in. steel wire, the outer diameter of the coils being $3\frac{1}{8}$ in. The free length of the spring is $7\frac{1}{4}$ in. and when in place with the clutch engaged, its length is $5\frac{1}{8}$ in. Assuming that there are 10 pairs of friction surfaces in contact, that the motor runs at 1,200 rpm, and that the coefficient of friction is 0.15, determine the following: (*a*) the axial force produced by the spring, and (*b*) the horsepower that the clutch will transmit.

17-7. Assuming that the normal wear of the surfaces in contact is proportional to the work of friction, derive an expression for the torque T that can be resisted by a single block of a flat-rim block brake for a given radial force on the block P.

Let b = width of block
f = coefficient of friction
D = diameter of wheel
θ = half the angle of contact of block and wheel

17-8. Determine the horsepower transmitted by the double-block clutch shown in Fig. P 17-8, assuming the following data: The drum is 10 in. in diameter and rotates

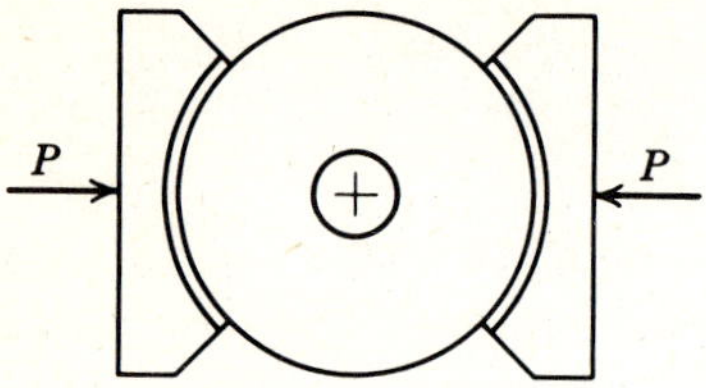

Fig. P 17-8

at 300 rpm. The contact angle for each shoe is 90°, and the coefficient of friction is 0.35. Assume the force on each block $P = 285$ lb.

17-9. Determine the torque that may be resisted by the single-block brake shown in Fig. P 17-9, assuming the following: The brake drum is 10 in. in diameter. $a = 4$ in.; $R = 140$ lb. The contact angle is 90°. Coefficient of friction between drum and lining is 0.35.

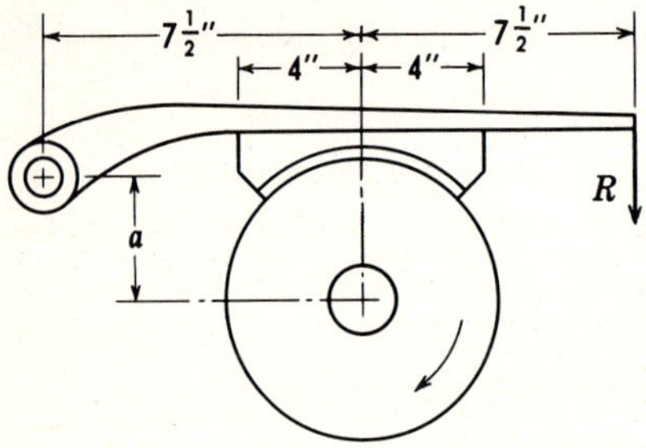

Fig. P 17-9

17-10. Solve Prob. 17-9, assuming $a = 6$ in.

17-11. The block brake shown in Fig. P 17-11 is set by a spring that produces a force S on each arm equal to 682 lb. The wheel diameter is 14 in., and the angle of contact for each block is 110°. Using 0.40 as a value of the coefficient of friction between the drum and lining, determine the maximum torque in foot-pounds that the brake is capable of absorbing.

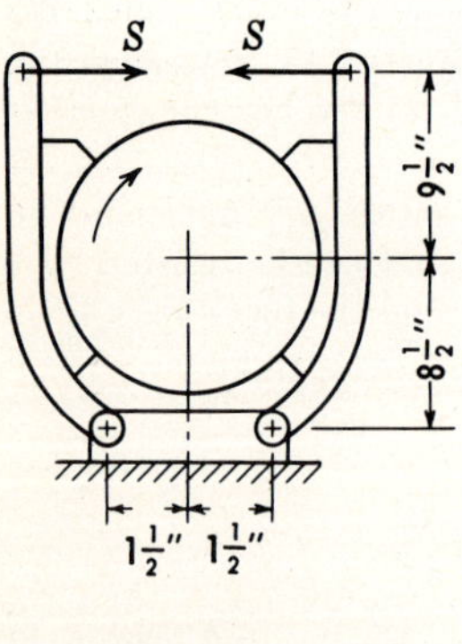

Fig. P 17-11

17-12. Solve Prob. 17-11, assuming counterclockwise rotation of the wheel.

17-13. Solve Prob. 17-11, assuming that the $1\frac{1}{2}$ in. dimensions are changed to 7 in.

17-14. The wheel of the double-block spring-set brake shown in Fig. P 17-14 is 12 in. in diameter and the dimensions are $a = 8$ in. and $b = 10$ in. (*a*) Assuming that there is a coefficient of friction of 0.4 and that the angle of contact for each block is 110°, determine the spring force that is required for the brake to resist a torque of 600 lb-ft. (*b*) Assuming that the value of the design factor pV is 28,000 ft-lb per min per in.^{2} of projected bearing area and that the normal operating speed of the drum is 300 rpm, determine the width of brake shoes required.

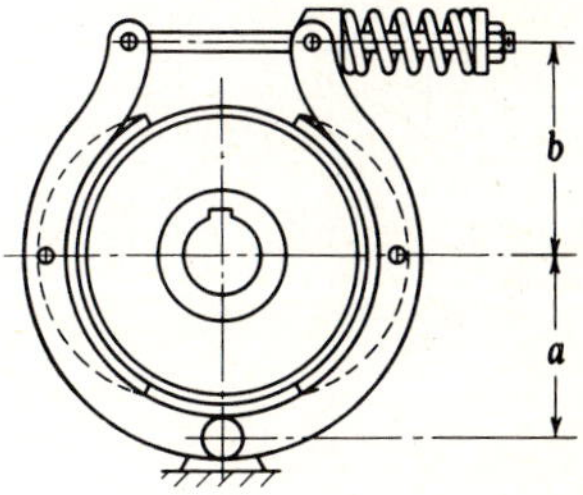

Fig. P 17-14

17-15. The layout and dimensions of a block brake are shown in Fig. P 17-15. (*a*) Assuming that the coefficient of friction for the brake lining and wheel is 0.4 and that the contact angle for each block is 90°, determine the force P on the operating arm required to set the brake for counterclockwise rotation of the wheel, which is 12 in. in diameter. The torque on the wheel is 667 lb-ft. (*b*) What type and size of chain would you select?

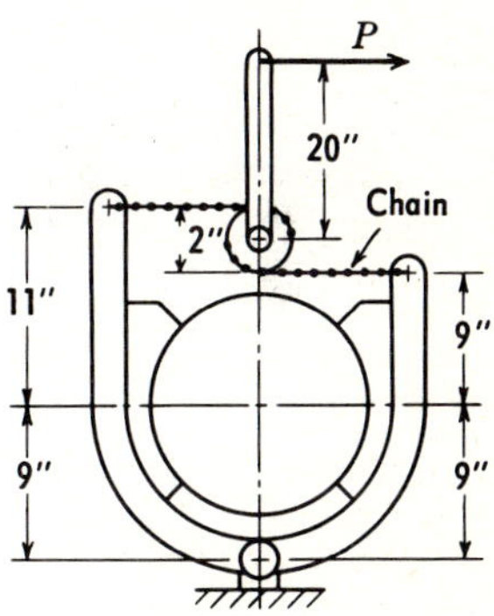

Fig. P 17-15

17-16. The arrangement of a transmission brake is shown in Fig. P 17-16. The arms are pivoted at O and, when force is applied at the end of the hand lever, the screw AB will rotate. The left- and right-hand threads working in nuts on the ends of the arms will move the arms together and thus apply the brake.

The force on the hand lever is applied 15 in. from the axis of the screw. The drum is 7 in. in diameter, and the angle subtended by each block is 90°. The screw has six square threads with a mean diameter of $\frac{3}{4}$ in. and a lead of $2\frac{1}{4}$ in.

Assuming a coefficient of friction for the braking surfaces that is equal to 0.30 and for the threads equal to 0.15, determine the force on the hand lever required to set the brake when the torque on the drum is 150 lb-ft.

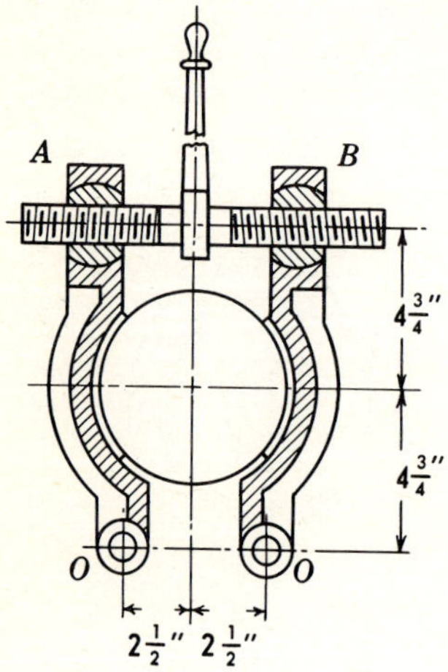

Fig. P 17-16

17-17. Determine the horsepower that the block brake shown in Fig. P 17-17 will absorb when the drum is rotating clockwise at 60 rpm and the operating force S is 100 lb. The coefficient of friction for the materials in contact is 0.3, and the angle of contact for each shoe is 90°.

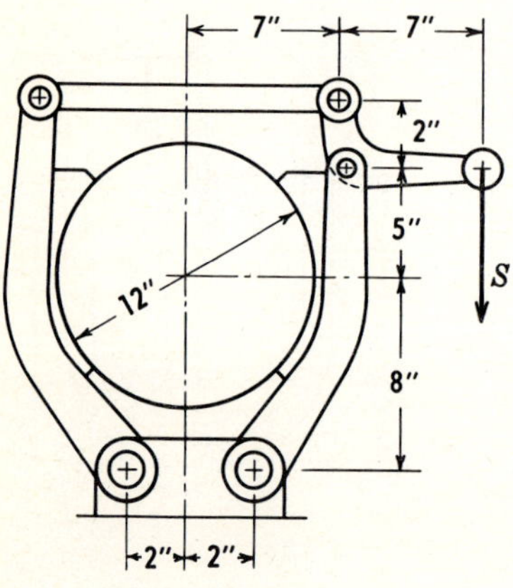

Fig. P 17-17

17-18. Figure P 17-18 shows the arrangement and dimensions for the preliminary design of double-block brake to be set by a spring providing a force S on the bell crank a. The brake is to have a torque rating of 2,400 lb-ft at 250 rpm. Assuming that the angle of contact for each shoe is 120° and that the coefficient of friction for the materials in contact is 0.35, determine the following: (a) the direction of rotation that requires the largest spring force for the rated torque and the value of the spring force S for that direction; (*b*) the width of shoes required, assuming $pV = 40{,}000$ ft-lb per min per in.^{2} of projected area; (*c*) the ratio of shoe width to drum diameter. Is this a satisfactory ratio?

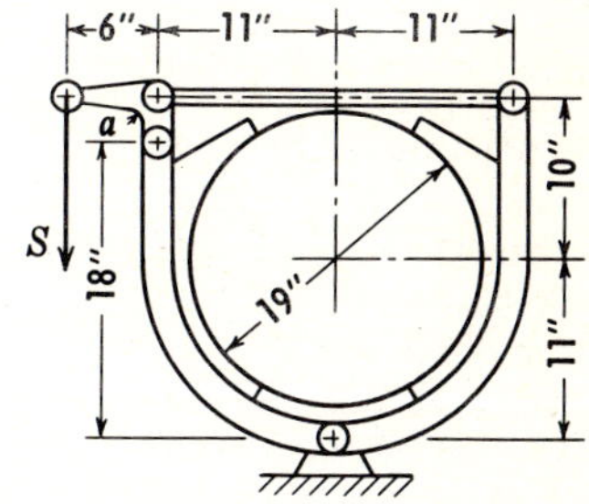

Fig. P 17-18

17-19. A 5-hp 850-rpm a-c motor is to be fitted with a centrifugal starting clutch. This type of clutch, as shown schematically in Fig. P 17-19, is composed of the arms *a*, connected to the rotating armature of the motor, and two rotating shoes *b*. These shoes, which are identical, are faced with friction lining, and they move outward by centrifugal force when the arm rotates, and, when a certain speed is reached, the force between the shoes and the ring *c* will produce sufficient friction to rotate the ring *c* against a resisting torque.

Determine the weight of each shoe required for picking up the load at the following conditions:

Torque at load pickup	80 percent of torque at rated hp and speed
Speed of motor at pickup	600 rpm
Coefficient of friction	0.3
Inner diameter of ring, *d*	8 in.
Radius to center of gravity of shoe, *r*	3 in.
Arc of contact of each shoe	90°

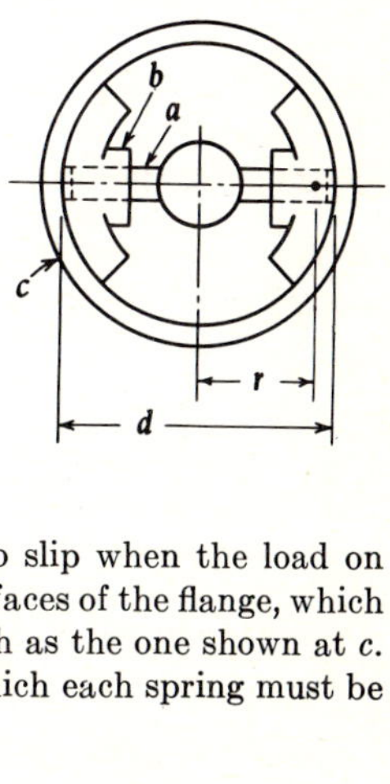

Fig. P 17-19

17-20. A slip coupling, as shown in Fig. 16-11, is designed to slip when the load on the driven member becomes excessive. The pressure on both faces of the flange, which is part of the hub *a*, is produced by a number of springs, such as the one shown at *c*. Using the data given below, determine the deflection with which each spring must be set up in order that the coupling will slip at a torque corresponding to 20 percent overload on a 250-hp motor running at 300 rpm. Use the equation based on the assumption that the normal wear is proportional to the work of friction.

Outer diameter of friction disks	31 in.
Inner diameter of friction disks	19 in.
Number of bolts and springs	12
Coefficient of friction (assumed)	0.14
Spring rate for each spring per inch	2,500 lb

17-21. The backstop shown schematically in Fig. P 17-21 is used to prevent backward rotation of the shaft. A sector is pivoted at O, and one end of the band is attached to it and operates at a radius $r_2 = 2\frac{1}{4}$ in. The other end of the band is attached at point a so that $Oa = r_1 = 1$ in. The diameter of the wheel is $8\frac{1}{4}$ in., the angle of wrap is 270°, and the width of the band is $2\frac{1}{8}$ in. The torque on the wheel is 300 lb-ft. Assuming a coefficient of friction between the band and wheel equal to 0.2, determine the following: (*a*) the maximum band tension; (*b*) the maximum pressure between the band and wheel; and (*c*) whether the backstop is self-locking.

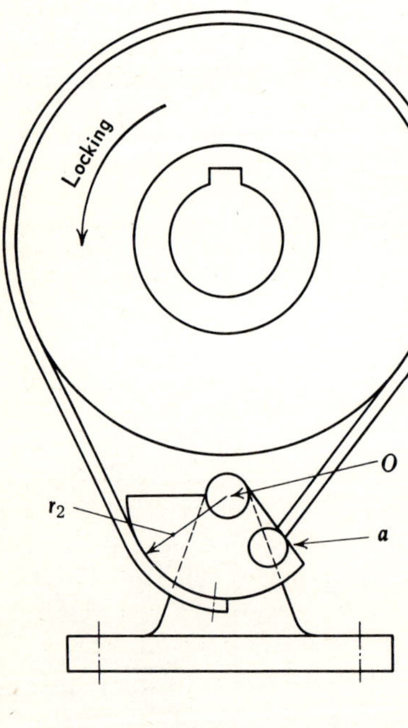

Fig. P 17-21

17-22. For a band brake similar to that shown in Fig. 17-15, determine (*a*) the torque in foot-pounds on the wheel for a force A on the operating lever equal to 20 lb (assume that the wheel is 9 in. in diameter, a = 8 in., b = 10 in., angle of contact θ = 210°, and the coefficient of friction f = 0.2) and (*b*) the time in seconds required to bring to rest a drum having WR^2 equal to 225 lb-ft² rotating initially at 200 rpm.

CHAPTER 18. SPUR AND PARALLEL HELICAL GEARS

18-1. The layout of a drum hoist driven by a 1,160-rpm motor through double-reduction gears is shown in Fig. P 18-1. The speed reduction between the motor shaft *A* and the intermediate shaft *B* is 4:1, and between shaft *B* and the drum shaft *C* is 5:1. The drum is 24 in. in diameter and the maximum cable load is 5 tons.

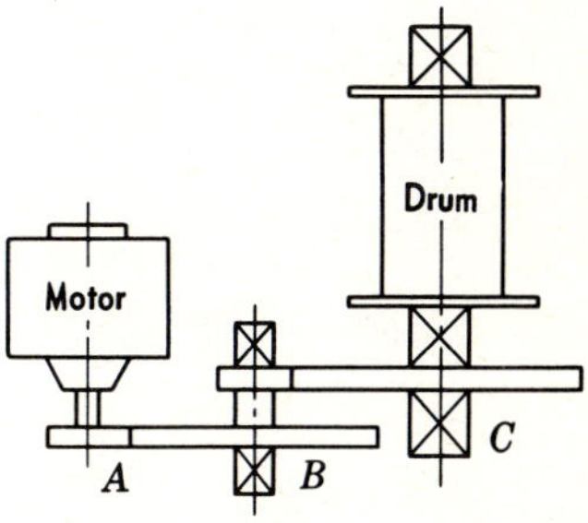

Fig. P 18-1

(*a*) Determine the torque in foot-pounds and the speed in rpm for shafts *A*, *B*, and *C*, and enter the results in the table below. The efficiency of each gear reduction is 95 percent.

Shaft	*A*	*B*	*C*
Torque, ft-lb			
Speed, rpm			

(*b*) Complete the table below.

	A to B (4:1)		*B to C* (5:1)	
	Pinion	*Gear*	*Pinion*	*Gear*
Diametral pitch				5
Number of teeth	24			
Pitch diameter	4			30

(*c*) What horsepower of motor would you recommend?

(*d*) If a spring-set brake is used to control the load, on which shaft, *A*, *B*, or *C*, would you attach the brake?

18-2. The two-speed hoisting winch shown in Fig. P 18-2 is to be rated at 3.5 tons on the drum cable at 100 fpm with the low-speed gears engaged. Data regarding the gears are shown in the table. (*a*) Copy the table, and enter values to complete it;

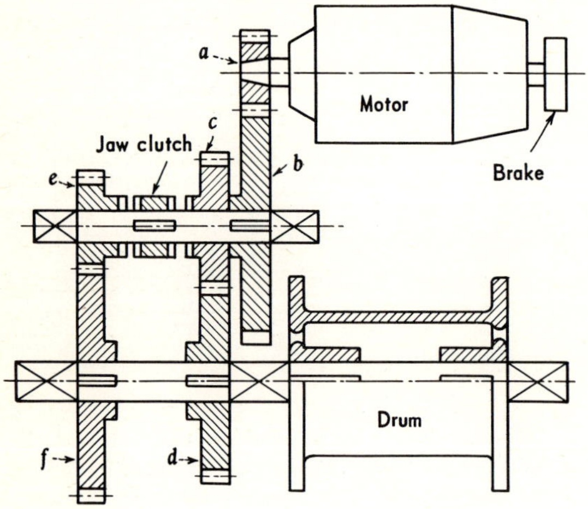

Fig. P 18-2

(*b*) determine the speed of the motor, assuming that the pitch diameter of the drum is 21¾ in.; (*c*) determine the horsepower of the motor required, assuming a 50 percent overload capacity and a 97 percent mechanical efficiency for each pair of mating gears.

Gear	*a*	*b*	*c*	*d*	*e*	*f*
Number of teeth	15	75	32		13	71
Diametral pitch	2½		2		2	
Pitch diameter						

18-3. (*a*) Make a neat sketch of a loaded gear tooth, showing the tangential load W, the inscribed parabola, and the dimensions of the tooth necessary for the derivation of the Lewis equation. (*b*) Assuming that a gear tooth is equivalent to a cantilever beam of uniform strength, derive an equation for the tangential load W in terms of the face width F, the circular pitch p, maximum stress s, and the Lewis form factor. (*c*) How is the allowable stress related to the static stress s_o?

18-4. It is required to determine the proportions of a pair of spur gears to transmit 40 hp from a shaft running at 250 rpm to a parallel shaft with a speed reduction of 3:1. The center-to-center distance of the shafts is 15 in. (*a*) Determine the pitch diameters of the pinion and the gear. (*b*) Assuming that the pinion is made of untreated SAE 1030 steel, the gear is cast steel, and the teeth are full-depth 20° involute teeth, determine by the use of the Lewis equation the diametral pitch, the face width, and the number of teeth for the gears.

18-5. Two parallel shafts are to be connected by means of spur gears to transmit 7.5 hp from 1,150 rpm of the high-speed shaft to the low-speed shaft with a reduction in speed of 5:1. The following assumptions may be made:

Pinion	SAE 1045 (hardened, 21 teeth)
Gear	cast steel
Teeth	AGMA 20° involute stub

By the use of the Lewis equation, determine the diametral pitch, pitch diameters, and face width.

18-6. A cargo winch is driven by a 35-hp 480-rpm motor through the medium of a single reduction of spur gearing so that the speed of the drum is approximately 57 rpm.

Using the Lewis equation, determine the diametral pitch, pitch diameters, and face width.

Assume the following:

Teeth	AGMA 20° stub
Pinion	SAE 1040, forged steel
Gear	cast steel
Number of pinion teeth	22
Maximum load on motor	150 percent rating

18-7. A straight spur pinion made of SAE 1040 steel is to be keyed to the shaft of a 15-hp 1,140-rpm motor and is to drive a gear made of ordinary cast iron. The speed reduction is to be 6:1. Assuming 20° involute stub teeth and 24 teeth on the pinion, determine the following: (*a*) which of the gears has the weaker teeth; (*b*) the value of the constant in the equation for the induced stress, that is, $s_{ind} = (\quad)P^3$; (*c*) the diametral pitch and pitch diameters; (*d*) the face width.

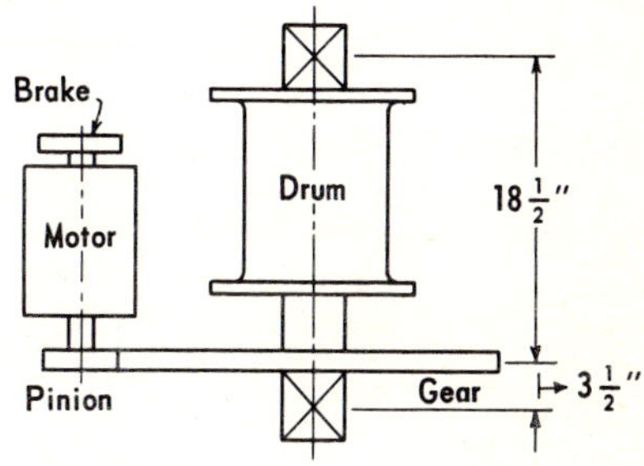

Fig. P 18-6

18-8. Using the data and results for Prob. 18-7, determine the following: (*a*) the dynamic load on the teeth, assuming class 2 cut; (*b*) the maximum stress corresponding to the dynamic load and the required Bhn for the gears, assuming steady load conditions; (*c*) the value of the load-stress factor and the required Bhn for the gears.

18-9. A 15-hp 1,140-rpm motor is to be connected to a drive shaft by means of a pair of spur gears having a speed ratio of pinion to gear equal to 6. The gear teeth are to be of 20° involute stub form. The pinion is to have 24 teeth made of SAE 1040 steel and the gear of medium-grade cast iron. (*a*) Determine the diametral pitch, the pitch diameters, and face width of the gears by the use of the Lewis equation. (*b*) Assuming steady-load conditions of service, determine the class of gears, and state if the cast iron as assumed will be of satisfactory grade. (*c*) Determine the Bhn required for the gears.

18-10. Values may be assigned or adopted for *A*, *B*, *C*, and *D* in this problem.

The following data apply to a single-reduction speed reducer having a pair of straight spur gears:

Rated horsepower	*A*______
Pinion speed	*B*______ rpm
Velocity ratio	*C*______ to 1
Form of teeth	20° stub

(*a*) Assuming SAE 2320 pinion, SAE 1040 gear, and a minimum of *D*______ teeth on the pinion, determine by the use of the Lewis equation the following: (1) diametral pitch of the teeth; (2) pitch diameters of the gears; (3) face width.

(*b*) Assuming steady-load conditions and class 2 gears, determine the Bhn required for the pinion and gear on the basis of dynamic loading.

(*c*) Determine the Bhn required for the pinion and gear on the basis of wear.

(*d*) Enter recommended values for the speed reducer in the table below.

	Diametral pitch	*Pitch diameter*	*Face*	*Bhn*
Pinion				
Gear				

18-11. Values may be assigned or adopted for *A*, *B*, *C*, *D*, *E*, and *F* in this problem.

The following data apply to a single-reduction speed reducer having a pair of straight spur gears:

Rated horsepower	*A*______
Pinion speed	*B*______ rpm
Velocity ratio	*C*______ to 1
Form of teeth	20° stub

(*a*) On the basis of 150 percent rating and using SAE numbers for pinion as *D*______ and gear as *E*______ and a minimum of *F*______ pinion teeth, determine by the use of the Lewis equation the following: (1) diametral pitch of the teeth, (2) pitch diameters of pinion and gear, and (3) face width of the gears. (*b*) On the basis of 150 percent rating and for steady-load conditions and class 2 cut, determine the Bhn required for the pinion and gear on the basis of dynamic loading. (*c*) On the basis of 100 percent rating, determine the Bhn required for the pinion and gear on the basis of wear. (*d*) Enter recommended values for the speed reducer in a table similar to that given below:

	Diametral pitch	*Pitch diameter*	*Face*	*Bhn*
Pinion				
Gear				

18-12. A parallel helical gear 10 in. in diameter having 20° involute stub teeth and a helix angle of 30° carries a torque of 3,000 lb-in. Determine the tangential and radial components of the load on the teeth in the plane of rotation, the axial component of the load on the teeth, the angle of obliquity in a plane normal to the pitch element of a tooth, and the resultant load on the teeth.

18-13. A 15-hp motor runs at 1,170 rpm and drives a ventilating fan through a gear reduction having a velocity ratio of 4:1. The gears are to have 20° involute stub teeth. The pinion is to be made of Micarta and the gear is to be made of cast iron. Assuming 24 teeth on the pinion, determine the diametral pitch, pitch diameters, and face width of the gears using the Lewis equation.

18-14. The shaft of a wire-pulling machine rotates at 1,800 rpm and is required to transmit a torque of 800 lb-in. from a gear located at one end of the shaft to a flexible coupling at the other end. The bearings that support the shaft are 30 in. apart, and the gear is located 4 in. outside its adjacent bearing, center to center. The gear is driven by a driving gear from a motor with a 1:1 velocity ratio. (*a*) Determine the horsepower of motor to be specified. (*b*) Assuming that the gears have 18 teeth of 20° full-depth form and made of SAE 1035 steel, determine, by the use of the Lewis equation, the diametral pitch, pitch diameters, and face width of the gears you would recommend. (*c*) Using an allowable shearing stress of 5,000 psi, determine the shaft diameter you would recommend.

18-15. A hotel-restaurant belt-conveyor system is to be driven by a 2-hp 870-rpm motor through a gear train having a velocity ratio of 4:1. Assuming a starting overload of 20 percent, a Micarta pinion having at least 26 teeth, cast-iron gear, and 20° full-depth teeth, determine by the use of the Lewis equation the diametral pitch, pitch diameters and face width you would recommend.

18-16. A reciprocating compressor system is to be driven by an 870-rpm motor through a pair of straight spur gears. The compressor operates at approximately 220 rpm at a running torque of 1,300 lb-in. Assume a starting overload of 20 percent, 20° full-depth teeth, 30 teeth on the pinion, and SAE 1045 and 1030 for the pinion and gear, respectively, and determine your recommendations for the horsepower of the motor, diametral pitch, pitch diameters, and face width for the gears.

18-17. In the line shaft shown in Fig. P 18-17, a belt runs over the pulley A which is 36 in. in diameter and rotates at 320 rpm. The belt tensions are $F_1 = 1{,}500$ lb and $F_2 = 1{,}000$ lb. (*a*) Assuming the pinion B is to be made of SAE 1020 steel and with eighteen 14.5° teeth determine the diametral pitch, face width, and diameter required by the Lewis equation. (*b*) Assuming an allowable shearing stress equal to 6,000 psi, determine the diameter of shaft to be specified.

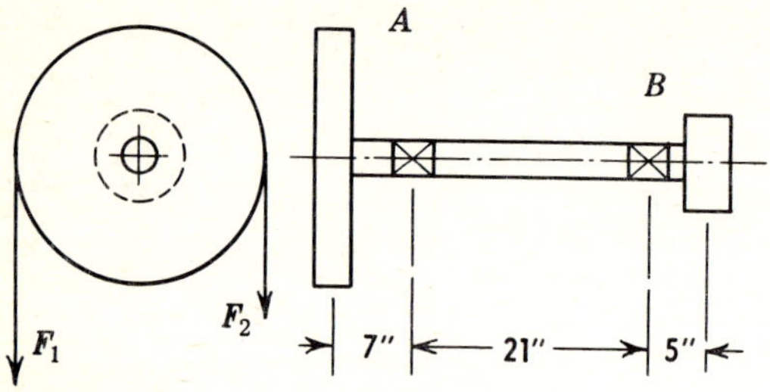

Fig. P 18-17

18-18. Refer to data for Prob. 14-11. (*a*) Assuming that the motor pinion and its mating gear have a face width equal to 3½ in. and are made of SAE 1025 steel, would you regard the gears as satisfactory, based on the Lewis equation? (*b*) Determine the diameter of the intermediate-speed shaft that is required (see Fig. P 18-18). (*c*) How would you anchor the cable to the drum? Use sketch.

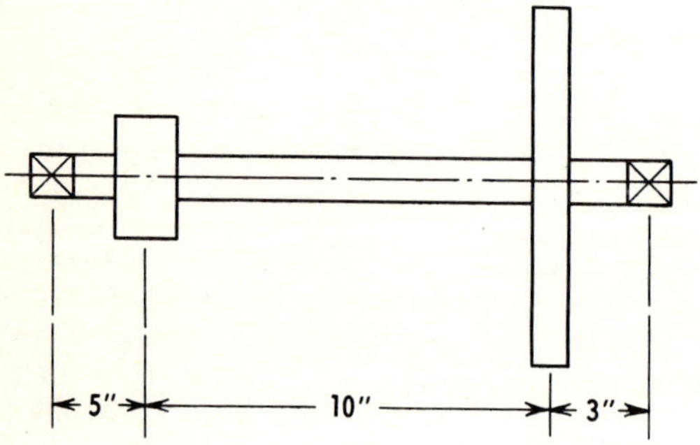

Fig. P 18-18

18-19. Assume that you are planning the design of a drive to connect the rotating drum of a cement kiln to a motor. The drum is to rotate at approximately 20 rpm and the speed of the motor is 1,160 rpm. Assume that it is reasonable to consider V-belt drive, roller chains, and/or gears in any combination of any one, two, or all three for the drive. (*a*) Make a neat sketch showing the layout of the drive you would propose. (*b*) Specify the speed ratios, belt-sheave diameters, numbers of sprocket teeth, and/or numbers of gear teeth in the combinations you propose in the drive. (*c*) Assuming that the loaded drum has a high WR^2, how would you provide for starting?

18-20. A bucket conveyor as shown in Fig. P 18-20 raises a total maximum load equal to 800 lb at an approximate velocity V of 100 fpm. The buckets are tripped and unloaded at station A just before they pass over the upper set of sprockets B, which are 30 in. in diameter. The geared-head motor C rotates its driving pinion E at 60 rpm. The mechanical efficiency of the conveyor is 85 percent.

(*a*) Determine the maximum horsepower required for the drive.

(*b*) Assuming SAE 1040 for the pinion, SAE 1020 for the gear, 24 teeth on the pinion, and 20° stub teeth, determine the diametral pitch, pitch diameters, and face width for the gears.

(*c*) Assuming an allowable shear stress s_s equal to 6,000 psi, determine the diameter of shaft required.

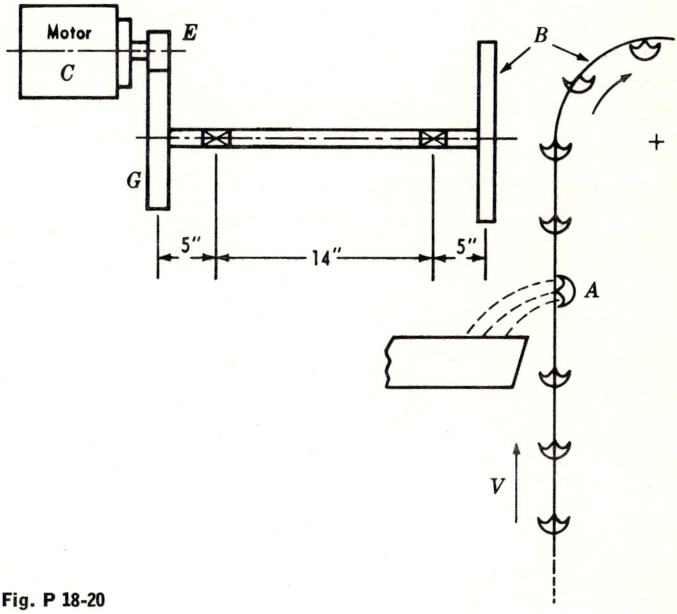

Fig. P 18-20

18-21. The shaft AB as shown in Fig. P 18-21 is driven by an 1,170-rpm squirrel-cage induction motor. The shaft is connected to a processing unit G through a flexible coupling E which transmits torque but no bending moment. The shaft AB is driven

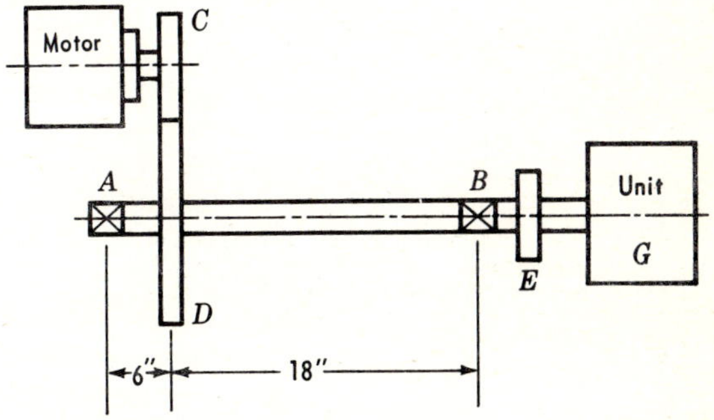

Fig. P 18-21

through spur gears C and D which have a speed ratio of 3:1. The unit G is to operate at 400 rpm and requires a torque of 600 in.-lb. The starting overload is negligible.

(*a*) Assuming that the mechanical efficiency of the drive, motor to unit G, is 90 percent, determine the maximum horsepower required for the motor.

(*b*) Determine the rated horsepower of motor to be recommended, assuming that the duty cycle for the unit G is as follows: full load for 2 min., half load for 1 min., running at no load for 1 min., and standstill for 2 min.

(*c*) Determine the diametral pitch P, face width F, and pitch diameters for the gears, assuming the following data:

Material for both gears, SAE 1030
Tooth form, 20°, full depth
Pinion, 20 teeth

(*d*) Assuming an allowable stress in shear $s_{s,\text{all}}$ equal to 6,000 psi, determine the recommended diameter for the shaft AE.

18-22. A 15-hp 1,170-rpm motor has a Micarta pinion keyed to its shaft, which meshes with a cast-iron gear. The velocity ratio is 4:1, the pinion has 24 teeth, tooth form is 20° involute, diametral pitch $P = 6$, and the race width of the gears $F = 2$ in. Determine whether the teeth have sufficient strength, using the Lewis equation, and whether they are well proportioned.

18-23. Figure P 18-23 shows the layout of a processing drive. Power is taken off equally at gears C and E, each delivering $1\frac{1}{4}$ hp. The data are as follows:

Motor speed, 1,200 rpm
Motor pinion, $2\frac{1}{4}$ in. diameter
Gear A, $11\frac{1}{4}$ in. diameter
Gears C and E:
  30 teeth, $P = 14$
  $F = \frac{3}{4}$ in.
  20° stub teeth
  SAE 1020
Duty cycle = $2\frac{1}{2}$ hp for 2 min, 1 hp for 1 min, no load for 1 min, rest for $\frac{1}{2}$ min

(*a*) Select the motor size.

(*b*) Determine the shaft diameter, assuming the shaft is uniform in diameter and using an allowable stress in shear of 5,500 psi.

(*c*) Are the gears satisfactory for the Lewis equation?

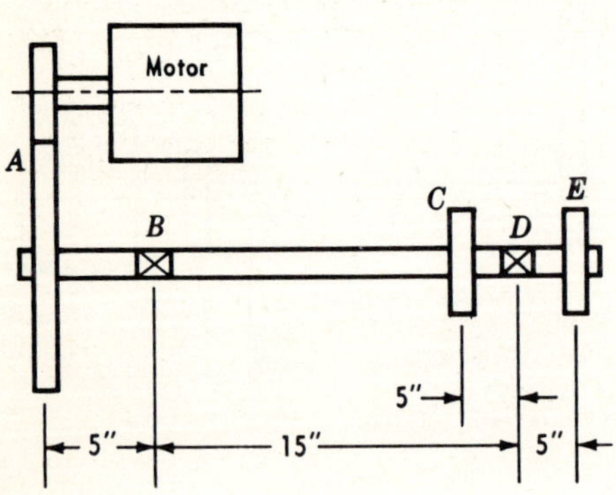

Fig. P 18-23

18-24. The following data apply to a motor and gear drive for a centrifugal pump. The shaft is supported by bearings A and B. A flexible coupling C connects the pump to the shaft.

Pump output, 500 gpm
Total head, 60 ft
Pump speed, 500 rpm
Efficiency of pump, 64 percent
Motor speed, 1,800 rpm
Efficiency of gears, 97 percent
Power cost, 3 cents per kwhr

Determine the following: (*a*) maximum horsepower required of motor; (*b*) rated horsepower of motor to be recommended, assuming a duty cycle of full load for 15 min and rest 15 min; (*c*) diametral pitch, pitch diameters, and face width of gears, assuming 14.5° teeth, 20 teeth for pinion, SAE 1045 for pinion and SAE 1030 for gear; (*d*) loads on bearing A and B; (*e*) size of shaft to be recommended, assuming allowable stress in shear, s_s = 5,500 psi; (*f*) power cost of operating the pump, dollars per day. The electrical efficiency of the motor may be assumed as 87 percent.

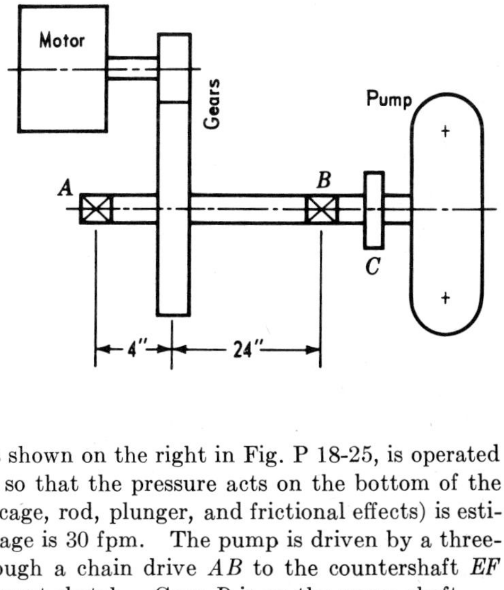

Fig. P 18-24

18-25. A hydraulic freight elevator, as shown on the right in Fig. P 18-25, is operated by oil pressure produced by a pump so that the pressure acts on the bottom of the plunger. The weight lifted (loaded cage, rod, plunger, and frictional effects) is estimated as 5 tons. The speed of the cage is 30 fpm. The pump is driven by a three-phase 220-volt induction motor through a chain drive AB to the countershaft EF and through a gear drive CD. See insert sketch. Gear D is on the pump shaft.

The data are as follows:

Motor speed, 1,750 rpm
Motor sprocket, 23 teeth
Chain pitch, ¾ in.
Chain reduction, 4:1
Pinion teeth, 20
Diametral pitch, 5
Gear reduction, 4:1

(*b*) Assuming an allowable shear stress $s_s = 6{,}000$ psi, determine your recommended size of countershaft d at bearing D.

(*c*) Assuming that the countershaft is made of SAE 1035 (machined finish), radius r of fillet is approximately $\frac{1}{8}d$ (see detail at the right in Fig. P 18-27) and that the torque is steady (static), determine the factor of safety at the critical section at the fillet r.

CHAPTER 19. GEARS FOR NONPARALLEL SHAFTS

19-1. Make a neat sketch of a bevel-gear tooth having large pitch radius R, pitch-cone element L, and face width F. Derive an equation for the tangential load W in terms of the maximum stress s, circular pitch p, Lewis form factor y, and F and L.

19-2. Derive an equation for the resultant tooth load W_0 and the radius at which it acts.

19-3. Design the full-depth 14.5° teeth of a pair of straight bevel gears to transmit 20 hp at 1,250 rpm of the pinion. The velocity ratio should be about 3.5 and the pinion should have 18 teeth. Determine the diametral pitch, face width, pitch diameters, and pitch-cone angles for the gears.

CHAPTER 20. SURFACE FINISH, FRICTION, AND WEAR

20-1. For a roller supporting a load W and rolling on a surface, derive the equation for the coefficient of rolling friction, $a = Fr/W$, where r is the roller radius and F is the rolling force applied to the axis of the roller.

Note: Draw a free-body diagram for the roller. What are the units of the coefficient of rolling friction?

20-2. A load of 4,000 lb rests on cast-iron rollers 8 in. in diameter that roll over a wooden floor. The load is supported by steel rails that rest on the rollers. Assuming the coefficients of rolling friction between the cast iron and steel and the cast iron and wood as 0.02 and 0.10, respectively, determine the force required to move the load.

Note: Draw a free-body diagram for a roller. Determine the work expended for each revolution of the rollers.

CHAPTER 21. SLIDING-CONTACT BEARINGS AND LUBRICATION

21-1. The following data apply to a lubricated bearing:

Load on bearing	785 lb
Speed of journal	360 rpm
Diameter of journal	3 in.
Length of journal	6 in.

Assume a clearance of 0.001 in. per in. of diameter, oil SAE 10 in Fig. 21-16, oil temperature _____°F, and ambient temperature 60°F. Determine the following: (*a*) the actual value ZN/p; (*b*) coefficient of friction; (*c*) heat generated in foot-pounds per minute; (*d*) heat dissipated in foot-pounds per minute; (*e*) assumed temperature to give thermal equilibrium.

21-2. A 3-in. shaft rotates at 1,000 rpm. One of the bearings carries a load of 1,000 lb. (*a*) If the bearing is an average industrial journal bearing 4 in. long, lubricated with SAE 40 oil, maintained at a film temperature of 160°F, and having a diametral clearance of 0.003 in., determine the friction horsepower. (*b*) If the bearing is a ball bearing, with a coefficient of friction of 0.002, and if it is assumed that the frictional force acts at the radius of the shaft, determine the friction horsepower.

21-3. The following data apply to an average industrial unventilated journal bearing for a generator:

Load on journal	500 lb
Speed of journal	870 rpm
Diameter of journal	1½ in.
Length of journal	2½ in.

Assume a clearance ratio of 0.001, SAE 20 oil, and a room temperature of 60°F. The maximum operating temperature t_o of the oil film is expected to be 140°F. (Note that the value for absolute viscosity Z at 140°F has been determined as 23.4 centipoises.) Determine whether fluid friction may be expected and whether artificial cooling is necessary, and if so, the heat to be removed in units of foot-pounds per minute.

21-4. A 15-hp 300-rpm centrifugal pump has two identical journal bearings, described below:

Journal diameter	3 in.
Bearing length	3½ in.
Load on journal	250 lb
Clearance ratio c/d	0.001
Ambient temperature	60°F
Average industrial unventilated bearing	

Assuming SAE 10 oil, determine whether the bearings may be expected to operate under conditions of thick-film lubrication. If so, determine the probable operating temperature and the power loss for the two bearings expressed in percent of the total horsepower.

Note: To obtain the approximate equilibrium temperature, the heat-generation and heat-dissipation curves may be assumed to be straight lines.

21-5. The following data apply to an average industrial unventilated journal bearing for a generator: load on bearing = 800 lb, speed = 870 rpm, diameter = 1½ in., length = 2¼ in.

Assume a clearance ratio of 0.001 in. per in., SAE 20 oil, and a room temperature of 60°F. The maximum operating temperature t_o of the oil film is expected to be 140°F. The value for the absolute viscosity Z at 140°F has been determined as 23.4 centipoises. Determine whether fluid lubrication may be expected in the bearing, whether artificial cooling is necessary, and, if so, the heat to be removed in units of foot-pounds per minute.

21-6. In a printing-press drive, the roll gear C is driven by gear A through the idler gear B. Each gear has 25 teeth of 5 diametral pitch. The centerline AC is vertical, and the centerlines AB and BC are 45° with the horizontal, as shown in Fig. P 21-6.

A torque equal to 2,920 in.-lb is transmitted from A to C. The idler gear B is supported midway between two identical bearings. It is required to investigate the operating characteristics of the two bearings. Assume the following:

Speed of gears, 870 rpm
Diameter of each bearing 1½ in. and length 2¼ in.

Clearance ratio, 0.001 in. per in.
Room temperature, $t_a = 60°F$

It is desired that the maximum operating temperature of the oil film be 140°F, at which temperature its absolute viscosity Z is 23.4 centipoises. Determine the following: (*a*) load on each of the idler-pulley bearings, considering only tangential components of tooth loads; (*b*) whether fluid lubrication is expected; (*c*) whether artificial cooling is expected; (*d*) if answer to part *c* is "Yes," determine the heat to be removed in units of foot-pounds per minute.

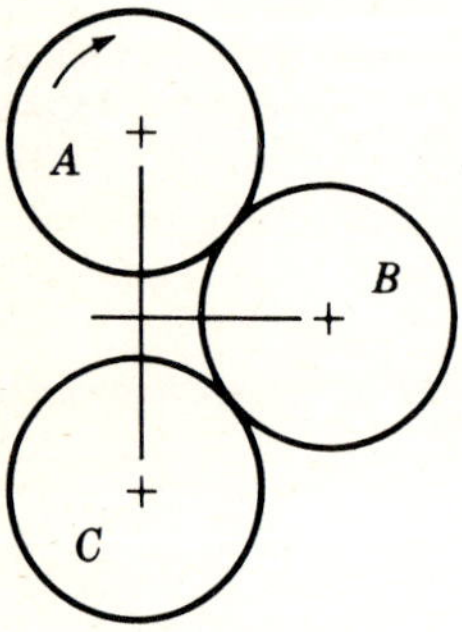

Fig. P 21-6

21-7. The following data apply to an average unventilated journal bearing for a 10-kw alternator:

Load on journal, 500 lb
Speed, 870 rpm
Diameter, 1½ in.
Length, 2½ in.

Assume a clearance ratio of 0.001, SAE 20 oil, and a room temperature of 60°F. The operating temperature t_o of the oil film is set to be 150°F. Determine: (*a*) whether fluid lubrication may be expected; (*b*) whether artificial cooling is necessary and, if so, the heat to be removed in units of foot-pounds per minute, and in Btu per minute; (*c*) the power loss as a percentage of generator rating, assuming two bearings as described above.

CHAPTER 22. ROLLING-CONTACT BEARINGS

22-1. A bearing is to carry a radial load of 500 lb and a thrust of 300 lb. The service imposes light shock, and the bearing will be in use 40 hr per week for 3 years. The speed of the shaft is 1,000 rpm. Determine the size of medium-series ball bearing that should be used.

CHAPTER 23. METAL FITS AND TOLERANCES

23-1. A cast-steel crank having a hub whose outer diameter and length are 7 in. and 6 in., respectively, is to be shrunk on the end of a steel shaft. The bore in the crank is 4 in. and the shaft is 4.004 in. in diameter. Neglecting the effect of the crank arm on the stresses, determine the following: (*a*) the radial pressure that is acting between the shaft and hub; (*b*) the tangential stress that is induced in the crank; (*c*) the temperature to which the crank must be heated in order that it may be slipped on the shaft. Assume a room temperature of 70°F, a coefficient of expansion for steel of 0.0000063, and a diametral clearance of 0.002 in.; (*d*) the maximum force required to press the crank off the shaft and the corresponding work in foot-pounds. Assume a coefficient of friction of 0.15.

CHAPTER 24. VIBRATION AND VIBRATION CONTROL

24-1. A 2-hp 1,200-rpm motor weighing 108 lb is mounted on four springs, each of which has a spring constant of 1,000 lb per in. The motor is mounted so that it can move only in a vertical direction. Determine the natural frequency of the system.

24-2. A 1,750-rpm motor and its support weighs 640 lb and is supported on four springs, each having a spring rate of 15,000 lb per in. A damping device is used that has a damping constant of 200 lb-sec per in. The rotor has become unbalanced to the extent that a periodic force for which $P_0 = 150$ lb caused the motor to vibrate. Assuming that the unbalanced force acts through the center of gravity of the motor and that the motor vibrates only in the vertical direction, determine the following: ω_n, x_0, and F for undamped and damped vibration, and enter the results in a table as shown below.

	Units	*Undamped*	*Damped*
ω	rad per sec		
ω_n	rad per sec		
x_0	in.		
$F = kx_0$	lb		

24-3. A 5-hp motor weighing 228 lb operates at 1,760 rpm. The motor is to be mounted on four springs so that 90 percent reduction in transmitted force (that is, 10 percent transmissibility) will be secured. Assuming that the effect of damping at the operating speed may be neglected, determine (*a*) the required natural frequency of the system in cycles per minute, (*b*) the static deflection of the motor in inches, and (*c*) the spring rate of each spring in pounds per inch. If an equivalent unbalanced weight of 1 oz is located centrally on the rotor at a radius of 8 in., determine (*d*) the maximum force transmitted to the base at operating speed.

24-4. A motor which weighs 30 lb and rotates at 1,150 rpm has a vertical shaking force equal to 5 lb due to an unbalanced rotor. The motor is mounted on two steel cantilevers placed side by side of depth 1½ in. and width ½ in. The distance between the motor centerline and the support is 24 in. Determine the maximum stress in the cantilevers.

SOME TYPICAL DESIGN-ROOM PROBLEMS

26-1. Cotter joint. Figure P 26-1 shows a common form of cotter joint used to connect rods that are subjected to axial loads. This joint is rugged, is dependable, and may be easily connected and disconnected, and on reconnection the rods assume their original alignment. The joint is used where ruggedness is essential and where the bulkiness of the joint is not objectionable.

1. Specifications: Materials. The materials of the rod end, the socket, and the cotter and the allowable stresses (pounds per square inch) are as follows:

	Rod end, mild steel	*Socket, wrought iron*	*Cotter, machinery steel*
Tension or compression	5,000	4,000	6,000
Shear	2,500	2,000	4,000
Bearing	10,000	10,000	12,000

2. Load. The joint is to be designed to support a tensile or compressive load equal to ______lb. (Assign 2,000, 3,000 or 4,000 lb.)

3. Assumptions. The thickness of the cotter may be taken as one-fourth the diameter of the rod end. The taper of the cotter should be ⅜ in. per ft. Allow clearances at the cotter equal to $\frac{1}{16}$ in., and clearance between the rod end and the bottom of the hole in the socket equal to ⅛ in.

4. Procedure. Determine the dimensions as follows:

Steel rod, d_1
Wrought-iron rod, d_2
Socket in tension across slot, d_4
Socket and cotter in bearing, d_5

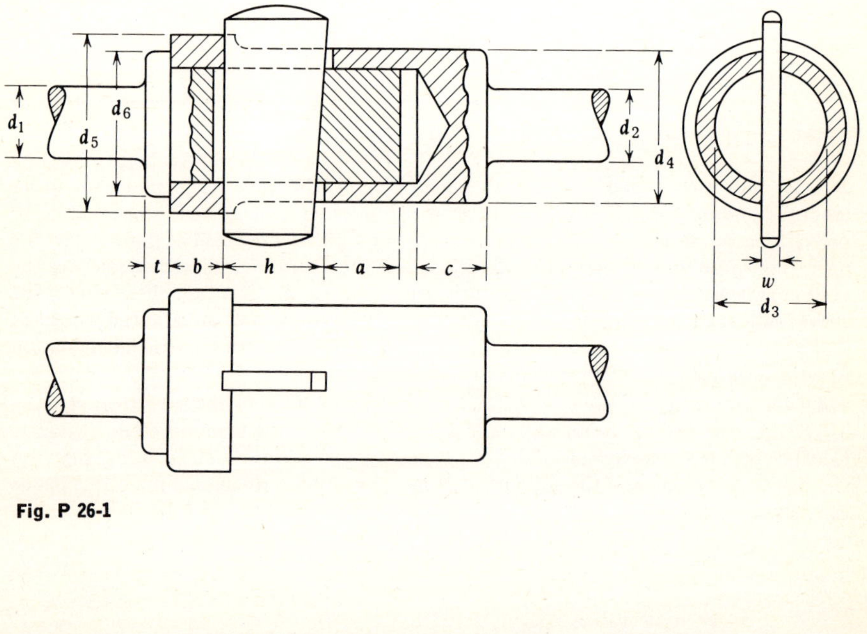

Fig. P 26-1

Rod end in tension, d_3
Rod end in bearing, d_3
Thickness of cotter, w
Width of cotter, h
Rod end in shear, a
Socket end in shear, b
Socket in shear, c
Collar in bearing, d_6
Collar in shear, t
Length of key (from sketch)

5. Sketch. Make a neat undimensioned freehand sketch approximately to scale (full size) of a side view of the joint in section and a plan view showing the outside.

26-2. Flanged shaft coupling. Specifications: This problem will consist of the design of a two-piece, flanged shaft coupling of cast iron for a given diameter of shaft. Starting with the diameter of the shaft, the design will be developed by working from the inside outward.

1. Size. The diameter of the shaft is $2\frac{1}{4}$, $2\frac{1}{2}$, $2\frac{3}{4}$, 3, $3\frac{1}{4}$, $3\frac{1}{2}$, $3\frac{3}{4}$, or 4 in. (Assign one.)

2. Proportions of hub and key. It is impossible to proportion the hubs of gears, pulleys, clutches, and couplings rationally. They are invariably proportioned empirically. The following equation is frequently used in determining the outside diameter of hubs of gears, pulleys, couplings, etc. (see Fig. P 26-2).

$$D_1 = 1\tfrac{3}{4}D + \tfrac{1}{4}$$

where D_1 = outside diameter of hub, in.
D = diameter of shaft, in.

The length of the hub is fixed by the length of key. Select the size of key required and specify its dimensions.

The key is to be made without taper, and a cup-pointed setscrew is to be used. The diameter of the setscrew should be somewhat less than the width of the key. To

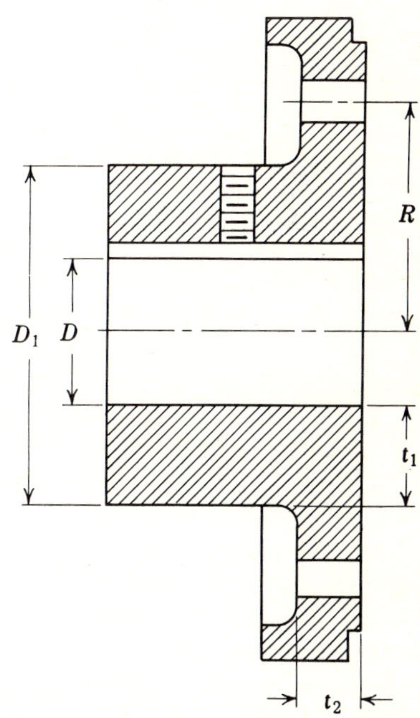

Fig. P 26-2

reduce hazard and to comply with safety requirements, a socket-head setscrew should be used. Give specifications for the setscrew and the tap drill size for the setscrew hole.

3. Coupling bolts. The number of bolts may be fixed empirically by the following formula:

$$n = \tfrac{2}{3}D + 3$$

where n is the number of bolts.

So that each bolt may take its share of the load, finished bolts fitted to reamed holes must be used. The size of the bolts will first be obtained by assuming that the shaft and bolts are of the same material and that the bolts are as strong in direct shear as the shaft in torsion, the friction between the halves of the coupling being neglected. To determine the size of the bolts, it is necessary to have the radius R of the bolt circle. This radius depends on the diameter of the head of the socket wrench, which, in turn, depends on the unknown diameter of the bolts. The diameter D_2 of the head of the socket wrench in terms of the diameter d of the bolt is

$$D_2 = 1.85d + \tfrac{5}{16}$$

and the radius of the bolt circle is

$$R = \frac{D_1}{2} + \frac{D_2}{2} + \text{clearance}$$

The clearance between the head of the socket wrench and the hub may be taken as $\tfrac{3}{16}$ in. On substituting R in terms of d in the expression for the direct shear of the bolts and placing this in terms of the torque on the shaft, the diameter of the bolts may be determined by trial. Having the diameter of the bolts, the radius of the bolt circle may be determined.

The bolts must also be of such size that they will not be overstressed when setting up the nuts. The equation in Art. 9-6 may be used with a value of $k = 10{,}000$ to 12,000 lb per in.

It may be assumed that the bolts are of SAE 1025 to 1030 steel. The setting-up stress may be assumed as three-fourths the yield strength. If the diameter found should exceed that obtained by the first method, the radius of the bolt circle must be altered accordingly.

Before reaming, a hole is drilled undersize to provide a small amount of material for sizing and hole alignment. Specify the drill size, reamer size, and spot-face diameter. In order that the holes may register perfectly, they should be reamed with the halves of the coupling assembled.

4. Coupling flanges. For good casting design, the change in thickness from the hub to the radial flange should not be too great. A satisfactory finished thickness t_2 for the radial flange in terms of thickness t_1 of the hub is

$$t_2 = \frac{t_1}{2} + \frac{1}{4}$$

For safety in operation and to comply with the safety code, the halves of the coupling are to be provided with circumferential flanges to function as guards for the boltheads and nuts. The thickness of the circumferential flange should be not less than one-half that of the radial flange. The inside diameter of the circumferential flange should be determined so that the boltholes are located midway between the flange and the hub. Determine the outside diameter of the coupling.

To provide for automatic alignment, one-half of the coupling is to be counterbored and the other turned to fit. The depth of the counterbore should be sufficient for the machinist to establish a fit without difficulty, and therefore should not be less than $\frac{3}{16}$ in. The height of the male boss should be $\frac{1}{32}$ in. more than the depth of the counterbore to ensure contact of the faces. Provision should be made for "one-position" assembly of the coupling halves.

Before beginning the drawing, the following dimensions should be computed: (*a*) spot-face thickness, (*b*) length of bolts, and (*c*) length of thread.

5. Drawing. Title: "(Size) Flanged Shaft Coupling." A complete detail of the halves of the coupling is to be made on 17- by 22-in. sheet. A bill of material is to be placed above the title.

26-3. Bench-model press. Hand-operated presses of this type are used for general-purpose work, including pressing for force fits of gears, pulleys, sprockets, couplings, bushings, and dowels and also for straightening or bending shafts, rods, and tubes. The press should be rugged and trouble free.

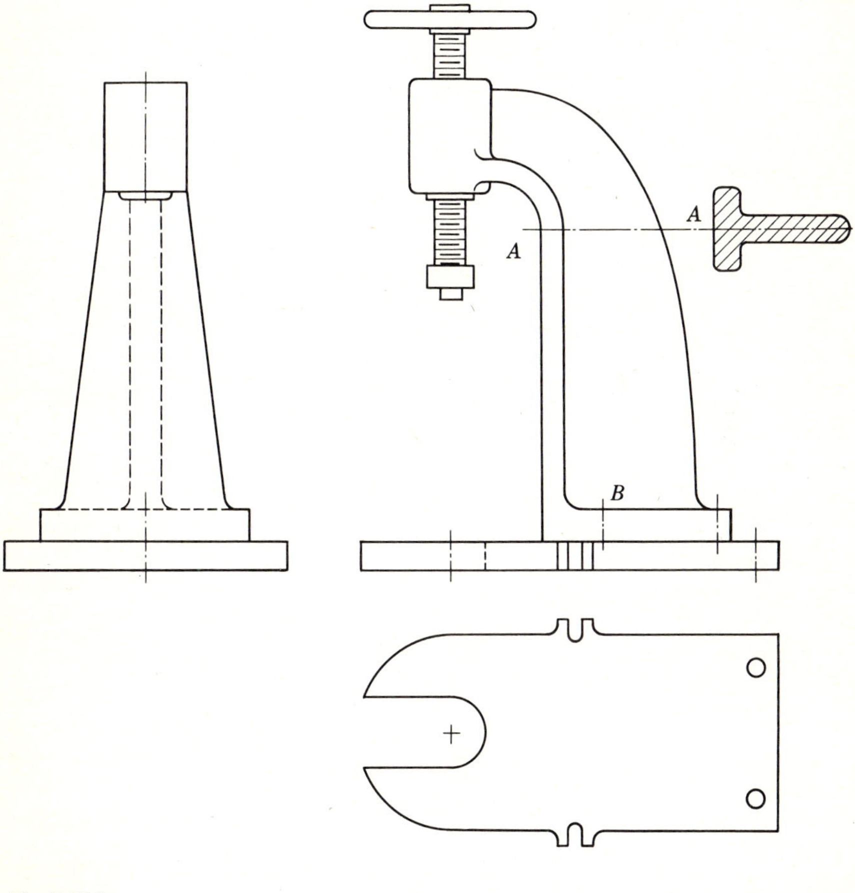

Fig. P 26-3

Specifications:

Maximum load = 2½ tons
Clearance, ram to base plate, maximum = 20 in., minimum = 6 in.
Throat = 10 in.
Diameter of handwheel = 10 to 18 in.
Maximum tangential force on handwheel at ¾ load = 60 lb (30 lb for each hand)
Materials are as follows: frame, cast iron; screw, SAE 1030 steel; bushing, bronze; base plate, 1015 steel.

Procedure. 1. Screw. (*a*) Determine the root area of the screw, select type of thread, and determine its size. A compressive stress of 10,000 psi may be used for the screw. (*b*) Design the collar, using a removable collet. (*c*) Check the force on the handwheel. The coefficient of thread and collar friction may be assumed as 0.1.

2. Determine the length of the nut, considering an allowable bearing pressure of 1,500 psi and a shearing stress for bronze of 3,000 psi. Provide for lubrication.

3. Make a detail sketch of the head including the screw, bushing, collet, and handwheel.

4. Frame. (*a*) It will be sufficient to calculate the dimensions of the frame at section *AA* in Fig. P 26-3. Assume dimensions for the frame at this section, and check the induced tensile and compressive stresses against allowable values in tension of 8,000 psi and in compression 32,000 psi. (*b*) Determine the size of bolts required at *B* to hold the frame against the base plate. For rebound, use two small bolts at the back of frame.

5. Base plate. Determine the thickness required for bending and provide for hold-down bolts.

6. Sketch. Make a neat, undimensioned, freehand sketch to scale of the assembly on an 8½- by 11-in. sheet, showing a side view and a front view of the press.

7. Determine the approximate weight of the press.

26-4. Capstan. A motor-driven capstan, as shown in Fig. P 26-4, is composed of a drum driven by a motor through a worm drive. A rope is wound around the drum so that when the drum rotates, a load F_1 may be moved by the application of a com-

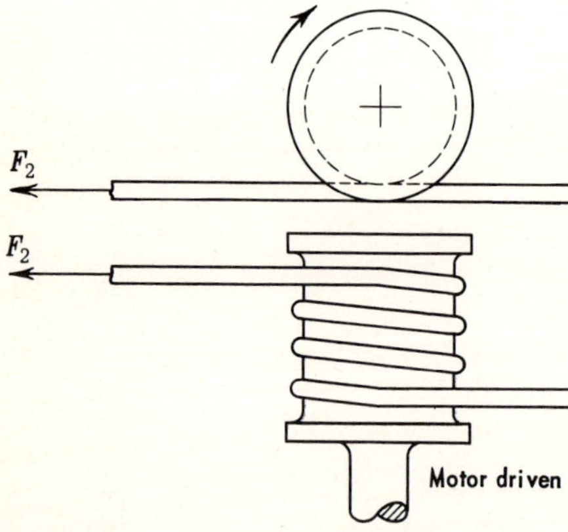

Fig. P 26-4

paratively small force F_2 produced manually on the free end of the rope. The connected end of the rope may be used to spot a railroad car, haul an anchor, dock a ship, for dragging operations in the field, etc.

The operation of the rope around the drum is similar to a belt drive where contact is several turns around the drum.

In the capstan shown, the maximum load F_1 is to be 3,000 lb and the manually applied force F_2 is limited to 40 lb.

(*a*) Assuming a coefficient of friction for the rope against the drum equal to 0.3, determine the number of turns or wraps of the rope required.

(*b*) Assuming that the capstan is rated at a load of 2,500 lb at a velocity of 50 fpm, determine the horsepower of motor to be specified. Assume a duty cycle.

(*c*) If the diameter of the drum is 10 in. and it is desired to use an 1,160-rpm motor, determine the overall speed reduction between the motor and drum.

(*d*) What approximate percentage of the work to move the load is provided by the operator?

26-5. Gears and brake. A motor-driven geared drum hoist, as shown in Fig. P 26-5(*a*), has a spring-set double-block brake attached to the motor shaft. The arrangement of the brake and its dimensions are shown at (*b*).

Data:

Rated load on hoisting cable	1,000 lb
Diameter of hoisting drum	12 in.
Number of teeth on pinion	24
Number of teeth on gear	72
Form of teeth	20° stub
Material of pinion	SAE 1030
Material for gear	Untreated 20 points C cast steel

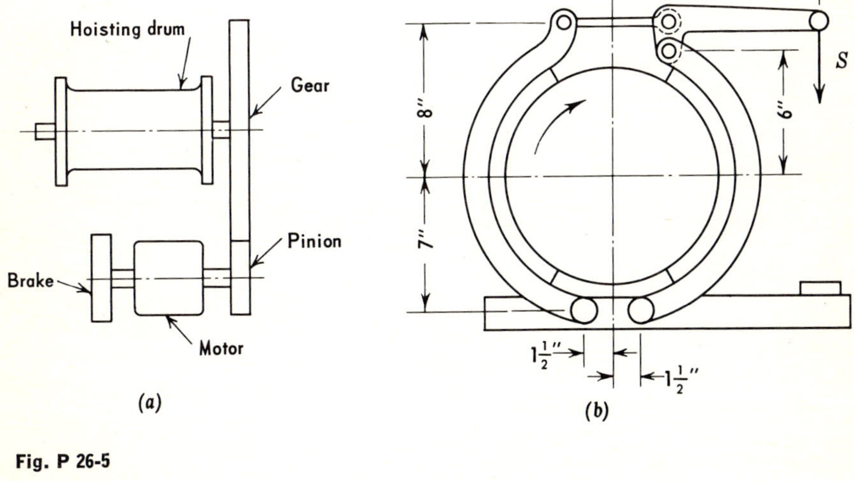

Fig. P 26-5

Motor speed	570 rpm
Brake-drum diameter	10 in.
Angle of contact for each shoe	120°
Coefficient of friction for brake	0.40

Determine the following:

(*a*) The size of motor required, using class D, squirrel-cage induction motor (see Fig. 25-1). For duty cycle: full load for 6 sec, rest for 1 min.

(*b*) The diametral pitch, pitch diameters, and face width for the gears.

(*c*) The spring force required to set the brake for a torque on the drum of 1 lb-in.

(*d*) The spring force required to set the brake for rated load on the cable.

(*e*) The width of brake shoes required for a pV value of 30,000 ft-lb per min. per in.^{2} of projected area.

26-6. Belt conveyor. A 20-in. belt conveyor in which the belt is supported by 2-in.-diameter steel rollers is 100 ft long and runs at 200 fpm of the belt. The conveyor is horizontal except for an inclined run (20-ft horizontal projection) to raise the belt from the loading level 2 ft above the floor to the 7-ft unloading level. At the unloading end is a metal platform with a 45° baffle to unload the material sideways.

The capacity of the hoist is to be six thousand 100-lb bags or boxes per hour. The friction for the rollers may be assumed as 10 percent of the weight on the belt.

(*a*) Make a sketch of the layout of the conveyor.

(*b*) Assuming an 1,160-rpm induction motor and 24-in. belt sheaves, determine the overall speed reduction for the drive and indicate the various reductions, i.e., belt, chain, gears, etc., and speed ratio of each.

(*c*) Assuming the efficiency of each reduction is 95 percent, determine the size of motor to be recommended, using a class C squirrel-cage induction motor for continuous rating.

26-7. Traveling stairs. In Fig. P 26-7 is shown a sketch of the elevation.

Data:

Capacity = 8,000 persons per hour
Width of stairs = 42 in.
Speed of stairs = 125 fpm
Rise H = 20 ft
Angle of incline = 30°
Number of steps = 44 on incline and 2 at each landing
Tread = 11 in.
Diameter of sprocket = 36 in.
Speed reductions:

1. V belt (efficiency 97 percent)
2. Helical gears (efficiency 98 percent)
3. Spur gears (efficiency 97 percent)

Brake is located on shaft for large V-belt pulley.

Assumptions:

Average weight of each passenger including luggage is 175 lb.
One person per step at full load
Coefficient of friction for stairs = 0.05
Motor speed = 1,170 rpm

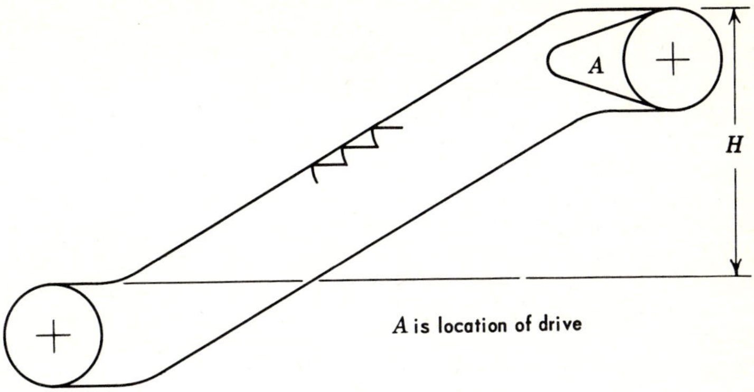

Fig. P 26-7

Determine the following:

Horsepowers

(*a*) For raising passengers (potential energy)
(*b*) For accelerating passengers (kinetic energy)
(*c*) For friction
(*d*) For motor selection at continuous rating

Speed reductions

(*e*) Overall, motor to sprocket
(*f*) Recommended for belt and gears

Required torque capacity for brake or backstop in foot-pounds at 50 percent overload

26-8. Single-reduction parallel helical-gear speed reducer.

1. Specifications. It is required to design a single-reduction, parallel helical-gear speed reducer having a 3:1 speed ratio and capable of transmitting the full-load rating of a ______-hp 1,150-rpm motor. (Assign 10, 20, or 30.)

(*a*) *Gearing.* The gears are to be of the single helical type and made of the following materials:

Pinion	SAE 3140, chrome-nickel steel
Gear	SAE 1040, steel

(*b*) *Shafting.* The shafts are to be made of SAE 1045 forged steel in the as-rolled condition.

(*c*) *Bearings.* The bearings must be of the rolling-contact type and must be protected against dust and leakage of oil.

(*d*) *Lubrication.* Adequate lubrication of the gears and bearings must be provided by a simple and positive splash system which will circulate the oil from a reservoir located in the housing.

(*e*) *Housing.* The housing is to be made of a good grade of cast iron and designed to enclose all moving parts properly. The upper part of the housing must be removable without disturbing the bearing alignment, and provision must be made for interchanging the shaft extensions from left to right, or vice versa. A suitable cover plate must be provided to facilitate casual inspection of the gears and for replenishing the lubricant in the reservoir. An air vent in the top of the housing and an oil drain and level indicator at the bottom must be provided.

(*f*) *General.* In designing the speed reducer, the following factors should be given consideration: rigidity, weight, accessibility, compactness, appearance, and cost of manufacture. The problem will consist of a set of computations and a 17- by 22-in. drawing.

2. Preliminary sketch. Designers find that their thinking is greatly stimulated if when beginning a new design a freehand sketch is made of the proposed machine. The sketch will frequently indicate the procedure to be followed in making the computations as well as provide a place to record any ideas that may come to mind. Many helpful suggestions may be secured by studying the literature of manufacturing companies of similar equipment. With the aid of any available literature and any original ideas, draw a neat freehand sketch of your proposed design.

3. Gears. The following data may be assumed:

Helix angle	23°
Form of teeth	20° involute stub
Minimum number of pinion teeth	24
Face width	3.5 times circular pitch

Electric motors will deliver under starting conditions around 200 percent rated torque. In the design of the gears for strength the overload torque may be used. For wear of the gears, the rated torque may be used because for the major portion of the operating life the motor operates at rated conditions rather than at starting conditions.

(*a*) Determine the diametral pitch, pitch diameters, and face width, using the Lewis equation.

(*b*) Assuming steady load conditions, determine the minimum required Brinell hardness numbers for pinion and gear.

(*c*) Determine the load-stress factor and the minimum surface hardness for the pinion and gear based on wear of the teeth.

(*d*) Specify the core and surface hardness of pinion and gear and required heat treatment.

4. High-speed shaft and bearings. The shaft and bearings are subjected to thrust and radial loads caused by the gears, to overhanging loads caused by the driven pulleys or gears, and to torsion due to the power being transmitted.

It is reasonable to make the tentative assumption that the maximum bending moment on the shaft due to the overhanging load is 150 percent of the maximum torque on the shaft.

The bearings should be selected with an expected life of 5 years operating 7 days each week at 10 hr per day.

(*a*) Make a neat sketch of the shaft showing the gear, bearing mountings, bearing cover, and oil seals.

(*b*) Compute the minimum size of the shaft at the bearing adjacent to the shaft extension.

(*c*) Tentatively select bearings.

(*d*) Determine the bearing span. Allow generous clearance between the pinion and the housing, and locate the bearing so that it is fully supported in the housing.

(*e*) Compute the length of the shaft extension measured from the bearing centers, allowing 2 in. for the cover, seals, etc., and record all known dimensions on the sketch.

(*f*) Compute the tangential, radial, and axial forces exerted by the pinion and the overhanging load.

(*g*) Determine the force on the most heavily loaded bearing due to the pinion tooth loads and the overhanging load. Assume that the reaction due to the latter is in the same direction as the resultant reaction due to pinion tooth loads.

(*h*) Determine the required bearing capacity, and check whether the tentative bearing selection is satisfactory.

(*i*) Make a sketch of the bearing showing its principal dimensions.

(*j*) Select an oil seal and record the catalog number, manufacturer, and its principal dimensions.

(*k*) Assuming the bore of the pinion to be the nearest nominal dimension above the bearing bore, determine the size of key.

5. Low-speed shaft and bearings. Because the torque and bending moments vary inversely as the speed, the low-speed shaft may be designed by simple proportion. Develop the equation for this proportion in terms of the speed ratio, and determine the minimum shaft size. Follow the same procedure for the low-speed shaft, as outlined for the high-speed shaft.

6. Housing. (*a*) Design the housing, observing the various factors and limitations referred to in the specifications.

(*b*) Suggestions. Thickness of housing wall $\frac{5}{8}$ to $\frac{3}{4}$ in. Provide ample clearance inside of housing for all moving parts, not less than $\frac{3}{4}$ in. around the gears. Provide suitable floor flanges. Provide suitable stiffening ribs on the housing walls. Use $\frac{1}{2}$- to $\frac{3}{4}$-in. screws for fastening housing top. Use $\frac{3}{8}$- to $\frac{1}{2}$-in. screws for fastening bearing retainers.

7. Drawing. Make a one-half size undimensioned assembly drawing of the speed reducer. Show two views, one side view and a top view. The top view should show the cover removed with the gears and one bearing mounting on each shaft in section. The title of the drawing should be

______ Horsepower
Single-reduction Speed Reducer
Assembly

26-9. Shaft deflection. Lay out the shaft and loading force vectors for the shaft shown in Fig. P 26-9 and, by graphical methods, determine the following: (*a*) bending-moment diagram, (*b*) M/EI curve, (*c*) dy/dx curve, (*d*) y/x curve, (*e*) maximum deflection, (*f*) critical speed, and (*g*) deflection within the bearings.

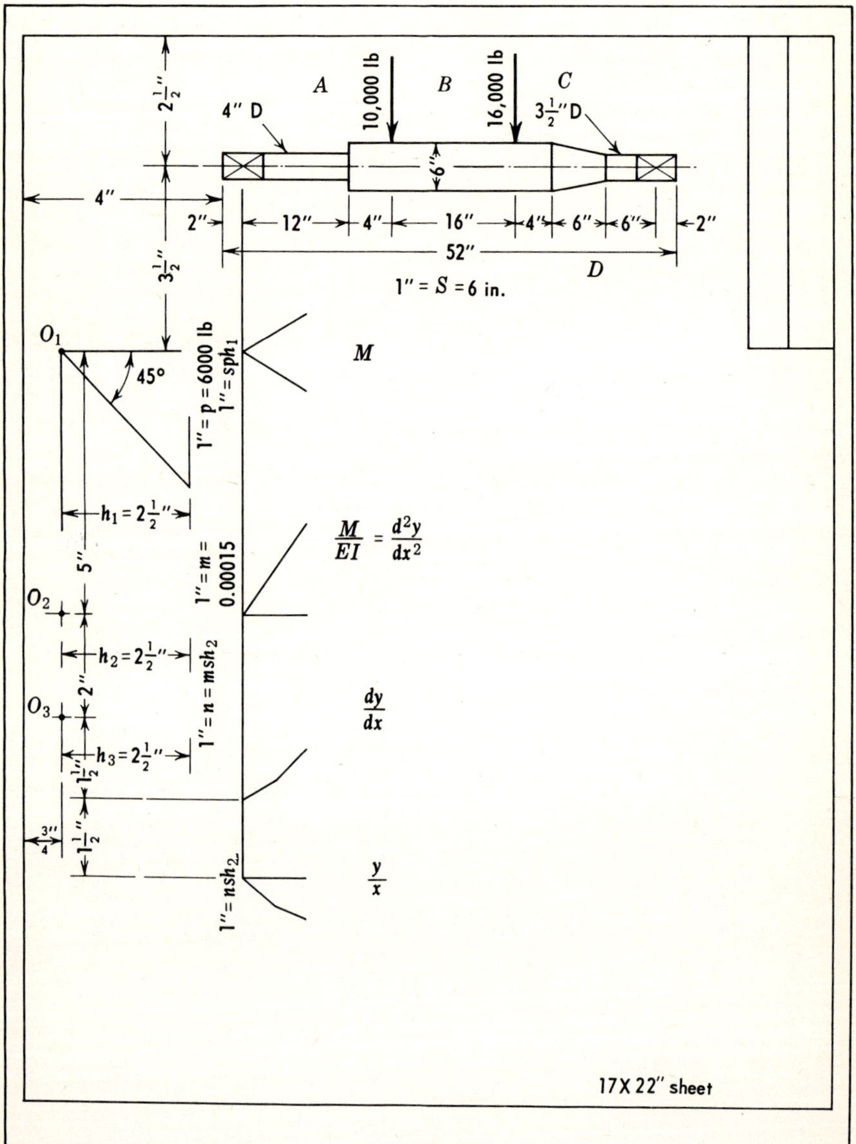

Fig. P 26-9

Indexes

Name Index

Albert, C. D., 25, 360, 367
Allan, R. K., 260
Almen, J. O., 72, 90, 110–121, 192–193, 216, 226, 232–233, 370, 467, 474–478
Amstead, B. H., 175
Archbutt, L., 414
Arkin, H., 493, 495
Atwood, G. H., 251

Baker, P. J., 467
Barr, J. H., 190, 415
Barth, C., 279
Barton, M. V., 72
Baud, R. V., 64
Baumgartner, T. C., 192
Begeman, M. L., 175
Benner, R. E., 17, 82
Berlin, D. R., 152
Bevin, T., 507
Bidwell, J. B., 474
Binder, R. C., 301
Blodgett, O. W., 163
Bolz, F. W., 482
Bowden, F. P., 408
Boyd, J., 429, 447
Brewer, G. A., 77
Broghamer, E. L., 368
Brown, J. S., 118
Brownell, L. E., 242
Bruce, C. S., 413
Buckingham, E., 358, 361, 369, 551
Buckwalter, T. V., 72
Buhl, H. R., 2
Burrell, C. R., 364
Burwell, J. F., Jr., 410

Church, A. H., 506
Clark, H. H., 223
Clarke, P. C., 230
Clayton, D., 439
Colbert, T. P., 390
Colton, R. R., 493, 495
Colwell, L. V., 131, 409, 410
Connors, H. J., 456
Cook, W. J., 230, 261
Crede, C. E., 46–47, 231, 506

Dally, J. W., 74
Davis, F. W., 261
Dean, P. M., Jr., 364
Deeley, R. M., 414
Den Hartog, J. P., 197, 499
Dieter, G. E., Jr., 34–35, 87–91, 93–95, 102–104, 124
Dimitry, D. L., 14
Dolan, T. J., 368
Drucker, D. G., 299
Duck, J. T., 413
Dudley, D. W., 346, 361, 366, 368, 385
Durland, M. A., 551

Edgerton, C. T., 222

Faires, V. M., 45, 47, 393
Faupel, J. H., 18, 47, 52, 487, 528
Feigenbaum, A. V., 496
Feng, I-Ming, 409
Finnie, Ian, 103, 105–106, 560
Frankland, J. M., 47
Frocht, M. M., 74, 536

Gatcombe, E. K., 373
Gere, J. M., 42, 101–102, 235
Germain, C. B., 14
Glassco, R. B., 77
Goodier, J. N., 20, 28, 43, 51, 60–61, 182, 189, 235, 237, 243, 411
Griffel, W., 41
Gyrna, F. M., Jr., 495

Hall, A. S., 432
Hall, S. G., 190

Ham, C. W., 256
Harris, C. M., 46–47, 231, 506
Hartenberg, R. S., 76
Hartman, J. B., 17, 82, 165, 224
Heller, W. R., 103, 105–106, 560
Henwood, P. E., 190
Hersey, M. D., 23, 422, 429
Hetenyi, M., 244
Higdon, A., 11
Hinkle, R. T., 220, 334
Hoersch, V. A., 410
Hollister, S. C., 172
Holowenko, A. R., 432
Horger, O. J., 72, 126
Hunter, M. C., 416

Ilmer, L., 439
Inglis, C. E., 61

Jacobsen, L. S., 62
Jennings, C. H., 166
Johnson, J. B., 39, 118
Johnson, S. J., 351
Jurah, J. M., 496

Kalikow, I., 470
Kelvin, Lord, 134
Kent, C. H., 235
Kimball, D. S., 415
Kincaid, T. R., 351
Kobrin, C. L., 467
Kommers, J. B., 73
Krick, E. V., 2
Krouse, G. N., 70
Kull, F. R., 192

Laughlin, H. G., 432
Leeson, D. N., 14
Leutwiler, O. A., 390
Lewis, W., 349
Libby, C. C., 515
Lipson, C., 409–412, 414, 529–530
Love, A. E. H., 49
Lyman, T., 80

Mabie, H. H., 317
McAdam, D. J., Jr., 233
McCracken, D. D., 32
McCrory, R. J., 1
MacCullough, G. H., 41
McKee, S. A., 14, 429, 431
McKee, T. R., 14, 429, 431
Macks, E. F., 416, 431
Maleev, V. L., 165, 224
Marin, J., 37, 84–85, 87–94, 106
Merritt, H. E., Jr., 355, 392, 395
Michael, D. M., 13
Mikelson, W., 407
Miner, M. A., 97
Moore, H. F., 70, 190
Morse, I. E., Jr., 220, 506
Murphy, G., 52, 64
Murray, W. M., 50, 116

Nádai, A., 79
Noll, G. S., 529–530
Norton, C. E., 467

Oberg, T. T., 118
Ohlsen, E. H., 41
Olchawa, J. M., 581

Palmgren, A., 416
Parker, E. R., 124
Pennington, R. H., 14
Pernick, A., 351
Peterson, R. E., 72, 77–78, 534
Polakowski, N. H., 84, 87, 88, 97, 104, 107
Polentz, L. M., 231
Poritsky, H., 351

Rabinowicz, E., 98, 414
Radzimovsky, E. I., 43, 45, 431
Raimondi, A. A., 31, 429, 447
Rand, C., 15
Reynolds, O., 428
Riley, W. F., 74
Ripling, E. J., 40, 84, 87–88, 97, 104, 107
Roark, R. J., 47, 76
Rossman, P. F., 152
Ryan, D. G., 256

Saunders, H. F., 235
Scarborough, J. B., 10

Schmitter, W. P., 367
Seely, F. B., 17, 20, 24–25, 43, 77, 237, 241–242
Shaw, M. C., 416, 431
Shewhart, W. A., 492
Shigley, J. E., 45, 94, 316, 336, 411
Siegel, M. J., 165, 224
Siegel, P., 13
Sines, G., 95
Slaymaker, R. R., 452
Smith, J. O., 17, 20, 24–25, 43, 77, 237, 241–242
Sommerfeld, A., 447
Southwell, R. V., 52
Spotts, M. F., 487, 493
Stiles, W. B., 41
Straub, F. G., 108
Sturm, R. G., 235
Sutton, A. D., 351
Sweeney, R. J., 182
Swigert, A. M., 402

Tabor, D., 408
Tachau, H., 299
Thomas, H. R., 411
Timoshenko, S., 20, 28, 41–43, 51, 61, 101–102, 111, 113, 216, 235, 237, 411, 487–488
Tse, F. S., 506

Uhlig, H. H., 107

Van Doren, H., 178
Van Vlack, L. H., 129
Vladimir, L., 165
von Heydenkampf, G. S., 124
von Mises, R., 79
Votta, F. A., 230

Wahl, A. M., 72, 77, 206, 209, 211–212, 214, 217, 223–228, 233
Walker, H., 393
Weese, J. A., 41
Weibel, E. E., 72
Weibull, W., 467
Wellauer, E. J., 364
Wescott, B. B., 107
White, A. E., 108
Williams, C. G., 118
Wilson, W. K., 513
Wulff, J., 403

Young, E. H., 242

Zimmerli, F. P., 213

Subject Index

Abbreviations, 523–524
Absolute errors, 9
Acme threads, 72, 184, 186, 250
Aero threads, 72, 186
Air springs, 230
Allowable stresses, 82
  in cyclic loading, 68
  for gears, 354
  in rivets, 148
  in shafts, 268
  for springs, 213
  in static loading, 83
  in welding, 163–166
Allowances, 483
American Standard threads, 184
Arc welding, 160
Axles, 264

Ball-bearing screw, 260
Ball bearings (*see* Bearings)
Band brakes, 337
Beam deflections, 51
Beams, curved, 24
Bearing design, lubrication systems, 457
  McKee and McKee, 431, 442
  Raimondi and Boyd, 448
  Reynolds' equation, 444
  Sommerfeld number, 446
Bearings, ball, 460
    comparison with roller bearings, 460
    dimensions of, 463
    installation of, 470
    lubrication of, 478
    radial, 463
    ratings of, 460
    selection of, 464
    series of, 463
    thrust, 471
    types of, 461
  collar, 421
  heat dissipated in, 441
  Hyatt, 472
  journal, 419
Bearings, journal, bearing pressure in, 432
    design of, 442
    friction in, 429
    heat generated in, 440
    imperfect lubrication in, 439
    Kingsbury, 426
    lubrication of, 422
    materials for, 435
    Michell, 426
    partial, 420
    pivoted-segment, 426
    ZN/p number for, 430
  roller, 472
    angular, 473
    compared with ball bearings, 459
Belleville springs, 228
Belt drives, American, 298
  angle of contact, 275
  initial tension in, 279
  materials for, 274
  power transmitted by, 278
  ratio of tensions in, 277
  Rockwood, 291
  service factors for, 283
Belts, coefficient of friction for, 280
  creep in, 273
  flat, 276
  timing, 289
  V (*see* V belts)
Bending of shafts, 23
Bending moments, equations for, 526
  by graphical integration, 552–554
Bevel gears (*see* Gears)
Birnie's equation, 241
Block brakes, 330
Block chains, 304
Block clutches, 330
Bolts, 180
  resilience of, 193
  shock loading, 192
  tap, 180
  tightening-up loads on, 191
  (*See also* Fastenings)

Brakes, 322–344
band, 337
block, force analysis of, 330
caliper, 328
differential band, 338
disk, 327
energy in, 339
heat dissipation of, 340
multidisk, 327
single-block, 330
torque resisted by, 332
Buckingham equations for gears, 358–359
Buckling, of columns, 99
of springs, 212
Buttress threads, 250

Carburizing, 115
Case histories, 573–588
Cast iron, 126
Castigliano, theorem of, 51–58
Casting and forging, 111
Casting design, 176
Castings, distortion in, 143
effect on, of cooling rates, 141
of mass, 142
fluidity in, 125
Caustic embrittlement, 107
Chain couplings, 316
Chains, power (*see* Power chains)
tension linkage, 311
Clavarino's equation, 241
Clutches, band, 338
block, 330
cone, 328
multidisk, 327
plate, 323
Columns, 38–42
beam, 40
buckling of, 38
Euler's formula, 38–39
factor of safety for, 41
J. B. Johnson equation for, 38–39
stability of elastic and plastic, 38
Combined stresses, 30
Computation errors, 10–12
Computations, form of, 8
importance of, 7
in revision, 12
Computers, 13
for shaft deflection, 556–559

Concepts, 1
Conversion factors, 563
Corrosion, caustic embrittlement, 107
fatigue, 107
Cotter joints, 202
Couplings (*see* Shaft couplings)
Creep, in belts, 273
in metal, 137
strength, 103
Critical speeds, 507
Curve, normal distribution, 491
Curved beams, stresses in, 24–26
Cyclic loading, 68
stress concentration in, 68
stress-concentration factors, 84
Cylinders, of brittle materials, 234, 240
Clavarino's equation for, 241
of ductile materials, 241
maximum energy of distortion, 241
maximum shear stress, 241
thermal stresses, 242
heads for, 243
Lamé's equation for, 237
thick-walled, 237
thin-walled, 236

Damping, in metal, 124
in vibration, 502
Decimal equivalents, 562
Deflections, beam, by Castigliano, 52
by computer, 556–559
as design criteria, 97
by graphics, 555
of springs, 210
Deformation, angular, 56
Design stresses (*see* Allowable stresses)
Diametral pitch, definition, 347
standard values for, 356
Die casting, 175
Differential screw, 259
Dimensional control, 481
Disk springs, 228
Distortion energy theory, 87
Ductility and stress concentration, 67
Dynamic loads in gears, 359

Eccentric loading, of bars, 30
in riveted joints, 149

Efficiency, of hoists, 294–295
of riveted joints, 156–157
of translation screws, 252
of worm gearing, 394
Elastic stability of columns, 99
Electric-resistance welding, 159
Electroplating, 117
Elements of sections, equations for, 525
Endurance limit, 70
chart for, 91
effect on, cold working, 72
of finish, 178, 406
of surface, 69
for range of stress, 46
Equations, compatibility, 20
generation, 14
Errors, absolute, 9
computation, 10–12
relative, 9
Euler's formula, 99

Fabrication, 144
Factors of safety, 82
in columns, 99
for machine parts, 82–83
for wire rope, 298
Failure, of gears, 371
methods of, fretting corrosion, 475
pitting, 475
in rolling bearings, 474
spalling, 475
subsurface corrosion, 476
subsurface defects, 476
wear, 408–413
by static loads, 84
distortion energy, 87
maximum principal stress, 85
maximum shear stress, 86
theories of, 84
Fastenings, bolts, 179
cap screws, 181
locking devices, 181–182
machine screws, 181
set screws, 181
studs, 180
tap bolts, 180
threads for (*see* Screw threads)
Fatigue, combined stresses, 95
corrosion, 107
cumulative damage, 96
Fatigue, design, 90
failures, 90
Miner's rule, 97
S-N curves, 92
Fatigue (cyclic) loading, 90
Feasibility, 2
Feather keys, 196
Finish (*see* Surface finish)
Finite life, 70
Fits, allowance in, 483
classes of, 484
force, 487
holding power of, 489
metal, 482
shrink, 487
assembly of, 489
tolerance in, 483
Flange coupling, 314
Flat keys, 194
Fluid coupling, 320
Fluid lubrication (*see* Lubrication)
Formulas, stress, 19
Frequency distribution, 491
Fretting, 411
Friction, clinging, 415
coefficient of, 415
Coulomb, 414
fluid, 418
rolling, 416
solid, 414
Functional design, 5
Fusion welding, 159

Galling, 409
in gears, 371
in screw threads, 187
Gas welding, 160
Gears, bevel, 380–385
mounting of, 385
resultant tooth force in, 383
strength of teeth, 380
terminology for, 380
types of, 378
failure of, 371
helical, crossed, 379
parallel, 366, 374
velocity factor for, 367
herringbone, 367
hypoid, 378
lubrication of, 373

Gears, nonmetallic, 370
  service factor, 375
  silent, 370
  spur, 346
    allowable stresses for, 354
    Buckingham equation for, 358
    design of, 355
    dynamic loads on, 359
    form factor for, 350
    Lewis equation for, 348
    maximum bore in, 356
    strength of, 348
    stress concentration in, 368
    terminology for, 346
    velocity factors for, 354, 370
    wear in, 362
  worm, 389
    design of, 389
    efficiency of, 394
    heat dissipation of, 393
    Hindley, 388
    mounting of, 395
    strength of teeth in, 392
    terminology for, 387
Goal, 1
Graphical integration, 552–554
Grinding and polishing, 113

Hardening, 116
Hardness, 125
Hardness numbers conversion chart, 533
Heat dissipation, in bearings, 441
  in brakes, 340
  in worm gear boxes, 392
Heat treatment, 73
Heli-coil thread, 186
Helical gears (*see* Gears)
Helical springs (*see* Springs)
Hoists, efficiency of, 294
  Manila rope for, 295
  relation of rope tensions in, 293
  safety factors for, 298
  wire rope for, 298
Hooke's coupling, 316
Hydraulic couplings, 320
Hydrodynamic theory of lubrication, 428
Hypoid gears, 378

Impact loads, 106
Imperfect lubrication, 439
Induction motors, 515
Instability, 38
Integration, graphical, 552–554
Iterative design, 14

Johnson, J. B., formula, 99
Joints, cotter, 202
  knuckle, 200
  riveted (*see* Riveted joints)
  single-weld lap, 167
Jominy hardenability specification, 132
Journal bearings (*see* Bearings)

Keys, feather, 196
  flat, 194
  Kennedy, 195
  length of, 198
  proportions of, 197
  round, 195
  square, 194
  standard sizes of, 199
  stresses in, 197
  Woodruff, 195
Keyways, 195
Kingsbury bearing, 426
Knuckle joints, 200

Lamé's equation, 237
Lapping, 405
Lay in surface finish, 400, 407
Leaf springs (*see* Springs)
Lewis equation, 349
Lewis form factors, 352
  method for determining, 350, 351
Loading cyclic, 18
  shock, 18
  static, 66
Lock nuts, 181
Lubricants, oiliness, 437
  viscosity, 437
Lubrication, boundary, 431
  converging film in, 422
  fluid, 422
  of gears, 373
  hydrodynamic theory of, 428
  imperfect, 439
  mechanism of film, 414, 422
  of rolling-contact bearings, 478
  (*See* also Bearings)

Machinability, 125, 134
Machine screws, 181
Machine welding, 161
Machining, built-up edge, 404
Manila rope, 295
Materials, nonferrous, 127–129
  properties of, 528
  selection of, 123
Mechanical pitting, 411
Metal fits, 481
Michell bearing, 426
Mohr's circle diagram, 35–36
Motors, application of, 515
  controls for, 517
  dimensions of, 517
  duty cycle for, 519
  induction, 515
  rating of, 514
    intermittent, 519
  selection of, 514, 519
  sizes of, 517
  types of, 515
  WR^2, in acceleration of, 520
Mountings, rubber, 231, 511

Needle bearings, 473
Nitriding, 115
Nonmetallic gears, 370
Nonmetallic springs, 230
Normal distribution curve, 91
Notch sensitivity, 80

Oiliness, 437
Oilless bearings, 437
Oldham coupling, 315
Optimum design, 220
Overhanging shaft loads, 373

Parallel helical gears, 366, 374
Peening, 73, 116
Permissible stresses (*see* Allowable stresses)
Photoelastic method, 74
Pins, 203
Pipe threads, 187
Pitting, 372
  of gear teeth, 372
  mechanical, 411
Plastic deformation, 110
Plastic stability of columns, 38
Power chains, 300
  block, 304
  cases for, 310
  horsepower for, 304
  inverted-tooth, 307
  length of, 309
  lubrication of, 310
  offset links in, 309
  roller, 304
    horsepower for, 304, 564–572
  service, factors for, 304
  silent, 307
  speed of, 300
  sprocket bore for, 308
Pressure, bearing, 435
Production design, 2
  marketing, 2
Profilograph, 400
Profilometer, 400
Properties of materials, 528

Quality control, 490
Quenching, 113

Range of stress, 46
  in springs, 217
Relative errors, 9
Residual stresses, 109–121
  classification of, 110
  measurement of, 120
  sources of, 110
  specification, 121
Resilience, 193
Rigidity of shafts, 267
Ring springs, 227
Riveted joints, allowable stresses in, 148
  ASME Boiler Code for, 155
  eccentrically loaded, 149
  efficiency of, 157
  failure of, 155
  machine-member, 147
  pressure vessel, 153
  structural, 147
  thin plates, 152
Riveting, dimpling in, 152
Rivets, Cherry, 152
  draw, 151
  explosive, 152
  stress concentration in, 146

Rolled threads, 188
Roller bearings (*see* Bearings)
Roller chains, 305
Rolling of metal, 174
  shafts, 116
Rolling friction, 416–418
  coefficients of, 418
  resistance, 416
Rope, Manila, 295
  wire (*see* Wire rope)
Round key, 195
Rubber mountings, 231, 511
Rubber springs, 231
Rupture, creep and stress, 102
  stainless steel, 560
  strength, 104
  stress diagram for, 560

SAE numbers for steel, 130
Safety factors (*see* Factors of safety)
Screw fastenings, types of, 180
Screw threads, Acme, 250
  Aero, 186
  American National, 184
  American Unified, 184
  buttress, 250
  designation of, 186
  endurance limit for, 187
  form of, 184
  Heli-coil, 186
  proportions of nut for, 188
  rolled, 188
  Sellers, 251
  sizes of, 185
  stress-concentration factors for, 190
  thread forms for translation screws, 250
  United Standard, 184
Screws (*see* Fastenings; Translation screws)
Section modulus, equations for, 525
  values for, 527
Service factors, for gears, 375
  for power chains, 304
  for speed reducers, 374
  for V belts, 283
Set screws, 182
Shaft couplings, 313
  Ajax, 317
  American, 315
  Amerigear, 316
  chain, 316
Shaft couplings, compression, 314
  Falk, 317
  flange, 314
  flexible disk, 319
  fluid, 320
  Hooke's, 316
  leather-link, 318
  Lord, 319
  Oldham, 315
  slip, 319
  Westinghouse-Nuttall, 318
Shaft seals, 479
Shafts, allowable stresses for, 268
  bending of, 266
    and torsion, 266
  deflection of, by digital computer, 556–559
    by graphics, 555
  rigidity of, 270
  seals for, 480
  in torsion, 266
    and bending, 266
    and compression, 266
  transmission, 271
Shear, 25
Shock factor, 46
Shock loads, effect of, on bolts, 193
Shot peening, 73, 116
  specifications, 121
Silent (nonmetallic) gears, 370
Simulation, 14
Sintering, 175
Snap rings, 203–204
Solid friction, 414
Speed reducers, 374
  service factors for, 374
  thermal rating of, 375
Splines, 195
Spring rates, 211
Springs, 206
  air, 230
  allowable stresses for, 213
  Belleville, 228
  conical, 226
  corrosion effects on, 233
  disk, 228
  helical, buckling of, 211–212
    concentric, 226
    deflection of, 210
    design stresses for, 218
    end connections for, 220

Springs, helical, index for, 208
natural frequency of, 223
stress factor, 209
stresses in, 208
surge in, 222
tension, 221
leaf, 224
deflections of, 224
stresses in, 224
Neg'ator, 228
nonmetallic, 230
range of stress in, 217
ring, 227
rubber, 231
shot peening of, 232
spiral, 227
torsion, 227
types, 206
volute, 226
Wahl factors for, 209
Spur gears (*see* Gears)
Square keys, 194
Statistical quality control, 490
Steel, hardness conversion chart for, 533
impact resistance of, 139
low temperature applications of, 140
machinability of, 134
properties of, 129
SAE numbers for, 130
wrought, applications of, 127
yield points for, 529, 531–532
Strain gauges, 75
Stress, due to axial loading, 21
due to bending, 23–24, 29
cyclic, 45
diagrams, 118
formulas for, 19
gradient, 80
due to impact loading, 46
relieving, 171
residual, 109
rupture, 102
due to shock, 46
Stress concentration, 59
brittle coatings for, 76
causes, 59
in cyclic loading, 68
effect on, of holes, 60
of notches, 61
electric strain gage for, 75
experimental methods for, 73
Stress concentration, methods for, brittle materials, 76
photoelastic, 74
short-gage length, 77
reduction of, 71
seriousness of, 66
in spur gears, 368
in static loading, 66
in tension, 60
visualization of, 62
Stress-concentration factors, 77
in cyclic loading, 18
determination of, 78
for gear teeth, 368
for screw threads, 190
in static loading, 209
values for, 534–549
Stresscoat, 76
Stresses, allowable (*see* Allowable stresses)
combined, 30
in keys, 197
in springs, 208
Structural welding, 161
Struts, 39
Superfinishing, 405
Surface finish, 178, 406
damage to, 402
effect of, on friction, 414
wear on, 408
irregularities in, 399
lay in, 400, 407
measurement of, 400
roughness in, 406
roughness-width cutoff, 407
specification of, 406
Synthesis, 2

Task specifications, 1
Thermit welding, 160
Threads, pipe, 187
screw (*see* Screw threads)
Thrust bearings, collar, 421
Kingsbury, 426
Michell, 426
Timing belt, 289
Timken bearing (*see* Bearings, roller, angular)
Tolerances, 483
Torrington bearing, 473

Torsion, moments, 26
  of rectangular bars, 28
  of shafts, 265
Torsion-bar springs, 227
Torsional vibration, 505
Translation screws, ball, 260
  bearing pressure in, 259
  coefficient of friction in, 260
  collar friction in, 260
  differential, 260
  efficiency of, 252
  Roton, 261
  stresses in, 259
  thread forms for, 250
Transmissibility, 508
Transmission shafting, 271

V belts, 280–289
  horsepower rating, 282
  service factors for, 283
  sheaves for, 283
  sizes for, 281
Vibration, 498
  control of, 510
  critical speeds in, 507
  dampers for, 511
  equation of motion for, 500
  forced, 502
  free, damped, 502
    undamped, 502
  lateral, 505
  mountings for absorbing, 511
  natural frequency in, 506
  torsional, 505
  transmissibility in, 508
Viscosity, chart for values, 438
  definitions of, 437
Volute springs, 226

Wahl factor for springs, 209
Waviness of surface, 399
Wear, measurement of, 413
  mechanical, 408
  types of, 408
    abrasive, 410
    adhesive, 408
    corrosive, 412
Welding, 157–174
  allowable stresses in, 164
  arc, 160
  cost comparison, 158
  electric-resistance, 159
  fillet welds in, 164
  fusion, 159
  gas, 160
  machine, 161
  parallel weld, 164
  precise calculations in, 166
  stress relieving in, 171
  structural, 161
  thermit, 160
Welding design, allowable loads, 164
Welds, eccentrically loaded, 168
  inspection of, 172–173
  moment of inertia for patterns in, 170
  single-lap joints, 167
  stress distribution in, 166
  stress flow in joints, 164–165
  stresses in, primary shear, 168
    secondary shear, 169
  testing in, 172
Westinghouse-Nuttall coupling, 318
Wire rope, bending stresses in, 297
  lay of, 296
  safety factors for, 298
  sizes of, 297
  wire sizes, table of, 561
Woodruff keys, 195
Worm gearing (*see* Gears)
WR^2, equations for, 550
  in motor acceleration, 520

Zerol bevel gears, 378
ZN/p number, 430